the Botanical World

second edition

David K. Northington
of National Wildflower Research Center

Edward L. Schneider
of Santa Barbara Botanic Garden

WCB Wm. C. Brown Publishers
Dubuque, IA Bogota Boston Buenos Aires Caracas Chicago
Guilford, CT London Madrid Mexico City Sydney Toronto

Book Team

Editor *Margaret J. Kemp*
Developmental Editor *Kathleen R. Loewnberg*
Production Editor *Jane E. Matthews*
Designer *Christopher E. Reese*
Art Editor *Mary E. Powers*
Photo Editor *Lori Hancock*
Permissions Coordinator *Mavis M. Oeth*

Wm. C. Brown Publishers

President and Chief Executive Officer *Beverly Kolz*
Vice President, Publisher *Kevin Kane*
Vice President, Director of Sales and Marketing *Virginia S. Moffat*
Vice President, Director of Production *Colleen A. Yonda*
National Sales Manager *Douglas J. DiNardo*
Marketing Manager *Thomas C. Lyon*
Advertising Manager *Janelle Keeffer*
Production Editorial Manager *Renée Menne*
Publishing Services Manager *Karen J. Slaght*
Royalty/Permissions Manager *Connie Allendorf*

A Times Mirror Company

Copyedited by Cathy DiPasquale

Cover Design: Jamie O'Neal

Cover photo: Spring maple leaves © Charles Krebs/The Stock Market

The credits section for this book begins on page 467 and
is considered an extension of the copyright page.

Library of Congress Catalog Card Number: 95–76844

ISBN 0–697–24279–X

Printed in the United States of America by Times Mirror Higher Education Group, Inc.,
2460 Kerper Boulevard, Dubuque, IA 52001

10 9 8 7 6 5 4 3 2 1

Brief Contents

Contents

Chapter Five
The Master Molecule at Work: DNA to New Cells

Part Three
The Plant Body

Chapter Six
Growth and Tissues

Chapter Seven
Roots

Chapter Eight
Stems

Chapter Nine
Leaves

Chapter Fourteen
The Control of Growth and Development

Part Five
Evolution and Diversity

Chapter Fifteen
Meiosis, Sexual Reproduction and Inheritance

Chapter Sixteen
Evolution and Taxonomy

Chapter Seventeen
Life's Origins and Prokaryotes

Chapter Eighteen
Diversity: Nonvascular Eukaryotes

Chapter Nineteen
Vascular Plants

Part Six
Plants and Society

Chapter Twenty
The Roots of Culture and Modern Agriculture

Chapter Twenty-one
Plants of Medicine and Culture

Chapter Twenty-two

Our Precarious Habitat

Chapter Twenty-three

Now, Why Botany?

Dedication

The second edition of The Botanical World is *dedicated to Dr. J. R. Goodin who coauthored the first edition and worked with the two of us on the second edition until his death in 1991. Joe Goodin was an outstanding scholar who was extensively published and was known worldwide for his expertise in arid land plant physiology.*

Joe was also an excellent teacher who dedicated over twenty years to improving the learning experience for the nonmajor student enrolled in his introductory botany course. His contributions to The Botanical World *and, therefore, to the students who use this text, are considerable. Although his name is not listed as a coauthor per publisher policy, his mark on this book is very real and an important part of the philosophy and organization of the second edition.*

With thanks and respect, we dedicate this to Joe Goodin, great teacher and good friend.

ANNOUNCING

JOE R. GOODIN MEMORIAL SCHOLARSHIP

$500 scholarships will be awarded twice a year!

Applicants should follow these easy guidelines:

- complete a general botany course using the text *The Botanical World* by Northington/Schneider.
- be enrolled as a full-time college or university student in '96, '97, or '98.
- submit a 500-word typed essay, describing why they feel qualified to receive the scholarship (for example: financial need, enjoy the study of botany, or will be seeking a career in the field of botany, etc.).
- accompany each essay with an endorsement from the class instructor (limit of two endorsements by one professor). Application should include the students' full name, social security number, permanent mailing address, and the name/address of the professor.
- mail completed application to: MARGE KEMP, EDITOR
WM. C. BROWN PUBLISHERS
2460 KERPER BLVD.
DUBUQUE, IA 52001

DUE BY	AWARDED BY
December 1, 1995	March 1, 1996
May 1, 1996	September 1, 1996
December 1, 1996	March 1, 1997
May 1, 1997	September 1, 1997
December 1, 1997	March 1, 1998
May 1, 1998	September 1, 1998

Note: Funds for the Joe R. Goodin Memorial Scholarship have been generated from royalty on *The Botanical World*. Awards are not guaranteed and are subject to change based on availability of funds. Taxes and fees are the sole responsibility of recipients. Wm. C. Brown Publishers reserves the right to change or cancel the scholarship at any time and has the right to revoke the winning entry if the winner is found to be ineligible.

Scholarship offer is void where prohibited or restricted by law and is subject to federal/state/local regulations.

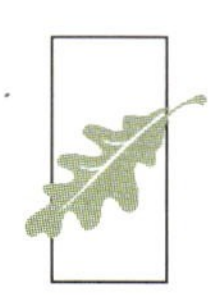

Preface

As with the first edition of *The Botanical World,* this second edition is written for the student. Our goal is to incorporate the strengths of the first edition with needed changes and updates to produce an introductory text that provides the best balance and depth of coverage with the most applied and readable writing style. It has been our experience that the typical student taking an introductory level botany or plant biology course can develop a real understanding of and excitement for the world of plants if the day-to-day value of that world is an ongoing part of each subject in the book.

We consider the natural environment and the ecological balances of nature to be of at least equal importance as the daily application of plants to human society. Therefore, we incorporate an ecological message throughout each topic as well as stress the applications of plants to our daily lives. We want the student to learn the vocabulary and basic grammatical structure of the study of plants as a necessary activity to fully appreciate the poetry and depth of the language of the natural world and our place in that world. Simply memorizing the traditional botanical structure, function, and diversity without developing a deeper understanding of the integration and interdependency of nature and especially humans in the botanical world is not only less enjoyable but far less pertinent and useful—and, therefore, less worth learning. Our commitment to awaken students to the excitement and importance of botany in their daily lives, and to maintain their involvement from the first class remain prime objectives.

APPROACH AND ORGANIZATION

Just as science majors would be intimidated if marched into a voice class and asked to sing something on the first day, many nonscience students are apprehensive about speaking about or even listening to lectures on science, biology, or botany. Introductory students appreciate a challenging but manageable menu in a course. Although chapters may be rearranged to tailor a course to the objectives of an instructor, there is much to be said for a front-to-back approach in an introductory course. Such a text is more likely to provide a logical, building sequence for the students as concepts gradually unfold. We have written *The Botanical World* in such a sequence. We have developed this text based on the approach we have found most successful in teaching introductory botany. That approach considers three basic elements: organization, balance, and depth of information.

Organization

We introduce the student to the most familiar topics first, and approach the subject by beginning not with the molecular level and building up to cells, tissues, organisms, diversity, and finally ecology (the traditional building block approach), but by starting with the most familiar subjects first. To the nonmajor, plant ecology, climatology, plant communities, and plant distribution are familiar topics that are often mentioned in the newspaper, on the evening news, and in other courses. There is also a greater comfort level with the macro versus the micro level of approaching any subject. With this approach, the student can develop an interest in the subject by concentrating on the most familiar topics first and address the concept of their place in nature early on.

In the second edition, we have then taken a more traditional approach to the organization of topics with Parts II through V. Throughout these subjects, however, we regularly refer to ecological principals and bring out the application of the subject being covered to our daily lives.

Part VI, Plants and Society, provides a logical conclusion to the organization of the book, and students respond well to pulling it all together both ecologically and in terms of human society's interdependence with (not use of) the natural world.

Balance and Depth of Information

Providing an approximately equal level of coverage on each topic allows for a greater breadth of information, including the integration of applied information throughout and the four chapters in Part VI, Plants and Society.

In the traditional coverage of the diversity of plants, the amount of space and the depth of detail are usually at that of a text designed for botany majors. In addition, depending on the particular interest of the author(s), other traditional subjects such as genetics, physiology, or morphology and anatomy often receive accordingly exhaustive coverage. This depth of detail and the number of pages dedicated to such coverage prevent the inclusion of applied information such as climatology, plant distribution, industrial plant products, the history of agriculture, horticulture, plants of medicine and culture, and a thorough coverage of environmental and ecological subjects as they relate to the successful balance of human needs and the protection of the natural world.

By balancing the depth of coverage, especially in the area of diversity, we have tried to treat each topic with the student in mind. We feel there is thorough coverage, even for a student who might decide to take upper-level courses in botany, but without the overabundance found in most texts.

Summary

The organization of topics, the depth and breadth of coverage, and the incorporation of applied material throughout the book are all aimed at helping the student learn. We have found in our courses that both majors and nonmajors are more successful with this approach than with a hierarchical organization: molecules, cells, tissues, organisms, ecology.

We do not subscribe to the philosophy which states that the capable students will learn the information and our only job is to cover everything thoroughly and carefully. Rather, we feel an obligation to help every student develop an understanding of how plants interact, grow, reproduce, and function. That understanding will promote informed decision making concerning the role of plants in a functional world. Our goal is to present the basic botanical information in a depth appropriate for beginning college students, while providing enough application to keep the students interested. We have tried to find a balance of these two components.

The overall theme of this book encompasses botany's ecological and applied components. The need for the enlightened management of plant resources is a large part of our message. Another is that humans need to function *within* the natural framework of all biological species.

DISTINCTIVE FEATURES

Although the first edition of *The Botanical World* was well received for its ecological and applied emphasis, several students, colleagues, reviewers, and adopters felt, and we concurred, that the second edition would benefit from a strengthening of the structural, evolutionary, and biodiversity coverage, while retaining our commitment to the original goals and objectives. It is our intent in this second edition to provide an enhanced balance of coverage, yet still retain our focus on informing students of the importance of plants in the world by providing a readable, stimulating, and compelling text. The outstanding production efforts made by our publisher have made this possible. Some of these enhancements are listed below.

REORGANIZATION

With the rewriting of any new edition comes reorganization of the existing chapters and the creation of new chapters. The second edition of *The Botanical World* is no exception. Part III, The Plant Body, includes three new chapters (6, 7, and 8), and all the others have been updated with new information since the appearance of the first edition ten years ago. In addition, many new photographs have been added, accompanied with new legends. Several of the previous photographs have also been upgraded together with their legends. These are important tools that allow instructors to teach complex concepts and enhance student interest and learning.

The addition of new chapters, photographs, and line art has resulted in an entirely new design and illustration program for the second edition of *The Botanical World*. The new design elements include:

Color

The generous use of full-color photographs throughout the text adds significantly to the clarity of the line drawings and provides vivid examples of the many specific topics discussed. A conscientious effort was made to include full-color photographs that would add instructional value.

New Photos

Numerous new color photographs have been selected with the purpose of enhancing the students desire to learn more about the structure, function, evolution, and importance of plants.

Original Drawings

Almost all of the line art in the second edition is original. Done under the close supervision of the authors, each drawing was planned to illustrate specific points in the text.

Chapter Outlines

Presented at the beginning of each chapter, these outlines provide an overall perspective of the major topics to be covered.

Chapter Overview

Each chapter opens with a brief summary of its content.

Concept Checks

Appropriately placed throughout all chapters are short topic and discussion questions. Students will benefit immensely from these, ensuring their understanding of the major concepts.

Cross-references

One of the pedagogical design aids in this text is cross-referencing where topics are discussed more than once. The intent is to indicate purposeful repetition of certain subjects and to allow the student to turn quickly to another explanation of the topic if desired.

Key Terms

Key terms are printed in **boldface type** the first time they occur in the text. Most of these terms can be found in the glossary. The student should use the glossary as often as needed, since not all terms are thoroughly defined the first time they are used.

Discussion Questions

At the end of each chapter, a list of discussion questions test student knowledge of the chapter's key content.

Boxed Essays

Examples of interesting botanical topics include the importance of seed banks (chapter 11), mycorrhizae (chapter 7), and a special boxed essay written by Dr. Sherwin Carlquist on the evolutionary history of island plants (chapter 16).

Enumerated Summaries

At the end of each chapter, these summaries provide a quick review of major concepts and serve as a study aid.

New Selected Readings

At the end of each chapter, we have provided an updated list of additional resource materials that may be assigned by the instructor or selected by the student when additional information on the subject material in the chapter is needed.

Glossary

A comprehensive glossary defines most boldface key terms found in the text, as well as several additional terms.

Ancillaries

A comprehensive package of supplementary materials has been developed for the second edition of *The Botanical World* to aid student learning and facilitate course management by the instructor. This package includes:

The Botanical World Laboratory Manual
The Botanical World Student Study Guide
The Botanical World Instructor's Manual and Test Item File
Overhead Transparency Acetates
Art Study Notebook
Color Slides (additional images not found in the text)
Computerized Testing Software

ABOUT THE AUTHORS

Professors Northington and Schneider combine over forty years of teaching experience in introductory botany.

David K. Northington, Ph.D.

Dr. Northington received his Ph.D. in systematic botany at the University of Texas, Austin in 1971 and accepted a faculty position in the Department of Biological Sciences at Texas Tech University in Lubbock, Texas. During thirteen years at Texas Tech, he received several teaching awards and served as curator of the E. L. Reed Herbarium, associate chairman in the Department of Biological Sciences, and director of the Texas Tech Center at Junction, Texas, a 400-acre campus and biology field station. During his academic career at Texas Tech, he published in regional, national, and international journals, edited two books on arid land plant resources and, with Dr. J. R. Goodin, wrote the first edition of *The Botanical World*. In 1984, Dr. Northington returned to Austin as executive director of the National Wildflower Research Center, a nonprofit research and educational organization dedicated to the preservation and reestablishment of the native flora of North America. He is also an adjunct associate professor in the Department of Botany of the University of Texas, Austin and the Biology Department of Southwest Texas State University. Dr. Northington has studied and traveled professionally for cooperative projects in Egypt, Australia, Mexico, and England.

Edward L. Schneider, Ph.D.

After earning a Ph.D. in structural and evolutionary botany at the University of California, Santa Barbara in 1974, he began an academic career at Southwest Texas State University, serving as professor, chairman of the Biology Department and dean of the School of Science before returning in 1992 to Santa Barbara to assume a new role as director of the Santa Barbara Botanic Garden, a scientific and educational institution dedicated to the study, display, and conservation of native California flora. Dr. Schneider is an adjunct professor of botany at the University of California, Santa Barbara. He is widely published and has presented numerous invited lectures on the structure, reproduction, and evolution of aquatic plants. His research, supported by the National Science Foundation, and cooperative projects have taken him to Japan, China, Malaysia, Australia, Mexico, and throughout much of Europe.

Scholarship

Because education was so important to Joe Goodin, a memorial scholarship has been established in his name and will be awarded to students who are engaged in botanical study. Complete information about the scholarship is available on page xii.

ACKNOWLEDGMENTS

Many people help make a text such as *The Botanical World* possible. We appreciate the valuable input, patience, and support of family, friends, colleagues, reviewers, and especially students. We enjoy teaching the introductory course and feel it should be an important part of any college student's education. We sincerely hope this book will convey the excitement and importance of the botanical world.

We would like to acknowledge the continued patience and support of Wm. C. Brown Publishers, especially that of our editors, the production staff, and the marketing coordinators. In particular, the incredible patience and encouragement of Marge Kemp, the Project Editor, and the guidance of Kathy Loewenberg, Lori Hancock, and Jane Matthews. They have been involved in every aspect of bringing this second edition to fruition, improving its quality, and ensuring its accuracy.

We would also like to acknowledge the endless hours of our copy editor, Cathy DiPasquale. Her care and talent have contributed an immeasurable amount to the style, flow, accuracy, and clarity of the text. If a misspelled word or dangling participle has slipped through, we apologize to her; she has displayed boundless talent in keeping us in line.

Obviously, the talents of our illustrator, Yevon Wilson-Ramsey, have added a great deal to enhancing the text. As with the first edition, it is a true delight to be able to work with a talented, patient, and dedicated person.

We would like to give particular thanks to the many people who provided the outstanding photographs, photomicrographs, and micrographs in this book. Each of these contributors has been cited in the photo credit listing, individually crediting their skill and collectively emphasizing that the science of botany grows through the application of the talents of many.

Those who contributed indirectly, although substantially, to this book include the botany professors who introduced us to the fascinating botanical world and the many, many students that have not only challenged us to excel as teachers, but who have instilled in us the rewards that come with quality instruction and the excitement of fostering successful student learning.

We could not have completed this text without the assistance of many colleagues. We wish to thank especially Dr. Paula S. Williamson and Dr. Arthur Elliot, who read portions of the manuscript and have offered many suggestions for improvement.

To our wives, Pat Northington and Sandy Schneider, we owe a tremendous debt. Without their understanding, support, and forgiveness of time spent preparing the text, it would not have been possible to complete this revision.

Reviewers

The authors and publisher would like to express appreciation to the botanists who provided critical reviews during the development of the second edition. Their support, advice, opinions, and criticisms concerning the content, organization, and philosophy of the book were instrumental in its revision. Since not all suggestions were incorporated, the authors accept full responsibility for any factual errors that might exist.

Neal M. Barnett
University of Maryland at College Park

Bruce M. Bennett
Community College of Rhode Island

Michael F. Gross
Georgian Court College

Richard E. Hall
Cypress College

Laszlo Hanzely
Northern Illinois University

William M. Harris
University of Arkansas

Richard W. McGuire, Jr.
San Juan College

L. Maynard Moe
California State University, Bakersfield

Clark G. Schaack

Daniel C. Scheirer
Northeastern University

Calvin P. Viator
Nicholls State University

Richard J. Weilminster
The Pennsylvania College of Technology

Part One

Plants and Our Environment

An oak leaf starting to show its autumn colors.

Chapter One

Why Botany?

Plants and Nature

Economic Importance of Plants

Aesthetic and Recreational Significance of Plants

Plants in Science and Technology

The Scientific Method
Plants and Societal Needs

The red hummingbird-pollinated flowers of *Aquilegia*.

When told that introductory botany is a part of their curriculum, it is perfectly fair for students to ask, "Why Botany?" This chapter addresses that question by providing an overview of the importance and relevance of plants. Their basic and critical role in the ecological balance of our earth is matched by the many specific, applied uses of plants to human society. Students are also first introduced to "science" as a discipline and as a method of investigation with emphasis on botany—the study of plants—and its several subdisciplines.

"Prairie Birthday"

Every July I watch eagerly a certain country graveyard that I pass in driving to and from my farm. It is time for a prairie birthday, and in one corner of this graveyard lives a surviving celebrant of that once important event.

It is an ordinary graveyard, bordered by the usual spruces, and studded with the usual pink granite or white marble headstones, each with the usual Sunday bouquet of red or pink geraniums. It is extraordinary only in being triangular instead of square, and in harboring, within the sharp angle of its fence, a pin-point remnant of the native prairie on which the graveyard was established in the 1840's. Heretofore unreachable by scythe or mower, this yard-square relic of original Wisconsin gives birth, each July, to a man-high stalk of compass plant or cutleaf Silphium, spangled with saucer-sized yellow blooms resembling sunflowers. It is the sole remnant of this plant along this highway, perhaps the sole remnant in the western half of our county. What a thousand acres of Silphiums looked like when they tickled the bellies of the buffalo is a question never again to be answered, and perhaps not even asked.

This year I found the Silphium in first bloom on 24 July, a week later than usual; during the last six years the average date was 15 July.

When I passed the graveyard again on 3 August, the fence had been removed by a road crew, and the Silphium cut. It is easy now to predict the future; for a few years my Silphium will try in vain to rise above the mowing machine, and then it will die. With it will die the prairie epoch.

The Highway Department says that 100,000 cars pass yearly over this route during the three summer months when the Silphium is in bloom. In them must ride at least 100,000 people who have "taken" what is called history, and perhaps 25,000 people who have "taken" what is called botany. Yet I doubt whether a dozen have seen the Silphium, and of these hardly one will notice its demise. If I were to tell a preacher of the adjoining church that the road crew has been burning history books in his cemetery, under the guise of mowing weeds, he would be amazed and uncomprehending. How could a weed be a book?

This is one little episode in the funeral of the native flora, which in turn is one episode in the funeral of the floras of the world. Mechanized man, oblivious of floras, is proud of his progress in cleaning up the landscape on which, willy-nilly, he must live out his days. It might be wise to prohibit at once all teaching of real botany and real history, lest some future citizen suffer qualms about the floristic price of his good life.*

—Aldo Leopold

*From *A Sand County Almanac, with other essays on conservation from Round River* by Aldo Leopold.

Aldo Leopold, one of history's great conservationists and a cofounder of the Wilderness Society, wrote the preceding essay out of concern not only for *Silphium* (figure 1.1), but for all plants and their importance to our world. Likewise, this book is concerned with helping you to develop an appreciation for **botany**—the study of plants—and the impact of plants on your life. Leopold's tongue-in-cheek statement that the teaching of botany should perhaps be prohibited ironically conveys that studying botany fosters an awareness of the crucial role of plants to the functioning of the world.

We appreciate plants for their contributions both to natural beauty and to our recreational pursuits in homes, gardens, parks, and wilderness. They enhance our enjoyment of life. However, other roles of plants come to mind; for example, plants are important for shelter, clothing, and diet. Wood is the world's most common building material; cotton is one of our basic fabrics; and fruit and vegetables are part of a balanced diet. The utility of plants in everyday life should not be underrated.

The *most* fundamental value of plants, however, lies in yet another dimension. They are vital members of our **ecosystem** (see chapter 2, p. 19)—the natural interrelationships among all the living organisms in a given area and with the environmental factors of that area (figure 1.2). Plants produce the oxygen in the atmosphere that we breathe. Also, by the process of photosynthesis, they convert light energy to food energy. This benefits not only human food consumption and sustenance, but also

Figure 1.1
Silphium.

Figure 1.2
As seen from outer space the planet Earth may look like a blue world, but in truth it is a green world powered by energy from the sun.

other members of the animal kingdom, who in turn are food sources for each other and for us. Just how crucial is the role of photosynthesis in the functioning of the world? *No other organisms, including humans, would exist on earth were it not for plants.*

PLANTS AND NATURE

The role of a single species—let alone a single plant—in the balanced functioning of any natural habitat can be easily overlooked. Leopold's awareness that his lone *Silphium* plant was a remnant of the native prairie and a relic of original Wisconsin demonstrates his understanding of natural **communities**—all the living organisms sharing a given area. Before clearing and plowing for cultivation were undertaken, the prairie species provided food for the wild animals of the region. From the buffalo to the smallest field mouse the diverse and numerous plants and animals found there coexisted in balance. Even ranching activities placed an unnatural burden on parts of the native flora, resulting in permanent changes in the numbers of species and their densities. There are precious few native prairies, or even small plots of them, still intact in North America. Their use to human society as valuable pasture or farmland has irretrievably altered their floristic composition. Unfortunately, the same process is occurring in all the major prairies of the world. Forests too, and even deserts, are now being irreversibly changed in response to the short-term needs of human society. The importance of plants in the balanced functioning of the biological world cannot be overemphasized. Even though plants may have applied uses and cultural importance in the development and maintenance of human populations, they act most significantly in the natural balance of all biological energy and of atmospheric oxygen and carbon dioxide.

The source for all energy available to living organisms is the sun. The warmth provided by the sun's rays makes the earth habitable, and in plant tissues sunlight energy is converted to food in the form of carbohydrates. Green plants, then, are the **primary producers** of food for the rest of the biological world, food that is subsequently converted to other forms of usable energy by those living organisms.

Plants combine sunlight energy with water from the soil and **carbon dioxide** from the atmosphere in a process called **photosynthesis** (see chapter 13, p. 219). One of the end products of this process is plant tissue, which can be used as food by animals. Another end product of this energy conversion is the production of **oxygen.**

Therefore, the two most important atmospheric components for life on earth are oxygen and carbon dioxide. Animals must have a constant supply of oxygen to carry out metabolic processes that convert food tissues into a form of energy that allows muscle contraction, brain functioning, new growth, and ultimately the perpetuation of the species. Plants, too, need oxygen to carry on their life processes. The source for all atmospheric oxygen is the same plant process that converts light energy to carbohydrates—the process of photosynthesis.

Since **respiration** (see chapter 13, p. 230)—the utilization of food for producing a form of energy to do biological work—results in the release of carbon dioxide into the atmosphere, the balance of these two essential gases is maintained. Plants carry out both processes, releasing both oxygen and carbon dioxide into the atmosphere, whereas animals are unable to produce oxygen. Water is crucial in both processes, cycling through plants, animals, the atmosphere, and the soil.

Thus, if the natural functioning of plants is disrupted, the flow of energy, the exchange of oxygen and carbon dioxide, and the availability of freshwater are all affected. These balances are essential; disturbing them ultimately affects the very organisms that have brought about changes in that natural balance—humans.

In his foreword to *A Sand County Almanac* Leopold summarized the cause for these changes. "We abuse land because we regard it as a commodity belonging to us. When we see land as a community to which we belong, we may begin to use it with love and respect." A true appreciation of what our responsibilities are within the natural community can come only from an understanding of the **botanical world.**

✔ Concept Checks

1. What is the most fundamental value of plants to human society? For what other reasons are plants important?
2. How do plants function in the natural balance of all biological energy on earth?
3. Explain how carbon dioxide and oxygen are involved in the processes of photosynthesis and respiration.

ECONOMIC IMPORTANCE OF PLANTS

Although Leopold made no direct allusion to any potential economic value of *Silphium,* his description of "saucer-sized yellow blooms resembling sunflowers" brings to mind an economically valuable relative. Prior

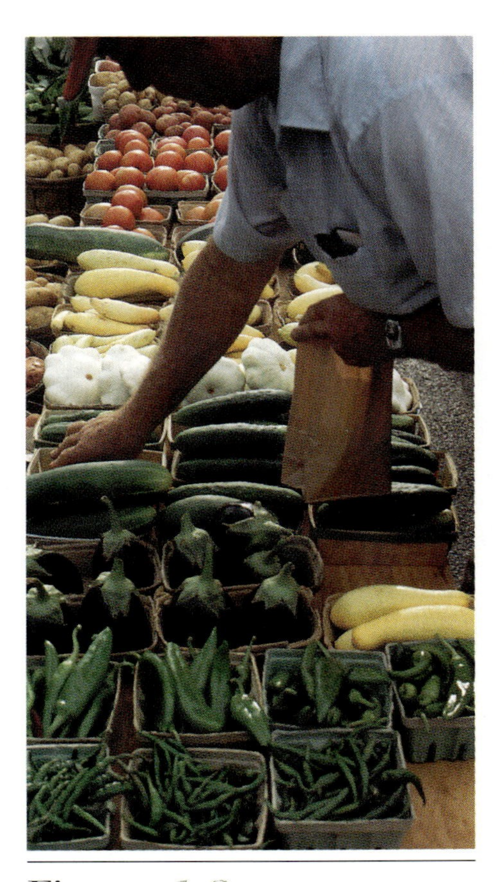

Figure 1.3
As population levels increase, food supply becomes more precarious. Food must be produced on a constantly larger scale, and localized food supplies fail to feed the multitudes.

TABLE 1.1

Top Food Crops. These seven crop species produce approximately 74% of all the food consumed by humans worldwide.

CROP	SCIENTIFIC NAME	ANNUAL PRODUCTION IN MILLIONS OF METRIC TONS
Wheat	*Triticum aestivum*	360
Rice	*Oryza sativa*	320
Corn	*Zea mays*	300
Potato	*Solanum tuberosum*	300
Barley	*Horedum vulgaris*	170
Sweet potato	*Ipomoea batatas*	130
Cassava	*Manihot ultissima*	100

to the commercial development of the Giant Russian sunflower for seed and oil, it too was only a common prairie weed (*Helianthus annuus*). This, of course, is true for all economically valuable plants; at one time the native plants that gave rise to today's crops were of no greater value to humans than the *Silphium*. The development of agricultural crops is the foundation on which civilizations were built and one of the key factors in the continuation of successful human populations. Modern agriculture produces an incredible amount of food annually, but with continued growth of the world population, further increases are necessary (figure 1.3). From approximately 500,000 plant species known to exist, over 93% of all food consumed by humans comes from less than twenty species, and over 70% of the food consumed is from only seven plant species (table 1.1). How many potentially valuable plant species are yet to be discovered and developed? And how many are no longer available to fulfill such potential?

The natural world still contains a vast array of different kinds of organisms, yet each day sees the extinction of additional species. No one can accurately predict the consequences of that loss to the world. It is quite possible that a plant with the genetic potential for becoming a major world food crop or a cure for cancer has just been destroyed by some other mowing machine in some other graveyard. It is equally possible that there are many plants still plentiful that have undiscovered potential. These plants need to be found and studied. The uses for many such plants are part of the lore of the native people of many primitive cultures around the world.

Plants influence both our conscious and unconscious decisions, even though most of us are unaware of their presence until we really need them, as at mealtime. The entire economic stability of many nations depends heavily on importing and exporting plant products. As the world's human population continues to grow, the economic importance of plants and plant products becomes even more significant.

Every time we write a check to pay for the groceries we buy, the economic importance of plants for food is clear—sometimes painfully so. From the produce aisle with its array of fresh fruit and vegetables through the canned goods to the bakery items, essentially everything we buy is directly or indirectly produced from plants or plant extracts (figures 1.4 and 1.5). The meats and dairy products would be unavailable without adequate feed for the domesticated animals from which they come. Even the cash register tape, the boxes or sacks in which our groceries are placed, and the check or cash used to pay for these purchases are plant products.

In the winter many tons of firewood provide heat throughout the world. Additional energy comes from coal, natural gas, and other **petroleum products** formed from plants that lived many millions of years ago (figure 1.6). Thus even plants from the past are available to us in the form of synthetic fibers, plastics, and other materials manufactured from petroleum by-products.

Many medicines and drugs are derived from plants (figure 1.7); synthetic pharmaceuticals are based on natural compounds first found in plants, bacteria, or fungi. The two most widely used pain relievers, morphine and aspirin (salicylic acid), are

Figure 1.4
Shopping at the supermarket is a constant reminder of the economic and ecological importance of botany in our lives.

Figure 1.5
The produce aisle of a modern supermarket is an excellent example of the economic value of plants in human society.

Figure 1.6
The fossil fuels so important to the technological world come from plants that lived several hundred million years ago.

such compounds. Plant poisons have great historical significance. Socrates was put to death with juice from the hemlock plant, and the Roman emperor Nero ruled after his father, Claudius, was assassinated with poison mushrooms and monkshood juice. Curare-tipped arrows and blowgun darts have aided many native tribes in the Amazon jungles in centuries of successful hunting.

Cannabis, the genus that produces hemp fibers for rope and the hallucinogen marijuana, is a well-known member of the plant world.

Many plants have industrial application as well. Their impact on the historical development of modern civilizations is vast. Natural rubber is from a tropical tree, and from another tropical species comes the only source of carnauba wax, the hardest of the natural waxes for polishes and industrial uses. Many of the finest lubricating oils for industrial machinery are found in plants, and various other plant extracts are important to a broad spectrum of industrial applications. Seeds from jojoba, a desert shrub, yield a substitute for sperm whale oil, with which it shares almost identical chemical properties (figure 1.8). Fibers for clothing, string, twine, and canvas come from plant sources; cotton is the most widely used. Dyes, essential oils for perfumes, and other oils and resins valuable in paint, varnish, linoleum, plastics, soaps, and many other products are also derivatives of plants. Thus for food, forage, fiber, fuels, medicine, and varied industrial uses plants have by far the greatest economic impact of any group of living organisms.

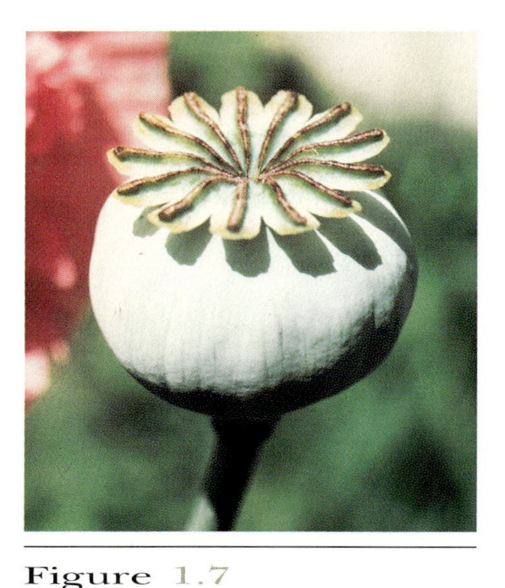

Figure 1.7
Some plants produce drugs that have been used and misused by humans throughout recorded history. The opium poppy, *Papaver somniferum,* contains a mixture of opiates from which are produced one of the world's most important pain relievers, morphine, and the world's most seriously abused illegal drug, heroin.

Figure 1.8
Plants with promising industrial potential include the desert shrub jojoba (*Simmondsia chinensis*). Oil from jojoba is an excellent substitute for the oil produced by the endangered sperm whale. The properties of this unusual oil make it particularly valuable as an industrial lubricant and as a cosmetic moisturizer.

Figure 1.9
Scenic view of Eldorado National Forest, California. The wilderness defines a unique natural beauty.

AESTHETIC AND RECREATIONAL SIGNIFICANCE OF PLANTS

Aldo Leopold's aesthetic sense must have been acute indeed. In the introductory essay his finely tuned appreciation for such a simple scene as the decorative yellow *Silphium* blossom against the backdrop of a country graveyard rings out. Leopold imagined that scene in another era, when millions of *Silphium* plants must have created the illusion of large, yellow inland seas. In our parks and botanical gardens plants provide panoramic beauty. On a somewhat smaller scale they provide shade and allow us to bring nature into our homes, businesses, and schools.

Sociobiologists contend that humans are genetically adapted to be comfortable in "natural" settings, but their immediate environment has evolved to include ever-increasing surroundings of metal, concrete, and plastic. Many people have learned to spend most of their time in such "unnatural" surroundings without any outward signs of physical or psychological discomfort. Others have not. In any case, a great many find enjoyment and relaxation when they are able to "return to nature" by backpacking, camping, picnicking, sitting under a tree, or walking barefoot in the grass or along a beach. But many urban dwellers enjoy such opportunities infrequently, if at all. Thus state and national parks and wilderness areas understandably are chosen frequently as vacation spots. The natural beauty and integrity of such areas must be preserved if they are to continue to provide the relaxation and recreation people seek there (figure 1.9).

Among the multitude of recreational activities humans have devised, one of the most consistently popular is golf. The proximity of the trees, hills, water, and grass of a golf course gives the feeling of being in a natural setting without having to invest hours of travel to reach it. Walking and jogging through city parks are also popular forms of recreation that can bring the participant in closer contact with plants and nature. Visiting botanical gardens and even weekend yardwork and gardening are, for many, welcome involvement in the plant world (figures 1.10 and 1.11).

Figure 1.10
Landscaping at the Nijo Castle Garden in Kyoto, Japan. Oriental gardens portray the grandeur of nature in confined spaces.

We humans also attempt to bring nature into our plastic worlds by "greening ourselves in" with houseplants and by landscaping our living spaces. The popularity of houseplants in working and living environments supports the theory that humans are genetically adapted to an environment in which we coexist with other living

Figure 1.11
Azaleas in bloom at the National Arboretum, Washington, D.C. Aesthetic qualities are important to us all. Landscaping provides relief from the hurried and disorderly world.

Figure 1.12
Gardening can be therapeutic. Horticultural therapy is recognized as an important tool in treating mental illness and emotional problems of the aged.

components of nature. There is even considerable evidence that mixing soil, pruning, watering, fertilizing, and tending to living plants are excellent forms of **horticultural therapy,** a subtle means of communing with nature by substituting cultivated plants for the larger and less accessible natural world (figure 1.12).

An understanding of how biological, geological, and environmental factors combine to produce the beauty of the natural world maximizes our appreciation and enjoyment of them, but it also helps us to understand how such beauty can best be maintained and preserved.

> ✔ *Concept Checks*
>
> 1. What seven species of plants have been used to produce food for over 70% of the world's human population?
> 2. Name one example each of plant products for fiber, forage, medicine, fuel, and industrial use.
> 3. List five recreational activities in which the aesthetic value of plants is important.

PLANTS IN SCIENCE AND TECHNOLOGY

The term *science* permeates every aspect of modern society. Food, consumer products, and even ideologies are scientifically designed, produced, and developed. There are natural, physical, political, and even social sciences. The scientific label has become a symbol of credibility, and yet a large segment of the human population still perceives science as a secretive or even mystical activity. Scientists are often depicted as rather strange white-coated people with exceptional IQs, who converse only in five-syllable words and mathematical formulas, often with beeping, blinking computers. Scientists are a diverse group of people observing, experimenting, and looking for new technologies and solutions to world problems. Thus **science** in general can be defined as a systematic search for knowledge and understanding of the natural and physical world. *A* science, in particular, is a branch of knowledge or study such as botany or physics.

The Scientific Method

Heredity and environment combine to shape the unique course of each of our lives. Similarly, our unique circumstances determine the kinds of associations we have with other parts of the biological world. Whatever our life's pattern, each of us will always have one or more relationships with the plant kingdom, be it as a consumer of shelter, clothing, and food, a gardener, a wilderness lover, an agriculturist, or perhaps a biologist. In any event, an understanding of individual plants and plant groups and their relationship to the rest of the biosphere affords us a viewpoint that will enhance our associations with plants.

But how does one understand plants? How does one know about any aspect of the physical and biological world? Most of our understanding is based on previous observations and discoveries of others. This text, for example, contains information based on a systematic assemblage of information from prehistoric to present times, organized into facts and theories. A *fact* is information that is thought to be known and specific. A **theory** is an idea that is well-supported by data but is not currently known to be entirely or universally true. A **hypothesis** is an idea or explanation that is a basis for further experimentation and inquiry but is currently without adequate evidence of support. The investigative process that incorporates reasoning and observation, with the objective of reaching theory or strongly supported understanding (fact), is called the **scientific method.**

Leopold's careful observation of the flowering dates for *Silphium* would have been missed by most of us. Who cares if flowering occurred on July 24, a week later than usual? Why did he make such an observation? Aldo Leopold was practicing science, that is, he was increasing his understanding of an event by using the scientific method. To this purpose he probably kept a diary describing personal observations in nature.

The scientific method always starts with a certain base of what is known. In observing the flowering dates of *Silphium* Leopold may have been collecting an information base from which he then could have generated a theory about the general flowering time for *Silphium.* Or, perhaps already having a hypothesis or testable assumption about why the flowering time for *Silphium* was later than usual, he might have been seeking out a cause and effect relationship for flowering time. The goal of the scientist is to observe with objectivity and accuracy so that ideas about what has been observed have validity. In science both reasoning and observation come into play. Science is the systematic accumulation of knowledge through the use of logic based on factual evidence. It is both an organized body of facts and a method of problem solving. The scientific method is the process for acquiring factual knowledge and for answering questions (solving problems) (figure 1.13). Ideally, the scientific method is

Figure 1.13
The goals in scientific investigation are to establish a hypothesis, record accurate observations, and make objective interpretations. In this experiment, up to 900 seeds can be tested with varying temperatures. Germinated seeds are counted and their location is scored to determine the temperature pattern most suitable for germination.

a repeated sequence of events that includes the following steps:

1. Recognition of a question (problem) or observed phenomenon
2. Establishment of a hypothesis to answer the question or solve the problem
3. Careful observations to gather factual information relative to the hypothesis
4. Reconsideration of the hypothesis in light of the accumulated data
5. Testing of the hypothesis to establish repeatability and thus validity of the knowledge

The scientific method is not to be regarded as inflexible or infallible. The important parts of an attempt to approach anything scientifically are accurate observation and objective interpretation. The number of observations made, the number of different observers involved, the analysis of the accumulated data, the questions asked, and the sequence in which all of this transpires are not the same from one situation to the next. The basic goals, however, are the same; **observations, hypothesis formation, testing,** and **repeatability to confirm validity** all must be present for science to function properly.

For example, we might *observe* that lemons taste tart. From this observation we might *hypothesize* that all yellow fruit taste tart. We would then *test* this hypothesis by tasting different yellow fruit. Discovering that grapefruit are also tart but that squash are not, we could then modify our hypothesis to say that some yellow fruit taste tart. To carry the process to the next level, we might then hypothesize that all yellow fruit in the citrus family taste tart or that all citrus fruit, regardless of color, taste tart. Continued testing and modification of our hypotheses ultimately provides sufficient evidence to formulate a theory.

When a theory has been repeatedly tested many thousands of times with the same results, it can be said to be a **natural law.** Few theories are considered laws, since there is a lack of sufficient experimental data to *prove* that there are no exceptions to the predicted results. One of the laws of nature describes the effect of gravity on objects. On the earth we can safely predict that every time we drop an object, it will move toward the center of the earth. To date there are no exceptions to this law, even though it has been tested in countless different ways.

Science does not attempt to determine what is "good" or "evil"; it does not have a moral purpose, nor does it attempt to convince anyone of anything. Science is simply an approach to establish what is and what is not supportable. Science is not a world only for the scientist; everybody can and does practice the scientific method.

Approaching an everyday situation scientifically is a matter of being objective and following a logical sequence through to the end. For example, comparison shopping for a car to obtain the best value for your dollar is essentially a scientific process of deciding among many possibilities that are closest to satisfying one's needs. Science in this context is basically an attempt to remove rumor and hearsay from the process, replacing them with supportable evidence and objective evaluations of reliability, performance, and durability.

Purchasing an item is actually a continuation of the process; using the item in question can be viewed as an experiment. The owner makes observations and comparisons of actual and claimed performance levels, which establish scientifically the validity of the manufacturer's claims.

So science is not mystical or limited to a unique intellect or personality. Science is basic and understandable in design. It is the framework in which knowledge is accumulated. **Biology** is the study of all living organisms; botany is the study of plants. Many more specialized subdisciplines delve into greater understanding of organisms, furthering the accumulation of knowledge (table 1.2).

Plants and Societal Needs

The twentieth century has suddenly thrust on both developed and underdeveloped countries complications not even dreamed of by our forebears. Events that now affect our daily lives would have sounded like science fiction in 1940. Technological advancements have developed at an incredible rate. As developed countries take advantage of the luxury of goods and services, all at the expense of tremendous energy and resource consumption, underdeveloped countries, having seen the splendor, suddenly rush to share it. The world's transportation and communication systems have increased at a phenomenal rate; we can know "what's going on" virtually anywhere on the globe.

What does all this have to do with botany? The realization that conventional food, fiber, forage, and energy resources are not inexhaustible has forced world leaders to consider where it all came from in the first place. Fortunately, the sun continues to supply sufficient radiation to keep the earth warm, but that is not enough to guarantee future human success. As long as the sun does provide sufficient heat and light energy, the role of plants to the very existence of humans on this planet remains critical. The interdependence of plant and animal life, especially human life, is and always will be of primary ecological importance. It is now crucial for average citizens to understand botanical implications so they can make intelligent decisions concerning the future. These are not easy decisions, either for the politician or for the voter. Some of today's social problems are **overpopulation,** inadequate food supply, and a reduced quality of life (figure 1.14). Does one nation with food have the right to dictate philosophy to another nation and use that food as a political weapon? Is it worth the environmental risks necessary to develop a supply of nuclear energy? Does one state have the right to take water from another state (or nation) that has an abundant supply but might need it in the future? Are you willing to pay a great deal more for environmental monitoring to ensure

Figure 1.14
Overcrowding in human society detracts from the quality of life. Social problems can increase when population densities reach high levels.

TABLE 1.2

Major Botanical Subdisciplines

PLANT TAXONOMY
Also referred to as plant systematics, this is the oldest botanical discipline. Taxonomists study plant relationships, identify and classify plants into groups based on genetic similarity, and name them according to these groups.

PLANT ANATOMY
The study of plant internal structure at cell, tissue, and organ levels of complexity.

PLANT PHYSIOLOGY
The study of plant function including biochemistry, hormonal control, and metabolic functions including photosynthesis and respiration.

PLANT MORPHOLOGY
The study of the form and structure of plants including vegetative and reproductive structures.

PLANT ECOLOGY
The study of interrelationships between plants, other organisms, and their environments.

PLANT GENETICS
The study of plant heredity and genetic control of structure and function.

CELL BIOLOGY
Once called cytology, this is the study of cells.

that the water you drink, the food you eat, and the air you breathe are safe?

In a botanical sense there exists a serious set of questions: How can we grow enough food to feed a world population two, three, or four times as large as the current one? What do we do if the quality of both drinking water and irrigation water becomes so poor that animals and plants no longer produce effectively? What do we do if the air becomes so polluted that plants will no longer grow?

These are not hypothetical questions but rather real concerns faced by today's leaders. In a democracy those decisions can be made only by a qualified electorate. As part of that electorate, voters with the understanding that knowledge provides play a serious part in how the botanical world and everything that depends on it are to be treated.

✔ Concept Checks

1. Define *science.*
2. List the repeated sequence of events that is involved in the *scientific method.* Explain the importance of each step.
3. How can an understanding of the botanical world help us take part in the decision-making process? What kinds of global questions need to be answered?

Summary

1. Plants are critical to continued human existence for aesthetic, economic, and ecological reasons.
2. Plants are the producer organisms for the world, converting sunshine into chemical energy for all organisms while releasing oxygen into the atmosphere. Ecosystems function properly only as long as nature is in balance.
3. Plants are the producers for all living organisms. Those which grew millions of years ago provide the world with fossil fuels, and modern plants provide human and animal food, shelter, medicine, fibers for clothing, and thousands of industrial products.
4. Plants permeate our conscious and subconscious lives in the form of nature, from which we have never totally separated ourselves. Parks, wilderness, and both interior and exterior landscaping remain at the center of human attempts to bring tranquility and beauty into a hurried, competitive, synthetic world.
5. Scientific inquiry begins with a question, followed by establishment of a hypothesis, observation, reevaluation of the hypothesis, and, finally, testing of the hypothesis for repeatability and validity.
6. National and worldwide food, energy, and water shortages are problems that both governmental leaders and citizens must face. As informed voters, those of us living in a democracy have an opportunity to help protect our botanical world.

Key Terms

biology 10
botanical world 4
botany 3
carbon dioxide 4
community 4
ecosystem 3
Helianthus annuus 5
horticultural therapy 8
hypothesis 9
natural law 9
observations 9
Oryza sativa 5
overpopulation 10
oxygen 4
petroleum products 5
photosynthesis 4
primary producers 4
respiration 4
science 8
scientific method 9
sociobiology 7
testing 9
theory 9
Triticum aestivum 5
Zea mays 5

Discussion Questions

1. In the past 200 years, has human society made any serious long-term mistakes that a stronger understanding of the botanical world might have prevented or altered?
2. Are plants more important to humans as food sources or in industrial and other nonfood uses?
3. How does world overpopulation affect the botanical world?

Suggested Readings

Brooks, R. R., and D. Johannes. 1990. *Phytoarchaeology.* Dioscorides Press, Portland.

Brucher, Heinz. 1989. *Useful plants of neotropical origin and their wild relatives.* Springer-Verlag, New York.

Davis, S. M., and J. C. Ogden, eds. 1994. *Everglades: The ecosystem and its restoration.* St. Lucie Press, Delray Beach.

Heiser, C. B. 1985. *Of plants and people.* Univ. of Oklahoma Press, Norman.

Lewington, Ana. 1990. *Plants for people.* Oxford University Press, New York.

Nepstad, D. C., and S. Schwartzman, eds. 1992. Non-timber products from tropical forests: Evaluation of a conservation and development strategy. In: *Advances in economic botany.* New York Botanical Gardens, Bronx.

Plotkin, M., and L. Famolare, eds. 1992. *Sustainable harvest and marketing of rain forest products.* Island Press, Washington, D.C.

Thomas, C., and D. Howlett. 1993. *Resource politics: Freshwater and regional relations.* Open University Press, Philadelphia.

Chapter Two

Plants and Ecology

Climatology

- Precipitation
- Temperature
- Light
- Limiting Factors

The Biosphere

Oxygen–Carbon Dioxide Balance

Cycling In the Ecosystem

- The Water Cycle
- The Carbon Cycle
- The Nitrogen Cycle

Trophic Levels

- Food Chain
- Food Web
- Food Pyramid

Ecological Succession

- Recolonization

Cereus, the saguaro or giant cactus, is characteristic of southeastern U.S. deserts.

In chapter 1, the relevance of studying about plants included their direct uses by society, the aesthetic and recreational significance of plants, and their role in nature. Understanding the natural interrelationships of plants and the other components of their environment is crucial to our understanding of our role in and impact on nature and the long-term consequences of our actions. This chapter provides a general background of where certain kinds of plants do and do not grow and why. Worldwide climate patterns, water and nutrient cycles important to plant growth, and the balances responsible for plant distribution are considered. All these topics are essential to understanding life and how living organisms interrelate. These are also concepts that develop the study of botany from general, familiar information early in the book to more detailed topics in later chapters.

One conclusion that can be drawn about living organisms is that the earth is the correct size and distance from the sun to allow life, as we know it, to exist. The size of the earth is important because it provides a gravitational force that holds an atmospheric sheath comprising a unique mixture of gases, including water vapor. Perhaps the most critical factor is the 150-million-kilometer distance from the sun, which allows an optimum amount of heat and light energy to reach the earth's surface. The possibility that these conditions exist on other planets has been investigated within the field of exobiology. This combination of heat, light energy, and the gases found in our atmospheric sheath provide the basic resources for life to exist on earth.

In recent years conflicts between resource exploiters and conservationists have led to a public awareness of **ecology,** even though many of the vocal advocates of both points of view are poorly informed of what ecology really means. The term is derived from the Greek *oikos,* meaning "home." Thus in a broad sense the home is simply the habitat for all living organisms, and ecology is a study of the factors that allow organisms to grow, compete, reproduce, and perpetuate the species. It is a study of the total environment and interrelates with many traditional fields in biology, such as morphology, anatomy, taxonomy, genetics, physiology, and biochemistry. Therefore a true ecologist needs to be a well-trained biologist, mathematician, chemist, physicist, and philosopher.

Ecologists study broad, functional systems in the living world. The systems in ecology follow an order of complexity based on the central question of why organisms live where they do. Part of the answer is that all organisms that depend on plants as a food source (and essentially all of them do) must live near their food. Therefore animals live only in those areas where they find sufficient food to survive.

A more specific question can be asked: Why do the producer organisms—the green plants—live there? Here the answer is much more complicated. This question can be answered by analyzing the abiotic (nonliving) environment. One plant may live where it does because the light, water, temperature, nutrients, soils, relative humidity, wind, and other factors are satisfactory for growth and survival. But mere survival is not enough; the plant must have a sufficient quality of growth to allow *reproduction and perpetuation of the species.* Sometimes humans settle for less. For example, the cotton plant grows from year to year in the warm environment of the tropics, but agriculturists have seen fit to move the cotton plant to a more temperate climate where it cannot survive the winter (figure 2.1). The farmer's only concern is with an economic yield (fibers and seeds from the fruit). It makes no difference that frost in late fall kills the plant, provided that the fruit (cotton bolls) have matured. The farmer will simply plant cotton seed again the next spring.

Figure 2.1
Mature cotton boll. Cotton is a major crop plant in temperate climates throughout the world.

Buying a tropical foliage plant for home or office, on the other hand, is a completely different situation. With varying degrees of success, you will attempt to properly care for your plant year round. Those of us who have had our houseplants die anyway may feel that we are not blessed with the innate ability to grow plants; we lack a "green thumb." In fact, we may actually feel that we are afflicted with the brown thumb syndrome; somehow, mysteriously, the ultimate death of our houseplants is ensured. Fear not; the brown thumb syndrome is not incurable. In fact, the trick of successfully growing plants is no trick at all; rather, it is an understanding of the combined effects of water, nutrition, light, humidity, temperature, and pest control. Houseplants are not specifically bred for inside growth but are really outdoor plants well adapted to their native environment, usually a tropical one. When plants are introduced into areas that do not naturally provide the environmental conditions to which they are genetically adapted, their needs go unsatisfied. We must therefore artificially compensate for those requirements for the plant to thrive or, in some instances, even survive. The better we understand the requirements of houseplants, the better we can modify their immediate environment and successfully ensure their healthy growth. The more we know, the greener our thumbs will become.

CLIMATOLOGY

The factors of precipitation, temperature, and light combine to provide most of the abiotic environment that controls worldwide plant distributions. Collectively these represent the climate to which a given plant species must be adapted to survive. The study of climate, climatology, is essential to a basic understanding of where plants do and do not grow. Since plants cannot move to secure water, light, optimum temperatures, or nutrients, the natural distribution of each species is controlled by what is provided.

Precipitation

Of the climatological factors, water is undoubtedly the most important factor in plant distribution. **Precipitation,** which includes all forms of moisture deposited on the earth's surface, is the source of essentially all freshwater.

Water vapor, although always present to some extent in the lower layers of the atmosphere, must exist in adequate amounts if condensation into water droplets or ice crystals is to occur. The droplets or crystals then must attain a sufficient weight to fall to earth as precipitation. For condensation to occur, the air containing the water vapor must cool to a temperature below its condensation point, the temperature at which water vapor condenses into a liquid state. The condensation point varies with the amount of water vapor present and the air temperature at which the vapor is being held.

Air is generally warmed by radiant heat from the sun, particularly near the equator and at lower elevations. Warm air is capable of holding more water vapor than is cold air; thus at and near the equator and at low elevations the air is warm and humid (full of water vapor). Warm air rises, cooling as it does so, and condensation occurs.

A complex set of events serves to produce high elevation air currents moving north and south from the equator toward either pole. These **circulation cells** do not reach the poles, however, but descend at about 30° N and S latitude (figure 2.2). This descending air has already lost most of its moisture as rain near the equator and has cooled considerably while circulating at higher elevations in the atmosphere. This cold, dry air warms as it nears the earth's surface. As it warms, its capacity to hold moisture increases, enabling it to absorb water vapor near the ground and produce a drying effect on the earth. This cell completes its circulation by moving back across the earth's surface to the equator and outward toward either pole. As the air travels, it continues to warm (especially air moving from 30° latitude to the equator) and absorb water vapor from the soil, ponds and streams (evaporation), and plants (transpiration) (see chapter 12).

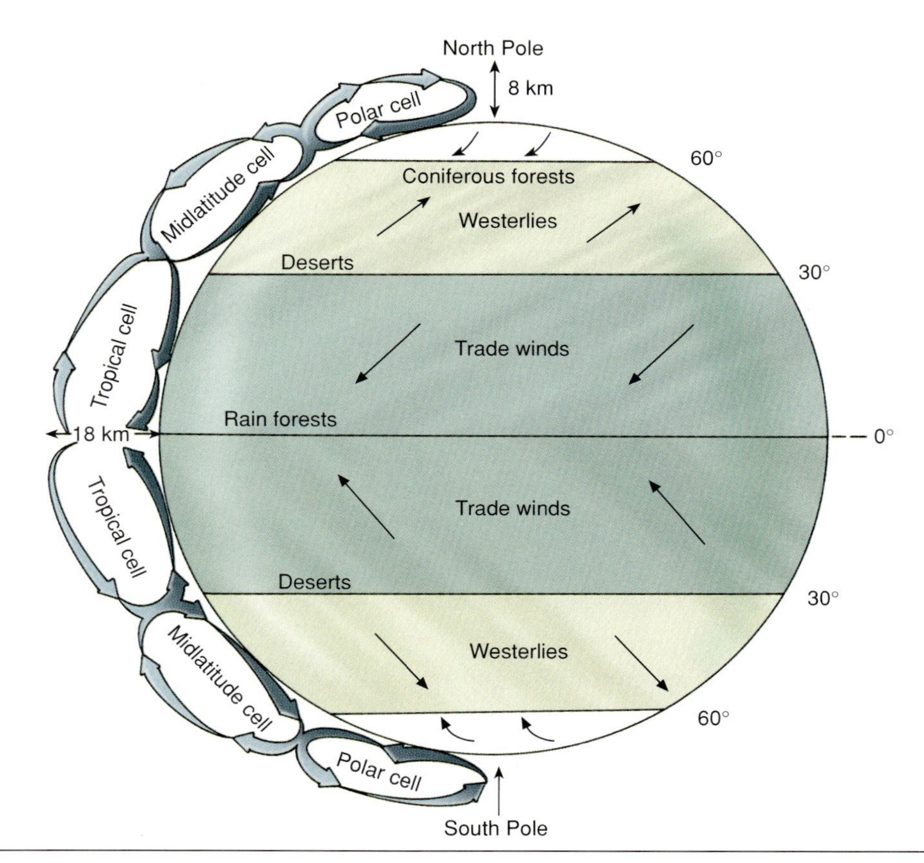

Figure 2.2
The circulation of air cells toward and away from the earth's surface produces bands of precipitation and dryness between the equator and approximately 60°. North and south of 60° latitude, cold has more effect on precipitation patterns than do air cells.

As a result of the movement of these circulation cells and their effects on precipitation, most of the earth's natural deserts occur between 20° and 30° N and S latitudes. Not all land between these two latitudinal belts is desert, and there are desert areas outside these belts. These anomalies occur because the precipitation effects of air circulation patterns in these cells are modified by topography, elevation, proximity to the coast, and ocean-current temperatures.

A coastal mountain range has a significant effect on precipitation patterns. As depicted in figure 2.3, as warm, moist sea air moves inland and is forced upward in elevation by a mountain range, it cools as it rises, causing condensation of water vapor into rainfall. The ocean (windward) side of the mountain therefore receives considerable precipitation as a result of the same physical phenomenon described for air-cell patterns. As this now cold, dry air moves over the range and descends the inland (leeward) side of the mountains, it begins warming as it descends, and its capacity to hold moisture increases. As it moves farther inland, the dry air takes moisture from the land and the plants, producing a **rain-shadow desert.** The Great Basin Desert of Nevada and Utah is formed by this rain-shadow effect, as is the Patagonian Desert on the leeward side of the Andes Mountains in Argentina (figure 2.4).

Generally, large continental landmasses are dry in the interior because

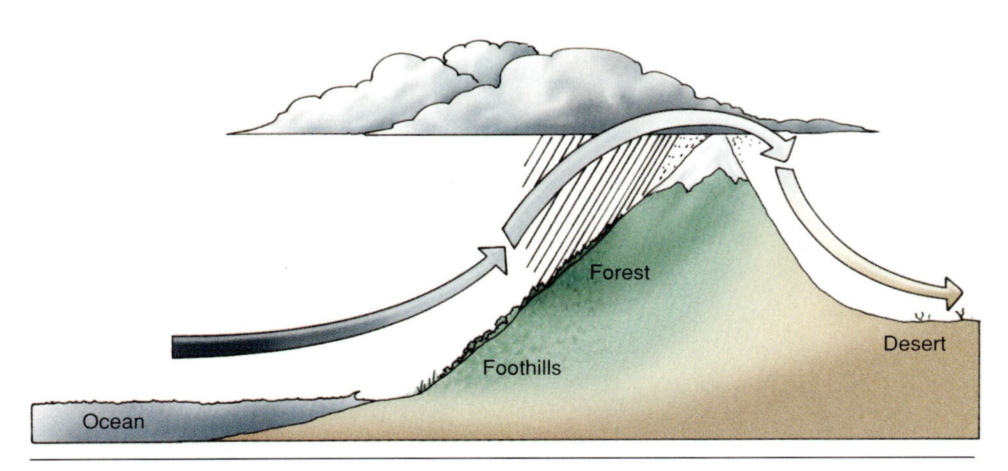

Figure 2.3

A rain-shadow desert is produced by the movement of warm, moist air passing over a coastal mountain range. Precipitation occurs on the ocean side of the mountains as the moisture condenses at higher elevations, passing dry air on to the other side of the mountains, where it warms and removes moisture from the plants and soil.

Figure 2.4

The Patagonian Desert in Argentina is formed by the rain-shadow effect of the Andes Mountains.

cold air masses collide with the warm, moist air from the sea. This forces the warm air to rise, and the condensation causes precipitation to occur before the moisture can reach very far inland. Even though no mountains may block the flow, the effect is similar to a mountain-induced rain shadow. Also, cold temperatures greatly modify precipitation patterns, so higher elevations tend to receive less rainfall (figure 2.5).

Although the effects of air-cell circulation patterns do produce increased precipitation near 60° N and S latitude, where air again rises away from the earth's surface (figure 2.2), as one moves farther north and south from 60°, the effects of temperature, especially cold, become increasingly important. North and south of approximately 60°, temperature is in fact the most significant factor influencing precipitation.

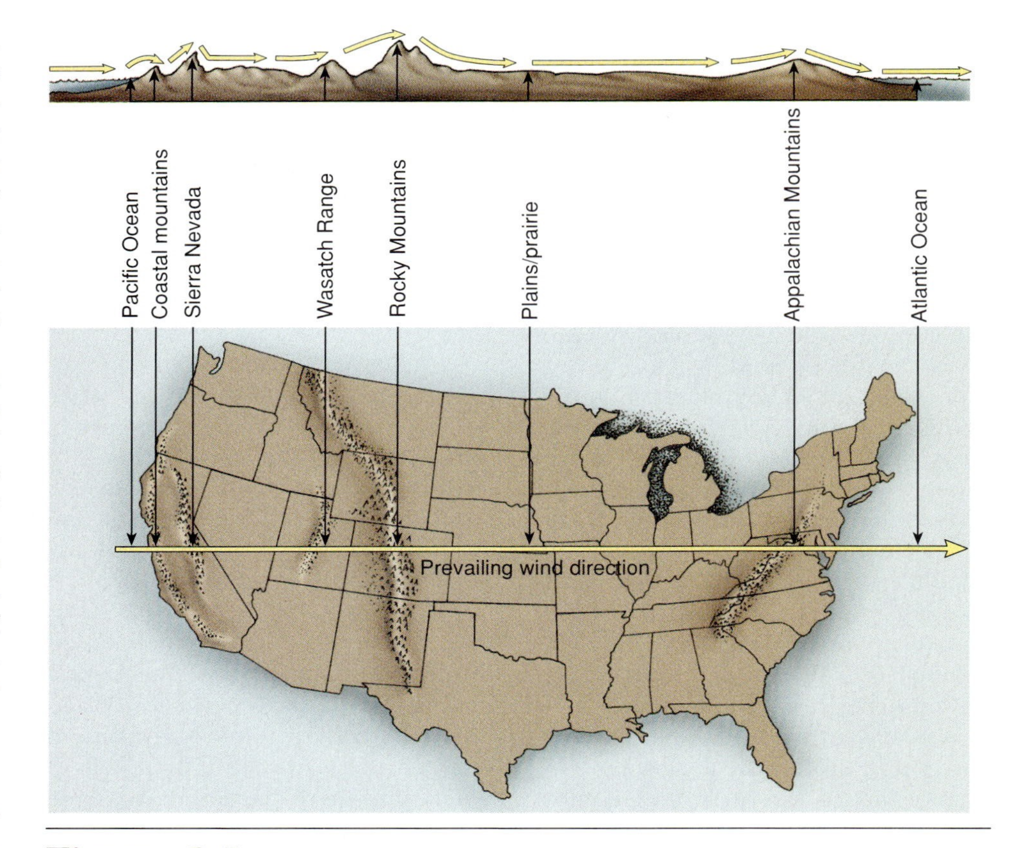

Figure 2.5

The annual prevailing winds are from west to east across the United States.

Temperature

Maximum and minimum annual ranges and **diurnal** (day/night) **fluctuations** in temperature produce another important set of climatic conditions to which plants must be adapted if they are to survive in a given area. Because of the relative position of the earth to the sun, the angle of the incident sunlight and the daylength result in the equator being warmer year-round than any other zone. As one moves north and south away from the equator, the annual temperature fluctuations become progressively more extreme (figure 2.6).

The sun's rays that strike the earth's surface most perpendicularly do not necessarily result in higher temperatures. This is because of another unique property of water. Water (including water vapor) gains and loses heat very slowly. An area of high humidity therefore has less extreme temperature fluctuations than do areas with low humidity. The maximum temperature for a particular tropical rain forest might be 30°C with an average diurnal fluctuation range of 5°C. In a desert, on the other hand, the maximum temperature might be 42°C and the diurnal fluctuation range 30°C, even though the equator gets the more direct rays of the sun.

Light

There are two important considerations when discussing light as a component of the climate affecting plant distribution and activity: both light intensity and daylength can have strong regulatory effects on plants (see chapter 14, p. 242).

Of the total radiation produced by the sun, light is only that portion known as the **visible spectrum** (figure 2.7). The intensity of this light varies with latitude and season, and it is greatest at the point of the earth most perpendicular to the rays of the sun at any given moment. Light intensity is progressively reduced from that point as the rays hit the earth's surface at greater angles. In addition, not all of the light emitted by the sun reaches the earth (figure 2.8).

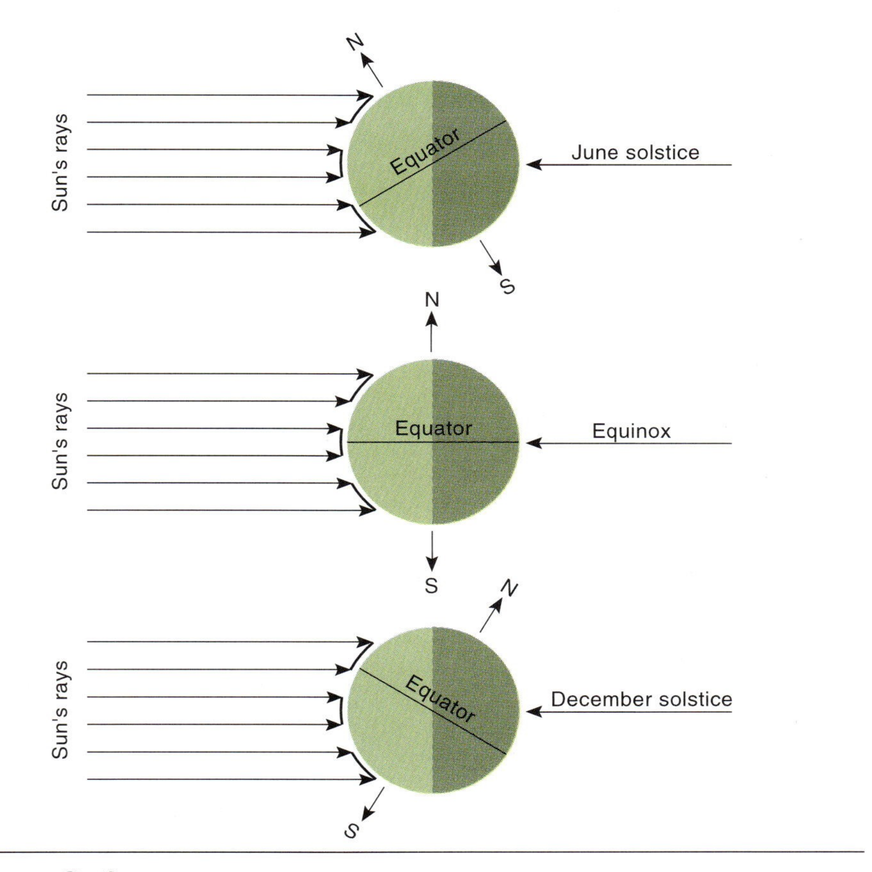

Figure 2.6
Since the sun is so far away, its rays can be considered essentially parallel. A unit of this radiant energy is dispersed over a greater area of the earth's surface whenever it strikes the earth at an oblique angle.

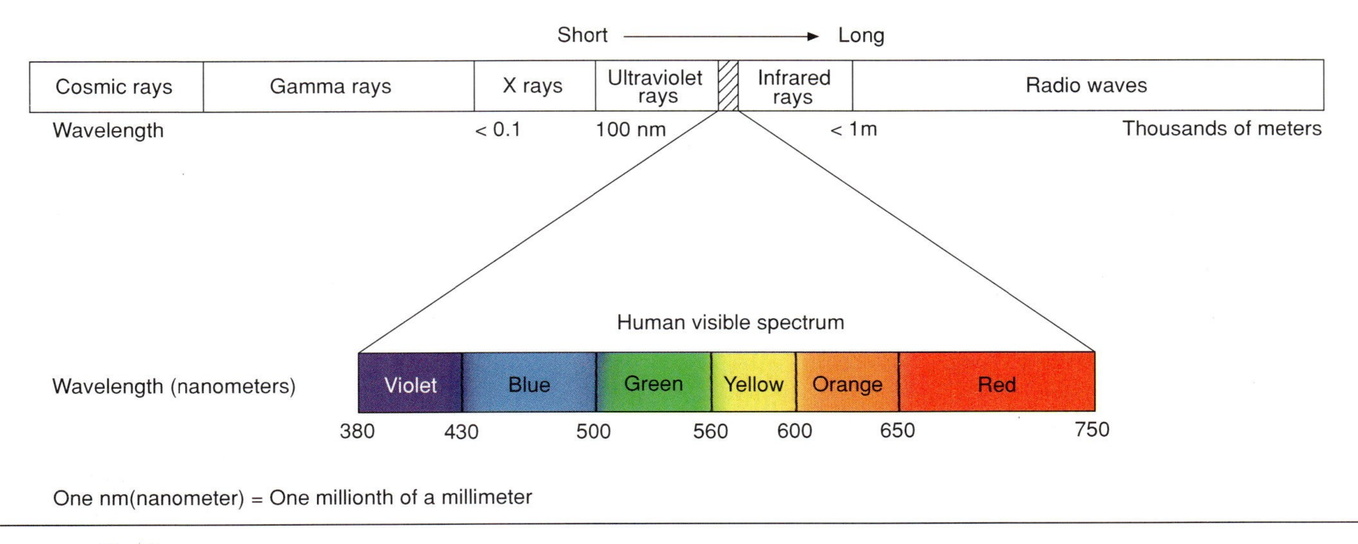

Figure 2.7
The portion of the total spectrum of light visible to humans is only a small range of wavelengths.

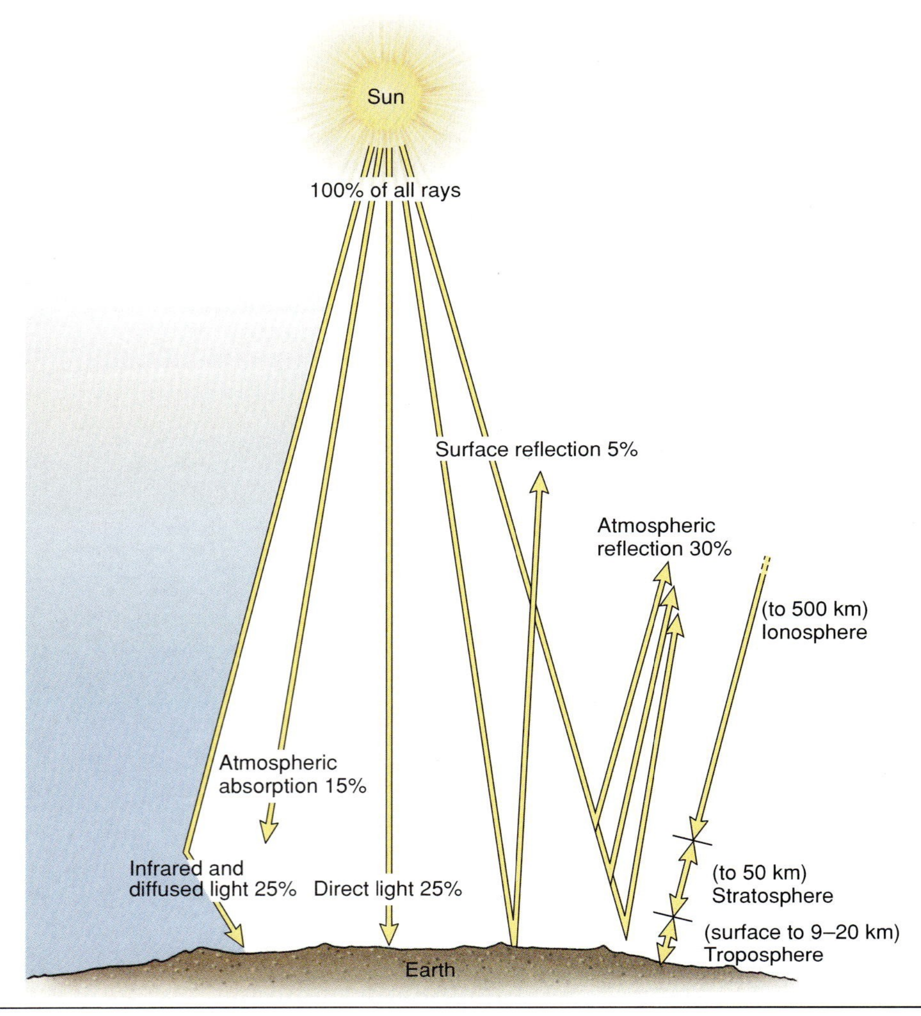

Figure 2.8
Only part of all the sun's rays reach the earth's surface, and some of that is reflected back into space.

Of the radiation that arrives at the outer limits of our atmosphere, only about half manages to get past the reflection and absorption by dust, clouds, water, and other gases, including the important ozone layer, which absorbs most of the harmful ultraviolet-B radiation. Some of the light reaching the earth's surface is reflected from rocks, water, and even plants themselves, whereas some is absorbed as heat or, in the case of green plants, converted to the chemical energy of carbohydrates (the process of photosynthesis, see chapter 13, p. 219).

More than half is in the infrared region of the spectrum and warms the earth. About 7% of the ultraviolet radiation manages to get past the ozone. The human eye sees light beginning at the short wavelength violet and then in successively longer wavelengths as blue, green, yellow, orange, red, and finally fading at the beginning of the infrared region (figure 2.7). Our eyes perceive sunlight as white light, simply because it is a mixture of all the colors of the visible spectrum—the colors of the rainbow. Yet specific wavelengths are absorbed and reflected differently; shorter wavelengths have more energy. Leaves are green because they have chlorophyll molecules that reflect green light while absorbing red and blue light. Thus any color appears to the human eye as that color, because light of that wavelength is being reflected. If all wavelengths are being absorbed, the object appears black; if all wavelengths are being reflected, the object appears white.

Limiting Factors

For a given plant species to successfully inhabit an area, it must have present in appropriate quantities all the necessary components of its environment. If any single requirement is insufficient, then the species in question will not be able to survive. Whichever environmental component causes the organism to fail is said to be the **limiting factor.**

Even though any of the requirements could theoretically be the limiting factor for a given species, only the one that actually prevents the species from surviving is so designated. Usually one of the abiotic factors, especially water or temperature, is the most likely candidate. For example, many plant species are limited from growing in a desert because there is not enough water. In fact, if irrigated, many of these plants could survive well in a desert. Others, however, would not be able to survive even with supplemental water because of the high summer temperatures. For them excessive heat would be the limiting factor. Other plants are limited from growing in more northern latitudes because of freezing winter temperatures that they cannot endure.

Many other abiotic and biotic (biological) components can also be limiting factors. Nutrients, space, soil characteristics, and interactions with other living organisms sharing the area are among the possibilities controlling plant success or failure. Other components of the total environment contribute to the success or failure of given organisms in various settings; they are considered in following sections.

THE BIOSPHERE

Our planet has often been referred to as "spaceship Earth," traveling through space with a critical atmosphere, critical energy input, and finite resources. Ecologists think of our entire earth as a giant system—the **biosphere.**

✔ Concept Checks

1. How do air circulation patterns and the concept of the condensation point of water produce worldwide patterns of precipitation?
2. How much of the sun's light rays reach the earth's surface, and of that, what do plants use?
3. List all of the things plants must have to be successful and decide which of those needs might be the limiting factor in your region of the country.

Any portion of the biosphere that represents a relatively closed system in terms of nutrient cycling, energy input, and therefore a definable set of plants, animals, soils, and climate may be defined as an **ecosystem.** It may be very small, as in a backyard pond with only a few organisms, or it may be very large, as in a deciduous forest or a grassland. The size of the ecosystem is in the eye of the beholder, and its limits must always be defined. It should be a relatively homogeneous area made up of living and nonliving components that have something in common; then the system can be studied as a unit. The ecosystem concept is useful for all biological study.

Within an ecosystem each species has a unique set of requirements, a spot in which it is better adapted than competing organisms. This set of specific requirements is termed the **niche** of that species. The niche of an organism should not be confused with the habitat in which the organism lives. The physical habitat is much more general and can include many different species, each with its own unique niche. In other words, the concept of niche includes every aspect of an organism's successful position within the habitat. The niche would include all the physical habitat characteristics of temperature, water, soil, and light plus all possible interactions with other organisms in the ecosystem. When two individuals of different species have some of the same specific requirements, they are sharing a part of the same niche. Such coexistence can theoretically occur for awhile if there are abundant resources available, but ultimately they will be in **competition** with each other for this specific need. The more similar the niches of these two species, the sooner they will find themselves in competition with one another.

It is easier to visualize animals than plants competing for the same resource, but the term is still appropriate. Plants compete for water, light, nutrients, space, and other components in their habitat by being better designed to acquire those needs. The depth or total volume of the respective root system, the vertical growth rate and leaf production, and the abilities to withstand climatic and herbivore stresses all enable one species to compete more successfully than another for common needs. Other interactions between individuals of different species are not competitive. In such situations each organism is still a component of the other's niche because it affects that organism's functioning in some way. If organism *A* benefits from its interaction with organism *B* but in the process affects *B* negatively, it is a parasitic relationship. For example, dodder (*Cuscuta*) is a parasitic vine that derives its nutrition from whatever host plant it lives on. The host plant ultimately dies in such a relationship.

Commensalism is an interrelationship in which one organism benefits and the other is unaffected. Spanish moss (*Tilandsia usnioides,* a bromeliad) is commensal on oak trees. When both individuals benefit from an interrelationship, they are said to have a mutualistic relationship. A classic example of mutualism is the lichen, an organism composed of a fungus and an alga. Any two or more dissimilar organisms having a close association, whether parasitic, commensal, or mutualistic, are said to be **symbiotic.**

In addition to plant-plant interactions, there are many plant-fungus, plant-insect, and plant-animal interrelationships that exemplify these systems. Some plant-herbivore (plant-eating animals) interactions significantly affect the plant's total environment. There are also plant-animal interactions in which the plant exerts a negative effect on the animal: there are toxic plants, plants with spines or thorns, and plants that provide shelter to one animal that in turn attacks herbivores that could damage its home. Some plants have more complex interrelationships and interdependencies with a wide variety of other organisms.

The likelihood of understanding all the possible niche interactions of any single species is improbable, so true comprehension of total ecological balance within even a single habitat therefore is a major undertaking. Studying selected components shared by the many species within a given area, however, is worthwhile and provides ecologists with some basis for comparison of general relationships. Even though the earth is extremely diverse in terms of numbers of species and numbers of individuals within a species, there are good reasons to consider the planet as a whole. Sometimes broad-scale conclusions about changes in gas concentrations in the atmosphere, increases or decreases in temperature, changes in precipitation patterns, and other serious problems can be viewed only on a global basis.

Ecologists have chosen to divide the world into a few major vegetation types based primarily on effective precipitation, temperature, and soils. These specific groups of ecosystems, representing recognizable types of vegetation that are remarkably stable with time (sometimes over hundreds or thousands of years) are called biomes (see chapter 3, p. 32).

Within a particular area in an ecosystem all the living components are collectively referred to as a **community.** The term may encompass a bit more than what we refer to as a human community, since from an ecological standpoint the community represents all the plants, animals, and microorganisms living together in an area. Within that community there may be groups of organisms, all of the same species, which constitute a **population.** A community might have a

Forest Islands and the Blue Jay

Loud, aggressive, nest-robbing avian huns are how the common blue jay is perceived by most of the general public. Although many of their qualities support this view, recent studies on large seed dispersal have shed new light on another side of these raucous birds. Actually, 150 years ago naturalist William Bartram wrote, "The jay is one of the most useful agents in the economy of nature, for disseminating forest trees. . . . " The jay's redeeming attribute of transporting and burying large nuts and tree seed is important in two ways.

During the last ice age, the North American glaciers pushed hardwood forest tree occurrence far to the south. Biogeographers have identified glacial refuges of trees such as oaks, beeches, and chestnuts in the southeastern United States and, until recently, assumed that the heavy-seeded trees revegetated slowly to the north as the glaciers receded. However, these scientists now know that oaks migrated quickly northward, even faster than some trees with light, wind-dispersed seed. Jays are credited as the agents of this rapid seed dispersal.

In the study of species diversity of fragmented habitats—landscape ecology—the number and range of species within a general habitat is often dependent on how isolated the fragments are, especially where the growth of cities and farms has led to the clearing of formerly vast hardwood forests resulting in small, isolated patches of trees. The biological connection between these "islands" of woodland is an important part of modern ecology. In such settings, ecologists now must study the sum of many separate parts rather than the whole. The dispersal by blue jays of large numbers of heavy seed, such as beechnuts and acorns, among and between these separate forest patches helps maintain plant diversity by mitigating the effects of isolation.

Amazingly, blue jays have an expandable throat and esophagus that allow them to transport as many as three white oak acorns or five pin oak acorns or fourteen beechnuts at one time. In the fall, jays make many foraging flights of up to five miles, caching many of these seed for later retrieval. Because of their generalist tastes, at one time or another, jays probably disperse all of the thirty plus species of oaks in the deciduous forests of eastern North America as well as native pecans, beechnuts, willows, and others. Jays disgorge this hefty load of nuts into a pile on the ground, then individually cache each one by pushing it into soft soil or working it down under grass. These caches are then covered with plant debris, which reduces the chances that small mammals will discover and eat them. Of course, migratory jays returning in the spring will not find all of their caches, and these buried and covered seed have an enhanced chance of germination and early seedling growth and establishment.

One recent study documented 50 jays transporting and caching 150,000 acorns in only 28 days. In addition, blue jays select the most viable nuts. In one sample, the germination of beechnuts cached by jays was 88% while a random collection from the same trees measured only 10% germination.

Thanks to the common blue jay, the fragmented remains of eastern hardwood forests have a much-improved chance of maintaining their genetic biodiversity. Conversely, blue jays benefit from having a high-energy food source in the form of large, oil-rich acorns and other hardwood tree seed. This is an excellent example of ecological interaction and mutualism in action.

population of bluegrass, a population of rabbits, a population of foxes, or a population of grasshoppers. Each population could have only the number of individuals that its **trophic,** or feeding, level and food supply would support.

This limit on the number of organisms that a given area can support without causing degeneration of the area is termed the **carrying capacity.** For example, the carrying capacity of a specific area of prairie may be 1000 field mice, 20 deer, and one coyote. In practical use, ranchers need to know how many sheep or cattle a given section can support without overgrazing the land, which could cause a smaller carrying capacity in subsequent years.

Many of the specific components of an ecosystem cycle through set phases within the normal functioning of the system. These cycles are all in balance at a worldwide level, but it is possible to throw them out of balance if the natural habitat is significantly altered. Overgrazing is only one of many such negative alterations possible.

OXYGEN–CARBON DIOXIDE BALANCE

The earth's atmosphere is made up of a mixture of nitrogen (N_2), oxygen (O_2), carbon dioxide (CO_2), water vapor (H_2O), and a number of other gases of lesser importance. Practically all these substances are critical for life processes, and the recycling of both oxygen and carbon dioxide is absolutely essential to almost all living organisms. The air around us contains only about 21% oxygen and 0.034% carbon dioxide. Essentially all of the remainder, about 78%, is nitrogen. The opposing processes of **photosynthesis,** which produces sugars, and **respiration,** which allows those sugars to be used for energy, also involve oxygen and carbon dioxide exchange (figure 2.9). As the sugars are being produced by the photosynthetic process from water, carbon dioxide, and sunlight, oxygen is given off into the atmosphere. At the same time, the respiration process consumes oxygen to "burn" sugars and release carbon dioxide and water. The reactions can be simply depicted as follows:

Photosynthesis

$$CO_2 + H_2O \xrightarrow{\text{sunlight}} \text{sugars} + O_2$$

(energy is stored in the sugars)

Respiration

$$\text{sugars} + O_2 \rightarrow CO_2 + H_2O + \text{ATP*}$$

*(a form of energy that carries out biological work)

The relative rates of photosynthesis and respiration, then, determine the amount of oxygen and carbon dioxide

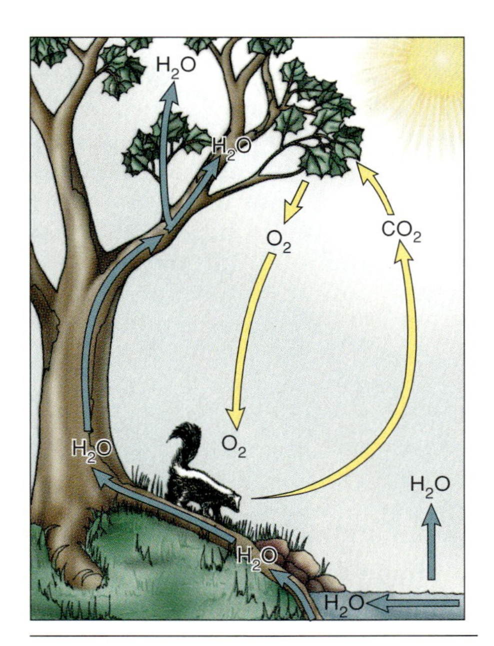

Figure 2.9
Oxygen–carbon dioxide balance. Animals consume O_2 and give off CO_2 through respiration; plants use water, CO_2, and light energy in photosynthesis, releasing O_2 into the atmosphere. Plants also respire.

in the atmosphere. All organisms carry on respiration and give off carbon dioxide, but only the green plants carry on photosynthesis and produce oxygen (figure 2.9). Green plants must therefore reach an acceptable balance between the two processes. If the amount of carbon dioxide released is exactly equal to the amount of carbon dioxide being consumed, the plant is said to be at the compensation point. Such plants cannot accumulate material and thus do not grow. In agriculture, for example, it is important that photosynthesis far exceed respiration if crops are to be productive. The balance of atmospheric oxygen and carbon dioxide is a critical interrelationship in which plants (which photosynthesize and respire) play the central role.

Photosynthesis, respiration, and other organismal interactions are not isolated events. Their interrelationships and their responses to environmental factors are complex, so a broad understanding of the entire system is necessary.

By 1772 Joseph Priestly had discovered that if an animal, such as a mouse, were placed in a closed container, it would die after a period of time. If a green plant were placed in the same container, and even if the container were glass so that sunlight could enter, the plant would also die. But if both the mouse and plant were placed in the container at the same time, they would coexist. If the gas balance (O_2 and CO_2) were correct, they could theoretically live in this condition indefinitely. A closed terrarium represents a similar system. Green plants within it consume carbon dioxide and synthesize sugars, giving off oxygen; the microorganisms, worms, and other animals living in the soil (or perhaps on the plants), and the plants themselves carry on respiration, consuming oxygen and releasing carbon dioxide back to the atmosphere of the terrarium. In effect, a balanced terrarium is an ecosystem.

CYCLING IN THE ECOSYSTEM

The ecological success of an ecosystem depends on its efficiency and stability. A great deal of that efficiency is related to the **cycling** of nutrients and water, which are essential components for all living organisms. If a single factor is missing, the entire system loses efficiency and slows down, and in a small system the balance may be permanently altered.

Organisms survive, grow, and reproduce because they have the energy, water, and nutrients to do so. *All energy for our entire biosphere is derived from the sun* (figure 2.10). This energy does not cycle, but it is converted from one form to another, eventually ending up as heat, which has little value in the overall functioning of the system. Because energy is "lost" as heat, there must be new input every day.

> ✔ **Concept Checks**
>
> 1. Define and distinguish between ecosystem, habitat, and niche.
> 2. Give examples of community interactions that are parasitic versus mutualistic.
> 3. What are the relative amounts of nitrogen, oxygen, and carbon dioxide in our atmosphere?
> 4. How do the overall processes of photosynthesis and respiration differ?

Figure 2.10
Sunlight penetrating a forest of redwoods in Redwood National Forest, California.

On the other hand, our supply of water and nutrients is finite. At any given time a portion of the water and nutrients is tied up in various parts of the system—in the air, the soil, the oceans, or the living or dead organic matter. As water and nutrients are transferred from one part of the system to another, a cycle is eventually completed and begins anew (figure 2.11). We derive part of our understanding of an ecosystem by knowing that water and nutrients cycle within certain physical boundaries. It makes no difference whether we are discussing a desert or a tropical rain forest; the manner in which nutrients cycle is exactly the same, although the rates may differ.

The Water Cycle

On the earth the total amount of water is enormous and essentially constant, but 99.4% of it is composed of

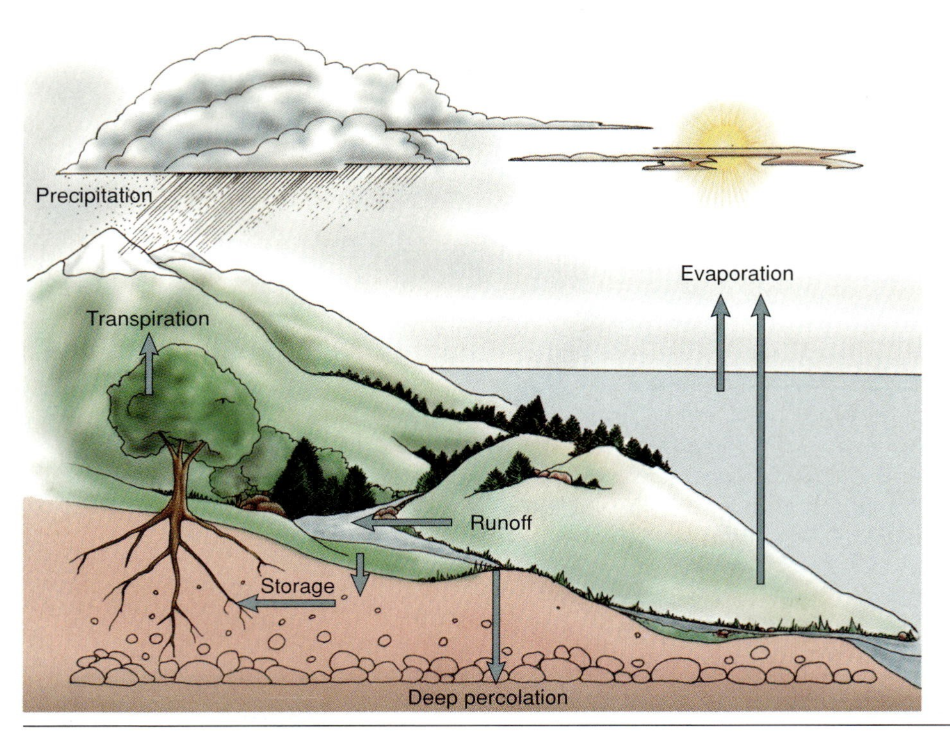

Figure 2.11
The water cycle. Water falls as precipitation and either evaporates directly, runs off, or percolates into the soil for plant use. About 99% of all water taken in by plant root systems is released into the atmosphere by transpiration (evaporation from plants). Some of the percolated water penetrates past root systems to replenish groundwater supplies.

salt water and ice found in oceans, inland seas, glaciers, and polar ice caps (table 2.1). The salt water provides a saline habitat for organisms adapted to those conditions, and it provides a reservoir from which pure water molecules can evaporate. The energy for this process is derived from the sun. As the surface waters of the oceans are warmed, evaporation occurs and water is moved into the atmosphere. When physical conditions are proper for condensation, clouds occur, and eventually precipitation is produced: rain, snow, hail, or sleet. As described earlier, air currents cause these cells of moist air to move around the earth, and precipitation may fall thousands of kilometers away from where the water was released by the ocean or landmass.

Some precipitation falls on land, a portion of it is evaporated back into the atmosphere, some goes to surface runoff (figure 2.12), some **percolates** through the soil to recharge groundwater supplies, and a portion is stored in the soil as a water source for plants and ultimately animals. The plants absorb the water from the soil, transport it through their cells, and eventually release it back into the atmosphere through transpiration, the evaporation of water from plant surfaces. About 99% of all water absorbed by the root system is given up this way. Only a small portion is stored in cells to be used for metabolism and maintaining water pressure, or turgidity. At any given time a very small fraction of the total precipitation is tied up in the living tissues of plants and animals.

Both surface and underground runoff water eventually returns to the oceans. Thus the problems in water management are not a matter of total supply but simply of having enough freshwater at the right place at the right time.

The Carbon Cycle

All **organic matter** contains carbon and hydrogen. That carbon is also cycled through the living/nonliving system in an orderly flow that allows for organic molecules to be constructed (figure 2.13). Carbon comes directly from the atmosphere as carbon dioxide. Earth's atmosphere contains about 0.034%, or 340 parts per million (ppm), CO_2. In comparison with the amount of nitrogen and oxygen, which dominate the air we breathe, this is a small percentage. It is remarkable that a gas so important should be present in such a small concentration.

In the process of photosynthesis CO_2 is taken into the green plant and first incorporated into sugars and then later metabolized into all of the organic

TABLE 2.1

World Water Budget

	TOTAL WATER IN BIOSPHERE (%)
Freshwater lakes	0.009
Saline lakes and inland seas	0.008
Rivers	0.0001
Soil moisture and surface runoff	0.005
Groundwater to 4000 m	0.61
Ice caps and glaciers	2.14
Atmosphere	0.001
Oceans	97.2279

Data from R. L. Nace, 1967, "Are We Running Out of Water?" in Geol Surv Circ. 536.

Figure 2.12
Runoff of water can produce scenic phenomena, as in Silver Falls State Park, Oregon.

Figure 2.13
The carbon cycle. Carbon in the form of atmospheric carbon dioxide (CO_2) is produced as a by product of respiration in plants and animals, the burning of fossil fuels and firewood, and even volcanic action. Both wild and cultivated plants use CO_2 for photosynthesis.

molecules important to life. In the normal process of respiration some CO_2 is cycled directly back into the atmosphere. Most of it is stored in living tissue, portions of it are consumed by animals in normal food chains, and some occurs in dead plant parts such as leaves, flowers, fruit, and seeds.

The level of CO_2 in the atmosphere is relatively constant, although there is considerable concern about its fluctuation. For example, there is good evidence that the level in the atmosphere has risen considerably during the past 100 years, primarily from the additional input from the burning of fossil fuels. In that period of time the level has increased from under 300 ppm to the present 340 ppm. At the current rate of increase a National Academy of Sciences study predicts that the concentration of CO_2 in our atmosphere will double by the year 2020. The oceans act as an effective buffer by absorbing most excess CO_2 as carbonates, but there is considerable controversy on how well the ocean's buffering capacity can counter the rate at which atmospheric CO_2 is increasing. If the CO_2 in the atmosphere were to ever rise dramatically, it would cause the sun's rays to be trapped at the earth's surface (the so-called **greenhouse effect**) and cause the temperature of the earth to rise. Studies indicate that only a 2° or 3°C increase in the earth's temperature would cause the polar ice caps to melt and result in serious ecological and economic damage. Such melting would raise the level of the oceans by as much as 5 m, according to some estimates, and significantly change the outline of our continents by submerging many millions of hectares of coastal land and many coastal cities.

Other sources of carbon released into the atmosphere include volcanic eruptions and the weathering of rocks, processes that are relatively stable over geologic time and about which we can do little (figure 2.13).

From the standpoint of the photosynthetic process an increase of CO_2 concentration in the atmosphere should not be a problem. In fact, CO_2 enrichment of the environment in greenhouses and other confined spaces is now practiced commercially to improve yields. The ecological consequences of doubled atmospheric CO_2 levels worldwide, however, would far override any benefit derived from additional photosynthesis in nature.

The Nitrogen Cycle

The air we breathe is about 78% nitrogen gas (N_2). As we shall see later, nitrogen is an extremely important element in the building of organic molecules, but plants and animals have absolutely no way of directly incorporating atmospheric nitrogen gas. Instead, plants obtain nitrogen as nitrate (NO_3) or ammonia (NH_3) from the soil. Once nitrate or ammonia nitrogen has been taken up by the plant, it is converted into organic matter and becomes part of living cells. If the plant is eaten by an animal, the organic nitrogen is partially converted into new organic matter in the animal. Whenever the animal or the plant dies, the organic matter is broken down by the **decomposers** into inorganic nitrogen in the soil. Most of this nitrogen will be recycled in the form of nitrate or ammonia; some will be volatilized into the atmosphere in a form that produces the pungent odor of decay. Bacteria present in the soil are capable of converting nitrate to ammonia, and vice versa. The cycle begins again as plants take up nitrogen from the soil (figure 2.14).

One way to provide these nutrients is to make them synthetically, usually involving methane, a fossil fuel. There was a time not too many years ago when fossil fuels were so inexpensive we felt that essentially all of our nitrogen fertilizer needs could be met this way. Now the energy crisis has forced us to look more seriously at natural **nitrogen fixation.** Many bacteria and cyanobacteria are capable of converting atmospheric nitrogen into inorganic nutrients in the soil that plants can use directly or that can be converted by other microorganisms into usable forms. They may do so as part of their independent, natural metabolism as free-living, nitrogen-fixing microorganisms, or in a mutually beneficial relationship in the roots of a higher plant as symbiotic, nitrogen-fixing bacteria. It has been known for many years that many legume (bean family) plants become infected with certain kinds of bacteria that cause nodules (tumors) on their roots (figure 2.15). The nodules become filled with bacteria, and the bacteria have the ability to fix nitrogen from the air into NH_3. In return for this "free lunch" the plant provides the bacteria with certain organic molecules necessary for the bacterial growth. These tumors do not harm the plant and, in fact, are considered highly desirable. Farmers routinely pull up alfalfa or soybean plants to see how many nodules occur on the root system and thereby judge the nitrogen nutrition of the plant.

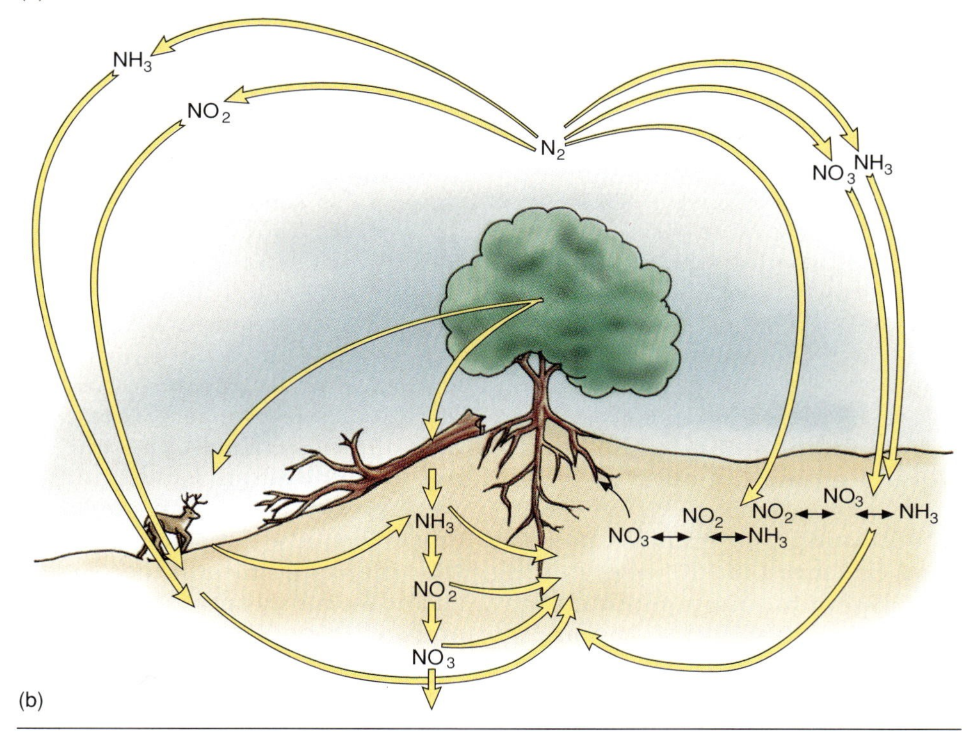

Figure 2.14

(*a*) The nitrogen cycle. Nitrogen occurs as atmospheric nitrogen gas (N_2), in the bodies of plants and animals, and in the soil as organic molecules from the decomposing action of soil microorganisms. (*b*) The cycling of the various compounds is shown schematically in relation to where each compound is located in (*a*).

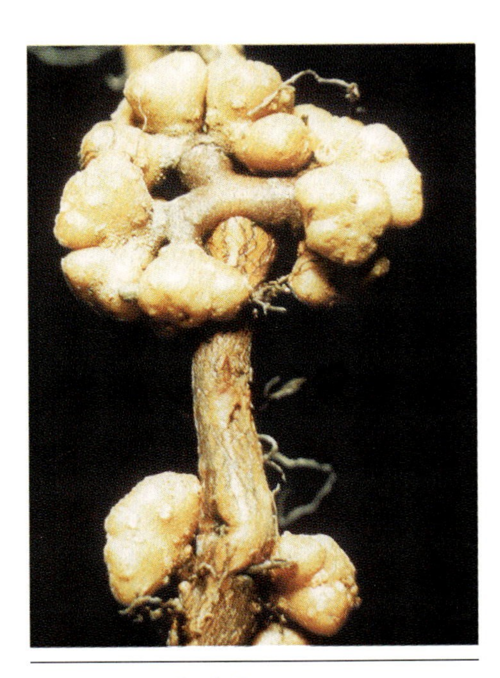

Figure 2.15
When members of the legume family are symbiotically infected by bacteria, root nodules form. In the nodules the bacteria convert atmospheric nitrogen to a form that can be taken up by plants. This is the process of nitrogen fixation.

In recent years it has been discovered that many nonlegume plant roots are infected by a different type of bacteria that do exactly the same thing. Such free-living bacteria and cyanobacteria appear to be far more important in nitrogen cycling than was originally believed. For example, the cyanobacteria that inhabit the water and soil of rice paddies in Asia are responsible for much of the nitrogen fertility of land farmed continuously for centuries. Most of these organisms have the ability to become dormant during periods of drought; then they grow vigorously in a matter of hours after receiving moisture, fixing nitrogen at a remarkable rate until water again becomes the limiting factor.

One other source of nitrogen from the atmosphere is important in certain regions. Electrical discharges during thunderstorms are capable of putting nitrite (NO_2) into the air; rainfall carries it to the ground, and bacteria oxidize it to nitrate (figure 2.16). The burning of fossil fuels may also put some ammonia and nitrogen oxides into the air. Ammonia is soluble in water and may be brought to the ground with precipitation.

Thus nitrogen cycles from the atmosphere by several fixation schemes to be captured in the soil or directly by the plant. It is taken up by the plant, fixed into organic matter, and later broken down by decomposers to the inorganic form where the cycle begins again.

These cycling nutritional materials combine with the abiotic components of the environment and the energy of the sun to control what kinds of plants grow where. Understanding these interrelationships increases our ability to grow them in other regions.

TROPHIC LEVELS

We refer to the organic matter present in organisms as **biomass,** an ecological description of the weight of the matter itself. We measure our biomass by simply weighing ourselves, and the biomass of a plant can be determined by weighing both aboveground and belowground parts. Such information is important in determining how efficiently the energy from the sun is being converted into the chemical energy of organic molecules.

Organisms that make their own food directly from sunshine, carbon dioxide, and water are called **autotrophic** (*auto,* self; *trophic,* feeding). All green plants are autotrophic. Other organisms, including humans, are said to be **heterotrophic** (*hetero,* other). All heterotrophs lack the ability to make their own organic molecules from inorganic substances and light energy. Not only do plants produce their own food, but they also produce tissues that can be used as food by animals. Therefore autotrophs provide food for heterotrophs; plants are the producers.

Food Chain

A food chain is a hierarchy of organisms in which the producer organisms, the green plants, are at the base (figure

Figure 2.16
Lightning can put nitrite (NO_2) into the air; rainfall can then carry it to the ground.

2.17). They are consumed by plant-eating animals (herbivores), which are in turn consumed by meat-eating animals (carnivores), or those which eat both plants and animals (omnivores). An example of such a simple food chain is a grass plant (producer) that is eaten by a cow (herbivore) that is eaten by a man (omnivore). Consider that many grass plants are required to feed one cow, and the number of grams of grass required for conversion to 1 gm of beef is fairly large; in fact, the ratio is about 9:1 at best. Similarly, it takes at least 9 gm of beef to provide 1 gm of weight gain for an actively growing human child. (The relationship becomes much more complicated for organisms that are fully grown.) What is apparent in this food chain is that it is indeed oversimplified. Cows

✔ Concept Checks

1. What is the ultimate source for all energy in our biosphere?
2. When precipitation falls to earth, what are the different things that might happen to it?
3. All organic matter contains what elements?
4. What is the greenhouse effect?
5. Describe nitrogen fixation.

may choose many different kinds of plants for forage, and humans eat many foods other than beef. A great deal depends on availability. (Many humans in the world have few choices.)

Food Web

For most organisms there is a choice of foods, and therefore it makes more sense to think of a **food web,** a hierarchy of consumption in which alternate food sources are represented (figure 2.18). The number of organisms available becomes very important because the total energy loss at each **trophic level** (actually a feeding level) is approximately 90%. Thus only about 10% of the total energy is incorporated as biomass at each trophic level. Since the green plants are at the first trophic level and are called producers, the herbivores that feed on them at the second trophic level are called primary consumers. Animals that eat the primary consumers are at the third trophic level and are called secondary consumers. The hierarchy continues until one reaches the top trophic level, occupied by the top carnivore. Nothing feeds on the top carnivore except scavengers, which consume the flesh after the top carnivore dies. Finally, decomposers (mostly bacteria and fungi) break down the organic matter into simple inorganic nutrients, which are cycled back to the soil, later to be absorbed by new plants, which start the cycle all over again.

A complex food web (figure 2.18*a*) is one in which the organisms at each level eat many different species available to them. Such flexibility of food sources develops in nature out of necessity. In unstable or unpredictable environments, where a given plant species may not be sufficiently available (or even at all) in a particular year, animals that would normally feed on that plant must feed on another species or fail to survive. These interrelationships are true at the secondary and tertiary consumer levels of the food chain. A desert is an example of such an environment because the precipitation is so unpredictable, and thus plant growth is not always ensured (figure 2.19).

A simple food web, on the other hand (figure 2.18*b*), is one in which a high degree of specificity exists in food sources. In many cases given animals feed on only a single plant species. (For example, koalas feed only on *Eucalyptus* leaves.) In turn, given carnivores may depend exclusively on a single species of animal as their sole food source. Such extreme specialization could have evolved only in a very stable environment, where each plant and animal species is always present in essentially the same abundance. Because these animals have never needed to search for and successfully compete for other food sources, they have lost the ability to do so. Their continued existence, therefore, is linked to the future stable supply of their food source.

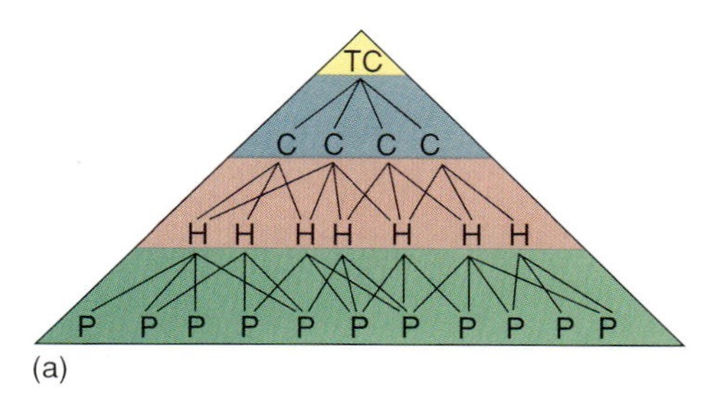

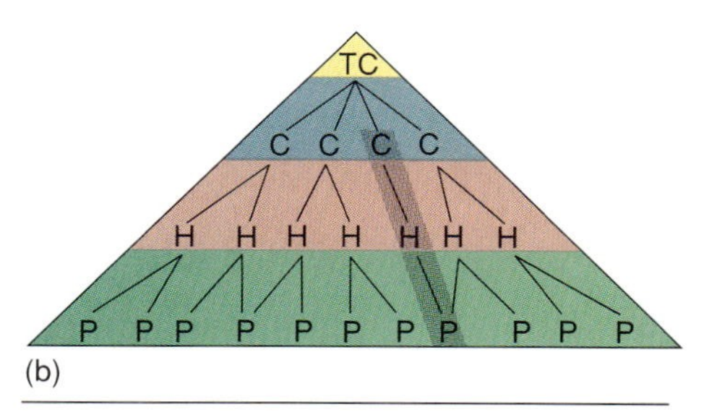

Figure 2.18
Food webs more accurately depict the feeding relationships between the producer plants (*P*), the primary consumer herbivores (*H*), the secondary consumer carnivores (*C*), and finally the top carnivore (*TC*). (*a*) Complex food web. (*b*) Simple food web. Shaded area represents the delicate balance and interdependence possible in some ecosystems.

Probably the most predictable and stable environments exist in the tropical rain forests. In these regions many highly specific ecological interrelationships exist—simple food webs. The elimination of any single species in such environments could well cause a chain reaction affecting several other species (figure 2.18*b*). That we are unaware of many of these specific interdependencies is one of the main reasons ecologists and conservationists concern themselves with protecting known endangered species and disappearing habitats.

Interestingly, another of the most predictable (stable) of the world's environments are the oceans, and yet a very complex food web exists there. The base of the aquatic food chain consists of single-celled algae generally referred to as **phytoplankton** (*phyto,* plant; *plankton,* small, free-floating aquatic organisms). These are in turn eaten by **zooplankton** (*zoo,* animal), single-celled animals that are then eaten by larger aquatic animals. The food source here is governed almost entirely by size and agility in avoiding

Figure 2.17
A food chain. Plants are the base of the chain, herbivores (mouse) eat plants, and carnivores eat animals. The snake is a first-level carnivore, the hawk a second-level, or top, carnivore.

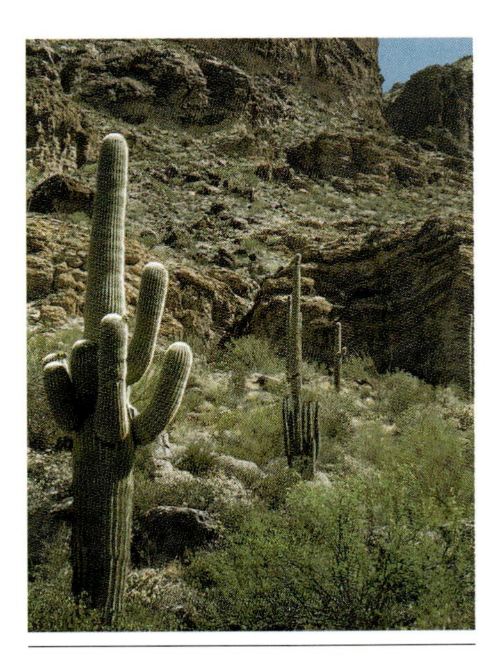

Figure 2.19
Deserts have unpredictable environments; thus, complex food webs develop. Saguaro cactus is a unique component of the Sonoran Desert.

capture. Flavor, texture, and other factors that we find important in foods apparently have little to do with fish diet. The little fish must be large enough to be an enticement to the larger fish, but it must not be so large that it cannot be swallowed.

Food Pyramid

The best way to depict these energy relationships is three-dimensionally in the form of a pyramid (figure 2.20). The broad base of the pyramid is made up of the first trophic level, the producer organisms with a very large biomass (generally, that means a very large number of individual plants). The second trophic level is represented by the primary consumers, the herbivores. Fewer individuals and less biomass occur here because the energy conversion efficiency is only about 10%. In other words, 90% of the plant tissue eaten by a herbivore does not go to make animal tissue. Much of the energy lost is in the form of heat, some as indigestible waste that goes to the decomposers. At the third trophic level, occupied by the carnivores, the biomass is again reduced by some 90%, and the process continues to the top carnivore. At any trophic level decomposers take their toll as some organisms die and are recycled to the soil and air (figure 2.21). Even the green plants lose leaves, flowers, and fruit during growth, and the decomposers break them down. A compost pile is an example of this principle (see chapter 12, p. 212).

The pyramid concept is used to demonstrate that very little energy and biomass are left after only three or four trophic levels. Thus food chains are never very long; the energy losses are simply too great. The lesson to be learned from this generalization is that more food is available closer to the base of the pyramid. Organisms that feed at the producer (green plant) level have access to a great deal more food than organisms that feed only on meat. The food supply for omnivores such as humans is greatly enhanced if we eat plants rather than animals. If we are willing to shift our diet so that plants constitute a greater portion of our total intake, we have access to more food, and the energy costs are much lower. High-priced meat is rapidly bringing this lesson home. This is not to say that all humans should become vegetarians, but modifications in diet toward more plant foods would mean additional food for a hungry world.

In summary, the food pyramid clearly shows the impossibility of having a larger top than can be supported by the base. The plant material on earth can support only a finite amount of animal life, including humans. It would be foolish, therefore, to presume that human populations can increase beyond a given size without proportionately increasing the available food base. This, of course, is what modern agriculture attempts to do. Continued pressure for ever-increasing productivity requires modifications in the habitats of the crop plants to allow production beyond what natural conditions would yield. Such modifications are not always in the best interests of the surrounding environment. In later chapters we will look more closely at energy flow and changes in energy form with plants as the all-important first step in the sequence.

ECOLOGICAL SUCCESSION

Ecological succession is the sequential replacement of one community type by another through a series of developmental stages. These are known as

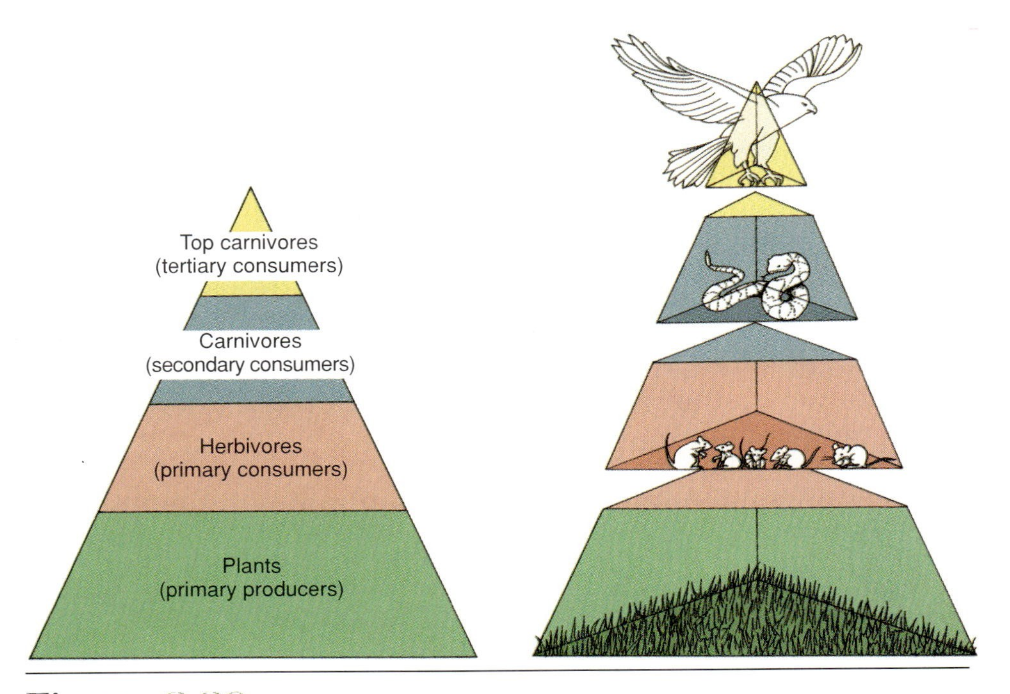

Figure 2.20
Food pyramid. The reduction in total biomass at each trophic level is depicted by a three-dimensional pyramid.

Figure 2.21
Decomposing log on the forest floor.

seral stages until the final community structure is reached. This last stage in succession is called the climax community because it is the optimum assemblage of species that the environment can support in that area. The measure to determine whether a climax community has been reached is stabilization of the dominant species; when those species begin replacing themselves rather than being replaced by a new species, the climax community has been achieved.

Although the environment determines the community structure as the process of change occurs, the environment does not *cause* that change; that is, succession occurs in spite of the fact that the climatic patterns remain the same. The change from one community to the next is actually brought on by the modifications produced by each temporary community. As the dominant species of each seral stage alter the area in which they are growing, they actually produce conditions less favorable for themselves and more favorable for a new assemblage of species.

When succession occurs in a new or pristine habitat or in one that has not previously had a similar community occurring there, it is called primary succession. When an area has at one time in the past already been occupied by a similar community, it is secondary succession. The classic example of primary succession is the normal transition of a pond to a bog and then to a woodland as it slowly fills with silt and organic material. The south shores of Lake Michigan have been undergoing primary succession for many years as the lake slowly retreats to the north.

Secondary succession occurs when land that has been cleared for pasture or for farmland is no longer maintained. Such abandoned fields slowly revegetate with plants native to the area. The actual species vary with the region, but the first year finds annual weeds dominating the site. In the second year perennial grasses and some herbaceous perennial broadleaf species join the annual weeds. For the next several years the grasses dominate, but an ever-increasing number of shrubs begin to appear, as do some tree seedlings. While the trees are slowly reaching maturity, the shrubby species and grasses codominate. The fast-growing tree species shade out the shrubs and grasses as they become large enough to form a dense forest, but they might be replaced in time by slower growing shade-tolerant deciduous tree species. If a deciduous forest is the dominant climax vegetation, a new group of shrubby plants take their place in the understory, maintained by the fact that deciduous trees do not form a shade canopy over them all year long.

Primary succession in ponds results from the slow filling of the body of water with silt and organic debris until plants can gradually invade from the banks toward the middle. As soon as submerged plants are able to root in the mud near the edge of the pond, the buildup of silt occurs much more rapidly, trapped and held by these plants. As the bottom continues to fill, floating surface plants such as water lilies become common near the banks, and the submerged plants move farther toward the middle of the pond. Next reeds, cattails, and similar rooted emergent plants become established in the deeper accumulation of sediment in the shallow water near the banks of the pond. The filling of the pond occurs much more rapidly now with plants occupying the entire area. Eventually true terrestrial plants establish along the original shallow zones of the pond, now dry land. The center of the rapidly filling pond becomes smaller and smaller, with an accompanying succession of seral vegetative zones, until it is a bog and finally dry land.

Generally, successional events share several common characteristics regardless of the wide variety of localities and plant species involved. Both the number of plant species and the rate of species replacement are higher in the initial stages, slowing and stabilizing in the older stages. In addition, the size of the plants and the total biomass increase through the seral stages until the climax community is reached. Finally, the food webs become more complex as the seral stages move toward climax.

Recolonization

Occasionally, a natural phenomenon totally destroys life in an area. The eruption of a volcano in a vegetated region and a devastating forest fire are examples of such phenomena. Given time, all of these areas will again become revegetated through the process of **recolonization** and succession.

An extreme example of recolonization occurred on the Pacific island of Krakatau. In 1883 a volcanic eruption destroyed almost half the island and covered the remainder in a thick layer of lava pumice and ash. All forms of plant and animal life were eradicated, but as soon as the substrate cooled, the process of recolonization began. Single-celled algae quickly established in pockets of rainwater caught by the lava folds. Slowly, organic materials combined with ash in these pockets to provide a shallow layer of soil sufficient for hardy species of vascular plants. Gradually, more and more plant species reestablished, and as they did they helped provide more humus for an ever-improving habitat for a greater diversity of life forms. By 1934, only 51 years later, over 270 species of plants occupied the island with a commensurate number of animal species also in full residence.

The eruption of Mount St. Helens on May 18, 1980, in southern Washington,

(a)

(b)

(c)

Figure 2.22
(*a*) Steam being released just prior to the eruption, May 18, 1980. (*b*) Mount St. Helens after the eruption on June 28, 1980. (*c*) The blast of the eruption devastated the forested slopes for large areas around Mount St. Helens.

produced large areas denuded by lava flow and ash (figure 2.22). The recolonization of this area is being monitored carefully by ecologists. Undoubtedly, succession to a climax forest community will take a long time, but the early seral stages of recolonization are already well established. This is also true of the recovery of Yellowstone National Park following the devastating fires that raged through the park in 1988. The formation of a new volcanic island would undergo the same sequence of biological habitation, but technically this would be an example of **colonization** rather than recolonization. The island of Surtsey was formed in the 1960s off the coast of Iceland by volcanic eruptions (figure 2.23). Within only a few months, living organisms had colonized the still warm lava.

Figure 2.23
The volcanic island of Surtsey, off the southern coast of Iceland, was born Nov. 14, 1963 and reached its final size of 2.8 km^2 in 1966. Named for Surt, the god of fire, this island is being studied by biologists interested in the colonization process.

✔ Concept Checks

1. What are the four basic trophic levels of the food chain?
2. Why is a food web a better way to depict feeding level interrelationships?
3. Why is there less biomass at each successively higher level of a food pyramid?
4. Describe the process of natural succession.

Summary

1. The term *ecology* implies a thorough understanding of the biotic and abiotic components of the environment. Organisms exist where they do in nature because the combination of these factors is appropriate for their existence.
2. Precipitation is the single most important factor in determining plant distribution. The annual distribution of precipitation is dictated by a complex pattern of air circulation cells, which distribute moist air to certain parts of the world. These cells are modified by elevational changes and collisions with other air masses.
3. Temperatures are modified by the amount of water in the atmosphere. Moisture-laden air heats and cools more slowly than dry air. The equator is warm year-round because of the incidence of angle of the sun's rays striking the earth.
4. Only a small portion of the entire radiation spectrum emanating from the sun is perceived by humans as visible light. Fortunately, most of the harmful ultraviolet radiation is absorbed by the ozone layer in the outer layer of the earth's atmosphere and does not reach the surface. Light in the red and blue portions of the spectrum excites the chlorophyll molecule to begin the energy capture process of photosynthesis.
5. Any of the abiotic or biotic factors critical to continued existence of an organism could also be the limiting factor for it.

6. Within the earth's total ecological system, the biosphere, balanced ecosystems include many different physical habitats. Plants occupy a specific niche within a given habitat, and competition for resources occurs when different species have similar requirements. Plants and animals also develop specific interrelationships with each other; parasitism, commensalism, and mutualism are very common associations.
7. The earth's atmosphere is primarily nitrogen, oxygen, and a small amount of carbon dioxide. Water vapor also occurs in varying amounts. Photosynthesis puts oxygen into the atmosphere and uses carbon dioxide, whereas respiration uses oxygen and produces carbon dioxide.
8. Water and nutrients important to plant and animal growth cycle through the ecosystem. At any given time they may occur in living or dead plant and animal tissue as organic matter, or they may exist in the soil or atmosphere as inorganic chemicals. Carbon and nitrogen both cycle through such stages. Carbon dioxide levels are critical to the carbon cycle; decomposition plus nitrogen fixation provide the necessary nitrogen for plant growth. The energy from the sun does not cycle but must be constantly replenished.
9. Organisms that obtain energy directly from the sun are called autotrophs or producers. These green plants are eaten by herbivores, which in turn are eaten by carnivores. A group of organisms that represents the producer-herbivore-carnivore sequence is called a food chain, or more commonly a food web. The biomass relationships are best represented in the form of a pyramid.
10. The natural change in community structure over time is termed *succession*. This process results in the climax community. Primary succession and secondary succession have different points of origin but proceed in a parallel manner. Recolonization is the successional replacement of organisms in a habitat denuded by a natural disaster such as a volcanic eruption.

Key Terms

Discussion Questions

1. How important are plants to the ecological stability of the earth?
2. Are humans part of the natural ecological balance of earth? Argue both answers.
3. How does an understanding of inefficiency in the food chain relate to world hunger?
4. Can planet Earth recover from the imbalances we have caused?

Suggested Readings

Alling, A., M. Nelson, and S. Silverstone. 1993. *Life under glass: The inside story of Biosphere 2.* Biosphere Press, Oracle.

Ambroggi, R. P. 1980. Water. *Sci. Am.* 243(3):100–117.

Carpenter, S. R., and J. F. Kitchell. 1993. *The trophic cascade in lakes.* Cambridge University Press, Cambridge.

Cole, J. A., and S. J. Ferguson. 1988. *The nitrogen and sulphur cycles.* Cambridge University Press, Cambridge.

Dunnett, D. A., and R. J. O'Brien, eds. 1992. *The science of global change: The impact of human activities on the environment.* American Chemical Society, Washington, D.C.

Golley, F. B. 1993. *A history of the ecosystem concept in ecology: More than the sum of the parts.* Yale University Press, New Haven.

Huggett, R. J. 1991. *Climate, earth processes, and earth history.* Springer-Verlag, New York.

Joseph, L. E. 1990. *Gaia: The growth of an idea.* St. Martin's Press, New York.

Lovelock, J. E. 1990. *The ages of Gaia: A biography of our living earth.* Bantam Books, New York.

Oelschlaeger, M., ed. 1992. *The wilderness condition: Essays on environment and civilization.* Sierra Club Books, San Francisco.

Peters, R. L., and T. E. Lovejoy, eds. 1992. *Global warming and biological diversity.* Yale University Press, New Haven.

Schopf, J. W., and C. Klein, eds. 1992. *The Proterozoic biosphere: A multidisciplinary study.* Cambridge University Press, Cambridge.

Smil, V. 1991. *General energetics: Energy in the biosphere and civilization.* Wiley, New York.

Walker, D. 1992. *Energy, plants and man.* 2nd ed. Oxygraphics, Brighton.

Chapter Three

Biomes

Terrestrial Biomes

Tropical Rain Forests
Savannas
Deserts
Grasslands
Temperate Deciduous Forests
Coniferous Forests
Tundra

Aquatic Biomes

Marine
Freshwater

Wetlands

With a base of ecological principles and the importance of abiotic factors in the distribution of vegetation on earth from chapter 2, the more specific groupings of plants into worldwide biomes is a logical next topic. While most of us recognize that different areas of the country are vegetatively different from each other, the natural grouping of those floristic areas with similar ones in other parts of the world is not a concept that is immediately apparent.

In this chapter, the major terrestrial and aquatic ecosystem types are presented as components of worldwide biomes. The climatic and geographical factors that control the distribution and floristic composition of each biome are presented and unique ecological associations are discussed. In addition, the effect of human activity in many of these areas is evaluated.

Lycopodium growing on the moist, nutrient-rich humus of the forest floor.

Biomes are worldwide groups of similar ecosystems. An ecosystem is a balanced and self-perpetuating assemblage of all the living organisms and the nonliving environmental factors in a given area. Ecosystems differ in size according to the geographic limits and climatic ranges that control the kinds of living organisms within them. In North America, for example, there are four ecologically distinct deserts, or four desert ecosystems (figure 3.1). Each occurs in a different geographical setting where latitudinal and elevational differences combine with proximity to mountain ranges and coastlines to produce four unique sets of indigenous plant and animal species. Although different, these organisms are all typical of deserts in general. The Chihuahuan, Sonoran, Mojave, and Great Basin Deserts of North America, the Sahara, Gobi, Namib, Patagonian, and all the other deserts of the world make up the desert biome. The plant types and the abiotic environmental factors common to all these otherwise unique ecosystems are the basis for describing this biome.

TERRESTRIAL BIOMES

Only one fourth of the earth's surface is land, and yet the vast majority of all plant and animal species are found there. Continents range from equatorial to polar latitudes and from sea level to more than 9000 m in elevation. Climatic and soil differences, combined with ranges in latitude and elevation, result in a phenomenal number of ecological settings for plants and animals to inhabit. Even though this variation results in many different plant types, reasonably accurate and useful assemblages or categories according to vegetation can still be recognized (figure 3.2). The number of distinct terrestrial biomes varies according to the authority, and boundaries between adjacent biomes are often variable and transitional, but the seven described in this text are the most common and fewest groupings normally used. In addition, there are several smaller subunits that can be validly recognized within many of these biomes.

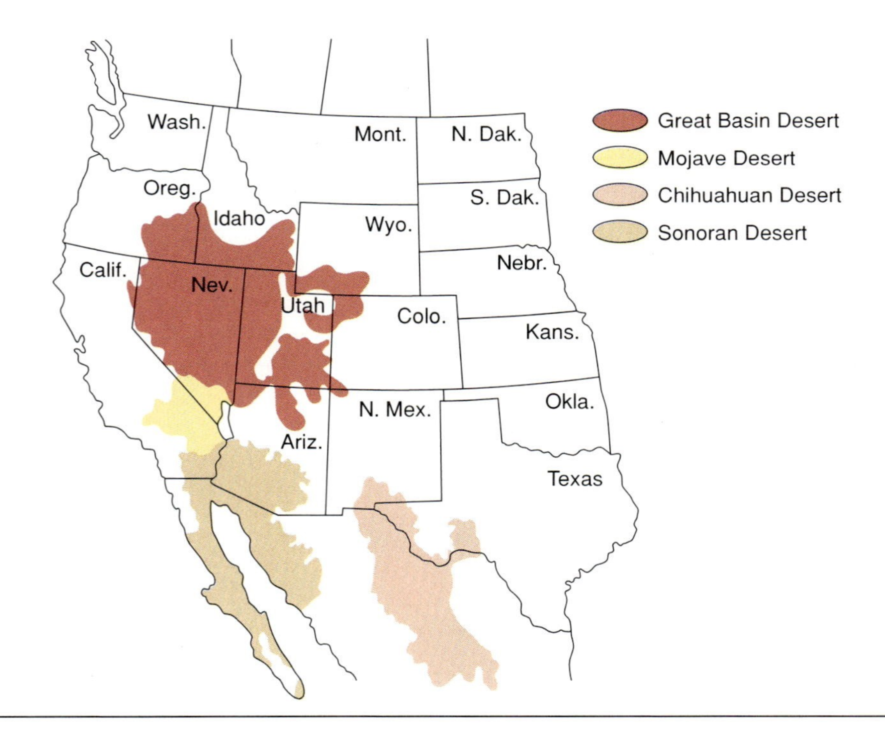

Figure 3.1
The deserts of North America.

Tropical Rain Forests

Found predominantly at or near the equator, **tropical rain forests** are characterized by having 200 to 500 cm of precipitation per year with some areas occasionally having over 1000 cm in a year (figures 3.3 and 3.4). Because of the equatorial location, there is no seasonality but rather a continual growing season with no cold periods. The daily temperature fluctuations are minimal because of the insulating effects of water, which gains and loses heat very slowly. Maximum daytime temperatures of only 30°C are typical, but the heat is oppressive due to the high humidity. Diurnal temperature fluctuations are on the order of only 5°C.

This consistently warm, moist climate with no extreme temperature fluctuations and no cold season provides a stable and favorable environment for plant growth. The vegetation is dominated by tall (50 to 60 m), broadleafed evergreen trees that branch near the crown, forming a solid layer, or **canopy,** of leaves. Because of the density of the trees in the forests, light availability is the primary limiting factor to plant growth below the canopy. Competition for light has resulted in tall, fast-growing trees that always have leaves on them.

The tree canopy shades the forest floor so that smaller trees and herbaceous plants cannot survive; therefore the floor is open, dark, and damp. Where the canopy is broken and light penetrates, a dense undergrowth of plants results. This may occur at the banks of a river, where an old and diseased tree falls, or in areas where the forest has been cleared.

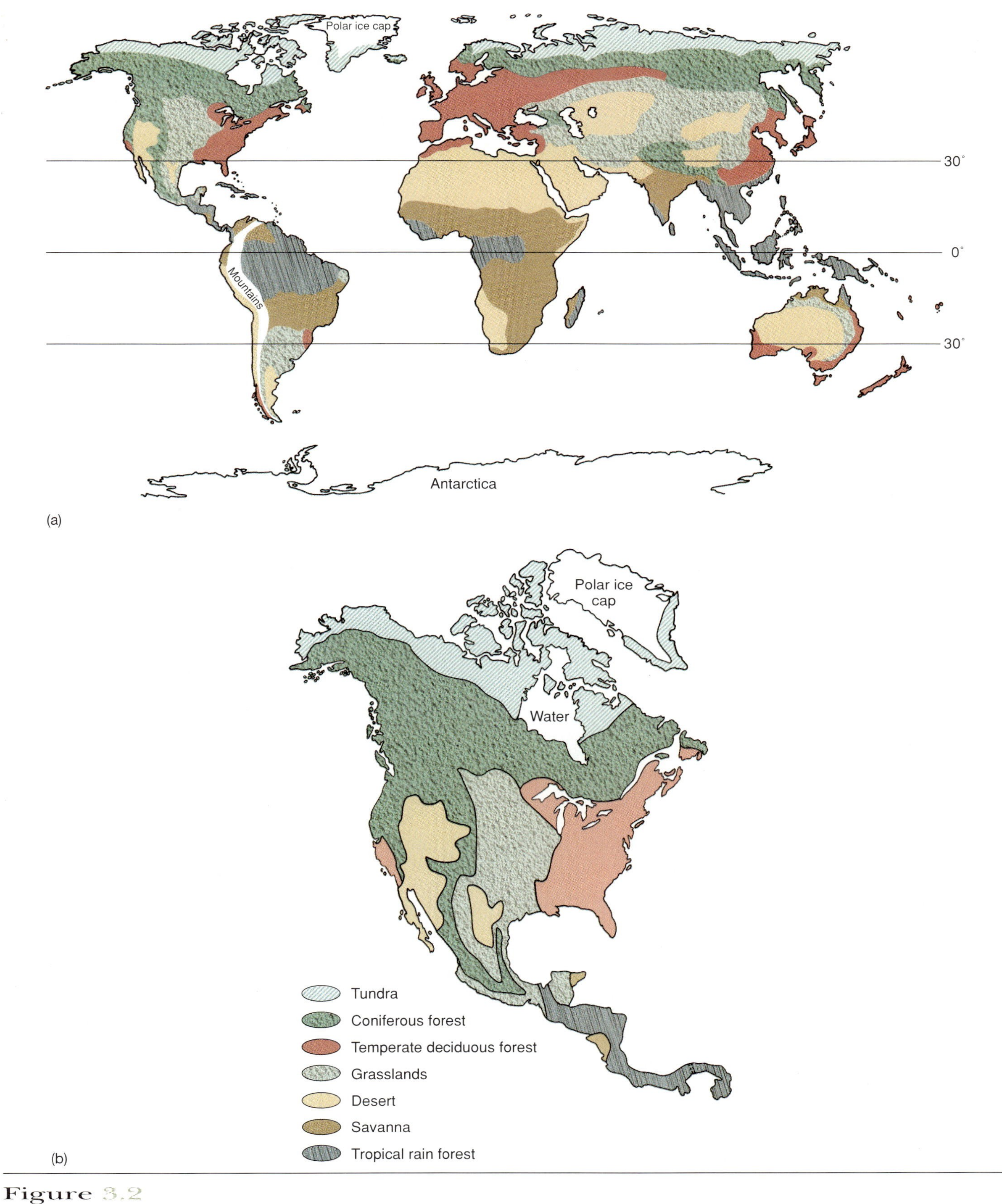

Figure 3.2
(*a*) The vegetative biomes, world view. (*b*) The vegetative biomes, North America.

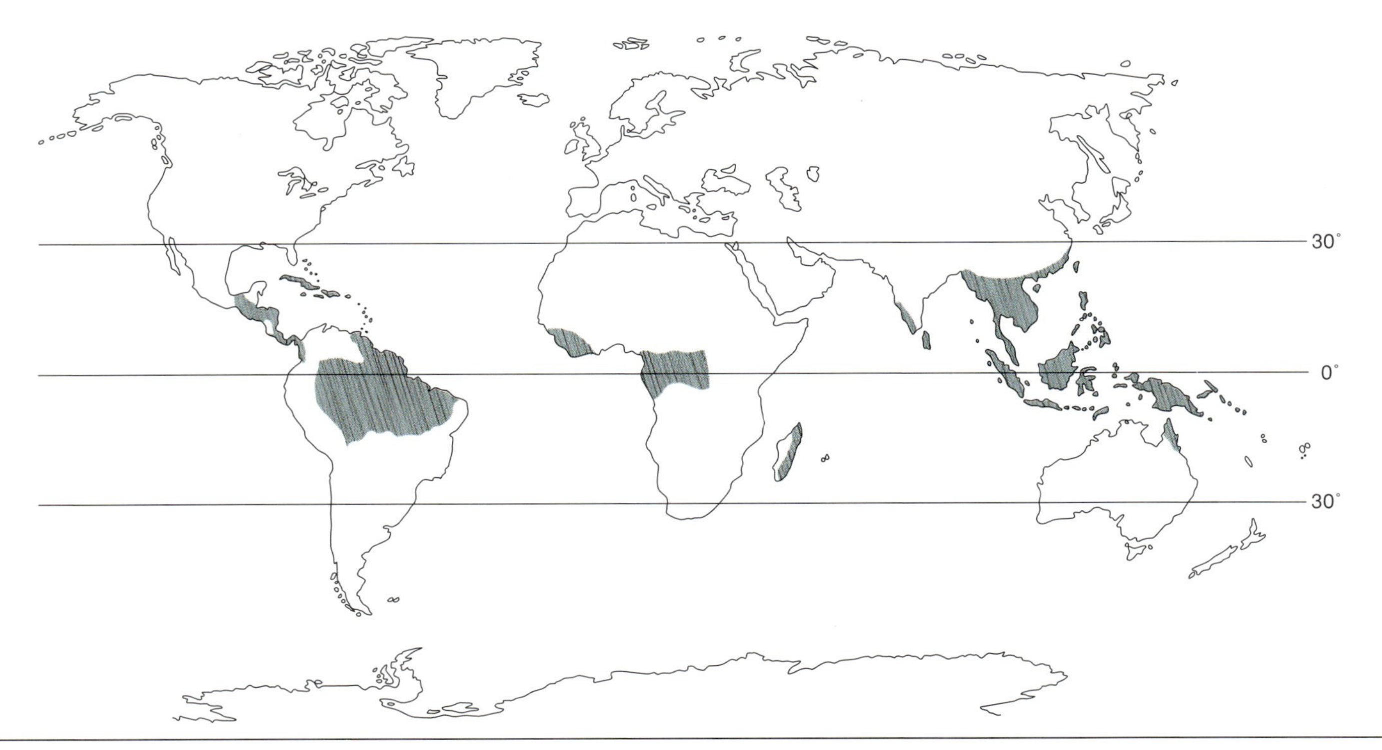

Figure 3.3
Tropical rain forest biome, world view.

Figure 3.4
Tropical rain forest biome. (*a*) Alakai Swamp, Kauai, Hawaii. (*b*) Oaxaca, Mexico. (*c*) Yucatan, Mexico.

The tropical rain forests are the oldest vegetative biome because their equatorial position has shielded them from the effects of past periods of glaciation. This great age (some 200 million years), combined with climatic stability and abundance of all resources, has produced phenomenal diversity in plants and animals native to these areas. Although tropical forests are relatively unexplored and poorly understood biologically, there are still more species of plants and animals described from this biome than in all the other biomes combined. It is estimated that two thirds of the world's approximately 265,000 plant species occur in the tropics—some 180,000 species.

Tropical diversity is exemplified in plant groups that have developed unusual methods of survival. **Epiphytes** are plants that grow with roots attached to another plant (figure 3.5*a*). These plants do not harm their benefactors in any way; they are not parasitic, but rather coexist with them. Bromeliads and orchids include many common epiphytes. Long **lianas** (hanging vines) use the trunks and branches of the tall forest trees to climb into the canopy, where they produce leaves for photosynthesis (figure 3.5*b*). Since these vines are rooted, they obtain water and nutrients from the soil. They often grow from one tree to the next, trailing long looping sections between trees. Many native primates travel throughout the forest by swinging from tree to tree without touching the ground. Entire communities of animals inhabit the forest canopy; exotic and beautiful plants, birds, insects, and other organisms are all part of this spectacular biome. Although the given species change from one tropical rain forest to the next, the general picture is very similar.

Nearly half the forested areas of the earth are tropical rain forests. The Amazon River Basin of South America, the Congo Basin in Africa, and the forests of Southeast Asia are the three largest areas; but, as shown in figure 3.3, tropical forests also exist in Australia, Central America, New Guinea, the Philippines, Malaysia, the East Indies, and many Pacific islands.

The warm, moist conditions of the tropics are also ideal for the bacterial and fungal processes that decompose dead plants and animals and return their rich inorganic nutrients to the soil. Because of the density of living plants needing these nutrients, deep topsoil rich in these decomposition products is not able to accumulate. The trees therefore have shallow root systems that quickly take water and nutrients out of the soil to be used for new growth.

In addition to having nutrients confined to a shallow topsoil, many tropical soils are lateritic—they have certain metals in combination with large amounts of clay that pack very tightly if not kept loose with organic materials. Annual plowing for cultivation and crop harvesting removes the source of organic matter. Without plants to take up the large amounts of water, rainfall leaches existing nutrients out of the topsoil. The lateritic soils compact and harden, so that after only a few years of cultivation such areas are as hard as concrete, and further cultivation or revegetation is impossible.

It is tragic to see this unparalleled natural resource being destroyed. Because of growing overpopulation in tropical countries, however, large-scale clearing and intensive cultivation is destroying these forests faster than they can be studied and understood. This situation has resulted in a race against time to classify and catalog tropical plants before they become extinct.

(a)

(b)

Figure 3.5

(*a*) Alakai Swamp, Kauai, Hawaii. Epiphytes are common in tropical forests, growing on the trunks and branches of the tall trees. Epiphytes are not parasites and thus do not harm the plant on which they grow. (*b*) Lianas in a Sumatran rain forest.

Clearing vast tropical forests for cultivation not only does nothing to solve food shortages, it can be an ecological disaster, resulting in the permanent loss of these areas. The loss of this vegetation affects not only ecological balances, but broad climatic patterns as well. In addition, the extinction of species eliminates their unique genetic potential.

Savannas

Savannas are usually found between the tropical rain forests and deserts (figure 3.2*a*). Their proximity to either of these two areas is greatly affected by the annual rainfall. The "normal" range often falls between 80 and 160 cm per year. Because of their latitudinal distance from the equator (figure 3.6), savannas do have seasonal temperature fluctuations even though they have no true cold period. Precipitation is also scattered; long dry periods, which are often very hot, are followed by heavy warm-season thunderstorms. There is little rain during the cool season. This lack of rainfall for a prolonged period apparently excludes many species that otherwise might occur there and gives the savanna its characteristic vegetative composition.

Savannas are primarily grassland with scattered deciduous trees. Since there is no true cold season, these trees generally lose their leaves during the long dry season each year, leafing out again when the rains come, and generally flowering while leafless. There are few annual plants in the savannas because of the density of the perennial grasses. A number of perennial herbs do thrive here, emerging from underground bulbs, rhizomes, and tubers after the rains begin. The trees have smaller leaves than do those in the tropical forests. Because it is necessary to reduce water loss in the dry season and because light is not a limiting factor, the increased surface areas characteristic of tropical forest species are not found. In savanna areas that border the drier desert regions the trees are smaller, denser, and often thorny. These areas are called **thorn forests.**

Some of the most extensive savannas are in central and eastern Africa (figure 3.7). In the African veld (pronounced "felt") a large variety of animal life, including many species of antelope, zebras, giraffes, elephants, and their predators depend on the grass species found there. Other savannas are found in Brazil, India, Southeast Asia, northern Australia, and North America. The maintenance of the grassland component requires periodic burning. Burns clear the dead dry grass so that new lush grass can grow when the rains come. Fire also keeps the trees thinned and scattered by removing young seedlings. The mature trees often have thick bark, and since they are leafless during the dry season, minor trunk scorching is usually the only damage done. For the proper balance of grassland for animal forage and scattered trees for shade and nesting sites, fires are essential. Lightning is a natural source of fire in the savannas as well as in other biomes.

For centuries, native inhabitants have set fire to grasslands, recognizing their role in the balance of plant and animal populations. In some parts of the world, including North America, controlled burns are used to bring about the same kinds of ecological manipulation that occur in nature

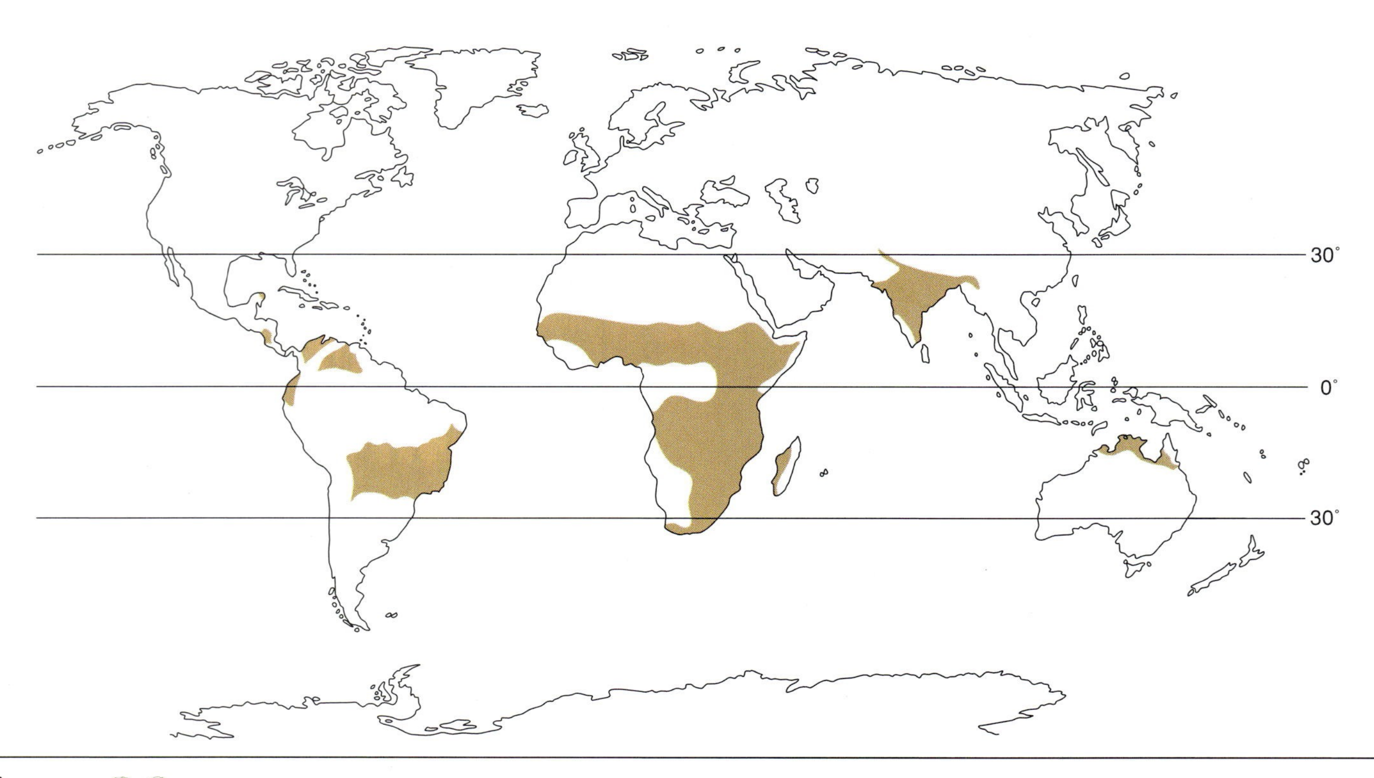

Figure 3.6
Savanna biome, world view.

(figure 3.8). If kept under control, these burns are not a misuse of fire. The study of fire ecology is an important part of overall management in several biomes.

Deserts

Most of the **desert** areas of the world are found in a belt from 20° to 30° N and S latitude with rain-shadow deserts at other latitudes (figure 3.9). A few deserts have no vegetation and lots of shifting sand dunes, such as parts of the Sahara, but most have scattered, low-growing vegetation.

Deserts receive 25 cm or less average annual precipitation. The driest deserts, including the Sahara, have less than 2 cm average annual rainfall, and all desert areas can have extended droughts with no rainfall for several years. In the Atacama Desert in northern Chile, for example, the total rainfall over a period of

> ✔ *Concept Checks*
>
> 1. What ecological conditions in the tropics have resulted in such incredible biological diversity?
> 2. Why is clearing the tropics for agricultural use destroying them?
> 3. What does fire do to help maintain the proper vegetative balance in a savanna?

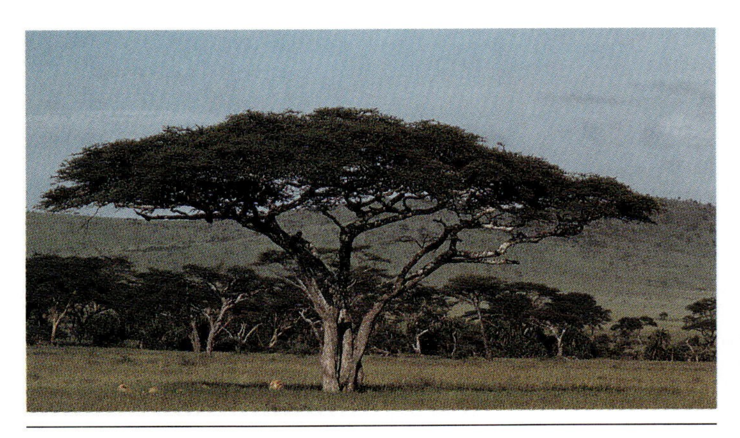

Figure 3.7
Savanna, the Serengeti Plain of Tanzania, Africa.

Figure 3.8
A controlled burn being used as a range management technique to improve grazing and to control weeds and shrubs in Santa Barbara County, California.

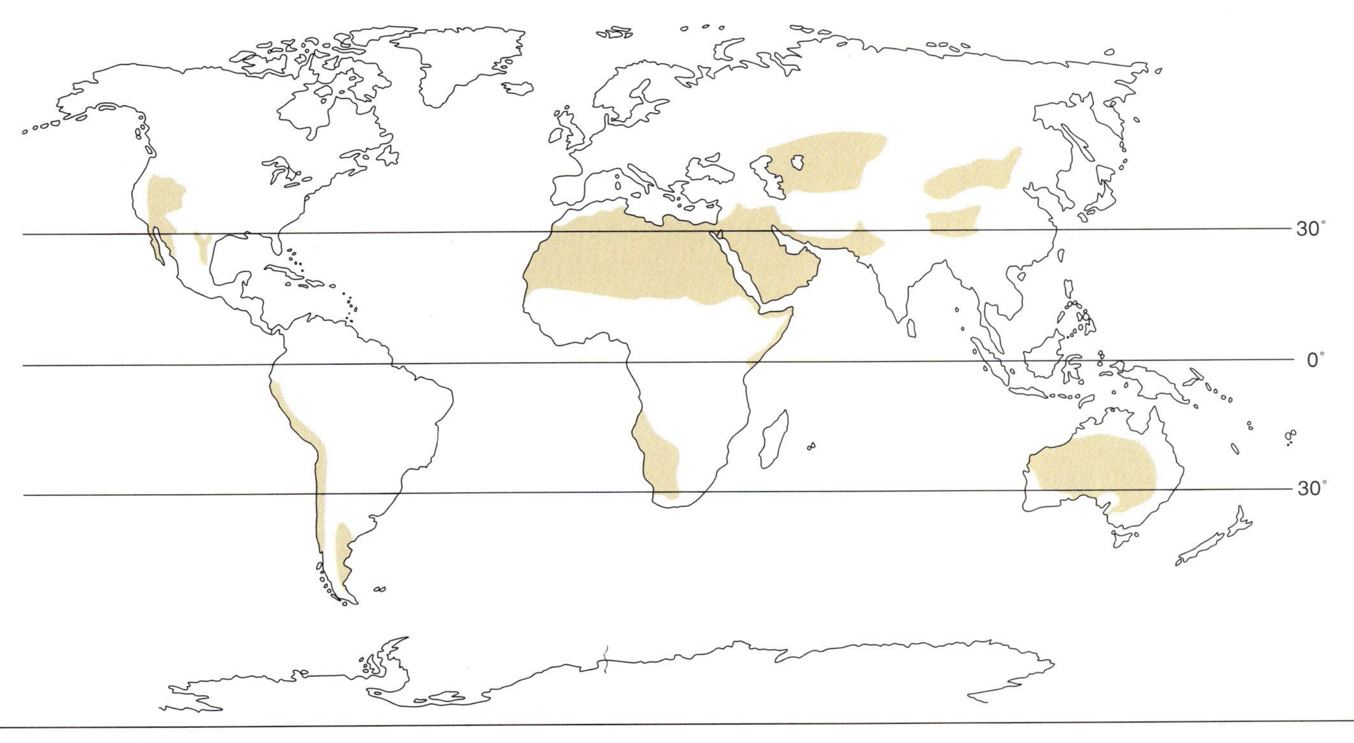

Figure 3.9
Desert biome, world view.

(a)

(b)

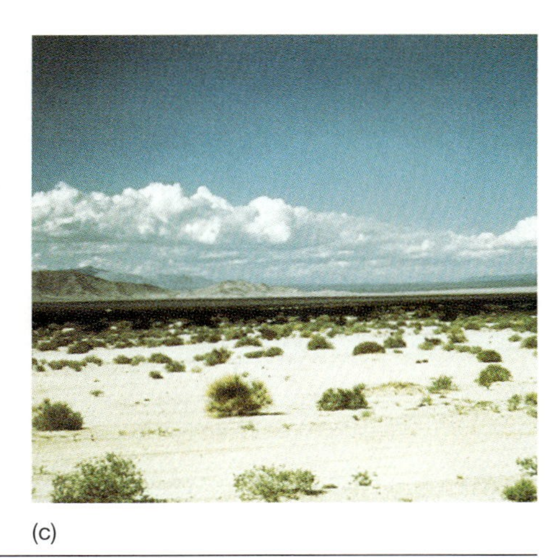

(c)

Figure 3.10

Desert biome. Relatively lush desert vegetation often can be found in a desert. (*a*) All three vegetation types, annuals, succulents, and small-leaved shrubs in the Saguaro National Monument, Tucson Mountains of Arizona (Sonoran Desert). (*b*) The red-flowered ocotillo is leafed out, indicating a significant recent rain (Chihuahuan Desert). (*c*) A more xeric desert setting found near Troy Lake, California (Mojave Desert).

17 years was 0.05 cm, with only three showers in that 17-year span that were heavy enough to measure. It is important, therefore, to note that the above rainfall figures are averages over many years, with considerable fluctuation possible. When the rains do come, they can be heavy enough to produce flash floods, or they can be light showers.

Because of the dryness, diurnal temperature fluctuations are great. It is not unusual to have a 25°C or greater drop in temperature at night, since there is little moisture to hold the heat produced during the day. Once the sun is down, the heat source is gone. Conversely, it warms up rapidly after sunrise. It is common for vacationers, camping in a North American desert in the summer months, to bring no blankets or sleeping bags. A thin sheet provides little warmth, and these campers usually spend a sleepless night in the car.

Desert temperatures also follow the seasons, with occasionally harsh winters, freezing temperatures, and snow. Elevation, proximity to the coast, and latitude also combine to produce some "warm" deserts, which seldom have freezing temperatures. Such is the case of the Sonoran Desert of North America, the only North American desert having the giant saguaro cactus, a plant that cannot tolerate the freezing temperatures found in the Chihuahuan, Mojave, and Great Basin Deserts.

Desert vegetation is amazingly diverse and uniquely beautiful, falling into three major categories based on the physical adaptations that allow their survival in times of drought.

1. **Succulents** are plants that store water in thickened leaves or stems and protect that supply with thorns and spines. Most of the cactus family are stem succulents with either jointed pads, as in the prickly pears and chollas, or barrels with a single thickened stem. The spines are modified leaves that protect the cacti from herbivores in search of moisture. Other plants, called leaf succulents, store water in modified leaves. The century plant (*Agave*), Spanish dagger (*Yucca*), and the low-growing *Sedum* and *Portulaca* are common examples, although many different plant groups have succulent members adapted to dry zones (figures 3.10 and 3.11).

2. Low-growing, small-leaved shrubs with spines or sharp branches survive in the deserts by conserving water in their woody stems and reducing water loss through reduced leaf surface area. In addition to or instead of spines, these shrubs often have foul-tasting compounds in their leaves to discourage herbivores from browsing. Creosote bush (*Larrea*) and tar bush (*Flourensia*) are such examples. Cat claw (*Acacia*) and mesquite (*Prosopis*) are protected by both spines and thorns.

3. A third group of desert plants survive in a totally different manner. Instead of storing water, they wait until an adequate supply is available. They exist as heat- and drought-resistant seed that will not germinate unless a heavy rain thoroughly washes off a self-produced chemical inhibitor to germination. The rain

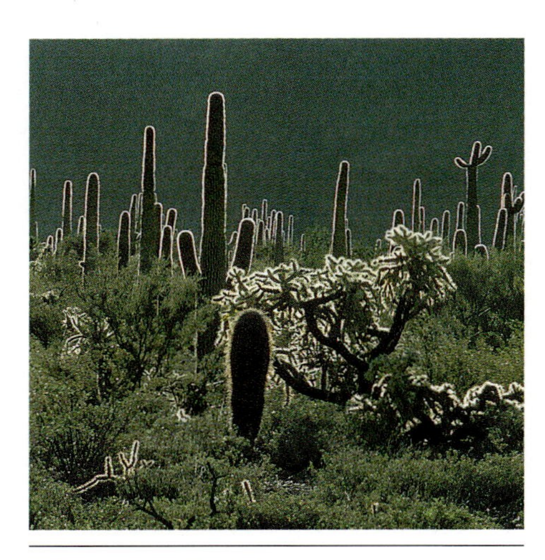

Figure 3.11

A desert area in Arizona. Note the two kinds of stem succulents in the jointed-stem cholla cactus and the barrel-stem saguaro.

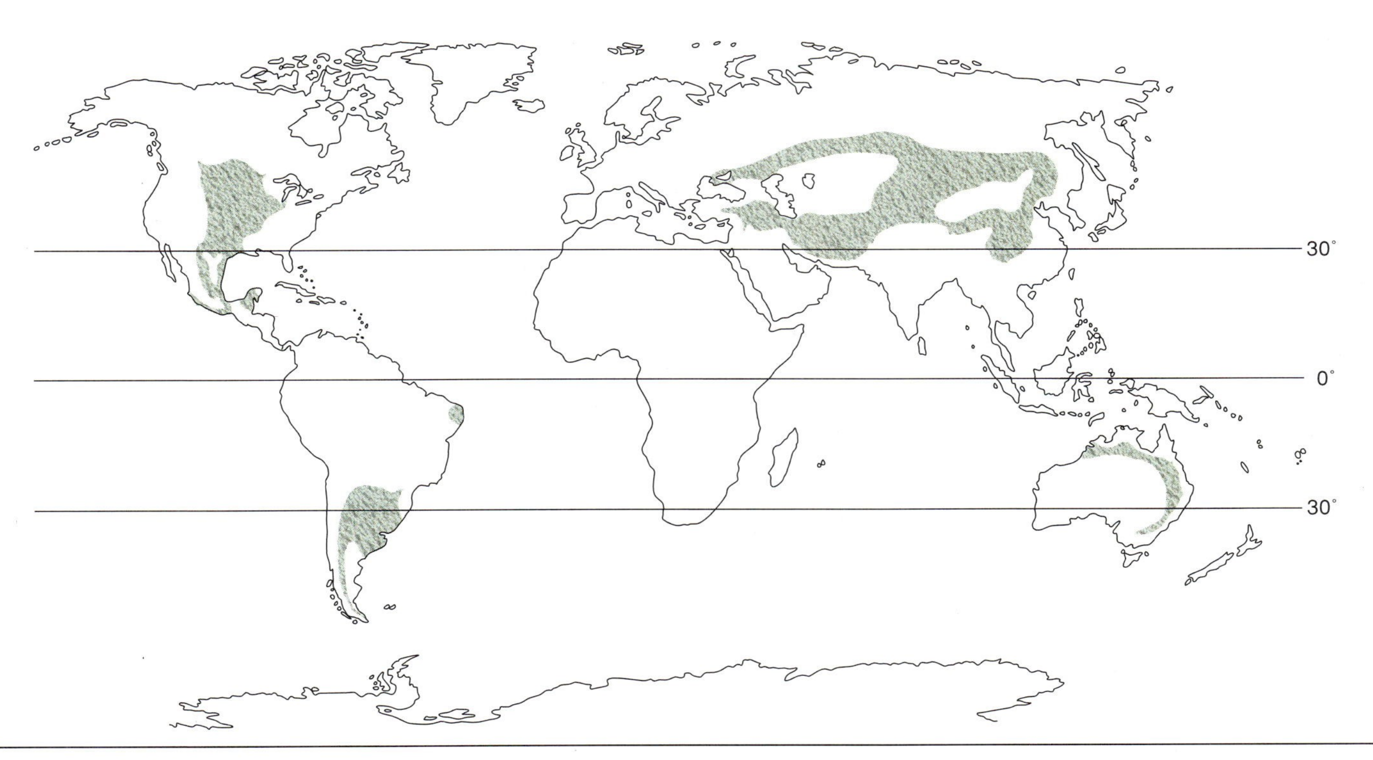

Figure 3.12
Grassland biome, world view.

also provides ample soil water for these plants to germinate, grow to maturity, flower, reproduce, and set seed before the supply of water is gone. This all happens in a short period of time—possibly a week or two, at most a few weeks, but certainly in one season. These desert annuals are sometimes called **ephemerals** because of their short life cycle. Their seed may have to wait as long as 20 or 30 years before conditions are right for germination and completion of their life cycle. Consider the fortitude of a seed able to survive so long in desert soils where surface temperatures are regularly over 60°C in the summer. If a year is wet enough, the flowering desert is as beautiful as any area.

The largest and driest desert in the world is the Sahara, followed in size by the Australian desert. Figure 3.9 shows where many of the other major deserts are located, including the four North American deserts, the Gobi of Mongolia, deserts in India, the Middle East, other parts of Africa, and in South America. Approximately one-third of the earth's land surface area is arid or semiarid. Deserts are recent in origin, some no older than 12,000 to 15,000 years and possibly none older than 5 to 6 million years. The relative youth of deserts is even more striking when compared with the age of the tropics, which have existed for some 200 million years.

Deserts have expanded due to worldwide drying trends since the last glaciation period, which ended some 15,000 years ago. Human activities are accelerating this process of **desertification** (conversion to deserts) in many areas through overgrazing, cultivation of marginal areas, and general removal of plant life and water for our own uses. It is estimated that the Sahara is advancing to the southwest at approximately 17 km per year. With wise use and controlled development, the deserts could provide much-needed living space and resources to help support the population. There are many developmental studies being conducted by government agencies, universities, and private groups in essentially every country containing desert lands. Some of the most interesting and promising research is being done in the area of desert agriculture. There have been attempts to develop strains of already existing crop plants that require less water, can withstand higher temperatures, and can tolerate a greater level of soil salinity.

There are also studies aimed at finding commercial uses for plants that naturally occur in desert regions and are therefore already well suited to desert climates. It is probable that this latter approach has the greatest potential with plants such as saltbush (*Atriplex*), prickly pear cactus (*Opuntia*), and ironweed (*Kochia*), which show great promise as forage plants. Guayule (*Parthenium argentatum*) for rubber, jojoba (*Simmondsia chinensis*) for oil, and several plant species for fuel biomass also have significant potential for commercial use.

Grasslands

This biome is dominated by areas of perennial grasses (figure 3.12). Predominately in temperate latitudes,

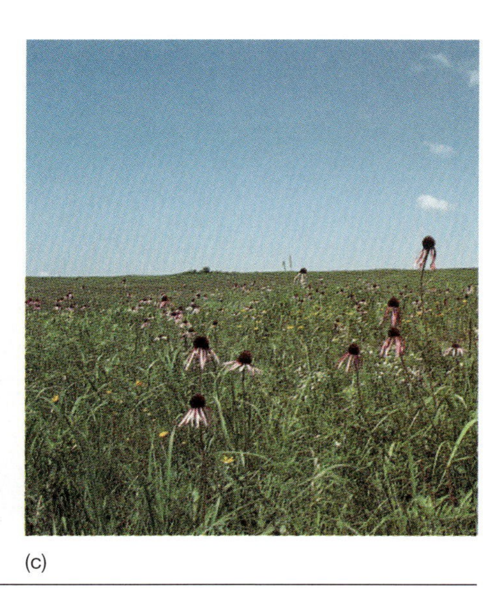

(a) (b) (c)

Figure 3.13
(*a*) Short grass prairie in the panhandle of Texas. (*b*) Close-up view of the density of grasses. (*c*) Tall grass prairie with purple coneflower (*Echinacea pallida*).

grasslands receive from 30 to 150 cm annual precipitation and have distinct seasons. Temperatures are often above 40°C in summer and far below freezing in winter, with occasional extremes. The annual precipitation is usually distributed throughout the year with occasional summer peaks. On the more mesic (wetter) end of the range, **grasslands** grade into savannas or temperate deciduous forests, whereas on the xeric (dry) end they grade toward deserts.

Because the matted turf of fibrous grass roots prevents seed from reaching the soil, there are very few annual plants in the grasslands. The herbaceous plants found there are mostly perennial with underground storage structures such as tubers, bulbs, and rhizomes. Occasional prairie fires help maintain the integrity of the grasslands (much as in savannas), but in the drier grasslands the invasion by desert species is common. Overgrazing of such areas has speeded up their desertification, allowing mesquite, cacti, and weedy annuals to become well established in the thinned turf.

There are few native grasslands (prairies) intact in the United States; most of these areas having been overgrazed by domestic animals or cleared for cultivation (figures 3.13 and 3.14). Other major grasslands are found in eastern Europe and Russia, central and western Asia, Argentina, and New Zealand.

Figure 3.14
Cattle on open grassland in New Brunswick, Canada.

The dust bowl of the 1930s in the Oklahoma and Texas panhandles was caused by an extreme drought that drastically reduced productivity; the few water wells were not adequate for irrigation. Without the cultivated plants to hold down the soil against the ever-present spring winds, the topsoil was literally blown away (figure 3.15).

Temperate Deciduous Forests

This biome is found in all the major continental areas of the northern hemisphere but is almost absent from the southern hemisphere (figure 3.16). The average precipitation of 75 to 225 cm per year is usually scattered throughout the year. Warm summers and cool to cold winters are typical.

(a)

(b)

Figure 3.15

Dust bowl. (*a*) Dust storm approaching from the west on June 4, 1937. Four minutes after this picture was taken, total darkness occurred from the dust blocking the sun. (*b*) An abandoned farm between Boise City, Oklahoma, and Dalhart, Texas, showing the dunes formed by the blowing dust. The inhabitants of the region, unable to make a living, fled mostly to California to search for jobs.

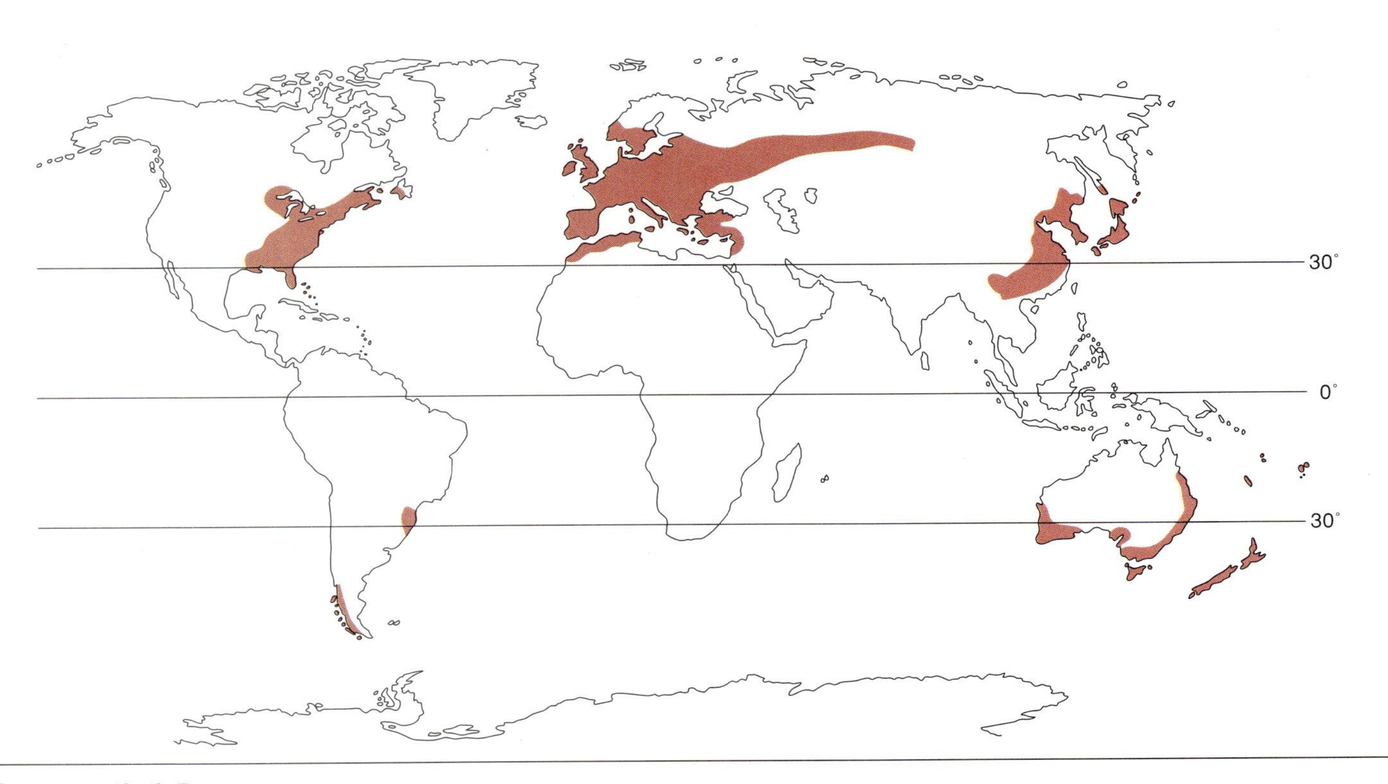

Figure 3.16

Temperate deciduous forest biome, world view.

Figure 3.17
Temperate deciduous forest in winter in Fairmount Park, Philadelphia.

These areas rarely have droughts and only limited periods of snow and subfreezing weather. The trees drop their leaves each fall; hence the name **temperate deciduous forests** (figure 3.17).

Because the forest leaf canopy is not intact year round, the ecology of these regions is unique. Low-growing, understory vegetation dominates in the early spring before the trees completely leaf out to form the shade-producing canopy. Once the canopy is formed, the forest trees are the dominant vegetation, followed again in the fall by a second understory assemblage that develops in response to the available light resulting from the leaf drop. Thus three distinct vegetational assemblages exist during the growing season, resulting in a complex ecological situation (figures 3.18 and 3.19). Also known as "hardwood forests," the diversity of tree species that are valuable for wood used in fine furniture includes oak, mahogany, walnut, and fruitwood, such as cherry. Much of the world's temperate deciduous forests have been cleared for cultivation (figure 3.20) as well as harvested for their wood, reducing the total biodiversity from this biome.

In drier areas of the biome, where the winters are cool and moist but the summers are hot and dry, a unique vegetational association called **chaparral** exists. Characterized by smaller, often thorny or roughly branched evergreen trees and shrubs and deciduous trees, the chaparral has a short spring growing season interrupted by the heat and drought of summer. Often referred to as a Mediterranean climate because of the winter rainfall and extensive chaparral areas along the shores of the Mediterranean Sea, such localized

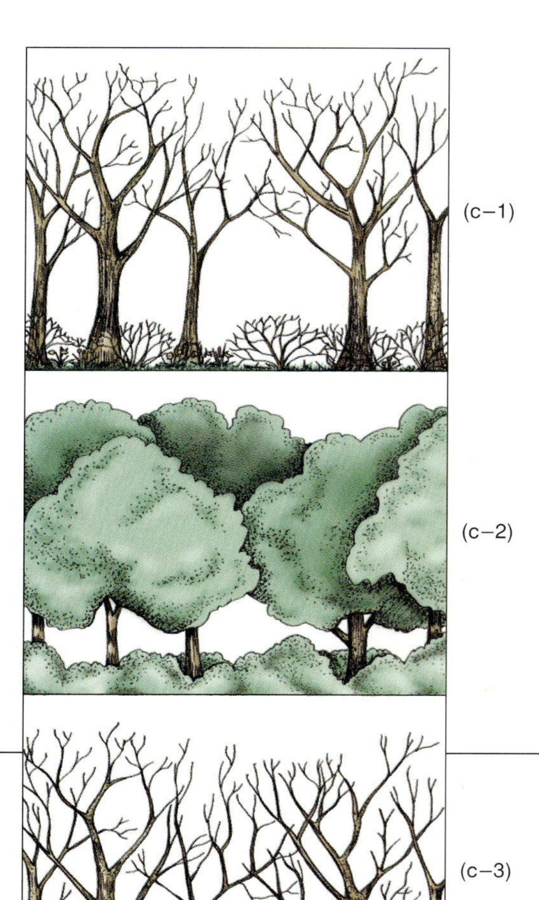

Figure 3.18
The three types of forest canopies. (*a*) The tall evergreen trees of the tropical rain forest. (*b*) The coniferous forest. (*c*) The temperate deciduous forest. (*c-1*) Early spring understory vegetation before the trees leaf out. (*c-2*) The trees become the dominant vegetative component. (*c-3*) A second understory vegetative complex becomes the major plant group after the fall leaf drop.

Figure 3.19
Temperate deciduous forest. (*a*) In the spring before the leaf canopy has developed. (*b*) In the summer the trees are fully leafed out. North Hardwood, Pennsylvania. (*c*) The leaves turn colors in the fall prior to dropping from the trees. Adirondack Mountains, New York.

Figure 3.20
Cultivation is a common practice in temperate deciduous forest areas because of the rich topsoil and climate.

areas are also found in southern California, southern Africa, coastal Chile, and coastal western and southwestern Australia. Although they are isolated from one another and contain different species of plants, their climatic conditions give these areas a similar appearance.

Like the grasslands, the temperate deciduous forest biome is represented in North America by only a fragment of what was once here. An additional crisis for all ecosystem balances is the introduction by human activity of plant species from other areas of the world. Many of these exotics escape cultivation and become overly successful in the wild, upsetting the natural balances of the native species.

Coniferous Forests

Coniferous trees are the cone-bearing members of the gymnosperms, a group of woody plants that lack flowers, having seed produced on the scales of cones or similar reproductive

Ecosystems under Alien Attack

It appears to be part of human nature to inquire of the unknown and covet the unusual, the exotic. Kings sent forth explorers to find what was out there in distant lands, and to bring back examples of the unique and beautiful plants they encountered. Gardeners and horticulturists were charged with learning how to keep these introductions alive and how to propagate more of them for the gardens of wealthy landowners.

This approach to horticulture remains the norm—select the unusual and learn how to propagate it and develop "tolerant" varieties for different climatic zones; or introduce species from similar climates in other countries that will be "hardy" and "adapted" in their new homes. Most of these introduced plant species were intended to be for the garden. Many of them, however, have escaped cultivation and now grow wild.

Other exotics were introduced as crop plants or as range grasses that were more tolerant of heavy grazing than the native grasses. And still other exotics hitchhiked in bags of agricultural crop seed or were part of the ballast dumped by visiting ships. Whatever their route of introduction to this country, many now have escaped and grow wild—in fact, some 3000 plant species from other parts of the world now grow wild in North America.

When an introduced species is added to a natural ecosystem, it displaces or at least changes the relative number of individuals of the native species there. By disrupting natural ecosystems in this way, some of the native species lose their natural habitat and openings are created for the more opportunistic species of introduced exotics, which can then become even more aggressive. Kept in check by predators, pathogens, or a balance of competition from associated plants in their natural habitats, these exotics can go unchecked when introduced into new habitats that lack these control factors. They spread so rapidly as to crowd out many native species.

One of the most serious examples of this problem is in Hawaii, where over 870 introduced plant species now grow wild—nearly ten times more than only a century ago. As a result of this change, nearly 200 native Hawaiian plant species have become extinct, and few, if any, of Hawaii's natural habitats exist unaffected by these introductions. With the loss of these native plant species there is a domino-effect loss of habitat for other plant and animal species that were dependent on the natural ecological balance that no longer exists. Florida is facing a similar problem because of the tropical climate and the number of introductions. Very likely no area of this country is absolutely unaffected by this invasion, and those few that are still natural are threatened with change.

Who are these alien invaders? We readily recognize many of the more visible and obvious—kudzu from Japan in the southeast, purple loostrife in the northeast; buckthorn from Europe now dominates areas of the eastern United States; Timothy grass, crabgrass, tumbleweed, salt spray rose, ryegrass, ox eye daisy, chicory, white and red clover, Queen Anne's lace, cocklebur, Japanese honeysuckle, Chinese ailanthus, Asian bittersweet, Norway maple, casurina, eucalyptus, and many hundreds of other examples either dominate or have become more common than the natives they have crowded out. Of the 18,000 native plant species in the 48 contiguous states, some 2000 are seriously threatened and 400 are in danger of extinction within the next ten years—mostly due to habitat disruption and competition from invasive introductions.

Obviously, continued introduction of exotic plant species is not only unnecessary, it has been extremely damaging to our ecosystem and should be curtailed. Further, use of existing cultivars and horticultural selections in our gardens and landscaped areas needs to be evaluated against the possibility of those exotics escaping cultivation and adding to an already critical situation of such plants establishing their presence in the wild. Of even greater concern is the use of these cultivars and introductions on roadsides, in parks, or in other areas more directly in contact with natural habitats.

The most desirable action that we can take is to try to reverse the process that has led to such serious ecological problems. Our hundreds of years of accumulated horticultural expertise needs to be directed at learning how to propagate and reestablish our own native plant species back into the habitats from which they are being crowded out. We need to learn how to manage habitats to encourage the reestablishment of the natives over the introduced exotics that are now so prevalent.

The introduction of most of these exotic species from around the world was innocent and, in many cases, unintentional. Only since our awareness of the ecological damage these introductions have caused is there an urgent need to change our well-established urge for the unusual, the exotic. We must reassess our priorities.

✔ *Concept Checks*

1. What are the three vegetation types found in the desert biome?
2. What has happened to the grasslands of North America?
3. What are the three seasonal vegetational assemblages in a temperate deciduous forest?
4. How does a chaparral differ from a temperate deciduous forest?

structures. Conifers, valuable for their use in the lumber industry, are evergreen, except for larch, bald cypress, and tamarack. Their needle-shaped leaves have a thick waxy covering (cuticle), which helps prevent water loss. Many conifers have shallow root systems that are well designed for the often rocky, shallow soils found commonly in mountainous areas. Mountains in Europe and Asia, as well as the Rocky Mountains and Appalachian Mountains of North America, have **coniferous forests** with similar climates and rainfall patterns (figures 3.21 and 3.22). The ability to thrive in thin, rocky or sandy soil that often contains little moisture explains the significant stands of coniferous forests in the southeastern United States (figure 3.23) and in the western coastal areas of California, where giant redwoods (*Sequoia*) grow (figure 3.24).

The far northern coniferous forests, found almost exclusively in the northern hemisphere north of

Figure 3.21
Coniferous forest of alpine spruce in Canada.

50° latitude, are referred to as **taiga** or snow forests (figure 3.25). The average precipitation ranges from about 35 to 100 cm per year, most of it falling in the summer. Winters are very long and cold and have a persistent snow cover. Although the winter air is dry, the ground remains moist because of the low evaporation rate resulting from the cold temperatures. These forests grade into either grasslands or temperate deciduous forests to the south, depending on precipitation levels (figure 3.26).

Tundra

At their northern limits, the coniferous forests gradually give way to **tundra** (figure 3.27). Found predominately in the northern hemisphere north of the Arctic Circle, tundra comprises approximately one tenth of the earth's total land surface. With less than 25 cm of annual precipitation and strong, dry winds, the subzero temperatures and long periods of winter darkness create an exceptionally harsh environment for plant growth. The ground is frozen solid for most of

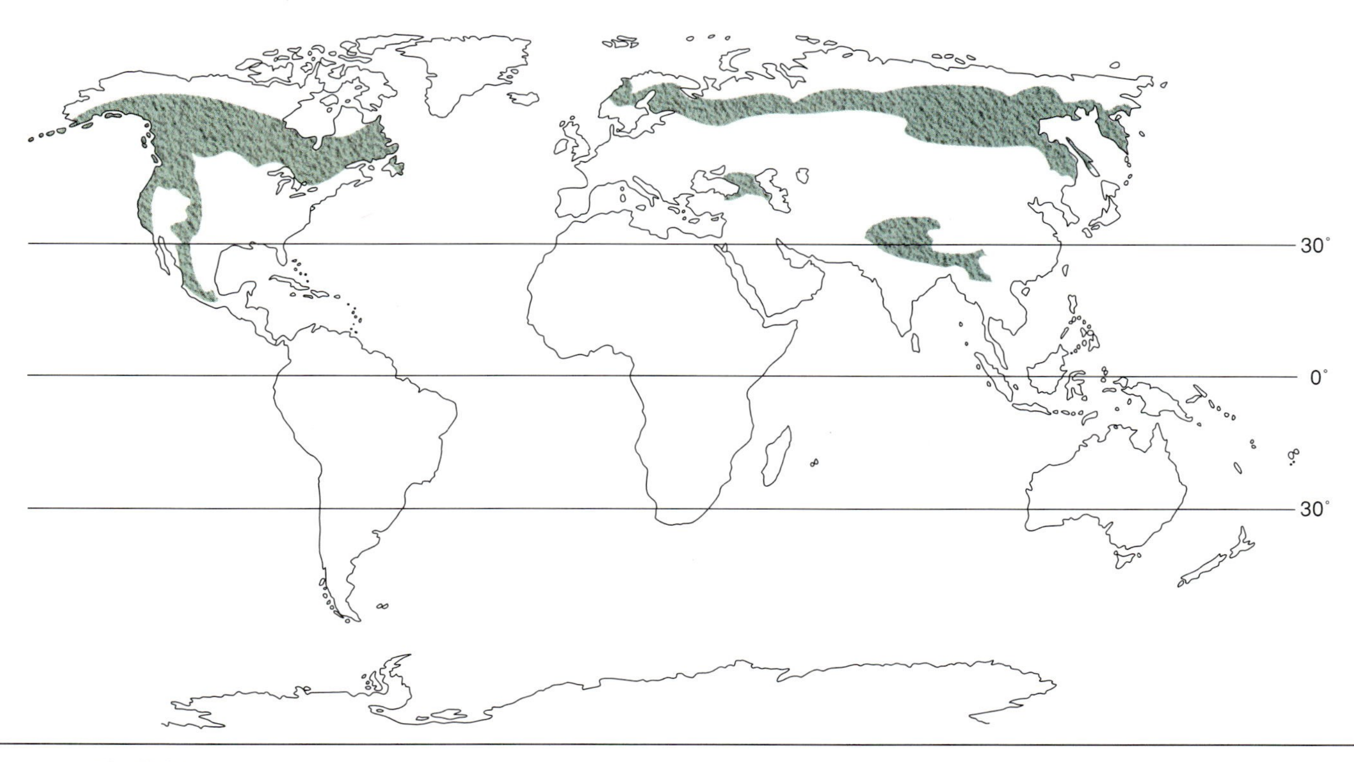

Figure 3.22
Coniferous forest biome, world view.

Figure 3.23
Longleaf pines in coniferous forest near Indian Pass, Florida.

Figure 3.24
Giant redwoods.

Figure 3.25
Taiga or snow forest area.

Figure 3.26
Taiga often grades into grasslands. Flathead Valley and the Mission Mountains, Montana.

the year, thawing only to a depth of about 1 m during the short summer. This frozen-soil zone, **permafrost,** causes plant root systems to be relatively shallow yet extensive. In the moist sedge-dominated communities, underground plant parts may be as much as ten times greater than aboveground biomass.

Tundra vegetation is typically composed of scattered, low-growing woody perennials that are well adapted to the drying winds and extreme cold. Certainly, tundra plants must survive long periods of moisture unavailability, and their morphological adaptations commonly include thick, waxy cuticle layers on the leaf surfaces and dense leaf pubescence (fuzziness, especially on the lower leaf surface). Additionally, their prostrate to shrubby woody trunks are often covered by a protective coat of lichens or moss that helps prevent desiccation as well as provide protection from the cold. Although no tundra vegetation grows taller than 1 m, and the harsh climate has limited

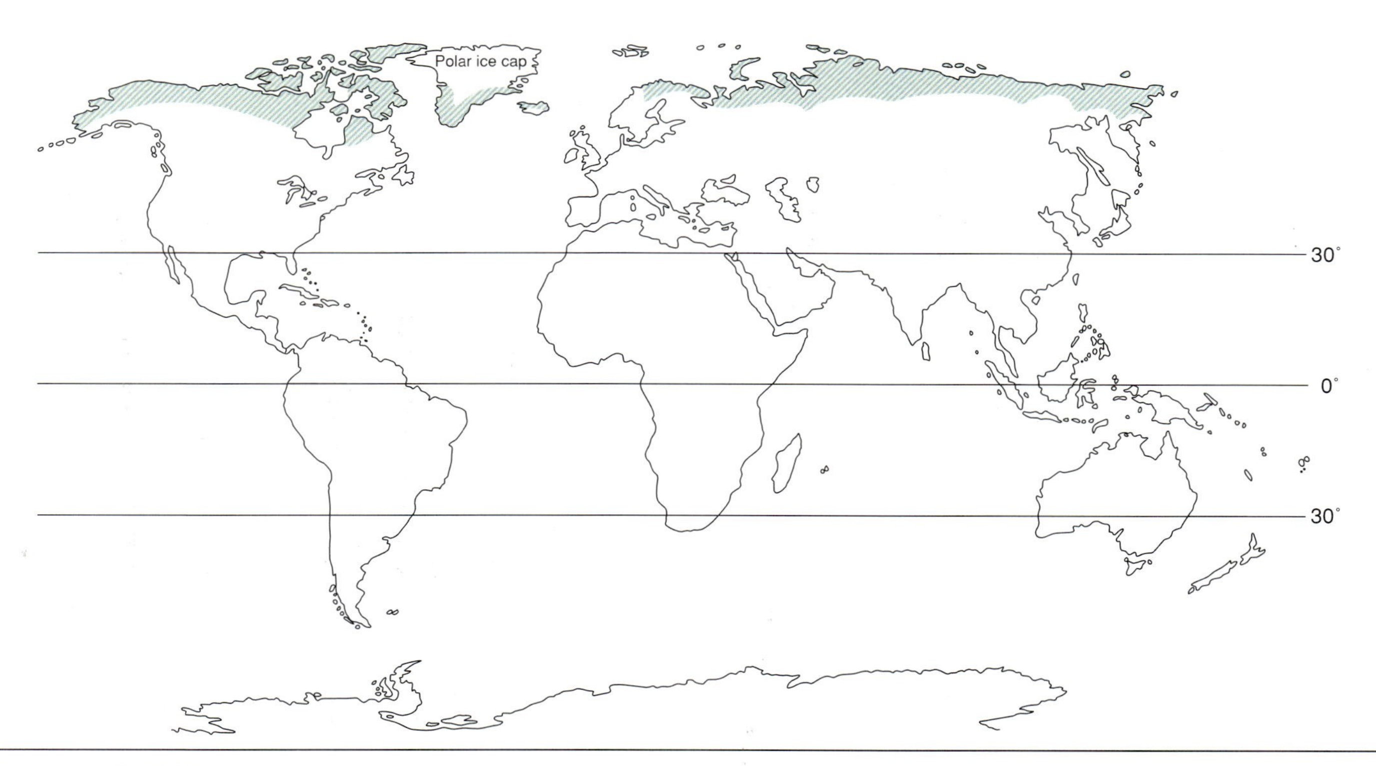

Figure 3.27
Tundra biome, world view.

the number of well-adapted species, the successful plants are abundant. Essentially every native plant community is dominated by only two or three species, but there are huge expanses of such areas (figure 3.28).

The mean daily temperature is above freezing for only about one month of the year, providing tundra plants with a very short growing season. During this period soil moisture is available because the ground thaws above the permafrost, and photosynthesis is possible for all 24 hours of summer daylight. Growth is generally minimal, however, because the replenishment of stored food reserves in the roots and woody stems is the primary plant function during this short period of favorable conditions. Some plants add only a few new leaves to each twig before the cold temperatures again become restrictive.

Sexual reproduction is an even more tenuous function than vegetative growth, since the plants only have four to six weeks to complete the entire cycle of flower development, pollination, fertilization, and fruit and seed maturation. Many species have developed dormant buds at the end of the previous summer; thus, when the snows melt, the mature flowers emerge within a few days. Many of these flowers are modified to concentrate the heat of the sun, thus increasing the maturation process by speeding up metabolic activity. The arctic poppy has a white cup-shaped flower that tracks the sun, focusing the sun's rays on its reproductive parts. In full sunlight the temperature inside the flower can be as much as 28°C higher than the air temperature around it. Insects attracted to this warmth effect pollination, and seed are mature about three weeks after flower opening. Not all the arctic plants have such sophisticated flowering adaptations; some depend on vegetative reproduction, managing to set mature seed only in unusually long and mild growing seasons, which may not occur for 50 years or more.

Once mature, seeds of arctic plants must be able to remain dormant for long periods. It is a rare year that provides a growing season sufficient in length to allow for seed germination and development to mature protective woody growth. The deep-freeze conditions of the arctic enable seed to retain their viability for an unusually long time, however. As an extreme example, a seed of an arctic lupine (genus *Lupinus,* family Leguminosae), found in a Yukon deposit dated at 10,000 years old, germinated after dry storage for 12 years at room temperature. Long dormancy enables seed to remain viable until the conditions are optimum for success.

Some mountainous areas have climatic conditions and vegetation similar to those of the arctic tundra (figure 3.29). These mountain tundra zones are at elevations above tree line in essentially all the world's large mountain ranges. The farther from the poles these ranges occur, the higher the elevation before tundra conditions can be found (figure 3.30). At extreme elevations and polar latitudes no vascular plants occur. The ice caps of both polar regions support a number of alga and fungus species but no higher or vascular plants. The tundra is therefore the farthest limit of plant growth.

(a) (b) (c) (d)

Figure 3.28

(*a*) Arctic tundra. (*b*) Arctic cotton (*Eriophorum*). (*c*) and (*d*) Arctic landscapes.

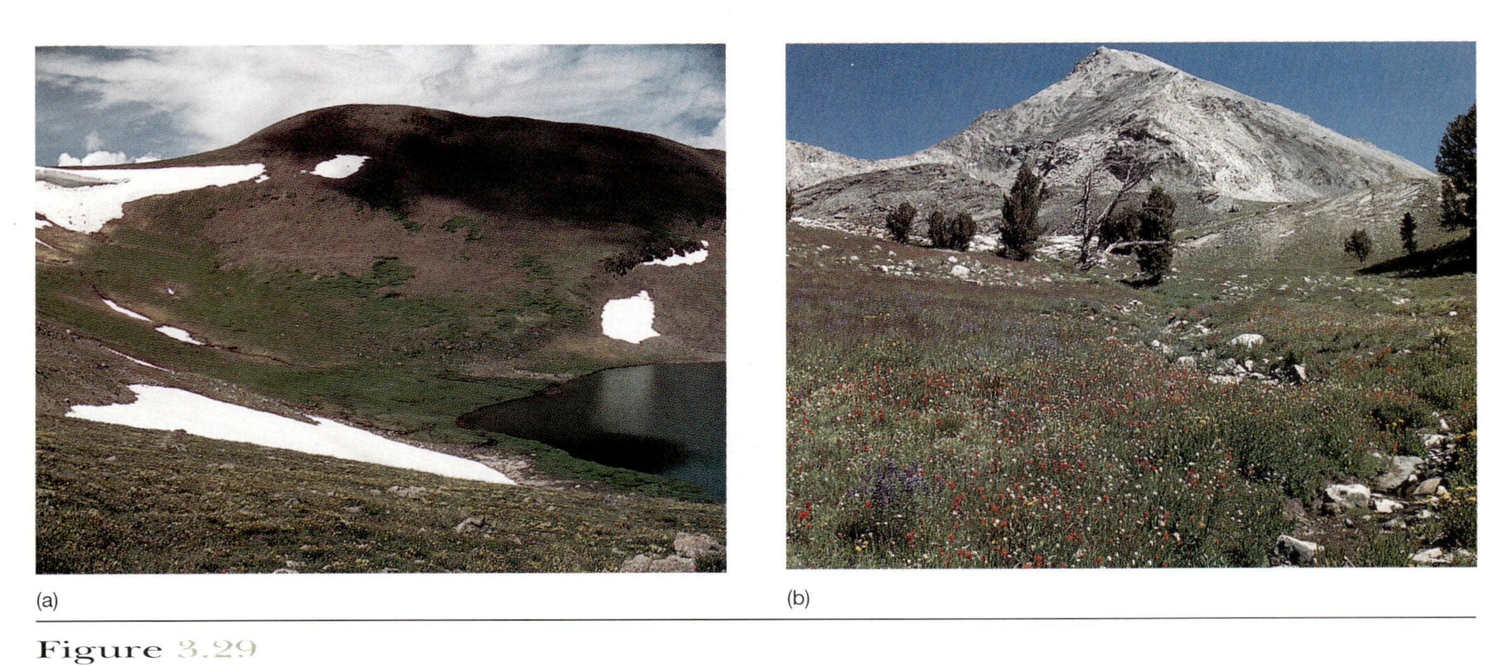

(a) (b)

Figure 3.29

Mountain tundra. At high elevations, above tree line, climatic conditions and vegetation are similar to those north of the Arctic Circle. (*a*) Tobacco Lake, South San Juan Wilderness, Colorado. (*b*) Hyndman Peak, Idaho.

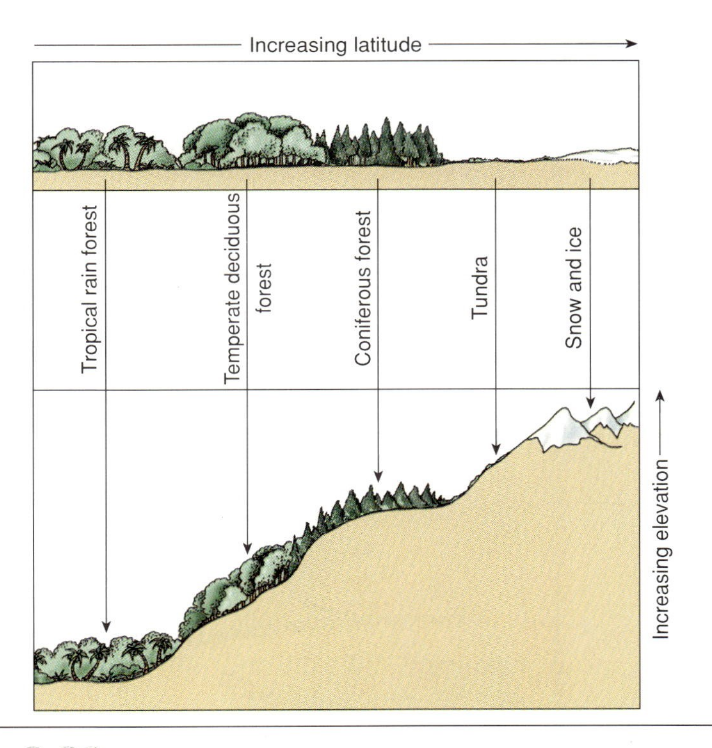

Figure 3.30
Elevational effects on vegetative associations parallel those produced by increased distance from the equator.

> **✔ Concept Checks**
>
> 1. Are there any differences between a coniferous forest and a taiga?
> 2. What characteristics of a tundra are the same as a desert and what are different?
> 3. How do latitude and elevation work to produce a tundra?

AQUATIC BIOMES

Marine

Covering nearly 75% of the earth's surface with an average depth of approximately 5 km, the marine biome is phenomenally large. The plant life there exists under unique circumstances. The most important is that light penetrates to an average effective depth of only a few meters. Below this shallow zone of adequate light only the short wavelength blue and green portions of the spectrum penetrate effectively, and even then it is essentially dark below 60 to 75 m. Thus the vast majority of plant life is limited to the lighted surface, with a few organisms capable of using shortwave light for photosynthesis at greater depths (figure 3.31). Bacterial, fungal, and animal forms inhabit the oceans even at their greatest depth, but, as on land, the base of their food chain is still plants. (See photosynthesis, chapter 13, p. 219.)

Water itself is obviously not in short supply; however, only organisms that can grow in salt water exist in the oceans and seas. These organisms are, for the most part, single-celled algae having no need for the complex vascular systems, supportive tissues, and reproductive organs of most terrestrial plants. Only in the shallow waters along the coastal shelves do more complex algal types exist. Simple transport systems, anchoring devices, and other multicellular modifications have been developed by these organisms in response to the constant wave action along the shores.

The climate for marine plants is to a great extent a function of the ocean currents. Plant distribution and temperature are especially current dependent. The most important factor in producing ocean currents is patterns of air circulation. In combination with

(a)

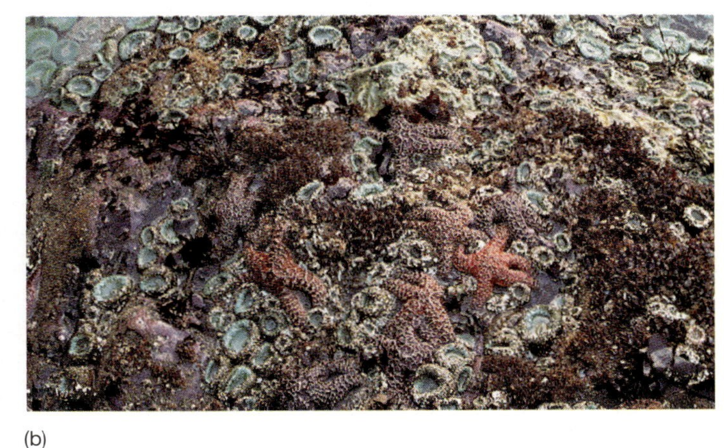
(b)

Figure 3.31
Tide pools. The shallow pools of water contain a beautiful variety of plant and animal life. (*a*) California. (*b*) Sea stars and anemones along Oregon coast.

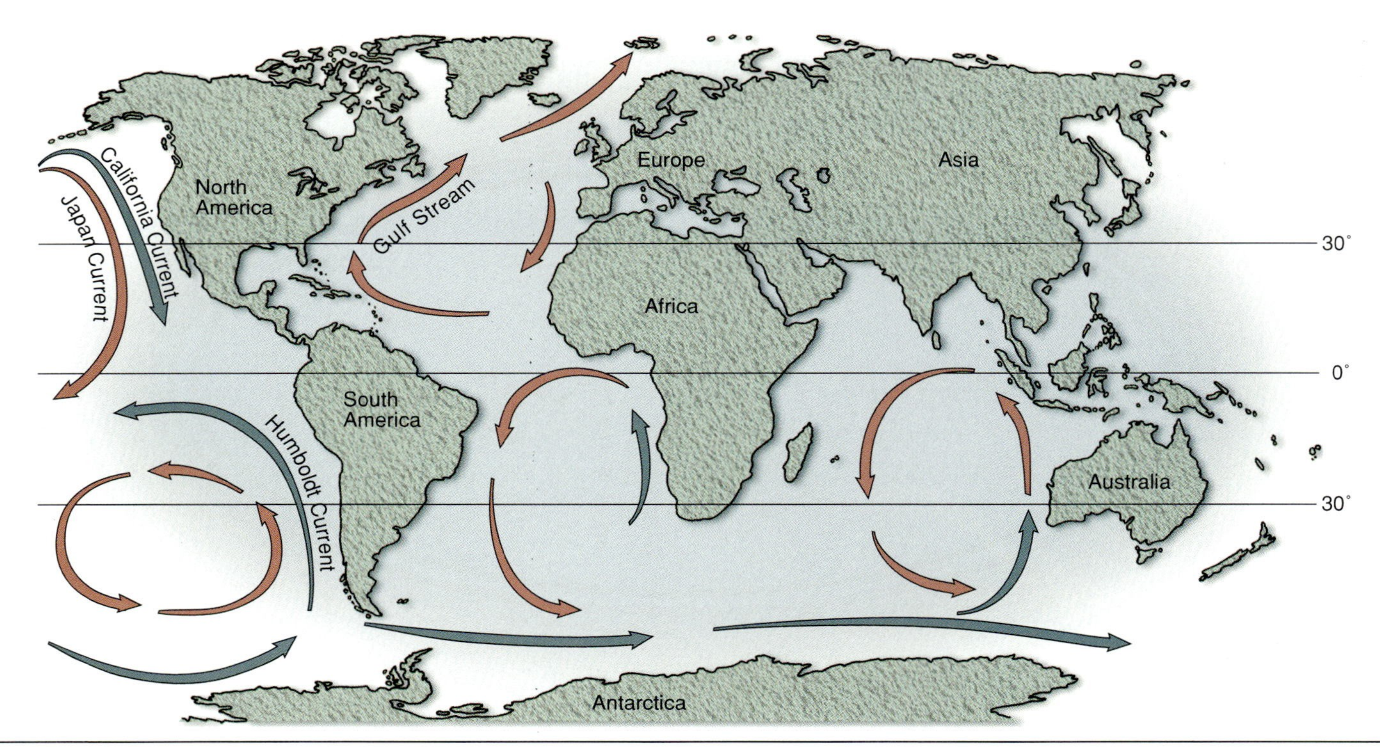

Figure 3.32
Worldwide ocean currents affect vegetative assemblages close to shorelines. Cold currents, blue arrows; warm currents, red arrows.

temperature, which affects water densities, and the deflection of currents off the continental landmasses, these predictable wind patterns create massive water movements around the world.

Ocean currents affect not only the distribution of plants and animals in the oceans, but some climatic patterns on land as well. The Gulf Stream, for instance, moves water warmed in tropical latitudes northward across the Atlantic Ocean to northern latitudes. When these warm ocean currents reach landmasses, they produce warmer climates and different precipitation patterns than would otherwise be normal for those latitudes. The British Isles are affected by these currents (figure 3.32), which change not only the climate, but also the terrestrial vegetation along the coasts and for some distance inland.

Marine organisms are sometimes classified as pelagic, or "of the open water," and benthic, or bottom dwelling. The free-floating organisms are primarily **phytoplankton** (single-celled plants) and **zooplankton** (single-celled animals). Phytoplankton are composed of single-celled algae, primarily diatoms. The open waters nurture millions of these organisms, plus eggs and larval forms of fish and invertebrates, to provide the early stages of the marine food chain. The larger pelagic animals are thus provided with a reliable and abundant food source. Animals of the benthic zone are usually sedentary or slow-moving clams, starfish, snails, worms, sea anemones, sponges, and the larger fish. Bacteria and fungi also inhabit this zone, thriving on organic debris that settles from the pelagic zone.

Not all ocean zones are considered productive; some regions are almost devoid of essential nutrients, and therefore few phytoplankton can survive. Since there is no food source for the zooplankton and larger marine animals, such regions have been referred to as ocean deserts. The commercial fishing industry is well aware of these unproductive zones.

One interesting and important part of the marine biome is the coral reef. Reefs have been formed only in warm, well-lighted waters of the world. The largest one is the Great Barrier Reef off the coast of northeastern Australia. Extending for a distance of about 2000 km, it, like other coral reefs, is composed of colonial coelenterates and encrusting algae. These organisms secrete calcium and become quite hard. Primary production (photosynthesis) is provided by symbiotic algae living among the coral. Coral exists in a variety of colors, and the reflection of light rays in tropical waters shows off a spectacular display of natural beauty.

Reefs provide both food and shelter for marine organisms, and the destruction of reefs is a major worldwide problem. Some of the more spectacular reefs have been plundered by skin divers and entrepreneurs; others are destroyed by massive storms or changes in populations of destructive organisms. It has been possible in some regions to construct artificial reefs from wrecked ships, automobile tires, and various other synthetic devices.

Coastal regions of islands and continents are generally rich in marine life, primarily because of the abundance of nutrients washed into the ocean from natural watersheds and raw sewage. In

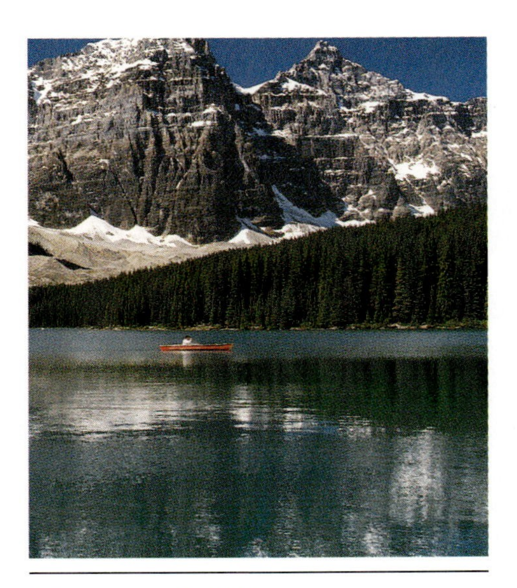

Figure 3.33
Freshwater biome. Human use of freshwater includes recreational activities, such as canoeing. Moraine Lake, Valley of the 10´ Peaks, Banff National Park, Alberta, Canada.

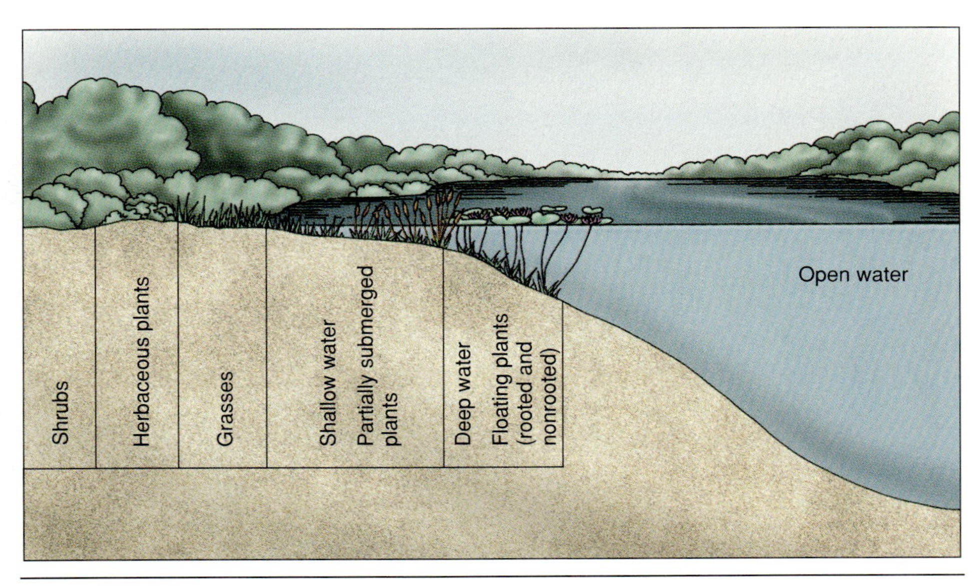

Figure 3.34
Freshwater lake ecology is complex. From completely terrestrial plants to rooted aquatics and nonrooted plants (especially algae) near the surface, a lake provides an array of different habitats for plant and animal growth.

temperate waters the giant kelps (brown algae) provide an exceedingly productive zone along the continental shelf, particularly along the western coast of North America. In warm tropical waters red algae are a primary food source that supports a great many species. Like the coral reefs, giant beds of seaweed provide food and shelter for large fish and benthic organisms. These regions are choice commercial and sports fishing grounds.

The macroalgae, or large multicellular algae, survive primarily along rocky coastlines where holdfasts attach the huge organisms to the rocks. Smooth sandy beaches provide no attachment surfaces for these organisms, and therefore they are seldom found in such regions. The habitat along the shoreline is particularly interesting because of the movement of the tides. Both plants and animals that survive there have been selected for a peculiar set of environmental conditions: alternately wet and dry, light and dark, and with high and low oxygen availability. Predators may have access only at high tide, but the drying, bright light of low tide may pose just as many obstacles for survival.

A similar set of circumstances exists in the salt marsh zone, regions where freshwater streams meet the sea. Organisms that survive the intertidal zone are subjected to alternating **osmotic shock**—the salts of seawater for a few hours followed by relatively dilute freshwater. (See osmosis, chapter 12, p. 203.) Few organisms could survive such treatment, and the marsh grasses and other producer organisms that grow there are special indeed. Interestingly, these are the same regions that receive rich nutrients from the land, and the productivity of these estuaries is among the greatest in the world. Estuaries are also prime breeding grounds for many forms of marine life.

Freshwater

Essentially all terrestrial organisms require a supply of freshwater. Plants obtain their moisture from the soil. Some animals are capable of securing all the water they need directly from eating plant tissues. The vast majority of animal life, however, needs additional freshwater supplies, and they find it in lakes, ponds, rivers, and streams. Humans are included among this number and, in fact, use a disproportionate share of the available freshwater in comparison with the remainder of the animal kingdom (figures 3.33 and 3.34). In addition to using more freshwater per unit of body weight, humans are the only animal group maintaining more individuals than the natural carrying capacity will support. The urban concentration, combined with continuing worldwide population growth, places an ever-increasing pressure on natural resources. Of the basic natural resources, humans require oxygen, food, fuel, freshwater, and raw materials for shelter, clothing, medicine, and industry. Loss of any one could theoretically limit future human growth. The one that first becomes inadequately available, however, is the limiting factor. Since only slightly more than 2% of the world's total land surface area is covered by standing or running freshwater, this resource is a prime candidate as the limiting factor for the human population.

The ecology of freshwater areas is complex. From the banks, where many vascular plants such as trees, shrubs, and herbaceous plants grow, to shallow water, where some specially adapted vascular plants such as cypress trees can exist, to progressively deeper water, where nonvascular plants, primarily algae, thrive, each slight change produces a new zone of plant species. Whether the water is

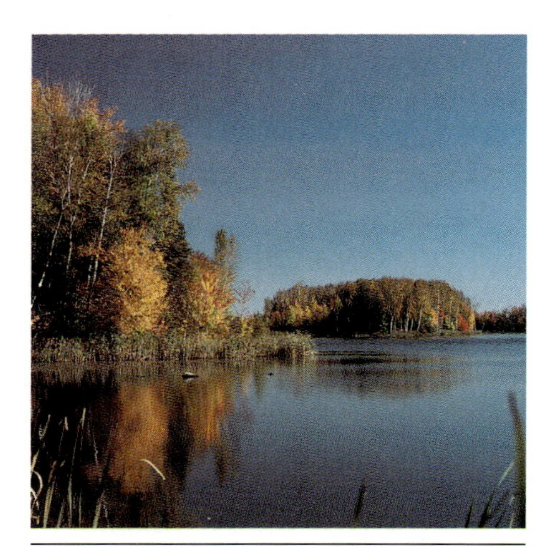

Figure 3.35
Freshwater lake—Spruce Lake in Taylor County, Wisconsin.

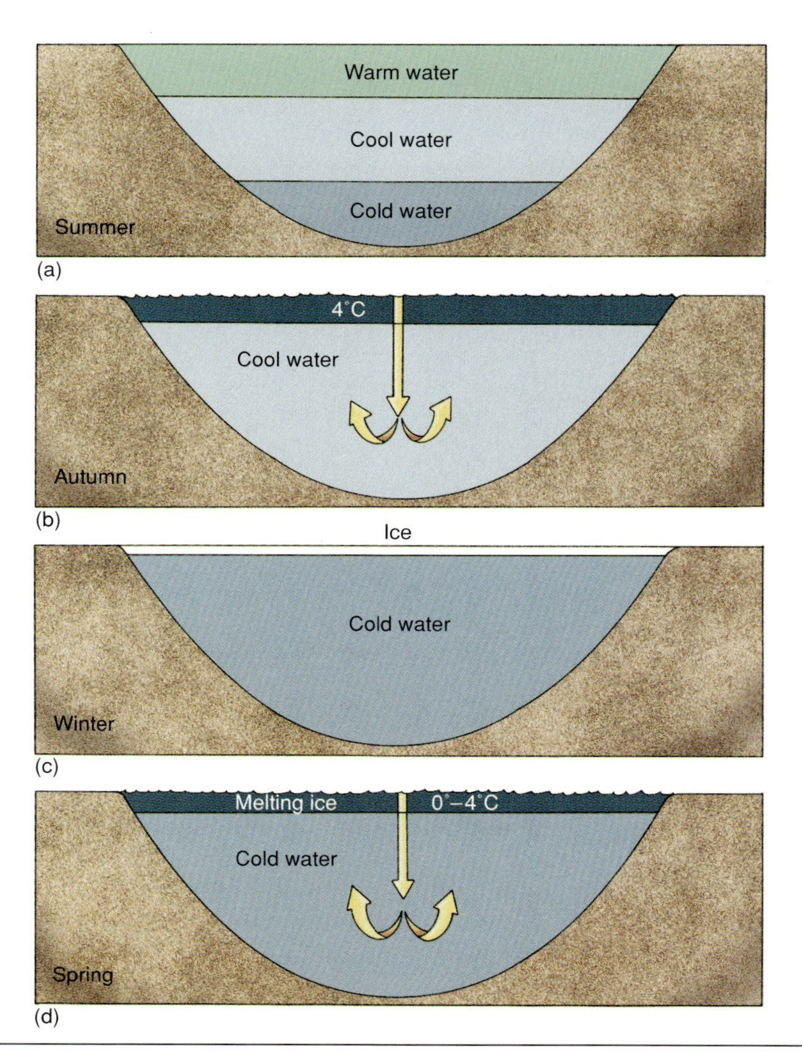

Figure 3.36
Thermal stratification. Water layers or stratifies at different depths, depending on temperature and resulting density. (*a*) During the summer the upper layers of the water become warmed by the sun while lower layers remain cool. (*b*) As the surface waters cool in the early winter to 4°C, the heavier water falls to the bottom, initiating fall turnover. (*c*) During the winter, ice forms at the surface, floats on the denser water below, and acts as an insulator. (*d*) As the surface ice melts in the spring and reaches 4°C, it becomes denser than the water below and sinks to the bottom, producing spring turnover.

still, as in ponds and lakes, or flowing, as in streams and rivers, it plays an essential ecological role.

Scientists who study freshwater biology are called **limnologists,** and they divide the biome into standing water and running water. In a sense, it is difficult to classify running water as an ecosystem, because the water and nutrients are not recycling within given boundaries. Standing-water lakes may be large or small, and the life zones are classified as the littoral zone, at the edge of the lake and quite productive (figure 3.35); the limnetic zone, a region of open water where phytoplankton are abundant in the upper layers, and the profundal zone, the region below the limnetic zone where there is no plant life. Principal occupants of this third region are the scavenging fish, fungi, and bacteria (figure 3.34).

The limnetic zone of lakes is similar in composition to the open water in the ocean (although the species may be different), but the littoral zone of the lake is unique. This shoreline may contain bottom-rooting aquatic angiosperms (flowering plants) such as cattails and rushes; water lilies and other rooted plants may extend farther out, and free-floating plants such as duckweed can extend out for a considerable distance.

Running water arises from melting ice or snow, from artesian water below the soil surface, or as an outlet from lakes. The flux or quantity of water transported per unit time determines to a large extent the kinds of organisms that prevail. Slow-moving streams may be rich in phytoplankton, whereas rapids and fast-moving water are not. Rapid water movement also precludes the attachment of angiosperms to the bottom of the river or stream. Under such conditions most productivity is confined to the quiet shallow areas, where algae and mosses can attach to rocks.

In areas with cold winters, **thermal stratification** occurs. This is the existence of layers of water at different temperatures and densities. One result of this is **turning over.** When water at the surface reaches its most dense state of 4°C it sinks to the bottom. As this layer descends, it oxygenates the water below and stirs up organic material on the bottom, resulting in a period of growth for freshwater organisms. Turning over occurs in the early winter as air temperatures cool the surface of the body of water, and again in the spring when the surface ice melts (figure 3.36).

Water is truly a unique liquid. If, as with other liquids, the freezing point produced a solid state more dense

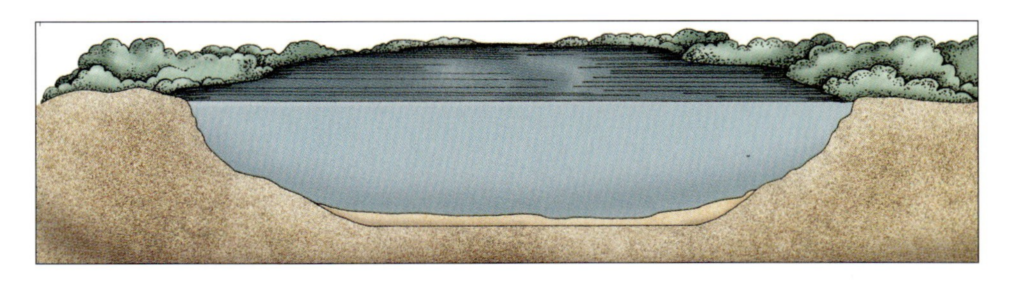

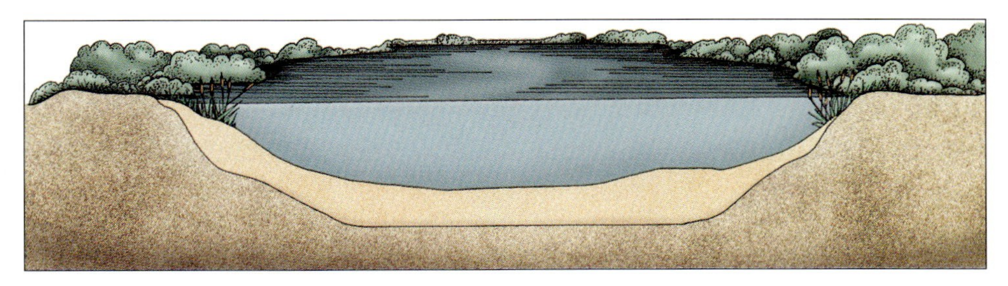

Figure 3.37
Eutrophication. Lakes naturally fill up with silt and organic debris from the plants and animals that have lived and died in the closed habitat. The filling is gradual and is accompanied by progressive encroachment of rooted shore plants.

than the liquid, ice would sink to the bottom. The pond or lake would be filled with ice, which would thaw slowly from the top down in the summer and probably never completely melt. This would eventually result in a solid block of ice in winter with only the upper portions ever being in the liquid state for part of the year. A vast majority of our deep freshwater bodies, then, would not support aquatic life, nor would it be available for human needs.

Because standing bodies of water have living organisms that reproduce, grow, and ultimately die, these bodies of water slowly fill with organic debris. During periods of eutrophic (nutrient-rich) conditions, bursts of growth and reproduction occur. As in the oceans, a majority of plant activity is at the surface where light is available. This often produces algal blooms, which cover the surface and initiate oxygen-poor conditions below. Excessive algal growth increases the animal populations that feed on them. The ultimate effect is a population crash in the pond or lake. This process of **eutrophication** is often artificially accelerated by human activities such as the dumping of raw sewage and the runoff of agricultural fertilizers, both of which provide rich nutrient supplies for the plant life growing there (figure 3.37).

Other human activities that not only disturb the ecological balances but also render the water unsafe for use include various forms of pollution. Chemicals, trash, and sewage are among the most serious water pollutants. (See pollution, chapter 22, p. 422.) Whatever the source, any negative use of the limited supplies of freshwater could have very serious consequences. We need to understand as best we can freshwater ecology and use this knowledge to wisely manage this critical resource.

WETLANDS

As a common feature of all terrestrial biomes, **wetlands** are variable and somewhat difficult to define; however, all definitions should include three basic components:

1. Wetlands are distinguished by the presence of water.
2. Wetlands often have unique, undrained hydric soil conditions that differ from surrounding upland areas.
3. Wetlands include hydrophytes, vegetation adapted to wet conditions.

Within these criteria, much variability exists from one wetland to another. Proximity to coastal areas, size, amount of time they are covered with water, and other influences affect the nature of each area. However, permanent waters of streams, deep lakes, and manmade reservoirs are not normally considered wetlands nor are areas that are so temporarily covered in water as to not have any effect on the development of moist-soil vegetation.

There are two main groups of wetlands: coastal and inland wetland ecosystems. Common names associated with wetlands include swamp, bog, marsh, slough, playa, moor (European), bottomland, fen, and others. They occur on every continent except Antarctica and only recently have scientists started to truly appreciate the importance of these ecosystems.

Historically, it was the swampy environment of the Carboniferous period that resulted in the fossil fuels on which we now depend for most of our energy needs. Of a more short-term importance, wetlands are known to cleanse polluted waters, act as buffers to help prevent floods and protect shorelines, and provide recharge for underground freshwater aquifers. These ecosystems also provide an important habitat for a wide variety of plants and animals including many migratory waterfowl.

Unfortunately, many of our country's wetlands have been drained, ditched, and filled to allow for human habitation and development. There are now laws that protect our remaining wetlands to allow scientists to adequately study and better understand these valuable areas.

Concept Checks

1. Where is the plant life in deep water of the marine biome?
2. In what part of the marine biome is species diversity greatest? Why?
3. What makes the ecology of freshwater areas so complex?
4. What are the basic criteria for an area to be a true wetland?

Summary

Biomes are worldwide groups of similar ecosystems. The terrestrial and aquatic biomes are summarized below.

1. **Tropical rain forests** Equatorial, large evergreen trees with a dense leaf canopy, little diurnal or annual temperature fluctuation, 200 to 500 cm of rainfall annually, shallow topsoils that are often lateritic, and many species of plants and animals.
2. **Savannas** At subtropical latitudes, there is some seasonality, 80 to 160 cm annual precipitation, perennial grasslands with scattered deciduous trees, maintained by periodic burning; the driest savanna areas are called thorn forests.
3. **Deserts** Primarily found in a latitudinal belt between 20° and 30° in both northern and southern hemispheres, only 25 cm or less annual precipitation, extreme daily temperature fluctuations, and annual seasons; low-growing, scattered vegetation is either small-leafed shrubs, succulents, or herbaceous annuals.
4. **Grasslands** Receive 30 to 150 cm of annual precipitation, distinct seasonality; constant coverage of perennial grasses is normal, but cultivation and overgrazing are permanently changing much of the world's native grasslands; in the more xeric grasslands, desertification is a common direction of change.
5. **Temperate deciduous forests** Found primarily in the northern hemisphere, 75 to 225 cm of annual precipitation is seasonal; deciduous trees (hardwoods) that dominate the vegetation in the summer produce a leaf canopy, but when the leaves drop in the fall, an autumn and spring understory of shrubbery and herbaceous plants becomes evident; the xeric end of the precipitation range produces what is termed chaparral or Mediterranean vegetation.
6. **Coniferous forests** Known as taiga in the northern latitude and high-elevation mountain areas; predominately northern hemisphere; annual precipitation of 35 to 100 cm limits vegetation to coniferous gymnosperms (softwoods), which are evergreen; at more southern latitudes this biome is produced by sandy soils.
7. **Tundra** Found mostly north of the Arctic Circle, about one tenth of the earth's land surface area; less than 25 cm of precipitation annually; tundra plants are scattered and low-growing perennials; permafrost soil conditions exist year-round and growing season is very short; mountain tundra occurs above tree line in high-elevation mountainous areas; vegetation and climatic conditions very similar to arctic tundra but at more southern latitudes.
8. **Aquatic biomes** The aquatic biomes include marine and freshwater systems; marine biome includes nearly 75% of the earth's total surface area and, although the oceans average 3 km in depth, most plant life is limited to the top few meters; most plant life is single celled, with the shallow water along the shores the only locality for the more complex kelps and other multicellular algae; ocean currents such as the Gulf Stream and Japan Current affect terrestrial vegetation by modifying the temperature of the area and thus the climate; freshwater biome includes flowing-water and standing-water ecosystem types; vegetation in both is more complex with more species than in the marine biome.
9. **Wetlands** Found in all terrestrial biomes, these areas are distinguished by the presence of shallow water, hydrophytic plants and unusual soil conditions. They include areas we know as swamps, bogs, marshes, and other names in both coastal and inland ecosystems. Our wetlands provide needed ecological resources and must be protected.

Key Terms

Discussion Questions

1. Are humans part of the natural world or not?
2. What are the long-term consequences of overpopulation in the tropics?
3. Do deserts have any potential for expanded human habitation and cultivation?
4. Can the marine biome solve our world food problems?
5. What should be our greatest concern about the world's supply of freshwater?

Suggested Readings

Chernov, I. I. 1985. *The living tundra.* Cambridge University Press, New York.

Colchester, M., and L. Lohmann, eds. 1993. *The struggle for land and the fate of the forests.* Zed Books, Atlantic Heights.

Cowell, A. 1991. *The decade of destruction: The crusade to save the Amazon rain forest.* Anchor Books, New York.

Gay, K. 1993. *Rainforests of the world: A reference handbook.* ABC-CLIO, Santa Barbara.

Hladik, C. M., ed. 1993. *Tropical forests, people and food: Biocultural interactions and applications to development.* Parthenon Publishing Group, Pearl River.

Kirk, R. 1992. *The Olympic rain forest: An ecological web.* University of Washington Press, Seattle.

Lovett, J. C., and S. K. Wasser, eds. 1993. *Biogeography and ecology of the rain forests of eastern Africa.* Cambridge University Press, New York.

Mabberly, D. J. 1992. *Tropical rain forest ecology.* 2nd ed. Chapman and Hall, New York.

Miller, K., and L. Tangley. 1991. *Trees of life: Saving tropical forests and their biological wealth.* Beacon Press, Boston.

Moffett, M. W. 1993. *The high frontier: Exploring the tropical rain forest canopy.* Harvard University Press, Cambridge.

Park, C. C. 1992. *Tropical rain forests.* Routledge, New York.

Peters, R. L., and T. E. Lovejoy, eds. 1992. *Global warming and biological diversity.* Yale University Press, New Haven.

Polis, G. A. 1991. *The ecology of desert communities.* University of Arizona Press, Tucson.

Torsten, A., and M. Diehl. 1992. *Deforestation of tropical rain forests: Economic causes and impact on development.* J. C. B. Mohr, Tubingen.

von Willert, D. J. 1992. *Life strategies of succulents in deserts with special reference to the Namib Desert.* Cambridge University Press, New York.

Part Two

The Inside Story

The common dandelion ensures seed dispersal by the wind with each seed having its own fluffy white "parachute" of bristles.

Molecules to Cells

Elements, Atoms, and Molecules

Molecular Bonding
Polar Molecules

The Molecules of Life

Water
Energy-Transfer Molecules
Macromolecules
Nucleotides and Nucleic Acids

Basic Cell Structure

Prokaryotic and Eukaryotic Cells
Primary Cell Wall
Secondary Cell Wall
Cell Membranes

Organelles and Other Inclusions

Nucleus
Mitochondrion
Plastids
Endoplasmic Reticulum and Ribosomes
Vacuole
Golgi Apparatus
Other Organelles
Microtubules

This chapter describes the chemistry of life—the molecules and the structure of the cell upon which all life depends. Beginning with atoms, one learns how molecules are constructed through chemical bonding, and then how these molecules become important to life processes. Water, energy-transfer molecules, simple carbohydrates, proteins, lipids, and nucleic acids are used to build macromolecules important to cell structure and function.

Cells function as they do because of compartmentalization of processes. Each organelle acts as an individual unit with specific metabolic functions. Membranes allow each organelle in the cell to specialize by separating groups of enzymes related to a particular function.

The raised venation on the underside of a giant water lily leaf.

As exciting as the study of ecology and the environment may be, it is impossible to understand the true organization and mechanisms of life without first understanding how the nature of life is developed from smaller and smaller building blocks. Different kinds of scientists look at organisms at different levels of organizations. Foresters (those who grow trees) may be concerned with the overall performance of a giant redwood tree, whereas a plant anatomist (a scientist who studies cells and tissues) might be concerned with the kinds of cells being produced in the trunk of the tree. A plant geneticist (a scientist who studies the mechanisms of inheritance) may be concerned with the cones and fertility of the pollen being produced, and a biochemist (a scientist who studies the chemistry of cells and tissues) may be concerned with the rate of photosynthesis being carried out by the leaves. Finally, there is a concern with the cells themselves and what is inside. This represents the field of cell biology (the structure and function of cells), an increasingly important component of life studies.

Anton van Leeuwenhoek (1632–1723) was among the first to examine tissue through a primitive optical system, the forerunner of the light microscope. About the same time, Robert Hooke saw boxlike compartments through a simple optical system that magnified what the naked eye could see. He described these structures as cells. Scientists later discovered that all living organisms except viruses are made up of these cells. It is true that cells come in all sizes and shapes, but the nature of cells is surprisingly similar in all organisms. They all have a **plasma membrane,** which encloses an aqueous solution called the **cytoplasm.** In the watery world enclosed by this membrane the mechanics of life occur: building large molecules from small molecules, making smaller molecules by tearing down larger ones, and always moving substances around. Constantly in motion, the cellular "factory" knows exactly where each "nut" and "bolt" goes.

Within this aqueous medium are even smaller structures, **organelles** (or, little organs) of various shapes and sizes. Each organelle has a specific job in the overall function of the cell.

Under a microscope, **cyclosis** (movement of the cytoplasm and many of its organelles) can be observed in the living cell; some go in one direction and others go in another. It resembles a network of freeways, with cars going in many directions. These movements are apparently not random, but very highly organized.

Chemistry is the center of cell structure and function. Certain chemical and physical laws govern how molecules are assembled from simple atoms, how the bonds are broken, and how they are re-formed in new molecules.

ELEMENTS, ATOMS, AND MOLECULES

All matter is made from various combinations of the **chemical elements,** substances which cannot be broken down by ordinary chemical means. The earth's crust contains 92 naturally occurring chemical elements; recently others have been made artificially. The basic particle of an element is the **atom,** and atoms are combined in various ways to form **molecules,** the building blocks of life.

Atoms are composed of three primary subparticles: proton (a positively-charged particle in the nucleus of the atom), neutron (a neutral particle that exists in the nucleus with the proton), and electron (a negatively charged particle that moves in specific tracks around the nucleus). Thus, the **nucleus** of the atom contains the protons and neutrons, and the electrons move in specified orbits around the nucleus. The proton has a specific mass and a positive charge; neutrons also have mass but no charge. Electrons have essentially no mass but have a net negative charge.

Various combinations of protons, neutrons, and electrons produce the chemical elements, and the structure of the elements is determined by the number of protons, the atomic number. The atomic weight is the sum of protons and neutrons. The simplest chemical element is hydrogen, with only one proton. Helium has two protons, **carbon** has six, and oxygen has eight (table 4.1). Each chemical element has its own symbol, often but not always indicated by the first letter. Some chemical symbols are derived from Greek or Latin terms. **C** is the symbol for carbon; **H** is the symbol for hydrogen, and **He** is the symbol for helium. But **K,** the chemical symbol for potassium, is derived from the word *kalium.* The symbol **P** is reserved for phosphorus; **Na** (*natrium*) stands for sodium; and **Fe** (*ferrium*) stands for iron.

Even though the atoms are often thought of as solid balls, they are actually more like hollow spheres. The nucleus, containing the protons and neutrons, occupies only a small amount of space. Likewise, the electrons themselves are small, but they move rapidly in orbits occupying a great deal of space. The distance from the electron to the nucleus is thousands of times the diameter of the nucleus. The number of electrons surrounding a nucleus is dictated by the number of protons in the nucleus. Different electrons of a particular atom have different amounts of energy, and they may occur at different distances from the nucleus. The greater the distance from the nucleus, the greater the energy of the electron. If energy is put into the atom, for example, when a photon (a discrete particle of light energy) strikes an atom, an electron may be displaced farther from the nucleus and therefore pick up additional energy. The energy levels are very discrete values, either high or low. Occupation of a position between these energy levels is not possible; changes in orbitals or

Smaller than a Breadbox

Since the beginning of time, there has been speculation about the nature of matter. The Greek philosophers Democritus and Leucippus suggested that if one were to cut a loaf of bread in half, and then in half again and again until there was nothing left to cut, then you would reach the ultimate building block of nature. This proposal was made some 2300 years ago, and the Greek philosophers called this "unit" an atom. The ancient Greek term is literally translated as "uncuttable."

We know now that slicing a molecule into its constituent atoms requires sophistication a bit greater than that of a bread knife, for the atom is so incredibly small that you hold a trillion trillion of them in your every breath. Atoms are virtually indestructible, and at any given time, you may be breathing in atoms once breathed out by Democritus, Julius Ceasar, Jesus Christ, or John F. Kennedy!

To grasp the scale of an atom, look at a period (.) printed on this page. If one uses the electron microscope to magnify that dot a million times, you will see a vast array of ink molecules. If you looked closely at one ink molecule, you would see the fuzzy outline of the largest atoms composing that molecule.

No scientist has ever seen the internal structure of the atom, but we know from thousands of years of experimentation and reasoning that atoms contain still tinier subatomic particles: protons, neutrons, and electrons. Electrons are so small that they carry essentially no mass, yet they soar at incredible speeds in orbits thousands of times the diameter of the nucleus of the atom.

It is interesting that exploring the smallest things in the universe requires the largest scientific equipment on earth. As physicists have probed deeper and deeper into the structure of the atom, the tools of the trade have become larger and larger linear accelerators. Stanford's accelerator near Palo Alto, California, is two miles long and it hurls electrons at 99.99% the speed of light. Other acclerators, like that of the European Laboratory for Particle Physics near Geneva, Switzerland, hurls protons instead of electrons—they are heavier and generate more collisions. It is more than four miles long and one of the largest pieces of equipment ever built. As subatomic particles are bombarded and split, they produce dozens of yet smaller particles, referred to in the 1960s as "a particle zoo." Many subatomic particles have been discovered, including at least six particles with the unlikely names of **quarks.**

Electric current is nothing more than flowing electrons, pointlike particles literally impossible to measure. They orbit an atom's nucleus governed by the principle of wave motion. Friction during the transmission process causes energy to be lost, and therefore electricity costs a great deal more than it might if the friction could be controlled. Recently the technology of superconductivity, which greatly reduces resistance to electric current at very low temperatures, has become feasible. New ceramic materials allow superconductivity to proceed even at higher temperatures. Plans were made for the construction of a superconducting supercollider (SSC) in the United States that would have had an orbital path of more than 50 miles. Unfortunately, due to cost overruns, construction of this facility was halted in the fall of 1993. If ever built, it is believed that this remarkable scientific achievement will hasten experiments to determine the true nature of matter.

TABLE 4.1

Selected Chemical Elements

CHEMICAL ELEMENT	PROTONS	NEUTRONS	ELECTRONS
Hydrogen (H)	1	0	1
Helium (He)	2	2	2
Carbon (C)	6	6	6
Nitrogen (N)	7	7	7
Oxygen (O)	8	8	8
Sodium (Na)	11	12	11
Chlorine (Cl)	17	18	17
Calcium (Ca)	20	20	20

**Note that the number of protons does not always correspond with the number of neutrons. On the other hand, the number of electrons always equals the number of protons—one negative charge for each positive charge. Currently, 109 chemical elements have been named, the newest termed Meitnerium.*

pathways of the electrons always follow exact patterns. For a given atom the electrons tend to seek the vacant spots in orbitals closest to the nucleus. But only two positions or "spaces" exist in each orbital. At the energy level closest to the nucleus, only two electrons can be positioned. In the next energy level there are four orbitals, each having a position for two electrons (total of eight). The third energy level can also accept eight electrons. Hydrogen has only one electron, and it occupies the first energy level. There is an additional, unfilled, position in that orbital. Obviously the larger and heavier atoms have electrons occupying many orbitals and energy levels.

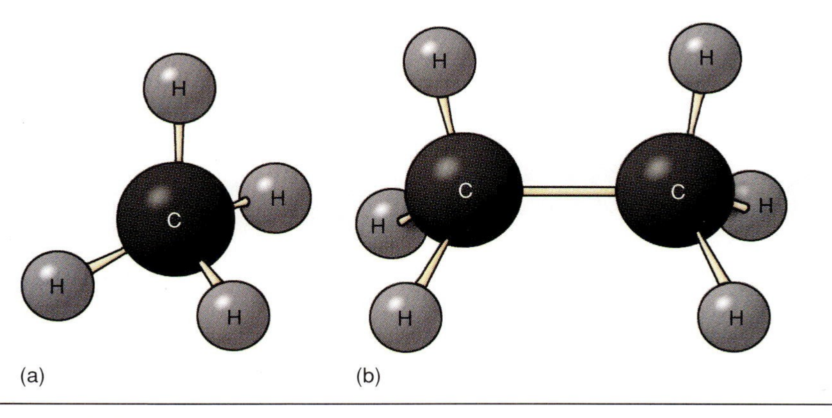

Figure 4.1
Various organic molecules are formed by combining atoms of carbon and hydrogen. (*a*) Methane has a single carbon, which shares electrons with four hydrogen atoms. (*b*) Ethane has two carbon atoms, which share electrons between themselves and three hydrogen atoms at each end of the molecule.

Atoms react chemically according to the number and arrangement of the electrons. Unfilled positions within an orbital tend to make the atom more reactive. Hydrogen, with only one electron, is extremely reactive; it will pair with other elements having unfilled orbitals. Helium, with two electrons in its first energy level, is a very inert gas and does not react readily, since both positions are filled in the first orbital. Neon, with an atomic number of 10, and argon, with an atomic number of 18, both have totally filled orbitals and thus are inert.

Molecular Bonding

Atoms are held together by various kinds of **chemical bonds** to form a relatively stable structure known as a molecule. Any two or more atoms that share a mutual attraction can form a molecule. For example, two atoms of hydrogen can form a molecule of hydrogen gas (H_2). Bonding of molecules may be due to ionic bonds. An **ion** is a particle with a negative or positive charge. In covalent bonding electrons are actually shared between atoms. Covalent bonds are particularly evident in the carbon-containing, organic molecules (figure 4.1). Thus atoms with unfilled orbitals can share electrons to form a new complete orbital and therefore remain stable. Sometimes atoms that share electrons are held together by two pairs of electrons, forming a **double bond,** or even three pairs of electrons, forming a triple bond. Molecules such as CO_2 are formed by oxygen atoms double-bonding to a carbon atom: O═C═O. Molecules such as nitrogen gas (N_2) are held together by the formation of three bonds (N≡N). This happens because the nitrogen atom (with an atomic number of 7) has two electrons in the first orbital, five in the second level of orbitals, and spaces for three more at that second energy level. Through the sharing of those electrons with another nitrogen atom, the triple bond is formed. Covalent bonds tend to be very strong and give stability to the molecule. Double and triple bonds tend to give additional rigidity to the molecule and make it even more stable. Such stability also imparts shape to the molecule, leading to a very specific three-dimensional structure. The presence of double bonds changes the physical properties dramatically, as seen in **lipids** (fats and oils) composed of carbon and hydrogen atoms. In oils (liquid at room temperature) some of the bonds are double, whereas in fats (solid at room temperature) all of the bonds are single. Lipids with a large number of single bonds are said to be saturated, and those with double bonds are said to be unsaturated. Lipids derived from plants tend to have a higher degree of unsaturation than those derived from animals and are thought to be more easily digested.

> ✔ *Concept Checks*
>
> 1. How are molecules formed from atoms?
> 2. How does polarity affect the orientation of one molecule toward another?

Polar Molecules

Sometimes the electrons in covalent bonds are not distributed symmetrically between the two atoms, but within the "electron cloud" they tend to form regions of more positive charge on one side of the molecule and more negative charge on the other side of the molecule. This uneven distribution of charge leads to polar molecules, and polarity influences the position of the molecules with respect to other adjacent molecules or ions (figure 4.2).

THE MOLECULES OF LIFE

Since all molecules are in motion and move faster as the temperature is raised, the chances become greater that two molecules will collide when the temperature increases. Likewise, increasing the concentration increases the chances of collision. Chemical reactions are nothing more than chance collisions between two particles capable of interacting in such a way that they share their electrons and perhaps form a new molecule or molecules. Chance collisions take place inside various parts of a living cell just as they do in a

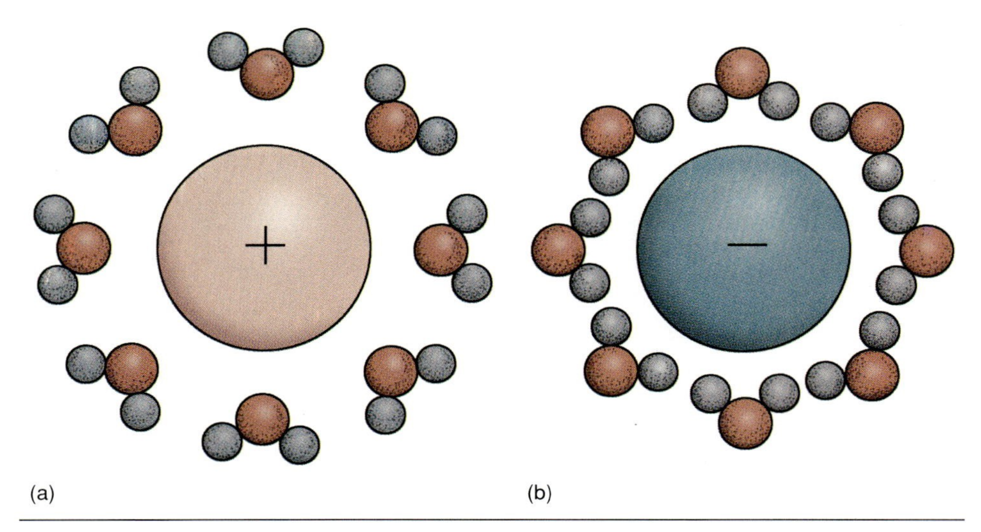

Figure 4.2
Water molecules are attracted to charged ions or to charged portions of polar molecules. (*a*) If a cation, or positively charged particle, is placed in an aqueous solution, water molecules surround the particle with the negatively charged portion of the water molecule oriented toward the ion. (*b*) If an anion, or negatively charged particle, is placed in an aqueous solution, water molecules surround the particle with the positively charged portion of the molecule oriented toward the ion.

chemistry lab test tube. The science concerned with describing the process in living cells is biochemistry. The only difference in biochemical and other chemical reactions is that the molecules of life are limited in number, and the chances that the reaction will occur are greatly enhanced by a very specific protein for each kind of chemical reaction. These proteins are called **enzymes,** and they act as a **catalyst** to make biochemical reactions proceed much faster than they would under chance conditions, without a catalyst.

Water

Most biochemical reactions take place in an aqueous medium. The water molecule is very special; similar molecules with respect to size, molecular weight, and shape simply do not perform the same functions. One might expect that a molecule with a single oxygen atom and two hydrogen atoms (H_2O) would be symmetrical, probably with one hydrogen jutting out of each side, exactly 180° from each other. Instead, the water molecule is arranged such that the hydrogen atoms form an angle of exactly 109°, with the oxygen at the center of the molecule. As the water molecule is formed, the electrons from all three atoms are shared and form an electron cloud around the molecule, which gives it an asymmetrical shape (figure 4.3). Even though the overall charge of the molecule is neutral, at the bulges created by the electron cloud around the two hydrogen atoms the molecule has a slight positive charge, whereas the other side of the molecule has a slight negative charge. Since negative charges attract positive charges, the water molecule will orient itself in specific directions toward any other charged site to which it is attracted, including other water molecules. Thus water has the property of hydrogen bonding, brought about by the attraction between a positive hydrogen atom and a slightly negative oxygen atom (figure 4.4). Hydrogen bonds are very weak when compared with covalent bonds, but they sometimes impart exceptional properties to molecules simply because of their great numbers. As one water molecule hydrogen-bonds to others, a specific pattern of molecules is determined, and a crystal-like lattice is formed, even in liquid water.

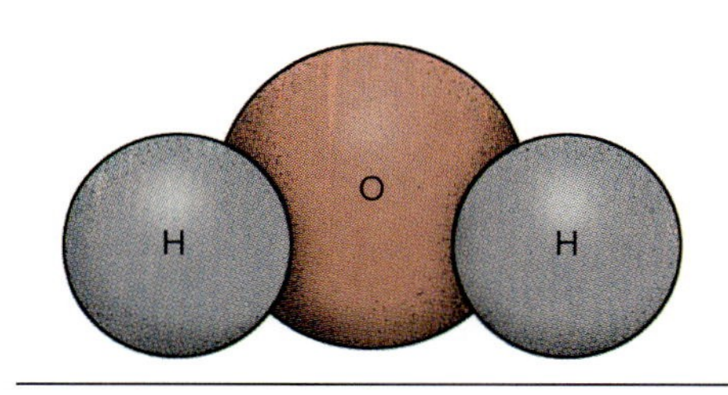

Figure 4.3
Water molecules consist of an oxygen atom adjoined to two hydrogen atoms. The angle between the hydrogen atoms is 109°, giving the overall electron cloud an asymmetrical shape. Such molecules are said to be polar.

Density

As the temperature becomes lower, the molecules move more slowly and become oriented more closely together. The molecules actually come closest together (water becomes more dense) at 4°C, not at the freezing point (0°C), where one might expect it to occur. The lattice actually begins to expand below 4°C, and as the water freezes its volume increases. Ice cubes have about a 10% increase in volume over liquid water. The ecological significance of this phenomenon is that because ice is less dense, it floats on water. Hence lakes and rivers freeze from the top downward, and seldom even in extreme latitudes does a lake freeze solid. Living organisms in the water are therefore protected from mechanical forces that would destroy them if all the water were to freeze.

Cohesion

The attraction between the positive side of one water molecule and the negative side of another is an example of a strong force called cohesion (taking place in perfectly pure water without dust, salts, or other contaminants). This attraction of similar particles is a major factor in the movement of water through confined spaces such as capillary tubes. The factor actually responsible for the attraction of one water molecule to another is the hydrogen bond. The strength of hydrogen bonding derives from the great number of bonds formed, even though individual bonds are quite weak.

Adhesion

Since hydrogen bonding can occur between hydrogen and unsatisfied charges of oxygen atoms of different

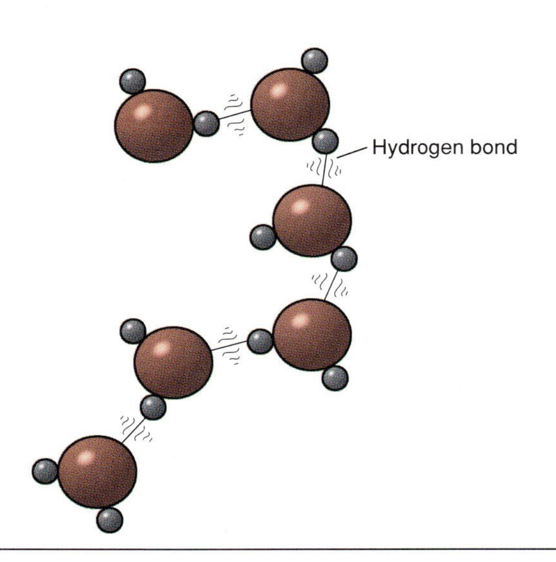

Figure 4.4
Water molecules are attracted to each other because of hydrogen bonding. The negatively charged portion of one molecule is attracted to the positively charged nucleus of a neighboring hydrogen molecule. Hydrogen bonds are very weak, but because of their large numbers they exert a tremendous force. This crystalline-like orientation of water molecules gives rise to cohesion.

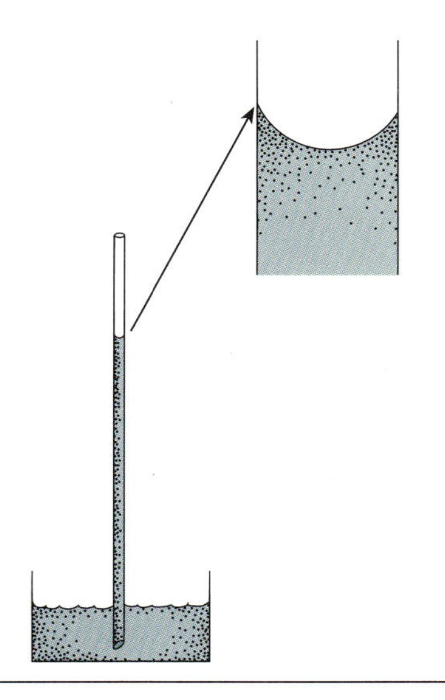

Figure 4.5
Cohesion and adhesion combine to produce capillary forces that allow water to rise in a glass tube. As the water molecules cling to exposed oxygen atoms in the glass, the molecules attempt to "climb" the walls of the tube and create a curved surface, or meniscus. Plastic tubes with no hydrogen bonding do not promote meniscus formation.

kinds of molecules, water also tends to be attracted to certain foreign substances. Adhesion is the attraction of unlike particles. In this case the hydrogen bond of water molecules is attracted to the oxygen of many substances, including wood, paper, and glass. Therefore water molecules tend to find certain substances hydrophilic (water loving) and other substances hydrophobic (water hating), such as waxed paper.

Consider a laboratory situation in which a small-diameter, glass tube is touched to the surface of water (figure 4.5). The liquid is drawn up into the tube to some height based on a number of physical factors, including the diameter of the tube. The attraction of the water to the sides of the glass tube is a function of the molecules of glass themselves, which have exposed, negatively charged oxygen atoms. Water molecules are attracted to these unsatisfied charges, and the orientation of the molecule is always toward the positively charged hydrogen portion of the molecule. Remember that water is attracted to certain materials simply because of the hydrogen bonding. This principle is important in understanding how water moves through plants.

Heat Gain and Loss

The standard unit of heat is called a **calorie,** the amount of heat required to raise the temperature of 1 gm of water 1°C. The amount of heat necessary to raise the temperature of water (specific heat) is about two times that for oil or alcohol. This means that water heats up and cools down very slowly, a physical property pertinent in climatology and ecology. Water is all around us—in the atmosphere, in soils, and in all living organisms. Because of the stability factor, all of the living organisms of our world tend to heat and cool less rapidly than if the water were replaced by some other molecule of similar shape and size. The greater the mass of water, the greater the stability; therefore landmasses located near oceans or large lakes tend to have a more stable temperature than landmasses located hundreds of miles from bodies of water. The atmosphere itself is a major factor in maintaining this temperature stability. If the relative humidity of a particular region is consistently high, then cold and warm cells of air are buffered in their effect, and the temperature changes slowly. On the other hand, if the relative humidity is consistently low, cold and hot air can change the temperature rapidly near the ground, and severe temperature extremes exist. This principle relates to the differences in organisms and their ability to withstand sudden temperature changes; aquatic organisms are usually highly sensitive to temperature changes. In all living organisms it is only within narrow temperature limits that chemical reactions will proceed.

Energy-Transfer Molecules

Unquestionably one of the most important molecules of life is the so-called energy-transfer molecule, **adenosine triphosphate (ATP)** (figure 4.6). ATP is present in all living cells, and it participates directly in providing the energy necessary for biochemical reactions. This molecule contains **adenine,** ribose, and three phosphate groups. It is possible to remove the third phosphate group from the second phosphate group by splitting the molecule and adding a molecule of water.

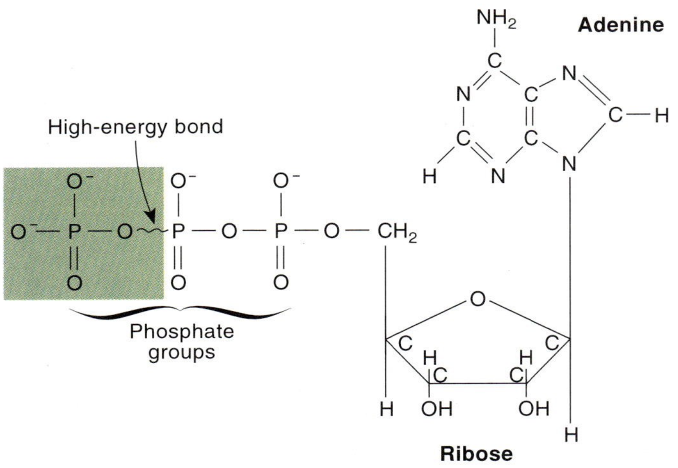

Figure 4.6
The energy currency molecule for all living organisms is adenosine triphosphate (ATP) or very similar molecules. The configuration consists of a molecule of adenine, plus ribose (a 5-carbon sugar), plus three phosphate groups connected by an ester linkage. Most of the energy is stored in the bond attaching the third phosphate group, accounting for about 7000 calories/mole of energy. ATP thus acts like a battery; the bond can form by the attachment of a phosphate group and the addition of 7000 calories/mole of energy. An enzyme catalyzes the making and breaking of the bond.

ATP is formed by adding a phosphate group to adenosine diphosphate (ADP) and releasing a molecule of water. The bond that connects the second and third phosphate groups on this molecule is a high-energy bond. The energy contained in this bond is more than 7000 calories/mole,* a value 100 to 1000 times greater than other chemical bonds. Obviously, when ATP is made, large amounts of energy must be put into the bond. When ATP gives up that energy, it becomes available to do metabolic work. Generally, as the energy is released for a particular reaction, the phosphate bond itself participates in the reaction and is transferred to the reactant molecule. Such molecules are said to be activated by the phosphate group. An example of this transfer is represented by the reaction:

$$\text{ATP} + \text{Glucose} \rightarrow \text{Glucose phosphate} + \text{ADP}$$

The energy contained in the new glucose phosphate molecule thus becomes available for further work.

*A mole is the molecular weight of a substance in grams.

All cells contain pools of ATP and ADP, and their relative concentration serves as a measure of the energy status of the cell.

Certain other energy-transfer molecules, such as nicotinamide adenine dinucleotide (NAD) and nicotinamide adenine dinucleotide phosphate (NADP), are also partially composed of nucleotides. These molecules are discussed in more detail with the processes of photosynthesis and respiration.

Macromolecules

By definition organic molecules are simply those which contain carbon and hydrogen. Most organic molecules also contain oxygen, but this component is not part of the definition. Methane, for example, is CH_4 and meets the definition of an organic molecule. Macromolecules simply refer to very large organic molecules formed by combining smaller molecules. Thus, all macromolecules are made of organic molecules, but not all organic molecules are macromolecules. Surprisingly few kinds of organic molecules make up the bulk of living tissue. The four principal types are carbohydrates, proteins, lipids, and nucleic acids. In addition, macromolecules include some of the more interesting secondary plant compounds, molecules found only in certain species or with some rarity in a number of species and not normally considered an essential part of a plant's metabolic function.

Carbohydrates

The most abundant organic molecules of life are the **carbohydrates,** in which the ratio of carbon:hydrogen:oxygen is 1:2:1. This means that regardless of molecular complexity, the common denominator for these atoms is the formula CH_2O. The simplest carbohydrates are the **monosaccharides,** such as 5- or 6-carbon sugars. They consist of a chain of carbon atoms to which hydrogen and oxygen are attached in various combinations. Essentially all organic simple sugars have 3, 4, 5, 6, or 7 carbons, sometimes referred to as the carbon skeleton or backbone of the molecule. Some of these simple sugars are shown in figure 4.7. Sometimes more than two monosaccharides can combine to form a **polysaccharide.** The principal polysaccharides in plants are **starch** and **cellulose.** Starch is primarily a storage carbohydrate, and cellulose is used exclusively for structure. In other organisms, different polysaccharides may be important. For example, **glycogen** is the primary storage carbohydrate in the human body, and chitin is the primary structural component of fungal cell walls and the exoskeletons (external surface layer that gives rigidity and structure) of insects and crustaceans. It is interesting to look at the chemical makeup of all these carbohydrates. They are surprisingly similar. All contain glucose or glucose derivatives, a simple hexose (6-carbon) monosaccharide hooked end to end for hundreds or thousands of units. Starch is made up of thousands of glucose units.

Polysaccharides are sometimes synthesized by an enzymatic reaction (essentially all biochemical reactions are brought about by enzymes) in which a molecule of water is split out between two glucose molecules. Later, when polysaccharides are broken down, as in the digestion of starch or glycogen, the molecule of water must be added back as the individual glucose molecules are re-formed.

Polysaccharides are relatively insoluble in water, and therefore serve storage or structural functions in cells. Their insolubility prohibits ready movement from one cell to another. Giant molecules cannot be readily transported across biological membranes. In animals, the transport of carbohydrates is most often accomplished as the simple sugar glucose; in plants, a molecule of glucose combines with a molecule of fructose, another simple hexose, and forms a disaccharide called **sucrose.** You will recognize this simple **disaccharide** as the same sugar obtained from sugarcane or sugar beets. It is extracted, concentrated, and crystallized to form table sugar.

Structural carbohydrates (cellulose in plants; chitin in insects, crustaceans, and fungi) are glucose-derived molecules linked end to end, but with slightly different chemical bonding such that the molecules become very rigid and relatively indestructible (figure 4.8). Cellulose is the primary component of plant cell walls and, in fact, is one of the most abundant molecules in nature. In primary cell-wall formation it is secreted across the plasma membrane to the outside of the cell, where it becomes the "box" in which the more fragile parts of the cell reside. Animal cells are not quite so fortunate; no wall ever forms, and the plasma membrane becomes the exterior of the cell.

Whenever cellulose becomes very thick in plant cells, the rigidity can be exceptional and lead to the formation of wood. Synthesis of cellulose in the secondary cell wall of the fiber of the cotton plant imparts its structure and the relative strength of its fibers. The greater the synthesis, the greater the strength and the longer your shirt or blouse will last. In structural woods

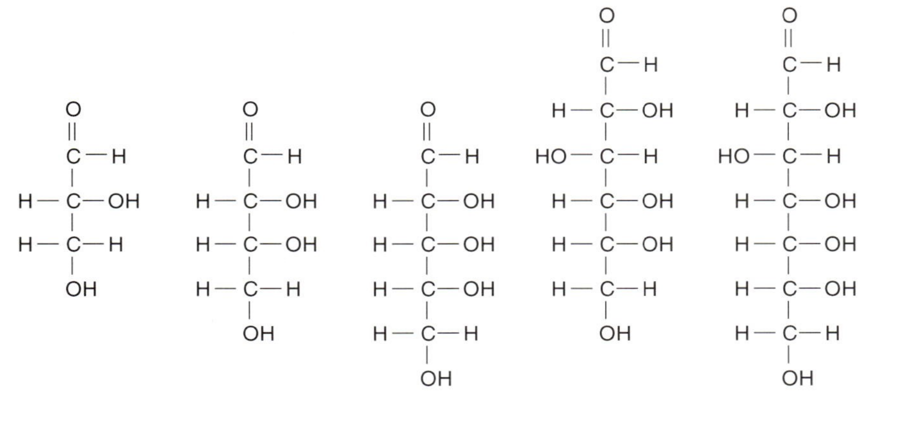

Triose (3-C)	Tetrose (4-C)	Pentose (5-C)	Hexose (6-C)	Heptose (7-C)
O=C—H	O=C—H	O=C—H	O=C—H	O=C—H
H—C—OH	H—C—OH	H—C—OH	H—C—OH	H—C—OH
H—C—H	H—C—OH	H—C—OH	HO—C—H	HO—C—H
OH	H—C—H	H—C—OH	H—C—OH	H—C—OH
	OH	H—C—H	H—C—OH	H—C—OH
		OH	H—C—H	H—C—OH
			OH	H—C—H
				OH
Glyceraldehyde ($C_3H_6O_3$)	**Erythrose** ($C_4H_8O_4$)	**Ribose** ($C_5H_{10}O_5$)	**Glucose** ($C_6H_{12}O_6$)	**Sedoheptulose** ($C_7H_{14}O_7$)

Figure 4.7
Most of the simple sugars contain from three to seven carbon atoms. The three-carbon sugars (trioses) and six-carbon sugars (hexoses) are among the most common in nature.

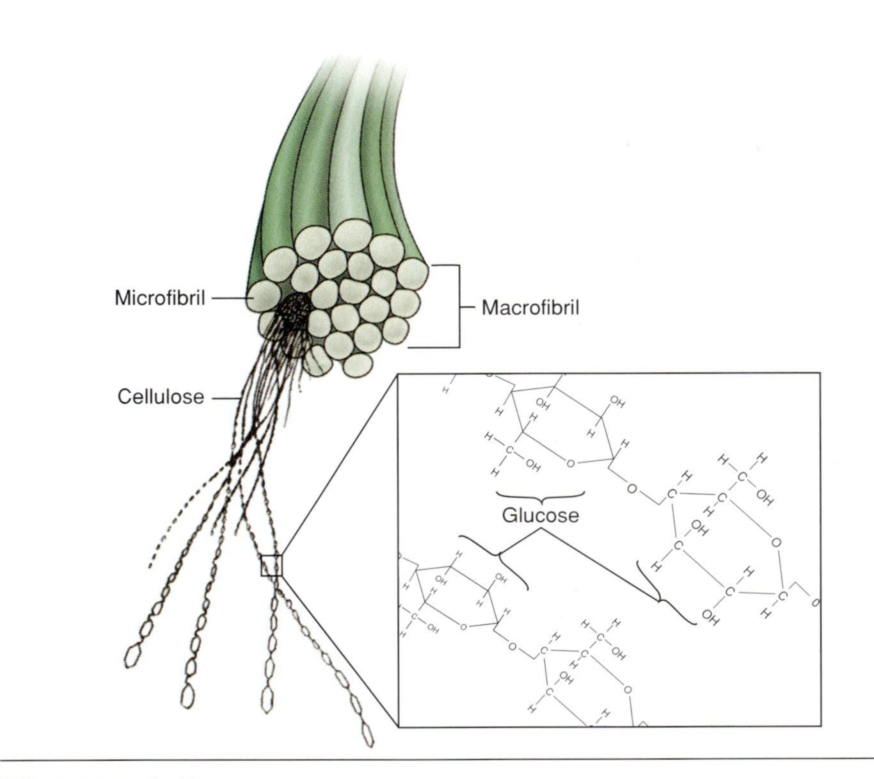

Figure 4.8
The cellulose molecule is formed by the linking of thousands of glucose molecules. These macromolecular strands are then cross-linked to form microfibrils, and groups of microfibrils are bound together to form macrofibrils. With a great deal of chemical cross-bonding, these structural carbohydrates give much strength to the cell wall.

(housing, furniture) and structural fibers (as in cotton) durability is brought about by the strength of the cell walls. Otherwise, a wooden house would constantly have to be rebuilt, and you would probably need a new shirt with each laundering. On the other hand, this permanence represents a problem in the recycling of organic matter in the ecosystem. Dead and decaying organic matter may be very slow to decompose, simply because the polysaccharides are so resistant to breakdown by the bacteria and fungi. Once cellulose has been formed, its function is strictly structural, and it is not an important energy source for the plant. Very few organisms have the ability to break down cellulose; certain bacteria and fungi, certain protozoans (such as those found in the gut of termites), and a few other higher organisms (such as silverfish) can do it. Otherwise, cellulose is resistant to decomposition, and the fibers made from it can last for hundreds of years, as in a well-maintained building. The enzyme responsible for the destruction of cellulose—cellulase—does not occur very frequently in nature.

Two other groups of polysaccharides are important constituents of plants: the **pectins** and the **lignins.** Pectic compounds are often found in association with cellulose and form the primary cementing substance between cell walls. Pectin is actually extracted from certain kinds of plants and used in making jellies and jams. During cooling, the polysaccharide structure forms a gel and gives the desired consistency to these products.

Lignin is also found in conjunction with cellulose in secondary cell walls. It is a very complicated, highly branched molecule made of many repeating units that gives rigidity and causes the wood to resist decomposers. Lignin has been a real problem in the pulping industry because, unless removed completely, it causes paper to yellow and lose quality. Strong chemicals must be used to remove the lignin, and the paper industry has long regarded this by-product a major cost and nuisance. New uses are being found for lignin, and certain chemical and industrial applications may make the cycling of wood and wood products more complete.

Proteins

All **proteins** are made up of nitrogen-containing molecules called **amino acids.** The characteristics of an amino acid are related to the name itself—an amino (—NH_2) group on one end and a carboxyl acid (—COOH) group at the other end. When the amino group of one amino acid is connected to the carboxyl group of another amino acid, a molecule of water is released, just as in the synthesis of polysaccharides, and the resulting bond is called a **peptide bond** (figure 4.9). Thus two amino acids connected by a peptide bond are called a dipeptide, three of them hooked together are called a tripeptide, and many of them constitute a **polypeptide chain.** Most enzymes are single polypeptide chains of several thousand amino acid molecules; some, however, consist of two or more polypeptide chains. The amazing procedure by which the genetic makeup of an organism tells the cell which amino acid sequence to use to make a particular protein is called **protein synthesis** (see chapter 5, page 66). Proteins are composed of up to 20 different amino acids (figure 4.10). Their basic structure is the same but, depending on side groups, they may be charged differently. Besides these 20 amino acids, certain species of plants may synthesize many other kinds of amino acids, all nonessential for protein synthesis. There are over 300 known different nonprotein amino acids, and some of them are so specific that they can be used for taxonomic identification.

Proteins are polymers of the 20 amino acids held together end to end, but the structure of the molecule is more complicated because the repeating units are different, and the sequence becomes critical. Whereas

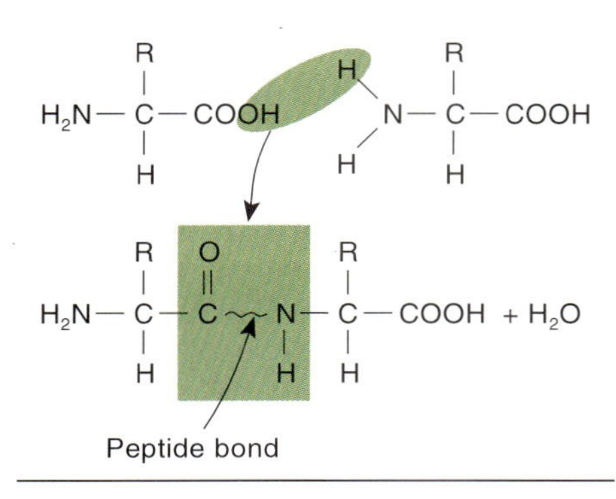

Figure 4.9
A peptide bond is formed by joining the amino (NH_2) group of one amino acid with the carboxyl (COOH) group of another one. A molecule of water is removed, and the peptide bond joins together the two amino acids.

starch is simply an array of repeating glucose units, specific proteins are made from thousands of a very specific sequence of the 20 different amino acids. This polypeptide chain (that is, a single linear array of amino acids) is referred to as the primary structure. Proteins do not occur as a straight thread but coil into a spiral shape, and the coil is held in position by the hydrogen bonds within the molecule. This coiling is referred to as a secondary structure. Finally, certain other types of chemical bonding may occur between other types of atoms, generally between the sulfur of one amino acid and the sulfur of another amino acid many units away to form an S—S (disulfide) bridge. This causes the molecule to loop back on itself and hold the overall structure in a particular three-dimensional configuration. Disulfide linking to form this three-dimensional image is referred to as tertiary structure (figure 4.11).

Based on tertiary structure there are two basic types of proteins—fibrous and **globular.** Fibrous proteins tend to be simpler and generally serve a more structural function. Fibrous, structural proteins give internal structure to both the plant and animal body. In globular proteins the tertiary structure is often more complex, and the protein function becomes much more specific. They are often involved in complex functions where the

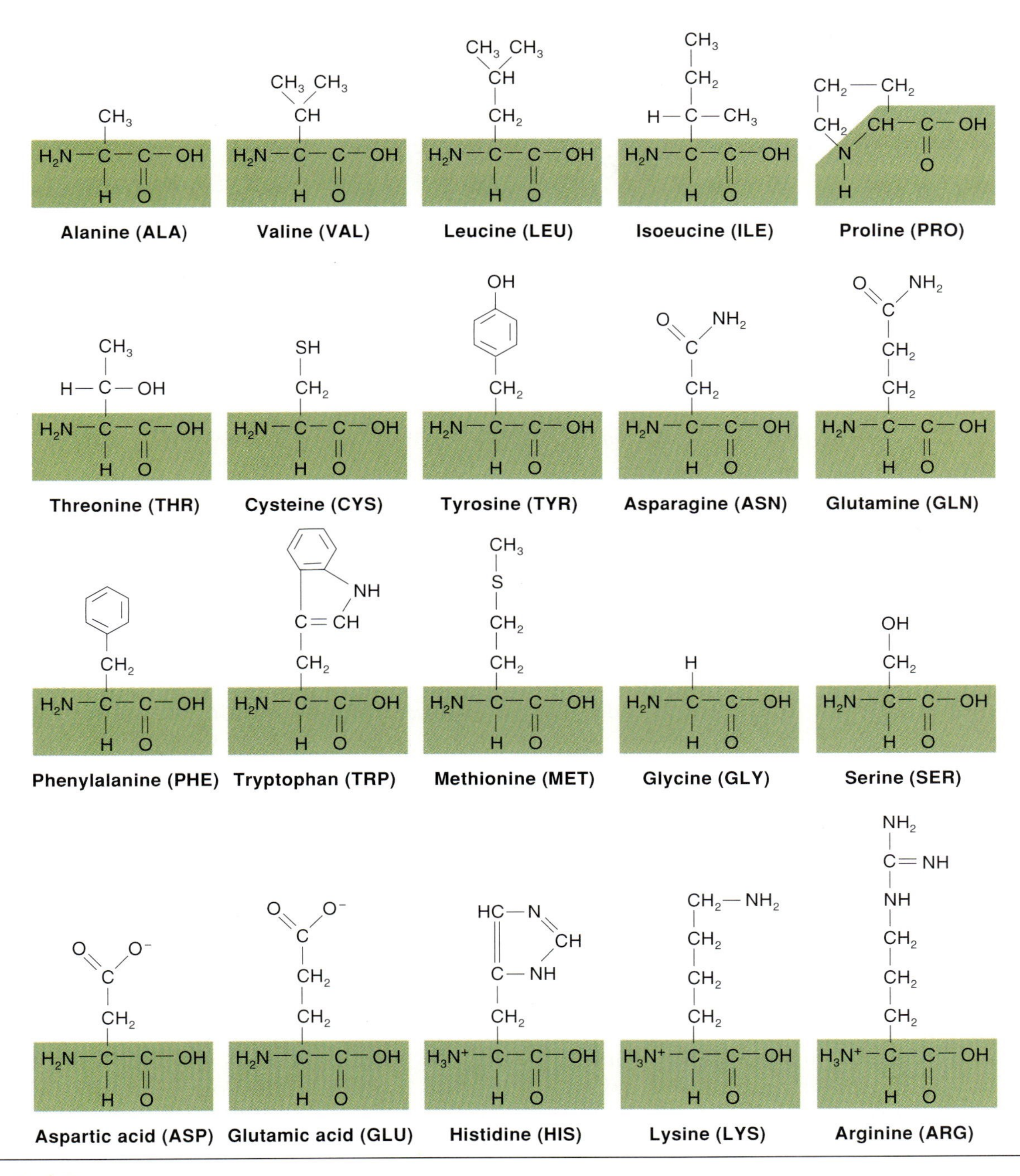

Figure 4.10

Twenty different amino acids, in various combinations, make up the protein molecule. Not all twenty may be found in any single protein. Even though the number of amino acids is relatively small, the possible combinations are great enough to account for the thousands of different kinds of protein. All have the basic structure: one or more amino groups (— NH_2) and one or more carboxyl groups (— COOH). The variability in amino acids is demonstrated by alanine and valine.

three-dimensional nature is all important. The most important functions are as enzymes, the catalysts that cause biochemical reactions to occur rapidly. Each biochemical reaction requires a specific, unique enzyme. Its specificity is dictated by the sequence of amino acids, secondary structure, and tertiary structure. In some cases different chains of the molecule are held together in a particular configuration, referred to as quaternary structure (figure 4.12). For an enzyme to function effectively, it is absolutely essential that the three-dimensional structure be correct. The shape of the molecule is important in its attachment to the substrate (the molecule on which the enzyme will effect a change). When this structure is lost—for example, when the white

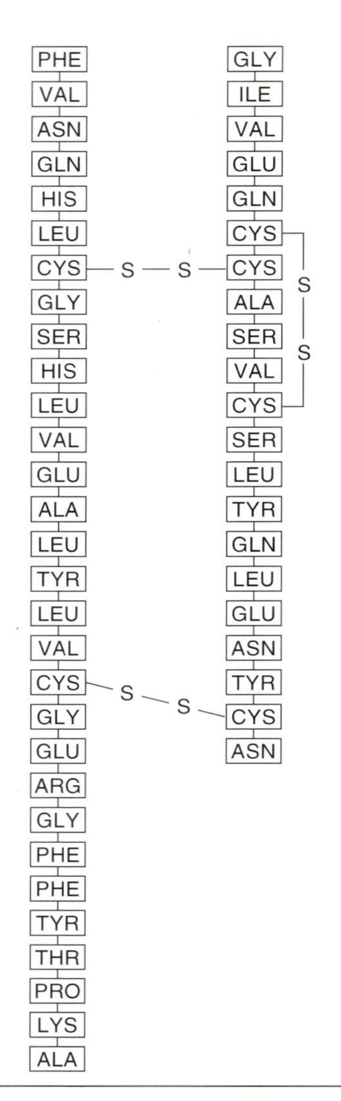

Figure 4.11
The structure of insulin. This protein is made of two amino acid chains, held together by sulfhydryl bonds (S—S) both between the chains and within one of the chains.

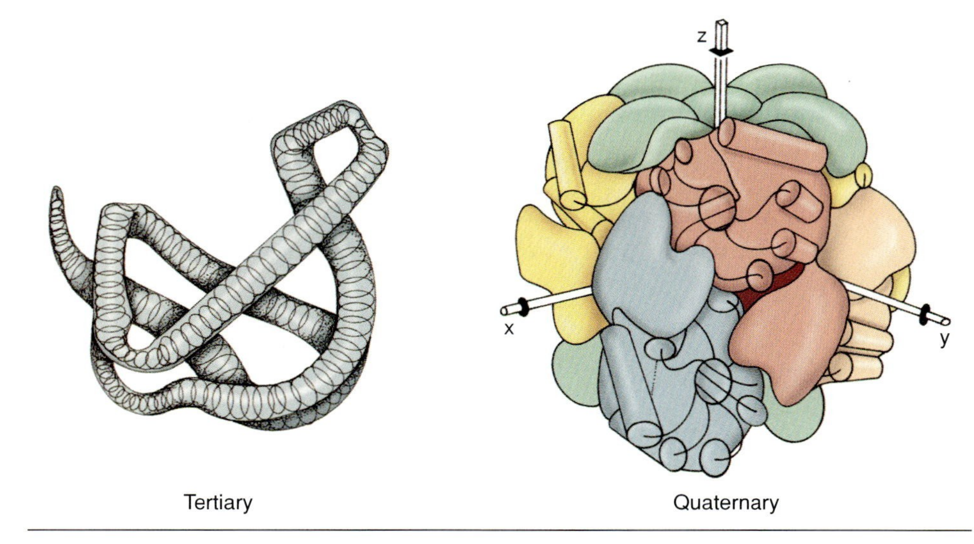

Figure 4.12
Globular protein structure. These models demonstrate how tertiary and quaternary structures hold the molecule in place and give rise to its three-dimensional shape.

of an egg is heated—the tertiary structure is destroyed and the loss of structure is not reversible. Such a protein is said to be denatured, a process caused by heating, heavy metal toxicity, and many organic molecules that promote pollution. Once specificity has been lost, that particular biochemical reaction usually cannot occur unless another enzyme molecule takes its place.

Proteins are sometimes used for energy storage, as in protein-rich seeds. Soybean, for example, stores much of its energy in the seed as protein. Whenever it is time for germination, that protein becomes hydrolyzed (digested), just as in the digestion of polysaccharides, and the end products are the original amino acids. These amino acids then become available for synthesizing new proteins.

Proteins are very large molecules. The smallest ones have molecular weights (sum of the atomic weights) of approximately 6000, and the largest ones may have molecular weights of hundreds of thousands. Compare this with a molecule of carbon dioxide (mol. wt. = 44) or even glucose (mol. wt. = 180). It is easy to calculate the approximate number of amino acids in a particular protein by using the average molecular weight of 100 for a single amino acid.

Lipids

Lipids are macromolecules that are oily or fatty in nature, are relatively insoluble in water, and contain a large number of C—H bonds capable of releasing a great deal of energy. A given amount of lipid yields much more energy on oxidation than do carbohydrates or even proteins. This is one reason we have trouble getting rid of a few extra pounds of fat. It takes a much larger energy output than intake to cause our bodies to burn up fat as an energy source, and a little fat produces lots of calories.

The general storage form of energy for lipids is as fats or oils. The only difference is that fats are solid at room temperature and oils are liquid at the same temperature. Cells are able to synthesize fats (as well as all other organic molecules) from sugars, the carbon skeleton having been produced in photosynthesis. A fat consists of three fatty acids hooked together to a molecule of **glycerol** (figure 4.13*a*). Fatty acids are long chains of carbon and hydrogen atoms that have a terminal carboxyl (—COOH) group. The terminal carboxyl group causes fatty acids to behave as very weak acids. In making the fat, the glycerol molecule links with the —COOH group, splitting out a molecule of water. You will recall that the same basic process is used in making polysaccharides and proteins. During digestion of fats a molecule of water must be reinserted when the molecule is broken down to its fatty acid and glycerol components.

In the investigation of human nutrition and heart disease there is concern about the degree of "saturation" of fatty acids. A saturated fatty acid is simply one in which each carbon is accompanied by as many hydrogen atoms as possible; that is, each carbon atom has four bonding sites (figure 4.13*b*).

On the other hand, an unsaturated fatty acid contains carbon atoms

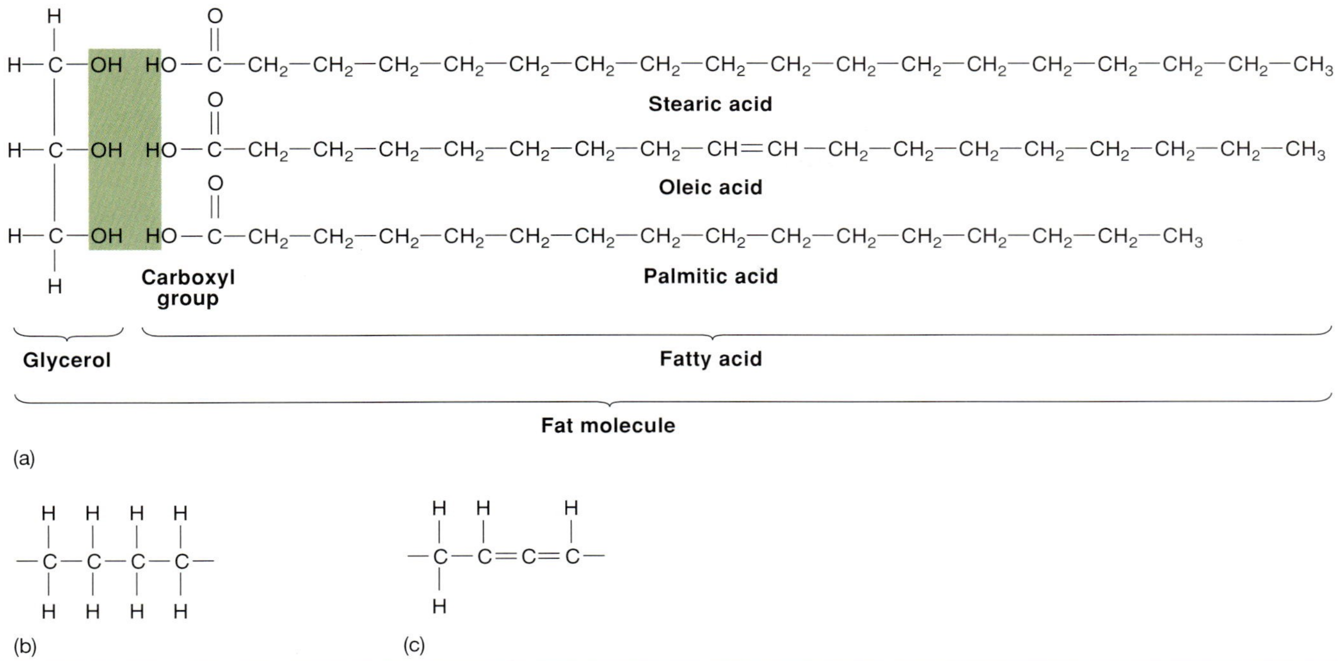

(a)

Stearic acid

Oleic acid

Palmitic acid

Carboxyl group

Glycerol

Fatty acid

Fat molecule

(b)

(c)

Figure 4.13
Fats are composed of three fatty acids attached to a molecule of glycerol. (*a*) The length of fatty acids varies greatly according to an even number of carbon atoms. Fats may be saturated (*b*) or unsaturated (*c*) depending on the number of hydrogen atoms.

joined by double bonds occurring at every position where a hydrogen atom is missing (figure 4.13*c*).

The number of double bonds indicates the degree of unsaturation, and unsaturated fatty acids are more common in plants; saturated fatty acids are common in animal tissues. Unsaturated fatty acids are more easily digested and are thought to contribute less to heart diseases. Typical unsaturated oils, liquid at room temperature, are corn oil, safflower oil, sunflower oil, and peanut oil. Commercial vegetable oils are usually combinations of some of these. Lard, an animal fat, tends to be solid at room temperature. Fatty acids in both plants and animals are synthesized in 2-carbon fragments, and therefore the total number of carbon atoms is always in even numbers. Typical fatty acids in plants contain 14, 16, and 18 carbon atoms.

Waxes Waxes are similar in structure to fats, except that a long-chain alcohol replaces the glycerol that becomes attached to the fatty acids. In waxes the length of the chain is longer, and therefore the melting point is higher. These wax molecules are resistant to degradation and provide an excellent barrier to water loss. A plant wax (cutin) is important as a covering (the cuticle) for the entire epidermis exposed aboveground. Another waxy layer that covers the walls of certain kinds of corky cells is called **suberin.**

Phospholipids Sometimes the glycerol molecule is attached to two rather than three molecules of fatty acid. The third position is occupied by a molecule containing phosphorus; hence the name **phospholipid.** Phospholipids are very important in the structure of **membranes.**

Even though the overall fatty molecule is not soluble in water, the phosphate portion is, so the molecule orients itself with the phosphate portion pointed toward the water layer. In the cytoplasm, phospholipids tend to line up in layers, with the insoluble fatty acids oriented toward one another and the phosphate ends pointed outward toward the water layers.

Nucleotides and Nucleic Acids

Nucleotides are macromolecules composed of a phosphate group, a pentose (5-carbon sugar), and a nitrogenous base. The nitrogen base is a ring structure containing carbon, hydrogen, nitrogen, and sometimes oxygen. The ring structures exist as the single-ring pyrimidines—**uracil, thymine,** and **cytosine,** and as the double-ring purines—adenine and **guanine.** Thus there are five kinds of nucleotides.

Nucleic acids are nothing more than polymers of nucleotides, and there are only two kinds: **DNA (deoxyribonucleic acid)** and **RNA (ribonucleic acid).** As the names imply, DNA contains a sugar called deoxyribose; RNA contains ribose. Both DNA and RNA contain the purines adenine and guanine, but a difference occurs in the pyrimidines. DNA contains thymine and cytosine, whereas RNA contains uracil and cytosine. In addition, the DNA molecule consists of two strands, whereas the RNA molecule is always a single strand; the only exception occurs in viruses.

✔ Concept Checks

1. Why is water so important to the chemistry of life?
2. How do energy-transfer molecules absorb and release large amounts of energy necessary for biochemical reactions?
3. What are the primary macromolecules important to life?

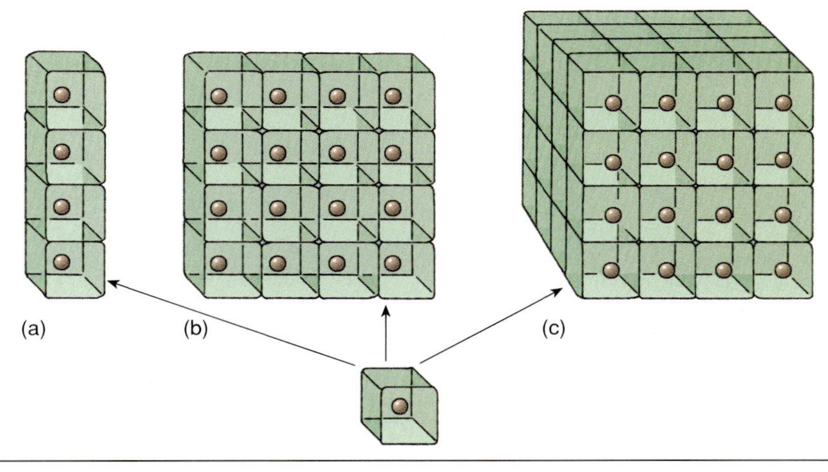

Figure 4.14
Whenever mother cells divide, they may do so in all three planes. (*a*) If the plane of division is the same every time, the result is a chain or filament of cells. (*b*) If the mother cell divides in two planes, a sheet of cells is the result. (*c*) If the mother cell divides in all three planes, the result is a three-dimensional block or sphere of cells.

DNA is the molecule responsible for storage of all hereditary information. Its significance in the biological world is almost beyond comprehension; this single molecule is responsible for the differences and similarities of all the organisms in the world. As far as we know, the method of hereditary information transfer is exactly the same today as it was when life first began with the single-celled organisms.

RNA exists in at least three different forms, each of which has a specific function. The largest one is called **messenger RNA (mRNA)** and carries the genetic code from the DNA. An intermediate-size molecule is the **ribosomal RNA (rRNA),** and it is a structural component, along with protein, for the ribosome. A third small molecule is **transfer RNA (tRNA),** which helps coordinate the sequence of amino acids at the time a protein is being formed (see chapter 5, page 84).

BASIC CELL STRUCTURE

Although plant and animal cells are very much alike, plant cells have a wall that becomes more or less rigid. Animal cells lack this **cell wall.** This is not to say that the plasma membrane surrounding an animal cell is "naked." Large molecules called glycoproteins (carbohydrate + protein) occur on the outside of the membrane and serve as a recognition surface. These glycoproteins form characteristic three-dimensional surfaces that allow cells sliding past each other to recognize similar surface features and then cling together. Such recognition may be less important in plant cells because the cellulose wall fixes the cells in place; there is little movement. Even in some plant cells, however, protein layers on the exterior surface of the cell wall are important in recognition. Plant cells are grouped together in tissues by virtue of their proximity at the time of division. The plane of division determines whether a group of plant cells will divide in only one plane to form a chain, in two planes to form a sheet, or in all three planes to form a cube (figure 4.14). If cell divisions were strictly random with respect to plane of division, a sphere of cells would result. Such is the case in tumors, which can develop in both plants and animals.

The basic shape of any tissue is brought about by the **turgor pressure** (the water pressure) inside each cell. Cells are literally blown up, and all push against each other. These cells are inflated by water pressure, rather than air pressure.

The large cellulose molecule in plant-cell walls is made from units of **glucose,** a 6-carbon sugar (hexose) (figure 4.15). The glucose units are connected end to end, forming a **macromolecule.** A cellulose molecule may be thousands of glucose molecules in length, and these long molecules are bundled together in packages to form a **microfibril** (figure 4.8). Groups of microfibrils are wound together much like a steel cable to form a **macrofibril.** The macrofibrils are a major component of the cell wall and are held together with other kinds of macromolecules, including hemicelluloses and pectic compounds. These substances glue the entire structure together in a sheet of fibers. The first microfibrils of the primary wall form a network with a predominantly transverse pattern. When turgor causes the cell to expand and the wall increases in surface area, the outer microfibrils become more parallel to the longitudinal axis of the cell. The overall effect is a cross-hatched appearance of the various layers (figure 4.16). Two adjacent plant cells are held together by the **middle lamella,** composed primarily of cementing pectic substances, and the **primary cell wall** on each side of the middle lamella. In many plants a **secondary cell wall** may be laid down at a later date, adding strength and rigidity to the tissue, particularly if lignin is present. Tree trunks, for example, have cells with very thick secondary cell walls.

Multicellular plants communicate by tiny passages that connect the cells—through one plasma membrane,

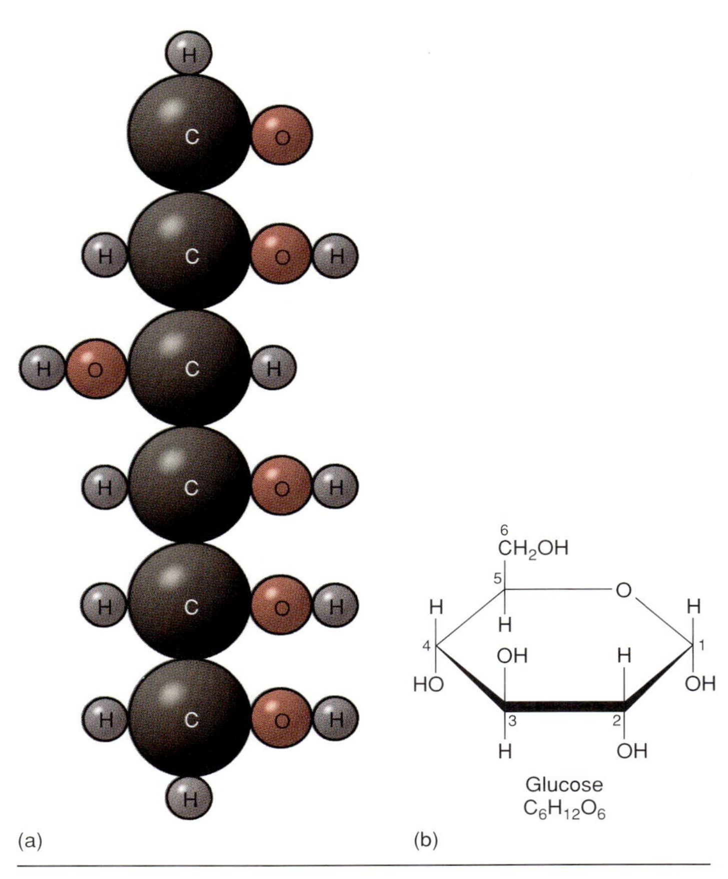

Figure 4.15

The glucose molecule is a simple hexose (6-carbon sugar). (*a*) The carbon atoms are shown schematically as a "backbone" or carbon skeleton. Oxygen and carbon atoms are attached at various positions. Note that electrons are shared between adjoining atoms; carbon must share four electrons, oxygen must share two, and hydrogen must share one. (*b*) Whenever a glucose molecule is placed in water, it assumes a ring configuration. The number of atoms is not changed, but the shape of the molecule changes drastically. By convention, carbon atoms are not shown in the ring forms.

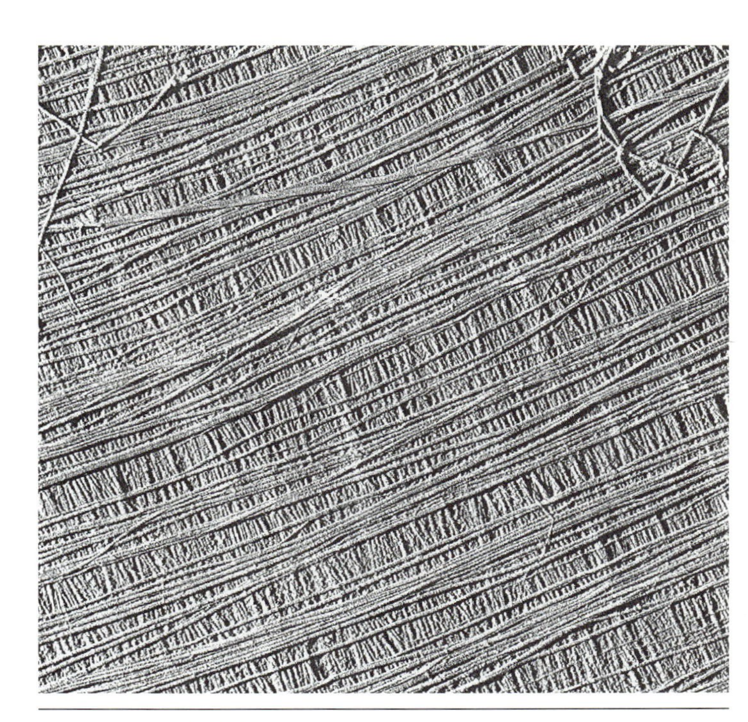

Figure 4.16

An electron micrograph of the cell wall of the green alga *Chaetomorpha melagonium* reveals two parallel layers of cross-linked macrofibrils. (× 15,000)

the primary cell wall, the middle lamella, the adjacent primary cell wall, and finally the adjacent plasma membrane. These connections occur in thin areas of the wall called **primary pit fields,** and each strand is called a **plasmodesma** (pl., **plasmodesmata**). Although too small for the interchange of most larger organelles, cytoplasmic connections apparently allow for the transfer of chemicals from one cell to another.

The wall itself is extremely permeable to all kinds of substances. Cross-linking of the molecules still allows water and various kinds of solutes to penetrate the wall; the barrier, determining what gets in and out of the cell, is the plasma membrane itself, just as in animal cells.

Prokaryotic and Eukaryotic Cells

Within the living world there are basically two types of cells: those which are **prokaryotic** and those which are **eukaryotic.** Prokaryotic means "before the nucleus"; therefore prokaryotic cells are those which do not have a well-defined nucleus. All other cells in the world, both plant and animal, do contain a nucleus and are therefore referred to as eukaryotic. Prokaryotes are the more primitive cells (figure 4.17). Probably the prokaryotic cells that exist today are similar to the very

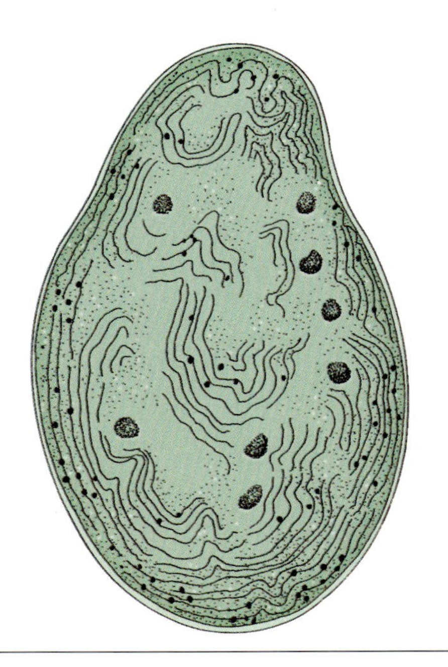

Figure 4.17

The prokaryotic cell has a simple structure; the organelles are either indistinct or not present. There is no nucleus, the chlorophyll molecules are not organized into chloroplasts, and there are no mitochondria. The cytoplasm is contained within a membrane, and other loosely organized membranes constitute the matrix. Various lipids and protein bodies may be seen.

first cells on earth. Prokaryotic cells first made their appearance in the oceans some 3.5 billion years ago and today are represented by two types of organisms, bacteria and cyanobacteria. Now, as then, all prokaryotic organisms were single celled. Some chains, groups, and colonies exist, held together within a gelatinous sheath, but no truly multicellular organisms with cellular differentiation occur. Prokaryotic cells are comparatively simple and small, rarely more than 1 or 2 μm in diameter. They consist of a rigid, noncellulose wall surrounding a plasma membrane that holds the components of the cytoplasm. They do contain ribosomes, but for the most part organic molecules are simply free in the cytoplasm. Packaging and partitioning in these cells is not nearly so dramatic as it is in the eukaryotic cells. The hereditary material within these cells, DNA, consists of a single long molecular thread. It is circular, but it is not organized into any sort of regular chromosome as occurs in the nucleus of eukaryotic cells.

Primary Cell Wall

In newly forming plant cells the wall that first surrounds the plasma membrane is referred to as the primary cell wall. Initially, the primary cell wall is relatively plastic, gradually becoming more rigid as the cell ages and enlarges. It finally stretches more and more, as new cell-wall material is synthesized, until the cell has reached its ultimate size. Certain cells grow more in certain directions than in others, leading to the elongation process. Elongation in some plant cells reaches many centimeters. The cotton fiber, for example, often reaches lengths of 4 to 6 cm. The size of certain cells is determined by the species in question, and genetics will ultimately determine how large a cell can become. Generally, both plant and animal cells are microscopic.

Secondary Cell Wall

Many plant cells develop a secondary cell wall, which is formed between the primary cell wall and the plasma membrane and may become much thicker than the primary wall. The secondary cell wall adds strength and rigidity to the cell. In some cases, for example, the flax fiber, the secondary cell wall gradually fills the space occupied by the cytoplasm, and the **lumen,** or living space inside the cell, becomes very small indeed. The same is true for many other types of fibers. By the time the secondary cell wall is complete, some plant cells die, and the cytoplasm is simply absorbed.

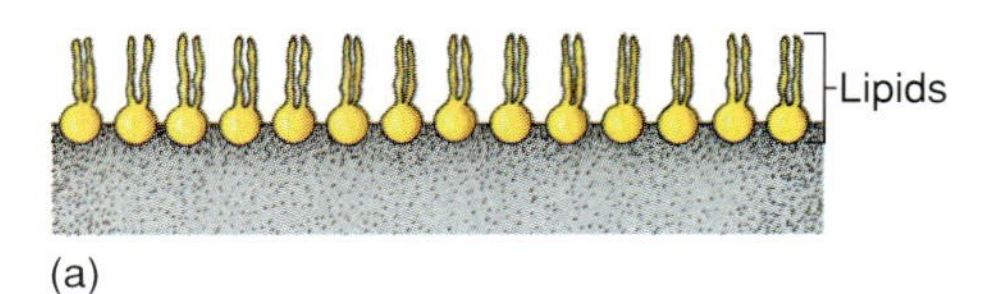

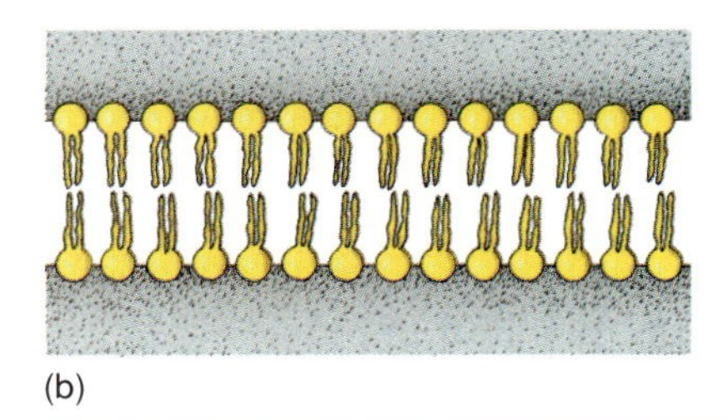

Figure 4.18
(*a*) If lipid molecules are placed in a container of water, the less dense lipid molecules form a layer at the top of the water, and the more hydrophilic (glycerol) portion of the molecule is oriented toward the water layer. (*b*) In biological membranes the lipid bilayers form with the more lipophilic (fat-loving) portion of the molecule oriented toward the center. The lipophobic portion of the molecule is oriented toward the outside, the region occupied by the protein.

Cell Membranes

Cell membranes are approximately half phospholipid and half protein. The three-layered structure of cell membranes consists of a double layer of phospholipids, with the insoluble portion in the center and the water-soluble phosphate portion oriented toward the outside in each direction (figure 4.18). Proteins are inserted on each side of the lipid layer; some of the protein molecules extend across the lipid layer and protrude out the other side; others are a component of only one side of the membrane. The proteins are laterally mobile in the double layer. In a biological membrane seen through an electron microscope, the fixation process causes the protein layers to appear as dense lines, but the phospholipid layers are transparent. Thus one sees two black lines with a space in between. This typical structure occurs in all biological membranes and is referred to as the **unit membrane.** This does not mean that all membranes are exactly the same; they have different permeability characteristics. Just because a substance can get across a chloroplast membrane does not mean that it can get across a mitochondrial membrane. **Membrane selectivity** suggests that each kind of membrane has subtle molecular characteristics that allow it to function in its own conditions. Scientists currently picture the biological membrane as a **fluid mosaic** in which large protein molecules float in a sea of lipid (figure 4.19).

Substances must pass across biological membranes to get into or out of a cell. These membranes tend to be very permeable to water and certain gases, including oxygen and carbon dioxide. Other kinds of molecules may have difficulty traversing the membrane because of size or polarity. Ions and polar molecules tend to move through the protein portion of the membrane. Many of these proteins are involved in the process called active transport in which **metabolic energy** (from ATP) is used to move substances across membranes against a **concentration gradient.** Membranes, then, serve as partitions to compartmentalize the

> **✔ Concept Checks**
>
> 1. What is the basic difference between prokaryotic and eukaryotic cells?
> 2. What is the significance of the secondary cell wall?
> 3. How do cell membranes compartmentalize and control the movement of molecules and ions?

cell into specific sections, each containing a specific group of enzymes and each responsible for a specific function or group of functions.

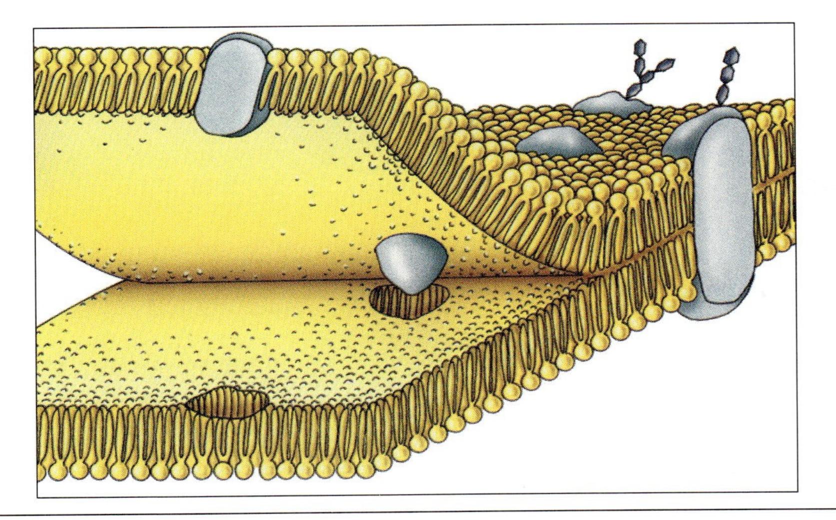

Figure 4.19
The fluid mosaic model of biological membranes demonstrates how large protein molecules are embedded in the lipid bilayer. Some protein molecules extend all the way through the bilayer, whereas others are embedded in it from either side.

ORGANELLES AND OTHER INCLUSIONS

There is some disagreement among cell biologists concerning the definition of an organelle. For the purposes of this book, an organelle is a distinct entity within a cell that performs a particular function as a **compartmentalization** of enzymes. Thus we include ribosomes and vacuoles as organelles, even though some scientists would classify only membrane-bound structures as organelles.

Typical organelles of a photosynthetic plant cell include the nucleus, **vacuole, plastids, mitochondria, ribosomes, Golgi apparatus, lysosomes,** glyoxysomes, and peroxisomes. A number of other cell inclusions, including **microtubules,** are important (figure 4.20).

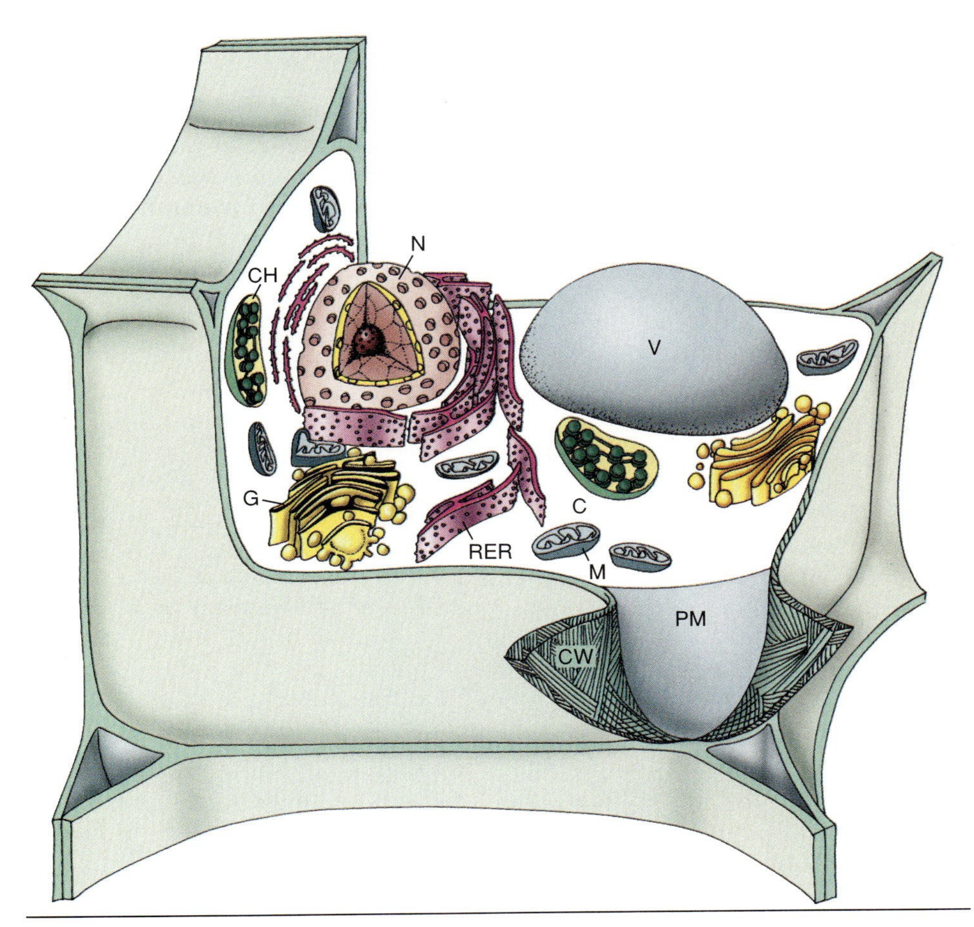

Figure 4.20
The typical green plant cell consists of a relatively thick cell wall (*CW*), a plasma membrane (*PM*), and a cytoplasmic matrix (*C*) filled with various organelles, including the nucleus (*N*), chloroplast (*CH*), mitochondrion (*M*), rough endoplasmic reticulum (*RER*), Golgi apparatus (*G*), and vacuole (*V*). This cell is not yet mature. The vacuole will continue to expand and dominate the interior of the cell.

Nucleus

The nucleus is a fairly conspicuous organelle within plant cells. During cyclosis it can be observed to remain in a relatively static position, attached by strands of membranes that form a network to suspend it in space; other organelles seem to slide by. In the process of cell division the nucleus undergoes dramatic changes as the hereditary material, DNA, is replicated and partitioned to daughter cells (figure 4.21).

A typical young plant cell is approximately 30 to 40 μm in diameter, whereas the nucleus itself is about 10 μm in diameter. It is enclosed by a double membrane system that makes up the so-called nuclear envelope. Viewed through the electron microscope the nuclear envelope is seen to contain relatively large holes, referred to as nuclear pores (figure 4.22), through which certain kinds of small and large molecules may pass. Many dramatic changes take place in the nucleus. A dense region in the nucleus is called the **nucleolus.** Some cells contain only one, others two or three, and others

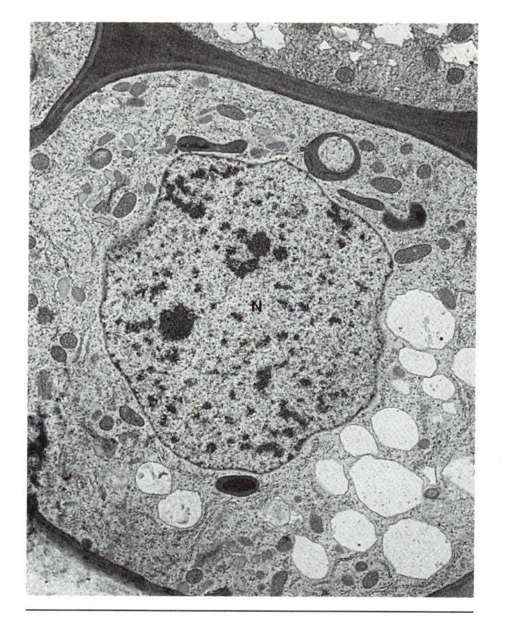

Figure 4.21
Electron micrograph of a corn (*Zea mays*) root tip cell. The primary cell wall is thick and quite distinct, and the nucleus is prominent in the center of the cell. The darkly stained matter within the nucleus (*N*) is chromatin, the hereditary material. The cytoplasm contains many organelles and vesicles, but the detail is missing in this low magnification.

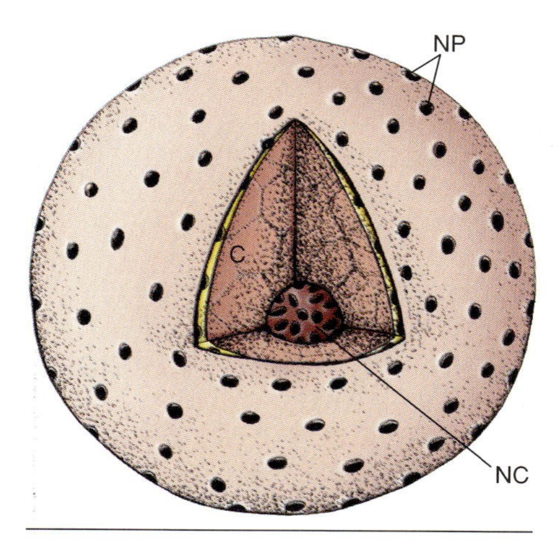

Figure 4.22
The nucleus is enclosed by a double membrane and is permeated by nuclear pores (*NP*). The central portion is chromatin (*C*) and one or more nucleoli (*NC*).

have literally hundreds of nucleoli. They appear to function in the synthesis of rRNA. During **interphase,** the so-called resting stage of cell division, the nucleoli can be observed in great detail, but as the cell begins the division process, they usually disappear at about the same time the nuclear envelope disappears. Since the nucleus contains the genetic information, it directs the framework of activity for the entire cell—when and how to divide.

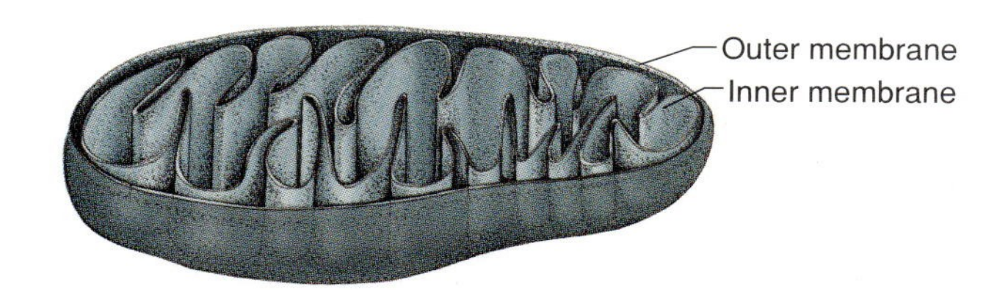

Figure 4.23
The mitochondrion is the site of aerobic respiration, consisting of an outer membrane and an inner membrane that invaginates and folds to form the cristae. The mitochondrial matrix is the site of enzymes of the Krebs cycle, and the surface of the cristae provides a surface for the enzymes of electron transport. The enzymes responsible for glycolysis occur outside the mitochondrion, free within the cytoplasm.

Mitochondrion

The mitochondrion is the organelle responsible for the process of **aerobic respiration** (*aerobic,* uses oxygen). It is capable of converting sugars into CO_2 and H_2O and releasing energy in the process. Energy from these molecules is produced as ATP, which is the main energy source for the cell. Mitochondria tend to be far more numerous than chloroplasts, with perhaps as many as a thousand per cell. They may be oblong, oval, or round and approximately 1 μm in diameter, about the same size as a bacterial cell. In fact, mitochondria have their own DNA similar in structure to the DNA found in bacterial cells. The structure of a mitochondrion consists of an outer membrane and an inner membrane that is involuted to form the **cristae** (figures 4.23 and 4.24). The involutions give a tremendous increase in the surface area of the cristae and provide a surface on which the enzymes of respiration occur. Plant cells that carry out a great deal of respiration and are required for producing a tremendous amount of ATP energy tend to have many mitochondria. Other cells that function primarily to provide some service other than respiration may have very few. It is important to remember that all cells carry on respiration, although some may do so under **anaerobic** (no free oxygen) conditions.

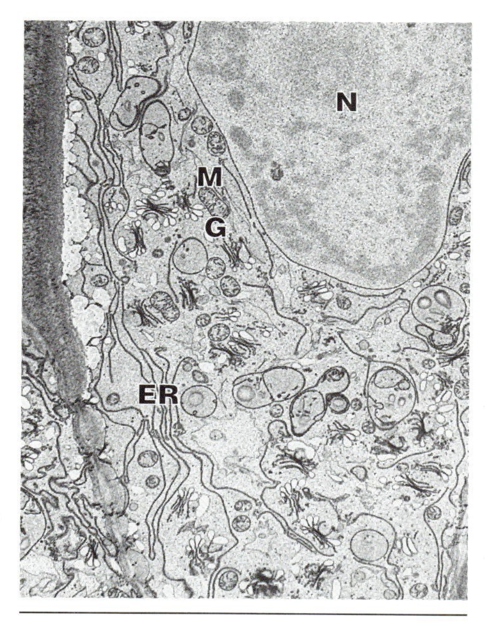

Figure 4.24
Electron micrograph of a corn (*Zea mays*) root tip cell. The details of a large number of organelles are apparent in this photograph. Note the large nucleus (*N*) in the center of the cell, and many mitochondria (*M*) and Golgi bodies (*G*). The endoplasmic reticulum (*ER*) is extensive in this cytoplasm, but the magnification is too low to emphasize the ribosomes.

Plastids

Other than the nucleus and vacuole, the plastids constitute the most conspicuous organelles of a plant cell. All plastids are bounded by a double membrane, just as the nucleus and mitochondria, and the internal structure is a system of membranes separated by a fairly homogeneous ground substance called the **stroma**. Like mitochondria, plastids also have DNA similar in structure to the DNA found in prokaryotic cells.

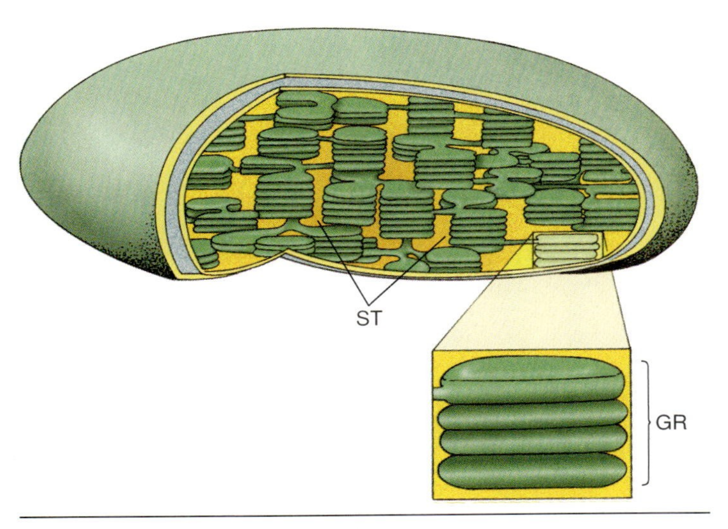

Figure 4.25
The chloroplast consists of a double membrane-bound structure enclosing dense stacks of membranes called grana (*GR*) surrounded by a fluid matrix called stroma (*ST*). The chlorophyll molecules occur in layers embedded in protein to make up the grana. Enzymes associated with the light reactions of photosynthesis occur in the grana; those responsible for the dark reactions of photosynthesis occur in the stroma.

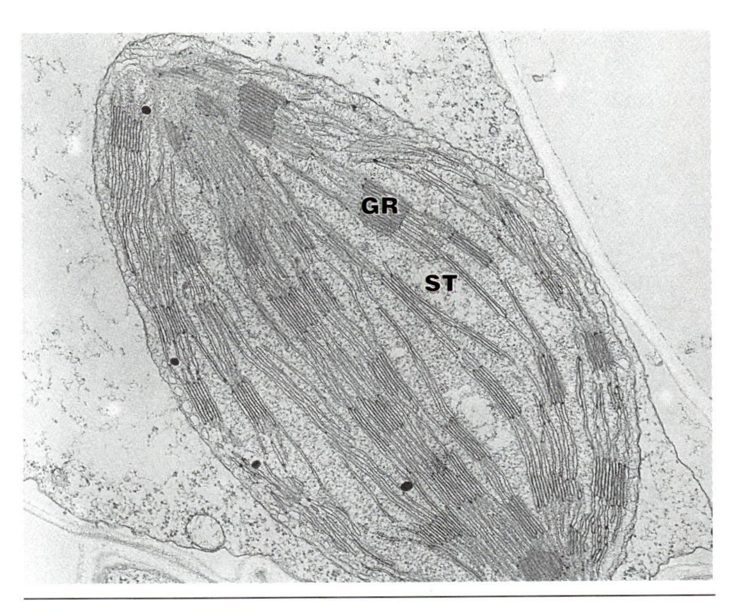

Figure 4.26
The schematic diagram of a chloroplast shown in figure 4.25 is reinforced by this electron micrograph of a corn chloroplast. The stacks of membranes are quite distinct and orderly.

There are three basic types of plastids. **Chromoplasts** are pigment organelles, as the name implies, but are specialized to synthesize and store carotenoid pigments (red, orange, and yellow) instead of **chlorophyll.** In the process of fruit ripening and in other pigmented tissues they accumulate large quantities of carotenoids to give the characteristic color to the tissue.

Leucoplasts are nonpigmented plastids but contain enzymes responsible for the synthesis of starch. Large starch grains may accumulate in plastids, as in a potato tuber.

Chloroplasts are the green plastids associated with the entire photosynthetic process, and they represent the functional unit in the transfer of light energy into the chemical energy of sugar production. All plastids begin as nonpigmented proplastids and then differentiate into one of the three basic types. Chromoplasts may begin as chloroplasts but lose chlorophyll and accumulate carotenoids to become a chromoplast during fruit ripening and other processes. Leucoplasts may be transformed into chloroplasts when exposed to light. They still retain the ability to store starch, as do all chloroplasts. Sometimes chloroplasts, following several hours of sunshine, will accumulate several large starch grains as products of photosynthesis, which distort the internal membrane structure.

As opposed to prokaryotic cells in which pigment molecules are attached to peripheral membranes of the cell, the chloroplast represents a highly organized arrangement of the chlorophyll and other pigment molecules. The molecules are arranged in specific double membrane layers called **thylakoids.** Stacks of thylakoids constitute a **granum** (pl., **grana**). The matrix in between the grana is called stroma, and the grana and stroma together make up the body of the chloroplast (figures 4.25 and 4.26).

Chloroplasts tend to be elliptical and 5 to 10 μm in diameter. In a green plant cell there might be 20 to 100 chloroplasts. During cyclosis they move freely throughout the cytoplasm. In carrying on the process of photosynthesis they respond directly to the energy from the sun by orienting themselves perpendicular to the rays of the light. In case the light energy becomes too great, they have the capability of moving away from the sun and orienting themselves at an oblique angle so that less light hits them.

Endoplasmic Reticulum and Ribosomes

The process of protein synthesis can occur only in the presence of ribosomes. The ribosomes are found on a series of interlacing membranes that traverse the cytoplasm and form the framework on which certain important functions are performed, including protein synthesis. This membrane system, the **endoplasmic reticulum,** provides the scaffolding to which the ribosomes are attached. Ribosomes are the site of protein construction from an ordered pattern of amino acid incorporation, although they need not be attached to the ER to do so (figures 4.27 and 4.28). Endoplasmic reticulum may have a group of ribosomes, much like buttons attached to a piece of cloth, in what is called rough endoplasmic reticulum; or it may simply consist of membrane without ribosomes, in which case it is referred to as smooth endoplasmic reticulum. The ribosomes themselves appear as dark round dots on the endoplasmic reticulum at low magnification, but as the magnification increases, it becomes apparent that they consist of two parts—a small spherical body and a larger concave body. This organelle is

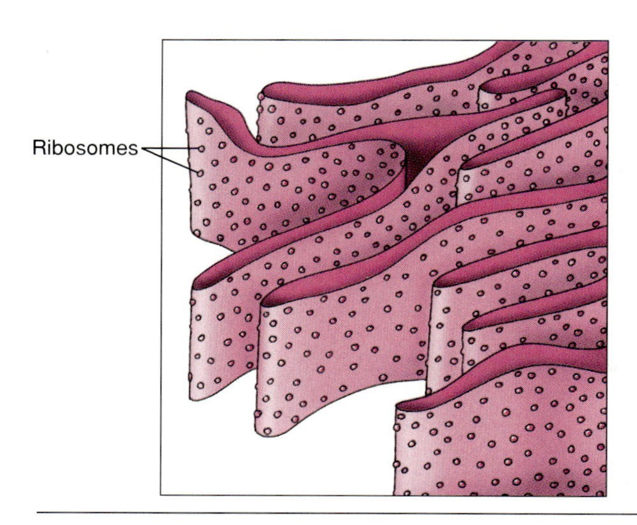

Figure 4.27
The membranes of the endoplasmic reticulum look much like a flattened rubber tube. The membrane may or may not be embedded with tiny beadlike ribosomes, the site of protein synthesis.

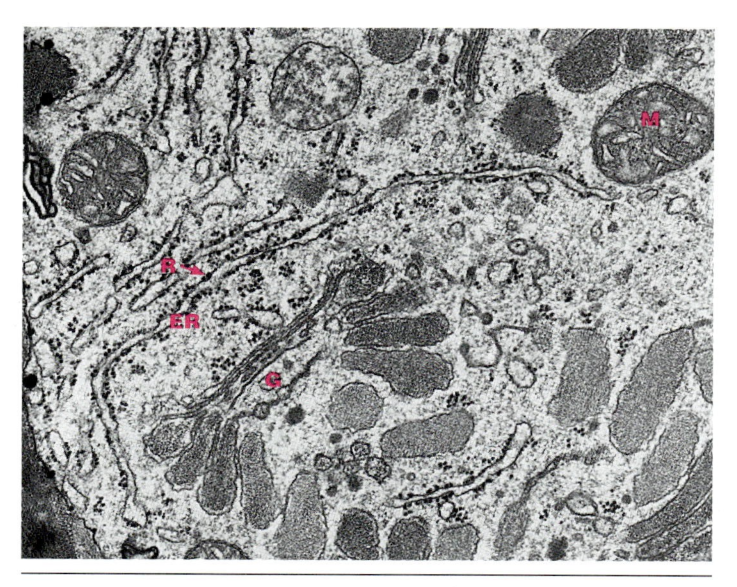

Figure 4.28
High magnification of an electron micrograph of a corn root tip cell reveals the detail of well-developed mitochondria (*M*) and Golgi bodies (*G*). At this magnification, ribosomes (*R*) are prominent on the endoplasmic reticulum (*ER*).

about 15 nm in diameter and therefore of much smaller dimensions than the other organelles already described. The ribosome is made of rRNA and protein. In this structure the amino acids are aligned in proper order for incorporation into the protein. In any given cell there might be many thousands of ribosomes. Thus, even though the process of protein synthesis may at first seem relatively slow, it is possible to make many molecules in a short period of time because each ribosome may be involved in the synthesis process (figure 4.29).

Vacuole

The vacuole begins as a very small organelle that eventually increases in size until, in a mature plant, it dominates the entire cell. As a matter of fact, the cytoplasm may be pushed to the outer limits adjacent to the cell wall because of increasing internal water (turgor) pressure. In some cases the nucleus is displaced into a "corner" of the cell, and the vacuole actually occupies most of the space within the cell. Vacuolar sap is mostly water and much less viscous than that of the cytoplasm proper. It is probably best to think in terms of the vacuole as the storage area of the cell, a place where nutrients and various solutes are maintained until they are needed in general metabolism or stored as waste materials. For the most part macromolecules are not part of the vacuolar system but are maintained within other organelles or directly in the cytoplasm. The membrane that surrounds the vacuole is the **tonoplast.** The tonoplast selectively acts to determine what gets in and out of the vacuole. There are many different kinds of relatively small molecules within the vacuolar sap, including the ions and small molecules such as sugars and amino acids.

Golgi Apparatus

Located throughout the cytoplasm is a group of organelles collectively called the Golgi apparatus. They appear as flattened membranes, much like a stack of pancakes (figures 4.30, 4.31 and 4.32). At the edges of these flattened membranes one can observe small pieces of membrane called vesicles being pinched off from the periphery of the "pancake." The vesicles contain the macromolecules used in constructing both membranes and primary cell wall. As the cell grows under the influence of turgor pressure against the plasma membrane, the membrane must enlarge and be strengthened by the deposition of new material. This packaging function is apparently performed by the Golgi apparatus, ensuring that, as the interior expands, the expanded membrane and wall will be able to take the additional stress imposed.

Other Organelles

Sometimes other small organelles are found in specific plant tissues. Lysosomes, glyoxysomes, and peroxisomes are organelles bounded by a single membrane and containing a package of enzymes for a specific task. Lysosomes contain acid hydrolytic enzymes capable of breaking down proteins and certain other macromolecules. Glyoxysomes are found primarily in fatty seeds such as cotton and peanut, and they provide enzymes for the conversion of fats to carbohydrates during the germination process. Peroxisomes provide a compartment for enzymes important in glycolic acid metabolism associated with photosynthesis.

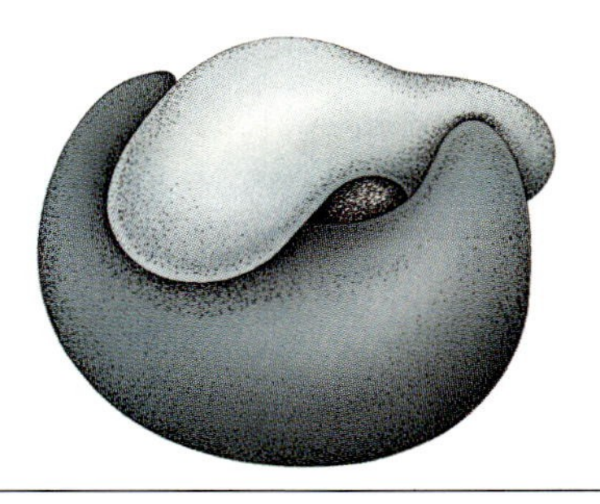

Figure 4.29
The typical eukaryotic ribosome occurs in two parts, one considerably larger than the other. The two parts curve to fit together with a small groove in the middle where the mRNA molecule is thought to pass.

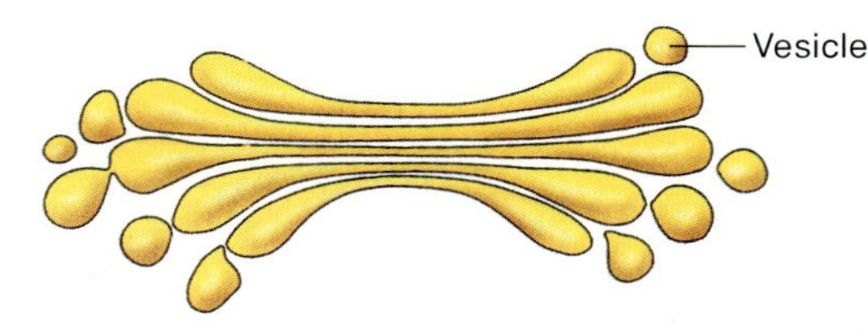

Figure 4.30
The Golgi apparatus as viewed with the electron microscope consists of stacks of membrane, like a layer of pancakes. At the margins, vesicles are pinched off and carry macromolecules to new plasma membranes or cell walls. Golgi bodies often may be found near the exterior of the cytoplasm, close to their site of transport.

Figure 4.31
The Golgi apparatus in three-dimensional view.

Microtubules

Microtubules are relatively small structures found in all eukaryotic cells and characterized by a tubular or rodlike appearance. For the most part they are found in the cytoplasm, but they may also be a part of cilia and flagella, whiplike projections on the surface of motile cells.

Cytoplasmic microtubules are rather uniform in size and remarkably straight. They are about 25 nm in outside diameter and several micrometers in length. The wall of the microtubule consists of individual linear or spiraling filamentous structures about 5 nm in diameter, and these are composed of about 13 subunits. There is a lumen (open area in the center), but occasionally dots or rods are observed in the center portion. Microtubules are composed of a special type of protein called tubulin.

Although the function of microtubules is still not perfectly clear, their orientation and distribution suggest that they form a framework which somehow shapes the cell and redistributes its contents. We will soon see how cell specialization brings about different shapes and functions, and cell differentiation begins to occur at the same time that numerous microtubules begin to appear. Microtubules may also transport macromolecules, possibly forming channels in the cytoplasm. In plant cells, microtubules occur inside the plasma membrane and are oriented tangentially to the cell wall. It has been suggested that they function in cell-wall deposition, and microtubules have been noted to underlie the points where secondary cell wall is being deposited in spiral or reticulate patterns. **Spindle fibers,** to be discussed with the cell division process, are microtubules composed of proteins.

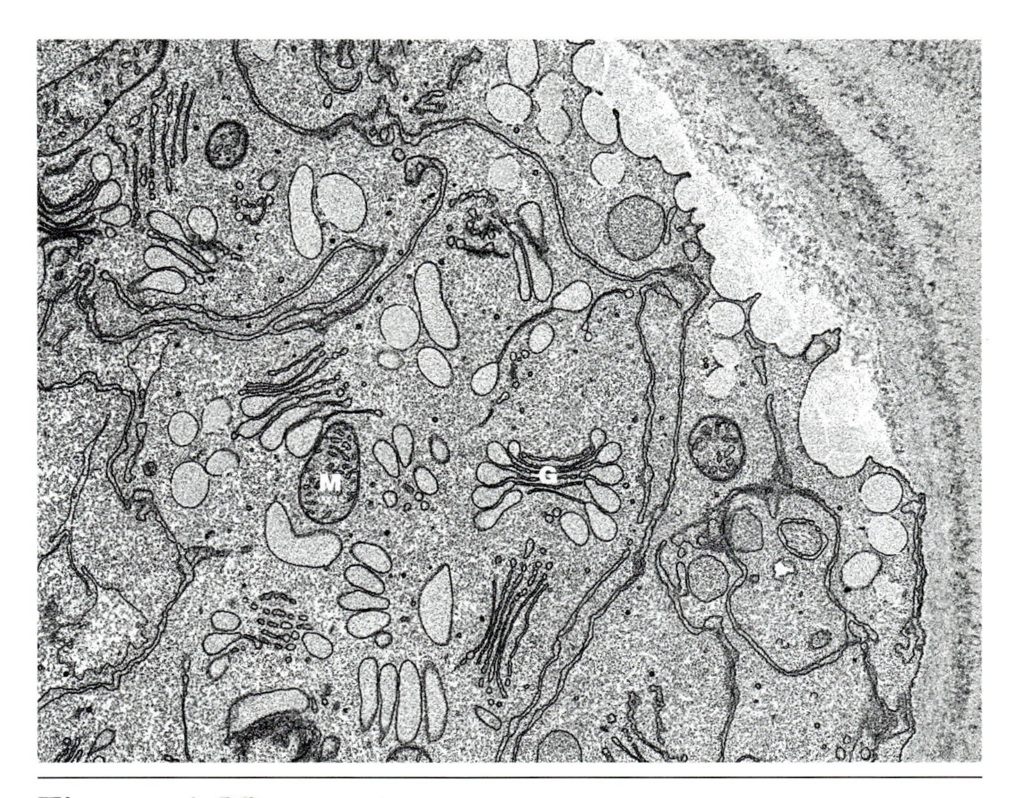

Figure 4.32
This electron micrograph of a portion of a corn root tip cell reveals excellent detail of organelles within the cytoplasm. The Golgi bodies (*G*) and mitochondria (*M*) are particularly clear, as are many small vesicles. Since this is nonphotosynthetic tissue, there are no chloroplasts.

✔ Concept Checks

1. What are the primary cell organelles, and what is the function of each?
2. How important is the vacuole in a mature plant cell?

Summary

1. The basic unit of life is the cell, and all organisms except viruses are composed of cells.
2. Chemistry is the science of atoms and molecules. Cells exist because of molecules, groups of atoms combined in particular ways so that electrons are shared to lend stability. Atoms thus combine by various kinds of bonding; hydrogen bonds are individually quite weak, whereas covalent and ionic bonds are quite strong. The stronger the bonds, the greater the energy required to break them.
3. Water is basic to all life as we know it, and its unique chemical and physical properties are related to hydrogen bonding. All the features of water density, cohesion, adhesion, and heat gain and loss can be attributed to this compound's special composition.
4. Energy is moved through living systems by a number of transfer molecules, including ATP, NAD, and NADP. These molecules allow for an orderly passage of energy from one chemical compound to another, ensuring that the efficiency of conversion is maintained.
5. The macromolecules of life are primarily carbohydrates, proteins, lipids, and nucleic acids. These large molecules, constructed from simpler molecules, provide the chemical framework of life.
6. Prokaryotic cells have no nucleus, whereas eukaryotic cells do have a distinct nucleus that can undergo division to produce new cells.
7. Plant cells have a distinct cellulose wall; animal cells do not.
8. Membranes are made of lipids and proteins, and substances move through biological membranes according to properties of size and solubility. Membranes provide a large surface area to allow for collisions of molecules responsible for biochemical reactions, and they compartmentalize functions by restricting certain enzyme combinations to certain organelles.
9. Within the green plant cell are found the nucleus, mitochondria, chloroplasts, endoplasmic reticulum with ribosomes, the Golgi apparatus, and microtubules. Many other smaller organelles are sometimes included.
10. Mature plant cells usually have a single, large vacuole dominating the water relationships within the cell.

Key Terms

adenine 63
adenosine triphosphate (ATP) 63
aerobic respiration 74
amino acids 66
anaerobic 74
atom 59
calorie 63
carbohydrate 64
carbon 59
catalyst 62
cellulose 64
cell wall 70
chemical bond 61
chemical element 59
chlorophyll 75
chloroplast 75
chromoplast 75
compartmentalization 73
concentration gradient 72
cristae 74
cyclosis 59
cytoplasm 59
cytosine 69
disaccharide 65
DNA (deoxyribonucleic acid) 69
double bond 61
endoplasmic reticulum 75
enzymes 62
eukaryotic 71
fluid mosaic 72
globular protein 66
glucose 70
glycerol 68
glycogen 64
Golgi apparatus 73
grana 75
guanine 69
interphase 74
ion 61
leucoplast 75
lignins 66
lipids 61
lumen 72
lysosome 73
macrofibril 70
macromolecule 70
membrane 69
membrane selectivity 72
messenger RNA (mRNA) 70
metabolic energy 72
microfibril 70
microtubule 73
middle lamella 70
mitochondria 73
molecule 59
monosaccharides 64
nucleic acid 69
nucleolus 73
nucleus 59
organelles 59
pectins 66
peptide bond 66
phospholipids 69
plasma membrane 59
plasmodesmata 71
plastid 73
polypeptide chain 66
polysaccharide 64
primary cell wall 70
primary pit field 71
prokaryotic 71
protein 66
protein synthesis 66
quarks 60
ribosomal RNA (rRNA) 70
ribosome 73
RNA (ribonucleic acid) 69
secondary cell wall 70
spindle fiber 77
starch 64
stroma 74
suberin 69
sucrose 65
thylakoids 75
thymine 69
tonoplast 76
transfer RNA (tRNA) 70
turgor pressure 70
unit membrane 72
uracil 69
vacuole 73

Discussion Questions

1. Discuss how the structure relates to the function of the following organelles: mitochondria, chloroplasts, endoplasmic reticulum, and Golgi apparatus.
2. A thin section is made through a portion of a plant. When examined microscopically, some of the cells are observed to be without nuclei. Suggest two possible explanations.
3. Imagine yourself to be a sugar molecule in a chloroplast. Describe two possible routes that would take you to a vacuole in a neighboring cell.

Suggested Readings

Becker, W. M., and D. W. Deamer. 1991. *The world of the cell.* 2nd ed. Benjamin/Cummings, Redwood City.

Blackburn, G., and M. J. Gait, eds. 1990. *Nucleic acids in chemistry and biology.* IRL Press, New York.

Darby, N. J., and T. E. Creighton. 1993. *Protein structure.* IRL Press, New York.

Dubyak, G. R., and J. S. Fedan, eds. 1990. *Biological actions of extracellular ATP.* New York Academy of Sciences, New York.

Erickson, R. P., and J. G. Izant, eds. 1992. *Gene regulation: Biology of antisense RNA and DNA.* Raven Press, New York.

Erlich, A., ed. 1989. *PCR technology: Principles and applications for DNA amplification.* Macmillan Publishers, New York.

Fry, S. C. 1988. *The growing plant cell wall.* Wiley, New York.

Hill, W. E., ed. 1990. *The ribosome: Structure, function, and evolution.* American Society of Microbiology, Washington, D.C.

Hyams, J. S., and C. W. Lloyd, eds. 1994. *Microtubules.* Wiley-Liss, New York.

Lewin, B. 1994. *Genes V.* Oxford University Press, New York.

Lewis, N. G., and M. G. Paice, eds. 1989. *Plant cell wall polymers: Biogenesis and biodegradation.* American Chemical Society, Washington, D.C.

Marin, B., ed. 1987. *Plant vacuoles: Their importance in solute compartmentation in cells and their applications in plant biotechnology.* Plenum Press, New York.

Marzuki, S., ed. 1989. *Molecular structure, function, and assembly of the ATP syntheses.* Plenum Press, New York.

McDonald, K. A. 1994. The last quark. *The Chronicle of Higher Education.* May 4, 1994.

Mohan, S., C. Dow, and J. A. Cole, eds. 1992. *Prokaryotic structure and function.* Cambridge University Press, New York.

Murray, A. W., and M. W. Kirschner. 1991. What controls the cell cycle. *Sci. Am.* 264(3):56–63. March 1991.

Murray, J. A. H., ed. 1992. *Antisense RNA and DNA.* Wiley-Liss, New York.

Osawa, S., and T. Honjo, eds. 1991. *Evolution of life: Fossils, molecules, and culture.* Springer-Verlag, New York.

Pavelka, M. 1987. *Functional morphology of the Golgi apparatus.* Springer-Verlag, New York.

Quinn, P. J., and R. J. Cherry, eds. 1992. *Structural and dynamic properties of lipids and membranes.* Portland Press, Chapel Hill.

Sillince, J. A. A., and M. Sillince. 1991. *Molecular databases for protein sequences and structure studies: An introduction.* Springer-Verlag, New York.

Watts, A., ed. 1993. *Protein-lipid interactions.* Elsevier, Amsterdam.

Westhof, E., ed. 1993. *Water and biological macromolecules.* CRC Press, Boca Raton.

Yeagle, P. 1993. *The membranes of cells.* 2nd ed. Academic Press, San Diego.

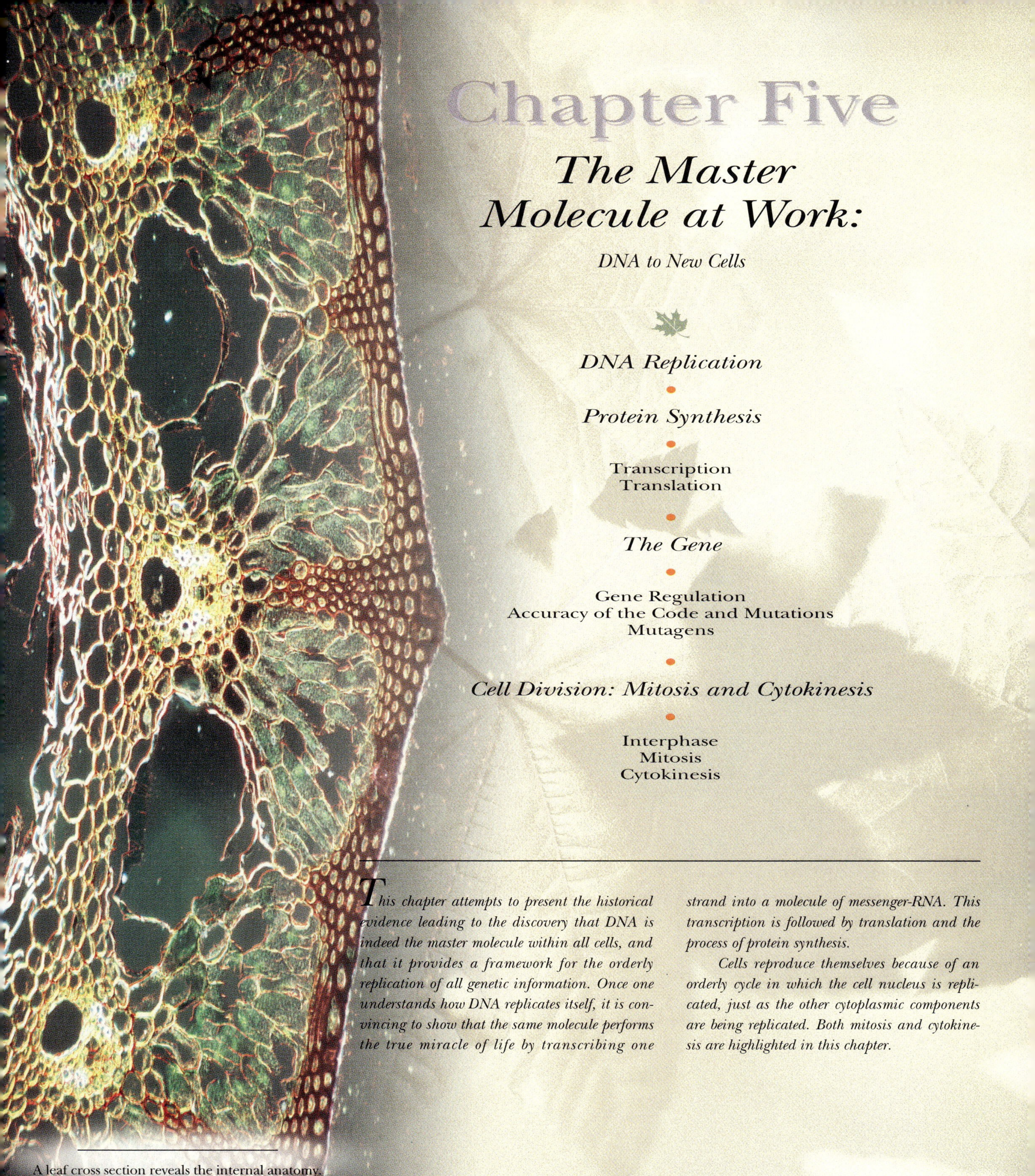

A leaf cross section reveals the internal anatomy.

Chapter Five

The Master Molecule at Work:

DNA to New Cells

DNA Replication

Protein Synthesis

Transcription
Translation

The Gene

Gene Regulation
Accuracy of the Code and Mutations
Mutagens

Cell Division: Mitosis and Cytokinesis

Interphase
Mitosis
Cytokinesis

This chapter attempts to present the historical evidence leading to the discovery that DNA is indeed the master molecule within all cells, and that it provides a framework for the orderly replication of all genetic information. Once one understands how DNA replicates itself, it is convincing to show that the same molecule performs the true miracle of life by transcribing one strand into a molecule of messenger-RNA. This transcription is followed by translation and the process of protein synthesis.

Cells reproduce themselves because of an orderly cycle in which the cell nucleus is replicated, just as the other cytoplasmic components are being replicated. Both mitosis and cytokinesis are highlighted in this chapter.

All of us, at some time during our lives, have wondered why there are so many different kinds of plants and animals. Why are they different? Do they ever change? These are commonly asked questions that have been answered only in recent years.

Understanding how plants grow, develop, and reproduce and do so with predictability is the key to understanding botany. Why do oak trees always produce acorns that in turn always grow into oak trees? Why do all other organisms display similar growth and reproductive sequences? Since this predictable repetition always includes a single-celled stage at some point in the process, it has become obvious to scientists that the controlling mechanism must be completely contained within a single cell and subsequently transmitted to resulting cells as growth proceeds. Therefore this awesome "responsibility" must surely be housed in some molecule that is present in all cells.

In addition, scientists have long been aware of the visibility and activity of chromosomes during cell division. It was only appropriate to suspect the **chromosomes** of being responsible for containing and passing on the information controlling the growth and development of an organism and the inheritance of that organism's traits in the next generation. But of what are chromosomes made?

One early rationale suggested that, to specify the complicated characteristics displayed by any single organism or species, surely the molecule of inheritance would have to be protein. Only here would there be the possibility of thousands and thousands of different combinations of amino acids. With 20 amino acids the number of possible sequences is enormous, and surely each sequence could specify the characteristics for a particular organism. The critical question was how the same molecules could be passed on from generation to generation. Carbohydrates or lipids, with their identical repeating units, could not specify complicated codes. The only possible solution seemed to lie in the proteins, but such was not the case. Beginning in 1951, James Watson, Francis Crick, and Maurice Wilkins began piecing together information about the other molecule that was a constituent of the chromosome—DNA.

"Monday morning quarterbacking" is easy when studying the history of the molecular basis of inheritance. The solutions may seem obvious to a student of modern biology, but the advances within the past century have been remarkable indeed, given the information at hand. It is particularly interesting that a German chemist isolated DNA in 1869, within a few years of Darwin's publication of *The Origin of Species* and Mendel's publication of his study of inheritance in peas. In the 20 years prior to Watson and Crick's elucidation of the DNA molecule, evidence began to accumulate concerning the importance of DNA:

1. In 1928 a public health official, studying bacterial pneumonia, found that something called a *transforming factor* can be passed from dead to live bacteria and cause a change in their hereditary characteristics. In 1943 scientists at Rockefeller University discovered that the transforming factor was DNA.
2. In 1941 George Beadle and Edward Tatum of the University of Chicago performed ingenious experiments with bread mold to show that a single gene is responsible for the synthesis of a single protein. The Nobel Prize winner Linus Pauling used this information to show that sickle-cell anemia is caused by a mutation in the blood protein hemoglobin, and the only difference in normal and sickle-celled hemoglobin is the substitution of two of the 600 amino acids.
3. At the same time as the Watson-Crick-Wilkins study, other scientists were able to show that a **bacteriophage,** a virus that attacks bacterial cells, can change the genetics of the bacterial cell by injecting its own hereditary material. Using radioisotopes, they supported the idea that the hereditary material was DNA and not protein.
4. Many different studies showed that the amount of DNA in body cells is two times the amount found in sex cells.
5. The convincing bit of evidence giving the basis for the Watson-Crick-Wilkins study was that DNA is composed of the 5-carbon sugar deoxyribose, some phosphate, two purines (adenine and guanine), and two pyrimidines (thymine and cytosine). The nitrogen bases (purines and pyrimidines) do not occur in equal proportions, but the proportions are fixed for a given species and are different between species. The amount of adenine always equals the amount of thymine, and the amount of guanine always equals the amount of cytosine.

When James Watson went to Cambridge to work at the Cavendish Laboratory with Francis Crick, they had access to a great deal of information concerning the basic chemistry of DNA. They knew not only that the molecule had to carry precise information that could be transferred from generation to generation, but also that the quantity of specific information needed would be enormous for complex eukaryotic organisms. Precision in information transfer was all-important, but if a mutation occurred, the new information would have to be transferred with the same precision as the original information.

When their studies began, they already knew from the work of Rosalind Franklin that the molecule consisted of a multiple helix with the bases inside; they also knew of Erwin Chargraff's experiments at Columbia University to show that the amount of guanine in an organism was always the same as the amount of cytosine, and the amount of adenine was equal to the amount of thymine. Considering all the chemical and x-ray crystallography data, they set about to construct a tin model of the molecule. They knew that if DNA was indeed the correct molecule, it would have to have the correct configuration, size, and complexity to code for the vast amount of information that had to be processed. Working within the physical dimensions imposed by purines, the sugar deoxyribose, and phosphate, they deduced that the molecule had to be not a single helical structure as in proteins, but a double helix.

In unraveling the complex structure, they discovered that the ends of the two strands were different. The phosphate group that joins two sugar molecules is attached to one sugar at the fifth carbon position (the 5´ position) and to the other sugar at the third carbon position (the 3´ position). This gave one strand with a 5´ end and the other with a 3´ end. In other words, the strands were opposite in direction, or antiparallel. It became apparent later on that decoding information could be done only from one direction, leading to the concept of a *sense* strand and a *nonsense* strand.

The four **nucleotides** of DNA are formed by a chemical bonding of the purine or pyrimidine to a molecule of deoxyribose and a phosphate group (figure 5.1). These nucleotides are paired with the purines and pyrimidines facing each other as the sugar and phosphate protrude toward the outside (figure 5.2). The purines adenine and guanine both have two rings and are thus the same size and shape (only a single ring substitution is different). The pyrimidines thymine and cytosine have single rings, and therefore the overall molecule is smaller. By pairing a purine with a pyrimidine, and thus a double ring with a single ring, all nitrogen base pairs are the same dimension and fit into the double-stranded molecule.

The complementary pairing of purines and pyrimidines is accomplished by hydrogen bonding between H—O and H—N. Only two such bonds occur between adenine and thymine, but three hydrogen bonds occur between guanine and cytosine. Therefore it is not possible for adenine to pair with cytosine, nor guanine with thymine.

The Watson-Crick model proposed two right-handed polynucleotide chains coiled in a helix around the same axis. This is often referred to as a **double helix.** The molecule might be thought of as a ladder, twisted into a coil. The purine and pyrimidine bases of each strand are stacked on the inside of the double helix with their planes parallel to each other and perpendicular to the long axis.

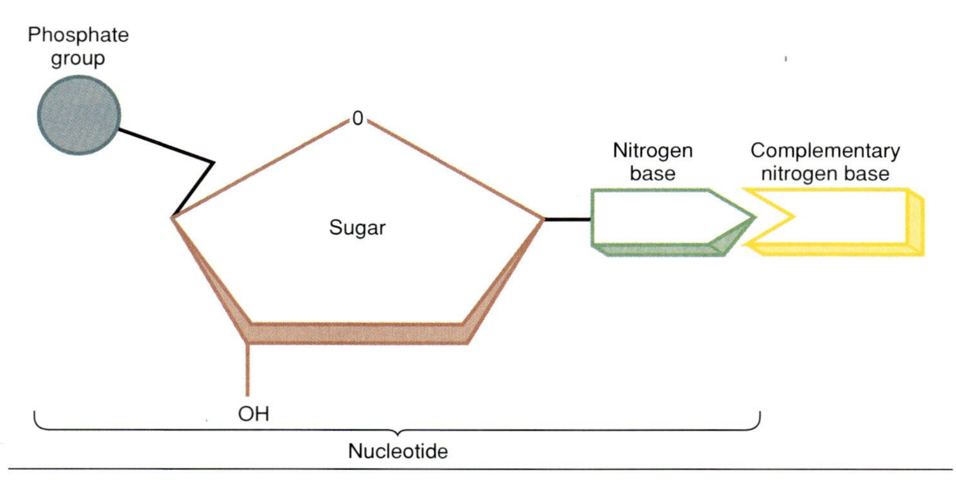

Figure 5.1
A nucleotide consists of a nitrogen base (either adenine [A], guanine [G], cytosine [C], thymine [T], or uracil [U]) connected to a 5-carbon sugar and a phosphate group. In DNA the sugar is deoxyribose and the bases are A, T, G, and C; in RNA the sugar is ribose and the bases are A, U, G, and C.

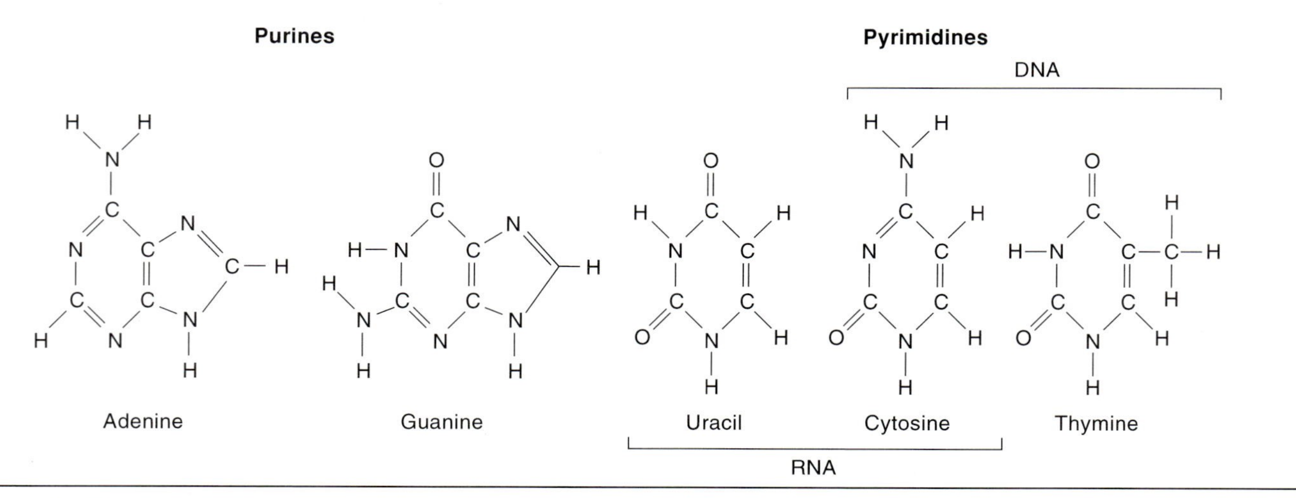

Figure 5.2
The nitrogen bases of nucleic acids are the purines adenine and guanine, and the pyrimidines uracil, cytosine, and thymine. Only the pyrimidines uracil and cytosine are found in RNA, whereas the pyrimidines cytosine and thymine are found in DNA.

The diameter of the entire helix is 2 nm. The space occupied by a purine is more than 1 nm, but the space occupied by a pyrimidine is less than 1 nm. Consequently, a purine-pyrimidine pair joined by hydrogen bonds occupies exactly 2 nm. The distance between nucleotide pairs in the coil is 0.34 nm, and there are ten nucleotide pairs in one complete coil of the helix ($0.34 \times 10 = 3.4$ nm). All these dimensions were revealed by x-ray crystallography. Thus constraints are imposed on the size and shape of the purines and pyrimidines themselves and the way they fit into the molecule. In a single strand of the DNA molecule one might encounter any possible sequence involving thousands of the four nitrogen bases, but in the matching strand complementary base-pairing would always occur.

DNA REPLICATION

All living organisms, and therefore each of their cells, must have one thing in common: the ability to replicate themselves exactly so that rose characteristics remain rose characteristics, cow characteristics remain cow characteristics, and all other species are perpetuated with exceptional accuracy. Knowing that the molecule of inheritance is DNA, one needs to know exactly how it manages to replicate itself, passing on its exact molecular structure to succeeding generations.

The total number of nitrogen base-pairings in a given chromosome varies, but for such a chromosome to replicate itself exactly and pass on this precise sequence to a new cell, it must "unzip" at the hydrogen bonds holding the two strands together and allow each of its two strands to build a new complementary strand. For example, the nucleotide containing guanine attracts another nucleotide containing cytosine in the nuclear sap to pair with it. The synthesis continues until each of the two original strands comprising the DNA molecule has added a new complementary strand and is bonded to it through the nitrogen bases.

Once the process is completed, the resulting chromosome is no longer one double-stranded helical molecule, but two identical such molecules. A cell in which all of its chromosomes have so replicated is ready to make two genetically identical cells (figure 5.3).

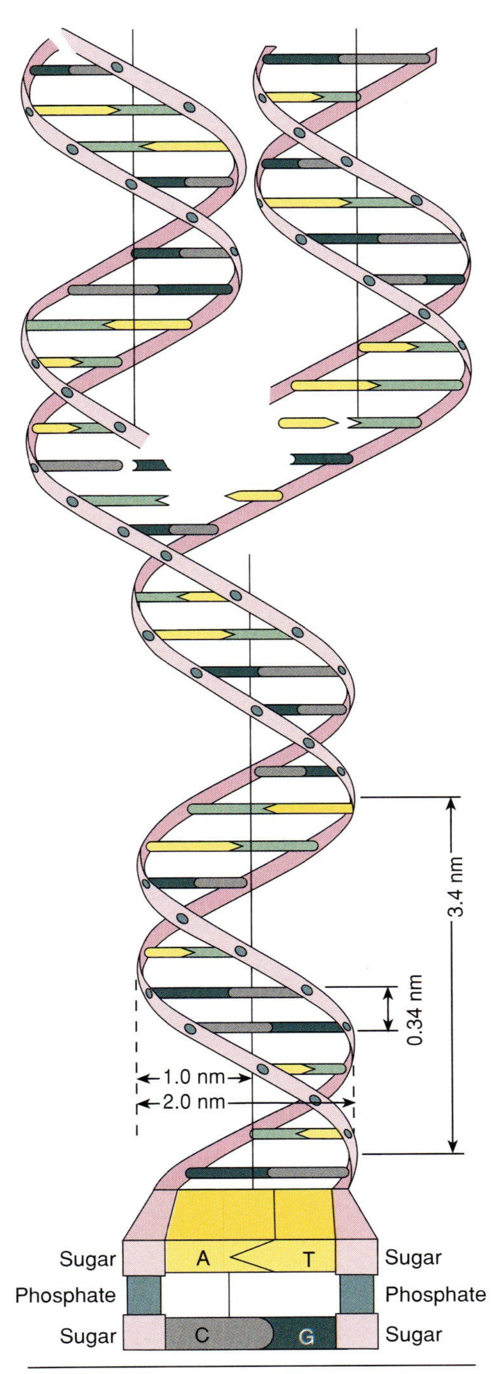

Figure 5.3
The double-stranded DNA molecule separates so that each strand may replicate with its complementary strand. At the completion of this process each chromosome consists of two identical chromatids with the same genetic information encoded.

> **✔ Concept Checks**
>
> 1. When was DNA first isolated?
> 2. Prior to 1951, what was the major evidence that DNA might be involved in cell replication?
> 3. What is meant by complementary pairing of purines and pyrimidines, and what is its significance to the structure of the DNA molecule?

PROTEIN SYNTHESIS

A typical chromosome has many thousands of subsections of its total nucleotide sequence, each controlling the production of a specific protein (figure 5.4). Each of these subsections of the chromosome is called a **gene.** Since a given gene might include a few hundred to several thousand nucleotides in sequence, and since proteins are composed of many amino acids linked together, the message contained in each gene must somehow be put to work building a specific protein with a set sequence of amino acids.

Both strands of the DNA do not function in this process; only one does so. Suppose, for example, one strand of the DNA molecule read in part: adenine-cytosine-guanine (A-C-G). Then the complementary pairing would have to read: thymine-guanine-cytosine (T-G-C). Such three-part sequences are referred to as **triplets,** and they specify

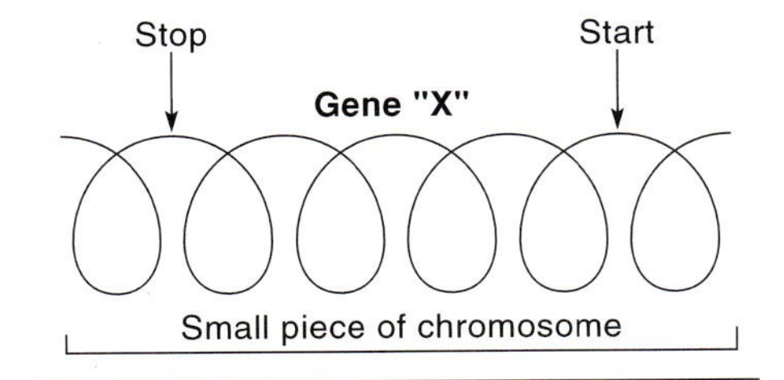

Figure 5.4
A chromosome consists of many thousands of genes linked end to end as a long, double-stranded molecule of DNA. At some point on the chromosome a codon START signal causes the message to be read until a codon STOP signal tells the system that the end of that gene has been reached. This piece of DNA between the START and STOP signals is the gene representing the genetic information for a single heritable characteristic.

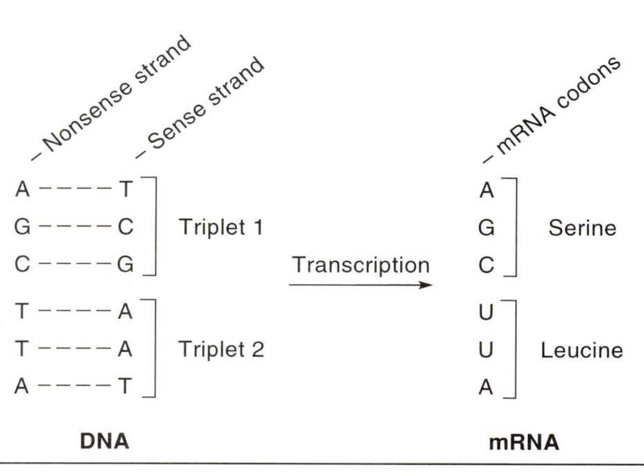

Figure 5.5
The double-stranded DNA molecule represents the "message center" from which the code is deciphered. The strand of nucleotides represented diagrammatically here on the left is a nonsense series. The nucleotides on the right represent a triplet message (group of three nucleotides) that can be transcribed into a codon of messenger RNA. In this example, triplet 1 specifies an mRNA codon that codes for the amino acid serine. Triplet 2 specifies an mRNA codon that codes for the amino acid leucine, and subsequent triplets of the gene are decoded in the same manner.

a portion of the genetic code, with each triplet specifying the location of a specific amino acid. Actually, the triplets of the DNA molecule do not translate directly into the code but serve as a template from which the code will be made. In the preceding example A-C-G might be the template for specifying a certain amino acid, but the complement T-G-C would not specify the correct message (figure 5.5).

The process by which proteins are synthesized is rather complicated, but it is now quite well understood by the scientific community. Since proteins are needed by the living cell for structure, storage, and, most important, for thousands of different kinds of enzymes with near-perfect specificity, it stands to reason that the process of forming proteins must be an exceptionally precise sequencing of amino acids. Enzymes, in particular, must have a primary structure that is accurate, predictable, and repeatable. The right kinds and amounts of enzymes must occur not only in the right place, but also at the right time. Sometimes the linkage of a complicated series of biochemical events fails if even one enzyme is missing. A cell that fails to make a particular enzyme at a particular time might not survive.

Protein synthesis is a sequential progression involving DNA, RNA, amino acids, and ATP. The "master plan" involves two basic processes: **transcription** and **translation.**

Transcription

In addition to the process of DNA replication, the DNA molecule within the nucleus provides the template for protein synthesis. Just as in the replication process, the double-stranded DNA molecule "unzips" (the bonds are actually broken by an enzyme) for a gene portion of its length, and one strand becomes the template for making a single-stranded messenger RNA (mRNA). Since the strands of the DNA molecule are antiparallel, the enzyme responsible for forming a new mRNA strand can read the message only from one direction, thus eliminating any confusion about which strand should be read. A particular triplet that becomes exposed by the unraveling and unzipping provides the genetic code for a new complementary nitrogen base-pairing (figure 5.6). A special kind of enzyme called RNA polymerase causes this triplet to pair with nucleotides previously synthesized and present in the nucleus. Returning to our original DNA triplet sequence: adenine-cytosine-guanine, the complementary pairing within the new mRNA would be: uracil-guanine-cytosine. Note that in RNA, uracil substitutes for the thymine found in DNA. All forms of RNA substitute uracil for thymine. This newly synthesized mRNA triplet is called a **codon** and is the code for a particular amino acid. Although triplets occur in both DNA and RNA, the term *codon* applies only to the triplets of mRNA. Subsequently, the DNA molecule unravels further, exposing more of the genetic code from the DNA. Each triplet accepts complementary pairing for making other codons, and the process proceeds until that entire DNA sequence (one gene) has been transcribed. Thus the single-stranded RNA is just as long as the specific segment of the double-stranded DNA carrying the code for the protein to be made—a very large molecule. Once an mRNA molecule has been made, the unzipped segment of the DNA molecule anneals by allowing the hydrogen bonds to reform. This process is a transcribing of the genetic code in the form of a series of many three-letter sequences. The codons can be read only from one direction; to do otherwise would specify a different amino acid.

In eukaryotic cells the newly formed mRNA molecule undergoes a modification or processing in which segments of the molecule are eliminated. Specific enzymes cut out or edit those portions of the message which are garbled or fail to specify the correct protein sequence. These intervening sequences may have some role in gene regulation or impart additional variability in genetic diversity.

Translation

Once the mRNA molecule has been formed and processed, it migrates out through a nuclear pore and becomes attached to a ribosome (figure 5.7). Even though the molecule may be very large, perhaps an average of some 500 codons, the nuclear pores are sufficiently large to allow movement from the nucleus to the cytoplasm.

At approximately the same time, the large mRNA molecule attaches itself to a ribosome. Many amino acid molecules are activated by an energy-dependent (requires ATP) attachment to rather small transfer RNA

Figure 5.6
In the process of transcription, one strand of the DNA molecule becomes the template for making mRNA. Each nitrogen base pairs with its complementary base. Within the new mRNA molecule, uracil substitutes for each thymine position. Transcription occurs in the nucleus, where intervening sequences of the message are edited prior to migration to the cytoplasm.

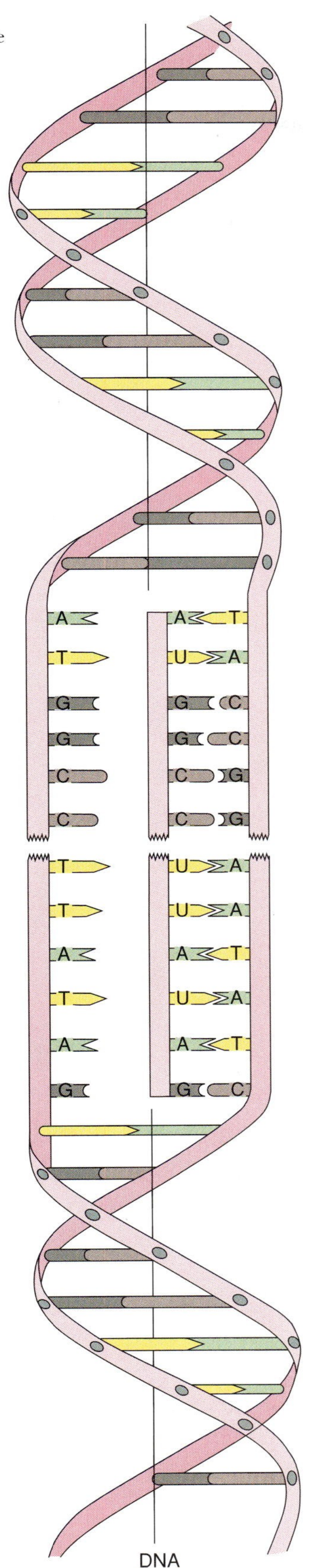

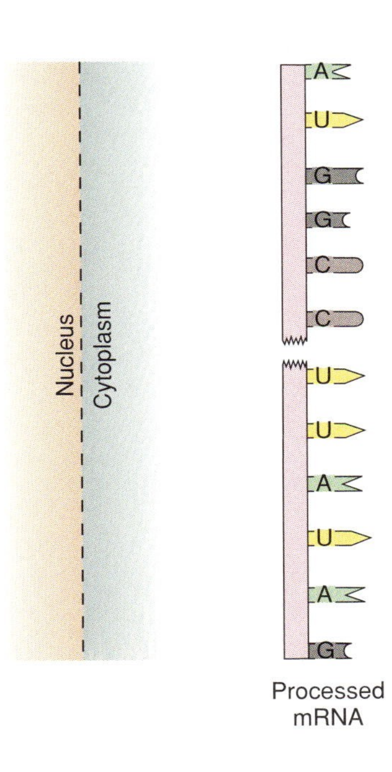

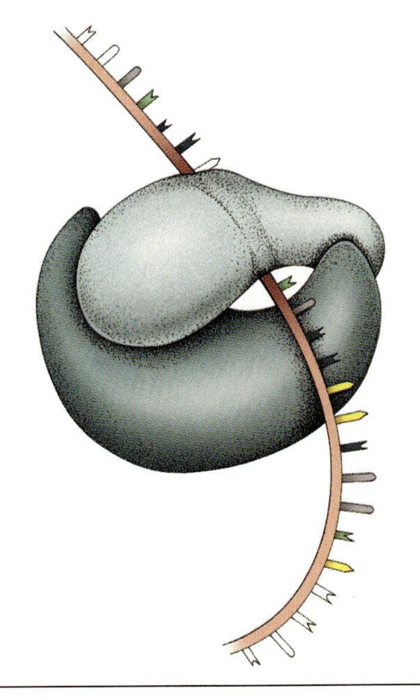

Figure 5.7
The two-segmented ribosome is attached to a strand of mRNA, and as the molecule is pulled through the channel, the message is translated into the proper amino acid sequence.

(tRNA) molecules. There is a separate tRNA molecule for each of the 20 amino acids, and each one is specific because of a particular **anticodon loop,** a sequence of three nitrogen bases that read as a complement to the codon at the site of the ribosome-mRNA. Thus a codon of mRNA aligns with the anticodon of tRNA and in the process positions the appropriate amino acid. Then the next codon is pulled across the surface of the ribosome, read, and matched with its appropriate anticodon loop, moving the second amino acid into the proper position. This process continues until the entire mRNA coding sequence has been translated (figure 5.8). As each amino acid is positioned in proper sequence, an enzyme causes a molecule of H_2O to be released

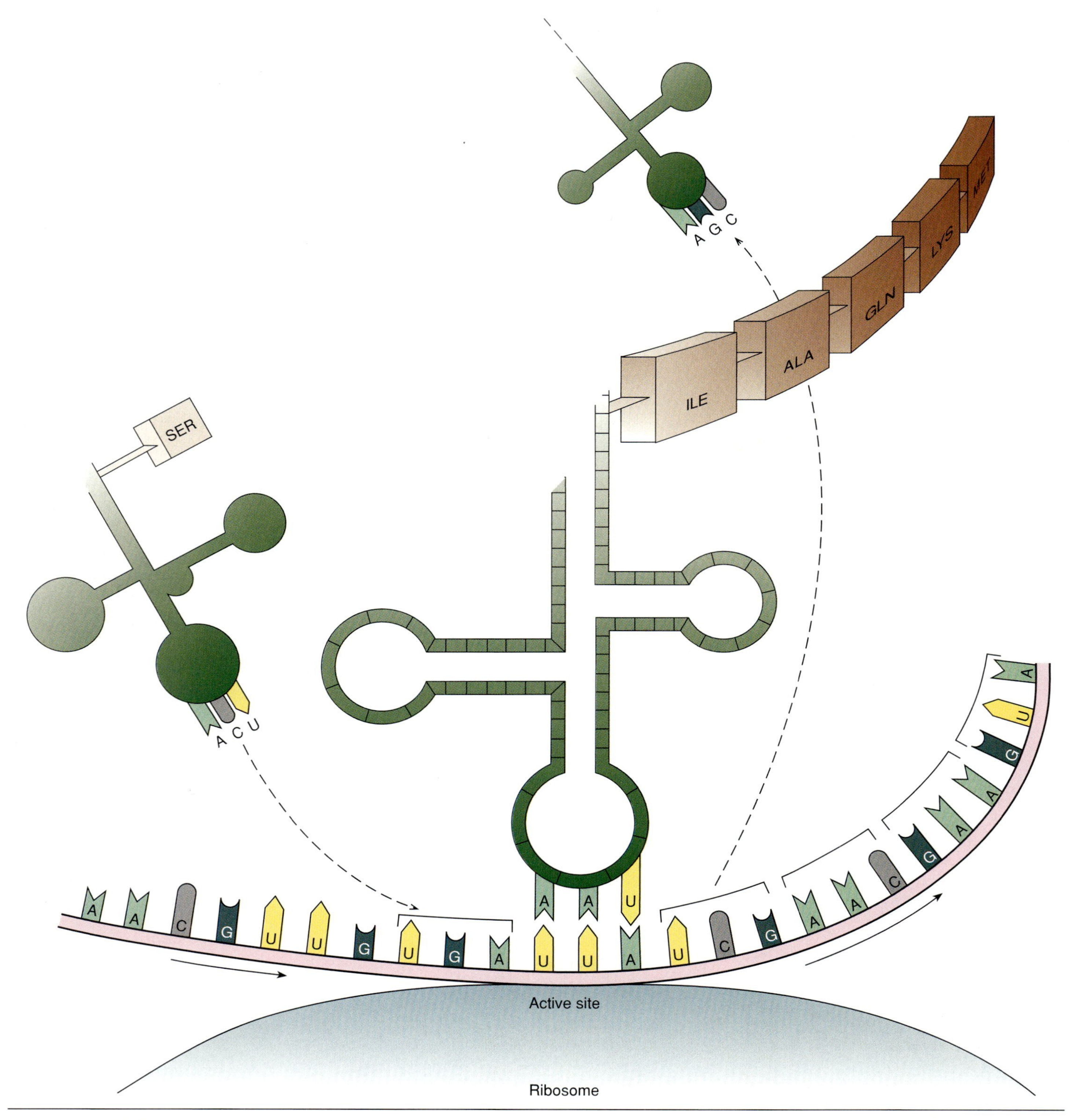

Figure 5.8

Translation occurs as the codon of the mRNA molecule is matched with the anticodon of the transfer RNA molecules. A previously activated tRNA-amino acid complex moves into position at the anticodon loop, which codes for the three complementary bases to those of the mRNA.

between the —COOH group of one amino acid and the $—NH_2$ group of the adjacent amino acid, forming a peptide bond. As each new amino acid in the chain is added, it is detached from its tRNA, which is then free to move back into the cytoplasm to activate and deliver another amino acid when "summoned" by an appropriate codon.

Thus, at the termination of the translation sequence, all amino acids are in the proper order and held together by peptide bonds. Proteins are also sometimes called polypeptides because of this structure. Technically a protein may consist of more than one polypeptide chain, but most proteins are but a single chain. In most cases the mRNA is degraded in a short time and is not used as a template for the production of more than a few identical proteins, possibly only one. Thus mRNA is usually thought to be short-lived.

Multiple copies of the same protein can be made at approximately the same time by directing the same mRNA molecule along the surface of several ribosomes. Such groups of ribosomes are called **polysomes** and apparently allow the message to be used repeatedly before it is degraded (figure 5.9).

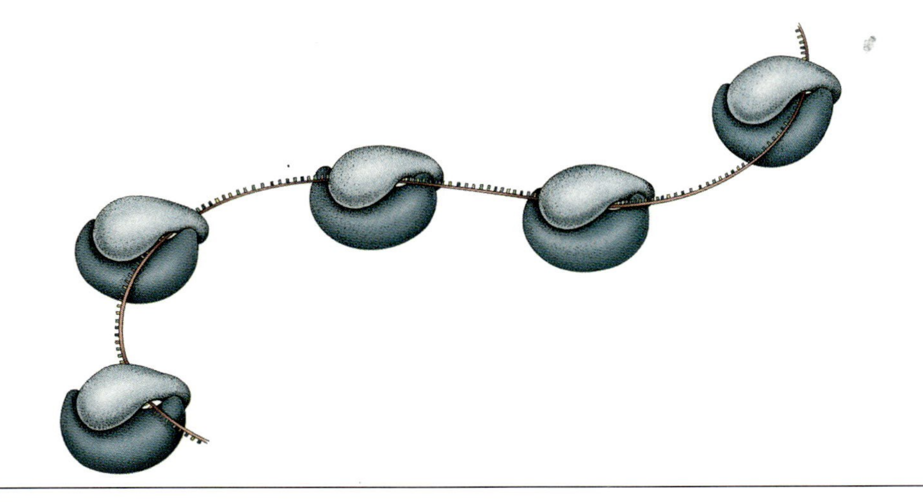

Figure 5.9
Protein synthesis usually occurs by the pulling of the mRNA molecule through a series of ribosomes known as polysomes. This allows the same protein to be made at each ribosome.

 ***Concept Checks***

1. Describe the processes of transcription and translation.
2. What is the function of the ribosome in protein synthesis?

THE GENE

In prokaryotic organisms the DNA is a single, circular molecule much like a very long rubber band. It is not associated with protein, it never condenses into a chromosome, and it is never visible under ordinary light microscopy. In eukaryotic organisms, however, each chromosome consists of a single linear thread of DNA associated with both basic and acidic nuclear proteins. During interphase the molecule is uncoiled and difficult to see, even with appropriate staining. During **mitosis,** however, the DNA condenses and coils so that the chromosomes do become visible under light microscopy.

Since cells are generally small and since the nucleus is even smaller, chromosomes are visible only under rather high magnification. Even when viewed with a light microscope at 1000 × magnification, some chromosomes are barely visible at their most highly condensed state. A study of their detailed structure requires further magnification.

It is important to realize that a chromosome is made up of some nuclear proteins plus the DNA molecules themselves. What one actually sees through the microscope is a stained preparation of protein and DNA, the so-called **chromatin.** Only during mitosis or nuclear division does the chromatin become visible as distinct chromosomes. Organisms vary in their number of chromosomes, and the number has absolutely nothing to do with the complexity of the organism. Each chromosome represents a large amount of DNA, and the molecules are coiled and doubled back on themselves much like a tangled kite string. The "string" is nothing more than a very long DNA molecule consisting of thousands and thousands of nucleotide pairs. From an inheritance point of view, a gene represents a heritable characteristic, one that can be passed on from generation to generation. From a molecular point of view, a gene is a piece of DNA molecule that carries the genetic message for a single polypeptide chain. Sometimes two or more polypeptide chains comprise the protein and two or more distinct genes are responsible for the entire code.

The overall problem of how an organism determines which genes to turn on and off is the basis of developmental biology, the combined field of study that synthesizes genetics, physiology, and morphology. Gene regulation is probably the least understood part of biology at the present time.

Gene Regulation

Most of the information available on how genes are turned on and off comes from studies with bacteria. Because the cell cycle is so short in such organisms, they lend themselves to elaborate experimentation in a short period of time.

In the organism *Escherichia coli,* a common bacterium in the digestive tract of healthy humans, the milk sugar lactose is split into glucose and galactose by the enzyme β-galactosidase. When lactose is present, the enzyme is present in large quantities; when lactose is absent, only traces of the enzyme can be found. Since lactose

Who Turns Them On and Off?

Not all genes are decoded or "turned on" at the same time. In fact, geneticists now believe that not more than 5% to 10% of the genome (the entire genetic complement) is turned on at any one time. The kinds of proteins needed at the time of seed germination—those associated with breaking down storage macromolecules, beginning rapid metabolism and the synthesis of a great deal of new structural material during meristem development, and the ultimate production of new leaves and roots—may be entirely different from the proteins needed after the plant is fully grown and ready for sexual reproduction. There are many examples in molecular biology to show that transcription occurs in genes only when they are needed, but the regulatory system must be controlled ultimately by a complex interaction of the environment and internal factors such as hormones. Even if a seed were placed in the soil in the presence of plenty of moisture, it might not germinate until the soil temperature reached some critical level (maybe higher or lower). A desert seed that responded to a sudden summer rain might never survive if the soil moisture were the only environmental cue. The seed needs to germinate during the cool season when more predictable rains are likely to occur. How can a plant (or an animal) possibly know all these things? How could a dry seed "know" what is best for its own survival?

Developmental biology is the synthesis of many specialized fields in an attempt to understand how an organism is able to turn on genes at different times during its life, turn some of them off at precisely the right time, and turn on others so that the organism enters into the next phase of its development. One of the intriguing questions, now that we know how the mechanics of protein synthesis occurs, is how does the cell find time or space to make those millions of enzymes necessary for carrying on life's processes?

If one knew just how long it takes for the translation process to occur at the surface of the ribosome, it would be possible to calculate the time required to make a particular protein, for example, one of 2000 amino acid units. By using precise radioisotope labeling experiments, it has been shown that only a few codons are translated per second, perhaps about ten. If that figure is accurate, then the calculation $2000 \div 10 = 200$ shows that it takes 200 seconds to make the single protein molecule. That may seem like a long time, but remember that each cell may contain thousands of ribosomes, and they could all be coding for the same protein at the same time. In such experiments protein synthesis usually occurs (that is, a polypeptide chain is completed) in about five to ten minutes.

One must also bear in mind that hundreds of different proteins could be synthesized in the same cell at the same time, provided that different genes were being transcribed from the chromosomes at the same time. When we put it all together, and consider all of those environmental cues like changing seasons (which changes length of day, temperature, and rainfall patterns), along with hormonal triggers and many other unnoticed and subtle factors, it's not so surprising that organisms do change with time, and somehow we make it through adolescence!

induces the enzyme to be formed, apparently turning on a gene, lactose is called the inducer and the enzyme is called an inducible enzyme. Some substances inhibit enzyme production and are called corepressors; such enzymes are called repressible enzymes.

The model developed for *E. coli* to explain gene regulation includes an **operon,** a group of structural genes that are functionally related and are aligned along a segment of DNA. These genes are all controlled as a single unit by an adjacent segment of DNA known as the promoter. The RNA polymerase attaches to the promoter at the beginning of transcription. An additional DNA segment called the operator acts as the on-off switch for the promoter. In the "on" position, mRNA can be transcribed; in the "off" position, no mRNA can be made. Yet another gene, the regulator, determines the position of the on-off switch. The regulator is controlled by a repressor, which may bind to the regulator and keep the entire system in the "off" position. It would appear that the repressor is controlled by inducers or corepressors. In the lactose operon, lactose controls the repressor.

In eukaryotic organisms, chromosomal proteins are believed to play an important role in gene regulation. Evidence suggests that each cell contains the same genetic information (i.e., that all genes are present in all chromosomes of all cells all the time). Complicated, multicellular organisms produce many different kinds of cells at different times during the life cycle and at different places in the organism. A majority of the genes are thus turned off or "silenced" most of the time.

Although the evidence is sketchy, it would appear that both acidic nuclear proteins and basic nuclear proteins, the histones, are important in turning genes on and off in eukaryotic organisms. Various hormones, vitamins, osmotic agents, and temperature shock treatments have been used to induce transcription in eukaryotic organisms under laboratory conditions. The complete elucidation of this fascinating part of developmental biology awaits further study.

You may be curious by now about the size of a gene, and where, if the DNA molecule is a continuous thread, one gene stops and another begins. The secret lies in the genetic code itself. As the result of hundreds of experiments to determine which nucleotide triplet codes for which amino acid, we now know that some triplets do not code for amino acids at all, but are STOP signals to indicate the end of the code for a particular gene. The STOP codons are U-A-A, U-A-G, and U-G-A. Since thousands of genes are all connected on the same chromosome, there must also be a START signal, and we know that codon is always A-U-G. When this codon is at the beginning of the gene, it will code for *N*-formylmethionine, which we now know must be the first amino acid in the polypeptide chain forming any protein. Thus the presence of *N*-formylmethionine, as the first amino acid coded for by a gene, allows for the start of the formation of the remainder of that protein. When A-U-G occurs anywhere else within the gene, it simply codes for methionine and does not function as a START codon (figure 5.10).

First Letter	Second letter: U	Second letter: C	Second letter: A	Second letter: G	Third Letter
U	UUU, UUC: PHE UUA, UUG: LEU	UCU, UCC, UCA, UCG: SER	UAU, UAC: TYR UAA: Stop UAG: Stop	UGU, UGC: CYS UGA: Stop UGG: TRP	U C A G
C	CUU, CUC, CUA, CUG: LEU	CCU, CCC, CCA, CCG: PRO	CAU, CAC: HIS CAA, CAG: GLN	CGU, CGC, CGA, CGG: ARG	U C A G
A	AUU, AUC, AUA: ILE AUG: MET	ACU, ACC, ACA, ACG: THR	AAU, AAC: ASN AAA, AAG: LYS	AGU, AGC: SER AGA, AGG: ARG	U C A G
G	GUU, GUC, GUA, GUG: VAL	GCU, GCC, GCA, GCG: ALA	GAU, GAC: ASP GAA, GAG: GLU	GGU, GGC, GGA, GGG: GLY	U C A G

Figure 5.10
There are 4^3 (64) possible triplet codons for specifying the 20 amino acids of protein. Some amino acids have more than one code (leucine has six), and three codons are STOP signals. Even though AUG codes for methionine, when it occurs at the beginning of a gene it codes for *N*-formylmethionine and is a START signal.

If one considers that there are only four possible nucleotides put together in groups of three, then the possible number of codons is 4^3 ($4 \times 4 \times 4$, or 64). That may seem like a surprisingly small number, but since there are only 20 amino acids in protein, 64 is more than sufficient. As a matter of fact, all possible combinations are used. Some amino acids are coded by more than one codon. Six separate codons, for example, code the amino acid leucine; only one codon translates for tryptophan. By analyzing duplicate codes, it becomes apparent that the first two nucleotides in the code are rather "fixed," whereas the third nucleotide is variable. (Geneticists refer to the third nucleotide as being "wobbly" or imprecise.)

Accuracy of the Code and Mutations

The 64 codons are the only possible codes for incorporating amino acids into protein. The grand scheme for passing inherited traits from one generation to another seems to follow a precise plan. Occasionally, however, mistakes are made, and the overall consequences of those mistakes (some good, some bad) are discussed in the section on evolution. It is important to emphasize here, however, that point mutations occur because changes are somehow incorporated into the primary structure or amino acid sequence during the synthesis of a protein. Following are types of point mutations that can occur as a result of miscoding:

Transitional mutants In the DNA molecule one purine-pyrimidine base pair may be replaced by another. For example, A-T may be replaced by G-C.

Transversional mutants A purine-pyrimidine pair may be replaced by a pyrimidine-purine pair. This type of point mutation is very common.

Insertion mutants This is sometimes called a frame shift, in which an extra base pair shifts the sequence out of phase by one pair. In reading all subsequent codons, the code is misread by one base in advance. This type of mutation may be caused by acridine dyes and of course has serious consequences.

Deletion mutants Here a base pair is deleted, which also causes a frame shift; the code is misread by one base delay. Deletions may also be quite serious and can be caused by high pH levels, high temperature, and by various chemical compounds, such as proflavin.

Not all mutations are lethal to the organism. Transitions and transversions are relatively benign; that is, they have little effect on the quality of the protein. They are often referred to as silent mutations. On the other hand, insertions and deletions are usually lethal. Occasionally, deletions will occur in multiples of three, so there is no frame shift, but an entire amino acid is deleted.

Mutagens

Any substance that causes a mutation to occur is called a **mutagen.** The effects of industrialization have caused concern about the number of possible mutagens in the environment. Sometimes human society continues to use them, even though the possible consequences are obvious. High concentrations of caffeine, for example, may act as a mutagen, and the drug LSD (lysergic acid diethylamide) can have mutagenic effects. Likewise, gamma

Concept Checks

1. What is a gene?
2. What causes genes to be turned on and off?
3. How can the genetic code for tens of thousands of different kinds of proteins be specified with only combinations of adenine, thymine, guanine, and cytosine?

rays, x rays, ultraviolet light, heavy metals, and pesticides are known to be highly mutagenic.

CELL DIVISION: MITOSIS AND CYTOKINESIS

Scientists must be concerned not only with what goes on in a single living cell, but they must also consider how those single cells grow to produce multicellular organisms. If all organisms maintained the cell uniformity found in most prokaryotes, there would be no opportunity for complicated, multicellular plants and animals with hundreds of different kinds of specialized cells in the same body. Sometimes a cell divides to produce two identical daughter cells that remain similar throughout their life span. Such similar cells make up a tissue. On other occasions the cells specialize to form new tissues.

The process of a cell dividing to produce two new cells occurs repeatedly in single-celled organisms and in specific tissues of multicellular organisms. In single-celled organisms the mother cell grows and then divides to produce two daughter cells that are essentially identical in size, structure, and genetic composition. They are, however, only about half the size of the original mother cell. Each of these two cells then begins a growth period until it reaches maximum size; then each repeats the division process.

The frequency of these cellular divisions varies from one species to the next; some divide every few minutes and others less often, up to once every day or so. *Escherichia coli* can replicate every 12.5 minutes. Given enough space and food, such organisms could produce incredible numbers in a short time. Available space and food do act as limiting factors to such growth, however, as they do for multicellular organisms. The balance of nature is kept intact by such controls.

All multicellular organisms contain eukaryotic cells with organized nuclei and chromosomes. The process of cell division involves two separate but interdependent steps. Mitosis is the process of nuclear division resulting in two nuclei, both genetically identical to each other and to the single nucleus that gave rise to them. **Cytokinesis** is the division of the parent cell into two new cells, each of which contains a full complement of all the different organelles and cytoplasmic constituents and one of the two newly formed nuclei.

Once initiated, the division of the nucleus is an essentially continuous process. Prior to the actual sequence of mitotic events is a period during which the cell makes extra cytoplasmic materials, including all the organelles. During this period the chromosomes are also replicated.

Interphase

The time when cells are not dividing is called interphase, an in-between stage (figures 5.11 and 5.15*a*). It actually represents a time of intense metabolic activity, even though the chromosome cannot be seen through a microscope.

Interphase can be divided into three stages: two gaps (G_1 and G_2) separated by a synthesis (S) stage. The sequence is $G_1 \rightarrow S \rightarrow G_2$. The first gap ($G_1$) is a time for growth of cytoplasmic contents, including numbers of organelles. At this time substances are synthesized that either promote or inhibit the subsequent stages of division. In some plant cells, as in those of nonmeristematic regions, specific levels of inhibitor apparently act as a permanent block to mitosis, and these cells never divide again.

During the synthesis (S) stage the DNA is replicated by the process described earlier. Purines, pyrimidines, sugars, and phosphate molecules that had been synthesized earlier are present in the nucleus and become the building blocks for the new DNA.

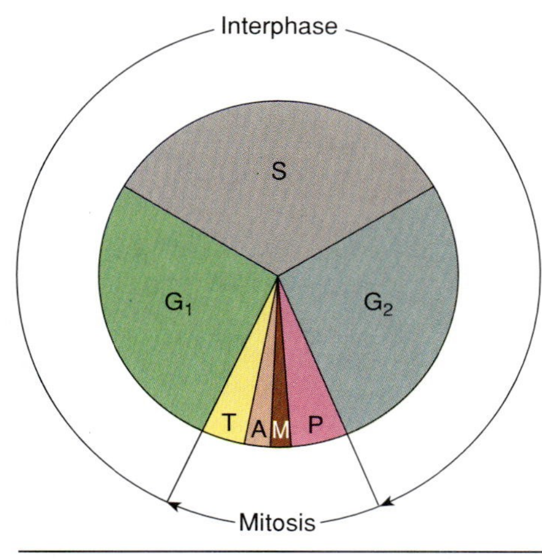

Figure 5.11
The processes of cell division and nuclear division are preceded by an interphase during which synthesis of materials, including replication of the DNA, occurs. G_1, or the first gap, is followed by *S* (synthesis phase), which in turn is followed by G_2, or the second gap. All these phases are actually periods of intense metabolic activity preceding mitosis. *P*, prophase; *M*, metaphase; *A*, anaphase; *T*, telophase.

During the G_2 stage the mitotic apparatus responsible for the division process itself is synthesized and assembled, including the spindle fibers.

Mitosis

Although a continuous process, mitosis has been divided into four phases of activity for easier comprehension. Since these phases are designated for convenience and do not reflect any true separation or pauses in the ongoing process, their application in the following description of mitosis is meant to be instructional only.

Mitosis begins with the chromosomes gradually changing from elongated, thin strands winding throughout the nucleus to shorter, thicker structures that finally become visible with the use of a microscope and appropriate stains. Beginning with DNA replication, each chromosome has been composed of two genetically identical, parallel double strands. As the chromosomes reach their most condensed and shortest condition, these double

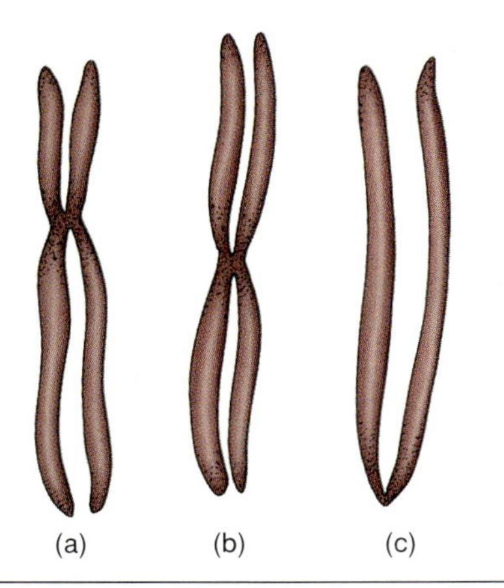

Figure 5.12
Chromosome structure may be partially described by the position of the centromere. Attachment can be submetacentric (*a*), metacentric (*b*), or telocentric (*c*).

TABLE 5.1

Chromosome Numbers of Different Organisms

Organism	Number
Haploppus gracilis (small desert plant)	4
Mosquito	6
Onion	16
Cabbage	18
Corn	20
Sunflower	34
Cat	38
Human	46
Plum	48
Dog	78
Goldfish	94
Ophioglossum reticulatum (fern)	1260

strands become visible as two **chromatids** connected by a noncondensed area called the **centromere.** The position of the centromere varies from the center of the chromosome in some (metacentric position), to off-center locations (submetacentric position), to end positions of the chromosome (telocentric position) (figure 5.12).

The position of the centromere aids in specific chromosome identification because of the relative lengths of the two chromosome arms on each side of it. Chromosome identification is important in many genetic studies and enables scientists to match up the two chromosomes in each nucleus that are identical in relative length, arm-length ratios, and most important in the traits genetically controlled by the genes of that chromosome pair. Even though the actual genetic message may be different on the two genes controlling each trait, these look-alike chromosomes are called **homologues.**

Essentially all plants have an even number of chromosomes in each cell because they exist in these homologous pairs. With the exception of cells that are produced in reproductive organs, all eukaryotic cells normally contain a **diploid** number of chromosomes. Diploid literally means "two sets" of chromosomes, each complete set coming from the two parents that originally gave rise to the new organism through sexual reproduction. The diploid (2n) number varies in the plant kingdom from only four chromosomes (2n = 4) to over 1200 chromosomes. A given species, however, normally has the same diploid number of chromosomes in all its individuals. Sometimes multiples of these sets of chromosomes are produced, leading to polyploidy. Table 5.1 gives a few representative examples of chromosome numbers for different species.

Prophase

The nuclear division begins with the condensation of the chromosomes to a shorter, thicker, and microscopically visible configuration. As the chromosomes are changing, the nucleolus or nucleoli gradually disappear from the nucleus, and the nuclear membrane (envelope) then begins to break up, becoming resorbed into the endoplasmic reticulum. Once the now visible chromosomes are fully condensed and free in the cytoplasm, long slender microtubules begin organizing into spindle fibers. The appearance of these structures signifies a transition from **prophase** to **metaphase** (figures 5.13 and 5.15*b*).

Metaphase

The spindle fibers continue to form from opposite ends (poles) of the cell, ultimately meeting in the middle of the cell. Three dimensionally, these fibers form a spindle- or football-shaped structure. Fully condensed chromosomes move to the center of the cell as the spindle is being formed and are attached at their centromeres to a pair of fibers, one from each pole. Not every fiber has a chromosome attached to it, but every chromosome, each appearing as a pair of chromatids, is attached to a spindle fiber. All the chromosomes are finally aligned in the center of the spindle and attached to fibers, forming what has been termed the metaphase plate. At this point in the process metaphase is complete, and the next sequence of events is grouped under **anaphase** (figure 5.15*c*).

Anaphase

Thus far each chromosome has been a single unit because there has been only one centromere holding two genetically identical chromatids of each together. As soon as all chromosomes are attached to spindle fibers in the center of the cell, their centromeres divide, allowing the two chromatids of each chromosome to separate (figure 5.14*a*). Current theory suggests that the spindle fibers attached to these centromeres begin to shorten and pull the newly divided centromeres and their chromatids

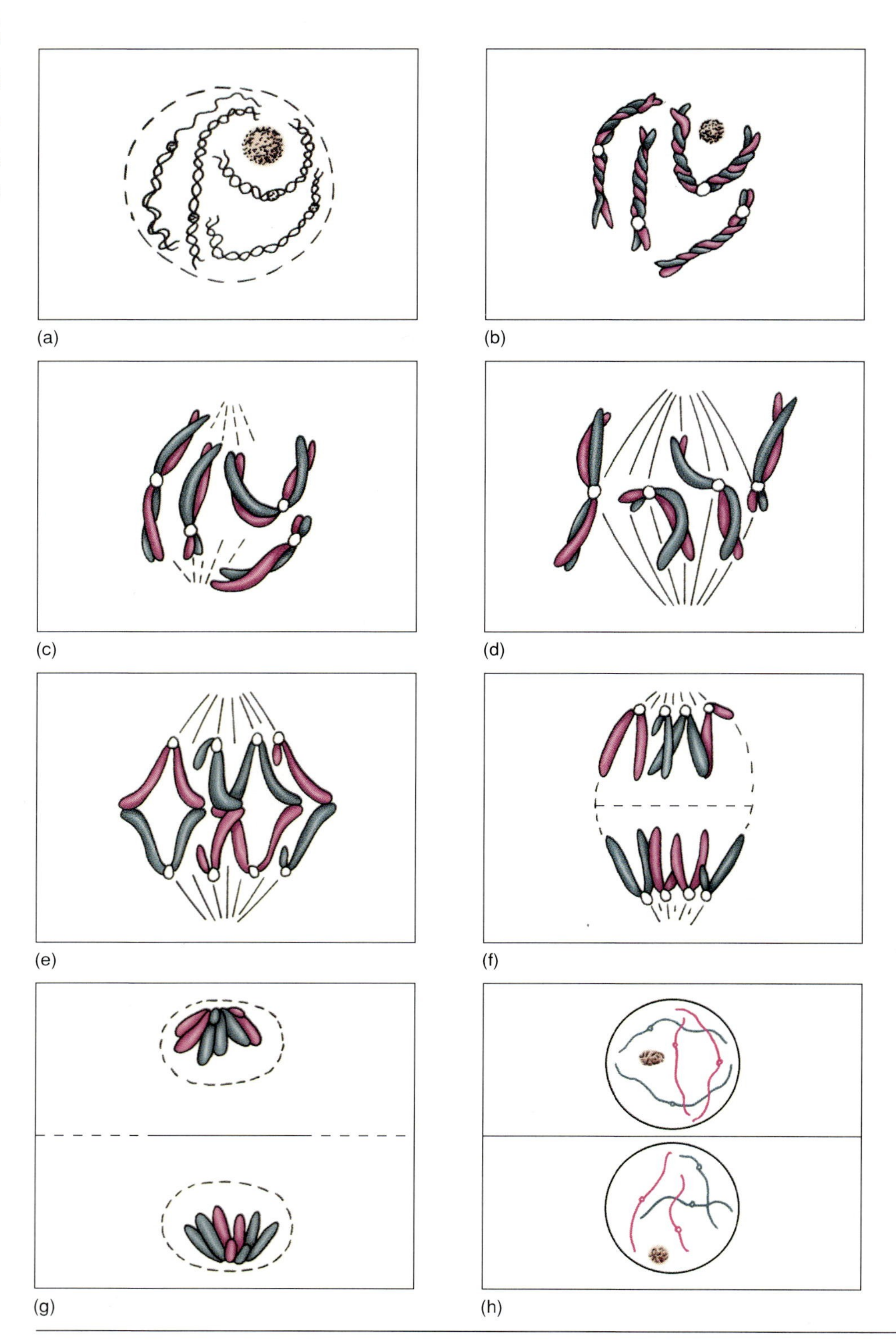

apart (figure 5.14*b*). The two chromatids of each chromosome are now called daughter chromosomes as they move apart through the cytoplasm to the opposite poles of the cell (figure 5.14*c*). Since the cytoplasm is an aqueous medium and the contracting spindle fibers pull on the centromeres, the arms of each daughter chromosome trail back from their centromere through the cytoplasm, forming an equal number of V-shaped strands aiming toward each pole (figures 5.14*d* and 5.15*d*). Anaphase thus consists of the short time from separation of the chromatids at the metaphase plate until the two groups of identical daughter chromosomes reach the opposite poles of the cell, then signifying the transition to **telophase.**

Telophase

As the two identical sets of chromosomes reach the opposite poles of the cell, several events begin to occur almost simultaneously. These events are essentially the reverse of prophase. A nuclear membrane begins forming around each of the two groups of chromosomes, and the spindle fibers completely disassociate and disappear (figure 5.15*e*). The chromosomes also become less distinct as they change into long, thin strands again. Nucleoli re-form, attaching to special nucleolar organizing areas of certain chromosomes. The end result of mitosis, then, is the formation of two complete nuclei, both genetically identical to each other and to the nucleus that gave rise to them. Since each nucleus is normally the control center for two new cells formed from the original cell (cytokinesis), the significance of their identical genetic makeup is clear. The growth, protein synthesis, and direction of all cellular functions in these two

Figure 5.13

Even though mitosis is an ongoing, integrated process, it is possible to recognize developmental stages. (*a*) Early prophase. The nuclear envelope is beginning to disappear, and the chromosomes are beginning to condense. (*b*) Late prophase. The nucleolus is beginning to disappear, and chromosomes have become tightly coiled. (*c*) Early metaphase. Spindle fibers begin to appear and attach to the centromeres. (*d*) Chromosomes move into a central position in the cell, with the centromeres aligned along the cell plate. (*e*) Early anaphase. Chromatids begin to pull apart by constriction of the spindle fibers. (*f*) Late anaphase. The chromatids have separated and have become new chromosomes. (*g*) Telophase. Chromosomes arrive at the poles, and the nuclear envelope begins to re-form. (*h*) Interphase. The nuclear envelope is complete, the nucleolus has re-formed, and the chromosomes are not visible.

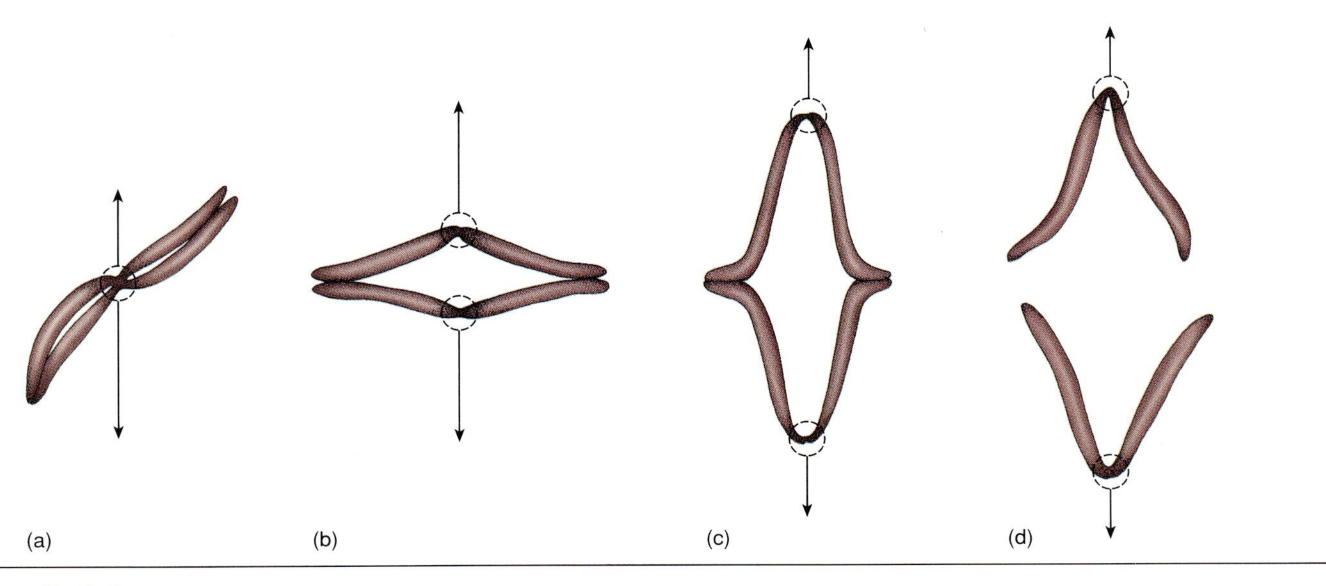

Figure 5.14

Chromosome separation at anaphase. The spindle is attached to the centromere (*a*), and separation begins at that point (*b*). The tips of the arms are the last to separate and trail along behind (*c–d*).

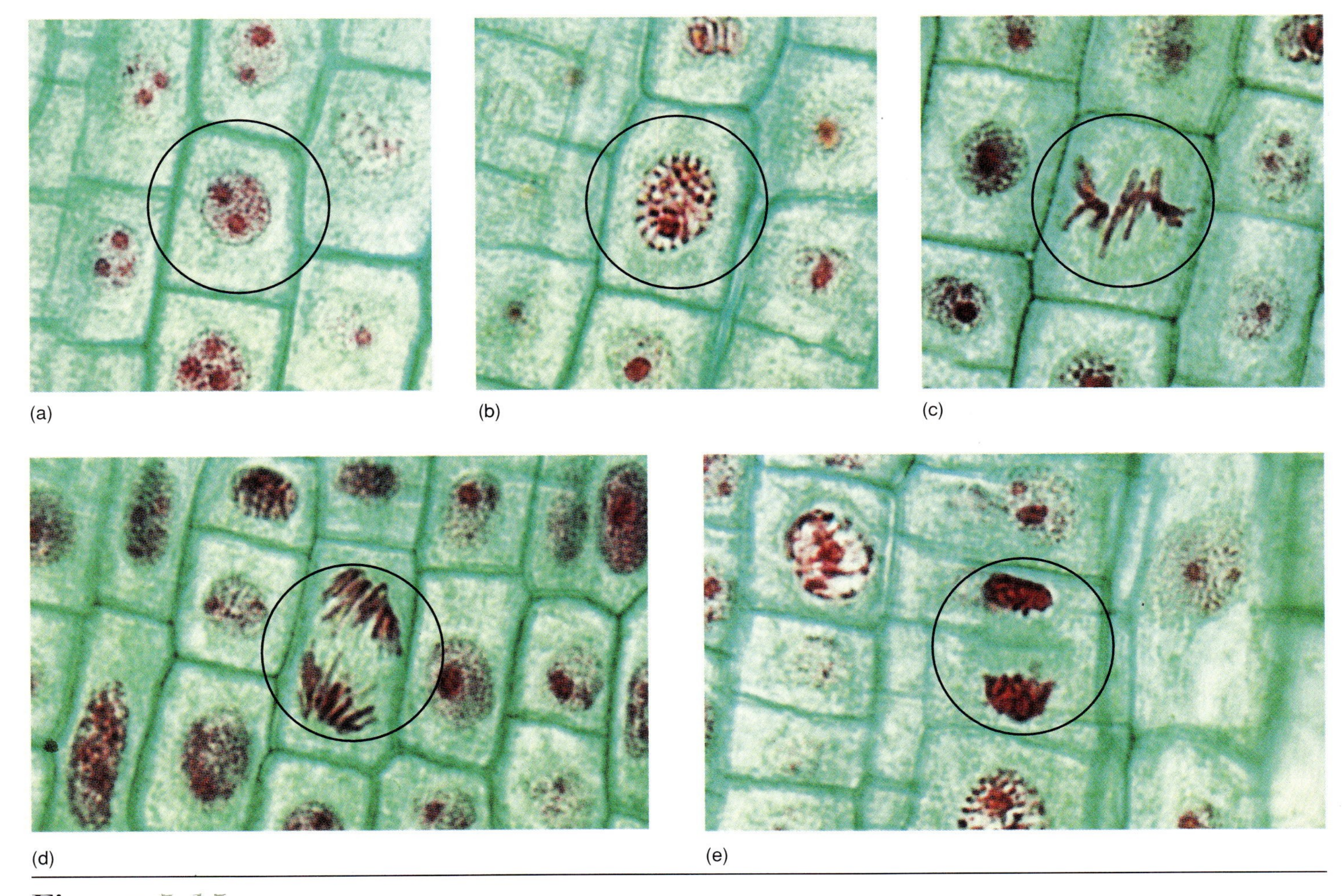

Figure 5.15

A stained thin section of *Allium* (onion) root tip with a clear interphase (*a*), prophase (*b*), metaphase (*c*), anaphase (*d*), and telophase (*e*).

new cells must be the same as in the parent cell, and it is the process of mitosis that ensures the identical genetic makeup of all new daughter cells.

Cytokinesis

Although discussed separately here, mitosis and cytokinesis usually occur simultaneously to produce two new daughter cells, each having genetically identical nuclei. Cytokinesis is the division of the original cell and its cytoplasm into two cells, each containing an approximately equal amount of all cytoplasmic materials. Occasionally, division results in daughter cells of unequal size and content. Cytokinesis begins during the early telophase activities of mitosis. As the newly formed daughter chromosomes pull apart and move away from the equatorial plate (metaphase plate) region of the cell, small vesicles migrate to that area and associate with microtubule fibrils found there. These fibrils may be derived from disassociated spindle fibers left in this region of the cell. The vesicles are produced by Golgi bodies and contain pectin, which forms the **middle lamella** between cells. The vesicle-fusing process begins at the middle of the cell and continues toward the outer walls, building the cell plate, which completely separates the original cell into two halves when completed.

Once the cell plate is formed, both newly formed daughter cells form a cell membrane across the plate to completely enclose their respective cellular contents. Primary walls are then laid down outside the cell membrane, and the two new daughter cells are complete although not yet full sized. The plasticity of the primary wall allows for cellular growth until each of the two daughter cells reaches maximum size; then secondary wall materials are added to provide rigidity and structural strength. It is during this period of cellular growth that the nuclei of each cell are said to be in mitotic interphase. During this period the cell synthesizes extra cytoplasmic materials and organelles, the chromosomes replicate, and finally specific materials necessary for the initiation of mitosis, such as extra microtubules, are produced. The two daughter cells are now mother cells in their own right and are ready to undergo mitosis and cytokinesis.

Not all cells follow this sequence of interphase activity because not all cells divide more than once. In plants new cells are produced only in apical meristems and lateral meristems. Those cells which do not remain in these meristematic areas are thought to go through the G_1 portion of interphase activities but are thought not to replicate their chromosomes or produce specific materials for mitosis (S and G_2 phases).

Those cells which do divide repeatedly vary in the amount of time required for mitosis and cytokinesis; some take a little more than an hour, and others take up to three or more hours. These times refer to eukaryotic cells only; as has been pointed out in discussions of *E. coli*, prokaryotic cells can divide much more rapidly, but they do not do so by the process of mitosis, since they do not have organized nuclei with true chromosomes.

Sometimes nuclei divide without cytokinesis taking place. The consequence is a cell with more than two sets of chromosomes, or a polyploid. The genetic implications of polyploidy are discussed in the section on genetics and evolution.

> ✔ ***Concept Checks***
>
> 1. What is the difference in mitosis and cytokinesis?
> 2. At what part of the cell cycle is DNA replicated?
> 3. What is the sequence of events leading to cell replication?

Summary

1. All living organisms are able to reproduce and maintain their characteristic morphology because of a precise replication system inherent in deoxyribonucleic acid (DNA). This macromolecule occurs in nearly all organisms and consists of a double helix of nucleotides.
2. DNA is replicated in a precise fashion as the double coil unwinds and separates, exposing nitrogen bases of the nucleotide that can attach to a complementary nucleotide.
3. In addition to replicating itself, DNA serves as the template for protein synthesis. This process occurs in two steps: transcription and translation. Transcription involves the uncoiling of the DNA molecule and exposing of the template, allowing it to synthesize a new single-stranded messenger RNA. The complementary pairing ensures that each nitrogen base of the DNA pairs with its complementary base in the mRNA. After detaching from the DNA, the mRNA is edited and migrates out of the nucleus and into the cytoplasm, where it attaches to a ribosome or polysome. As the mRNA is threaded through the ribosome, specific transfer RNAs for each amino acid match at the anticodon loop with the codons of the mRNA. This allows the ribosome to translate the message and place the tRNA-amino acid complex in the correct position. Finally, a peptide bond is formed between adjacent amino acids, ensuring the proper sequence for each specific protein.
4. A gene consists of a piece of DNA responsible for the coding of one particular characteristic. Genes are regulated by switches that turn them on and off, but the control mechanism is poorly understood.
5. Cell division in eukaryotes consists of mitosis, or nuclear division, and cytokinesis. The carefully controlled developmental sequence proceeds from prophase, to metaphase, then to anaphase, and finally telophase. After division the nucleus goes into interphase, which consists of two gaps separated by a synthesis phase in which the DNA is replicated. Interphase is a time of intense metabolic activity.

Key Terms

anaphase 91
anticodon loop 85
bacteriophage 81
centromere 91
chromatid 91
chromatin 87
chromosome 81
codon 84
cytokinesis 90
diploid 91
double helix 82
gene 83
homologue 91
metaphase 91
middle lamella 94
mitosis 87
mutagen 89
nucleotide 82
operon 88
polysome 87
prophase 91
telophase 92
transcription 84
translation 84
triplet 83

Discussion Questions

1. How did Watson, Crick, and Wilkins develop the DNA model based almost entirely upon a few bits of evidence?
2. What would be the minimum number of triplets needed to code for the amino acids essential for life? Why do you suppose that there are duplicate codons for specifying certain amino acids?
3. Why wouldn't one expect that the entire process of transcription and translation could be accomplished within the nucleus?
4. How might seemingly insignificant but metabolically perceptible mutations lead to a change in a species?
5. Just how important is interphase in the overall cell-cycle process?

Suggested Readings

Grierson, D., ed. 1991. *Developmental regulation of plant gene expression.* Chapman and Hall, New York.

Hames, B. H., and S. J. Higginns. 1993. *Gene transcription: A practical approach.* IRL Press, New York.

Hyams, J. S., and B. R. Brinkley, eds. 1989. *Mitosis: Molecules and mechanisms.* Academic Press, San Diego.

Nierhaus, K. H., ed. 1993. *The translational apparatus: Structure, function, regulation, evolution.* Plenum Press, New York.

Verma, D. P. S., ed. 1993. *Control of plant gene expression.* CRC Press, Boca Raton.

Wagner, R. P., M. P. Maguire, and R. L. Stallings. 1993. *Chromosomes: A synthesis.* Wiley-Liss, New York.

Warmbrodt, R. D. 1992. *Gene expression in horticultural crops.* National Agricultural Library, Beltsville.

Part Three

The Plant Body

Functional and beautiful are the pattern of veins providing transport of water and food throughout the millions of cells of this leaf.

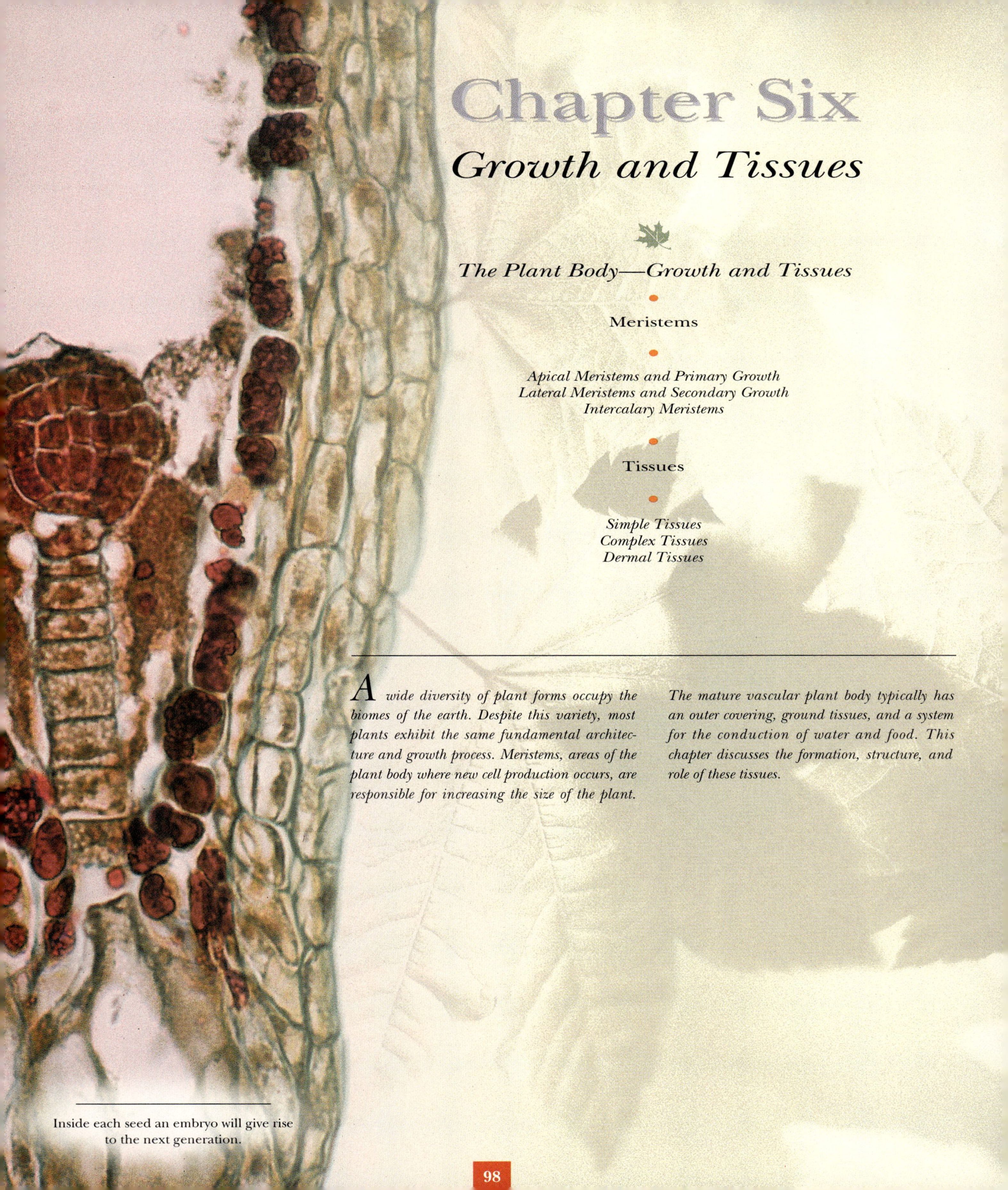

Chapter Six
Growth and Tissues

A wide diversity of plant forms occupy the biomes of the earth. Despite this variety, most plants exhibit the same fundamental architecture and growth process. Meristems, areas of the plant body where new cell production occurs, are responsible for increasing the size of the plant. The mature vascular plant body typically has an outer covering, ground tissues, and a system for the conduction of water and food. This chapter discusses the formation, structure, and role of these tissues.

Inside each seed an embryo will give rise to the next generation.

One of the consistent features common to plants is a lack of motility. Except for a rare example of motility in some aquatic plants, plants are stationary—fixed in their location and usually anchored by a root system. This immotility subjects plants to the dictates of nature. Like animals, the typical plant body is over 90% water. Thus, in addition to being able to withstand fluctuations of light, temperature, space, and nutrient availability, plants also must acquire water. Unlike animals, however, plants do not have the ability to move to a place with more water.

THE PLANT BODY—GROWTH AND TISSUES

Nonvascular plants (those that are less complex—typically single-celled or small multicellular organisms) get water and dissolved materials to each cell of the plant body by being in direct contact with moisture. Vascular plants (those that are more highly evolved and complex) have **vascular (conducting) systems** in which water and solutes are transported to the different parts of the plant body.

The two most visible, largest, and most complex groups of vascular plants are the gymnosperms, notably represented by the cone-bearing trees such as pine, fir, spruce, and juniper; and the angiosperms, or flowering plants (chapter 11). Since angiosperms dominate the landscape and are of utmost importance to human existence, our discussions center on them more than the gymnosperms.

Meristems

This chapter deals with the organization and growth of the flowering angiosperm plant and with the function and structure of its tissues and cell types. The body of an angiosperm (flowering plant) and other vascular plants may be divided into two parts,

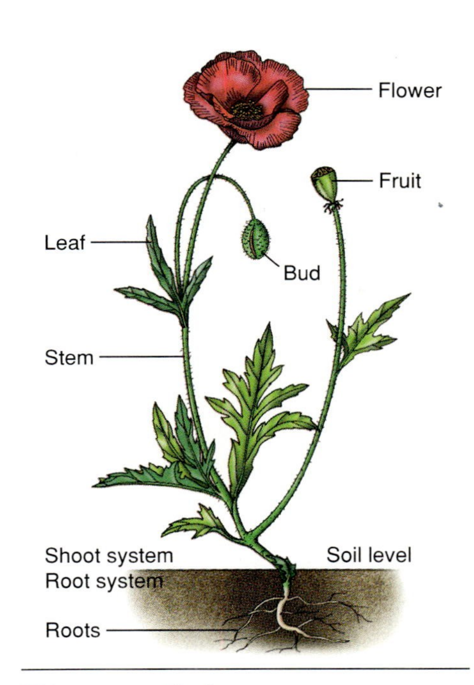

Figure 6.1
This poppy (*Papaver rhoeas*) exhibits all the normal vegetative and reproductive structures of an angiosperm.

the **shoot system** and the **root system** (figure 6.1). The shoot system consists of the stems and their attached appendages, the leaves and reproductive (flowers, fruit, seed) parts. The root system typically includes the subterranean parts of the plant body. Although the variability is great, the external morphology (overall form) of vascular plants is composed of four basic parts: roots, stems, leaves, and reproductive organs (figure 6.1). Each **organ** is composed of several different **tissues** grouped together into a structural and functional unit. The architecture and role of each organ will be discussed in subsequent chapters.

In multicellular organisms, actively dividing cells result in either new growth or replacement of old cells for overall tissue maintenance (chapter 5, p. 90). Growth regions in vascular plants are called meristematic regions, and they occur in the tips of growing shoots and roots. Woody plants also have two kinds of **lateral meristems:** the vascular cambium and cork cambium.

Apical Meristems and Primary Growth

The growth of the shoot and root systems can be attributed to the production of new cells in regions termed **meristems.** Meristems are composed of meristematic cells, whose main function is to divide and contribute new cells to the existing plant body. In vascular plants, meristems typically occur at the tip(s) of each stem and root (figure 6.2). These meristems are termed **apical meristems** because of their location at or near the apex. Every time a meristematic cell divides, one of the two resulting offspring remains a meristematic cell, commonly referred to as an initial, within the meristem. The rate of cell division of the other cell, referred to as a derivative, is gradually reduced. Along with this change the derivative begins to enlarge, pushing the meristem with its initials ahead, either through the soil as in the case of the root system, or higher into the air in the case of the stem. This overall increase in the length of the root and shoot system is termed **primary growth.** Tissues that are produced as a result of primary growth are termed primary tissues and collectively they comprise the primary body of a plant.

Lateral Meristems and Secondary Growth

Plants that grow for a relatively long time, and particularly those that survive the winter, do so as the result of lateral meristems, cylinders or sheaths of meristematic cells that cause a plant stem or root to increase in diameter and give rise to secondary tissue. By far the most important lateral meristem is the vascular cambium, a cylinder of cells in the roots and stems of woody plants that gives rise to secondary xylem and secondary phloem (chapter 8, p. 125). In addition, most woody stems develop a phellogen or cork cambium that gives rise to **periderm,** the outer bark of a tree. The production of secondary tissues which increase the diameter and circumference of the shoot and root system is

Figure 6.2
The higher plant body typically has at the apex of each stem and root a group of cells which continually undergoes cell division. These regions are the apical meristems.

termed **secondary growth.** Plants which exhibit secondary growth concurrently undergo primary growth. Secondary growth will be discussed in greater detail in chapter 8.

Intercalary Meristems

Although apical meristems and lateral meristems account for most growth, other pockets of meristematic activity sometimes persist. At the base of a leaf of grass, for example, an intercalary meristem provides continuing cell division long after the leaf is fully developed. This feature has adaptive significance because, if the leaf is cut off by a herbivore (or lawn mower), it continues to grow from the base and ensures survival of the plant.

Concept Checks

1. Distinguish among apical, lateral, and intercalary meristems.
2. Distinguish between primary growth and secondary growth.

Tissues

Since cell division in plants occurs only in meristematic regions—specific, localized pockets of cell division—cells adjacent to those regions begin to change and eventually become the specialized tissues that make up mature organs. Thus one finds a lineage of cells trailing away from the meristem of the shoot or the root. The meristematic region itself is quite small, ranging from a single cell in most nonvascular and non-seed-producing plants to hundreds of cells in gymnosperms and angiosperms. Directly behind the apex is a group of cells that have recently divided and are in the process of becoming mature. They do so by going through a process of **cell enlargement** in which the cell expands greatly through the uptake of water. The direction of expansion is dictated by the physical forces exerted on the cell wall and is primarily upward, hence the region immediately behind the shoot or root apex is referred to as the zone of elongation (figure 6.3). Small meristematic cells, originally only 20 to 30 micrometers (μm) in diameter, may suddenly elongate to perhaps several hundred micrometers.

Within the zone of elongation, three fairly distinct primary meristem areas derived from the tissue of the apical meristem can be recognized. These are the protoderm, ground meristem, and procambium (figure 6.3).

During the elongation phase and as the cells begin to reach their ultimate shape, subtle internal chemical changes begin to occur. This regional change in cell composition, thickness of the wall, and function is directly behind the zone of elongation and is called the **zone of differentiation.** The specialized nature of the cell, which contributes to the separation into tissues with specific functions, begins here.

Even though all cells start out appearing the same, as daughter cells produced in cell division, they begin changing in the zone of elongation and continue to do so in the zone of differentiation. Here they take on structural and functional characteristics that are to remain until they decompose as litter. Some plant cells die early in life, but they still function for a number of purposes. Vessel members, for example, specialize and die early. In this case the disappearance of their protoplast is essential to their function in water and mineral transport.

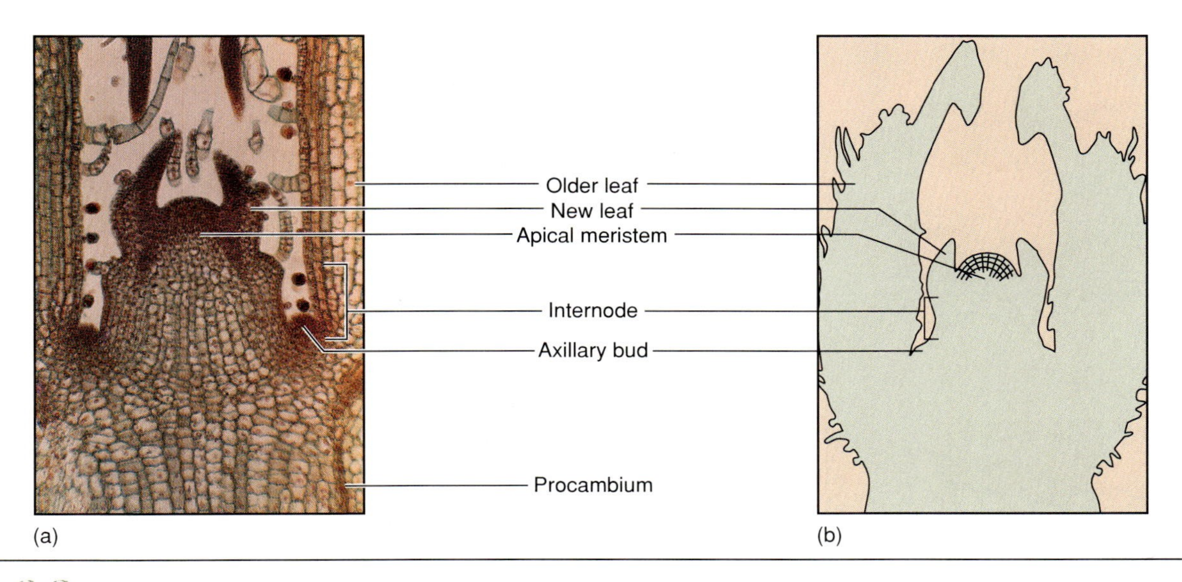

Figure 6.3

The shoot apex of *Coleus* as viewed in the light microscope (*a*) and diagrammatically (*b*). The dome of tissue at the top is the apical meristem with two newly formed leaves. At the level of the second (older) pair of leaves the dark areas indicate the location of axillary buds. The space between is the internode.

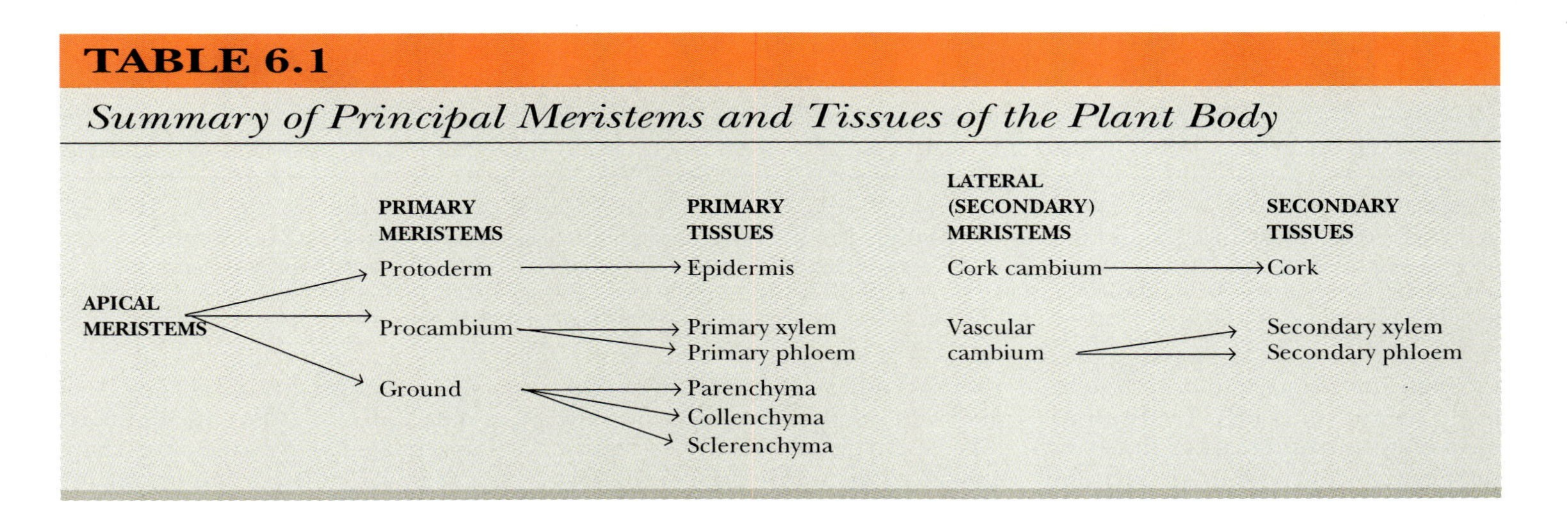

TABLE 6.1

Summary of Principal Meristems and Tissues of the Plant Body

	PRIMARY MERISTEMS	PRIMARY TISSUES	LATERAL (SECONDARY) MERISTEMS	SECONDARY TISSUES
APICAL MERISTEMS →	Protoderm →	Epidermis	Cork cambium →	Cork
	Procambium →	Primary xylem Primary phloem	Vascular cambium →	Secondary xylem Secondary phloem
	Ground →	Parenchyma Collenchyma Sclerenchyma		

Simple Tissues

Groups of cells that are structurally and/or functionally distinct are known as tissues. Tissues composed of one type of cell are termed simple tissues. Simple tissues differentiate from the ground meristem and (table 6.1) are of three types: **parenchyma, collenchyma,** and **sclerenchyma** (figure 6.4). Each of these tissues, often referred to as ground tissues, has specific functions. The primary functions of the ground tissues are synthesis, support, and storage.

Parenchyma Parenchyma tissue is composed of parenchyma cells. Parenchyma cells have relatively thin primary cell walls and are often isodiametric; that is, they have essentially the same diameter in all planes. They generally retain all membranes and organelles and appear very much like what one might expect a typical plant cell to be. Parenchyma cells have large vacuoles and are involved in storage. Parenchyma tissues remain alive and functional at maturity, and occur abundantly in stems, roots, leaves, and in fruit. Many parenchyma cells are capable of cell division (i.e., they are totipotent—see "Mighty Meristems") and are therefore important in wound healing and regeneration. They also function in various tissues for photosynthesis, storage, respiration, protein synthesis, and secretion. Spaces, termed intercellular spaces, are commonly found associated with parenchyma tissue and are important in the storage and movement of gases and water in the plant body (figure 6.4*a,b*). Numerous plasmodesmata (chapter 4, p. 71) occur

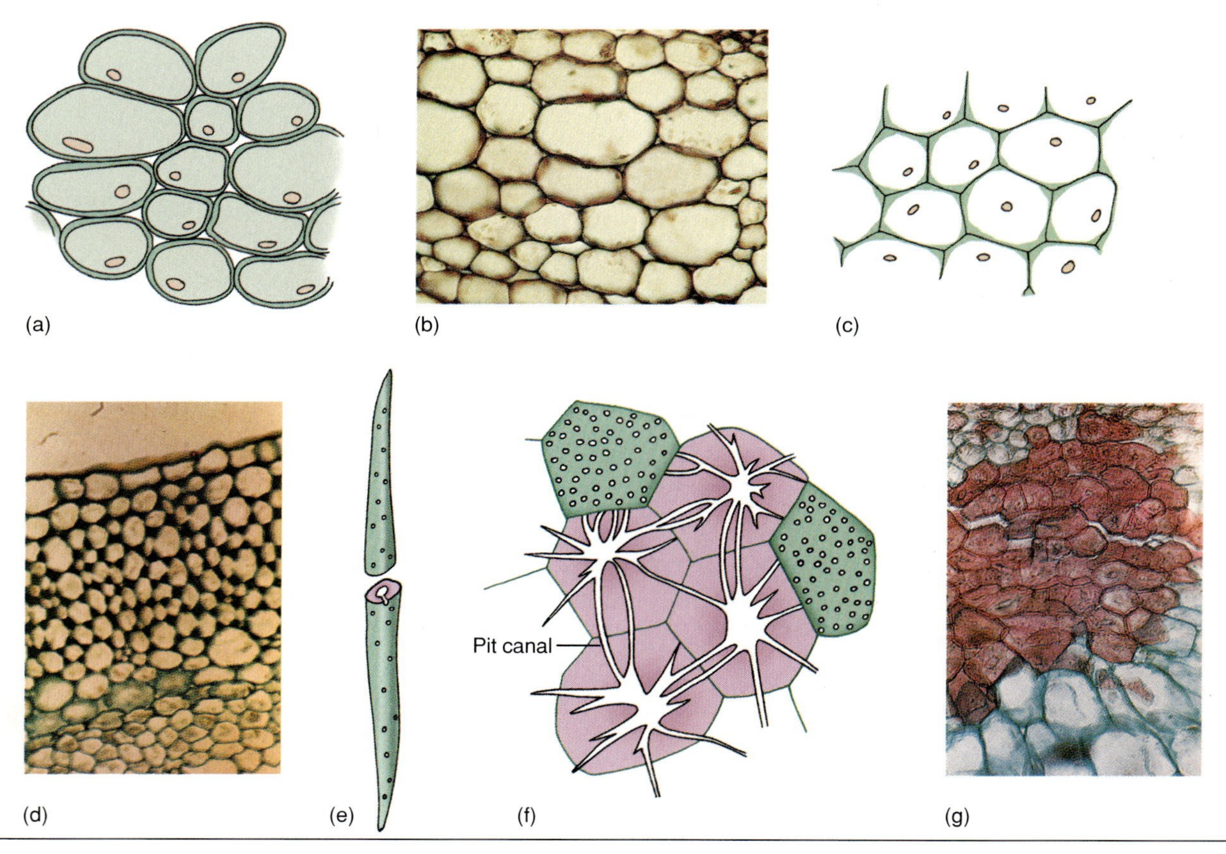

Figure 6.4

(*a, b*) The basic tissue type is parenchyma. The living cells are thin walled, isodiametric, and typically possess large storage vacuoles. Considerable intercellular space is usually present among the cells. (*c, d*) Collenchyma is much less common but easily recognized by the uneven thickening of the primary wall. Sclerenchyma may be either fibers (*e*) or sclereids (*f, g*). These thick-walled support cells are typically dead at maturity and have lignified secondary walls that provide support and protection to the plant. Note the presence of pit canals in the secondary walls.

through the primary wall of adjoining parenchyma cells, facilitating movement of solutes from the cytoplasm of any one cell to all adjacent living cells.

Collenchyma Collenchyma cells usually occur as discrete groups of cells that form a tissue directly beneath the epidermis in stems and leaf petioles. The cells are generally elongated and unevenly thickened in the primary walls (chapter 4, p. 72) and are capable of giving exceptional strength to young actively growing organs. The walls of these cells stretch readily and offer little resistance to elongation. Collenchyma cells, like parenchyma, are alive at maturity. The uneven wall thickening characteristically occurs at the junction with two other cells, forming a "corner" with thickened deposits of cellulose, pectin, and water (figure 6.4*c,d*). This additional thickness provides flexible mechanical support for young stems and leaves when they are subjected to a stress such as wind. One excellent place to observe collenchyma is in the leaf stalk of celery or beets. In windy regions celery may form a great deal of collenchyma, which tends to make the celery stalk (properly termed the petiole) tough.

Sclerenchyma Sclerenchyma is characterized by cells with very thick walls and that commonly lack protoplasts at maturity. Sclerenchyma generally falls into two categories of cells: fibers and sclereids. Fibers (figure 6.4*e*) are elongated and have secondary cell walls that are lignified (impregnated with lignin—see chapter 4, p. 72) and are important for strength and support in plant parts that have completed elongation. Fibers generally occur in strands or bundles composed of several cells, and they are very important in the consumer industry. Hemp and jute fibers are used to make twine; flax fibers are used in making linen; cotton fibers (although termed fibers, these cells are technically epidermal hairs of the seed and differ anatomically from sclerenchyma fibers) are converted to thread and woven into a myriad of products. Other plants provide fibers that are processed into twine and rope. Originally most kinds of rope were made from hemp, but

Mighty Meristems

The primary growth of young root and shoot systems is indeterminate. Cell division, enlargement, and differentiation can continue to produce the aerial and subterranean organs as long as these systems are well nourished and no environmental signals change the pattern of growth in the meristem. This is in contrast to the determinate growth process of animals, which limits the size of the adult, the number of organs that it will develop, and the time period over which the individual can grow. With the potential of unlimited growth concentrated in the meristems these regions and the cells that compose them have been the focus of continuing research. The cells of meristems and other cells with a nucleus such as parenchyma are considered to be totipotent, that is, they are capable of differentiating into any cell type of the plant body.

When small pieces or single cells from meristems are placed *in vitro* (in glass) under sterile (antiseptic) conditions on a defined nutrient-containing medium in a controlled environment the meristematic cells or tissue can be maintained indefinitely, or they can proliferate and be encouraged to grow into an unorganized mass of cells called callus. With the addition of hormones (chapter 14), in specific amounts and proportion, the callus can be induced to differentiate and develop into mature plants (figure 1*a*). This process, referred to as plant tissue or cell culture, is a highly efficient means of propagating plants, since literally millions of new plants can be initiated from a small piece of tissue or individual cells. As many as 10 million cell cultures can be produced in a single 250 ml flask (about 1 cup) in a very short time period. This can be compared to the space and time requirement needed if individual plants were rooted from stem cuttings by conventional methods or propagated by seeds. Greater genetic uniformity is also exhibited in cultured plants compared to seed-derived individuals.

Commercial seed and crop producers as well as foresters and horticulturalists are using culture techniques to produce plants with improved yields such as potatoes, tomatoes, corn, tobacco, wheat, strawberries, apples, peaches, asparagus, cabbage, citrus, sunflowers, carrots, orchids, pines, redwoods, and many other species of plants (figure 1*b*).

Successful *in vitro* cultivation of plant cells and tissues underlines the new biotechnologies in genetic engineering discussed in chapter 15.

(a)

(b)

Figure 1

(*a*) At the Oil Palm Research Station in Malaysia, callus (*left*) is supplied a nutrient-rich media resulting in the development of young oil palm plants (*right*). (*b*) Oil palm fruits (*Elaeis guineesis*) are harvested for two distinct oils; palm oil is extracted from the fruit pulp, and palm kernel oil is extracted from the seed.

since this plant also produces marijuana, its cultivation is no longer legal in the United States. Sclereids, often termed stone cells, are variable in shape, some are shaped like a bone, others relatively spherical or cubical (figure 6.4*f,g*), and others highly branched (figure 6.5). Compared with fibers, they are short cells, and they may occur singly or as groups of cells throughout the ground tissue. They are the primary component in seed coats, nut shells, and the hard pits of stone fruit, and they give a ripe pear its gritty texture.

As the thick secondary wall is formed, the protoplasm of a sclerenchyma cell degenerates. In dying they reach the ultimate specialization. The strength of a fiber or sclereid depends on the thickness of the secondary cell wall. Prior to death, that cell must synthesize a large number of cellulose and lignin molecules, which impart strength to the cell. The secondary wall thickness is one of the important quality considerations in the cotton fiber and can be a major factor in the farmer's price. The thickness and number of wood fibers determine the strength and quality of the lumber (chapter 10, p. 154).

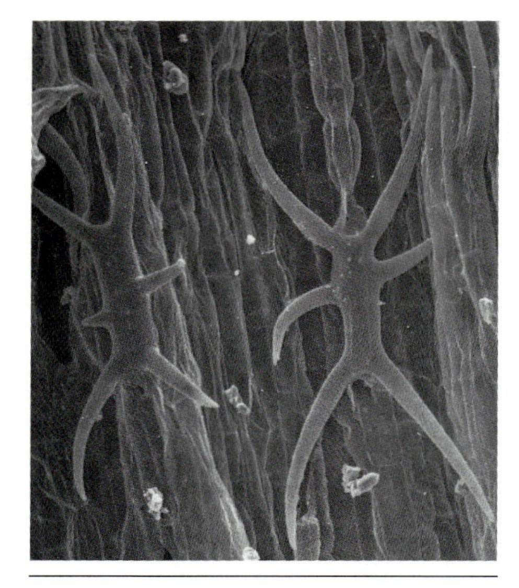

Figure 6.5
Scanning electron micrograph of an astrosclereid, a highly branched sclereid which prevents the collapse of the air canals in *Nuphar,* a common water lily.

> ### ✔ Concept Checks
>
> 1. List the three primary meristems. What tissue does each produce?
> 2. What is meant by the term differentiation? Outline the sequence of events that occur following cell division of a meristematic cell until a mature parenchyma cell is produced.
> 3. Compare parenchyma and sclerenchyma with regard to location in the plant body, cell wall structure, and function.
> 4. Why is collenchyma, rather than sclerenchyma, a supporting tissue in young growing regions?

Complex Tissues

In contrast to simple tissues, complex tissues consist of more than one cell type, and include the **xylem, phloem, epidermis,** and periderm.

The procambium gives rise to the vascular system of the primary plant body—the primary xylem and primary phloem. The cells of these tissues are highly specialized; xylem transports water and mineral nutrients, whereas phloem transports organic solutes produced within the plant. These pipelines are the support system for higher plants; without them the stature of land plants would be much smaller than it is.

Xylem Four basic cell types make up the primary xylem: **vessel members, tracheids, fibers,** and xylem parenchyma (figure 6.6*a–d*). Although they all have specialized functions, vessel members in angiosperms are the most important cells in water and mineral nutrient transport. Vessel members combine end to end, much like sections of stovepipe, making a vessel that may run throughout the body of the plant. Vessels traverse the plant from small roots up through the center of the root system, through the stem, out through branches, into petioles, and terminate near the margin of leaves. Vessels are very large in diameter, and therefore the flux of water through them can be enormous. As the cells destined to become vessel members reach maturity, they take on a characteristic shape and size, and the protoplast, or living part of the cell, dies. At the same time, the end walls are dissolved forming the perforation plate (figure 6.6*a*), and the vessel members are connected end to end to form a continuous pipe, much like a drinking straw.

Tracheids also function in water transport but are less efficient than vessel members in doing so because they are smaller in diameter and the end walls are not dissolved as the cell reaches maturity. Therefore resistance to the vertical movement of water is greater. Obviously, those plants that have more vessels and few tracheids are more efficient in water transport. In surveying all the plants that have a vascular system, the tracheophytes, it can be seen that most of the angiosperms have vessels; most gymnosperms have none, and neither do most ferns. Angiosperms have both vessels and tracheids, and each contributes to the overall transport.

In addition to these water-conducting cells, xylem consists of fibers that are dead, thick-walled, and

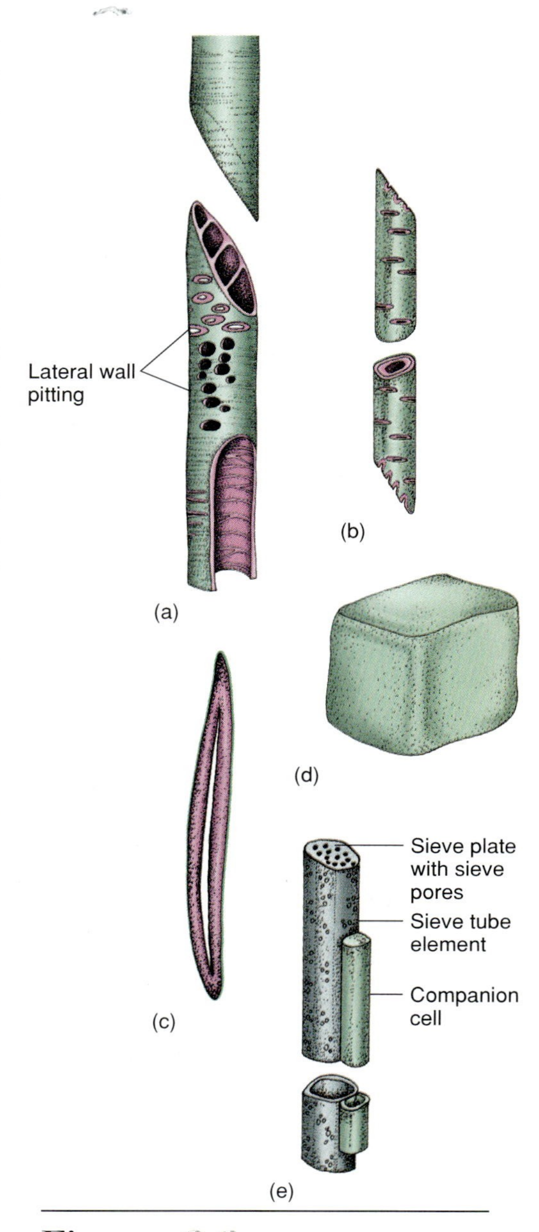

Figure 6.6
Xylem in angiosperms consists of vessels (*a*), tracheids (*b*), fibers (*c*), and parenchyma (*d*), whereas phloem (*e*) consists of sieve tubes, companion cells, and fibers and parenchyma similar to those in xylem.

nonfunctional for transport. Their function is in support, although occasionally fibers remain living and function for storage purposes. The parenchyma cells associated with primary xylem function for storage and synthesis of certain metabolites.

During the formation of the secondary cell wall in primary xylem the wall may be laid down in a variety of patterns including rings and spirals, much like the reinforcing fibers in a plastic garden hose (figure 6.7). These thickenings give strength to the vessel members and tracheids while allowing for expansion of the cell. The diversity in the secondary wall pattern is specific for each plant species and is therefore of taxonomic and evolutionary importance to botanists.

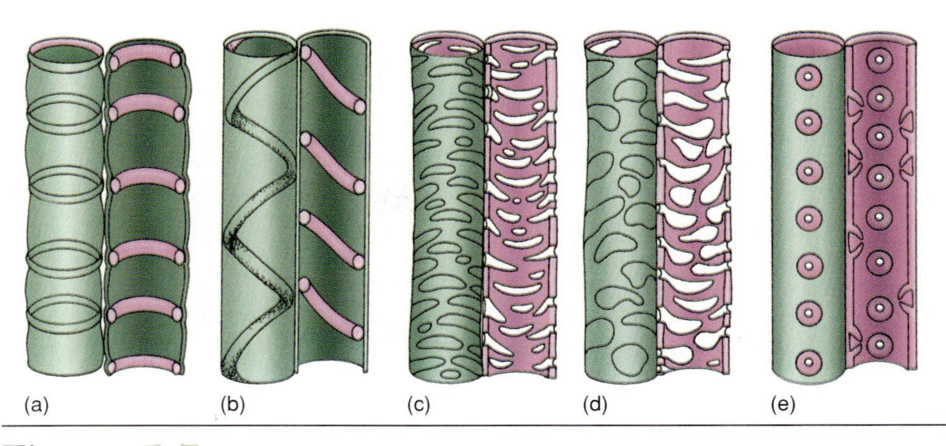

Figure 6.7

The secondary cell wall may be deposited in a variety of patterns in tracheids and vessel members. (*a*) Annular, (*b*) spiral, (*c*) scalariform, (*d*) reticulate, (*e*) circular-bordered pitted.

Phloem Primary phloem tissue is made up of **sieve-tube members, companion cells,** fibers, and phloem parenchyma (figure 6.6*e*). The phloem transports organic solutes, primarily sugars, over long distances within the plant. Students are often taught that xylem movement is *upward* from the soil, and phloem movement is *downward* from the leaves. Such comments are oversimplifications, since a great deal of transport is also lateral. Water moves to the point where the water potential (chapter 12, p. 203) is most negative, and sugars move from areas where they are synthesized or stored (source) to the point where the *sugar deficit,* or sink, is greatest. Transport of sugars and other organic solutes occurs through sieve tubes, formed by sieve-tube members connecting end to end (figures 6.6*e*, 6.8). The term sieve tube is appropriate because the end walls, termed **sieve plates,** are left with portions of the wall remaining to form a sieve, much like a spaghetti strainer. The protoplasts with dissolved organic solutes of the sieve tube are connected through holes in the sieve plate. These holes are termed **sieve pores** (figure 6.6*e*). Although the pores are large in terms of molecular size, they provide some resistance to movement of the viscous fluid, and transport through the phloem is considerably slower than water movement through the xylem. In gymnosperms a more primitive type of sugar-conducting cell, a sieve cell, performs essentially the same function, although the efficiency of transport is reduced.

When wounding of phloem occurs, the sieve tubes synthesize a polysaccharide called callose, which plugs the sieve pores and reduces the flow. Such a mechanism stops the loss of organic solutes from a wound and thus enhances the plant's chance for survival.

In the process of differentiation of sieve-tube members, a unique developmental process occurs. Generally, whenever a cell loses its nucleus, the regulatory processes are uncoordinated and the cell dies. In sieve-tube members, however, maturity leads to a cell that is enucleate (without a nucleus), but otherwise the cell is alive and functions normally.

How a cell could function without a nucleus has long puzzled scientists. We believe that the close association of a sieve-tube member with a companion cell through cytoplasmic connections allows for its regulation. The two cells have a great deal in common, having been derived from the same mother cell. They are functionally connected, and the companion cell appears to have a great deal to do with the loading and unloading of solutes into and out of the sieve tube. Although much smaller in size than the sieve-tube member, companion cells have all the normal cell components, including a nucleus. Much remains to be discovered about how the companion cell actually functions. The function of fibers and parenchyma is the same in phloem as in xylem.

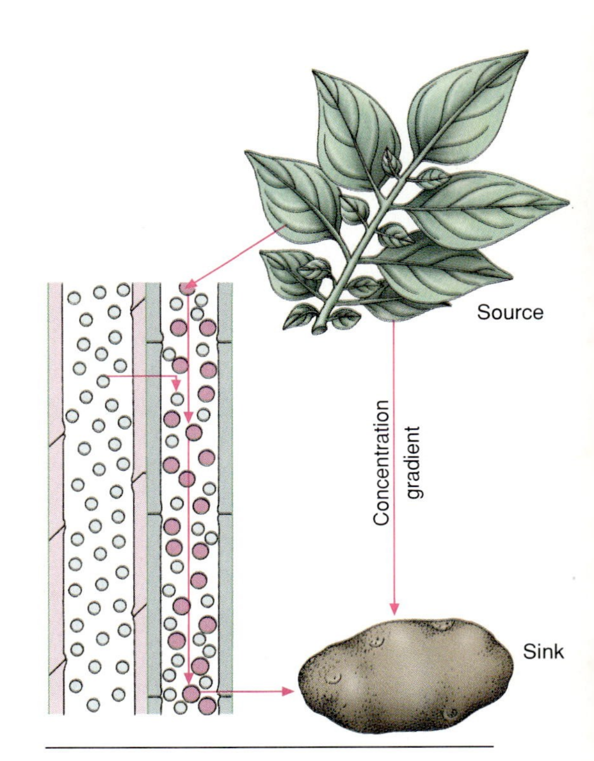

Figure 6.8

This diagrammatic potato plant can be used to explain the source/sink relationships in plants. Sugar molecules manufactured in the leaves are loaded into the phloem (sieve tube on right adjacent to xylem vessel on left). As the sugar concentration increases in the sieve tube, water moves in from the vessel by osmosis, pressure builds up in the sieve tube, and the sugar solution is forced down through the plant to the root system, where developing tubers accept the sugar as it is "unloaded" from the phloem. In potato tubers the sugar is rapidly converted to starch and stored in the vacuoles of parenchyma.

Dermal Tissues

Epidermis The protoderm, the primary meristem that gives rise to the epidermis, produces all the dermal or outermost tissue of the primary plant body. The epidermis produced by the shoot apex is made of a layer of cells that covers all parts of the plant, including the stems, both upper and lower surfaces of the leaves, all surfaces of flowers and fruit, and the root system. This covering functions as a protective barrier to reduce desiccation and prevent the entrance of harmful agents such as bacteria and fungi. The epidermal cells fit together tightly, sometimes looking like a jigsaw puzzle in face view (figure 6.9). The outer surface and the cell wall are coated with a waxy substance named **cutin.** As cutin builds up on the outer surface a layer called the **cuticle,** which reduces evaporation, is formed. Sometimes the cuticle can become very thick on old leaves, even though it may develop cracks through which water can escape. There is typically no cuticle covering the epidermis of the root system.

The epidermal layer may also have various types of appendages, called **trichomes,** which may be specialized for protection, secretion, and other functions (figures 6.10, 6.11, and 6.12). Trichomes come in many

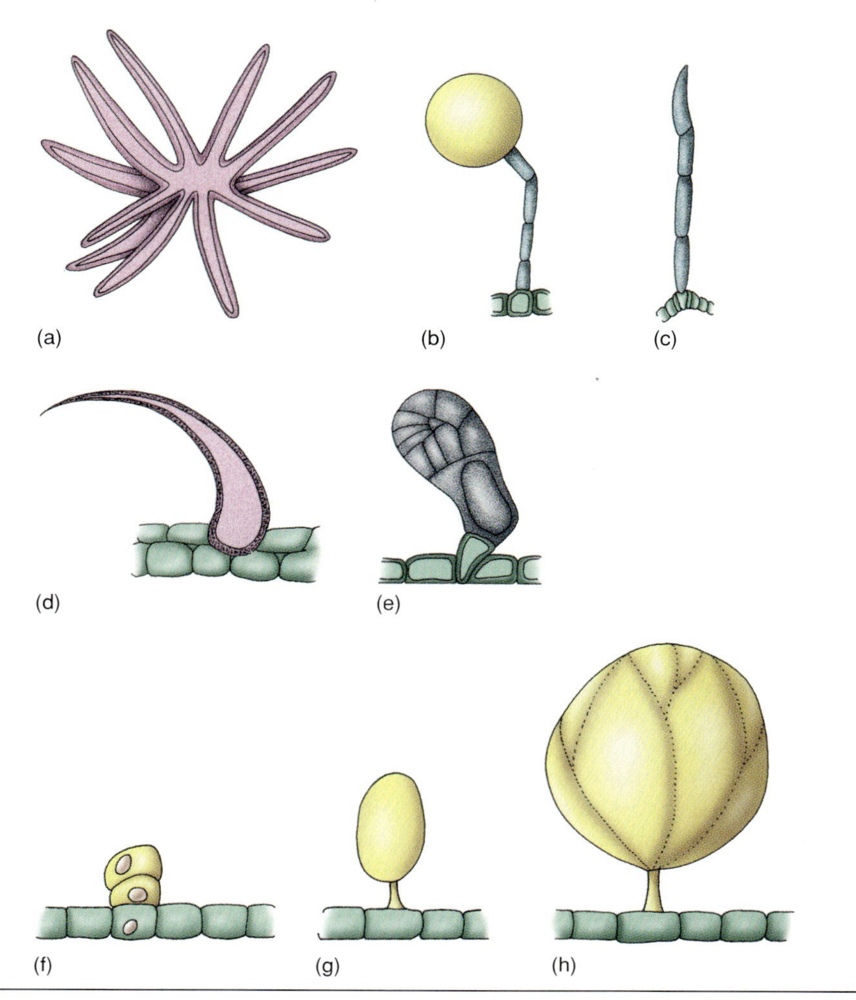

Figure 6.10

(*a* to *e*) The varied morphology of trichomes. They come in all shapes and sizes, but each species can have its own particular trichome type. Some are single celled; others are multicellular. (*f* to *h*) The development of a trichome (epidermal hair) in *Atriplex.* In the epidermal layer, cell division gives rise to a new cell that begins to enlarge and protrude above the epidermal layer. A second division produces the balloon cell. At maturity the huge balloon cell is held in position by the stalk cell, which attaches it to the epidermis. These trichomes are also involved in salt transport and accumulation.

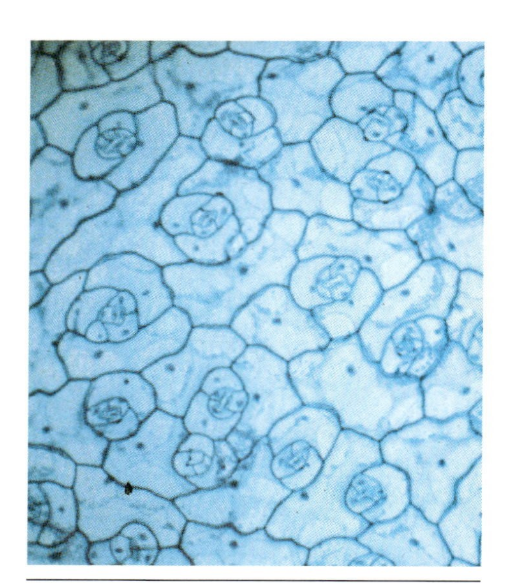

Figure 6.9

The epidermal layer looks like a jigsaw puzzle interspersed with stomata. This pattern is typical for dicotyledons.

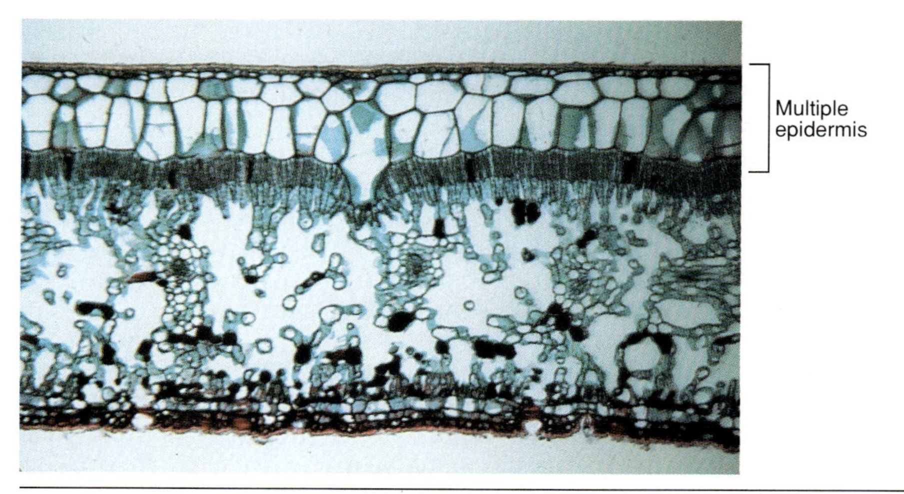

Figure 6.11

The epidermis in *Ficus elastica* (the rubber plant) is several layers thick and serves as a water storage tissue.

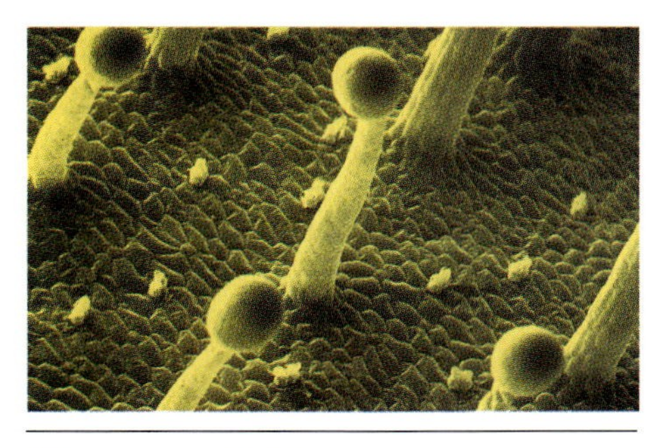

Figure 6.12
Scanning electron micrograph of the leaf surface of sundew (*Drosera*), an insectivorous plant, showing glandular trichomes. (×365)

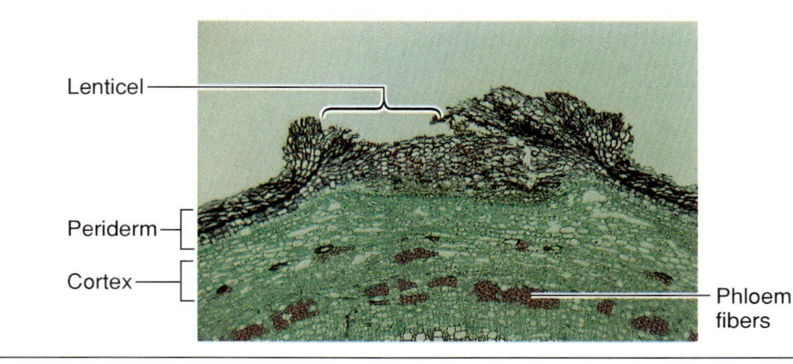

Figure 6.13
Cross section of the periderm, composed principally of cork cells produced by the cork cambium. A lenticel is also shown.

TABLE 6.2

Summary of Plant Tissues, Cell Types, and Cell Functions

TISSUE	CELL TYPES	PRINCIPAL FUNCTION
Parenchyma	Parenchyma cells	Storage; synthesis; wound healing
Collenchyma	Collenchyma cells	Flexible support
Sclerenchyma	Fibers or sclereids	Rigid support
Epidermis	Epidermal (parenchyma) cells; guard cells; trichomes	Protection; gas exchange
Periderm	Cork cells; parenchyma cells	Protection; gas exchange
Xylem	Tracheids; vessel members; sclerenchyma cells; parenchyma cells	Water and mineral conduction
Phloem	Sieve cells or sieve-tube members; companion cells; parenchyma cells; sclerenchyma cells	Conduction of organic solutes, principally sugars

shapes and sizes; they may be part of an epidermal cell, or they may be a series of cells attached to the epidermis. Generally, the epidermis is only a single cell-layer thick, but in some plants, such as *Ficus elastica* (rubber plant), the epidermis may be several layers thick and serve as a water storage tissue (see figure 6.11).

Interspersed throughout the epidermal cells are **guard cells,** two elongated cells capable of flexing at the ends to form an opening between them. The two guard cells, together with the pore, which they form between them, are called a **stoma** (pl., **stomata;** figure 6.9). Through regulation of the turgor pressure in these guard cells, the size of the pore may be controlled so that the pore is fully open, fully closed, or any stage in between. The stomata function in gas exchange, allowing water vapor to escape from the leaf surface and carbon dioxide and oxygen to enter and leave the leaf. Other gases also come and go, but CO_2 and O_2 are of primary importance (see chapter 13, p. 228).

Periderm In plants which undergo secondary growth, the epidermis is typically ruptured as the diameter of the stem is increased. In these plants the epidermis is replaced by another dermal cover, the periderm or outer bark (chapter 8, p. 126). The periderm is primarily composed of tightly arranged, rectangular **cork** cells. During differentiation the protoplast of a cork cell secretes **suberin,** a waxlike, impermeable substance which impregnates the cork cell wall. At maturity cork cells are dead, but the presence of suberin, like cutin on the epidermal surface, provides protection for the underlying tissues. Areas of loosely organized cork cells, termed **lenticels,** occur in the periderm and permit the exchange of gases between the interior of the stem and the atmosphere. The periderm will be considered in greater detail in chapter 8.

✔ *Concept Checks*

1. What cell types are found in the xylem? What is the general role of xylem in the plant body? How is the architecture of vessels suited for this role?

2. What cell types are found in the phloem? What is the general role of the phloem in the plant body? Define the following terms: sieve plate, sieve pores, sieve tube.

3. What are the dermal tissues of the plant body? From what meristem does each develop?

Summary

1. Despite the wide diversity of form in vascular plants, all possess similar growth processes, tissues, and cell types.
2. Meristems function in the production of new cells. The subsequent enlargement of these new cells results in the increase in size of the plant body.
3. Apical meristems of the shoot and root lead to primary growth, and lateral meristems lead to secondary growth.
4. Cells in the zone of elongation retain limited ability to continue cell division and produce the primary meristems: protoderm, ground meristem, and procambium.
5. The three simple tissues, parenchyma, collenchyma, and sclerenchyma, all arise from differentiation of the cells produced by the ground meristem.
6. Parenchyma cells constitute the bulk of the primary plant body. They have thin walls, large vacuoles for storage, and are active in synthesis reactions.
7. Collenchyma cells have unevenly thickened primary walls and occur in peripheral strands. They provide flexible support for appendages of the shoot.
8. Sclerenchyma cells have thick, lignified secondary cell walls that provide a rigid and/or protective framework.
9. Dermal tissues provide the outer protective coverings of the plant body. The epidermis, which arises from the protoderm, covers the primary plant body. The periderm, composed principally of cork cells, covers the older portions of the plant body where secondary growth occurs.
10. Primary xylem, the water and mineral conducting tissue of the primary plant body, differentiates from the procambium. The principal cells responsible for conduction are the hollow vessel members and tracheids. These cells are vertically aligned to facilitate the upward movement of water.
11. Organic substances, primarily carbohydrates produced in photosynthesis, are transported through the phloem in the plant body. Sieve-tube members are joined end to end to increase the efficiency of upward and downward conduction.

Key Terms

apical meristem 99
cell enlargement 100
collenchyma 101
companion cell 105
cork 107
cuticle 106
cutin 106
epidermis 104
fibers 104
guard cells 107
lateral meristem 99
lenticel 107
meristem 99
organ 99
parenchyma 101
periderm 99
phloem 104
primary growth 99
root system 99
sclerenchyma 101
secondary growth 100
shoot system 99
sieve plate 105
sieve pores 105
sieve-tube member 105
stoma 107
suberin 107
tissue 99
tracheids 104
trichomes 106
vascular system 99
vessel members 104
xylem 104
zone of differentiation 100

Discussion Questions

1. What are meristems? Explain how the activity of the apical meristem is responsible for the increasing length of the root and shoot system.
2. What is the difference between an organ and a tissue? How do simple tissues differ from complex tissues?
3. Imagine yourself to be an architectural engineer. Design a water-conducting system for a plant body that would function more efficiently than the vessels.
4. Imagine yourself to be a sugar molecule in a primary phloem parenchyma cell of a stem. Trace the pathway, listing all cell types, tissues, regions (or zones), and the general forces involved until you arrive into the zone of cell division in the apical meristem of a stem.

Suggested Readings

Baker, D. A., and J. A. Milburn, eds. 1989. *Transport of photoassimilates.* Wiley, New York.

Behnke, H. D., and R. D. Sjolund, eds. 1990. *Sieve elements: Comparative structure, induction and development.* Springer-Verlag, New York.

Esau, K. 1965. *Plant anatomy.* 2nd ed. John Wiley & Sons, New York.

Fahn, A. 1990. *Plant anatomy.* 4th ed. Pergamon Press, Elmsford.

Hillis, W. E. 1987. *Heartwood and tree exudates.* Springer-Verlag, New York.

Mauseth, J. D. 1988. *Plant anatomy.* Benjamin/Cummings, Menlo Park.

Palmer, P. G., and S. Gerbeth-Jones. 1988. *A scanning electron microscope survey of the epidermis of East African grasses.* Smithsonian Institution Press, Washington, D.C.

Raven, P. H., R. F. Evert, and S. E. Eichhorn. 1992. *Biology of plants.* 5th ed. Worth Publishers, New York.

Romberger, J. A., Z. Hejnowicz, and J. F. Hill. 1993. *Plant structure: Function and development: A treatise on anatomy and vegetative development, with special reference to woody plants.* Springer-Verlag, New York.

Sandved, K. B., G. T. Prance, and A. E. Prance. 1993. *Bark—the formation, characteristics and uses of bark around the world.* Timber Press, Inc., Portland, OR.

Saunders, P. T., ed. 1992. *Morphogenesis.* Elsevier Science Publishing Company, New York.

Steeves, T. A., and I. M. Sussex. 1989. *Patterns in plant development.* 2nd ed. Prentice Hall, Inc., Englewood Cliffs, NJ.

Roots

Root Functions
Root Morphology
Root Structure

Root Cap
Root Epidermis and Root Hairs
Cortex and Endodermis
Pericycle
Vascular Tissues
Secondary Growth in Roots

Adventitious Roots
Modified Roots

Anchorage
Water and Nutrient Absorption and Conduction
Storage
Nodulation
Mycorrhizae

A knowledge of the development, structure, and function of cells and tissues discussed in the preceding chapter is a prerequisite to the understanding of the biological roles of the plant organs. The root system, typically subterranean and hidden from the casual observer, can be many times larger than the aerial system of the plant body. In this chapter the functions, morphology, structure, and adaptations of roots are explored.

Prop roots provide extra support for this *Pandanus* tree in the Singapore Botanical Garden.

The five main functions of plant root systems are (1) anchorage, (2) storage of food materials, (3) absorption (uptake) of water and mineral nutrients, (4) conduction (movement) of that water and the minerals dissolved in it to the aboveground parts of the plants as well as the movement of food (sugars) through the root, and (5) growth.

ROOT FUNCTIONS

Anchorage holds the plant in position and supplies support for the shoot system. Winds, which can buffet a tree at high velocities, exert an enormous force. The flexibility of the tree allows it to bend rather than blow over, but it is the root system that ultimately anchors the tree. Although grasses have a relatively shallow root system, it is so extensively intertwined among the soil particles that it forms a dense mat of roots, which provides an effective anchorage against grazing animals (figure 7.1) and the erosion of the soil particles.

Some plants have extensive storage capabilities within their root systems. Sugar beets and carrots are examples of root adaptations for storage of water and carbohydrates. These materials are used by the plant for new shoot growth during the second growing season. Many such highly modified storage roots produce important food products (chapter 20, p. 377).

The amount of water (and dissolved minerals) absorbed by roots can be extensive. Water absorption and conduction in a typical corn plant, for example, can surpass two liters per day. This requires an enormous root surface area through which water can be absorbed.

The root system like the shoot system contains meristematic tissue, resulting in primary growth and in the case of woody dicotyledons, secondary growth.

ROOT MORPHOLOGY

There are two basic types of root systems: **taproots,** which have a primary root that is dominant and larger than other roots in the system, and **fibrous roots,** which have many roots of equal size. Each of these has extensive lateral branching, with the fibrous root system fashioning an interwoven mass of roots (figures 7.2 and 7.3).

These root systems originate from the embryonic root, the **radicle,** which emerges from a germinating seed (chapter 11, p. 184). As the radicle begins its growth into the soil it is termed the primary root. In gymnosperms and some dicotyledons, the primary root develops into a taproot.

The depth that taproots penetrate the soil varies from only a few centimeters to a reported 53 meters (mesquite). The taproot grows straight down, developing **lateral roots (branch roots)** as it grows. These in turn have tertiary roots, which have quaternary roots, which successively branch and lengthen to provide increased surface area for water absorption. As the root system develops, the aerial portion of the plant body grows, and the increased volume of tissues demands a constant water supply. Some trees are known to have taproots that extend far into the soil to ensure a supply of water even in periods of drought. Young pecan trees often have a taproot extending to a depth exceeding their height.

In fibrous root systems the primary root is either short-lived or loses

Figure 7.1
Grass roots form a mat of dense roots (turf) and soil that anchors the plant against being pulled up by grazing animals.

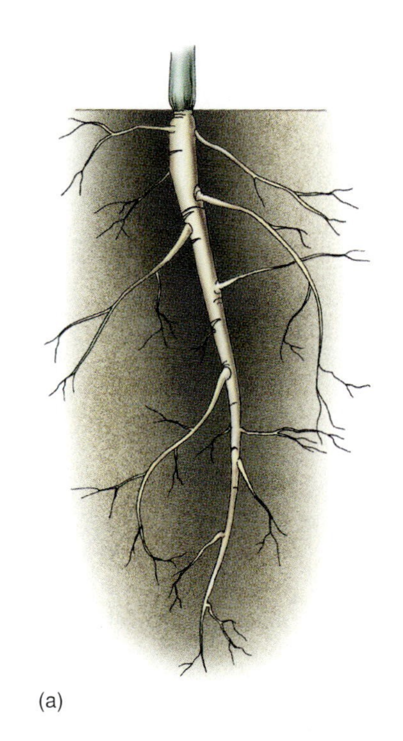

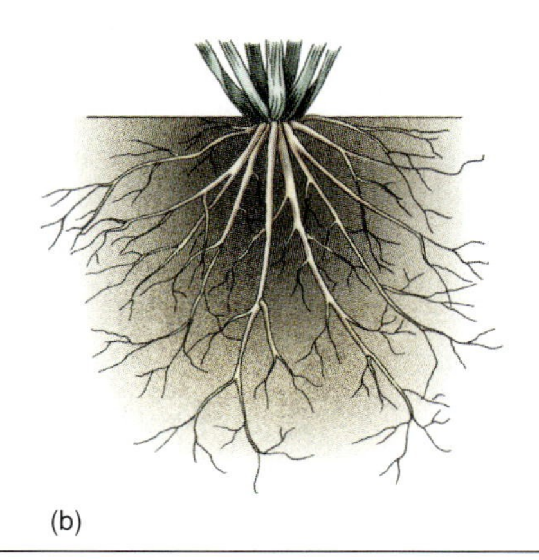

Figure 7.2
The two types of plant root systems. (*a*) Taproot. (*b*) Fibrous roots.

Figure 7.3
The length of the main taproot and the number of lateral roots vary considerably with type of plant, soil, and age. Shown is the taproot of *Taraxacum officinale,* the common dandelion.

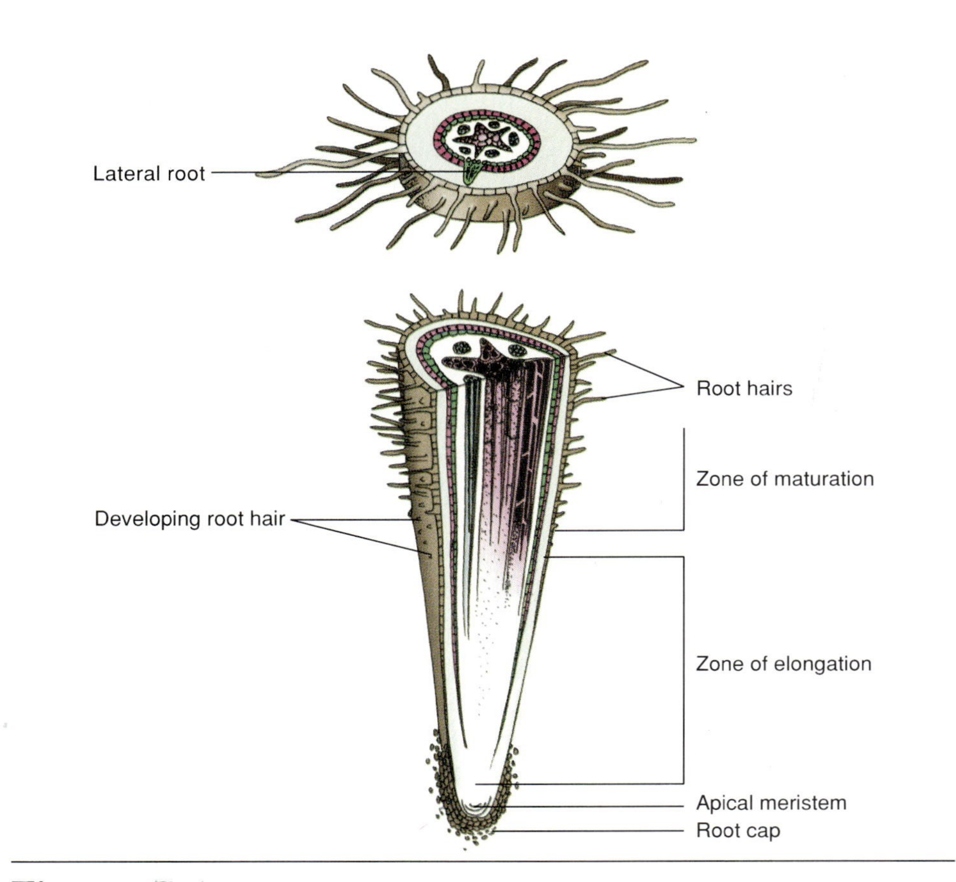

Figure 7.4
Diagram of a dicotyledonous root showing the root cap, apical meristem, zone of elongation, and zone of maturation. Note that root hairs occur only in the zone of maturation.

its dominance. If the primary root is short-lived, adventitious roots, roots that arise from tissues other than those of the root, produced from the stem together with their lateral roots form a fibrous root system.

Fibrous root systems are generally much more diffuse and closer to the surface than the taproot system. Each of the equal-size roots develops secondary and tertiary branch roots, as in the taproot system. This root network can effectively prevent any other plant from becoming established. The grassland biome can be solid grasses without many other plants, except where the grass turf has been thinned or removed (chapter 3, p. 39). Having such a large fibrous root system close to the soil surface is very important for plants in environments with relatively little rainfall. What little moisture there is can be more efficiently used. Conversely, in such an environment plants with taproot systems must develop the taproot very quickly to reach the soil moisture present below the fibrous root zone.

> **✔ Concept Checks**
>
> 1. List five basic functions of the root.
> 2. What is the radicle? How is the radicle involved in the formation of a taproot system? A fibrous root system?
> 3. What is the functional advantage of a taproot over a fibrous root system?

Early development for desert plants having a taproot system is much greater below the ground than above, and lateral roots do not develop as soon or as close to the surface in such plants.

ROOT STRUCTURE

Roots differ from stems in that they lack nodes and internodes, never produce leaves, and have a special covering, the **root cap,** which protects the apical meristem as the root apex is pushed through the soil. As with the shoot, the apical meristem of the root is the site of cell division. Directly behind the meristem is the zone of elongation. The elongation of cells in this region results in the increasing length or primary growth of the root. Differentiation of the enlarging cells results in a region of maturation in which the primary tissues can be distinguished (figure 7.4).

Root Cap

The delicate meristematic cells of the root apex are protected by the root cap, a thimble-shaped covering of parenchyma cells. As primary root growth occurs and the root tip is pushed through the soil, cells of the root cap are torn or rubbed off through abrasion with the soil particles, but the

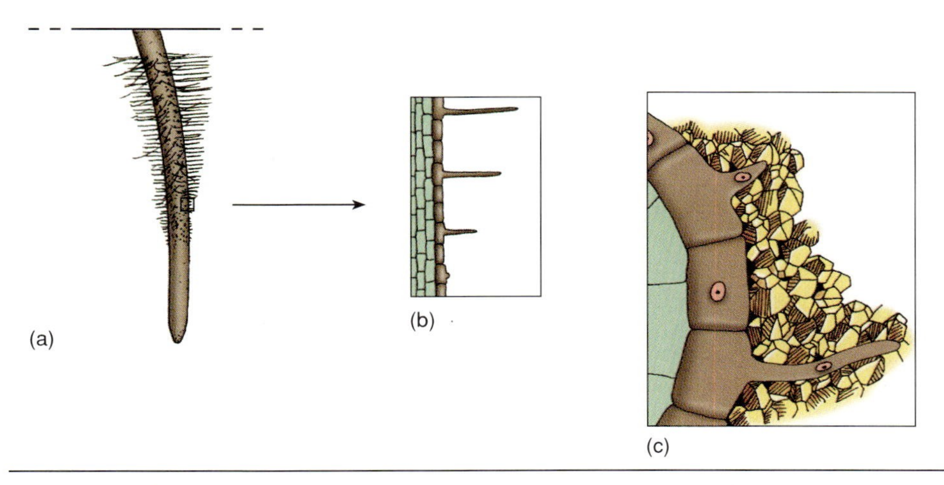

Figure 7.5
Above the growing apex of a young root (*a*) many of the epidermal cells elongate to form root hairs (*b*). These increase manyfold the total absorptive surface area as they penetrate the soil (*c*). Note that the root hairs are formed through the extension of the outer epidermal cell wall.

meristematic cells of the root apex remain undamaged. As cells are torn from the root cap, the underlying apical meristem contributes new cells to the root cap as well as new cells to the zone of elongation. The apical region of the root secretes a mucilaginous, slimy substance termed **mucigel,** which lubricates the growing root tip enhancing its penetration of the soil. In addition to lubricating the root apex, it has been suggested that mucigel may facilitate absorption of soil ions, such as phosphate and potassium, as well as serve as a culture medium for soil bacteria that aid in the growth of the root system. The positive response exhibited by most roots to gravity is also related to the presence of the root cap (chapter 14, p. 255).

Root Epidermis and Root Hairs

The root is typically bound by a single-layered epidermis, derived from the protoderm. The root epidermis has an extremely thin cuticle compared to the epidermal cells of the shoot system. In most plants many of the root epidermal cells in the youngest portion of the zone of maturation begin to expand outward, forming long slender **root hairs** that increase the total root surface area many times (figures 7.4 and 7.5). This increase allows much greater water absorption because these root hairs extend out among the soil particles not in direct contact with the root itself. On plants with actively growing root systems, new root hairs are continually being produced at the rate of many millions each day. Without root hairs most plants would not be able to absorb even a fraction of the water they need. A newly germinated seedling is an excellent subject for observation of root hairs. Root hairs are short-lived and are capable of secreting mucilage. In chronologically older regions within the zone of maturation, root hairs wither and cease to function. Each root, therefore, has a well-defined root-hair zone. When gardeners remove plants from the soil, care must be taken not to rip off the fragile root hairs and root tips. To minimize shock to the plant as much soil as possible should be maintained around the roots. After transplanting, a tree or shrub is typically pruned, decreasing the amount of evaporative surfaces, and watered thoroughly until the absorptive system has reestablished.

Cortex and Endodermis

Derived from the ground meristem, this region occupies by far the greatest volume of primary tissues in the root. Parenchyma is the principal tissue of the cortex and affords the plant an excellent region for the storage of excess food produced through the process of photosynthesis. Numerous starch grains may be observed in the cytoplasm of the parenchyma cells. The innermost layer of the cortex consists of a single layer of compactly arranged cells; no intercellular spaces occur in this layer, as commonly occur to the exterior. This cell layer is termed the **endodermis.**

Each cell of the endodermis has a characteristic band of suberin impregnated in the radial and transverse walls (figure 7.6). This band is called the **Casparian strip.** Since the Casparian strip is impermeable to the passage of water through the otherwise permeable cell wall, all water passing from the cortex through the endodermis must do so by passing through the protoplast. (See chapter 12, p. 205; apoplastic and symplastic movement.)

The function of the Casparian strip therefore is to direct the flow of water, previously moving in the intercellular spaces of the cortex or along the cell walls, through the **differentially permeable membranes** of the endodermal cells. This action assists the root in regulating the passage of water and dissolved substances rather than allowing the indiscriminate movement of water and solutes into the vascular tissues.

Pericycle

Just inside the endodermis is the **pericycle,** a cylinder of cells one or two layers wide that, together with the vascular tissues, are derived from the procambium. The cells of the pericycle maintain a high degree of meristematic potential, are involved in the **endogenous** origin of lateral roots (figure 7.7), and participate in the formation of the vascular cambium and cork cambium in all roots that undergo secondary growth.

Vascular Tissues

To the interior of the pericycle are the vascular tissues, primary xylem and primary phloem. These tissues have the same basic cellular composition

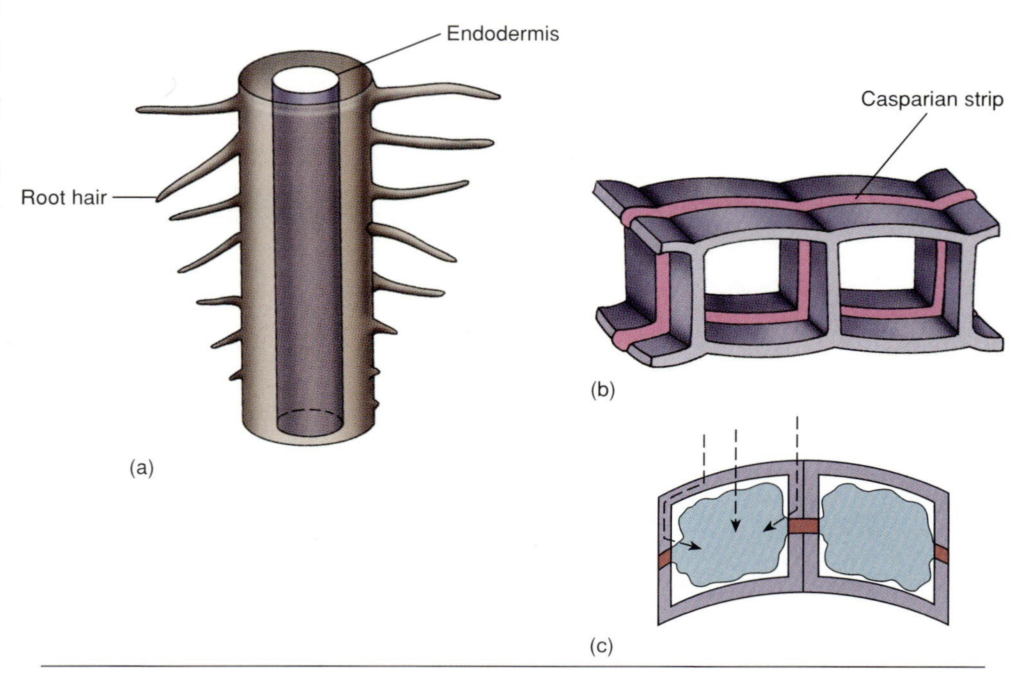

Figure 7.6

(*a*) The endodermis may be thought of as a cylinder beyond which substances may not pass except through living, differentially permeable membranes. (*b*) The Casparian strip represents a gasketlike layer of suberin that prohibits the movement of water and solutes between the cells and through the cell wall. (*c*) If viewed in cross section, the Casparian strip and plasma membrane are firmly attached. Arrows indicate pathway of water.

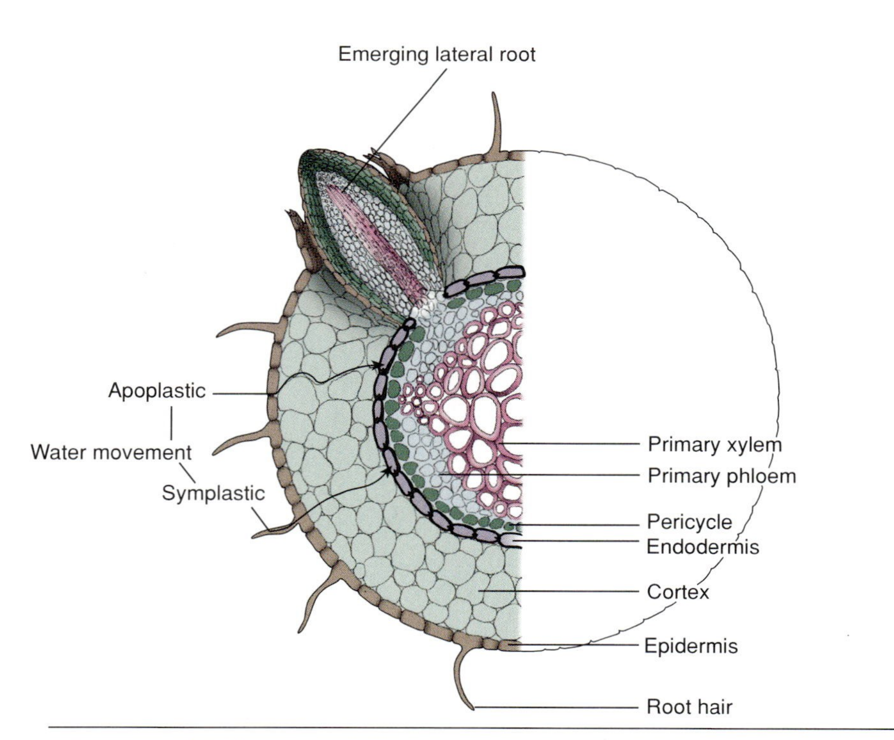

Figure 7.7

The layer of cells just inside the endodermis is the pericycle. Various points in the pericycle may become meristematically active and give rise to lateral roots. The new root breaks its way through the endodermis, cortex, and epidermis.

and functions as the vascular tissues elsewhere in the primary plant body. The arrangement of the primary xylem and phloem, however, is variable. As examples, we shall compare the dicotyledonous and monocotyledonous roots.

The Dicot Root

In dicots the vascular cylinder, or **stele,** arises in the center of the root (figure 7.8). The pattern typically has primary xylem in the center, radiating like spokes of a wheel as observed in cross section. The number of spokes varies according to species. A common form has four spokes and is called a tetrarch pattern; those with three spokes are called triarch; those with six spokes are called hexarch, and so on. The primary phloem occurs in pockets between the spokes.

The Monocot Root

Most monocot roots have a pith in the central position surrounded by several xylem points (figure 7.9). Most monocots have many xylem spokes and are referred to as polyarch. Alternating between each xylem spoke is primary phloem. Surrounding the xylem and phloem are the pericycle, endodermis, cortex, and epidermis.

Secondary Growth in Roots

As in dicot stems, a portion of the procambium remains undifferentiated between the primary xylem and primary phloem in the root. These undifferentiated cells constitute a portion of the vascular cambium. The cambium forms secondary xylem to the inside and secondary phloem to the outside. At the end of each xylem spoke, where the xylem lies next to the pericycle, the meristematic pericycle cells undergo cell division, producing cells that participate in the formation of the vascular cambium. Initially the outline of the cambium is lobed, conforming to the outline of the xylem spokes. As secondary growth continues, secondary xylem is produced more rapidly in the areas

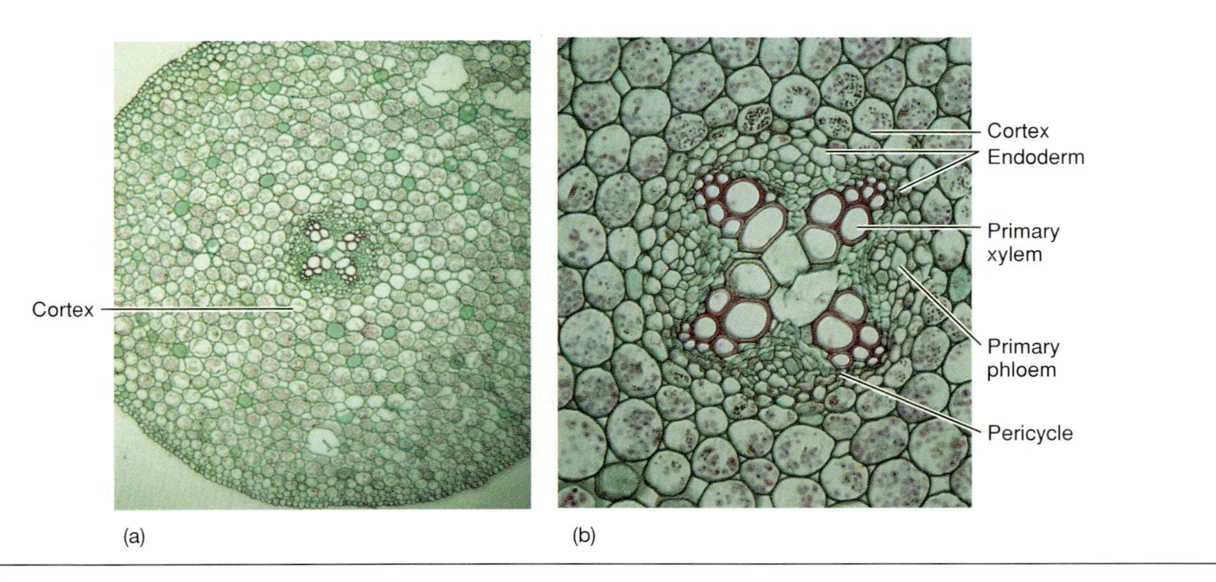

Figure 7.8

The dicot root (*Ranunculus*) consists of a great deal of cortex and a small vascular cylinder. In this species the pattern of the xylem is tetrarch. (*a*) Cross section of entire root. (*b*) Detail showing cortex, endodermis, primary xylem, primary phloem and pericycle.

underlying the position of the phloem. Continued activity of the vascular cambium eventually results in a cylindrical arrangement of secondary xylem and secondary phloem similar to that of the stem. The first cork layer in a dicot root is formed by a cork cambium that arises from cells of the pericycle. The production of the suberinized cork layer, the phellem, physiologically isolates the cortex and epidermis from the plant body. New cork layers are successively formed from the living parenchyma located in the secondary phloem.

Monocot roots, like most monocot stems, typically do not develop lateral meristems.

ADVENTITIOUS ROOTS

Some plants form roots on other parts of the plant body. These are termed **adventitious roots,** and they develop on stems (chapter 8, p. 134) and leaves. The ability of many ornamentals to root from leaves or stem pieces has facilitated commercial growing. Most begonias and African violets are propagated from leaf cuttings. Many new plants can be propagated from one adult *Dieffenbachia* (dumb cane) if its stem is cut into sections and rooted in moist soil or vermiculite (figure 7.10).

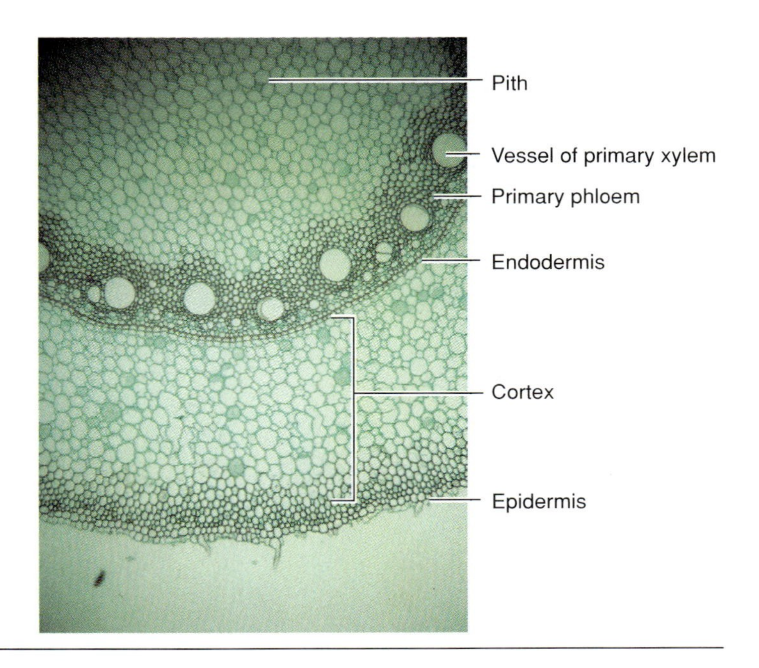

Figure 7.9

In the monocot root of corn the vascular cylinder occurs in the center of the ground tissue, delimiting pith toward the inside and cortex toward the outside.

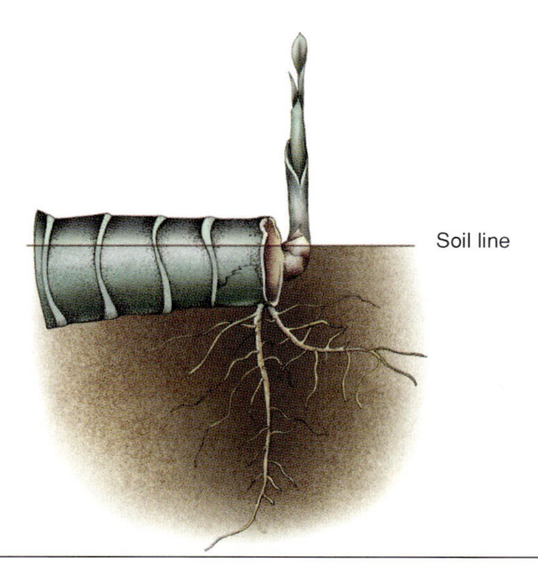

Figure 7.10
A stem section of *Dieffenbachia* will form adventitious roots below the soil line while developing a new shoot above. Such stem cuttings are important methods of propagation.

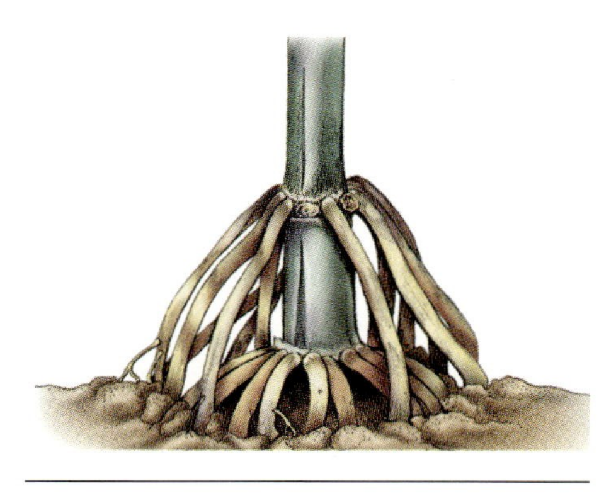

Figure 7.11
Corn plants develop adventitious prop roots for support.

MODIFIED ROOTS

Structural modification of roots can affect any of their basic functions.

Anchorage

Prop roots are modified adventitious roots providing anchorage and stability. Developing from the lowest nodes of the stem, they extend from the stem into the soil and help brace the plant. These aptly named prop roots are found most commonly on plants growing in marshes and mud flats. Prop roots can even grow out of the branches of some trees, penetrating the ground and thereby thickening and supporting the tree, as in the banyan tree (*Ficus bengalensis*) and the red mangrove (*Rhizophora mangle*). In the banyan tree they are so large that they form a "miniforest" under the main tree. Corn plants (*Zea mays*) also develop prop roots (figure 7.11).

Contractile roots, roots that pull the parent plant deeper into the soil, are produced by many plants. The action of contractile roots explains why many bulbs become more deeply seated in the soil year by year.

Plants that grow on other plants but that do not draw food from them are termed **epiphytes,** and usually have modified roots. Many of the tropical bromeliads and the more temperate ball moss (*Tillandsia recurvata*) and Spanish moss (*Tillandsia usneoides*) (also bromeliads, not mosses) are anchored by their roots to the trunks and branches of trees (figure 7.12).

✔ Concept Checks

1. Diagram a root as viewed from a longitudinal profile. Label the following: root cap, zone of cell division, zone of cell elongation, zone of maturation, root-hair zone, lateral roots.
2. What is the origin of the cells that compose the root cap? What is the primary function of the root cap?
3. Describe the appearance of a root hair. How are root hairs produced?
4. Explain how the Casparian strip directs the flow of water and mineral nutrients through the differentially permeable membrane of an endodermal cell.
5. What are the functions of the pericycle?

Figure 7.12
Spanish moss (*Tillandsia usneoides*), an epiphyte, growing on trees in North Carolina.

Water and Nutrient Absorption and Conduction

Parasitic plants, those that draw nutrients from other plants, have a detrimental effect on their hosts. Mistletoe (*Phoradendron*) and dodder (*Cuscuta*), for example, produce haustoria, threadlike cells, which not only anchor the parasite to its host but also penetrate the host's vascular tissue. In this way the parasite avails itself of water and/or food from the host plant.

Since roots require an oxygen supply to carry on respiration, some plants that grow in poorly drained soil or in soil covered by stagnant water develop root modifications that grow out of the water to aerate the plant. Roots of the black mangrove (*Avicennia nitida*) develop **pneumatophores,** or air roots, which stick straight out of the swampy water. The bald cypress (*Taxodium distichum*) has characteristic "knees," which may also provide oxygen to the root system, although there is some doubt about this function (figure 7.13).

Storage

Many plants have extensive root storage capacities. Starch and other molecules are stored for growth or flowering or as a reserve against periods of harsh environmental conditions. Beets and carrots are well-known examples (figure 7.14). The sweet potato (*Ipomoea batatas*) also has an enlarged storage root, as does the tropical cassava (*Manihot esculenta*), from which we obtain tapioca.

Nodulation

Some plants, notably members of the pea family (Leguminosae), are unique in their ability to increase the levels of soil nitrogen in a useful form. **Nitrogen fixation** involves the formation of a root nodule (chapter 2, p. 24) in response to an infection initiated by bacteria from the genus *Rhizobium*. The nodule, which is actually a tumor formed by parenchyma of the root cortex in response to the bacteria, provides a home for the bacterial colony. In return the *Rhizobium* bacteria convert, or fix, atmospheric nitrogen (N_2) that is found in the air spaces of the soil into ammonia (NH^+_4). Atmospheric nitrogen cannot be used by plants, but ammonia can. In this way the modified roots and their bacterial guests effectively fertilize the soil with nitrogen.

Mycorrhizae

These are usually short roots that form symbiotic, or mutually beneficial, associations with certain fungi found in the soil. The fungi actually enter the tissues of these modified roots and benefit nutritionally from them. The roots, in turn, become more efficient in the absorption of certain minerals needed by the plant (see "The Roots of Success"). **Mycorrhizae** are grouped into two categories: **ectomycorrhizae** and **endomycorrhizae.** In ectomycorrhizae the fungus forms a structure, termed a mantle or sheath, that encloses the root. The fungus also penetrates between root cells to form

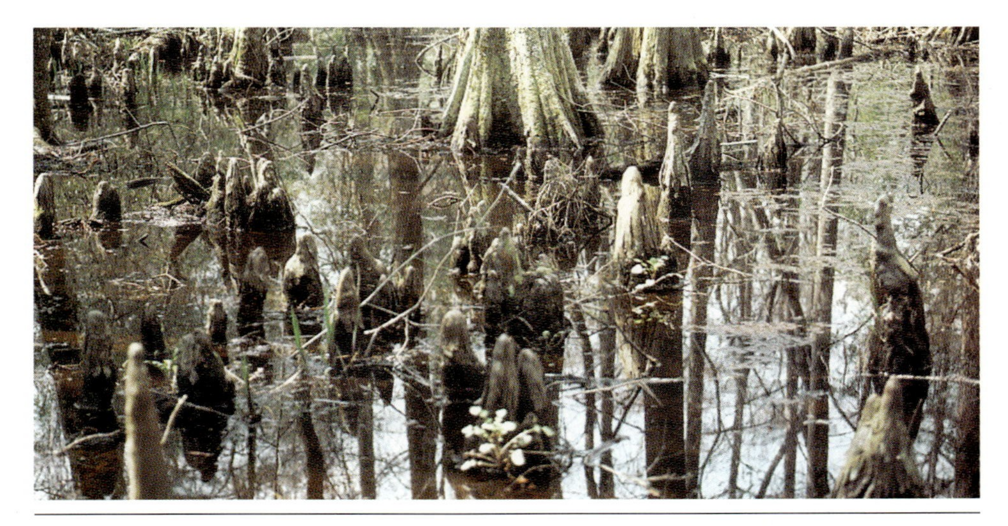

Figure 7.13
The "knees" of the bald cypress roots may function in oxygen uptake.

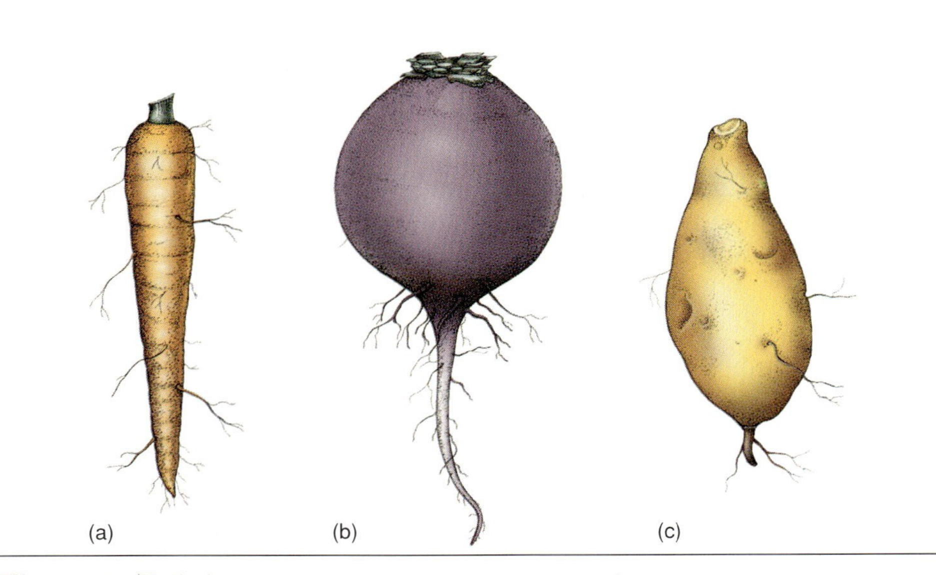

Figure 7.14
Roots modified for storage. (*a*) Carrot. (*b*) Beet. (*c*) Sweet potato.

an intercellular network known as the Hartig net, but there is no penetration into living cells. Plants in which ectomycorrhizal associations occur, such as species of the pine and willow families, have roots which appear thicker and more branched (figure 7.15). In endomycorrhizae, there is no discernible change in the appearance of the root. A fungus mantle is not present, but the fungal cells, termed **hyphae,** enter into the parenchyma cells of the root cortex. Once the fungus penetrates the cells, the hyphae repeatedly branch forming featherylobed arbuscules, which fill with phosphate granules, which then dissolve and are absorbed by the plant. Following arbuscule formation, vesicles appear. These are saclike swellings at the tip of the hyphae. They occur between cells and are lipid-containing storage organs of the fungus. Endomycorrhizae are extremely common, and because of the characteristic intercellular swellings and highly branched intracellular hyphae, they are termed VAM, or vesicular-arbuscular mycorrhizae.

✔ Concept Checks

1. What are adventitious roots? How does the origin of a lateral root differ from the origin of an adventitious root?
2. Explain each of the following terms: prop roots, epiphyte, haustorium, pneumatophore, nodulation, ectomycorrhizae, and endomycorrhizae.

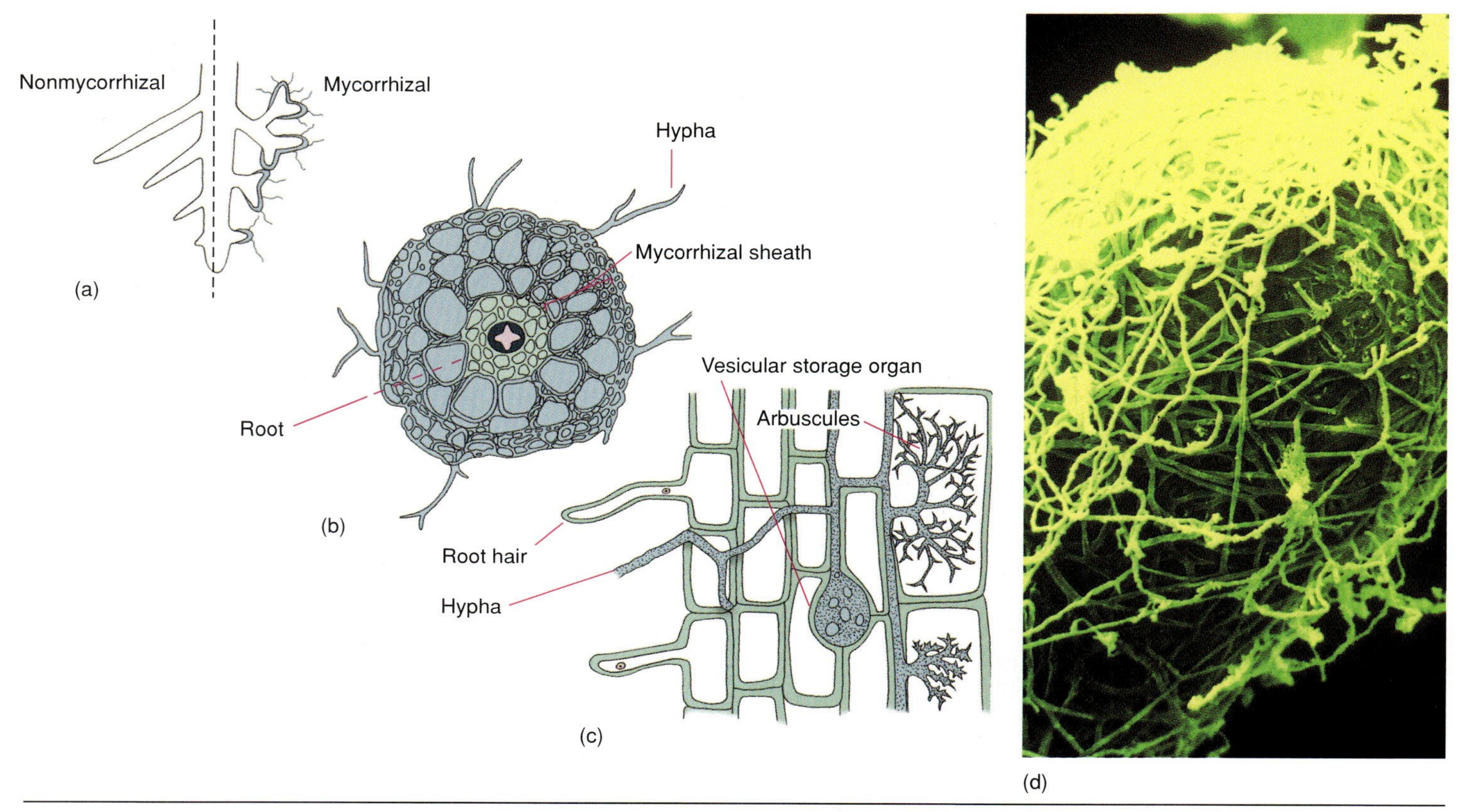

Figure 7.15

Mycorrhizae. (*a*) Comparison of nonmycorrhizal root with ectomycorrhizal root. (*b*) Ectomycorrhizal root in transverse section. (*c*) Endomycorrhizae illustrating vesicular storage organ and arbuscules. (*d*) Scanning electron micrograph revealing fungal hyphae on root of aspen. (×200)

The Roots of Success

A unique kind of association between the roots of vascular plants and fungi growing in the soil has been recognized for a long time, but only in recent years has the scientific community come to realize the widespread occurrence and significance of this relationship. These associations are termed mycorrhizae, or fungus roots. It now appears that in addition to most living vascular plants, many fossil plants—even the earliest vascular land plants dating to 400 million years in age—had mycorrhizae. The advantage to the plant is the increased surface area; thus, more of the soil mass and litter layer, the layer composed of decaying leaves, twigs, fruit, etc., is "mined" for water and mineral nutrients. The tubular cells of the fungus act much like additional root hairs would, by absorbing additional water and mineral nutrients. By using radioactive elements as tracers, scientists have found mycorrhizae are particularly important in the absorption of phosphorous and other nutrients such as copper, zinc, and manganese required for the maintenance of normal growth and development in plants. In turn, the plant supplies organic compounds, such as photosynthates, that promote the growth and development of the fungus. It is this mutualistic association that probably played a major role in the successful colonization and subsequent proliferation of plants onto the land—a land thought to be originally characterized by nutrient-deficient soils. In addition to enhancing phosphorous uptake, even in nutrient deficient soils, secondary benefits of mycorrhizal relationships also occur. These include increased resistance to disease transmitted by soil-borne pathogens, better water transport, enhanced nitrogen fixation in legumes, and increased tolerance to highly acidic soils.

Scientists are becoming increasingly aware that mycorrhizal associations are crucial to the proper maintenance and management of many ecosystems. Land disturbances such as strip mining, the indiscriminate use of fungicides, the deforestation of tropical rain forests, and the occurrence of acid rains all reduce populations of mycorrhizal fungi and result in decreased plant growth and even death.

Summary

1. Roots anchor plants, absorb and conduct water and minerals to the rest of the plant, store starch as a food reserve, and are capable of primary growth.
2. Both taproot and fibrous root systems increase the surface areas of the root in contact with soil particles, thereby increasing the anchorage and absorption roles of the root systems.
3. The apex of the root is covered by the root cap, an important structure for the protection of the meristematic cells against the abrasive action of primary growth through the soil.
4. Epidermal cells of the root may elongate outwardly into the soil, forming root hairs, which greatly increase the surface area of the root for absorption. Mucigel, a mucilaginous secretion, is produced by the root hairs and growing root tips.
5. The cortex is the principle storage area of the root and is composed principally of parenchyma. The innermost layer of the cortex is the endodermis which, with the aid of the Casparian strip, regulates the passage of water and minerals into the vascular tissues.
6. The pericycle, a cylinder of meristematic cells, is the site of lateral root formation as well as vascular cambium and cork cambium formation.
7. Lateral meristems, the vascular cambium, and cork cambium produce the same tissues and function as they do in stems. Roots which have undergone secondary growth are similar in appearance to secondary stems.
8. Adventitious roots are produced from leaves or stems and are important in the propagation of plants.
9. Root modifications include adventitious prop roots, which help support the stem. Haustoria, the modified roots of parasitic plants, drain nutrients from their hosts such that they may ultimately kill them. Roots also enlarge for storage. Nutrient uptake is improved through root nodulation for nitrogen fixation and by fungal mycorrhizae associations.

Key Terms

adventitious root 115
Casparian strip 113
contractile root 116
differentially permeable membrane 113
ectomycorrhizae 117
endodermis 113
endogenous 113
endomycorrhizae 117
epiphyte 116
fibrous root system 111
hyphae 118
lateral root 111
mucigel 113
mycorrhizae 117
nitrogen fixation 117
pericycle 113
pneumatophores 117
prop root 116
radicle 111
root cap 112
root hairs 113
stele 114
taproot system 111

Discussion Questions

1. Imagine yourself as a water molecule clinging to a soil particle. A root hair makes contact and absorbs you. List in correct sequence the tissues, regions, and type of cells through which you would pass until the time you enter into a tracheid or vessel member of the primary xylem.
2. Diagram two dicot roots; one illustrating only primary tissues, the other showing secondary tissues with at least three annual rings.
3. Perhaps you have heard the statement "roots grow in search of water." Why is this statement botanically incorrect?

Suggested Readings

Davis, T. D., and B. E. Haissig, eds. 1994. *Biology of adventitious root formation.* Plenum Press, New York.

Epstein, E. 1974. Roots. *Sci. Am.* 228(5):48–58.

Evans, M., R. Moore, and K. H. Hasenstein. 1986. How roots respond to gravity. *Sci. Am.* 255(6):112–119.

Fahn, A. 1990. *Plant anatomy.* 4th ed. Pergamon Press, Elmsford.

Glinski, J., and J. Lipiec. 1990. *Soil physical conditions and plant roots.* CRC Press, Boca Raton.

Kolek, J., and V. Kozinka, eds. 1992. *Physiology of the plant root system.* Kluwer Academic Publishers, Boston.

Nadkarni, N. 1985. Roots that go out on a limb. *Natural History* 2:43–48.

Raven, P. H., R. F. Evert, and S. E. Eichhorn. 1992. *Biology of plants.* 5th ed. Worth Publishers, New York.

Saether, N., and T. H. Iversen. 1991. Gravitropism and starch statoliths in an *Arabidopsis* mutant. *Planta* 184:491–497.

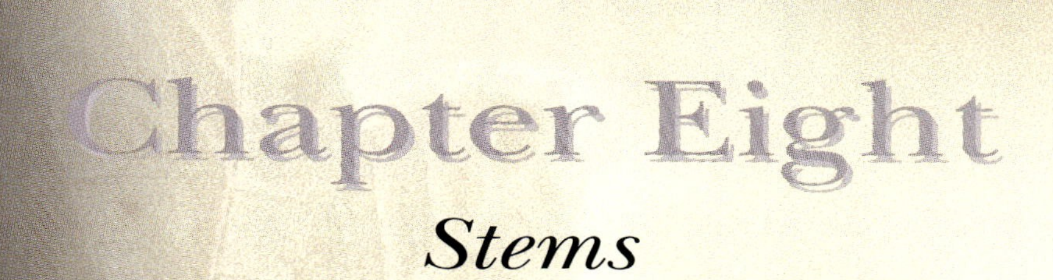

Chapter Eight

Stems

Stem Functions
Stem Morphology
Primary Growth of the Stem
Stem Anatomy

The Dicot Stem
The Monocot Stem

Secondary Growth of the Stem

Vascular Cambium
Secondary Xylem
Secondary Phloem
Bark

Stem Adaptations

Rhizomes
Stolons
Tubers
Corms
Bulbs
Succulents
Thorns
Tendrils

Budding and Grafting
Stem Cuttings
Layering

Aside from the underground root system, there exists the visible shoot or aerial system. The stem is the aerial axis of the plant body. Its growth and tissues are similar to those of the root system. This chapter deals with the anatomy, morphology, modifications, and functions of the stem.

This modified stem is well protected from herbivores by its spine.

Plant stems come in a great array of shapes and sizes (figure 8.1), but they all share the same basic functions: (1) the attachment and support of leaves, flowers, and fruit, (2) the conduction of materials, (3) storage, and (4) growth.

STEM FUNCTIONS

Stems are the place of attachment for leaves, flowers, and fruit. Erect stems provide structural support for leaves, raising them to allow for adequate exposure to light. Similarly, flowers are elevated to allow greater visibility to pollinators, and the resulting fruit and seeds are better dispersed from their positions of attachment on the stem.

A second function of the stem, in combination with the roots, is conduction. The stem conducts water and minerals from the roots to all the aboveground parts, carries food produced by photosynthesis in the leaves to roots, stems, flowers, and fruit, and transports hormones from the tissues in which they are synthesized to those areas where the effects are produced.

Third, most stems store nutrients, organic molecules, water, and byproducts. A large amount of storage occurs in certain modified stems, but even unspecialized stems contain some storage tissues.

Finally, the stem contains meristematic tissue (chapter 6, p. 99), which results in new cell production, elongation of stem tips, increases in stem diameter, and production of tissues and organs such as leaves and flowers.

STEM MORPHOLOGY

Regardless of whether stems are erect, prostrate, or in some other position, all stems have areas of leaf and bud attachment called **nodes.** The portion of stem between one node and the next is termed the **internode.** Normally the leaf is attached to the stem at the node, with a **bud** (embryonic shoot) arising in the leaf **axil** (the angle above where the leaf connects to the stem). Because

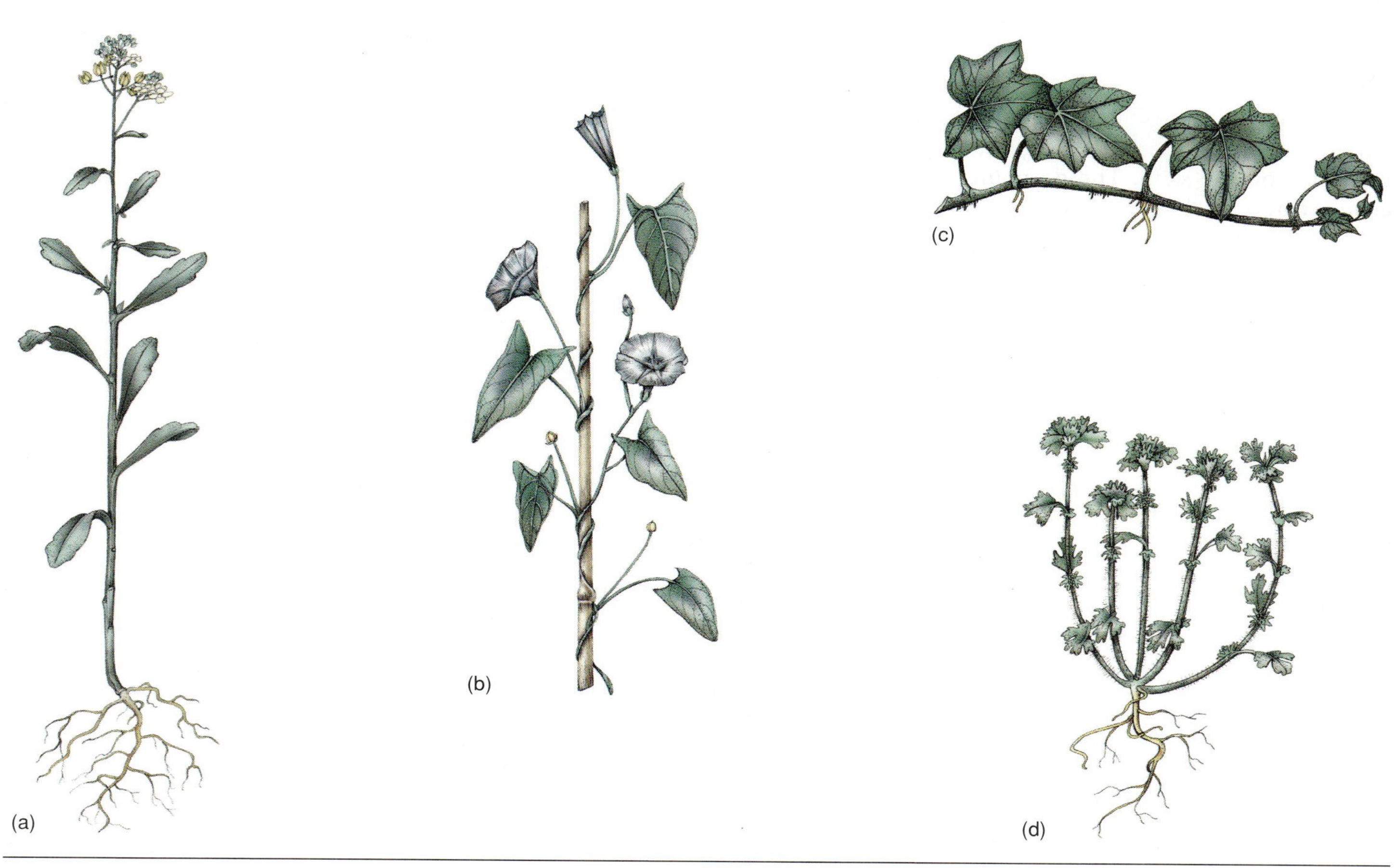

Figure 8.1

Plant stems come in a variety of shapes and sizes. Erect (*Iberis*) (*a*), twining (*Convolvulus*) (*b*), prostrate (Boston ivy) (*c*), and ascending stems (*Alchemilla*) (*d*) are some of the more common types.

of their location, such buds are called lateral or **axillary buds** (figures 8.2; 8.3). Plants also have terminal buds at the end of each shoot.

Since buds are undeveloped shoots, new leaves and/or flowers may be initiated from them when and if they become active. Not all axillary buds develop, and the hormonally controlled pattern of bud growth and inhibition, as well as the length of the internodes, is responsible for the vast array of plant shapes (chapter 14, p. 249).

Herbaceous stems are so called because they are typically green, soft, and normally live only one season. They may develop lateral meristems that yield only a minimal quantity of secondary wood or cork (bark). Woody stems, on the other hand, produce new growth each season, overwintering in a dormant state. Dormant lateral buds on most woody plants are protected by a number of overlapping **bud scales** (figure 8.3). As the bud begins new growth, these scales fall off, leaving **bud scale scars** on the stem. The dormant **terminal bud** is also protected by a series of scales that leave a complete ring of scars around the stem when they fall off at the onset of new growth in the spring. The length of a year's growth can be determined by measuring the distance between any two rings of terminal bud scale scars. The total annual increase in length (new growth) of a stem varies from one species to the next; but it can also vary from year to year on the same plant in response to environmental conditions such as water availability and length of frost-free weather (figure 8.3).

On the stems of deciduous trees and shrubs **leaf scars** may also be observed, indicating the former positions of leaf attachment. Within each leaf scar, vascular bundle scars mark the location where the vascular tissues (i.e., xylem and phloem) supplied water and food to the leaf. The number of vascular bundle scars and their arrangement within the leaf scar, as well as the shape of the leaf scar are characteristic for each species. In fact, these patterns are characteristic enough to identify deciduous trees in winter by looking only at the twigs.

✔ *Concept Checks*

1. List four functions of stems. Briefly explain each.
2. What is a node? An internode? How can terminal bud scales be used to determine the age of a woody stem?

PRIMARY GROWTH OF THE STEM

As discussed in chapter 6, stems arise from cells produced in the shoot apical meristem. A longitudinal section through a typical stem reveals a dome-shaped apical meristem flanked by developing leaves (see figure 6.3).

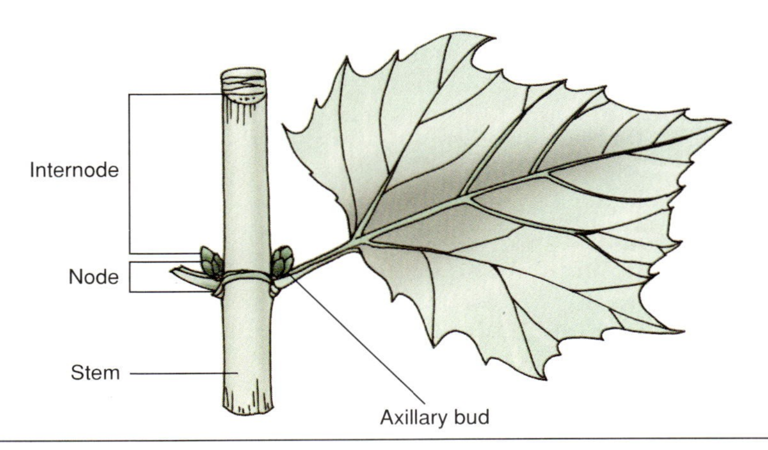

Figure 8.2
Leaves attach to the stem at nodes and have axillary or lateral buds arising in the axil of the leaf-stem union.

Below the meristematic region, within the zone of elongation, three primary meristems, the protoderm, ground meristem, and procambium, are located and are collectively responsible for production of the primary tissues.

STEM ANATOMY

One of the impressive developmental phenomena of angiosperms is that the two major subgroups, Monocotyledons (monocots) and Dicotyledons (dicots), are easily distinguished on the basis of stem structure and other structural aspects of the plant (chapter 11, p. 186). Although considerable variations exist in the arrangement of the primary tissues in stems, it has become traditional for students of introductory botany to learn the topographic patterns of the primary tissues as they differentiate in each angiosperm subgroup.

The Dicot Stem

If the young dicot stem, such as a sunflower, is cut in cross section (i.e., across the long axis) in the region of maturation, stained properly, and viewed under a microscope, the outside of the stem is covered by the epidermis with its waxy cuticle. Directly inside is the **cortex,** a region composed of simple tissues, principally parenchyma and collenchyma. Separating the cortex from the **pith,** the central region of the stem (also composed principally of parenchyma), is a ring of vascular bundles (figure 8.4). Each vascular bundle repeats the pattern of primary xylem toward the inside and primary phloem toward the outside of the bundle. Both vascular tissues may include large numbers of fibers that lend additional support to the stem. The area between adjacent vascular bundles is termed the interfascicular region or **pith ray,** and is composed of parenchyma.

The Monocot Stem

Monocots include such plants as true lilies, orchids, palms, and grasses. As in herbaceous dicots, the surface of a monocot stem is covered by an epidermis. The vascular bundles produced by the procambium, however,

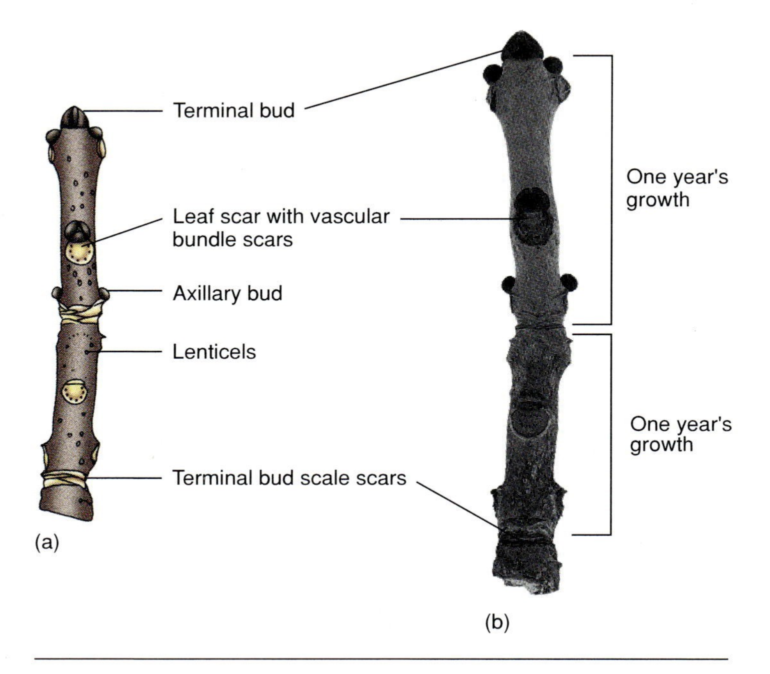

Figure 8.3

(*a*) Woody stems produce a terminal bud at the end of each growing season. This bud overwinters wrapped in protective bud scales, which fall away in the spring when the terminal bud initiates new growth. (*b*) The distance between two sets of terminal bud scale scars indicates the total growth for the stem in that year. The scars form a circle around the stem.

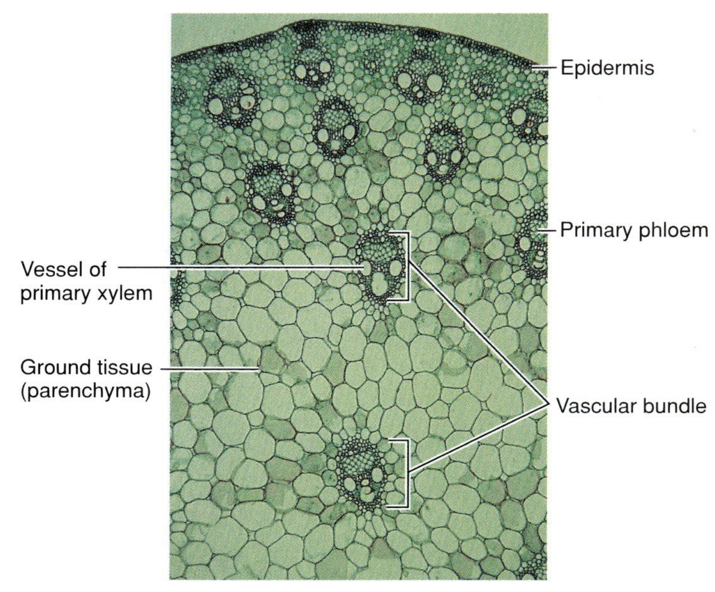

Figure 8.5

In cross section the vascular bundles of a typical monocot, such as *Zea mays* (corn), are scattered throughout the ground tissue.

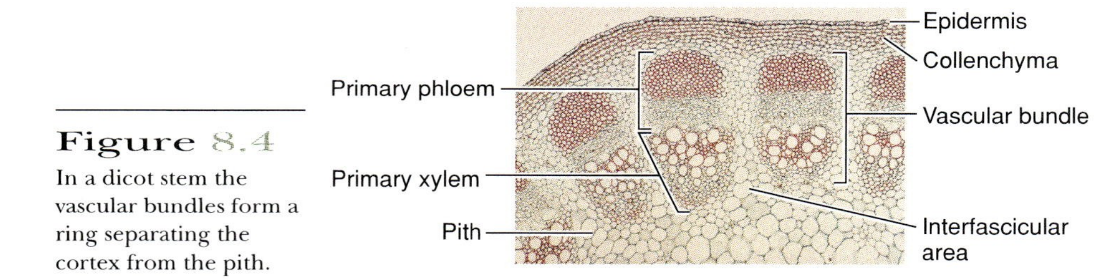

Figure 8.4

In a dicot stem the vascular bundles form a ring separating the cortex from the pith.

Concept Checks

1. Draw a dicot stem in cross section. Label all primary tissues.
2. Illustrate a monocot stem in the region of maturation. Label the following: epidermis, vascular bundles, primary xylem, primary phloem, ground tissue.

when viewed in cross section, appear scattered throughout the parenchyma tissue of the stem, as compared to the ring of bundles in a dicot (figure 8.5). Since the bundles are more or less randomly scattered and do not separate a cortex region from a central pith, the tissue around the bundles is simply called ground tissue or ground parenchyma. Some monocots have hollow stems, however, and their vascular bundles may occur in a ring, as in the dicots. Each bundle is oriented such that the primary xylem is closest to the center of the stem, and the primary phloem is closest to the epidermis. Each vascular bundle has large vessel elements, some of which collapse under the stress of early growth and rapid elongation. Each vascular bundle is surrounded by a sheath of sclerenchyma fibers (chapter 6, p. 102), making several monocots a commercially important source of fibers, in the production of ropes and twines.

SECONDARY GROWTH OF THE STEM

If one were to categorize the basic shape of plants, as compared with the shape of animals, the similarities would tend to be greater among plants. Most of us would think of field crops and trees of forests and orchards as being relatively tall and slender, although many tree trunks reach enormous diameters through the activity of lateral meristems. This thickening around the middle occurs in woody stems and roots and is termed secondary growth. The tissue produced is termed secondary tissue.

In plants, elongation is restricted to the tips of stems and roots, whereas in animals growth is not limited to specific tissues. Once a woody stem has completed a year's growth, no new elongation will take place in that section. This can be illustrated by driving a nail into a stem at a measured distance above the ground. The nail will remain at that height for the life of the tree, although the tree will certainly become taller and increase in diameter.

Vascular Cambium

As the dicot stem becomes older, a **vascular cambium** forms. The vascular cambium is a single-layered cylindrical

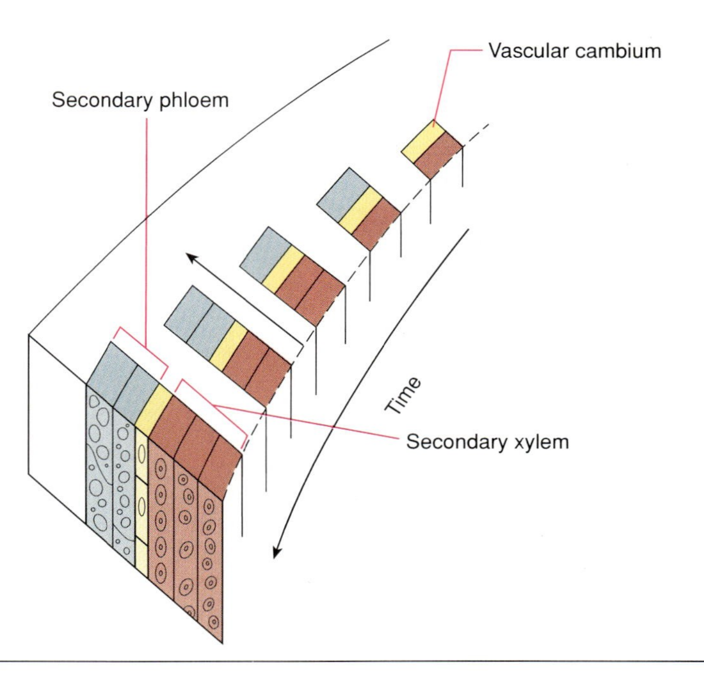

Figure 8.6
Cell divisions of the vascular cambium result in the formation of new water conducting tissue (secondary xylem) to the interior of the vascular cambium and food conducting tissue (secondary phloem) to the exterior. The outward enlargement of the new cells is responsible for the expansion in the diameter (secondary growth) in stems and roots.

sheath of meristematic cells derived from portions of the procambium that remain undifferentiated, as well as through the reversion of parenchyma in the pith ray into meristematic cells. The portion of the cambium that develops within the vascular bundle between the primary xylem and primary phloem is referred to as the fascicular cambium (figure 8.4), and the portion that develops between the bundles from parenchyma of the pith ray is termed the interfascicular cambium.

When a cell of the vascular cambium divides, one cell remains meristematic (an initial) while the other, the derivative, begins enlargement and subsequently differentiates. Cellular enlargement is primarily outward, such that all cells and tissues to the exterior of the enlarging cell are pushed outward, hence increasing the diameter of the axis. All cells that differentiate to the interior of the vascular cambium become cell types of the secondary xylem. All cells that differentiate to the exterior of the cambium form the secondary phloem. Over time, much more secondary xylem, or wood, is formed than is secondary phloem (figure 8.6).

Secondary Xylem

The basic cell types that comprise the secondary xylem are the same as those of the primary xylem. Cells that differentiate to form vessel members, tracheids, fibers, and parenchyma are produced toward the inside of the stem by the vascular cambium. **Xylem rays,** radial rows of parenchyma cells, are distinguishing features of the secondary xylem (figure 8.7). They appear as lines or streaks in the wood when viewed in cross section. In longitudinal section, the width and height of the rays may be ascertained. Rays function in the lateral conduction of water, minerals, and organic solutes (carried from the secondary phloem) as well as storage.

When a tree is cut down, concentric rings, termed growth rings or **annual rings** can be observed (figure 8.7) in the secondary xylem (wood). A single ring typically develops each year as a response of the derivatives produced by the vascular cambium to environmental or seasonal fluctuations. Chapter 10 provides a discussion on annual rings, the aging of trees, and the economic importance of wood and other plant tissues.

Secondary Phloem

The cells comprising secondary phloem are the same types as those of primary phloem. Again, the vascular cambium produces new sieve-tube members, companion cells, fibers, and parenchyma toward the outside of the stem (figure 8.6). Secondary phloem is important in long-distance transport of organic solutes from the "source" to the "sink," but the thickness of the tissue is never very great. Old primary phloem and secondary phloem are gradually crushed and sloughed off as part of the bark. Only the newly formed secondary phloem nearest the vascular cambium remains functional.

Phloem rays, radially aligned rows of parenchyma located in (figure 8.7) the secondary phloem, together with the xylem rays constitute the vascular rays characteristic of the secondary vascular tissues. As the size of the stem becomes greater, the epidermis and cortex become stretched and crack at irregular intervals. At approximately the same time that the outer tissues begin to crack, new regions of meristematic activity occur inside the epidermis. This new lateral meristem is called the **phellogen** or cork cambium and leads to the production of cork, a dead tissue that protects the inner tissues from desiccation, mechanical injury, insects, and disease.

Bark

The production of a rough, thick, mostly dead protective outer layer is common to most trees. **Bark** is the general term for all the tissue located outside the secondary xylem or wood. Bark results from the activities of two lateral meristems: the vascular cambium, which produces secondary phloem toward the outside, and the phellogen or cork cambium, which produces an outer cork layer, the **phellem,** and an inner

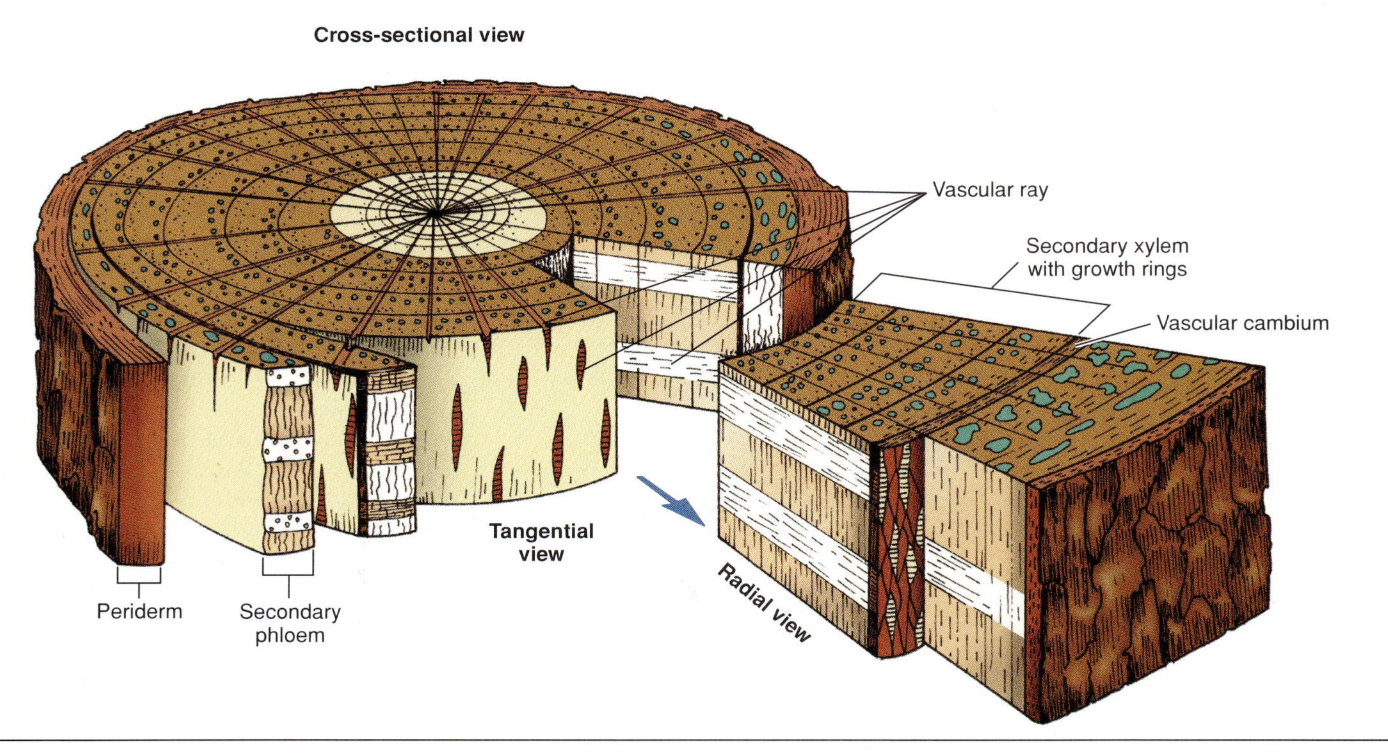

Figure 8.7
Three-dimensional view of a woody dicot stem illustrating relationships of tissues. Annual rings and vascular rays, characteristic of secondary tissues, are shown.

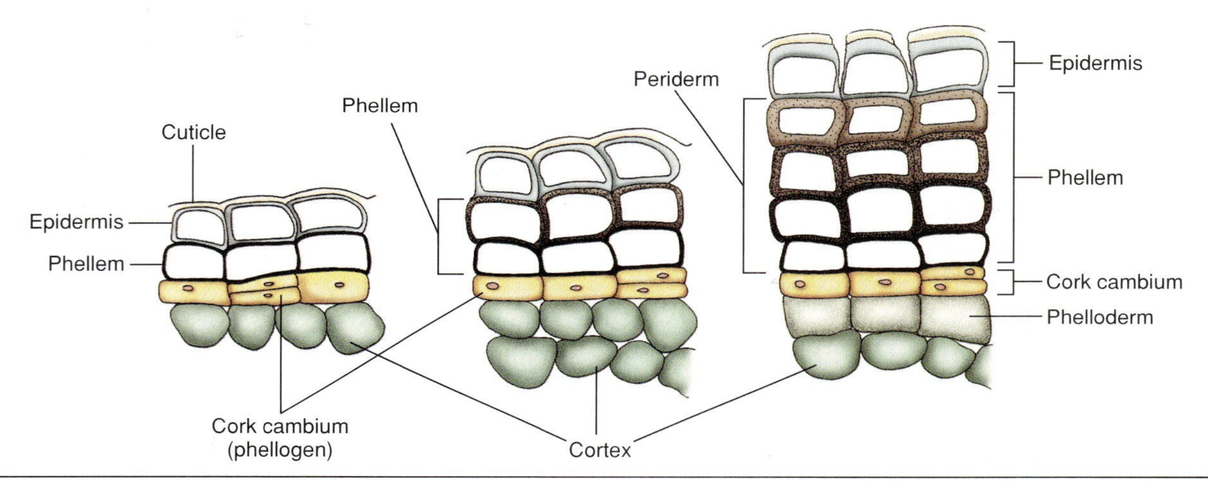

Figure 8.8
Structure of the periderm. Several periderms typically occur in the outer bark of a tree.

parenchymatous layer called **phelloderm** (figure 8.8). Collectively the cork cambium and the cells it produces—cork cells and phelloderm—make up the **periderm,** a tissue that replaces the epidermis as the protective outer covering. The tissues of the periderm are collectively called outer bark, and tissues to the inside are termed inner bark.

During the first year's growth, before any secondary phloem has been produced by the vascular cambium, bark consists only of primary tissues. At the end of the first year of growth the bark includes (from inside to outside) the vascular cambium, secondary phloem, primary phloem, cortex, periderm, and the epidermis. During each subsequent year more secondary phloem is produced. Since some cells of the phloem are soft walled, the older secondary phloem becomes crushed by the expanding woody core pushing outward, and these cells cease to function in conduction of organic materials. Only the most recently produced phloem cells are functional in this capacity.

(a)

(b)

(c)

Figure 8.9
Rough bark on white pine (*Pinus alba*) (*a*) and cottonwood (*Populus deltoides*) (*b*). Smooth bark on gum tree (*Eucalyptus deglupta*) (*c*).

Not all cork cambium activity is uniform, and the cambial layer usually is not a continuous cylinder around the stem. During the life span of a tree, new cork cambia arise from the living parenchyma cells interior to the existing periderm(s). All tissue exterior to the most recent periderm becomes physiologically isolated from food and water supplies, owing to suberin and wax deposited in the cork cell wall. These substances retard lateral conduction and provide the floatability, oil and fire resistance properties of cork. In barks where new periderms form to the inside of the earlier formed periderm(s), entrapping phloem between the layers, the result is an overlapping series of scales or plates. These rough barks (elm, maple, pine) are much more common than smooth barks (birch, sycamore), which are formed by periderms that occur as essentially continuous cylinders and form concentric rings of new periderms every few years. The combination of periderm formation patterns and continued increase in the girth of the trunk produces a variety of bark appearances, from an essentially smooth surface to deeply furrowed or ridged bark (figure 8.9).

Some species of woody plants have a great deal of phellogen activity, giving rise to very thick barks. Others are only a few cells thick, and therefore the bark is very thin. The most extreme example of cork thickness is *Quercus suber,* the cork oak (figure 8.10) widely cultivated in regions of Spain. This tree produces essentially all of the cork for wine bottles, and most of the trees are grown in Portugal. The periderm(s) of the cork oak have an abundance of lenticels that enhance gas exchange allowing the wines to "breathe" during storage and aging.

✔ Concept Checks

1. Draw a three-year-old dicot stem in cross section. Label all primary and secondary tissue.
2. How does secondary xylem differ from primary xylem?
3. What is bark? Distinguish between the periderm, phellogen, phellem, and phelloderm. How does bark form on a tree?

Figure 8.10
The bark of *Quercus suber* is used commercially in the production of corks for wine bottles. Care is taken not to disrupt the vascular cambium by leaving the most recently produced secondary phloem and periderm intact.

Those Amazing Australian Boabs

The boab tree (*Adansonia gregorii*) is a large deciduous tree occurring on sandy plains, creek beds, and stony ridges throughout the Kimberley region of Western Australia, and in the Victoria and Fitzmaurice River basins of the Northern Territory (figure 1*a*).

The name "boab" is probably a mispronunciation derived from the "baobab" tree (*Adansonia digitata*) of the tropical African mainland.

(a)

The distinctive, immense trunk varies from bottle-shaped to unusual, even grotesque shapes. The boab can have a girth of 20 meters and a canopy over 25 meters high. During the dry season leaf deciduousness occurs, enhancing the boab's drought tolerance. The wood of the boab is soft and spongy, owing to its water storage capabilities. Moisture can be extracted by chewing the parenchymatous stems or roots, or it can become entrapped in the hollows that develop inside the trunk and at the bases of branches. The boab bark varies from dull gray-brown to bronze in color, is smooth in texture, stores water, and is unique in that it is not sloughed (continually replaced) as is the bark of many tree species.

Australian aboriginals place great importance on boab trees. The fruit, resembling pendulous gourds, contain a juicy pulp in which are malic, tartaric, and ascorbic acids (vitamin C). The pulp is eaten dry or mixed with water as a beverage. Aborigines carved marks in the bark of boabs, apparently to record the number of fruit taken. Another mark, signifying the encroachment of civilization, was carved into a boab on the banks of the Victoria River in 1856. Close examination of the boab photograph (figure 1*a*) will reveal the inscription "July 2nd 1856," placed there by the explorer Augustus Charles Gregory. Immediately below the month and date is the inscription "Letter in Oven." This inscription indicates the location of a letter that gave details of the explorers' plan in case they were delayed or perished on their return overland trip to Brisbane (figure 1*b*).

Several boabs have exceptionally large, swollen hollow stems. Aborigines and early settlers used these as shelters or as temporary prisons during the transportation of criminals (figure 1*c*). Ferdinand von Mueller, a famous botanist and member of the North Australian Expedition, collected specimens of the boab tree, and the species was published as *Adansonia gregorii* in honor of Gregory.

(b)

(c)

Figure 1

(*a* and *b*) Gregory's Boab on the banks of the Victoria River near Timber Creek, Northern Territory, Australia. (*c*) An Australian Boab, *Adansonia gregorii,* prisoner tree in Derby, Western Australia.

STEM ADAPTATIONS

Stem adaptations include a variety of changes in both external morphology and internal anatomy. Minor adaptations include production of thicker bark, greater height, more flexible stems, basal buttressing (figure 8.11), and increased photosynthesis. All enable the plant to adjust to some environmental stress. Major adaptations significantly alter the form of the stem. These specialized stems are still recognizable, however, in that they retain some or all of the typical stem structures, such as nodes, internodes, buds, and leaves.

Figure 8.11
Cypress trees develop buttressing at the base of their trunks for added support in swampy areas.

Rhizomes

These are horizontal stems usually located beneath the soil surface. Superficially they resemble roots, but like typical aerial stems they have nodes, internodes, buds, and often leaves.

Adventitious roots, roots that arise from stems or leaves, develop at the bottom side of the **rhizome** in the area of the node, and shoots emerge aboveground from the same location (figure 8.12*a*). The chief functions of rhizomes are food storage and vegetative reproduction. As a stem, they also function in water conduction and food translocation. Most rhizomes are **perennial,** living year after year, increasing in length each year and sending up new plants at their nodes.

Stolons

These also are horizontal stems that produce roots and shoots at the nodes, but they form aboveground. Bermudagrass (*Cynodon dactylon*) and a number of other grasses spread vegetatively by **stolons** (figure 8.12*b*). The stolons of strawberries are unusual in that they produce new growth at every other node (figure 8.12*c*).

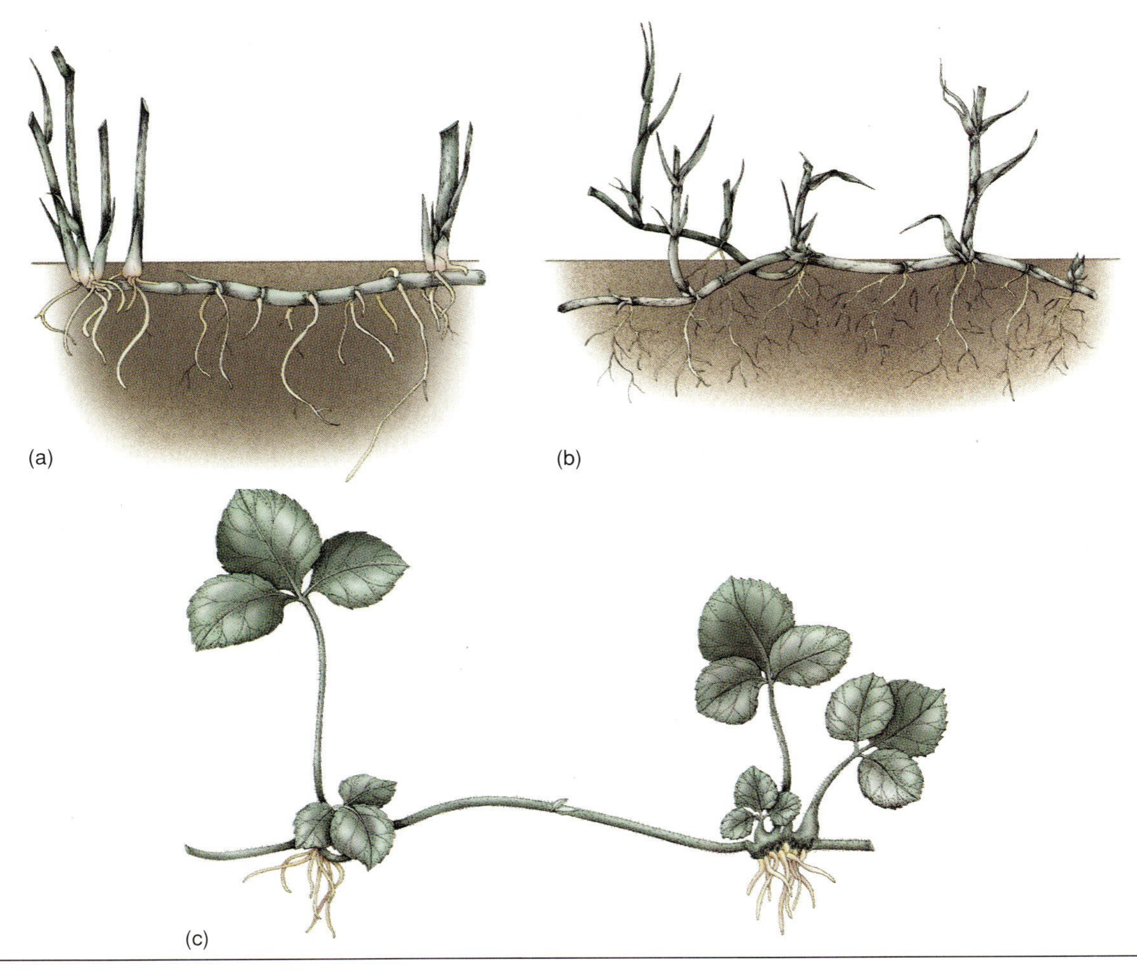

Figure 8.12
Stems modified for vegetative reproduction. (*a*) Rhizomes of Johnsongrass. (*b*) Stolons of Bermudagrass (often called runners). (*c*) Stolons of strawberries.

Tubers

These are enlarged storage stems that develop at the ends of slender rhizomes. The Irish potato (*Solanum tuberosum*) is a well-known **tuber.** The "eyes" of the potato are actually groups of lateral buds. The eyes are used for vegetative propagation (figure 8.13). Potatoes are of tremendous food value due to the enormous amounts of starch stored in the parenchyma.

Corms

These are short, thickened, vertically oriented, underground stems with dry, papery, scalelike leaves. *Gladiolus* is an example of a **corm** from which a single shoot develops by using the stored food (figure 8.14). Once leaves are produced photosynthesis provides the food necessary to continue growth, produce a flower, and develop a new corm with the next year's food store. The Hawaiian dish poi, a staple in Polynesia and Southeast Asia, is made from crushed and fermented taro corms (*Colocasia esculenta*).

Bulbs

These are cone-shaped stems surrounded by many scalelike leaves that are modified for food storage. The conical stem produces a single aboveground shoot from the terminal bud and a new **bulb** from a lateral bud. Unlike corms, the food reserves of the bulb are in the modified leaves. These reserves are exhausted in the production of the leafy aboveground shoot, and food for storage in new bulbs is produced in the leaves by photosynthesis. Both onion and daffodils (figure 8.15) have bulbs. The food storage leaves of the onion bulb constitute the edible part.

Succulents

Plants that have extensive parenchyma tissue for the storage of water (figures 8.16, 8.17, and 8.18) are referred to as **succulents.** Cacti are easily recognizable examples. Succulent stems are found not only in the Cactaceae (cactus family) of North and South American deserts but also in members of the Euphorbiaceae native to the African deserts. Members of these two families are often remarkably similar in external morphology because of the analogous habitats to which they have adapted. This phenomenon of similar appearance in unrelated plants resulting from adaptation to similar habitats is termed convergent evolution.

Thorns

As modified stem branches, **thorns** arise from the axils of leaves as do regular branches. The honey locust (*Gleditsia triacanthos*), hawthorn (*Crataegus*), and firethorn (*Pyracantha*) are thorn-bearing plants. Rose (*Rosa*) "thorns" are not true thorns but are stem surface outgrowths called prickles. Spines and thorns are terms often used synonymously; however, most

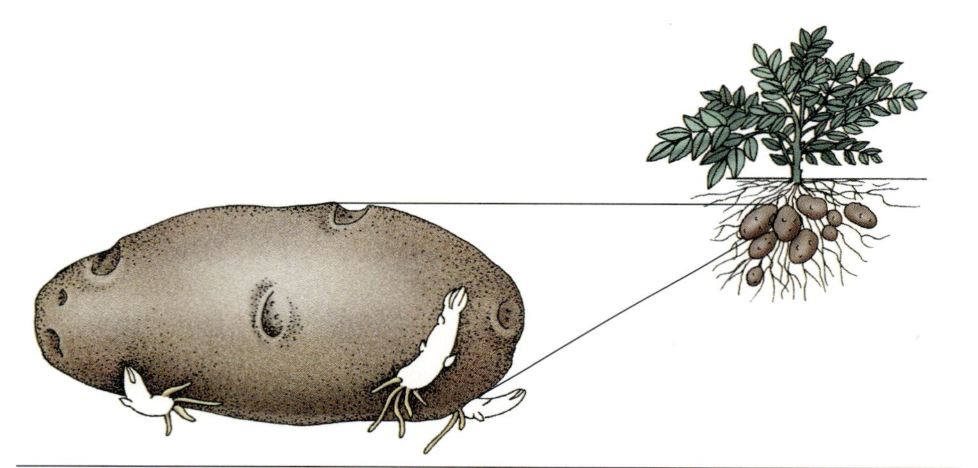

Figure 8.13
The tuber of the Irish potato is a stem modified for storage. The "eyes" are axillary buds capable of forming new stems.

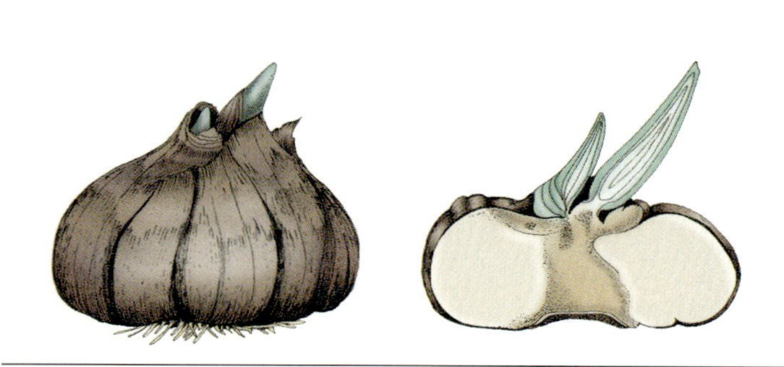

Figure 8.14
Corms are short, thickened underground stems. Food stored in the corm of a *Gladiolus* enables it to produce an aboveground shoot.

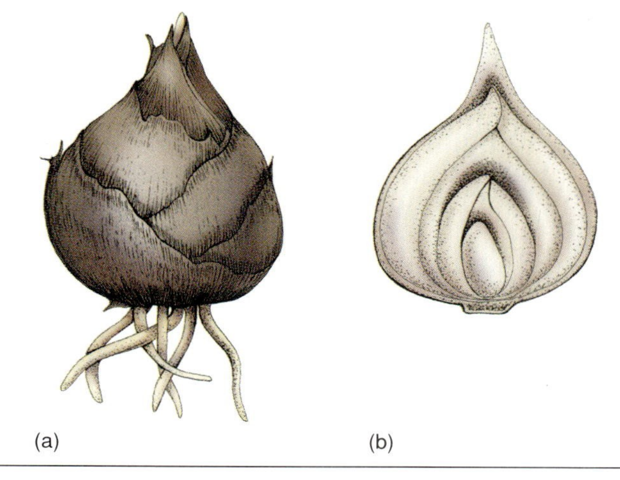

Figure 8.15
Bulbs are actually buds on small cone-shaped stems surrounded by modified leaves, which act as storage structures. Shown is a whole daffodil bulb (*a*), and a longitudinal cut through the bulb (*b*).

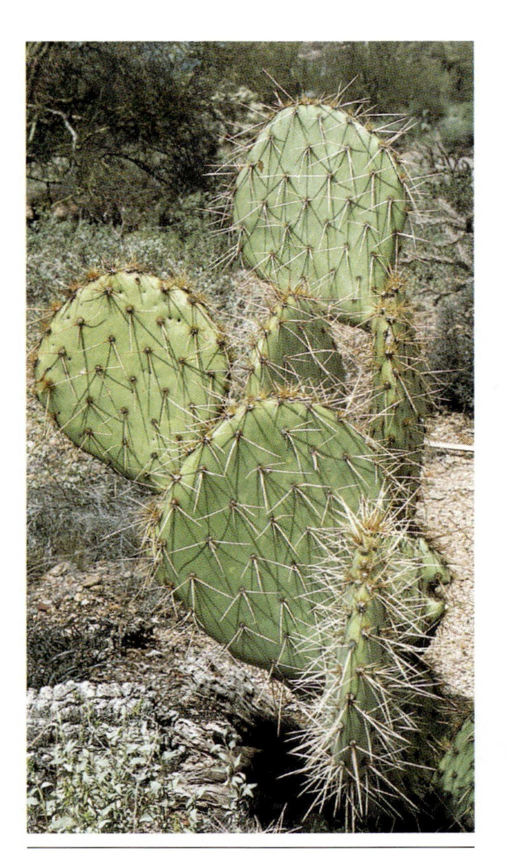

Figure 8.16
Succulent stems of a prickly pear cactus (*Opuntia*). Note spines that protect the plant from herbivores.

Figure 8.17
(*a*) *Ariocarpus* has no spines, but it has a succulent stem and is also well camouflaged as are (*b*) stone plants, *Lithops marmorata*.

Figure 8.18
(*a*) These barrel cacti are well protected by stiff spines. (*b*) (*Echinocactus*).

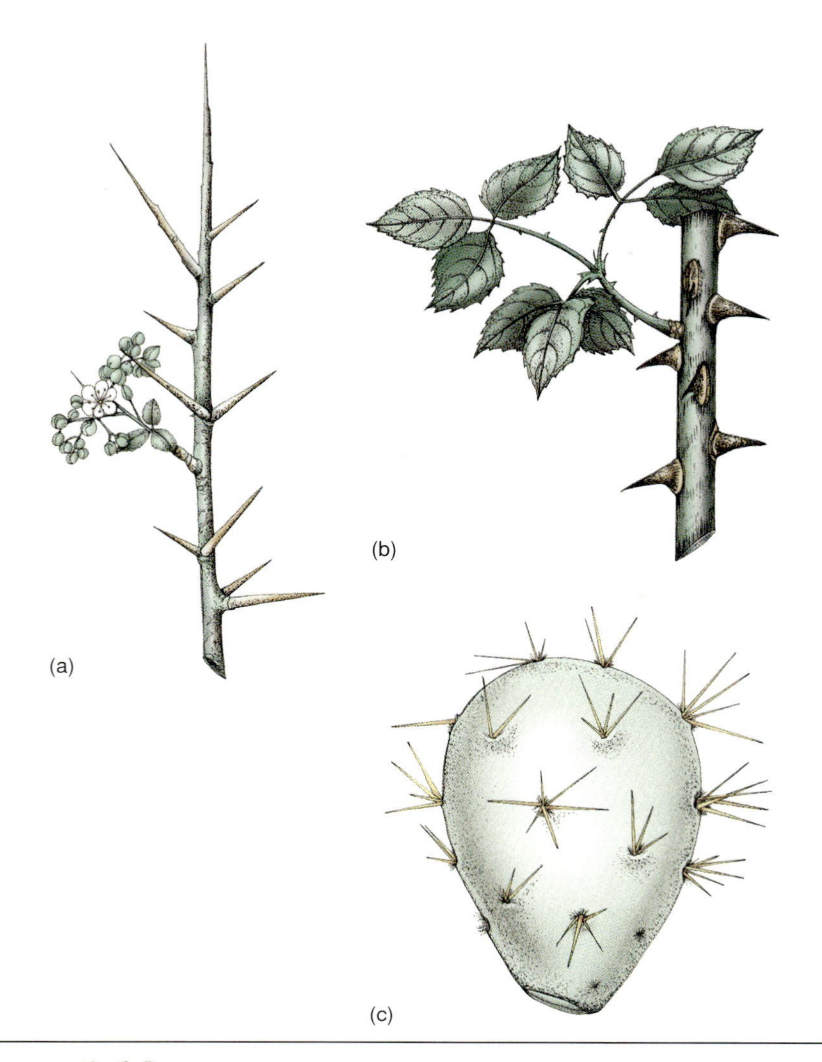

Figure 8.19
(*a*) Thorns of *Pyracantha.* (*b*) Roses have prickles, and (*c*) cacti have spines, not thorns.

spines are modified leaves arising from below the epidermis. Both are stiff, sharp-pointed, woody structures that can be equally painful if encountered inadvertently, and thus the technical difference is understandably unimportant to the victim. Both are equally effective in reducing predation by herbivores (figure 8.19).

Tendrils

These can be modified leaves or stems. Grape plants (*Vitis*) and Virginia creeper (*Parthenocissus*) are two well-known examples of plants with stem **tendrils.** Grape tendrils twine around a support structure; Virginia creeper tendrils anchor the plant with a sticky substance so tightly that, when they are removed, mortar and even a building's facing can break away. Once anchored, these tendrils coil or contract to pull the plant closer to its anchoring surface (figure 8.20).

BUDDING AND GRAFTING

Specific adaptations of organs allow asexual propagation from essentially all vegetative parts of the plant—leaves, stems, buds, roots, and even single cells. Sometimes it is possible to transplant an organ from one plant to another, and this process is called budding or grafting, depending on the organ used for transplant (figures 8.21 and 8.22). The plant receiving the new organ is referred to as the rootstock, or simply the stock. The donor organ is referred to as the scion. Just as in organ transplants in animals, genetic compatibility is essential, although it is not quite as specific. Often two species within the same genus can be grafted, and they are perfectly compatible. The critical factor in ensuring that the graft or bud "takes" is the matching of the vascular cambia, the lateral meristematic regions from which new cells arise. If this tissue fails to join, the scion will die. For a graft, an entire stem piece is placed on a stock, according to the size of the stem, time of year, and various other factors. The art (as well as the science) of grafting ensures a high degree of success. Commercial apples, peaches, pecans, etc. are grafted plants in which the scions of desirable fruit are grafted onto stocks with vigorous roots. In budding, a single bud removed from the donor plant is placed on a stock. Essentially all commercial roses are propagated in this manner.

STEM CUTTINGS

Many ornamental and horticulturally important species are propagated by stem cuttings (figure 8.23). When stem cuttings are placed in a suitable propagating medium (sand, peat moss, vermiculite), adventitious roots form at the base of the cutting, after which the cutting can be potted and grown in a normal manner. Certain hormones promote the rooting process in most cases, and their use is standard practice in commercial operations (chapter 14, p. 248). A few ornamental species are propagated by single leaves (African violets and *Begonia* are often propagated this way). The base of the leaf or the base of the petiole will root in a suitable medium and produce a new shoot from the leaf (figure 8.24).

LAYERING

This is a method of propagating plants with long flexible stems, such as grapes or berries. The stem is pinned to the ground at a node and covered with soil, and adventitious roots form at the node. Once the roots have appeared, the stem may be cut from the parent plant, and the new plants are on their own. Air layering is the process of packing a stem node with moist peat moss and wrapping it in plastic (figures 8.25 and 8.26). Once adventitious roots form at the node, the top of the plant can be cut off just below the rooted node and planted as a new individual. The apex of the air-layered plant will initiate new apical growth.

Figure 8.20
In Virginia creeper (*Parthenocissus*) padlike appendages arising at the nodes secrete a sticky substance, which anchors the stem to surfaces.

✔ Concept Checks

1. Why is a rhizome a stem rather than a root?
2. Compare bulbs and corms.
3. Define the term succulence. What is the adaptational significance of succulence?
4. Explain how grafting is accomplished.

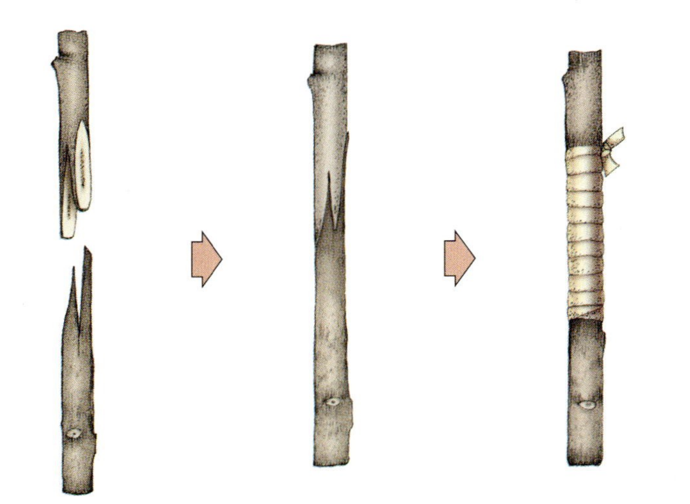

Figure 8.21
A whip-and-tongue graft is used to join stems or branches of different plants.

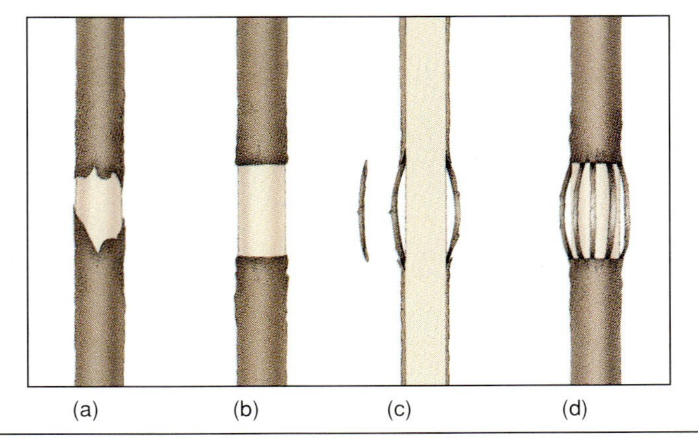

Figure 8.22
A bridge graft is used to repair damage caused by girdling, the removal of bark from the circumference of a tree. (*a*) The damaged area. (*b*) The damage is trimmed. Twigs from the same tree are cut with tapered ends (*c*) and inserted under the bark all around the wound (*d*).

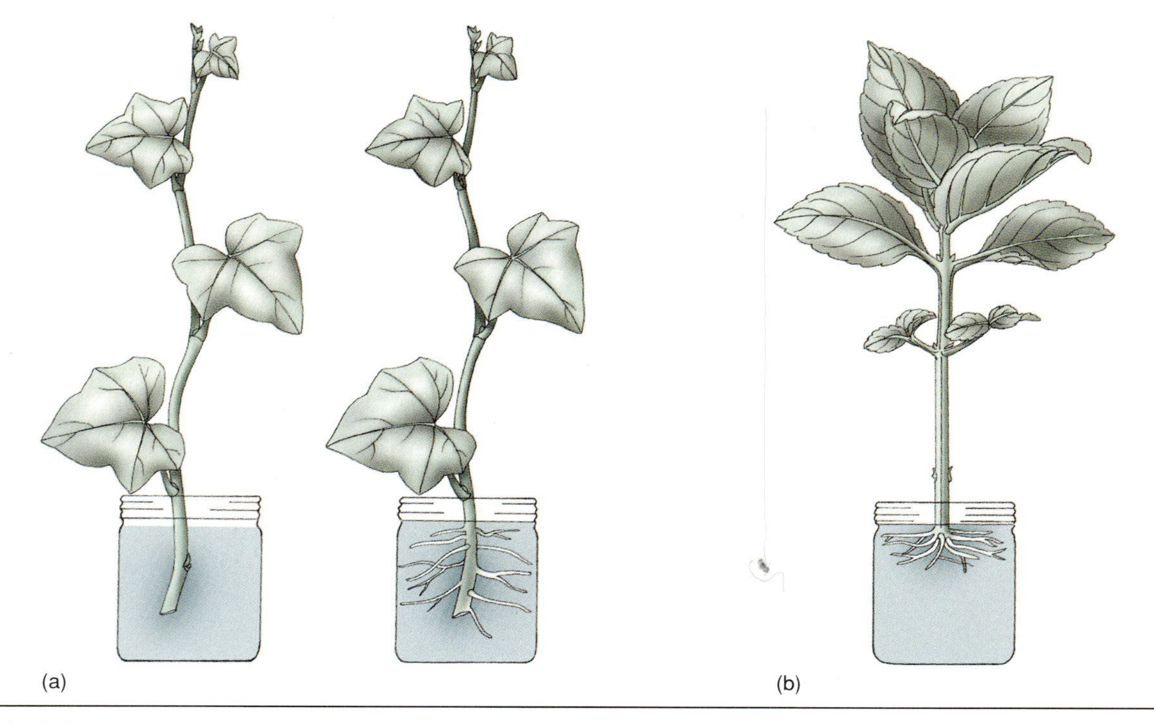

Figure 8.23

Stem cuttings form adventitious roots at the cut. (*a*) *Hedera*. (*b*) *Coleus*.

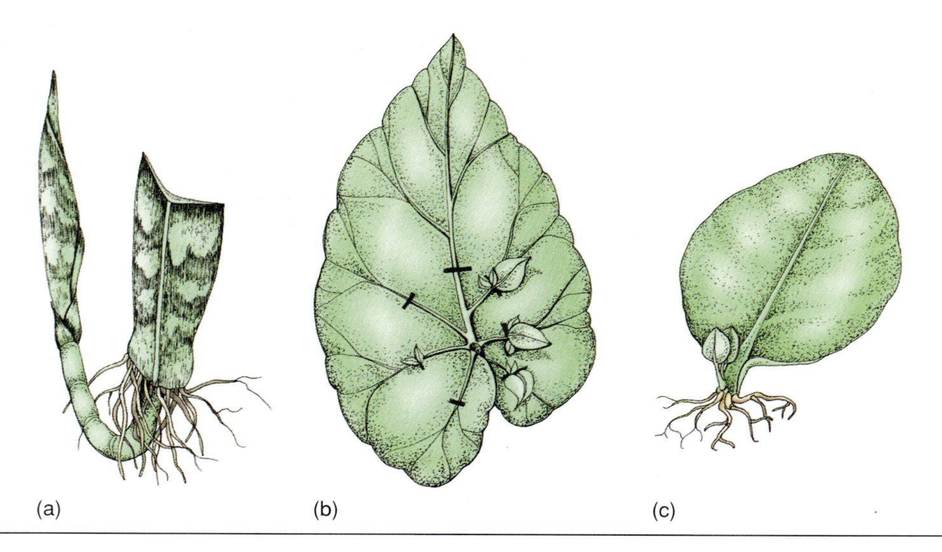

Figure 8.24

Leaf cuttings. (*a*) Mother-in-law tongue (*Sansevieria*). (*b*) *Begonia*. (*c*) African violet (*Saintpaulia*).

Figure 8.25
Commercial air layering of *Sorbus* trunks.

Figure 8.26
Some plants can be propagated by air layering. A section of bark is removed (*a*), the area is packed in damp sphagnum moss (*b*) and wrapped in plastic (*c*). Roots form at the cut surface.

Summary

1. Stems conduct organic solutes, water, and nutrients throughout the plant body, produce new growth, support leaves, flowers, and fruit, and store materials. Stems can be prostrate, various upright shapes, or even twining on other structures. Stems are either herbaceous or woody.
2. Stems have nodes and internodes. Buds, embryonic shoots, are attached to the stem and are responsible for branching.
3. The apical meristem of the stem leads to primary growth. Lateral meristems of the vascular cambium and cork cambium lead to secondary growth.
4. The herbaceous dicot stem has a vascular pattern of a single ring of vascular bundles. As the stem becomes older and if it becomes woody, the fascicular cambium between the xylem and phloem of the bundles links with a newly formed interfascicular cambium to produce a ring of dividing cells called the vascular cambium. All secondary vascular tissues arise from this region, and the commercial product referred to as wood is technically secondary xylem.
5. The monocot stem is characterized by vascular bundles scattered throughout the ground tissue, and consequently there is no distinction between cortex and pith. Except in rare cases, there is no secondary growth in monocots.

6. In secondary growth, the vascular cambium produces secondary xylem toward the center of the stem and secondary phloem toward the outside of the stem. The cork cambium, or phellogen, produces an outer layer(s) of cork cells that serve a protective role.
7. Higher plants, especially angiosperms, are adapted through their external morphology to survive the stresses of the environment.
8. Stem modifications include rhizomes and stolons, both producing roots and shoots at the nodes for vegetative propagation. Enlarged storage tubers and reproductive corms and bulbs are all underground stems. Succulent stems are modified for water storage. Thorns protect the plant from herbivores.
9. Vegetative (asexual) reproduction by rhizomes, stolons, and other underground modified stems provides mechanisms for commercial plant propagation. There are a number of advantages to cloning plants, although the variability resulting from sexual reproduction is lost. Budding and grafting are very common techniques for vegetative plant propagation, as are stem cuttings and layering.

Key Terms

adventitious roots 129
annual ring 125
axil 122
axillary bud 123
bark 125
bud 122
bud scales 123
bud scale scar 123
bulb 129
corm 129
cortex 123
herbaceous 123
internode 122
leaf scars 123
node 122
perennial 129
periderm 126
phellem 125
phelloderm 126
phellogen 125
phloem rays 125
pith 123
pith ray 123
rhizome 129
stolon 129
succulent 130
tendrils 132
terminal bud 123
thorns 130
tuber 129
vascular cambium 124
xylem rays 125

Discussion Questions

1. Explain the origin of the fascicular and interfascicular cambia.
2. Describe how the diameter of a tree increases from year to year.
3. If the bark is removed in a complete ring from around the stem axis, how would this affect the flow of water or food through the vascular tissues?
4. Explain why a nail driven into a five-year-old tree, three feet above the ground, does not change position as the plant grows.
5. Describe how certain stems overcome the following adverse environmental factors: shortage of water in the soil, herbivory, lack of internal supporting tissues, continued, year-to-year growth without the production of lateral meristems.

Suggested Readings

Bell, A. D. 1991. *Plant form: An illustrated guide to flowering plant morphology.* Oxford University Press, New York.

Dahlgren, R., H. Cliffore, and P. Yeo. 1985. *The families of the monocotyledons: Structure, evolution, and taxonomy.* Springer-Verlag, New York.

Fahn, A. 1990. *Plant anatomy.* 4th ed. Pergamon Press, Elmsford.

Fuller, D., and S. Fitzgerald, eds. 1987. *Conservation and commerce of cacti and other succulents.* World Wildlife Fund, Washington, D.C.

Raven, P. H., R. F. Evert, and S. E. Eichhorn. 1992. *Biology of plants.* 5th ed. Worth Publishers, New York.

Rees, A. R. 1992. *Ornamental bulbs, corms and tubers.* C. A. B. International, Oxon, UK.

Willert, D. J. von. 1992. *Life strategies of succulents in deserts: With special reference to the Namib Desert.* Cambridge University Press, New York..

Chapter Nine

Leaves

Leaf of *Victoria,* the giant Amazon water lily.

Leaves are typically the most prominent organ of the shoot. As outgrowths of the shoot apex, they are the light-capturing, solar-energy collectors of the plant body. Although tremendous variations occur in the size and shape of leaves, most are thin and flat with the internal arrangement of tissues well suited for the primary function of photosynthesis. Like the root and stem, adaptations among leaves are numerous. This chapter explores the morphological variations, internal structure, and adaptations of leaves.

Leaves are figuratively the heart of the green plant. One may think of roots and stems as organs for support, absorption, and transport, but leaves contain the majority of the chloroplasts and are responsible for essentially all photosynthesis, with the exception of a few plants with green stems and few or no leaves. Plants are green because of the light-trapping pigment chlorophyll. In the leaf, light energy is converted into food energy in the form of carbohydrates. Although many herbaceous stems are green and photosynthetic, the greatest amount of photosynthesis (chapter 13, p. 219) carried on by terrestrial plants occurs in their leaves. Leaves, then, can be thought of as energy factories providing food energy necessary to sustain all life on earth.

LEAF FUNCTIONS

The process of photosynthesis involves the exchange of carbon dioxide (CO_2) and oxygen (O_2) with the atmosphere. Carbon dioxide is taken into the leaf, and oxygen is released into the atmosphere. During this gas exchange at the surface of the leaf, water is evaporated from the leaves. This water loss, the process of transpiration, is the end of the water movement through the plant, which began with the roots absorbing water from soil. Thus the interrelated processes of photosynthesis, gas exchange, and transpiration take place in the leaves.

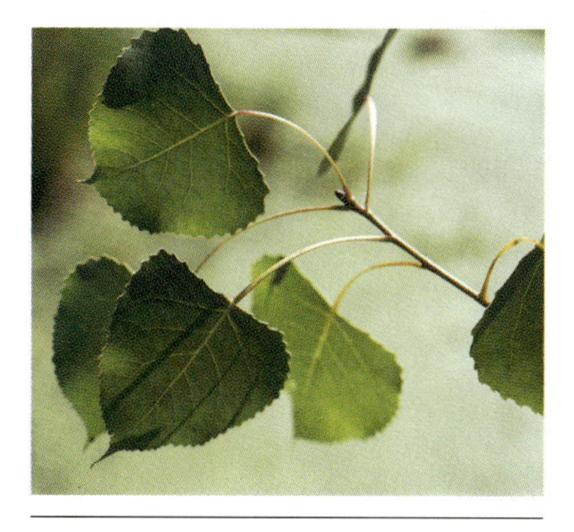

Figure 9.1
The leaves of the eastern cottonwood (*Populus deltoides*), showing the petiole and blade.

LEAF MORPHOLOGY

Leaves are well designed to carry on their functions, usually having broad, flat, and thin **blades** that are connected to the plant stem by a stalk called a **petiole** (figure 9.1). The larger the leaf blade, the greater the photosynthetic surface for trapping sunlight. The leaf blade can be either **simple,** having one blade attached to a petiole or stem node, or **compound,** having two or more separate bladelike subunits (leaflets) making up the blade (figure 9.2). Even though some compound leaves are large enough that the individual leaflets might appear to be separate leaves, confusion can be avoided by remembering that buds always occur in the axil of leaves; leaflets do not have buds in their axils. The arrangement of the leaflets of a compound leaf can be either **pinnately compound** (with the leaflets arranged along the length of a central stalk, the rachis) or **palmately compound** (with the leaflets attached to the end of the petiole like the fingers radiating from the palm of the hand).

Leaf blades come in an incredible array of sizes and shapes, and although the morphology is sometimes consistent within a given plant group, the amount of variability displayed is often so great as to prevent the use of leaf morphology as a diagnostic technique for the identification of plant groups.

Innumerable combinations of overall leaf shape, apex, **margin,** and

Figure 9.2
(*a*) The simple leaf of a quaking aspen (*Populus tremuloides*). (*b*) A pinnately compound leaf of a rose (the structure at the base of the petiole is a stipule). (*c*) The twice pinnately compound leaf of mesquite (*Prosopis*). (*d*) A palmately compound leaf of *Schefflera.*

base modifications are possible. Leaf margins are generally said to be entire (smooth), dentate (with sharp "teeth"), or lobed (with rounded intrusions). The apex (tip) and base, where the petiole attaches, can be variously rounded, angled, pointed, or indented. In addition, leaves can be smooth, rough, thin, thick, leathery, or **succulent** (fleshy and juicy with stored water) (figure 9.3).

The petiole can also add to the overall leaf variability by being short or long, attached to the middle of the blade (peltate) or to the edge, as is normal, or even absent. When the petiole is absent, the leaf is said to be **sessile.** Sessile leaves can be further modified to wrap around the stem (clasping) or even form a **sheath** (very common in grasses). Additionally, the presence or absence of **pubescence** (hairs or epidermal trichomes), which can be simple, branched, glandular, barbed, long, short, fine, coarse, dense, sparse, and specifically localized or generally distributed, makes it evident how complex and variable a thorough discussion of leaf morphology would be (figure 9.4).

Leaf and Bud Attachments

The leaves on a stem are produced by the shoot apex usually in a definite pattern. The most common pattern of leaf and axillary bud attachment is an alternating sequence of one leaf and bud per node; this is termed an alternate or spiral arrangement. Two leaves (and their axillary buds) attached at the same node directly across the stem from each other produces an opposite arrangement. Three or more leaves attached at each node produces a whorled arrangement. Plants such as dandelion (*Taraxacum officinale*) have all leaves grouped in a basal rosette at ground level in a tight spiral (alternate) arrangement (figure 9.5).

Although most plants have only one bud in each leaf axil, some have several. In such cases the central bud is the true axillary bud that will develop into a new lateral branch; the others are called accessory buds. Buds that develop on the plant at positions other than the leaf axil and apices are termed adventitious buds. These buds may develop on stems, roots, or leaves, producing new shoots from any of these positions. Adventitious buds are often produced on stems as a response to pruning or injury. As with axillary buds, they initiate new growth in response to hormonal controls, which in turn are linked to growth at the apex of each stem.

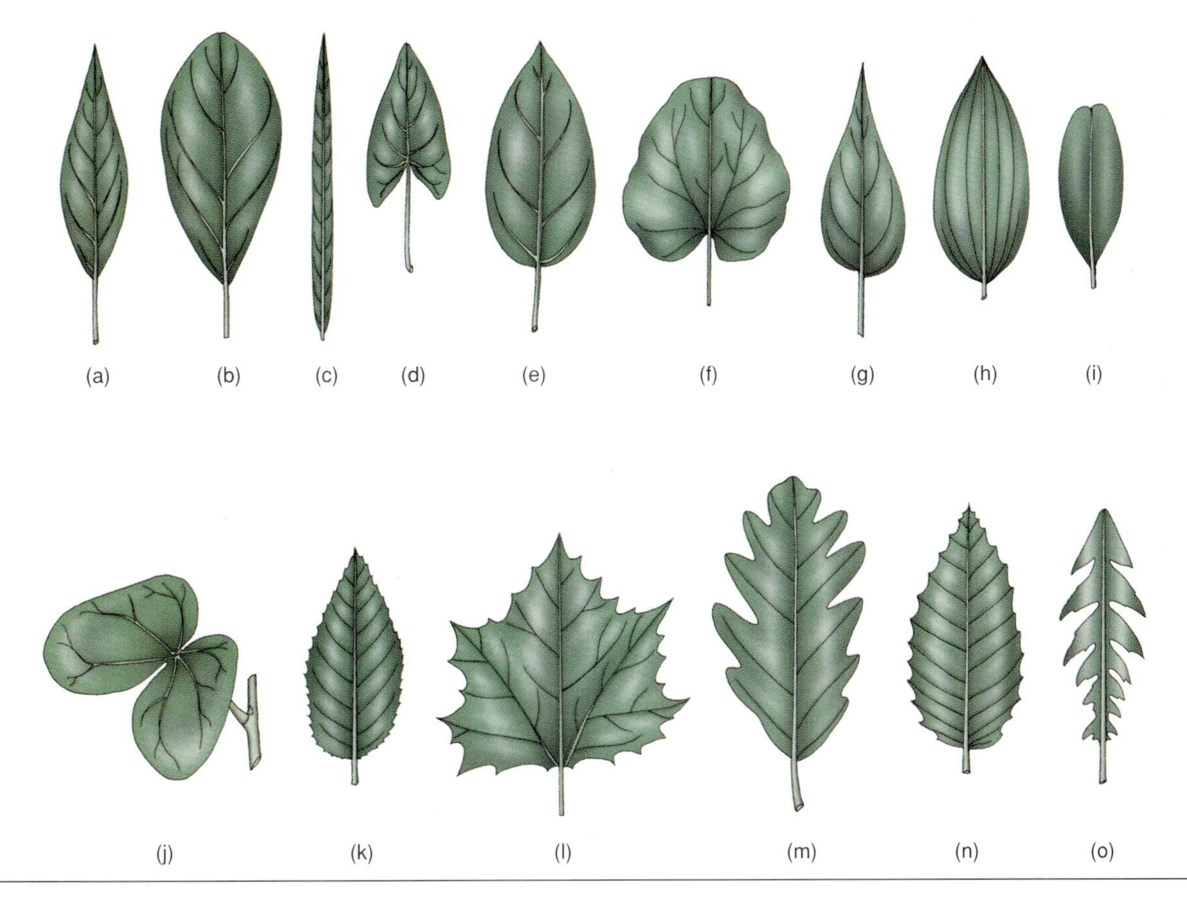

Figure 9.3

Small sampling of the enormous variety of leaf shapes, margins, apices, and bases. (*a*) Lanceolate shape. (*b*) Oblanceolate shape. (*c*) Linear shape. (*d*) Sagittate shape. (*e*) Rounded base. (*f*) Cordate base. (*g*) Oblique base. (*h*) Acute apex. (*i*) Emarginate apex. (*j*) Bilobed. (*k*) Serrate margin. (*l*) Incised margin. (*m*) Lobed margin. (*n*) Dentate margin. (*o*) Divided margin.

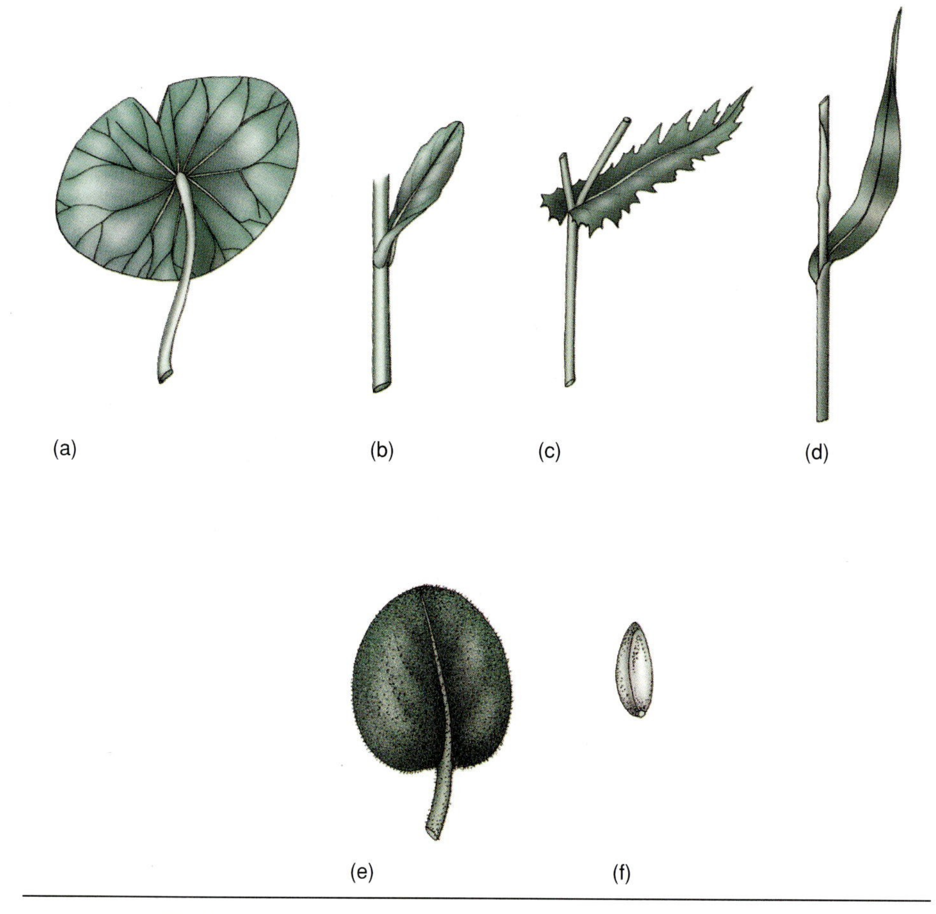

Figure 9.4

A leaf petiole may be attached to the edge of the blade or to the middle, as in the water lily's peltate leaf (*a*). Leaves without a petiole are sessile (*b*), clasping (*c*), or sheathing (*d*). Surface pubescence (*e*) and leaf succulence (*f*) also add to total leaf morphological variability.

Venation

The vascular bundles or veins of the blade occur in specific patterns or arrangements. Leaf venation can be either pinnate, **parallel,** or palmate. Pinnate vein arrangement, common in dicotyledonous leaves, has a main central midrib, or primary vein, with secondary veins branching from it (figure 9.2*a*). Tertiary and then quaternary lateral leaf venation arises from the secondary and tertiary veins, respectively, with each becoming progressively smaller. The smaller veins commonly merge, forming an intricate network of veins. Because of this, many botanists have called this type of venation **netted.** The midrib, then, is the most visible of all these levels of vein branching.

Parallel venation characteristic of monocotyledons commonly lacks a central midrib. In addition, several equal-sized veins run the length of the leaf blade parallel to each other and converge near the apex of the blade (figure 9.3*h*). Between these major veins is a network of smaller venation. Palmate venation is similar to parallel venation in that several equal-sized veins exist, one going to each palmately arranged lobe of the blade (figure 9.3*f*).

LEAF DEVELOPMENT

A leaf begins as a **leaf primordium,** a small bulge or protrusion of meristematic cells produced by the activity of the shoot meristem. As the primordium expands, two longitudinal ridges appear, one on each side of the developing leaf. These are the **marginal meristems,** which ultimately control the size and shape of the leaf. If the marginal meristems undergo cell division at a uniform rate, the blade develops a smooth, even margin. Differential rates of cell division along the marginal meristem, however, could produce lobed, dentate, or other margin types.

In contrast to roots and stems, leaves are of determinate size and have a limited life span. Plants whose leaves fall off at the end of the growing season leaving the stem temporarily barren until the following spring are termed **deciduous.** Plants whose leaves persist for several seasons and are gradually shed as new leaves are produced are called evergreen.

LEAF STRUCTURE

Despite the variety of leaf shapes, the typical green foliage leaf consists of dermal, ground, and vascular tissues that function in concert for the primary functions of photosynthesis, gas exchange, and transpiration. Anatomical and morphological adaptations of these tissues occur as a result of the limiting and retarding environmental factors present in the habitat.

Epidermis and Stomata

The leaf is bounded by both an upper and lower epidermis. Each epidermis is typically composed of a single layer of cells, compactly arranged and covered with a cuticle that retards transpiration. As discussed in chapter 6, a variety of epidermal hairs or trichomes may be present on the epidermis. In some plants these hairs may aid in reflecting light from the surface of the leaf, thereby reducing heat absorption

✔ Concept Checks

1. List three main functions of the leaf.
2. Distinguish between simple and compound leaves. Between alternate and opposite leaf attachment. Between netted and parallel venation.
3. Where are leaves produced? How is the size and shape of leaves controlled?

Figure 9.5
Leaf attachment to the stem is normally alternate (*a*), opposite (*b*), whorled (*c*), or basal (*d*), as in the dandelion.

and minimizing transpiration. Epidermal hairs may also deter potential predators through the secretion of noxious chemicals or by the formation of barbs or hooks, resulting in physical injury or even death, especially to small, visiting insects.

Although the cuticle is responsible for retarding transpiration, it also impedes gas exchange, the movement of carbon dioxide and oxygen between the internal atmosphere of the leaf and the external atmosphere. The abundant occurrence of stomata (singular, stoma), typically several thousand per square centimeter, however, provides a pathway for this exchange. Each stoma is a pore between a pair of kidney-shaped guard cells. As will be discussed in chapter 12, the size of the pores can be regulated by the entrance or exit of water into the guard cells and their uneven cell-wall thickness. When a stoma is open, carbon dioxide enters the leaf and oxygen may depart. The wider the pore the greater the rate of gaseous exchange. In most plants the stomata are open during periods of light and closed in the dark. Although the entrance of carbon dioxide is critical for the photosynthetic process, stomatal opening, especially during the warmer daylight hours imposes increased transpiration demands upon the plant. Several hundred water molecules exit through an open stoma for every molecule of carbon dioxide that enters.

Because the typical leaf is flat, the upper epidermis is subjected to greater light intensity and heat absorption than the lower epidermis. In response to such environmental differences, the upper epidermis characteristically develops a thicker cuticle and fewer stomata compared to the lower epidermis. Although the lower epidermis typically contains a larger number of stomata than does the upper epidermis, the positions are sometimes reversed. The floating leaf of a water lily, for example, has functional stomata only on the upper surface of its pads. Additional adaptations of the leaf epidermis and transpiration rates are discussed in chapter 12.

Mesophyll

Between the epidermal layers of the leaf are the veins and the photosynthetic tissue, collectively termed the **mesophyll.** The mesophyll is composed of photosynthetic parenchyma or **chlorenchyma,** characterized by an abundance of chloroplasts and intercellular spaces. In many leaves, two well-differentiated regions can be discerned within the mesophyll. Below the upper epidermis are one or more layers of column-shaped cells, oriented perpendicular to the epidermis. This region is the **palisade parenchyma** (figure 9.6) and is the primary photosynthetic region of the mesophyll. Below the palisade layer, and making up the bulk of most leaves, is the **spongy mesophyll,** a layer of loosely packed parenchyma cells also containing chloroplasts and a network of many intercellular spaces. These intercellular spaces facilitate the movement of gases within the leaf as well as through the stomata, which are abundant in the lower epidermis.

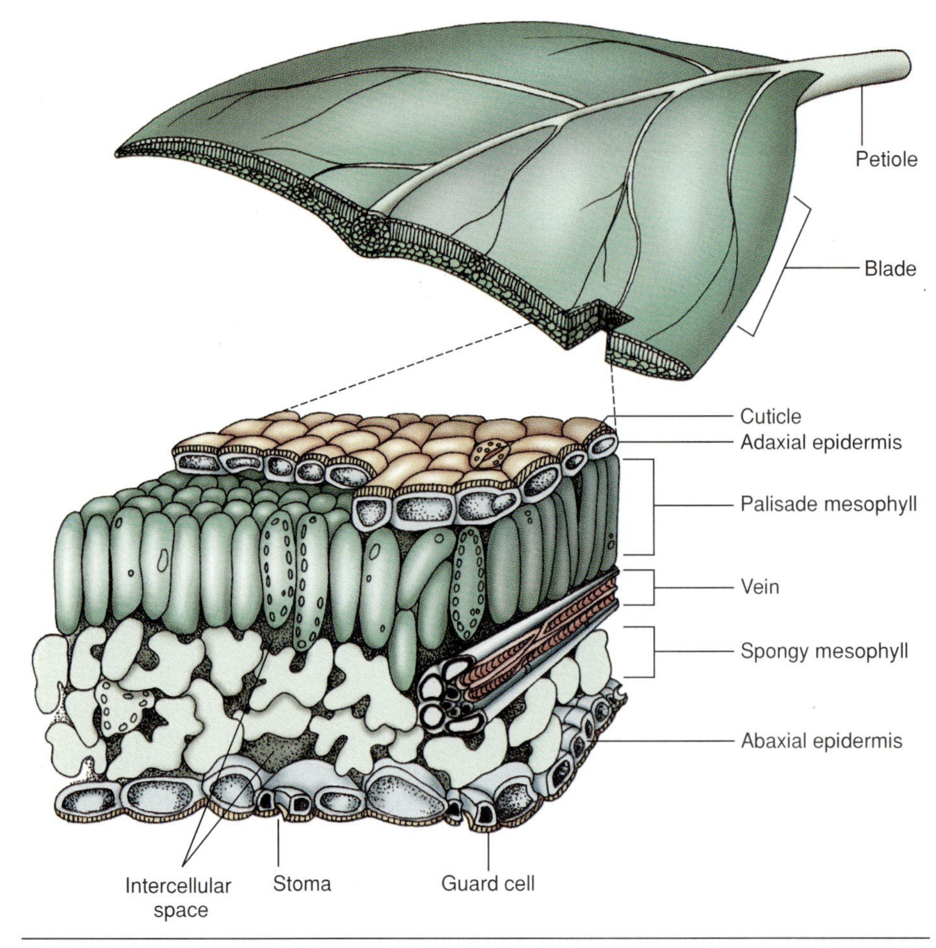

Figure 9.6
Three-dimensional diagram of a dicotyledonous leaf illustrating the morphology and spatial relationships of tissues.

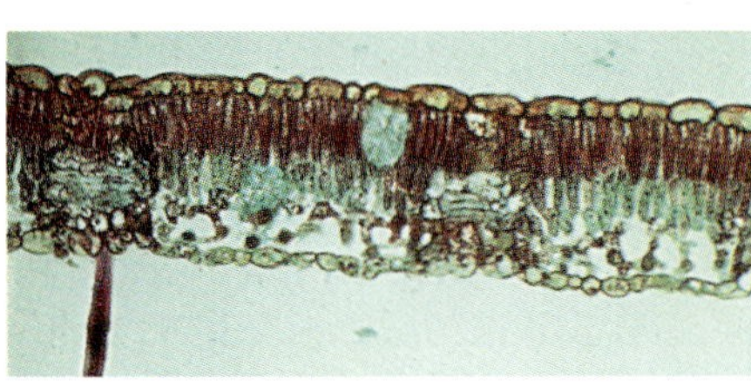
(a)

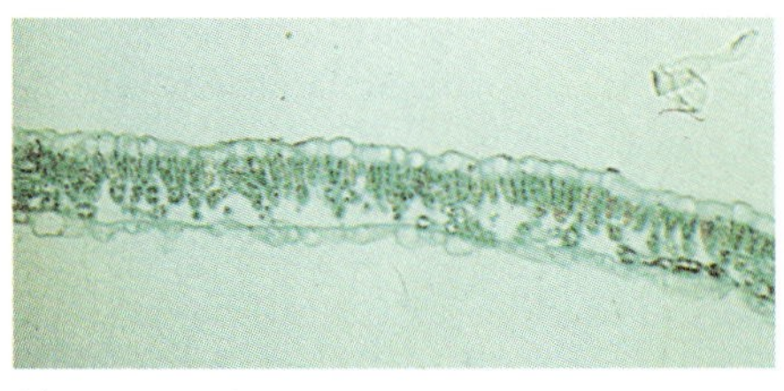
(b)

Figure 9.7
Cross section of two leaves from the same plant. The thicker one (*a*) was grown in full sunlight, and the thin one (*b*) was grown in dense shade. Palisade parenchyma development is related to light intensity.

Leaves taken from the same individual plant can be used to illustrate the variation that is possible in the development of mesophyll and epidermises to environmental parameters. Leaves exposed to full sunlight during development (sun leaves) for example, will exhibit more palisade parenchyma, a thicker mesophyll, and thicker cell walls than will leaves that develop in the shade (shade leaves) (figure 9.7).

Vascular Bundles

Embedded in the palisade or spongy layer are the veins or vascular bundles. Viewed in cross section, they appear similar to a vascular bundle in a stem with the primary xylem oriented toward the upper epidermis and the primary phloem toward the lower epidermis. The midrib may be a very large vascular bundle and in some evergreen dicotyledons and gymnosperms, secondary vascular tissue may also be present. In smaller veins the number of cells become smaller as one progresses from the midrib to the edge of the leaf. Each lateral vein branches again and again, finally terminating with a single vessel, tracheid, or sieve-tube member in the mesophyll tissue. Water entering the tiniest root must make its way through a massive pipeline system to the main root, through the main stem, out through branches, and finally from a stem into a petiole and its blade. All parts of the system must be highly integrated and connected; there can be no breaks, or the tissue will die. Likewise, the last sieve tube terminating in a leaf is a site of "loading" of sugars produced in the process of photosynthesis. Once loaded into the phloem pipeline, those sugars may be transported to any region of the plant in which they are needed.

The veins are enclosed by a tight-fitting sheath of parenchyma cells, termed the **bundle sheath** (figure 9.8). These parenchyma cells facilitate the rapid loading of sugars.

In some leaves the veins and bundle sheaths are in turn surrounded by a wreathlike layer of mesophyll. Such plants are said to have **Kranz leaf anatomy,** or the Kranz syndrome (figure 9.9). This anatomical feature is associated with C_4 dicarboxylic acid metabolism (chapter 13, p. 226).

Many veins are also surrounded by supporting cells, such as collenchyma and sclerenchyma fibers. In perennial **xerophytes,** plants that are adapted to grow in arid habitats, the leaves may be extremely tough and leathery, owing to extensive fiber development that minimizes leaf damage that occurs during wilting (figure 9.10).

LEAF ADAPTATIONS

The morphology and anatomy of a typical leaf are well suited to its major functions: absorbing light, maximizing gas exchange and the exchange of water and food with the mesophyll, and minimizing water loss through transpiration. Clearly, however, all

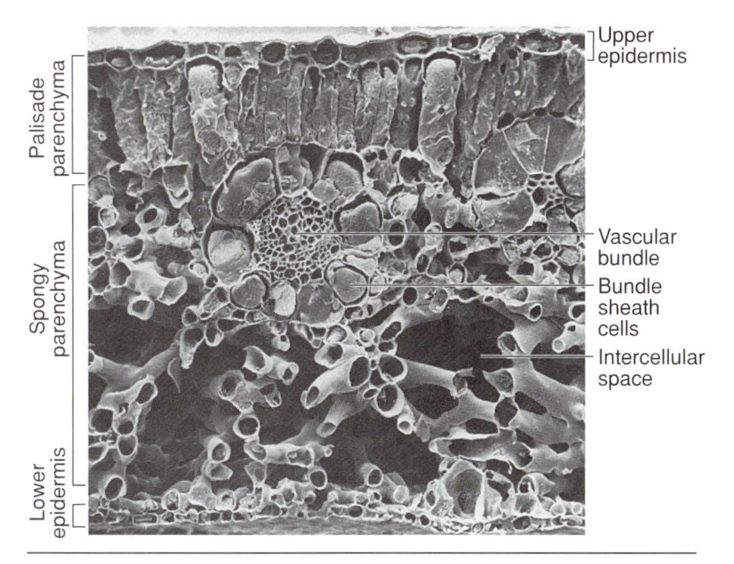

Figure 9.8
Cross section of a *Euphorbia* leaf as viewed with a scanning electron microscope reveals the upper and lower epidermis, palisade parenchyma, spongy parenchyma, and the vascular bundle (vein) with bundle sheath cells.

From R. G. S. Bidwell, Plant Physiology *2nd ed. © 1979 Simon & Schuster.*

Figure 9.10
Some leaves, such as those of *Phormium*, are very thick, and the vascular bundles are protected by massive caps of fibers, which form bundle extensions or support beams within the leaf.

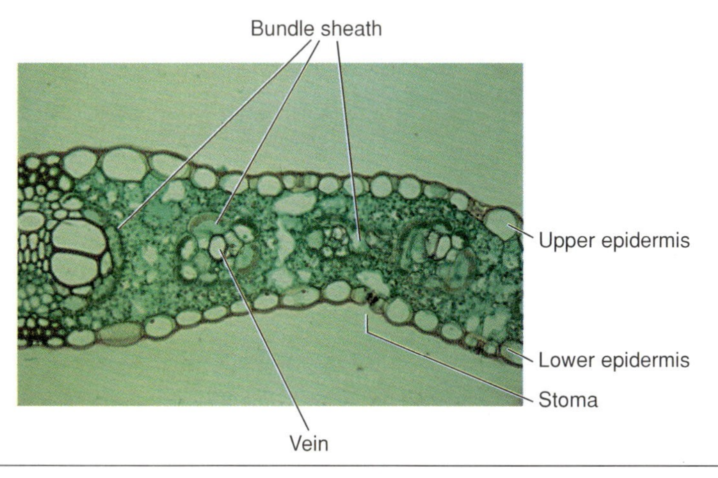

Figure 9.9
Plants with Kranz anatomy have essentially no palisade parenchyma but do have prominent sheaths that surround the vascular bundles. Such plants have the C_4 pathway of photosynthesis.

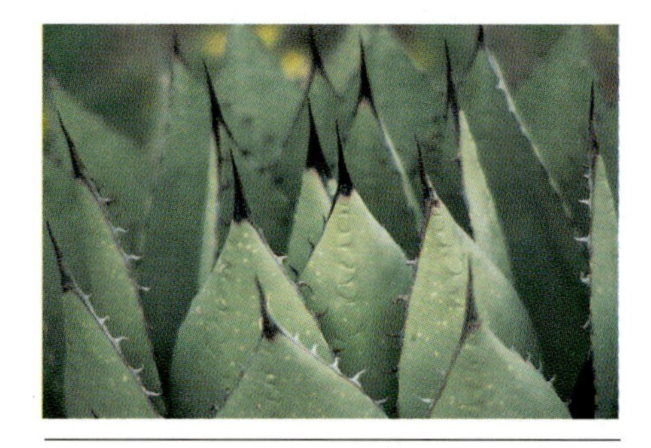

Figure 9.11
Agave leaves are thick, succulent, and well protected by a sharp tip and marginal teeth.

Concept Checks

1. In a typical leaf how does the upper epidermis differ from the lower epidermis?
2. Does gas exchange occur more rapidly during daylight hours or at night? Explain.
3. Distinguish between palisade and spongy mesophyll. What are bundle sheaths? What is their function?

leaves exhibit modifications to the habitat in which they develop, survive, grow, and reproduce.

Leaves of Desert Plants

Leaf surface-to-volume ratio is a factor in water loss. The greater the surface area from which water can evaporate (transpiration), the greater the amount of water that is lost. Since the total leaf volume controls the amount of water available, a large surface area with a small volume can rapidly result in wilting. A large volume and a small surface area conserve water against loss. Desert plants (xerophytes) usually have small, often thicker leaves with a small surface-to-volume ratio. Plants that grow in wetter areas, **mesophytes,** often have much larger and thinner leaves, resulting in a larger surface area for the leaf volume. Some desert plants have large, thick leaves with heavy wax cuticles. These succulent leaves have greater water storage capacities and are well designed to allow only minimal water loss. The *Agave* (century plant) and *Yucca* have sharp teeth on the leaf margins (figures 9.11 and 9.12). Other desert species have marginal "armament"; some

Victoria—The Queen of Leaves

The Amazon water lily (*Victoria amazonica*) is an incredible plant and ranks as one of the great botanical curiosities. Discovered by European botanists in the early 1800s in the backwaters of the Amazon tropical rain forest river basin, it was named in honor of Victoria, Queen of England (1837–1901). This beautiful giant water lily has been celebrated by all who have observed it, either in its natural habitat or in botanical gardens.

The plant is characterized by huge, round peltate floating leaves of up to 7 feet in diameter with upturned margins (see figure 1*a*), and large, fragrant nocturnal flowers, frequently visited by beetles. The rate of leaf growth is phenomenal, expanding at the rate of 1 inch per hour and as much as 4–5 square feet in 24 hours. Over a growing period of 21–25 weeks, 600–700 square feet of leaf surface may be produced. The petioles and abaxial (lower) surface of the leaves are covered with stout prickles and have an intricate pattern of girderlike, air-filled ribs surrounding the veins that branch, rebranch, and crisscross with symmetrical regularity (see figure 1*b*). This framework of support ribs allows a single leaf of about 1 millimeter in thickness to sustain nearly 300 pounds of sand before collapsing. The mechanical strength afforded by the arrangement of ribs has had a major influence on architectural design. In 1849 Joseph Paxton realized that to cultivate *Victoria* properly a greenhouse was needed that would admit far more light than that permitted by the clumsy, small-paned, thick timber-supported greenhouses of the day. Using the *Victoria* ribs as a pattern, Paxton designed a greenhouse using slender beams of iron to support long panes of glass, ribbed at regular intervals. Paxton's *Victoria* house design proved so useful that it was eventually widely adopted and set the pattern for greenhouse design for over a century. The architecture of *Victoria* was incorporated into the roof design of the famous Crystal Palace, erected for the Great Exhibition of 1851 in Hyde Park, London (see figure 1*c*).

(a)

(b)

(c)

Figure 1
Gigantic leaf of *Victoria* showing upper surface with upturned margin (*a*) and lower surface (*b*) growing in Longwood Gardens, Kennett Square, Pennsylvania. (*c*) London's Crystal Palace showing architecture gleaned from the ribbed patterns of the *Victoria* water lily.

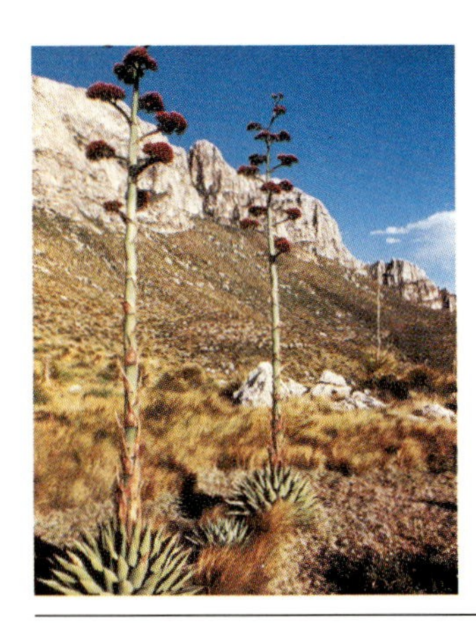

Figure 9.12
The water and food stored in *Agave* leaves (and thick roots) provide the materials for producing a large flowering stalk.

exhibit extreme indentation, and others display sharp, stiff teeth, which are neither thorns nor spines. These effectively discourage predators. Regulation of water loss (transpiration) in xerophytes is discussed in chapter 12.

Spines

As modified leaves, **spines** are technically different from thorns, which are modified stem tissue. The most familiar examples of plants bearing true spines are the members of the Cactaceae, whose stems are modified for water storage. The ocotillo (*Fouquieria splendens*) is another desert plant with true spines that are modified petiole and midrib of the first season of growth for that stem (figure 9.13). In subsequent years true leaves are produced in the axils of the spines after a rain.

Tendrils

All **tendrils** twine around solid objects, thus providing a means of support for the plant. Some plants have twining stem tendrils (grapes), whereas true leaf tendrils are found on sweet pea (*Lathyrus*) and other members of the pea family (Fabaceae) and on the trumpet flower (*Bignonia*). Leaf

Figure 9.13
The ocotillo has spines along its stems to help prevent herbivory.

tendrils in these plants are actually modified leaflets of compound leaves instead of a modified simple leaf, as in other plants (figure 9.14).

Bulbs

The modified leaves of a bulb, such as onion, store food in their thickened bases, allowing for growth until aerial leaves are developed adequately to begin photosynthesis.

Insectivorous Plants

Insectivorous plants display some of the most unusual adaptations of leaves to specialized functions. In general, the leaves are modified to entrap insects but occasionally trap small frogs and rodents. Enzymes secreted by the plant then digest the insects, providing necessary nutrients, especially nitrogen. **Insectivorous plants** grow in nitrogen-poor soils such as bogs and swamps. Inadequate soil oxygen in these habitats results in little decomposition of organic material, the normal source of nitrogen for plants.

The manner of entrapment varies among the species (figure 9.15). The Venus fly trap (*Dionaea muscipula*) has a folding leaf, the sundew (*Drosera*) a sticky surface, and pitcher plants (*Darlingtonia*) columnar tubes. Pitcher plants differ from one species to the next in their insect-trapping modifications. Some have only water at the bottom of the column in which the insect drowns; others have stiff hairs on the inside of the tube pointing down, which prevent the insect from crawling back out; and in others, sticky or gummy surfaces hold the insect.

The misleading term *carnivorous plant* implies a general meat-eating capability. This occasionally has produced an astounding science fiction vision of giant man-eating plants with tendrils or vinelike stems that can reach out, entangle their helpless victim, and draw them into the plant's dark digestive interior. Such imaginary activities are, of course, just that. Insectivorous plants are hardly to be feared (unless you are a fly), but are

Figure 9.14
The stem tendrils of grapes (*a*) and the leaf tendrils of peas (*b*) both serve the same function.

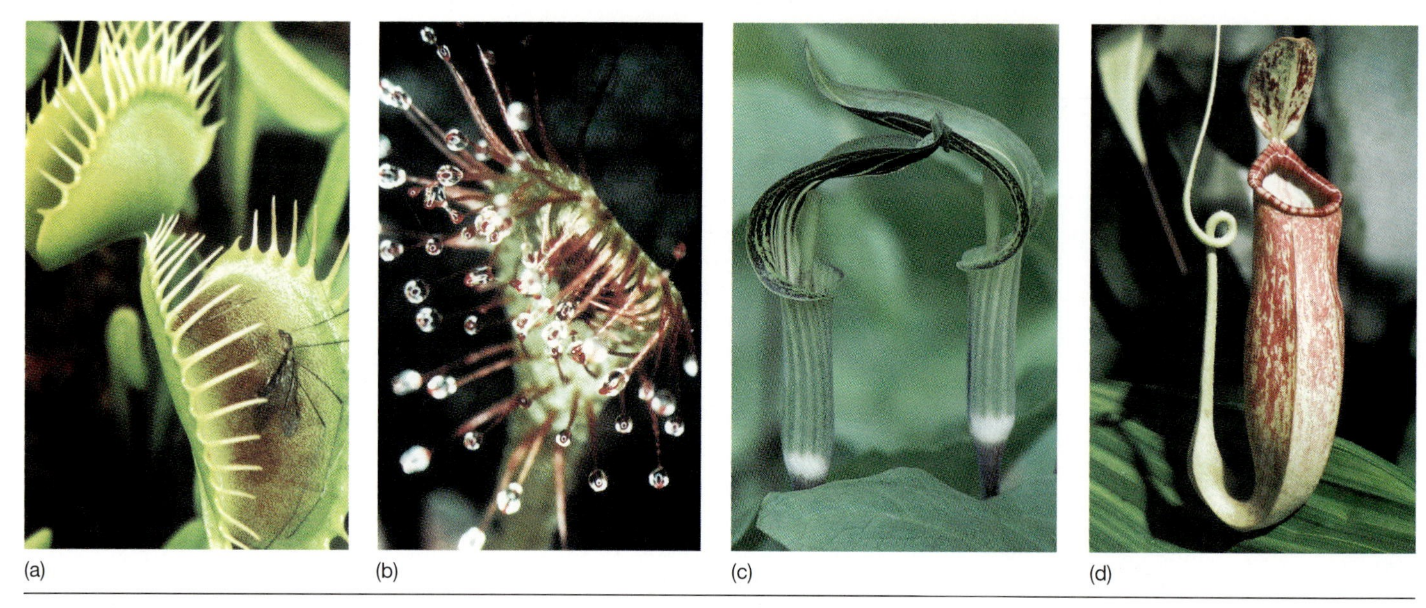

Figure 9.15

Insectivorous plants entrap their prey with various leaf modifications. (*a*) The Venus fly trap has a folding leaf. (*b*) The sundew has a sticky leaf surface. (*c*) and (*d*), Jack-in-the-pulpit (*Arisaema*) and *Nepenthes*, each have a tubular trap.

remarkable examples of modified plant parts that result in increased adaptability to an otherwise inhospitable environment.

VEGETATIVE PROPAGATION

Although most vegetative reproduction occurs by budding from the parent stem or from rhizomes or stolons, some plants are capable of vegetative (asexually) reproducing via modified leaves (figure 9.16). *Bryophyllum* leaves develop small plants at the notches along the dentate margins. Complete with leaves, short stems, and roots, these baby plants fall from the still healthy and active parent leaf, root, and begin growing. *Begonia* species also produce new plants from the upper leaf surface once it is in contact with the soil.

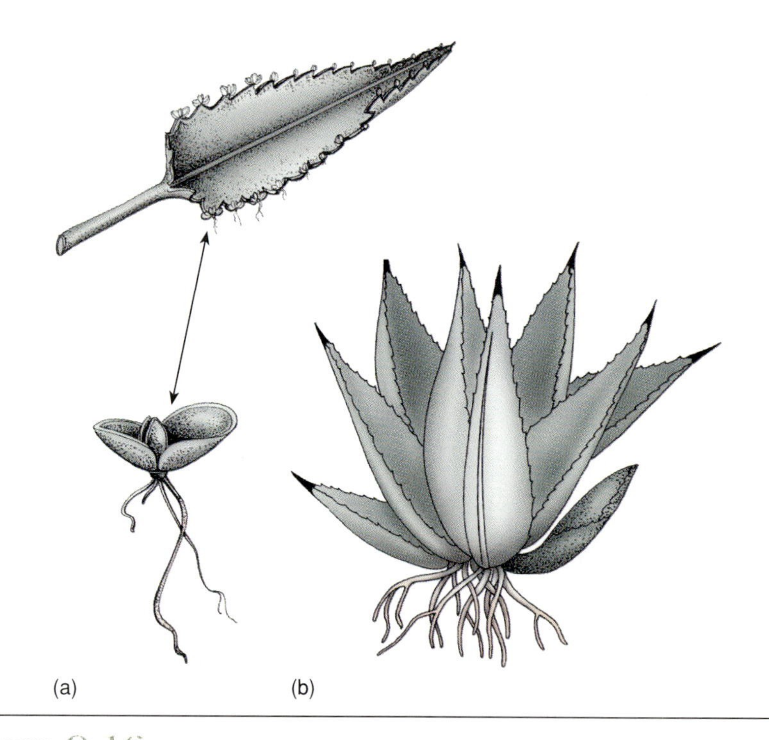

Figure 9.16

Asexual reproduction by budding. (*a*) The leaf margin of *Bryophyllum* produces buds. (*b*) Century plants (*Agave*) can produce buds.

COMMERCIAL PLANT PROPAGATION

The process of propagating plants asexually is important in ornamental plant commercial operations, fruit tree and grapevine production, and in other horticultural production.

Reproduction in plants is, for the most part, a completion of a sexual life cycle. The advantage of sexual reproduction is that it produces variation within the species (chapter 15). Should the environment change, at least some members of the population might be adapted to the new condition. These individuals would survive, allowing for the successful continuation of the species.

On the other hand, it is sometimes advantageous to have an asexual reproductive system that produces genetically identical individuals. Individuals produced from a single plant are referred to as a member of a clone. Although there might be short-term advantages, even in nature, to

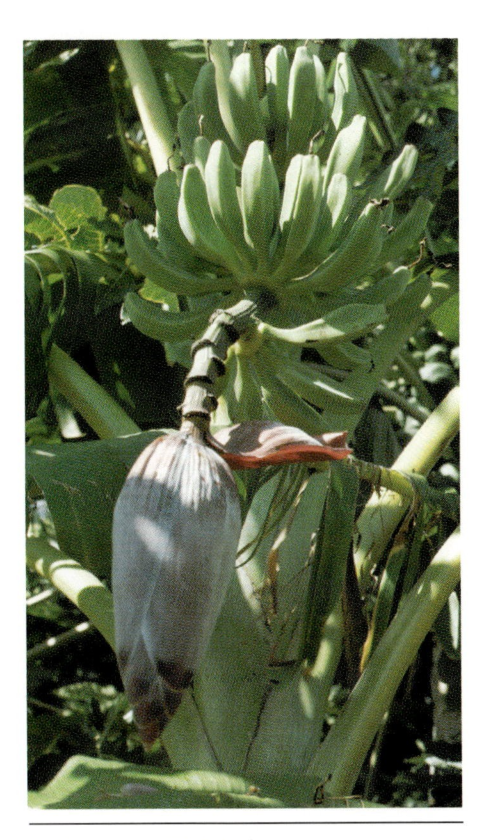

Figure 9.17
The commercial banana does not reproduce sexually.

producing individuals that are identical and perfectly suited to the environment, it is difficult to see how long-term survival could be served by a species with no means of sexual reproduction. The advantages of genetically identical and uniform plants in agriculture are obvious. Mechanical harvesting, the timing of market crops, and reducing perishability all depend on reliable and constant factors. Much the same results are achieved in agriculture with sexually reproducing plants by selecting and breeding specifically for that uniformity. Individuals with too much variability are simply eliminated.

Although many plants reproduce asexually in nature, few of them have totally lost the ability for sexual reproduction. One plant that reproduces exclusively asexually is the commercially grown banana (figure 9.17). As far as is known, there is no record of this plant having reproduced from seeds. Small, vestigial seeds are found in bananas, but they are not fertile. Thus all commercial banana propagation is accomplished by offshoots (buds) from the mother plant. As the central stalk flowers and produces bananas, that stalk dies, but offshoots at the base create new plants, which can simply be pulled off and planted. Commercial banana plantations generate their crops in this manner.

✔ *Concept Checks*

1. Explain why leaves with a large volume and small surface area are found in arid regions rather than in the wet tropics.
2. How do leaf spines and tendrils differ in function and structure compared to stem thorns and tendrils?
3. Cite three examples of insectivorous plants and explain for each how the leaf is adapted for capturing insects.
4. Explain what is meant by leaf budding. What is vegetative (asexual) reproduction? What are the short-term vs. long-term advantages to this form of reproduction?

Summary

1. The main functions of the leaf are photosynthesis and gas exchange; huge quantities of water can be lost during the latter activity.
2. Most leaves consist of two basic parts: the thin, flattened blade, which possesses a large surface area for the collection of solar (light) energy, and the petiole, which attaches the blade to the stem.
3. Leaf structure is incredibly variable, with blade shape ranging from needlelike to flat and broad. Leaves can be simple or variously compound, and their margins can be smooth to deeply lobed or divided. Leaf surfaces are smooth, pubescent, prickly, or many other textures.
4. Leaves are produced by the apical shoot meristem. The differential activity of the marginal meristem determines the shape and margin form. Leaves are produced one at a time at the apex (alternate attachment) or two at a time (opposite attachment) or several at the same time (whorled leaf attachment).
5. The vascular tissue within the leaf blade forms characteristic patterns (venation). Monocots have parallel venation. Dicots have netted venation.
6. The blade is composed of an upper and lower epidermis, enclosing a palisade parenchyma, spongy parenchyma, and veins of xylem and phloem. These veins unload water through terminal vessels and load sugars through terminal sieve tubes for transport to other parts of the plant. Some leaves have Kranz anatomy, a specialized leaf structure in which the palisade layer is not well developed, and a sheath of chlorophyllous cells surround the vein. This type of anatomy is associated with the C_4 dicarboxylic acid form of photosynthesis.
7. Leaf modifications include succulence for water storage, bulb leaves for storage, spines, and climbing tendrils. Insectivorous plants have leaves modified for insect entrapment and digestion to supplement nitrogen intake.

Key Terms

blade 138
bundle sheath 142
chlorenchyma 141
compound leaf 138
deciduous 140
insectivorous plants 145
Kranz leaf anatomy 142
leaf margin 138
leaf primordium 140
leaf sheath 139
marginal meristem 140
mesophyll 141
mesophyte 143
netted venation 140
palisade parenchyma 141
palmately compound leaf 138
parallel venation 140
petiole 138
pinnately compound leaf 138
pubescence 139
sessile 139
simple leaf 138
spines 144
spongy mesophyll 141
succulent 139
tendrils 144
xerophyte 142

Discussion Questions

1. Prepare a diagram of a dicotyledonous leaf blade cut perpendicular to the surface. Label all regions, tissues, and cells.
2. Assume you were given a leaf and asked to determine which was the upper surface. Explain how you might reach a determination comparing only the epidermises. Only mesophyll. Only the veins.
3. Discuss how the activity of the marginal meristem results in the following leaf (blade) forms:
   a. simple leaf with smooth margins
   b. simple leaf with lobed margins
   c. pinnately compound leaf, each blade with a smooth margin
4. Leaf structure is closely associated with the habitat of the plant. In what type of environment (biome) would you expect to find the following:
   a. an extremely thick cuticle
   b. an extremely thin cuticle
   c. several layers of palisade parenchyma
   d. little or no palisade parenchyma, mostly spongy parenchyma

Suggested Readings

Andrews, J. H., and S. S. Hirano, eds. 1991. *Microbial ecology of leaves.* Springer-Verlag, New York.

Cheers, G. 1992. *A guide to carnivorous plants of the world.* Harper-Collins Publishers, New York.

Fahn, A., and D. F. Cutler. 1992. *Xerophytes.* Gebr. Borntraeger, Stuttgart 1, Federal Republic of Germany.

Klucking, E. P. 1986. *Leaf venation patterns.* J. Cramer, Berlin.

Prance, G. T. 1985. *Leaves: The formation, characteristics and uses of hundreds of leaves found in all parts of the world.* Crown Publishers, Inc., New York.

Raven, P. H., R. F. Evert, and S. E. Eichhorn. 1992. *Biology of plants.* 5th ed. Worth Publishers, New York.

Roth, I. 1990. Leaf structure of a Venezuelan cloud forest. In *Encyclopedia of plant anatomy.* Vol. 14. Pages 234–239. Borntraeger, Berlin.

Widnels, C. E., and S. E. Lindow, eds. 1985. *Biological control on the phylloplane.* American Phytopathological Society, St. Paul.

von Willert, D. J. 1992. *Life strategies of succulents in deserts with special reference to the Namib Desert.* Cambridge University Press, New York.

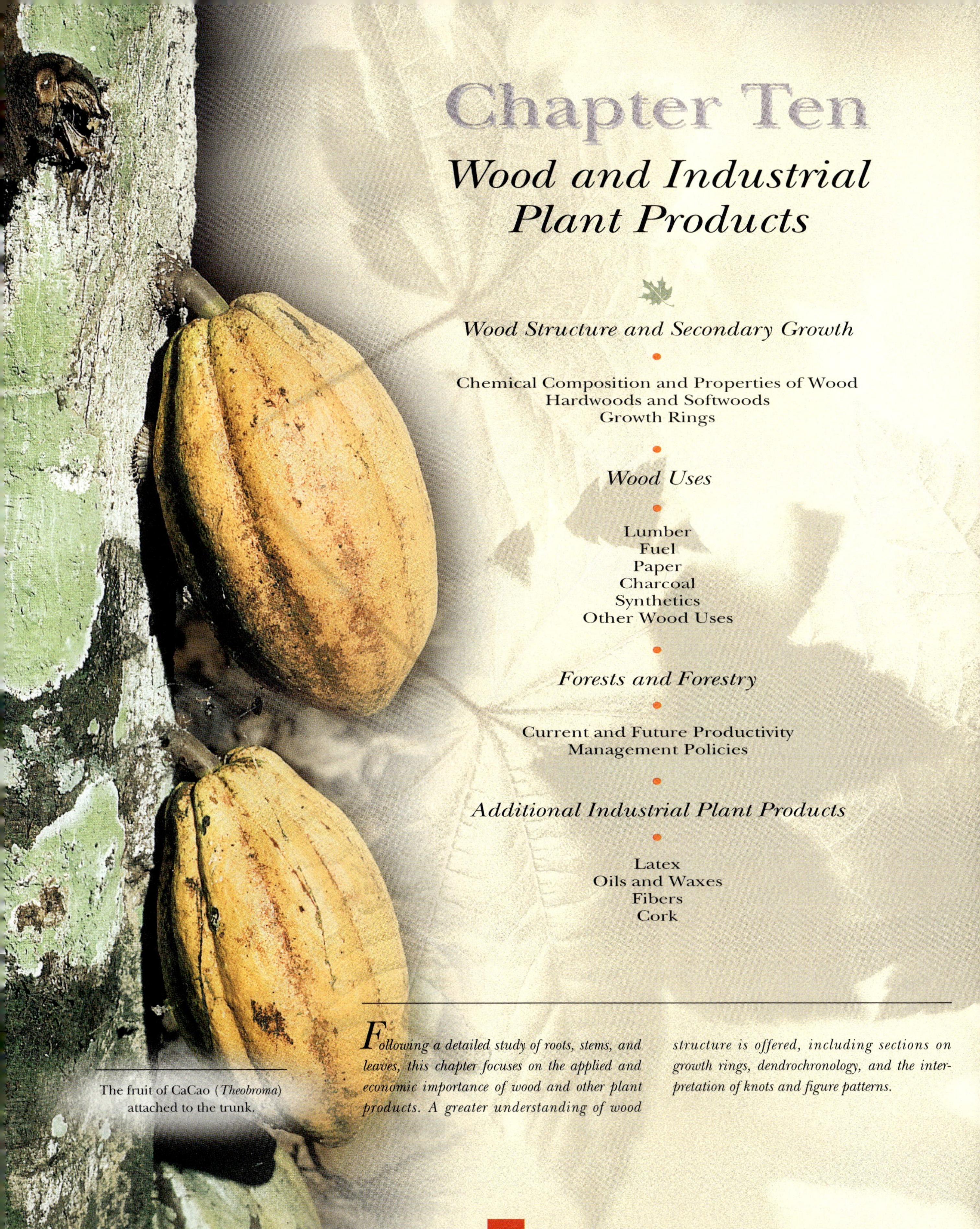

Chapter Ten

Wood and Industrial Plant Products

The fruit of CaCao (*Theobroma*) attached to the trunk.

Following a detailed study of roots, stems, and leaves, this chapter focuses on the applied and economic importance of wood and other plant products. A greater understanding of wood structure is offered, including sections on growth rings, dendrochronology, and the interpretation of knots and figure patterns.

In overall botanical importance to human existence, only food plants rank above wood and wood products. In early human history wood had even greater importance than the food plants, as a fuel and for weapons and tools. Today there are over 4000 products that come wholly or in part from the secondary xylem (wood) of trees. Wood is used for housing, furniture, fuel, paper, charcoal, distillation by-products, and synthetic materials such as rayon, cellophane, and acetate plastics.

Trees come in all shapes, sizes, and ages. The world's tallest tree is reputed to be a giant redwood (*Sequoia sempervirens*) (figure 10.1) from Humboldt County, California, measured in 1970 at 111.6 m. Although giant redwoods are generally considered to be the tallest tree species, there are claims of *Eucalyptus reganus* in Australia measuring over 140 m. Certainly there are a number of eucalyptus trees validated to be taller than 91 m, so they are easily the second tallest tree species in the world.

A tule tree (*Taxodium mucronatum*) found outside of Oaxaca City, Mexico, is believed to have the world's largest trunk circumference (42 m) (figure 10.2). Only 40 m tall, this tree is over 2000 years old and was visited by Cortez after he heard stories of its size. The tule tree is a youngster compared with the bristlecone pine trees (*Pinus longaeva*) found in the White Mountains of California, and in Nevada, Utah, and Colorado (figure 10.3). Several of these remarkable trees are known to be over 3500 years old; one of the oldest living ones, named Methuselah, is approximately 4900 years old.

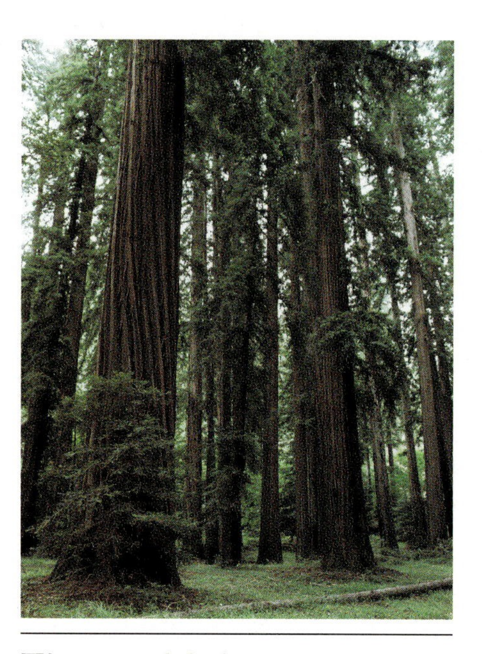

Figure 10.1
The coastal redwoods (*Sequoia sempervirens*) are considered to be among the world's tallest trees, reaching 345 feet.

Figure 10.2
This tule tree (*Taxodium mucronatum*) is believed to have the world's largest trunk circumference (42 m). Cortez is reported to have traveled to see this tree, located just outside of Oaxaca City, Mexico.

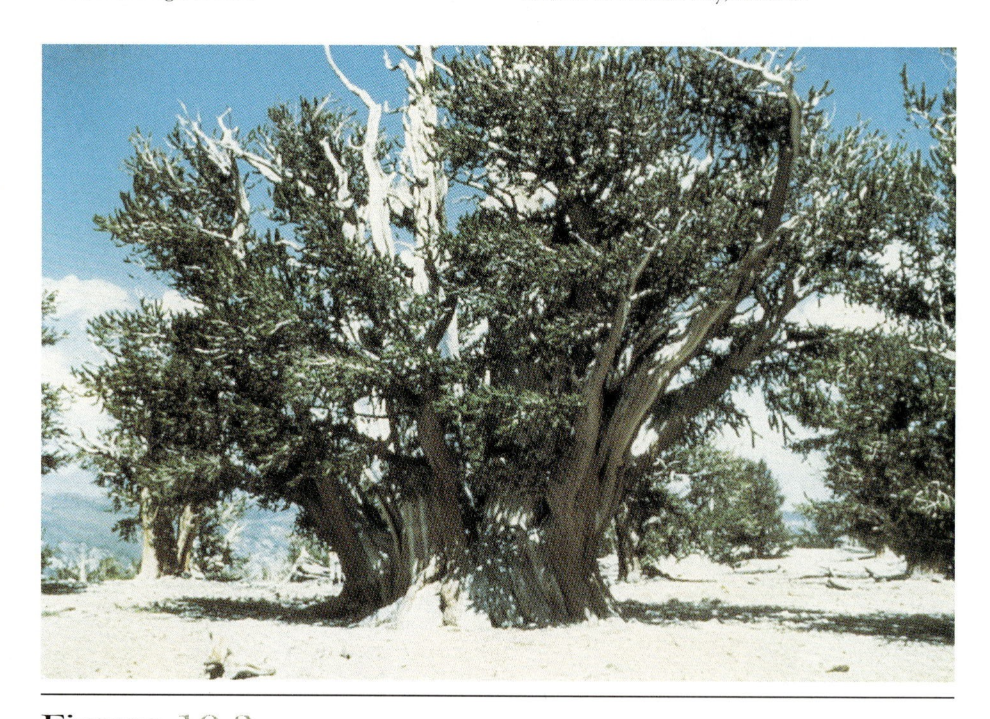

Figure 10.3
Bristlecone pines (*Pinus longaeva*) are the oldest living organisms on earth; several are more than 4500 years old.

WOOD STRUCTURE AND SECONDARY GROWTH

Recall that wood is composed of the secondary xylem formed by the vascular cambium (chapter 8, p. 125). Generally, the vessels and tracheids function in water and mineral transport, and fibers are support cells. Additionally, xylem rays provide for lateral transport throughout wood tissue.

Chemical Composition and Properties of Wood

Many of the structural properties of wood depend on the arrangement of the component cells and on the chemical composition of their secondary cell walls. The three major cell-wall constituents are **cellulose** and **hemicellulose** (both polysaccharides) and **lignin,** a complex polymer, which cements together the cellulose and hemicellulose increasing the mechanical strength of the secondary cell wall. Cellulose is the most abundant naturally occurring organic compound. It is a long compound of several thousand glucose molecules linked end to end, which provides for the physical organization of the cell wall. Hemicelluloses are the other polysaccharides found in cell walls.

Lignin is the second most abundant component in cells. Lignin is not found in all plant cell walls, but when present it is especially important and abundant in cells having a supporting function. The deposition of lignin, a process termed **lignification,** normally occurs after the cell has reached its maximum size.

The relative amounts of each of these materials control some of the physical properties of wood. The greater lignin content in gymnosperm wood makes it more stable and less prone to warping. The commercial value of wood depends on a combination of characteristics that make certain woods better suited to different uses. The specific gravity, figure, grain, cuts, and knots of wood are some of these properties (figure 10.4). The following properties are also considerations for the best use of woods.

Durability

The resistance of wood to decay, wear, and insect damage is especially desirable in wood that is to be used structurally or that comes in contact with moisture. Since fungal decay is the most common form of wood destruction, and since fungi thrive in warm, moist conditions, wood in contact with damp soil or subject to frequent rain and high humidity is more likely to decay. Fence posts, railroad ties, telephone poles, greenhouse tables, mine timbers, and coastal or tropical structures require wood of exceptional durability. The natural preservatives found in many trees are often toxic or unpalatable to decay organisms and insect pests. This is especially true of

Figure 10.4
The many different wood products require adequate supplies of various species of trees.

tannin, which is found in amounts up to 30% in some woods. The most durable and resistant woods include redwood, cedars, black walnut, junipers, chestnut, bald cypress, black locust, and catalpa.

Color, Luster, and Polish

In wood used for furniture or cabinetry, the color of the wood is important. In addition, the natural luster, or ability to reflect light, and the ability to take a polish are functions of cell-wall structure and the types of cut. Some woods known to polish well are cedar, white pine, cherry, maple, walnut, holly, and some oaks.

Moisture and Shrinkage

The amount of moisture in the wood of a freshly cut tree varies from less than 10% of green weight (i.e., the weight prior to air or kiln drying) in some species to over 75% in others. Some moisture is found in the **lumens,** or hollow centers of the vessel elements, and evaporates readily without causing any shrinkage in the wood. However, the cell walls of green wood comprise approximately 25% to 30% water, which is removed with more difficulty. As this water is removed from the cell walls, the wood shrinks. Most woods shrink between 10% and 20% in volume if all the water is removed by oven drying.

To prevent uneven shrinkage and resulting warping, most commercial lumber is "seasoned" or "cured" by drying the wood under controlled conditions. This is especially important in hardwoods because of their use in furniture. In addition to preventing warping, proper drying reduces shipping weight and costs, increases strength, and improves the wood's ability to be glued, painted, stained, and polished.

Acoustical Properties

For wood used in musical instruments the resonance depends on a combination of elasticity, density, thickness, and cut. The soundboard in a piano, responsible for resonance and tonal quality, is best when made of spruce; laminated hard maple holds the metal tuning pegs tightly. The various woodwind instruments, such as clarinet, oboe, and bassoon, and the string instruments, such as violin, guitar, mandolin, and bass, all depend on the acoustical resonance properties of the woods used in their construction. A master instrument craftsman must know woods as well as music. In addition, the reeds of the clarinet, saxophone, and oboe are tapered strips of the woody cane from a large tropical grass, *Arundo donax.* Reeds from this plant have been used for woodwind instruments since at least 3000 BC.

Hardwoods and Softwoods

One of the most common distinctions made when discussing wood is the categorization of all woody dicots as **hardwoods** and all coniferous gymnosperms as **softwoods.** Although there is some justification for such an oversimplification, there are a number of exceptions. One of the softest (lightest) of woods is balsa (*Ochroma lagopus*), a dicot, whereas slash pine (*Pinus elliottii*), a conifer, is harder than many hardwoods. What is actually being measured is relative density, the mass of wood per unit of volume. Generally, the less dense or lighter a wood, the softer and weaker it is.

Relative density is determined by measuring the specific gravity of the wood, which depends on cell size, cell-wall thickness, and the number of different kinds of cells. For example, fibers can be thick walled and small lumened and can be packed closely together, providing for a very dense (high specific gravity) wood with little air space. Conversely, fibers with thin walls and large lumens produce a wood with a lower specific gravity. Vessel elements have relatively thin walls and large lumens, so a high vessel volume results in decreased specific gravity.

Specific gravity is determined by weighing a paraffin-coated block of wood (to prevent water absorption), immersing it in water, and weighing the displaced volume of water (1 cubic centimeter of water = 1 gram).

$$\text{Specific gravity} = \frac{\text{Dry weight of wood}}{\text{Weight of displaced volume of water}}$$

Table 10.1 includes the specific gravities of wood for a number of tree species.

Generally, a specific gravity of 0.41 or less is considered softwood, and above 0.41 hardwood. Several woods are above a specific gravity of 1.0, which makes them more dense than water, so they sink.

Growth Rings

Although in a cross section of a log the phloem does not normally appear as discernible rings (because of the compacting of the soft cells), the hard-walled xylem cells form a new **growth ring** of wood each time the vascular cambium becomes active. In temperate regions such activity usually occurs only once during the growing season each calendar year, and thus these are termed annual rings. Annual rings are also formed in tropical trees where a seasonal dry period occurs. The rings are a cross-sectional view of the annual growth layer that forms the full length of the tree (figure 10.5).

Since water availability is one of the most important environmental factors controlling plant growth, a drought may cause early cessation of growth followed by a second burst of growth after subsequent rainfall. Infrequently, then, false annual rings may form, resulting in two or more apparent growth rings in one year. Such sporadic growth patterns are generally restricted to areas with unpredictable climatic patterns, such as arid and semiarid regions.

Woody growth appears as visible concentric rings because of the contrast in cell diameter and cell-wall thickness from the early to the late part of the growing season. In a typical year in temperate regions growth begins in the early spring, while plenty of soil moisture is available from winter rains and snow. The cells produced are large and thin walled, making them less dense than the xylem produced in the summer. The less-dense **earlywood** therefore appears

TABLE 10.1

Specific Gravity of Some North American Trees

NAME	SPECIFIC GRAVITY	NAME	SPECIFIC GRAVITY
Softwoods (coniferous gymnosperms)			
Arborvitae, American (*Thuja occidentalis*)	0.29	Pine, lodgepole (*Pinus contorta*)	0.38
Cedar, eastern red (*Juniperus virginiana*)	0.44	Pine, northern white (*Pinus strobus*)	0.34
Cypress, bald (*Taxodium distichum*)	0.42	Pine, slash (*Pinus elliottii*)	0.64
Fir, Douglas (*Pseudotsuga menziesii*)	0.45	Pine, western yellow (*Pinus ponderosa*)	0.38
Hemlock, western (*Tsuga heterophylla*)	0.38	Redwood (*Sequoia sempervirens*)	0.41
Larch, western (*Larix occidentalis*)	0.48	Spruce, Englemann (*Picea engelmannii*)	0.31
Hardwoods (angiosperm dicots)			
Ash, white (*Fraxinus americana*)	0.55	Hickory, bitternut (*Carya cordiformis*)	0.64
Aspen (*Populus tremuloides*)	0.35	Ironwood, black (*Krugiodendron ferreum*)	1.30
Balsa (*Ochroma lagopus*)	0.12	Lignum vitae (*Guaiacum officinale*)	1.25
Beech (*Fagus grandifolia*)	0.56	Locust, black (*Robinia pseudoacacia*)	0.66
Catalpa (*Catalpa speciosa*)	0.38	Maple, sugar (*Acer saccharum*)	0.57
Cherry, black (*Prunus serotina*)	0.47	Oak, white (*Quercus alba*)	0.60
Chestnut (*Castanea dentata*)	0.40	Osage orange (*Maclura pomifera*)	0.76
Corkwood (*Leitneria floridana*)	0.21	Persimmon (*Diospyros virginiana*)	0.64
Cottonwood (*Populus deltoides*)	0.37	Sycamore (*Platanus occidentalis*)	0.46
Elm, American (*Ulmus americana*)	0.46	Walnut, black (*Juglans nigra*)	0.51
Gum, sweet (*Liquidambar styraciflua*)	0.44	Willow, black (*Salix nigra*)	0.34
Hackberry (*Celtis occidentalis*)	0.49		

Figure 10.5
This cross section of an Acacia tree has generally uniform growth rings, indicating consistently favorable growing conditions.

lighter than the smaller, thick-walled **latewood** (figure 10.6). Whereas the morphology of cells formed from the vascular cambium changes gradually with the growing season and does not present a sharp contrast from earlywood to latewood, the interface between the latewood of one year's ring and the earlywood of the next year's ring is clearly delineated. The annual rings are therefore distinctly visible.

The width of the growth rings is affected by a complex of environmental factors besides total water availability. Temperature, length of growing season, time of precipitation, disease, competition from surrounding trees, and soil fertility are among the variables that work together to control the patterns of lateral growth from year to year. A wide ring generally reflects a long, wet, and moderate growing season, whereas narrow rings usually reflect some kind of environmental stress. An extremely dry, cold winter followed by a hot, dry summer could even prevent any lateral growth for that year.

Wood Anatomy

Some woody plants produce larger vessel elements in the early spring growth than later in the season, and this pattern is termed **ring-porous** wood. Trees with ring-porous wood are easily recognized in cross section because the growth rings are quite distinct. Other trees produce the same size vessel elements throughout the growing season, and their wood is termed **diffuse-porous.** In other woods such as gymnosperms, no vessels are produced. These woods are termed nonporous (figure 10.7).

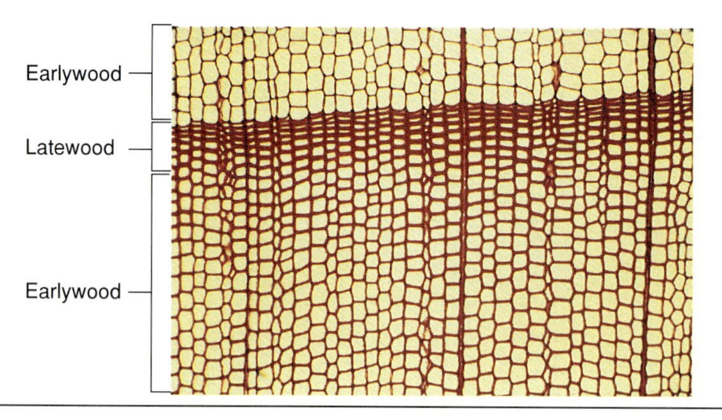

Figure 10.6
At the beginning of the season, cells laid down in secondary xylem are very large and thin walled, but they become smaller and thicker walled as the season progresses. Growth finally stops in winter. Large cells are again produced the following spring. This pattern of cell size gives rise to the annual rings (growth rings) that are used in determining the age of the tree.

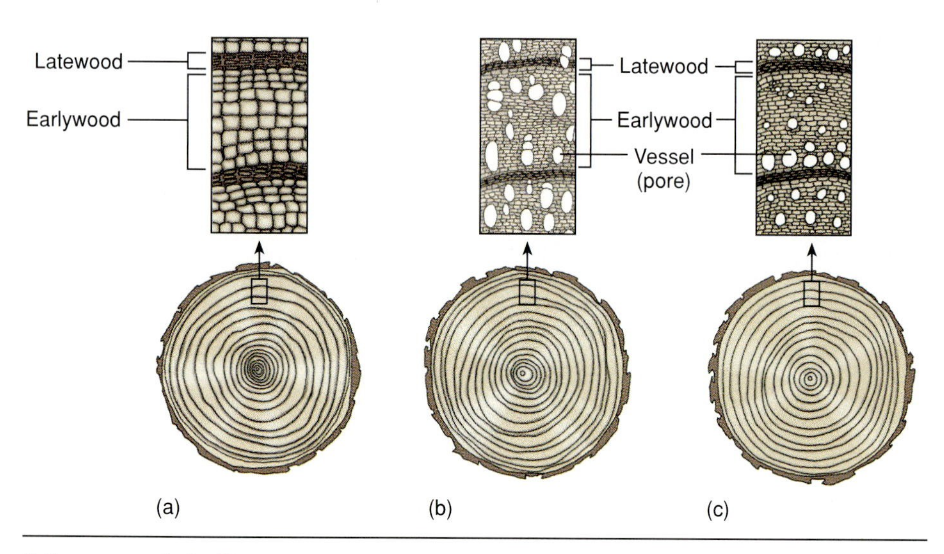

Figure 10.7
(*a*) Growth rings are visible because the secondary xylem produced at the end of a growing season is composed of smaller cells. (*b*) In some species the size of the vessels, known as pores, is barely distinguishable at the beginning and end of the growing season. The wood is said to be diffuse-porous. (*c*) In other species the size of the vessels is larger in the earlywood. The wood is said to be ring-porous. Wood that lacks vessels, such as the one illustrated in (*a*), is said to be nonporous.

There are several structural differences between coniferous wood (softwood) and dicot wood (hardwood). Conifers are more homogeneous and less complex than dicots. They contain no vessels, limited fibers, and approximately 90% of the woody tissue is composed of tracheids with little parenchyma. Some of the parenchyma present is associated with resin canals, long intercellular spaces present in the longitudinal system of cells and in some of the rays. Resin canals are found in pine, spruce, larch, and Douglas fir. Parenchyma cells surround the resin canal and produce the resin, which is secreted into the canal. Most parenchyma cells in coniferous woods are in the xylem rays, which are normally 1 cell wide and 1 to 20 cells high.

Dicot wood has not only more and larger xylem rays, but also a greater variety of cell types in the longitudinal (axial) system. In general the presence of vessel elements, which are the primary water-conducting cells, is the main difference. Dicot woods also contain fibers and parenchyma cells, and different species contain varying amounts and kinds of fibers. Thus the kinds of cells, the relative amounts of each kind, and, most important, the presence of vessels distinguish the more complex structure of dicot wood from conifer wood.

Sapwood and Heartwood

With increased age the center of the trunk ceases to function in the transport of water and minerals because it accumulates resins, tannins, oils, gums, and other metabolic by-products. Since trees do not have the ability to remove these compounds, they remain in the plant body. The centralization of all these materials allows the outer wood to remain unclogged and functional. These central rings of wood are often darker in color, more resistant to decay, and sometimes aromatic. This central **heartwood** is usually visibly distinguishable from the lighter **sapwood** in fresh-cut or cured and processed wood (figures 10.8 and 10.9). The natural preservative properties of some heartwoods, such as cedar, cypress, and redwood, lend exceptional resistance to decay. The color of other heartwoods, such as black walnut and mahogany, makes them especially valuable for fine furniture. The ratio of heartwood to sapwood varies from one species to the next, as does the degree of visible difference between the two woods. Some characteristics of heartwood increase its commercial applications.

✔ *Concept Checks*

1. Identify the three major cell-wall chemical constituents. What is the role of lignification?
2. List at least three properties of wood that increase its commercial value.
3. Briefly discuss the difference between hardwood and softwood. Between ring-porous and diffuse-porous wood.

Figure 10.8

This cross section of a 28-year-old redwood (*Sequoia sempervirens*) trunk clearly shows the darker heartwood (inner 19 or 20 rings) and lighter sapwood.

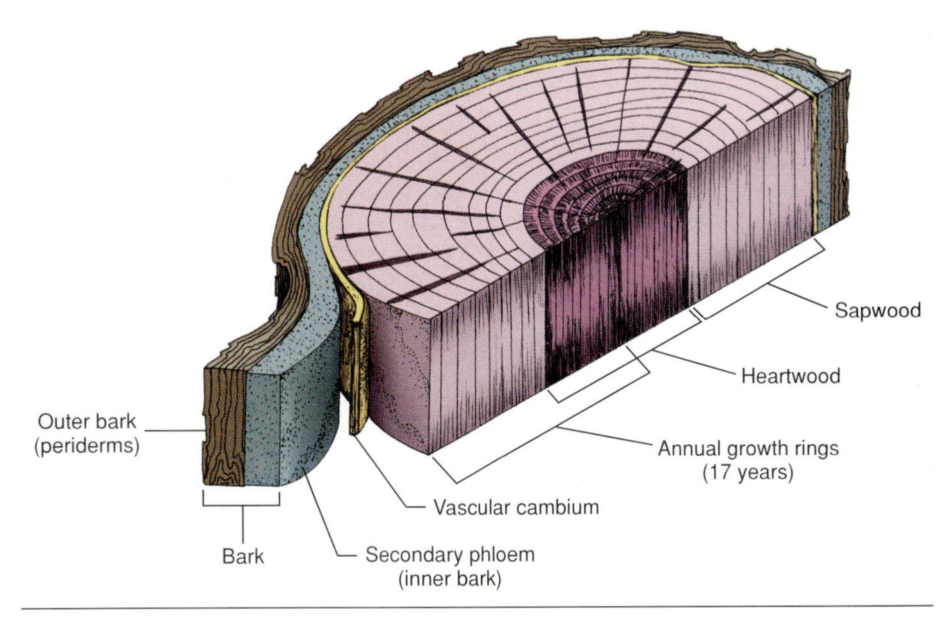

Figure 10.9

From the center of the log section outward, heartwood, sapwood, vascular cambium, phloem, and outer bark can be seen.

WOOD USES

There are many uses of wood, some obvious and some less well known. The uses of wood and wood products are expanding because trees are a renewable resource. It is important to note that while individual trees or stands of trees may be thought of as renewable, this concept does not apply to forest ecosystems. With proper management and use the forests of the world will be able to provide this valuable raw material in more than sufficient quantities as long as it is needed.

Lumber

One of the most obvious uses for trees is the production of lumber for building and furniture (figure 10.10). Many millions of board feet of certain softwoods are used each year for home construction because of their physical properties (a board foot is 1 foot long, 1 foot wide, and 1 inch thick or the equivalent volume). Trees such as white pine have a soft, uniform texture and an even grain that can be machined easily, does not shrink, swell, or warp significantly, is strong, and holds nails well. Not all gymnosperms have equally desirable qualities, although several other species are used for construction because of the ease with which nails can be driven into them. Framing a house out of oak, walnut, maple, or hickory, on the other hand, would be quite a task because their wood is so hard that boards would have to be drilled and screwed together instead of nailed.

Because of the grains, colors, and durability of hardwoods, they are most often used in furniture making. Some softwoods, especially certain pines, are also used for furniture. The uses of different woods depend on several properties, which are partially a function of the type of cut and the part of the tree used.

Cuts and Grains

In addition to the kinds of cells and the width of annual rings in a given wood, the appearance of a finished wood surface is a function of how the log has been cut. In a **transverse cut** or **cross section** (figure 10.11) the annual rings appear as concentric circles. This cut is not commonly used to produce a commercial piece of lumber, although transverse slices that include the bark can make a unique and beautiful tabletop.

Figure 10.10

One of the most important commercial uses for trees is lumber.

The Secrets of Secondary Xylem

Aside from their commercial utilization, trees are living archives, carrying within the structure of their wood a record of their age, past precipitation, and temperature patterns as well as evolutionary and systematic clues. **Dendrochronology** is the science of interpreting past conditions or historical events by studying the growth rings of trees. Since climatic conditions, especially water availability, influence the width of growth rings, and since there is normally only one ring produced each year, patterns of annual ring production directly reflect past climates. By counting the number of annual rings a given tree has, one can determine the age of that tree. The oldest trees are the bristlecone pines (*Pinus longaeva*) found in the White Mountains of California at above 3000 m elevation (see figure 10.3). Because of the short growing season and small amount of rainfall at higher elevations, bristlecone pines grow very slowly and have narrow rings, as many as 1000 or more in less than 13 cm of lateral growth. Because of this slow growth, bristlecone pines are much smaller in both trunk diameter and height than the massive giant redwoods (*Sequoiadendron gigantea*) (figure 1).

By matching ring patterns from the wood of a living tree (figure 2) with the wood of an older tree containing a partially overlapping growth-ring sequence and continuing this with progressively older woods from trees long dead, a continuous chronology of climatic patterns can be established for a particular region. Such a chronology has been established for the White Mountains, extending back over 8000 years. Because bristlecone pine wood, dating back about 9000 years, has been found, it is probable that this chronology can be extended even further back. Interestingly, by using radiocarbon dating techniques on wood

Figure 1

Sequoiadendron gigantea (the Big Tree) in Yosemite National Park. Under United Nations auspices a World Heritage Convention was adopted in 1972, its purpose to help preserve great landmarks of the earth, both natural and man-made. *Sequoiadendron* groves have been proposed as designated World Heritage sites since they represent botanical treasures that transcend national boundaries.

Figure 2

Dendrochronologists use a borer to remove a thin core of wood. This allows dating and ring study without sacrificing the tree.

Figure 10.11

A log has three basic planes of sectioning. In a cross or transverse section the growth rings appear as concentric circles. In a radial section, cut longitudinally along the radius, the growth rings appear as parallel lines whereas in a tangential section, a longitudinal cut perpendicular to the radius, the growth rings appear as wavy bands.

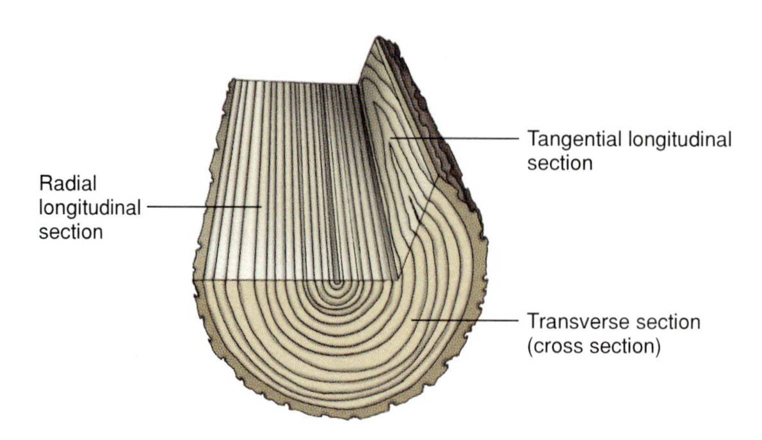

that has a known age through ring sequencing, it has been determined that radiocarbon dating is increasingly inaccurate for material over 1000 years old. Carbon 14 dates of 1600 and 3300 years were found for bristlecone pine specimens having 2000 and 4000 annual rings, respectively. With continued cross-dating, a more accurate age estimate for archaeological studies can be made, thus improving our understanding of our cultural beginnings. Wooden utensils, tools, ornaments, and building timbers from archaeological digs can help provide information about the climate as well as the age in which a particular culture existed.

The study of tree rings also can tell a much more complete story about an individual tree's history than just its age and the general climatic conditions during its lifetime. Figure 3 shows what kinds of events can be read from the annual rings of a single tree, especially when ring patterns from other trees in that region tell a slightly different story for the area in general. Damage due to fire, landslide, insects, and other natural occurrences, such as earthquakes and volcanic eruptions, can be seen in the patterns of annual ring formation.

Figure 3

The study of tree rings can reveal much more than the age of a tree. Past climates, injury, and historical information can be interpreted from these rings.

A **radial cut** (figure 10.12*a*) is made longitudinally through the center of the log. In a radial section of wood the annual rings appear as parallel lines running the length of the flat board surface with the rays running at right angles across them. Since this cut is made through the center of the log, only a few boards can be cut from each log. Radial cuts are also said to be quartersawed.

The most common board surface is a log cut longitudinally but not through the center. This **tangential cut** (figure 10.12*b*) results in the annual rings appearing as wavy bands with the ends of the rays scattered throughout. Tangentially cut lumber is also said to be flat-cut, slab-cut, or plainsawed, and it is the most common cut because so many more boards can be sawed from a log than with the radial cut. The design resulting from these different cuts is the figure of the wood. The grain of a wood technically refers to the direction of the fibers, tracheids, and vessel members.

Knots

Another textural and structural consideration in woods is the presence of **knots,** which are the bases or shafts of dead branches that subsequent lateral

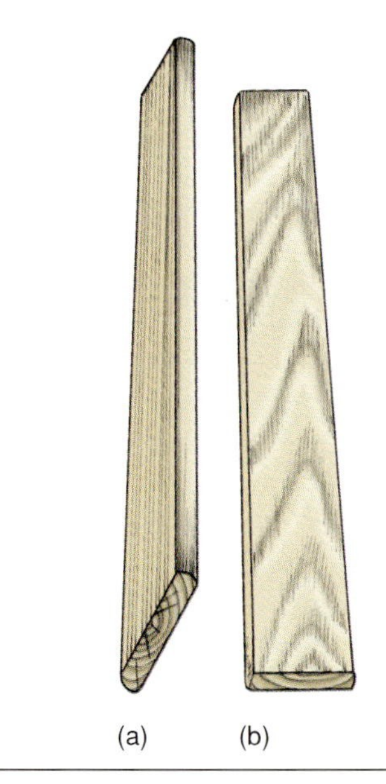

Figure 10.12
The radial (*a*) and tangential (*b*) cuts produce different figures on the board surfaces. The radial cut produces parallel lines down the length of the board; the tangential cut produces the wavy bands.

Figure 10.13
Longitudinal cuts through these ponderosa pine logs from Idaho show how lateral branches appear as knots when the trunk grows over them.

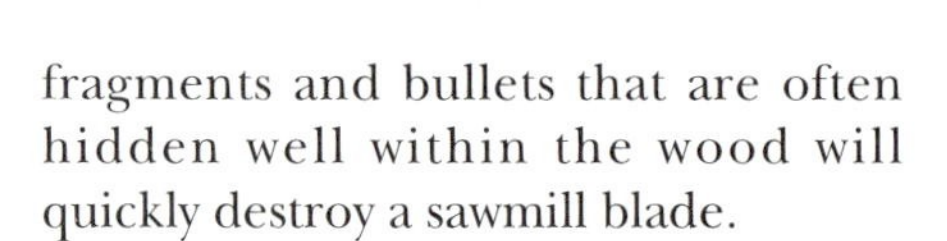

Figure 10.14
A sheet of plywood produced from the oldest (innermost) wood of Douglas fir. A sheet cut from the pruned outer (youngest) portion of the same trunk would not have knots, making it a higher grade plywood.

growth has covered over (figures 10.13 and 10.14). There is a higher proportion of knots in the center of the trunk because the tree was younger and smaller with more lower branches then. As the tree grows and ages, lower branches cease to be formed, and the old ones die and eventually break off. Just as the lateral growth will eventually cover the base of old branches, so will it cover scars from fire and other injury. You may have noticed that a nail or eyescrew put in a tree trunk to hold a clothesline or bird feeder many years ago is now only partially, if at all, showing. It is the same lateral growth that ultimately turns two closely adjacent young trees into a "single" double-trunked tree. Occasionally two different species growing in proximity result in the faster growing one gradually enveloping the lower trunk of the other (figure 10.15). Tree trunks have even been found with rocks in them. In many of the hardwood forests of France no tree old enough to have been present during World War II is harvested, because the metal bomb fragments and bullets that are often hidden well within the wood will quickly destroy a sawmill blade.

Fuel

More than one third of the world's total population depends on wood for heating and cooking. The significance of wood as a fuel is greatest in the developing countries, where more than 86% of all wood consumption is for fuel.

Approximately 1.5 billion people derive at least 90% of their energy requirements for cooking and heating from wood and charcoal, and another billion people depend on wood for at least 50% of their needs. These are disquieting statistics, especially since these are also the countries with the fastest growing populations.

Because of the dependence on firewood in these countries, it is estimated that 50% of all wood used worldwide each year goes to fuel. In developed countries with stabilized population sizes and appropriate management and use practices, wood is a renewable resource that should never be in short supply. Theoretically the developing countries, which depend much more heavily on wood, could produce enough through replanting and appropriate management to meet their needs. Practically, however, this resource has already been depleted to a critical level. Many people are now

Figure 10.15
Trees can engulf rocks, nails, and even trunks of other trees growing beside them. This Oregon oak tree has "cannibalized" a smaller Douglas fir tree.

depending on dried animal dung and crop residues for fuel. This removes much-needed nutrients from the soil, which will in turn be less productive. The vicious cycle continues as new land is cleared for more crop production to feed the growing populations in these countries. Clearing without replanting removes even more of the wood that could have been used in the future for fuel.

In the United States, although only slightly over 10% of our total wood use is for fuel, over 1 million homes now use wood for their primary heat source. Since 1933 the wood fuel-use growth rate has been about 15% annually. More recent trends are even greater because of the renewed interest in wood-burning stoves.

Even at this rate of increase, wood use for fuel is critical only in the developing countries, where the "energy crisis" is currently much more serious than ours, even with our petroleum shortages. If the developed countries fail to solve the energy crisis, however, we could well join the rest of the world in our dependence on wood as a primary fuel.

Figure 10.16
Elongated fibers from wood are very strong. These stained and macerated fibers are typical of those found in paper.

Paper

Paper production accounts for over half of the wood that is pulped through mechanical or chemical means. The balance of the wood pulp produced each year goes into cardboard and fiberboard. Paper is basically composed of separate plant fiber cells (figure 10.16) that have been slurried and then spread into thick sheets and dried. True paper was first produced around 105 AD in China. Before the Chinese developed a way of making paper, they wrote on strips of wood with a stylus and later on woven cloth, especially silk. Also in pre-Christian times, Egyptians made a type of paper by beating laminated stems of papyrus until they were very thin. The only other nonpaper material used prior to the second century was parchment, which is made from animal skins. The Chinese guarded the secret of paper making for over 500 years, establishing themselves as the only supplier of paper. Paper making was a slow hand-labor process until the first machines were invented around 1800.

In 1865 the sulfite process for removing unwanted lignins from the pulp was developed. Lignins cause paper to turn yellow and become brittle with age. For economy, newsprint production does not include this lignin-removing process, and so newsprint eventually becomes yellow and brittle. To add weight and body to paper and to produce a smooth ink-impervious or "hard" surface, given fillers are mixed with the pulp while it is in the liquid slurry stage. Clay, alum, and talc are common fillers to add weight and stiffness to paper; certain resins, starch, and glues are used to produce the smooth, hard surfaces for writing purposes. To make colored paper, dyes are added at this pulp-slurry stage. Over 90% of all paper comes from wood; the rest is made from the fibers of other plants such as flax (paper money and cigarette paper), cotton, and hemp.

Charcoal

Charcoal is made by partially combusting hardwood blocks in the presence of very little air. Today charcoal is made by destructive distillation of hardwood, and the vapors are collected. From these distillation vapors wood alcohol (methanol), methane gas, wood tar, acetic acid, acetone, and hydrogen gas are separated. The distillation of softwoods, especially pines, yields turpentine and rosin, both valuable in the paint industry. Rosin is also used on musicians' bows, and baseball pitchers handle a rosin bag to gain a better grip on the ball. Boxers and ballerinas shuffle their feet in a low-sided square rosin box to improve their footing.

Synthetics

Not all synthetics are made from petroleum by-products. Rayon was the first commercial synthetic fiber and is made from dissolved cellulose materials of wood pulp. Other products from wood include cellophane, acetate plastics, photographic film, and other molded plastics used as handles for tools. Although synthetics from petrochemicals are much more common and popular now, because wood is a renewable resource, rayon and other wood-pulp synthetics will probably be revitalized in the future. In fact, rayon blends are increasingly common in clothing.

Other Wood Uses

The applications of wood as lumber include plywood, particleboard, veneers, and paneling as well as board lumber for construction and furniture. In fact, the use of plywood for covering large surface areas is now preferred over board lumber in construction framing. Particleboard is especially popular in Europe, as is fiberboard, made in a similar manner. The manufacture of these materials was not economically feasible until the development of modern wood-processing techniques and glues that would permanently bond wood and wood particles.

A veneer is a sheet of wood sliced from a log. Most commonly, the log is rotated against a stationary blade that moves inward, producing a continuous sheet of desired thickness. Plywood is made by cutting the continuous sheets to desired size and bonding them together under pressure. Alternate sheets are laid with their grains at right angles to each other to increase strength and reduce warping. When three or seven sheets are used, the center sheet is twice as thick as the outer ones so an equal amount of wood has the grain running in each direction. Different veneering techniques allow for thinner and smoother slices, which can be used as the "paneled" surfaces for finished walls or for the facing on furniture.

Veneered furniture has been produced by master craftsmen for hundreds of years, and veneer itself has been known since 1500 BC; evidence of this has been found in the tombs of Egyptian pharaohs. Undoubtedly the items were painstakingly made by hand and were highly prized because of the beautiful figures on the surface of the fine furniture. Although thicknesses of as little as 0.023 cm ($\frac{1}{110}$ inch) are possible, most veneer is about 0.084 cm ($\frac{1}{30}$ inch) thick.

Particleboard is made by gluing wood chips together under pressure. Fiberboard is made by the same treatment, but the wood fiber is obtained by various chemical or mechanical pulping methods. Their use is primarily as insulation board or for building containers. However, when faced with a finished veneer or vinyl overlay, they are used for paneling and furniture.

Raw wood treated with preservatives is used extensively for fence posts, telephone poles, mine support timbers, boat dock pilings, and timbers that are buried in unstable soils under building foundations to improve support. Most of the preservatives used are either creosote mixtures (i.e., an oily liquid obtained by the distillation of wood tar), inorganic chemicals toxic to decay organisms, or toxic organic oils that prevent bacteria and fungi from decomposing (rotting) the wood tissues. These chemicals may be introduced into the wood by pressure treatment or vacuum. If free from such decomposition, wooden posts, poles, and timbers are incredibly strong and durable.

FORESTS AND FORESTRY

The total U.S. land surface area of the 50 states is approximately 2.3 billion acres. About one third of that amount, 760 million acres, is forested, and almost two thirds of that forest land, 482 million acres, is classed as commercial timberland.

Current and Future Productivity

The commercial forest lands of the United States totaled approximately 800 billion cubic feet of standing timber in 1977. The 1978 production from this timberland was 12.2 billion board feet. This wood provided the raw materials for over 4000 products worth more than $25 billion annually. This industry employs nearly 1.5 million people and provides an average of over 200 board feet of lumber for every person in the United States. We are by far the largest consumers of wood in the world, using over 2 million tons of both newsprint and writing paper. In addition, more than 230,000 tons of napkins and hundreds of millions of fence posts, railroad ties, construction boards, paper bags, cardboard boxes, and gallons of turpentine and acetic acid are purchased annually.

The 1978 production of 12.2 billion board feet of lumber was for a U.S. population of approximately 220 million. The official 1980 population was 228,340,000, and it is predicted to be 280 million by 2010. It is predicted that our demands on the forest resources will significantly increase in the years to come despite conservation and recycling efforts. One of the critical factors in a successful effort to keep production up with need is proper management and efficient use of our forested areas nationwide.

Management Policies

Proper management of existing forests includes thinning inferior trees to allow for more rapid growth of healthy ones, removing litter and diseased trees, which increase fire hazards, and employing the most efficient harvesting and restocking methods (figures 10.17 to 10.19). In addition, studies show that some 168 million acres of commercial timberland would yield greater returns from the application of such intensive management techniques than other lands.

Usage

With cost-effective technologies developed, those parts of harvested trees relegated in the past to waste could be used for pulp, fuel biomass, or synthetic extraction (figure 10.20). Inferior, diseased, and damaged trees that are now culled from the harvestable stands could also become a resource instead of a liability.

Reducing Losses

An average of over 4 billion cubic feet of timber is lost each year to such destructive agents as insects, diseases, storms, and fires. Biological control of insect pests and fungal diseases may effectively supplement existing efforts of controlling these destructive organisms. Controlled burning to remove ground fuel in the form of dead trees, dry underbrush, and dead limbs is beginning to gain acceptance as a new

Figure 10.17
Reforestation of Apalachicola National Forest in Florida with slash pine.

Figure 10.18
Proper management practices include thinning out smaller trees and removing underbrush litter that might provide fuel for a fire. This results in a stand of more uniformly sized trees.

(a)

(b)

Figure 10.19
(*a*) Modern logging requires large and sophisticated equipment. (*b*) Compare earlier logging equipment.

management tool, and many federally controlled forest lands now have a "let burn" policy in some of their lands where the litter (ground fuel) is small enough to keep the fire from destroying the living trees and from spreading extensively or rapidly.

"Supertrees"

New "super strains" of trees are being produced that will increase productivity in many future forests (figure 10.21). By using cuttings, grafts, and, most recently, tissue culture to speed up the reproductive cycle, new strains are being developed that might be twice as productive as wild genetic strains.

ADDITIONAL INDUSTRIAL PLANT PRODUCTS

Latex

The story of natural rubber is one of the most interesting in the history of plant use. The latex sap of many plants contains small amounts of cis-polyisoprene, or natural rubber. Only one plant has ever been a significant source for this material; most of the others simply do not contain enough to be commercially exploitable. A tropical tree native to the Amazon Basin, *Hevea brasiliensis,* is still the only commercial source for rubber (figure 10.22).

First known to eighteenth-century Europe as a bouncy plaything of the Mexican Aztecs and the Amazonian tribes, it received its English name when Joseph Priestley discovered that this toy would rub out pencil marks. In 1823 a Scotsman named Macintosh

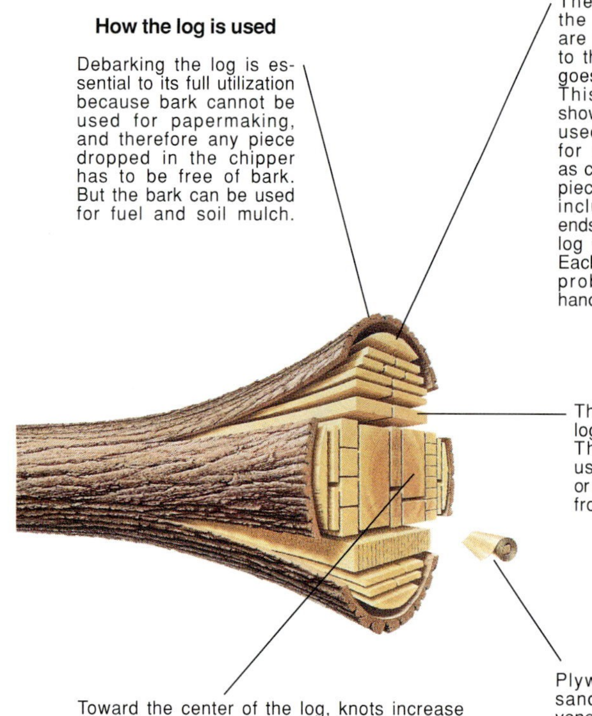

Figure 10.20

Maximum use of a log is a very important part of the forestry industry.

Figure 10.21

Research on new strains of pine seedlings may yet produce "supertrees" that will significantly increase forest productivity.

(a)

(b)

Figure 10.22

(*a*) *Hevea brasiliensis* is the main source for natural rubber. (*b*) Collection of latex from a rubber tree plantation in West Malaysia.

discovered that rubber was soluble in naphtha. When such a solution was used to impregnate cloth and the naphtha evaporated away, the result was a waterproof material suitable for protection from the rain. Even today, raincoats are still called mackintoshes in the British Isles.

In 1839 Charles Goodyear accidently heated raw rubber with some sulfur. The result was a tougher material that would not become tacky in hot weather or stiff and brittle in cold. The process of vulcanization made rubber a much more important product. Vulcanized rubber wears longer, is more resilient, and can be molded into any shape, including that of a tire.

Natural rubber shortages during World War II prompted the development of synthetic rubber, but it does not have all the properties of natural rubber. Many rubber products, including radial tires, require natural rubber. Since many of these items are being sold in increasing amounts, it is feasible that another rubber shortage could occur unless a new source can be developed for commercial rubber production.

Guayule (*Parthenium argentatum*), a shrubby member of the sunflower family, is found in the desert regions of the southwestern United States and northern Mexico (figure 10.23). It was originally developed as a viable source for natural rubber in the early 1900s, and its production was federally subsidized during World War II to help make up for the shortage of rubber. Guayule is still not a commercially profitable source of rubber. It is, however, a plant that is being actively researched, and it is expected to become commercially valuable in the future. Besides American tire companies and the Mexican government, Australia has initiated a major effort at guayule introduction and usage.

(a)

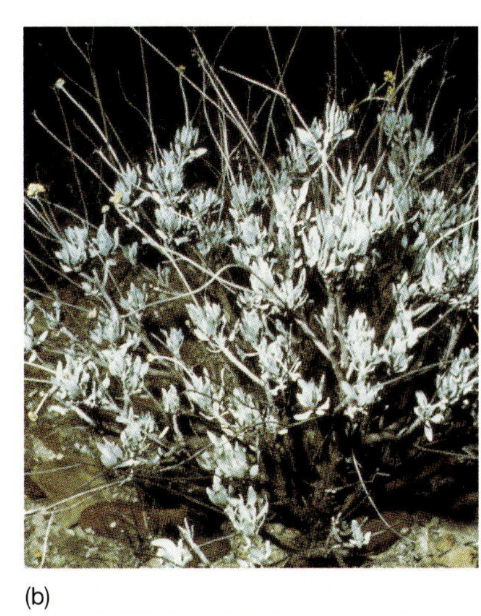

(b)

Figure 10.23

(*a*) Guayule (*Parthenium argentatum*) is an excellent source of natural rubber and is being studied for potential commercial development as a crop plant. (*b*) A low-growing desert shrub with gray leaves, guayule is native to the warm North American deserts.

Oils and Waxes

Oils contained in fruit, seeds, and other plant parts fall into several categories based on their chemical composition and use. Cooking oils, soaps, plastics, paints, linoleum, lubricants, and printing inks are a few examples of the range of plant-oil uses. Tung, castor, olive, palm, coconut, soybean, peanut, cottonseed, linseed, sunflower, safflower, and corn oils are some plant oils with a wide variety of commercial uses.

A plant with potentially valuable oil is the jojoba plant (*Simmondsia chinensis*) (figure 10.24). The seeds of this plant yield an oil essentially identical to sperm whale oil in structure, which makes it valuable in industrial lubrication. Already in small-volume

(a)

(b)

(c)

Figure 10.24

Jojoba (*Simmondsia chinensis*) is another plant native to the warm deserts of North America (*a*). Jojoba contains an excellent oil in its seeds (*b,c*), equivalent in quality to the oil from sperm whales.

commercial production, jojoba products include soaps, lotions, and cosmetics in addition to its machine lubricant uses. Like guayule, jojoba is native to the arid southwestern United States, where it is being harvested commercially by a number of Indian tribes. Its graduation to more intensive agriculture depends on the market value for the oil and improvements in establishment, yield, and harvesting techniques. Without doubt, it possesses an oil of excellent quality and many uses.

Waxes from plants play a fairly minor role in the world's economy, but several are important. Used in candles, textile sizing, and leather treatments and coatings, vegetable waxes yield to animal, mineral, and synthetic sources in economic importance. The most significant vegetable wax is undoubtedly carnauba wax, from a palm tree native to northeastern Brazil (*Copernicia cerifera*). One of the finest automobile waxes, carnauba is also used in the manufacture of carbon paper, phonograph records, lubricants, chalk, matches, plastics, films, cosmetics, and as an additive to other waxes. It is the hardest natural wax and has the highest melting point of any known wax. Candelilla wax comes from *Euphorbia antisyphilitica,* which grows in the desert areas of northern Mexico. Extracted in small quantities from wild plants, this wax is very similar to carnauba wax in most of its properties. In spite of being a high-quality wax, it will probably never become a cultivated plant producing a large annual yield because of its preferred habitat. It is used primarily in candle production. Other species of *Euphorbia,* such as *Euphorbia lathyrus* (figure 10.25), have been investigated as a source of hydrocarbons.

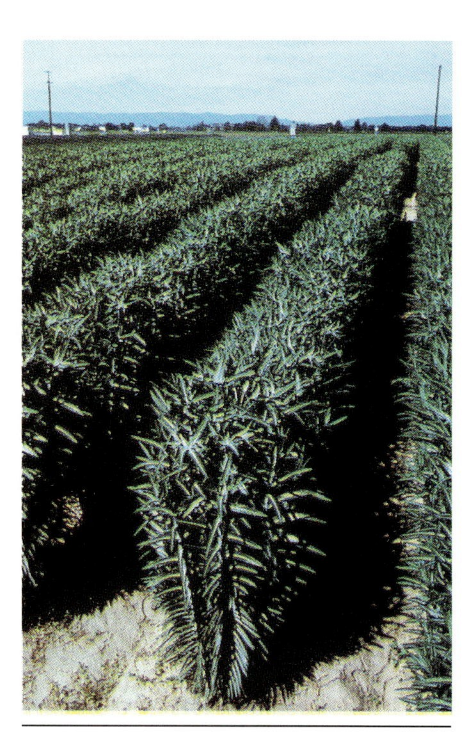

Figure 10.25
Many unexploited plants contain hydrocarbons that might be used directly as an energy source. *Euphorbia lathyrus* comes from the California deserts and holds promise as an alternative energy source.

Fibers

Cotton (*Gossypium hirsutum*) is probably the most widely used plant fiber. Cotton was first popularized as a replacement for wool in Europe during the Dark Ages. Wool garments harbored disease-carrying lice and fleas, could not be washed expediently, and were hot and scratchy for summer wear. Cotton was soft, washable, reasonably durable, and, once available commercially, inexpensive. Cotton fibers are not like the fiber cells found in stems; rather they are elongated seed epidermal cells.

Another fiber-producing plant is flax (*Linum*), from which linen, paper money, and cigarette paper are made (figure 10.26). Popular in ancient Egypt, fine linen was even used for wrapping mummies. Sisal is a coarse fiber from the leaves of the *Agave* and is primarily made into binding twine. Hemp rope is made from the fiber strands extracted from *Cannabis* stems. Hemp is not commercially produced in the United States. Russia is the leading producer of hemp materials. Manilla rope is made from the fibers extracted from *Musa* (frequently called manilla hemp).

Figure 10.26
Paper money is made from flax fibers (*Linum*), not wood.

Cork

Cork is a nonwood product of trees, specifically the "bark" of the cork oak, *Quercus suber.* Native to Europe, most cork is produced in Morocco, Portugal, and Spain because the climatic conditions there are most conducive to rapid outer periderm growth. The cork cells are naturally impregnated during growth with a wax, suberin, which makes them watertight and produces incredible buoyancy. A cork oak can be stripped of its outer bark after the first 20 years of growth and then every 10 years thereafter until the tree is approximately 150 years old.

✔ Concept Checks

1. Define the following terms: renewable resource, board foot, quartersawed, plainsawed, grain.
2. What is a knot? How are knots formed?
3. Describe four uses of wood.
4. Explain the differences among plywood, particleboard, veneer, fiberboard.
5. Explain the economic importance for each of the following plants: *Hevea brasiliensis, Parthenium argentatum, Simmondsia chinensis, Copernicia cerifera, Linum, Cannabis.*

Summary

1. There are over 4500 different commercial products that come from wood, or secondary xylem. The giant redwood is the world's tallest tree; the bristlecone pine is the oldest living organism. Wood is composed of cellulose, hemicellulose, and lignin. Properties of wood that depend on the relative composition of these materials include durability, color, luster, polishability, shrinkage, heat conduction, and acoustics.
2. The terms *softwood* and *hardwood,* although somewhat misleading, refer to coniferous gymnosperms and woody dicots, respectively. The hardness of wood actually refers to its specific gravity or density.
3. In most trees a single annual growing season results in the formation of a growth ring. These rings are xylem tissue, and as the tree ages, the inner (earliest) rings usually accumulate metabolic by-products and take on a different appearance. This heartwood no longer conducts water and minerals as the outer sapwood layers do.
4. The study of the past by using tree rings is called dendrochronology. The number of annual growth rings establishes the age, and the widths of the rings reveal the climatic conditions for that growing season. A continuous chronology for about 8000 years has been established using dead tree trunks of bristlecone pines.
5. Building lumber can be categorized according to density, texture, and the part of the tree used. Transverse, radial, and tangential cuts results in different grains. The presence of knots is also a consideration when selecting building lumber. Knots are old branches covered over by the expanding trunk girth or diameter.
6. Wood used for fuel is an essential natural resource in most of the developing countries. The use of wood as fuel is increasing in the United States and in several other developed countries as well.
7. Paper production is a significant commercial use for wood, as is charcoal. Rayon is a synthetic fiber produced from wood, not petroleum by-products.
8. Other uses for wood include veneer making and plywood manufacture. In addition, some woods have natural preservatives that allow them to be used as timbers and pilings in wet environments.
9. Although the United States has adequate wood resources, some countries are not properly harvesting and caring for their forests. Proper management policies mean adequate timber production in the United States for many years to come.
10. Plants with industrial importance include the latex producers, such as *Hevea* and guayule; those which contain oils, such as jojoba; fiber-containing plants such as cotton, flax, hemp, and sisal; and those that produce a high quantity of cork.

Key Terms

Discussion Questions

1. Explain how the study of wood can be used to determine the age of a tree as well as past climatic conditions.
2. Examine several boards found in either a local lumber yard or stored in your garage. For each, determine which cut surface most closely relates to a cross section, a radial section, and a tangential section.
3. How can improved management policies reduce losses in our annual timber production?

Suggested Readings

Biermann, C. J. 1993. *Essentials of pulping and papermaking.* Academic Press, San Diego.

Carlquist, S. H. 1988. *Comparative wood anatomy.* Springer-Verlag, New York.

Cook, E., and Kairiukstis. 1989. *Methods of dendrochronology: Applications in the environmental sciences.* Kluwer Academic Publishers, Boston.

Emrich, W. 1985. *Handbook of charcoal making: The traditional and industrial methods.* Kluwer Academic Publishers, Boston.

Goldstein, I. S. 1991. *Wood structure and composition.* M. Dekker, New York.

Haase, E. F., and W. G. McGinnies. 1972. *Jojoba and its uses.* Office of Arid Land Studies, University of Arizona, Tucson.

Keating, W. G. 1982. *Characteristics, properties, and uses of timbers.* Texas A&M University Press, College Station.

Schmidt, Karen F. 1991. Good vibrations. *Science News.* Vol. 140(24):392–394.

Schweingruber, F. H. 1988. *Tree rings: Basics and applications of dendrochronology.* D. Reidel Publishing Co., Boston.

Sherbrooke, W. C. 1978. *Jojoba: An annotated bibliographic update.* Office of Arid Land Studies, University of Arizona, Tucson.

United States Department of Agriculture. 1991. *The Bristlecone pine: Nature's oldest living thing.* USDA, Forest Service, Ogden.

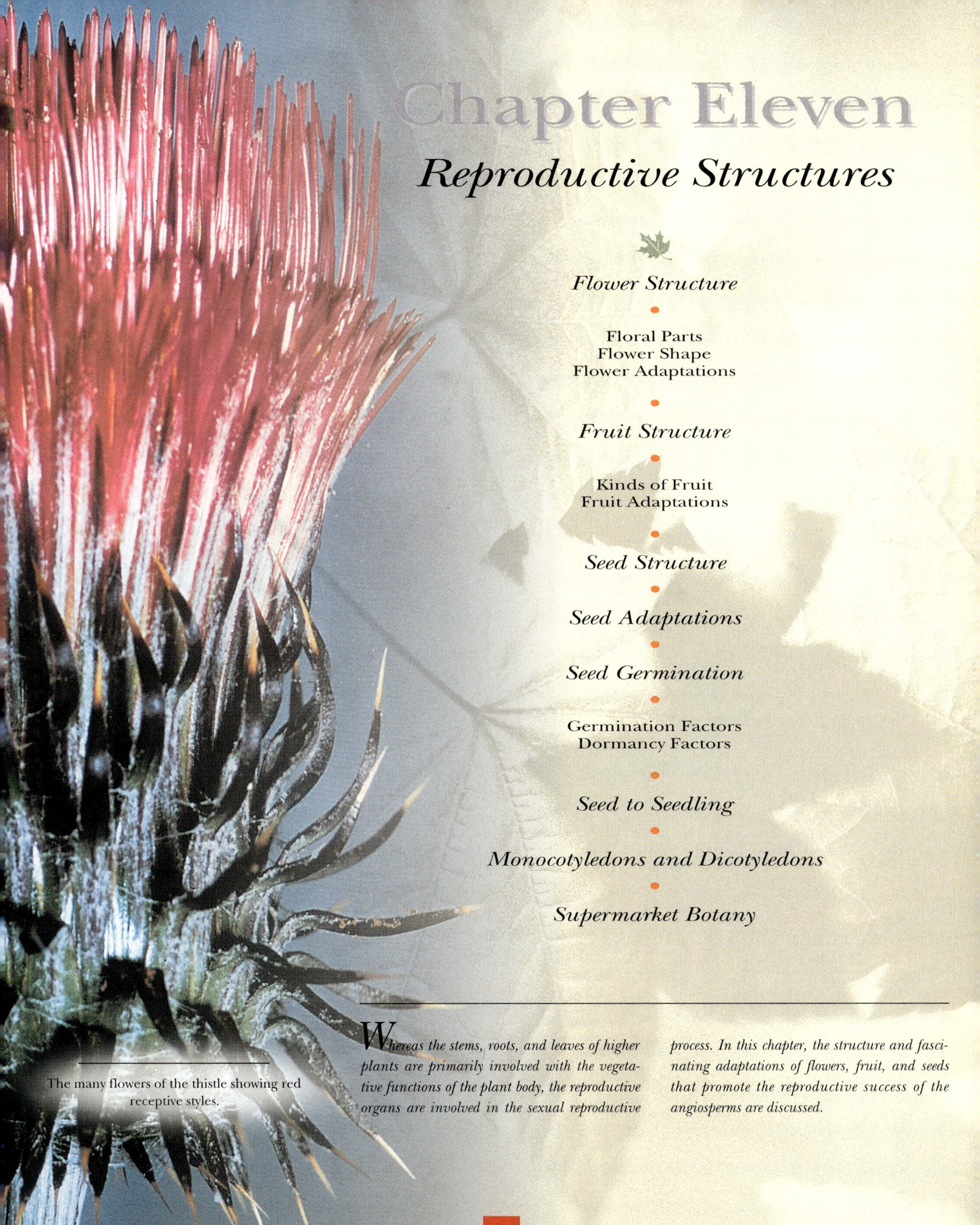

Chapter Eleven

Reproductive Structures

The many flowers of the thistle showing red receptive styles.

Whereas the stems, roots, and leaves of higher plants are primarily involved with the vegetative functions of the plant body, the reproductive organs are involved in the sexual reproductive process. In this chapter, the structure and fascinating adaptations of flowers, fruit, and seeds that promote the reproductive success of the angiosperms are discussed.

Sexual reproduction in vascular plants always takes place in specialized plant parts that develop during specific periods of the plant's life. In angiosperms sexual reproduction occurs in the **flower,** resulting in the production of **fruit** containing **seeds,** which house the new plant embryo. Flowers come in a vast array of sizes, shapes, colors, and arrangements, as do the resulting fruit and seeds.

FLOWER STRUCTURE

Although the flower has deservedly been made the object of poetry and often symbolizes beauty, love, peace, and happiness, the basic biological function of the flower is sexual reproduction. Exotic colors and shapes of flowers are actually devices to attract specific **pollinators.** This ensures their reproductive success, since **pollen** will be carried among flowers of the same species. Not all flowers are large or beautifully showy, but even the plainest flowers function. Some flowers are not conspicuous at first glance, but closer observation reveals remarkably complex, colorful, and beautiful design.

Floral Parts

A typical flower possesses four different floral parts that are attached to the **receptacle** in the following order, from outside to inside: **sepals, petals, stamens,** and **carpels** (figure 11.1). The sepals are collectively referred to as the **calyx,** which encloses and protects the developing flower. The **corolla,** the collective name for the petals, is frequently colorful and exhibits diverse sizes and shapes. Together the calyx and corolla are termed the **perianth.** The stamens are collectively termed the **androecium.** The collective term for all the carpels within a flower is the **gynoecium.**

A stamen is composed of pollen-producing **anthers,** attached to the end of the slender **filament.** The filament raises the anther to a position where the pollen, units responsible for the production of sperm, is more easily dispersed or accessible to visiting pollinators.

At the center of the flower is the carpel. The base of the carpel is termed the **ovary,** which contains one or more **ovules.** The ovules ultimately house the female sex cells or eggs. The slender necklike portion of the carpel is the **style,** which elevates

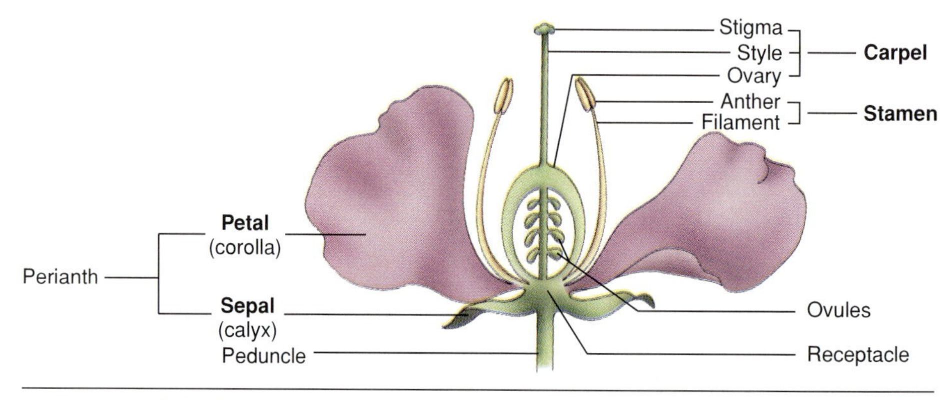

Figure 11.1
Typical flower, showing the four basic floral parts attached to the receptacle and cross section of pollen-producing anther.

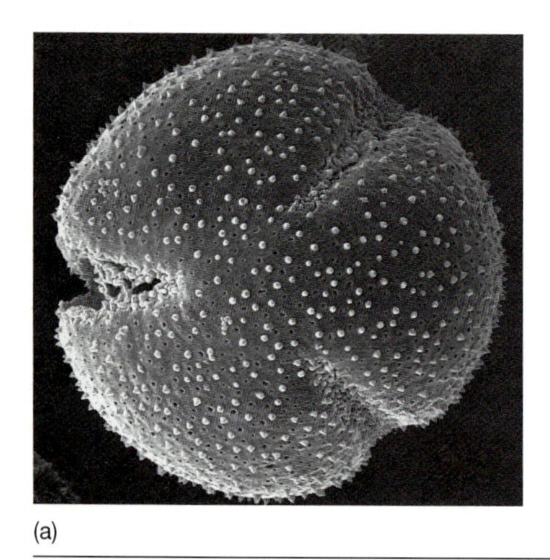
(a)

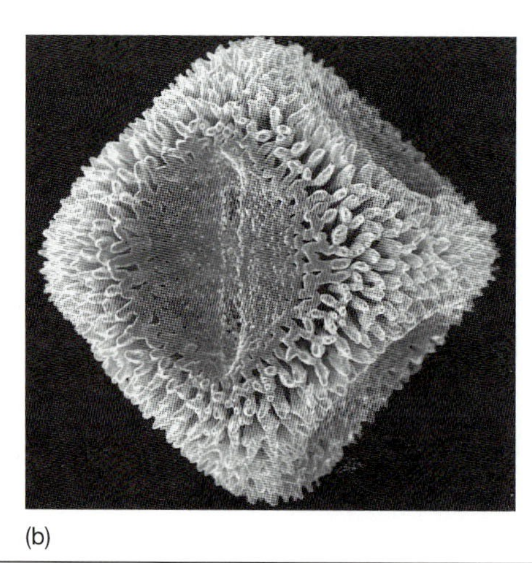
(b)

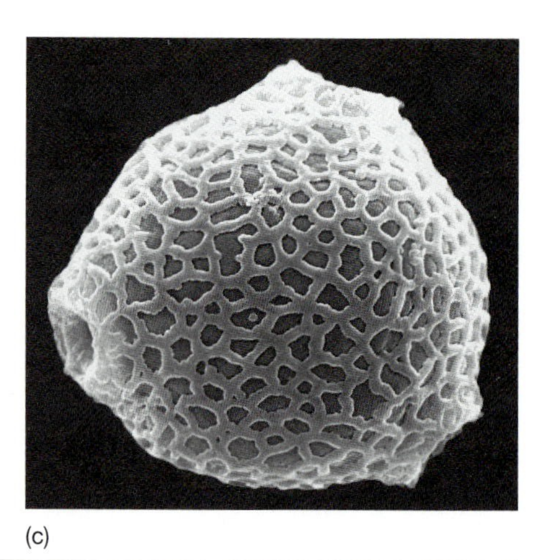
(c)

Figure 11.2
(*a*) Pollen of *Echinocereus chinerascens* has small spines. (×1785) (*b*) *Basella alba* has cubical pollen with long projections. (×2400) (*c*) Pollen of *Auroia hirsuta* has a pattern of ridges on its surface. (×2500)

the **stigma** to a favorable position relative to contact with pollen. Whether pollen is airborne or brought by a visiting pollinator, the process of pollen landing on the stigma is **pollination.** Many stigmatic surfaces are sticky, pubescent, or otherwise modified to help ensure pollen attachment. In addition, pollen grains, which vary in size from less than 20 µm to over 200 µm in diameter, may be "ornamented" with spines, ridges, and barbs that further aid pollination (figure 11.2).

An ovary has a placental surface to which the ovules are attached (figure 11.3). The ovary of a single carpel is composed of a placental surface and ovules. When a gynoecium is composed of a single carpel, as in the flowers of peas and beans, it is called a simple carpel. When there are multiple carpels, such as those found in tulips, lilies, grapefruit, and poppies, it is termed a compound carpel (figure 11.4). The sections of a cut grapefruit are carpels of the compound ovary. The ovules of each of these carpels are where the female **gametes,** or eggs, are produced and where sexual **fertilization** (fusion of the egg and sperm) takes place in flowering plants (chapter 15, p. 268). The resulting plant embryo is housed in the seed (mature ovule) within the fruit (mature ovary). Also within the seed, surrounding the embryo, is a highly nutritive tissue termed the **endosperm.**

In most flowers the carpel is attached to an enlarged apical portion of the stem called the receptacle. The attachment of the carpel to the receptacle is usually above the insertion of the other floral parts (sepals, petals, and stamens). When so positioned the flower is said to have a superior ovary (figure 11.5). In some flowers, however, the ovary has receptacle or perianth tissue surrounding it so that the other floral parts appear attached above the level of ovary attachment. Such flowers are

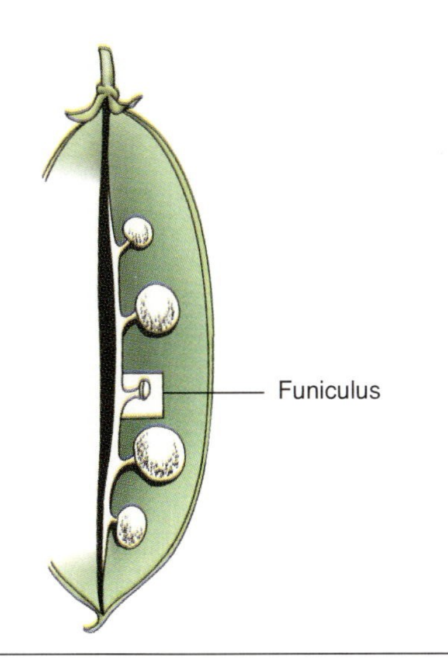

Figure 11.3
Ovules are attached to the placental wall of the ovary by a stalk termed the funiculus.

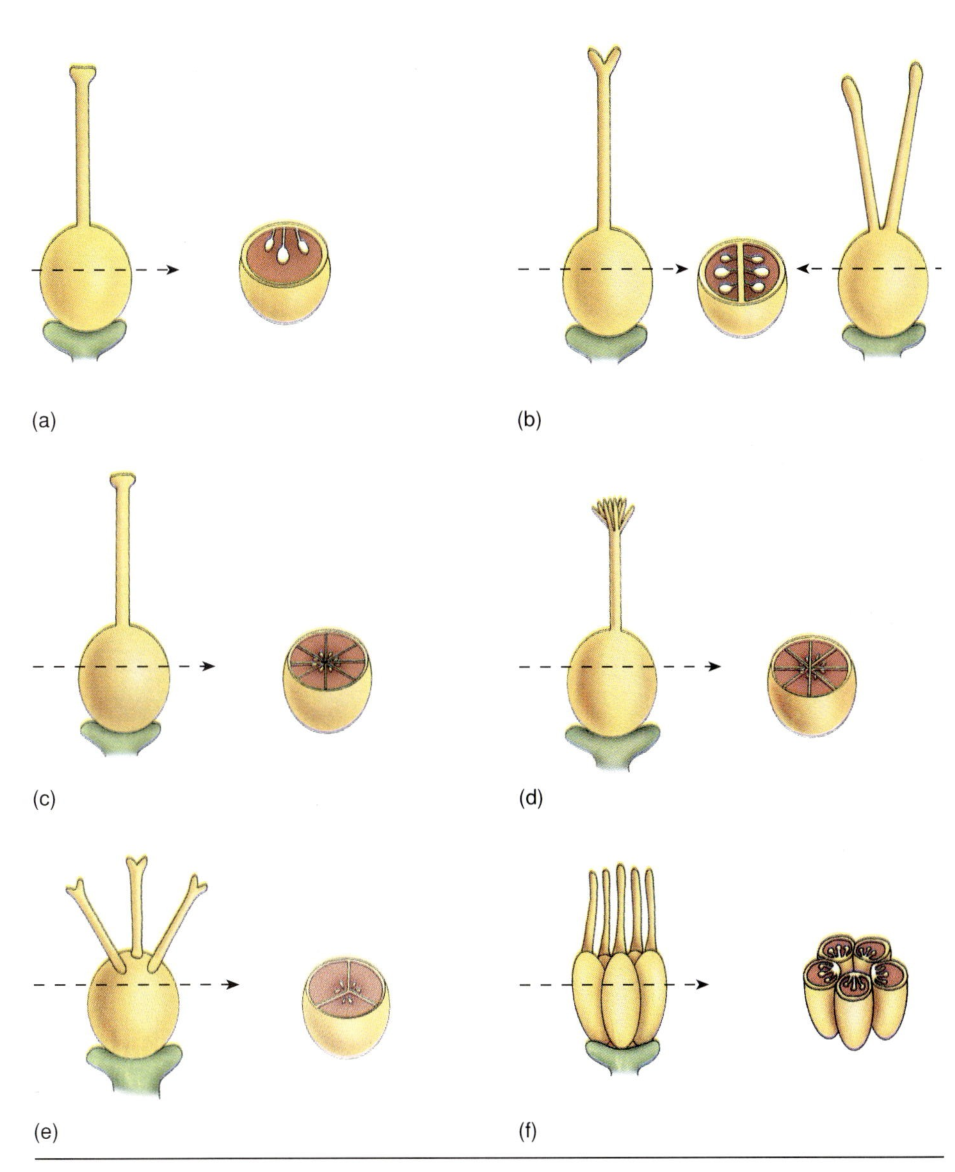

Figure 11.4
(*a*) A simple carpel. (*b* and *c*) Compound carpels composed of two and eight carpels, respectively. (*d*) A compound carpel often has an equal number of stigmas and carpels. (*e*) Others have an equal number of styles and carpels. (*f*) Some flowers have several simple carpels.

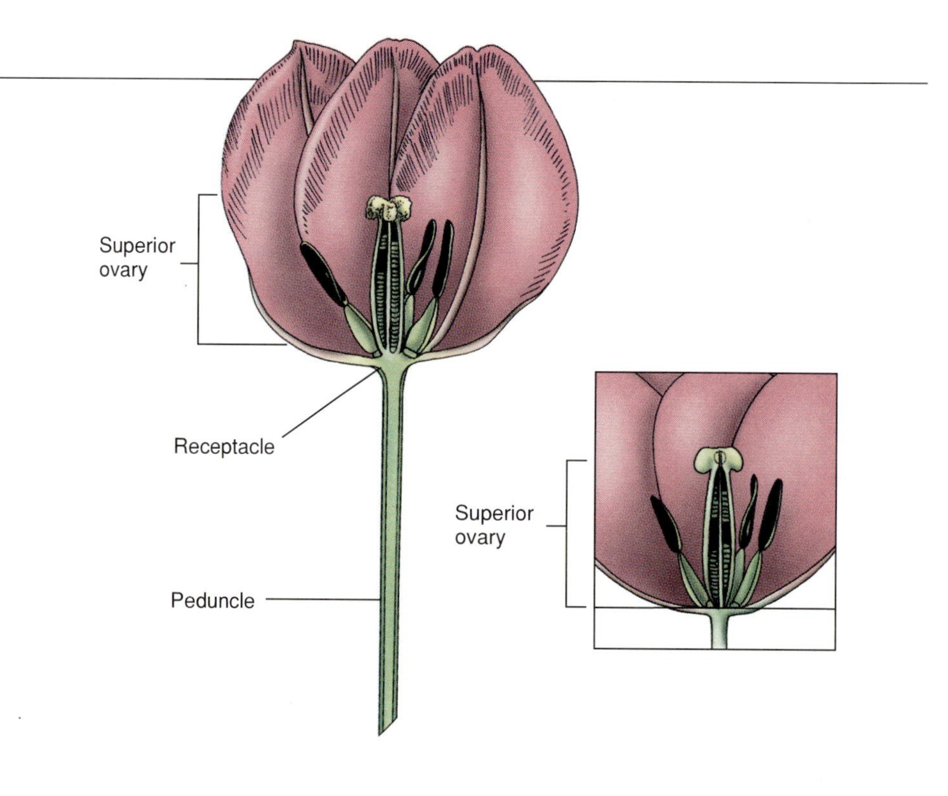

Figure 11.5
The superior ovary of a tulip, showing the attachment points of the other floral parts at the bottom of the ovary.

said to have an inferior ovary (figure 11.6). The position of the ovary is an important taxonomic characteristic.

Flowers can occur singly or in an **inflorescence** of several to many flowers. Inflorescences can be as simple as only a few flowers attached near one another on the flowering stem or as complex as the head of a sunflower, which is composed of hundreds of tightly grouped flowers with the head actually being in the shape of a single large flower. The range of inflorescence complexity and shape and the modifications of the component flowers within some inflorescences are great (figures 11.7 and 11.8). All these arrangements and modifications in some way facilitate pollination and fertilization and thus the reproductive success of the species. Flower appearance is usually consistent because pollinator-flower specificity often exists. If a pollinator fails to recognize the flowers of some individuals of that species, pollination, and therefore reproduction, cannot occur. Failure to reproduce sexually could ultimately result in fewer offspring, reduced genetic variability, less adaptability, and finally extinction.

A complete flower has sepals, petals, stamens, and carpels; all are present. Many species, however, lack one or more of these four basic floral parts. Such flowers are termed

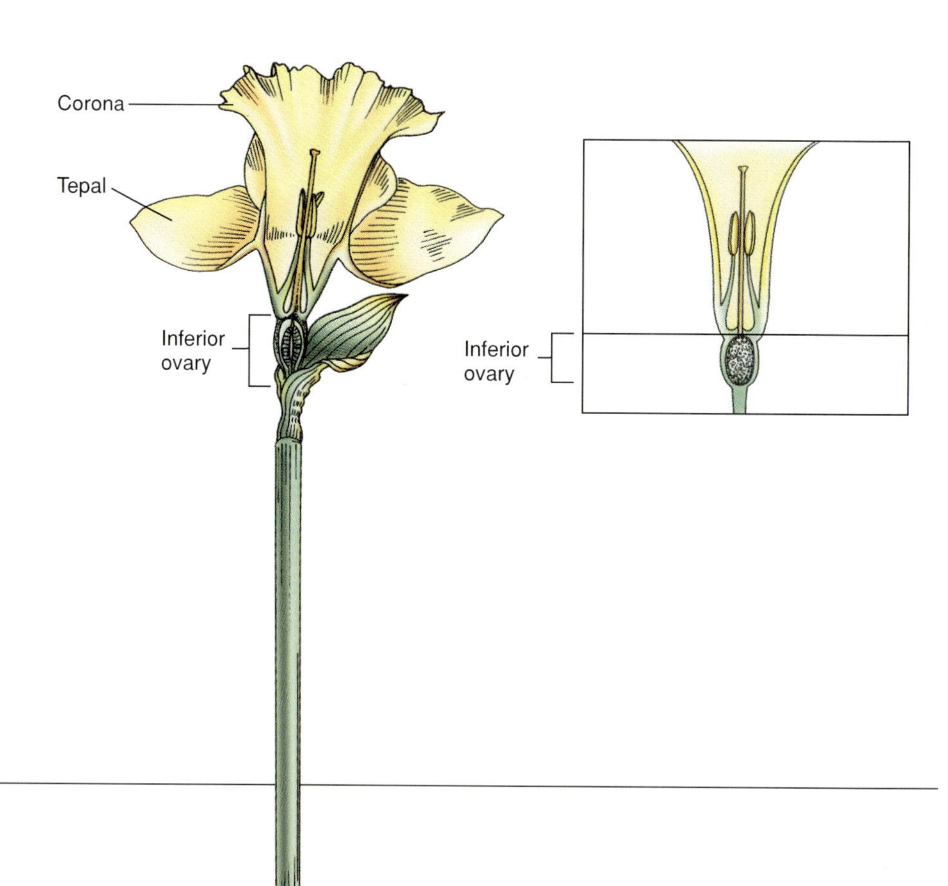

Figure 11.6
The daffodil has an inferior ovary with the other floral parts attached at the top of the ovary, and six tepals (two series of three each) with a secondary corona (small crown).

Figure 11.7

Inflorescence types. (*a*) Spike. (*b*) Raceme. (*c*) Panicle. (*d*) Corymb. (*e*) Umbel. (*f*) Compound umbel. (*g*) Catkin. (*h*) Head. The peduncle is the stem just below a solitary flower or an inflorescence; pedicels are the stalks to each flower in an inflorescence.

Figure 11.8

(*a*) The compound umbel of cowparsnip (*Heracleum lanatum*). (*b*) A raceme of the wild cherry (*Prunus serotina*).

incomplete flowers. Flowers having both stamens and carpels (that is, the capability to produce both types of gametes) are called perfect flowers, whereas an imperfect flower contains either stamens or carpels but not both. Those containing only stamens are logically called staminate flowers, and those lacking stamens but having carpels are termed carpellate flowers (figures 11.9 and 11.10).

Plants having both staminate and carpellate flowers on the same plant are **monoecious,** literally meaning "one house." **Dioecious** (two houses) plants have the staminate flowers on one plant and the carpellate flowers on another. Corn (*Zea mays*) and squash (*Cucurbita*) are monoecious; willow trees (*Salix*) are dioecious. Historically, the designations of "male" and "female" parts have been applied to the stamens and carpels, respectively, and thus with imperfect flowers one might see reference to the male or female flower. Such designations are not without some basis; however, this terminology should never be scientifically used. It is best to use the more descriptive terms staminate and carpellate.

Flower Shape

Flowers are either regular (radially symmetrical), as a daffodil or rose, or irregular (bilaterally symmetrical), as a snapdragon (figure 11.11). A regular flower can be cut through the center in more than one plane and result in identical halves. An irregular flower is bilateral, having only one plane through which it may be cut to result in mirror-image halves.

Figure 11.9
Staminate flowers of the honey locust (*Gleditsia triacanthos*).

Figure 11.10
The staminate and carpellate flowers grouped in catkins on paper birch (*Betula papyrifera*). Species with both staminate and carpellate flowers on the same plant are monoecious.

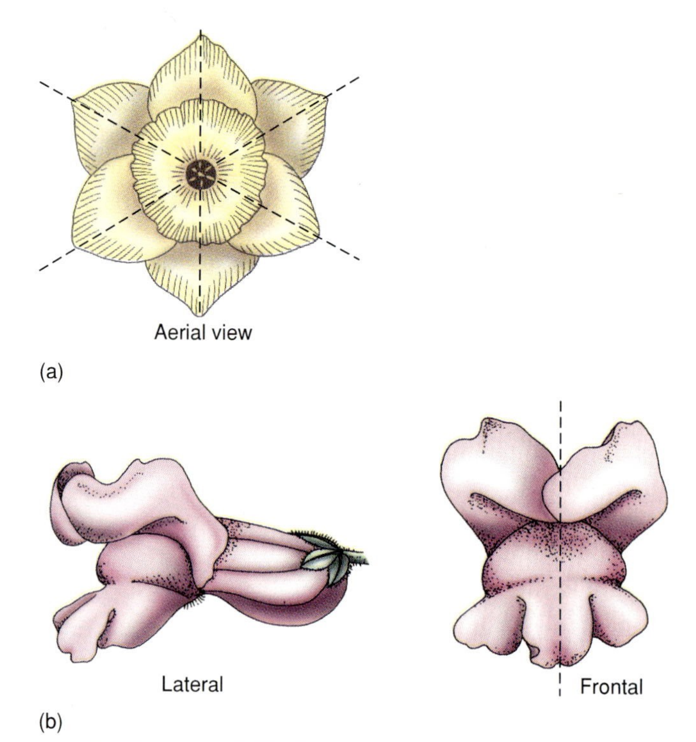

Figure 11.11
Regular flowers, such as the daffodil, have radial symmetry (*a*). Irregular flowers, such as the snapdragon, have bilateral symmetry (*b*). Dotted lines indicate the lines of symmetry.

Additionally, the petals can be either distinct (not fused to one another) or partially to wholly fused into a single structure. Sepals, stamens, and carpels also may be distinct or variously fused.

Flower Adaptations

The array of flower shapes, sizes, colors, and arrangements seems endless. The variability that exists, however, is not haphazard. Flower modifications heighten the likelihood that pollination, and subsequently fertilization, will occur. They help attract animals, primarily insects and birds, that physically transfer pollen from plant to plant. Pollination agents, or vectors, may be general visitors, or they may be attracted to specific flowers. They are enticed by a unique floral presentation that results from alteration of the flower parts. The changes include different sizes, shapes, fusions, colors, patterns, odors, and edible materials, as well as relocation of the stamens and stigma (figures 11.12 and 11.13). Thus flower adaptations promote reproductive success.

(a) (b) (c) (d)

Figure 11.12

Many flowers depend on insect pollinators. Flower modifications attract these visitors. (*a*) Blue curl (*Phacelia*). (*b*) *Lantana*. (*c*) *Albizzia*. (*d*) Thistle (*Cirsium*).

(a) (b) (c) (d) (e) (f)

Figure 11.13

Flower modifications. (*a*) Elephant head (*Pedicularis groenlandria*). (*b*) Adder's Tongue (*Erythronium*). (*c*) *Solenophora*. (*d*) Wild orchid (*Epipactis*). (*e*) *Eucalyptus*. (*f*) Indian Pipe (*Monotropa*).

Petal Modification

Petal color, size, shape, and fusion are the most common variables in flower morphology. Each such modification exists as a result of increased reproductive success through pollinator visitation. Some flowers are successful exclusively through visual attraction. Others must additionally produce nectar as a food reward for a visiting pollinator or have a strong, detectable odor. Odors can be sweet and perfumy or foul. An aroma similar to rotting meat attracts flies as the pollination vector in certain plants.

Some petal surfaces are marked to signal to appropriate insects which way to approach, where to land, and which way to enter the flower. These markings often change after pollination. This signals to other potential visitors that the flower has already been visited; significant pollinator efficiency results.

The human eye cannot appreciate all these petal markings, but certain insects can. Reflections of ultraviolet wavelengths from given portions of the petal produce a specific pattern for bees and wasps, which can see in the ultraviolet range. Such insects, however, do not discriminate colors at the opposite end of the visible spectrum and thus are essentially red color-blind.

Other floral and nonfloral parts display modifications involved in pollinator attraction. In some flowers the sepals appear petaloid (tepals, figure 11.6); in others, highly modified petal-like stamens termed staminodia are present. The stamens are visually attractive through elongated and colorful filaments or large, colorful, and showy anther sacs (*Caesalpinia* and *Tradescantia*). The bright red poinsettia "petals" are in fact colorful leafy bracts located below the cluster of small, relatively inconspicuous flowers. *Bougainvillea* also has petaloid bracts and less conspicuous flowers (figure 11.14).

These are only a few examples of modified flowers and flower parts, whose evolution has increased pollinator visitation and reproductive success. The complex mosaic of natural beauty that has resulted from these modifications is intellectually intriguing and enjoyable for us to simply observe and appreciate.

Figure 11.14
Bougainvillea. The three colorful bracts are much more showy than the three small tubular flowers.

Pollination Systems

There are two general categories of pollination mechanisms: plants that are **self-pollinated** have their own pollen land on their own stigma; those that are **cross-pollinated** receive pollen from the flowers of other individuals of the same species.

Self-pollination Most self-pollinating plants have perfect flowers; their flower design allows their own pollen to be shed onto their stigma (or a visiting insect can initiate this). Since all the flowers of a single plant have the same genetic origin, pollen transfer to any flower on the same plant is still considered to be a self-pollination event. It does not increase genetic variability.

Cross-pollination When genetic information from two different plants is mixed, the seeds have a larger pool of genetic information. The resulting progeny will display variations in form and function. This cross-pollination is also called outcrossing. Because greater genetic variability is desirable, outcrossing has been ensured in many plants by the development of genetic **self-incompatibility.**

Self-incompatibility This can be physical or chemical in nature. Some flowers are structurally designed so that the pollen of that flower cannot be shed onto its stigma. In some the style is elongated more than the stamen filaments, positioning the stigma well above the anther sacs. In others the pollen is not released from the anthers when the stigma is receptive; it is shed either before the stigma is ready or after the stigma has already been pollinated from a different plant.

Chemical incompatibility is less easily determined. Certain proteins in the outer layer of the pollen grain (termed the exine) are involved in a recognition reaction with the surface of the stigma, which ensures that pollen from the same plant will be rejected. This is similar to immunological reactions in animals. Pollen from a different plant of the same species will not produce this recognition reaction. It will be allowed to germinate and produce a tubular extension called the pollen tube, which grows through the style and ultimately reaches the ovule, where fertilization takes place.

Wind Pollination For a plant to be cross-pollinated, or outcrossed, there must be a dependable source of pollen from other plants of that species. Wind-pollinated plants generally have many small, inconspicuous flowers that produce large quantities of lightweight pollen grains. Unlike much of the pollen of insect-pollinated plants, the pollen grains are usually smooth and do not stick together. Since wind carries the pollen, colorful petals, edible tissues, attractive odors, and nectar production are generally lacking; these adaptations are necessary only for plants that attract insect visitors. Wind-pollinated flowers are usually modified, having long styles with well-exposed stigmas that are often feathery or branched and fairly large. Their stamens are also well exposed to the wind, sometimes hanging down away from the flower on long, slender, flexible filaments to better shed their pollen when the winds blow across them (figure 11.15). Since they are usually very small flowers, they most often occur in inflorescences. Their great number and density further increase the chances of pollination. Most of these flowers have only a single ovule and produce a single-seeded fruit, such as a grass grain, the winged fruit of the elm, or the acorn of an oak.

Because wind pollination is chancy, copious amounts of pollen are shed from each flower. The vast majority never lands on a stigma but ends up on the ground not far from its source. Many wind-pollinated species are imperfect (monoecious) with small staminate flowers grouped together on one part of the plant, and carpellate flowers together on a different part of the same plant (figure 11.16). Other wind-pollinated species, such as cottonwood and honey locust, are dioecious having staminate and carpellate flowers on different plants.

Since wind pollination is inefficient, it is most common where many plants of the same species grow close together in open areas. The tropics, therefore, have few wind-pollinated species. They are found mostly in temperate zones. Most grasses and deciduous hardwood trees are wind pollinated, the latter flowering primarily in the early spring before the leaves develop.

Gymnosperms are also wind pollinated. Most of the cone-bearing trees (conifers) have small pollen-producing cones and much larger seed- producing cones on the same tree (chapter 19, p. 353). It is probable that angiosperms evolved from gymnosperms; so, for a long time botanists theorized that since gymnosperms are all wind pollinated, wind-pollinated angiosperms must be relatively primitive. Now it is generally accepted that they are not primitive, but in many cases fairly advanced, having evolved from insect-pollinated groups.

During the flowering season many areas have so much pollen shed by wind-pollinated trees (both angiosperms and gymnosperms) and grasses, that the air appears hazy. Ponds and shorelines of lakes and still streams have a layer of billions of yellow pollen grains floating on the surface. Hay fever and many other allergic reactions are common ailments during these periods.

Insect Pollination By far the most common pollinators are insects. Insects are the only group of organisms that display as great a diversity in numbers of species, distributional range, and overall complexity as the flowering plants. The origins and development of such variability and numbers in these two groups have been processes of coevolution.

✔ Concept Checks

1. Diagram a complete flower, labeling all parts.
2. What is the distinction between pollen and pollination?
3. Distinguish between simple and compound carpels. Superior and inferior ovaries. A perfect and imperfect flower. Radial and bilateral symmetry. Self- and cross-pollinated flowers. What is a head inflorescence?
4. How does the structure of a wind-pollinated flower differ from one that is insect-pollinated?

(a)

(b)

Figure 11.15
(*a*) Grass flowers with the anthers exerted so that their pollen is picked up by the wind. (*b*) Cattails (*Typha*) have their flowers densely clustered into tight inflorescences for increased efficiency.

Figure 11.16
Corn is monoecious. *Bottom,* staminate flowers. *Upper right,* carpellate "silks."

FRUIT STRUCTURE

In most plants, after fertilization, the ovary of the flower matures into a fruit. Fertilization actually takes place in the ovules, which in turn develop into seeds. Fruit development, therefore, is normally triggered by pollination-fertilization-seed development, and if there is no fertilization or if the seeds fail to develop, the ovary will not mature into a fruit and commonly abscises from the plant.

As with most other "normal" events, there are exceptions. Some fruit (bananas, for example) develop without fertilization. This process is **parthenocarpy,** and parthenocarpic fruit are seedless. Crop scientists specialized in fruit production have selected from nature seedless grapes and oranges.

Fruit play an important role in the reproductive cycle of flowering plants, providing continued protection for the enclosed seed and aiding in their dissemination. The step from naked seeds (in gymnosperms) to enclosed seeds (in angiosperms) was a major one in the evolution of plants (chapter 19, p. 352).

As the ovules begin developing into seeds, the ovary wall matures into the fruit wall, which is then called the **pericarp.** The pericarp usually has an outer exocarp, a middle mesocarp layer, and an inner endocarp. The distinctiveness of these three subunits of the fruit wall varies from plant to plant. On the outside of the developing ovary the remains of the style and stigma may still be visible, as might the shriveled stamens, corolla, and calyx. The calyx, in fact, is often not only still visible, but in some species remains healthy and even enlarges as the fruit matures.

The most common kinds of fruit and examples for each are given in table 11.1. The following general terms are used in classifying these fruit.

TABLE 11.1

Fruit

TYPES	DEVELOPMENT	EXAMPLES
SIMPLE FRUIT		
Dry and dehiscent		
Follicle	Develops from single-carpel ovary; splits open down one side	Columbines, magnolia, milkweeds
Legume	Develops from single-carpel ovary; splits open along both sides	Pea family (Leguminosae); all peas and beans
Silique	Develops from two-carpel (bicarpellate) ovary; halves fall away, leaving seeds attached to persistent, central wall	Mustard family (Crucifereae)
Capsule	Develops from compound ovary with two or more carpels; capsules dehisce in many different ways	Cotton, poppy, primrose, pinks
Dry and indehiscent		
Achene	Small, one-seeded fruit; pericarp easily separable from seed coat, although closely encasing it	"Dry" fruit of strawberry, buckwheat, and sunflower family (Compositae). (Some authorities consider the sunflower fruit to be different because it is derived from a compound inferior ovary called a *cypsela.*)
Samara	Winged, one- or two-seeded achenelike fruit; wing(s) form from outgrowth of ovary wall	Elms, ash, one-seeded maples
Schizocarp	Two or more carpel ovary that, on maturity, splits into separate, one-seeded sections that fall away	Maples considered samaras or winged schizocarps
Caryopsis	One-seeded, usually small fruit with pericarp completely united to seed coat	"Grain" of all grass family (Gramineae); includes wheat, oats, rice, corn, barley, rye, and other important grasses
Nut	One-seeded fruit with hard pericarp (shell)	Walnut, hazelnut, chestnut, acorns

Kinds of Fruit

Simple Fruit

Fruit that develop from a single ovary, whether it is composed of only one or several fused carpels are **simple fruit** (figure 11.17). They can be further subgrouped as either fleshy or dry, depending on whether the pericarp is soft and juicy. Additionally, dry fruit can be either dehiscent, where the dry pericarp splits open at maturity, releasing the seeds, or indehiscent, where the seeds remain with the fruit and the entire fruit falls from the parent plant at maturity.

Compound Fruit

These are formed by the development of a group of simple fruit. There are two general kinds of **compound fruit,** aggregate and multiple.

Aggregate Fruit These form from many separate carpels (ovaries) of a single flower, as in the strawberry, raspberry, and blackberry (figure 11.18). Strawberries differ from the latter two in having small, dry, individual fruit or achenes attached to an enlarged juicy receptacle; blackberries and raspberries are an aggregation of separate fleshy fruit (drupes) attached to a common receptacle.

Multiple Fruit These result from the development of the ovaries of several separate flowers that have fused on the axis of the inflorescence. The pineapple and fig are both multiple fruit.

Accessory Fruit

These consist of simple or compound fruit together with additional tissues, usually the calyx and/or the receptacle, as in strawberries, pineapples, and apples.

Fruit Adaptations

As among flowers, the differences among fruit have evolved directly as a result of increased reproductive

TYPES	DEVELOPMENT	EXAMPLES
SIMPLE FRUIT—cont'd.		
Fleshy		
Berry	Two or more carpel ovary, each usually having many seeds; inner layer of pericarp (mesocarp and endocarp) is fleshy	Tomatoes, grapes, dates
Hesperidium	Berry with thick, leathery "peel" (exocarp and mesocarp) and juicy, pulpy endocarp arranged in sections	Oranges, grapefruit, lemons, limes; all citrus fruit; rind has oil glands
Pepo	Berry with outer wall or rind formed from receptacle tissue fused to exocarp; fleshy interior is mesocarp and endocarp	Gourd family (Cucurbitaceae), including cucumbers, watermelons, squash, pumpkin
Drupe	Usually only one-carpel ovary and with only one seed developing; endocarp is hard and stony, fitting closely around seed; mesocarp is fleshy, and fruit is thin skinned (thin, soft exocarp)	Many members of rose family (Rosaceae), including cherry, peach, plum, almond, apricot; not in the Rosaceae, olive and coconut are also drupes (coconut has fibrous outer coat rather than fleshy one)
Pome	From compound, inferior ovary (one embedded in surrounding receptacle or perianth tissue); fleshy edible part is ripened tissue surrounding ovary, which matures into "core" and contains seed	Apples and pears, both members of subfamily of Rosaceae
AGGREGATE AND MULTIPLE FRUIT		
Aggregate fruit	Development of numerous simple carpels from a single flower; some are dry fruit attached to fleshy receptacle, others an aggregation of simple fleshy fruit (drupes)	Strawberry, blackberry, raspberry
Multiple fruit	Individual ovaries of many separate flowers clustered together; usually nutlets on enlarged fleshy receptacle or group of berries	Mulberry, pineapple, fig

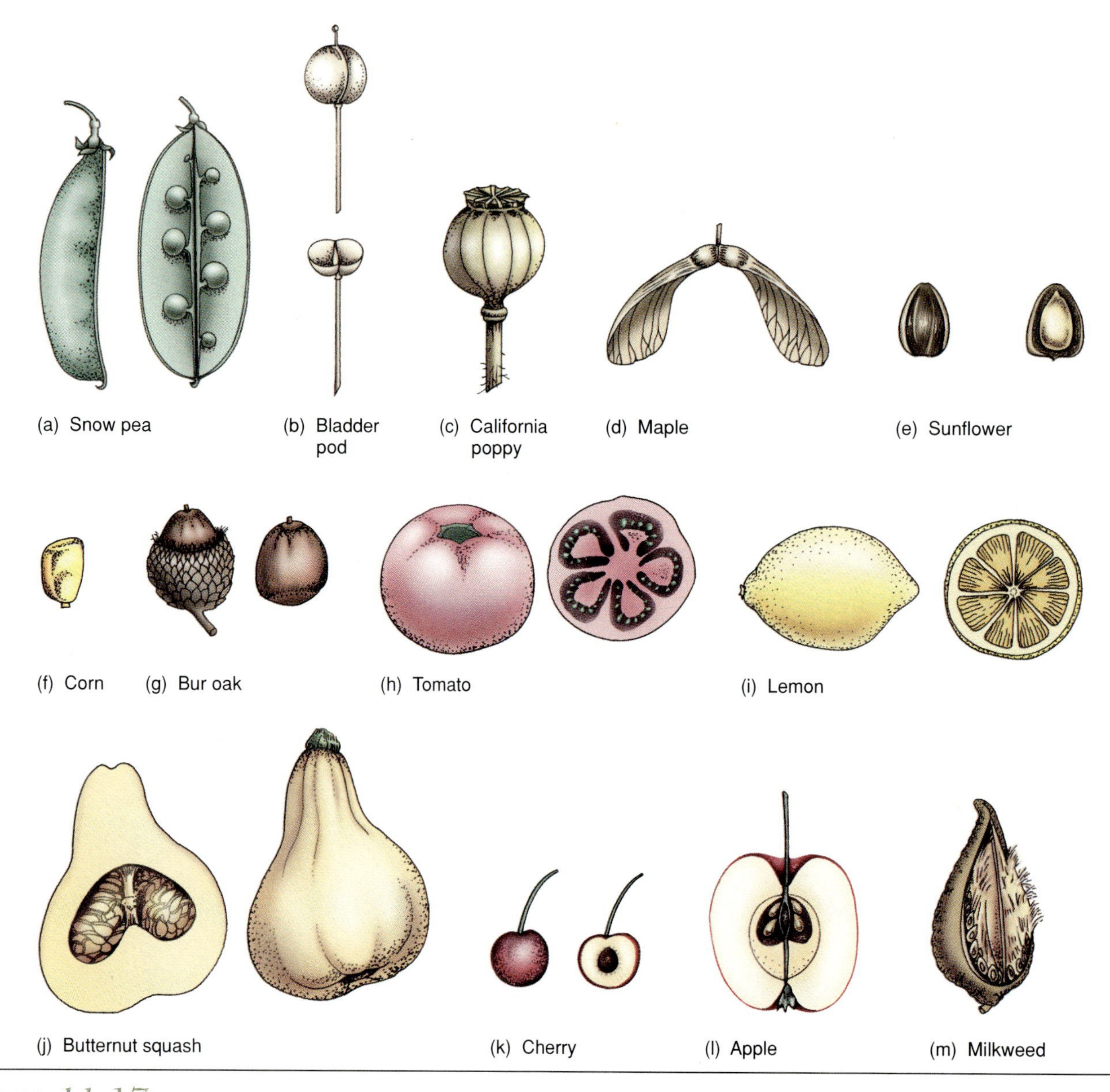

Figure 11.17

Examples of simple fruit. (*a*) Legume. (*b*) Silique. (*c*) Capsule. (*d*) Achene. (*e*) Samara. (*f*) Caryopsis. (*g*) Nut. (*h*) Berry. (*i*) Hesperidium. (*j*) Pepo. (*k*) Drupe. (*l*) Pome. (*m*) Follicle.

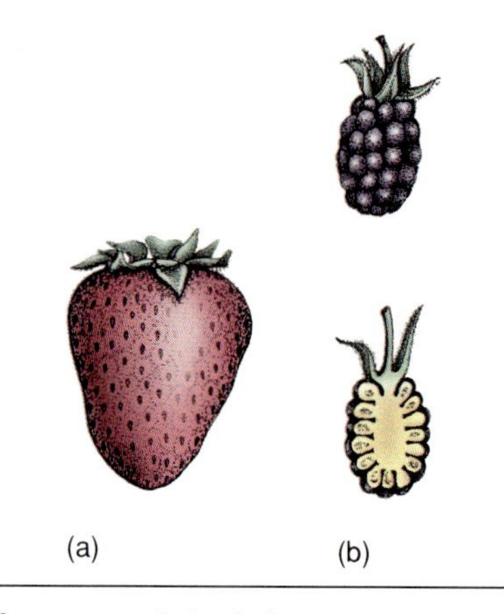

Figure 11.18

Compound fruit include the strawberry (*a*), which has accessory tissue, and the aggregate blackberry (*b*).

success. The role of fruit in the reproductive cycle of plants is seed dispersal. The modifications therefore reflect the methods by which the fruit is transported—by wind, water, or animal. If the fruit (and the seed within) is dispersed to areas away from the parent plant, each new plant seedling has a better chance for adequate space, water, and nutrients, which are needed to survive. Additionally, the species thereby has the opportunity to increase its total distributional range. The genetic variability of new generations contributes to their success in new environments.

Wind Dispersal

Very small, lightweight fruit can be carried some distance from the parent plant by the wind (figures 11.19 and 11.20). Dandelion fruit, with their parachute tufts of soft, fine bristles, and the winged fruit of maple trees are wind borne.

Water Dispersal

Some fruit have thick, fibrous outer coverings that provide buoyancy and protection from salt water. The coconut's husk has ensured its dispersal

Figure 11.19
The seeds of the cottonwood (*Populus deltoides*) are carried by the wind after release at fruit maturity.

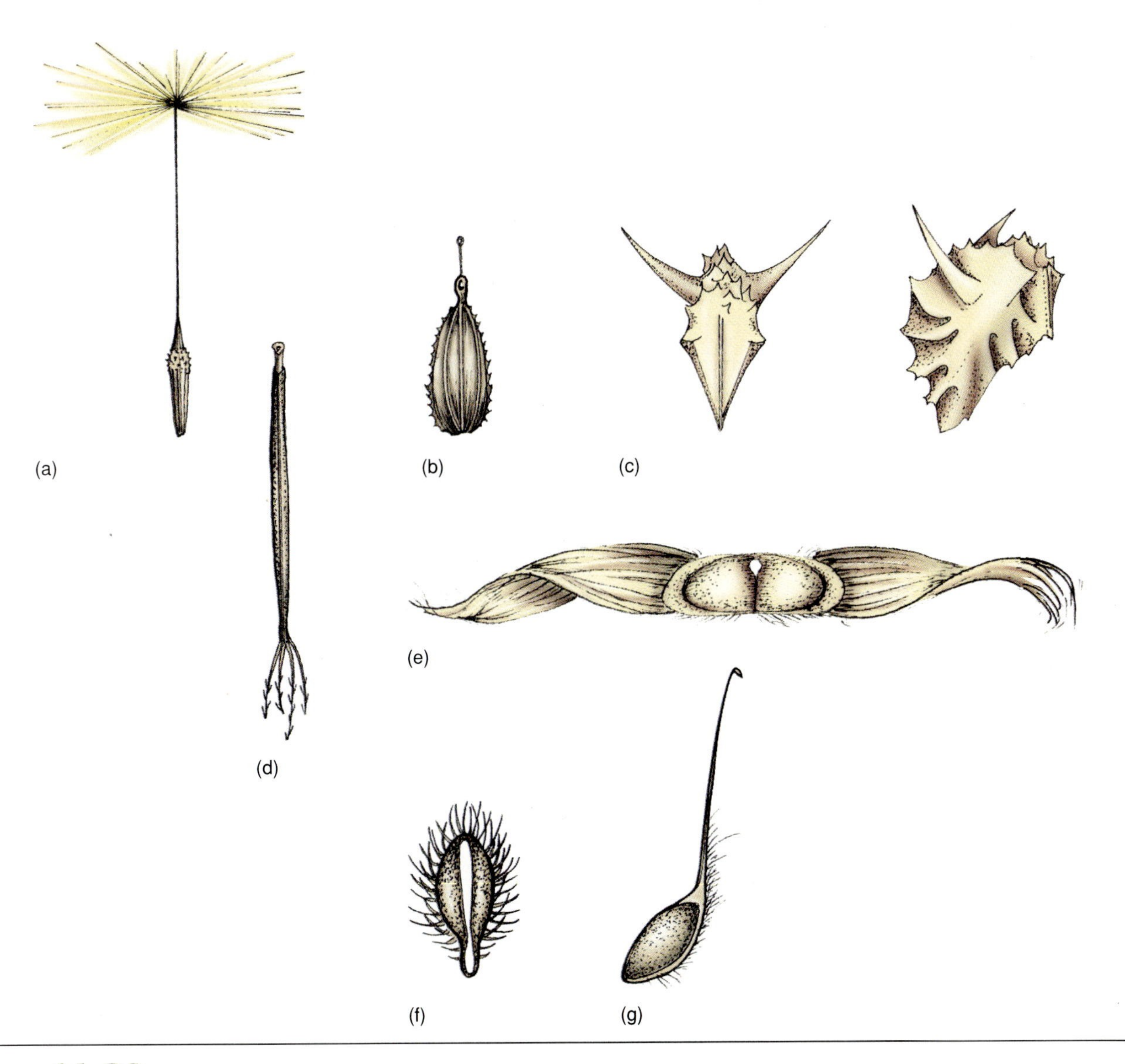

Figure 11.20
Fruit modifications allow for dispersal by several different mechanisms. (*a*) Dandelion (wind). (*b*) Carrot (animal fur). (*c*) Goathead (animal fur and bicycle tires). (*d*) *Bidens* (animal fur). (*e*) *Chilopsis* (wind). (*f*) *Sanicula* (animal). (*g*) *Geum* (animal).

to virtually every tropical sandy shoreline in the world. Freshwater streams and lakes also are dispersal agents for buoyant fruit of many of the plants found only in these habitats. Some fruit float because of low-density buoyant tissue, whereas others have air inside them. Other fruit are not particularly buoyant or water resistant but can float long enough to be carried at least a short distance from the source.

Rainwater can act as a dispersal agent. A reasonably heavy downpour will help knock some fruit from the parent plant and wash them away if there is any incline available to produce a runoff.

Animal Dispersal

When ripe, many fruit are brightly colored, thin walled, and juicy, especially those that are red. Although not attractive or even visible to most insects, red fruit are highly visible to birds and other animals. Many such fruit are also sweet, making them even more attractive to animals. When the fruit is ripe, the enclosed seeds are dry and hard enough to pass through the digestive tract of the animals. The animals distribute the seeds effectively throughout their range of movement.

Although the seeds pass through the animal's digestive tract intact, the seed coats are altered by the digestive acids acting on them. Many actually could not germinate without this scarification. Immature seeds cannot survive the action of these juices; thus most fruit are green when unripe, camouflaged against the green leaves of the plant. The level of sugar in these fruit remains low until maturity, further reducing their attractiveness to animals before the enclosed seeds are mature. Some fruit even taste bitter before completely ripening to discourage animals from eating them before they are mature.

Not all fruit that depend on animals for dispersal are juicy and sweet; many hard-shelled nuts and dry fruit such as grass grains are gathered and stored away for the winter by squirrels and other small mammals such as packrats and field mice. Since not all of these seeds are eaten, some may germinate when the conditions are appropriate.

Still other fruit are dispersed externally by hitching a ride on the animal. Such fruit are externally modified with barbs, hooks, bristles, or sticky surfaces by which they adhere to the fur, hair, feathers, or skin of animals (figures 11.20 and 11.21).

Figure 11.21
The devil's claw (*Proboscidea*) is one of the most unusual fruit. Its two long hooks catch on animals' legs, where it is carried until it splits open and releases its seed.

> **✔ Concept Checks**
>
> 1. Define the following terms: fruit, pericarp, accessory fruit, simple fruit, compound fruit.
> 2. Give a common example for each of the following fruit types: legume, achene, caryopsis, nut, berry, drupe, pome.
> 3. Discuss at least three mechanisms for fruit dispersal. Give an example for each.

SEED STRUCTURE

Sexual reproduction in flowering plants culminates when ovules develop into seeds. Some fruit have only one seed and others thousands; some seeds are very large (palm) and others microscopically small (orchids); some must germinate immediately, and others can remain dormant for many years. Seeds, however, have certain common characteristics, including a tough seed coat or testa, which encloses the embryo and its nutritive tissue, the endosperm. When the seed reaches maturity it often becomes dormant or inactive until the germination process begins.

SEED ADAPTATIONS

Seeds that display adaptive features do so for improved dispersal and reproductive success, and most of their modifications are similar to those found in fruit. Seeds released from their surrounding dehiscent fruit fall either to the ground or into water. Those which land in water are often as buoyant as fruit and are dispersed as readily. Small lightweight seeds can be carried away by the wind, whereas larger seeds develop wings, which allow them to spin or flutter through the air. It is possible for the seeds to land far away from the parent plant.

Some seeds are actually edible; some, like that of *Erythrina,* only appear to be. *Erythrina* seeds possess brightly colored coats that are displayed in open fruit to entice passing birds (figure 11.22).

Seeds also develop a full array of surface modifications to promote adherence to animals. Some others have been found economically useful, such as the long surface fibers on the cotton seed (figure 11.23).

One of the most interesting dispersal mechanisms is the physical ejection of seeds by the fruit. Several species are explosively dehiscent, propelling seeds up to several meters away from the parent plant. Some split open and eject violently when a certain degree of dryness is reached.

Figure 11.22
The seeds of *Erythrina* extruded from their fruit at maturity. The bright red color attracts birds, who carry them only a short distance and then spit them out because they are very hard.

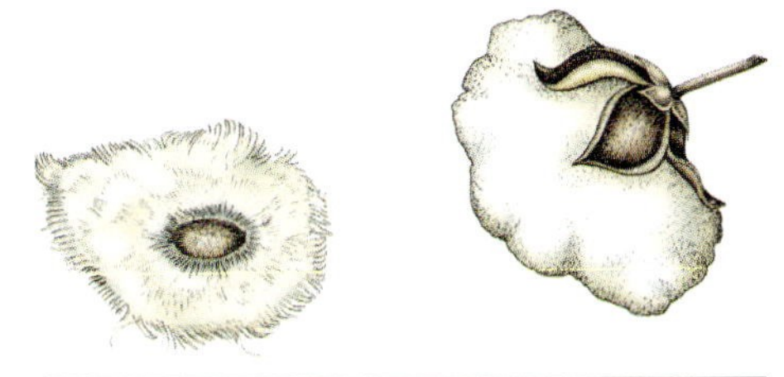

Figure 11.23
The seeds of commercially grown cotton have long fibers on their surface, which are economically desirable. In wild strains such a modification would be important in seed dispersal.

Other seeds actually germinate while still attached to the parent plant. These viviparous seeds later fall to root in the soft soil and mud. Still other seeds (figure 11.24) are capable of "planting" themselves.

Once dispersed, seeds next must germinate and grow to maturity. Although not normally considered modifications, the different strategies that have evolved controlling seed germination and seedling growth are responses to environmental factors.

SEED GERMINATION

The germination of the seed begins a new plant life. Even though the embryo is formed and has its primary tissues well developed, the mature seed may be stored for varying periods of time and still retain its viability—the ability to germinate. Only when the proper environmental conditions are provided does the seed revitalize and produce a seedling. For seeds such as *Acer* (maple) such longevity may be only one week; but other seeds, such as *Nelumbo* (lotus), may retain viability for hundreds of years under proper storage conditions. Most common cereal plants retain viability for about ten years.

The environmental requisites for seed germination include suitable oxygen concentrations, temperature, moisture, and in some cases light. Mature seeds are dry; for germination to begin, these dry tissues must be hydrated. It may be difficult to tell that a dry and dormant seed is really living, but respiration and metabolism continue throughout dormancy at a much reduced level.

If moisture enters the seed coat, a strictly physical process called imbibition causes the tissues to swell with enormous expansion forces. This process will occur even in dead seeds, or in sticks of wood that become wet. Dry seeds can be placed in plaster of paris and hardened into a block. When the block is wetted, the imbibing seeds will swell so dramatically that they will shatter the plaster. It is said that the Egyptian pyramid laborers drove pegs of wood into holes bored in rock, poured water around the pegs, and thereby split huge boulders with ease.

The amount of moisture required for seed germination varies greatly among species. Some seeds germinate in what might be considered a very dry soil. Modification of the seed coat allows some surfaces to act as a wick and absorb more water. Some seeds have mucilaginous coatings that attract water and enhance the imbibition process.

Figure 11.24
(*a*) Stork's bill (*Erodium*). The long, slender "beak" of the seed dries and twists like a corkscrew, literally screwing the seed into the ground (*b*) and (*c*).

To the naked eye the imbibition process produces a swollen seed, with a large increase in volume and weight simply from the uptake of water. The next observable step is usually the protrusion of the **radicle,** the embryonic root. It may protrude hours, or even days, before the first sign of the shoot.

The growing point of the shoot above the point of **cotyledon** (embryonic leaf) attachment is called the epicotyl, and the section of stem below the cotyledons is called the hypocotyl (figure 11.25). At the base of the hypocotyl the transition zone separates the shoot from the root. The epicotyl consists of a short stem apex and primordial primary leaves.

Some seedlings emerge with the cotyledons rising above the ground (epigeous germination) and participating for a short while in photosynthesis before they shrivel and fall off (figure 11.25). Others, such as the pea, emerge with the cotyledons still underground, in approximately the same place that the dry seed was placed (hypogeous germination). The difference is strictly a genetic one and is determined by whether the hypocotyl elongates sufficiently to elevate the cotyledons above the soil. Epigeous germination is an adaptation for exerting force at the soil surface to penetrate the soil crust while protecting the delicate shoot.

Investing in Our Future—Seed Banks

Aside from the high caloric value making them a significant food source, over 70% of the world's food supply is derived directly from seeds. Because plants often produce seed in large quantities that can be readily harvested they have traditionally been the most common, practical and inexpensive method for propagation. In addition, when seeds are kept cool, between –16°C and 5°C, and dry, they offer a convenient means of storing plants over long periods of time.

In 1956 Congress appropriated funds for the construction of the National Seed Storage Laboratory in Fort Collins, Colorado. This seed bank, also termed a gene bank, has a twofold function: (1) conduct research on ways to extend seed viability, and (2) preserve for posterity the valuable diversity of plant germplasm (tissue containing the hereditary information). At present over 230,000 germplasm collections are maintained in storage. Since seeds will not remain viable indefinitely, samples are periodically tested for germinability and if low, are germinated, grown to maturity, and self-pollinated with the resulting seeds collected for continued storage. In many cases plant pollen or tissue cultures can also be stored, and with the application of proper hormones and environmental conditions, they can be grown into new plants.

In an attempt to help alleviate worldwide hunger, plant breeders have improved the agricultural yield of crop plants by genetically manipulating for improved resistance to insects, diseases, and environmental stresses. During this process however, the genetic variability among crops has been reduced, making crops genetically vulnerable to epidemic (epiphytotic) losses. Tragic epiphytotics have occurred since biblical times: the Irish potato famine of the 1840s, the Ceylon coffee rust in 1870, the United States wheat rust in 1916, the Bengal rice problems in 1942, and the southern United States corn leaf blight of 1970. Epidemic losses can be minimized by retaining the original genetically varied stock from which plant breeders would be able to produce new crop strains. Such action might seem relatively easy to effect, but in fact, the wild relatives of crop plants are vanishing at an alarming rate due to destruction of natural habitats through urbanization, monoculture, and industrial development. The loss of genetically varied plant stock will place all humankind in a precarious position. Seed banks, genetic repositories of variability, are essential in conserving the genetic resources of the plant kingdom (see biodiversity, chapter 22) since extinction, the loss of species and their hereditary information, is permanent.

Germination Factors

Oxygen

As the ovule matures, the cell layers that make up the seed coat become rather impermeable to water, and this ensures that during the long period of dormancy and storage the seed will not lose too much water and that respiration will remain at a low rate. On the other hand, it is essential for the seed coat to retain sufficient permeability to begin the absorption of oxygen whenever renewed growth begins. Some seeds, such as many kinds of beans, become so impermeable to both water and oxygen that germination cannot proceed. These seeds must go through an abrasion of the seed coat (scarification) to allow penetration of water and gases.

Light

Light can influence germination in some species. Since most seeds germinate underground, the general trend is for germination to occur in darkness. However, some seeds, particularly those that are very small and that germinate right at the soil surface, may be light sensitive. Certain types of lettuce seeds, for example, will germinate only if they are exposed to light. Researchers studying this phenomenon discovered that red light at a wavelength of about 660 nm stimulated germination. Interestingly enough, if the same seeds were exposed to light of a slightly longer wavelength, about 730 nm (far red), germination was effectively inhibited.

Environmental Interactions

Most adaptive strategies, including those for germination, do not depend on a single environmental factor. More likely, several factors may interact to trigger the germination response. In the desert, for example, ephemerals complete their life cycle in a matter of a few weeks. Their chance of success is dictated almost entirely by an interaction of temperature and rainfall. Moisture is needed for imbibition to initiate germination, but a sudden summer thundershower could spell doom for a seed that began to develop at that time of the year. Fortunately, summer temperatures are too high for germination to begin; the seedling could never survive the intense summer heat. Germination must wait until early spring or fall. At that time there is adequate moisture and cooler temperatures.

The same situation holds true for species in alpine meadows, where the growing season is very short because of temperature restrictions. Moisture may be always plentiful, but germination too early might cause a young seedling to die because of frost. Even though a few warm days may occur in early spring, the moisture-temperature combination signals the delay of germination until later in the season, when all danger of frost is past. Once the seedling has been established, the life cycle must be completed rapidly before the onset of cold weather in late summer or early fall.

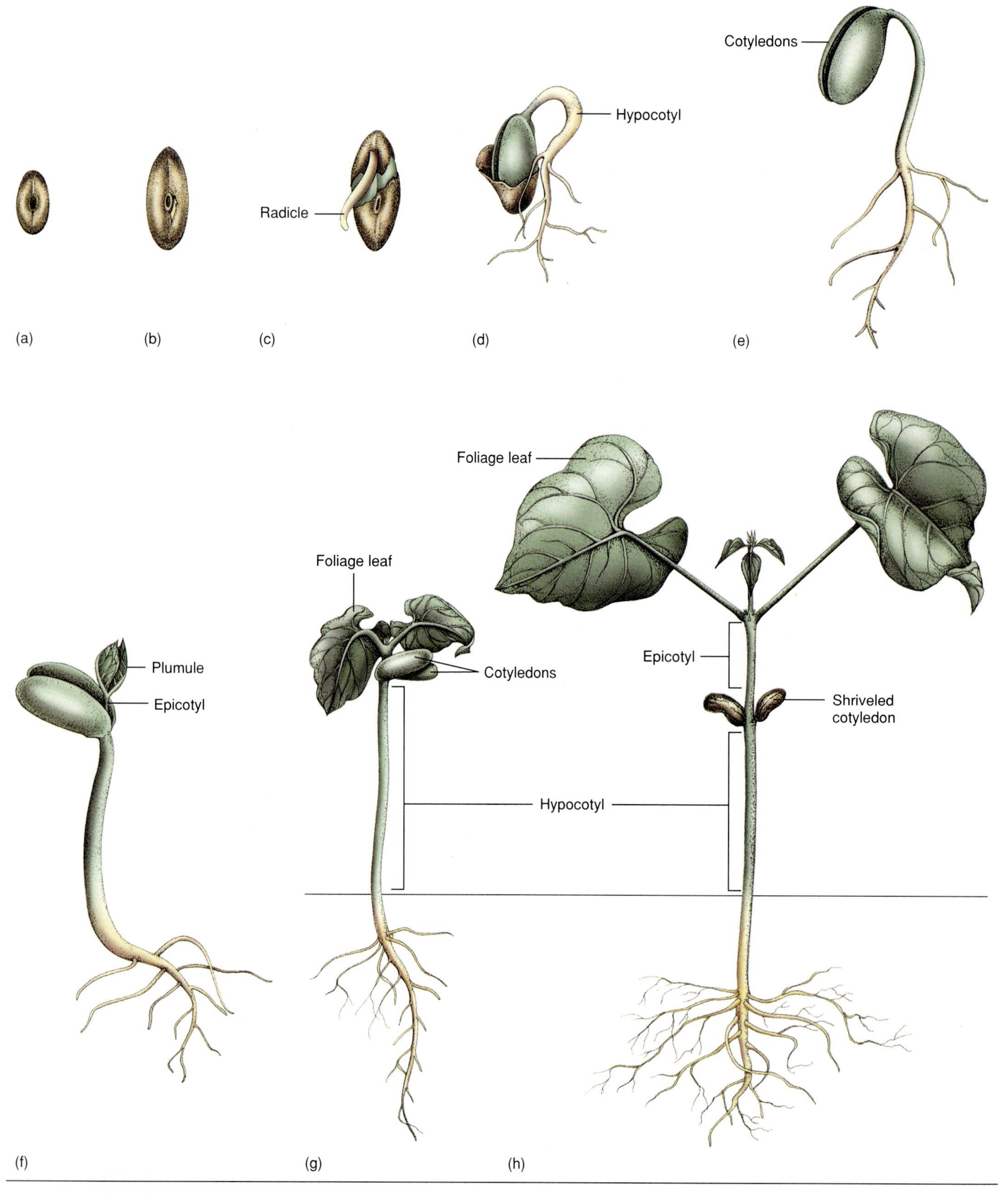

Figure 11.25

Epigeous seed germination through seedling establishment. The dry seed of the bean (*Phaseolus*) (*a*) swells when it imbibes or absorbs water (*b*). The radicle protrudes (*c*) and begins developing into the root system (*d*). The hypocotyl "hook" next emerges aboveground (*e*); then the first true leaves and shoot apex, the plumule, begins growing (*f*). As true leaves continue to develop (*g*), the nourishment provided by the cotyledons is replaced by the energy from photosynthesis, and the cotyledons shrivel (*h*) and fall off.

Dormancy Factors

Mechanical Barriers

Many times during ovule development the cells of the seed coat become so lignified (impregnated with lignin) or otherwise become so hard that uptake of water and oxygen is essentially impossible. As the seed ages, various chemical and physical forces gradually break down the seed coat so that it finally does become penetrable. This might come about by freezing and thawing, by the seed passing through the digestive tract of animals, or by wind abrasion and water erosion. Not all seeds from the same plant, even if produced in the same year, will have the same coat thickness. Some of them may be thin and germinate the first year, those of medium thickness may germinate the second year, and the very hard ones may germinate several years later. Such variability increases the chances for survival, even if no seeds at all are produced in certain years.

Physiological Factors

Occasionally seeds appear to be mature, but in fact the development of the embryo is not yet sufficiently advanced to allow germination to proceed. In species where this occurs it is necessary to delay germination until embryo maturity is completed. Even under the microscope the embryo may appear to be mature, but certain chemical adjustments are necessary before germination will proceed.

Chemical Factors

Sometimes germination is inhibited by the accumulation of chemicals. This appears to be an adaptive strategy to spread out the germination process over time. As the seed ages, certain chemicals may be broken down until the concentration is so low that germination can proceed. More often, the chemical inhibitors are water soluble, and with rainfall they are simply leached from the seed. Sometimes the concentration is such that very little leaching is needed; other seeds may have very high concentrations, which require years of leaching before the seed will germinate. Again, the adaptive strategy is clear: if all seeds were to germinate at one time, and it happened to be the wrong time, the species could become extinct.

SEED TO SEEDLING

The mature seed contains a new embryo, a miniature plant, complete in every detail. In the germination process the embryo expands its existing tissues with the aid of turgor pressure and, through the process of cell division, adds new tissue. The digestion and respiration of storage molecules (carbohydrates, proteins, or lipids), found in either the endosperm or cotyledon(s), provides the energy necessary for building these new parts.

Environmental factors are critical in the germination process and in the establishment of the seedling. It is not strictly by chance that the radicle begins to emerge and grow faster than the shoot. Although proper temperature and oxygen are obviously important in the process, proper moisture is probably the most critical factor in the success or failure of the new plant. An emerging root system must remain in contact with moist soil, or the embryo will die. It is absolutely essential that the root system grow rapidly enough to penetrate soil to depths that provide additional moisture. In many situations seeds may germinate on a moist soil surface but fail to reach deeper soil moisture before the surface layers dry. This is one of the critical factors in dryland farming (no supplemental irrigation). The same principle also holds in natural ecosystems. Not only is rainfall critical to germination per se, but stored soil moisture also is essential to seedling establishment. Some genetic strains are particularly adapted for seedling vigor and are thus better able to survive under stress conditions.

If the embryo does become established, it proceeds through its life cycle as a rapidly growing seedling, reaching vegetative maturity, passing into reproductive maturity (the ability to flower and set seed), and finally into senescence (literally, old age). Some plants complete this process in one growing season and are called annuals. Most crop plants, like corn, are annuals.

A second group of plants, the biennials, require two growing seasons to complete their life cycle. Many members of the cabbage family (Brassicaceae), among others, require one season of vegetative growth followed by a cold period to induce flowering the following year. During the first year, growth is strictly vegetative, and many leaves are produced on short internodes of stem forming the cabbage "head." In the spring of the second year the stem begins to elongate rapidly and produce flowers from which fruit and seeds will be formed.

The third group of plants are perennials. They continue vegetative growth and reproduction during their indefinite life span. All woody plants are perennials and many are specially adapted to survive harsh winters and revive in the spring. Many perennials go through a long period of vegetative growth before they begin to flower. It is not known why these plants require such lengthy leaf production. Citrus trees, for example, may need eight to ten years of vegetative growth prior to the onset of flowering. If farmers could hasten the beginning of reproductive activity, they could bring orchards into production sooner.

Some perennials, such as chrysanthemums, die back to the ground level each winter but produce new growth each spring from an underground crown of stem tissue. Such plants are called herbaceous perennials.

Botanists categorize plants as annuals, biennials, or perennials according to life cycle. However, there are complications where certain plants are concerned. Consider *Agave americana,* the century plant, so named because it remains vegetative for many years, flowers once, and dies. Is this plant truly perennial, or should it be considered an annual or some modified form of biennial? One way to simplify the problem is to classify plants only according to whether they flower once or more than once. Since botanists often refer to fruit as ripened ovaries or carpels, we can call all plants that flower once monocarpic plants. Those plants that flower over and over again are called polycarpic plants. The latter scheme is definitely simpler, but the terms annual, biennial, and perennial will probably continue to be used, since they are readily understood.

MONOCOTYLEDONS AND DICOTYLEDONS

One of the impressive developmental phenomena of angiosperms is that the two major subgroups, Monocotyledons (monocots) and Dicotyledons (dicots), are so easily distinguished throughout their life cycles. As seeds, the monocots (*mono,* one; *cot,* cotyledon) have only a single cotyledon, or embryonic leaf; the dicots (*di,* two) have two. As they develop into seedlings and adult flowering plants, their morphological distinctiveness continues to be evident. Table 11.2 summarizes the major differences between these two groups of flowering plants (see also figure 11.26).

SUPERMARKET BOTANY

Early in human history plants were gathered for food or medicine by small populations of predominantly nomadic people. In time, humans evolved from food gatherers to food growers. Members of the primitive farming communities recognized the value of some of the plant modifications and were able to capitalize on them. Wild strains that possessed beneficial variations could be selectively propagated. These adaptations might have included plants with larger edible roots, increased seed yields, and the capacity to withstand unusually adverse environmental conditions.

Once humans began domesticating plants, they were able to select individual plants and species systematically according to desirable traits. Not only could usefulness and yields be increased, but also new strains could be developed through hybridization, the mating of unlike parents, and plant breeding. These techniques are employed today in the development of many commercially valuable consumer products. Although these products have a wide variety of applications, certainly the most important is as a food source.

The vast majority of food for human consumption comes from the flowering plants (figure 11.27). Most of us are familiar with the broad selection available in the produce section and the canned fruit and

TABLE 11.2

Monocotyledon and Dicotyledon Characteristics

"MONOCOTS"	"DICOTS"
One cotyledon	Two cotyledons
Flower parts in threes or multiples of three	Flower parts in fours or fives or multiples of these numbers
Herbaceous, almost never woody	Can be woody or herbaceous
Usually linear leaves with parallel venation	Leaves with netted (reticulate) venation
Scattered vascular bundles in the stem	Vascular bundles in a ring
Fibrous, most adventitious root systems	Taproot system

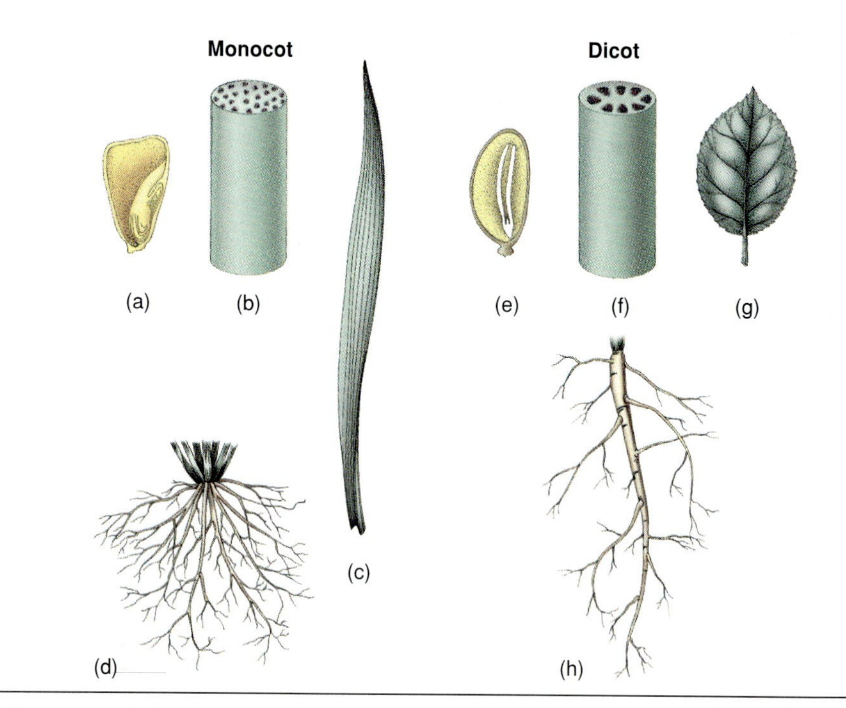

Figure 11.26

Monocot and dicot differences. Monocots have one cotyledon in the seed (*a*), scattered vascular bundles in the stem (*b*), parallel leaf venation (*c*), and fibrous root systems (*d*). Dicots have two cotyledons in the seed (*e*), stem vascular tissue in a single ring (*f*), netted leaf venation (*g*), and a taproot system (*h*).

vegetable aisles of the supermarket. The commercial designation of these plant products does not always reflect their botanical function. Squash, cucumber, pepper, tomato, and corn are commonly called vegetables ("Eat your vegetables, dear."), even though they are reproductive, not vegetative, parts of the plant. Many other grocery store "vegetables" are also misnamed. Beans and peas, for instance, are actually fruit (green beans) or seeds (lima, kidney, pinto, navy, soybeans, and green and black-eyed peas). In addition, many of these consumer items belong to the same plant family. Table 11.3 lists some of the more common edible plant materials by family and identifies the actual part of the plant that is eaten. Scientific names or binomials (genus and species names—chapter 16, p. 287) are also included. Note how frequently several species of a single genus (for example, *Brassica*) are commercially developed consumer products.

✔ Concept Checks

1. Distinguish between a fruit, a vegetable, and a seed. What characteristics do seeds share?
2. Describe the events that occur from the time that a seed is watered until it has produced photosynthetic foliage leaves.
3. Cite three factors that affect germination. What is dormancy? What causes seed dormancy?

Figure 11.27
Some commercially produced plants. (*a*) Radishes (roots). (*b*) Papaya (fruit). (*c*) Coconut (seed). (*d*) Peppers (fruit). (*e*) Strawberries (accessory fruit). (*f*) Cranberries (fruit).

TABLE 11.3

Common Food Plants

FAMILY NAME	COMMON NAME	SCIENTIFIC NAME	PLANT PART
Monocotyledons			
Gramineae (Poaceae) (grass family) All cereal grains belong to this family of plants, as does sugarcane, a great source of granulated sugar	Wheat	*Triticum aestivum*	Fruit
	Rice	*Oryza sativa*	Fruit
	Corn	*Zea mays*	Fruit
	Barley	*Hordeum vulgaris*	Fruit
	Oats	*Avena sativa*	Fruit
	Rye	*Secale cereale*	Fruit
	Sorghum	*Sorghum bicolor*	Fruit
	Sugarcane	*Saccharum officinarum*	Stem
Liliaceae (lily family)	Onion	*Allium cepa*	Bulb (leaves)
	Garlic	*Allium sativa*	Bulb (leaves)
	Asparagus	*Asparagus officinalis*	Young stem
Bromeliaceae (pineapple family) A family of mostly epiphytic plants; pineapples are rooted in soil	Pineapple	*Ananas comosus*	Fruit
Dicotyledons			
Cruciferae (Brassicaceae) (mustard family) Most food plants in this family have a tangy, sharp taste	Mustard	*Brassica alba*	Seeds
	Broccoli	*Brassica oleracea*	Inflorescence
	Cabbage	*Brassica oleracea*	Leaves
	Cauliflower	*Brassica oleracea*	Young inflorescence
	Brussels sprouts	*Brassica oleracea*	Lateral buds
	Turnip	*Brassica rapa*	Root
	Radish	*Raphanus sativus*	Root
	Watercress	*Nasturtium officinale*	Leaves
Leguminosae (Fabaceae) (bean family) A very large family having many members with high protein levels and nitrogen fixation in roots nodulated with the bacterium *Rhizobium*	Broad bean	*Vicia faba*	Seed
	Green bean	*Phaseolus vulgaris*	Fruit
	Pinto bean	*Phaseolus vulgaris*	Seed
	Navy bean	*Phaseolus vulgaris*	Seed
	Kidney bean	*Phaseolus vulgaris*	Seed
	Lima bean	*Phaseolus lunatus*	Seed
	Black bean	*Phaseolus mungo*	Seed
	Soybean	*Glycine max*	Seed
	Green pea	*Pisum sativum*	Seed
	Black-eyed pea	*Vigna sinensis*	Seed
	Peanut	*Arachis hypogaea*	Seed
Rosaceae (rose family) A large family with many different kinds of fruit	Cherry	*Prunus avium*	Fruit
	Apple	*Pyrus malus*	Fruit
	Pear	*Pyrus communis*	Fruit
	Peach	*Prunus persica*	Fruit
	Plum	*Prunus domestica*	Fruit
	Apricot	*Prunus armeniaca*	Fruit
	Blackberry	*Rubus canadensis*	Fruit
	Strawberry	*Fragaria virginiana*	Fruit (receptacle)

TABLE 11.3

Common Food Plants—Continued

FAMILY NAME	COMMON NAME	SCIENTIFIC NAME	PLANT PART
Dicotyledons, Continued			
Solanaceae (nightshade family) This family also contains other economically important species, including tobacco and a number of poisonous members	Tomato	*Lycopersicon esculentum*	Fruit
	Potato	*Solanum tuberosum*	Tuber (stem)
	Peppers (jalapeno, bell, cayenne, etc.)	*Capsicum*	Fruit
	Eggplant	*Solanum melongena*	Fruit
Cucurbitaceae (gourd family)	Squash	*Cucurbita*	Fruit
	Pumpkin	*Cucurbita pepo*	Fruit
	Cucumber	*Cucumis sativus*	Fruit
	Gherkin	*Cucumis anguria*	Fruit
	Watermelon	*Citrullus vulgaris*	Fruit
	Honeydew melon	*Cucumis melo*	Fruit
Chenopodiaceae (goosefoot family)	Beet	*Beta vulgaris*	Root
	Spinach	*Spinacia oleracea*	Leaves
Compositae (Asteraceae) (sunflower family) Although one of the largest families, it contains very few economically important plants	Sunflower	*Helianthus annuus*	Seeds (fruit)
	Artichoke	*Cynara scolymus*	Inflorescence (bracts)
	Lettuce	*Lactuca sativa*	Leaves
Umbelliferae (Apiaceae) (carrot family)	Carrot	*Daucus carota*	Root
	Celery	*Apium graveolens*	Petiole
	Parsley	*Petroselinum crispum*	Leaves and stem
Convolvulaceae (morning glory family)	Sweet potato	*Ipomoea batatas*	Root

Summary

1. The flower, the site of sexual reproduction in angiosperms, has four main parts: sepals, petals, stamens, and carpel. The stamens produce pollen; the ovary of the carpel contains the ovules. A simple carpel contains only a single carpel; a compound carpel is made up of two or more carpels. The ovary, which can have a superior or inferior position, matures into the fruit, and the ovules mature into the seeds. Flowers occur singly or in an inflorescence and can be perfect or imperfect. Imperfect flowers are either monoecious or dioecious.
2. Flower shape is highly variable, but is either regular (radially symmetrical) or irregular (bilaterally symmetrical). Floral parts can be fused or distinct and have a wide variety of colors, patterns, sizes, and other modifications. Flower adaptations exist for pollinator attraction. Most pollination is by animals, especially insects, but wind pollination is also common in nonshowy flowers.
3. Flowering plants are either self-pollinating or cross-pollinated. Self-incompatibility ensures outcrossing. Flower-insect specificity can be very highly developed and results from long periods of coevolution.
4. Fruit develop only if fertilization occurs in the ovules. Parthenocarpic fruit develop without fertilization and are normally seedless. A mature fruit usually has three layers to its pericarp or fruit wall. The pericarp can be fleshy or dry; when dry, it can be either dehiscent or indehiscent.
5. There are simple fruit and compound fruit, which include aggregate and multiple fruit, according to how they develop. Most fruit adaptations are dispersal mechanisms. Small light fruit are wind dispersed, whereas other fruit can be water dispersed or animal dispersed.
6. Seeds contain the embryo and nutritive tissue. Seed adaptations are also a result of developing dispersal mechanisms.
7. Seed germination can occur immediately following seed dispersal or up to many years later. The first step is imbibition, the absorption of water. The emergence of the radicle precedes the emergence of the cotyledons, epicotyl, hypocotyl, and plumule.
8. Several mechanical and environmental components affect seed germination. The thickness of the seed coat or the

presence of chemical inhibitors produces dormancy until scarification or washing breaks that dormancy. Oxygen, temperature, and light all can affect the germination of seeds, either individually or in combination.

9. The development of a seedling until it is an adult plant involves many physiological and anatomical changes over time. Annuals develop much more quickly than biennials and perennials, the latter being either woody or herbaceous. The terms *monocarpic* and *polycarpic* refer to the number of times a given plant flowers in its lifetime.

10. Angiosperms contain two natural subgroups, the monocots and dicots. Each has several sequentially less inclusive groups, including families, genera, and species. Scientific binomials always include the genus and species. The supermarket name for a plant part often differs from the botanical name for the part. Many commercially useful food plants reflect vegetative and reproductive modifications.

Key Terms

Discussion Questions

1. Summarize the structural differences between monocotyledons and dicotyledons with respect to roots, stems, leaves, and flowers.
2. Envision yourself to be a flowering plant well adapted and established to a particular biome (assume a freshwater biome for this example). Describe how fruit and seed dispersal can be accomplished. What difficulties might be encountered in seed germination? What adaptations could counter these difficulties?
3. Think of ten common vegetables or fruit that could be encountered at the produce department of your local grocery store. For each, determine whether it is fruit or vegetable. If fruit, determine if they are simple or compound. If vegetables, are they modified roots, stems, or leaves? Lastly, for each try to classify by family and give the binomial name using table 11.3.

Suggested Readings

Barrett, S. C. H. 1987. Mimicry in plants. *Sci. Am.* 257(3):76–85. September 1987.

Barrett, S. C. H. 1989. Waterweed invasions. *Sci. Am.* 261(4):90–97. October 1989.

Barth, F. G. 1991. *Insects and flowers: The biology of a partnership.* Princeton University Press, Princeton.

Bawa, K. S., and M. Hadley, eds. 1990. *Reproductive ecology of tropical forest plants.* Parthenon, Park Ridge.

Cox, P. A. 1993. Water-pollinated plants. *Sci. Am.* 269(4):68–74. October 1993.

Marshall, J. G., ed. 1992. *Fruit and seed production.* Cambridge University Press, New York.

Moore, P. D., J. A. Webb, and M. E. Collinson. 1991. *Pollen analysis.* Blackwell Scientific Publications, Boston.

Nandel, S. N., and A. J. Beattie. 1990. Seed dispersal by ants. *Sci. Am.* 263(2):76–83. August 1990.

Niklas, K. J. 1987. Aerodynamics of wind pollination. *Sci. Am.* 257(1):90–95. July 1987.

Ottavilano, E., ed. 1992. *Angiosperm pollen and ovules.* Springer-Verlag, New York.

Vogel, S. 1990. *The role of scent glands in pollination: On the structure and function of osmophores.* Smithsonian Institution Libraries and National Science Foundation, Washington, D.C.

Weberling, F. 1989. *Morphology of flowers and inflorescences.* Cambridge University Press, New York.

Willemstein, S. C., ed. 1988. The works of Charles Darwin. Vol. 17. "The various contrivances by which orchids are fertilized by insects." New York University Press, Washington Square.

Young, J. A. 1991. Tumbleweed. *Sci. Am.* 264(3):82–87. March 1991.

Part Four

Function and Control

The leaf of this dogwood carries on photosynthesis to provide food for the rest of the plant's tissues.

Chapter Twelve

Plant-Soil-Water Relationships

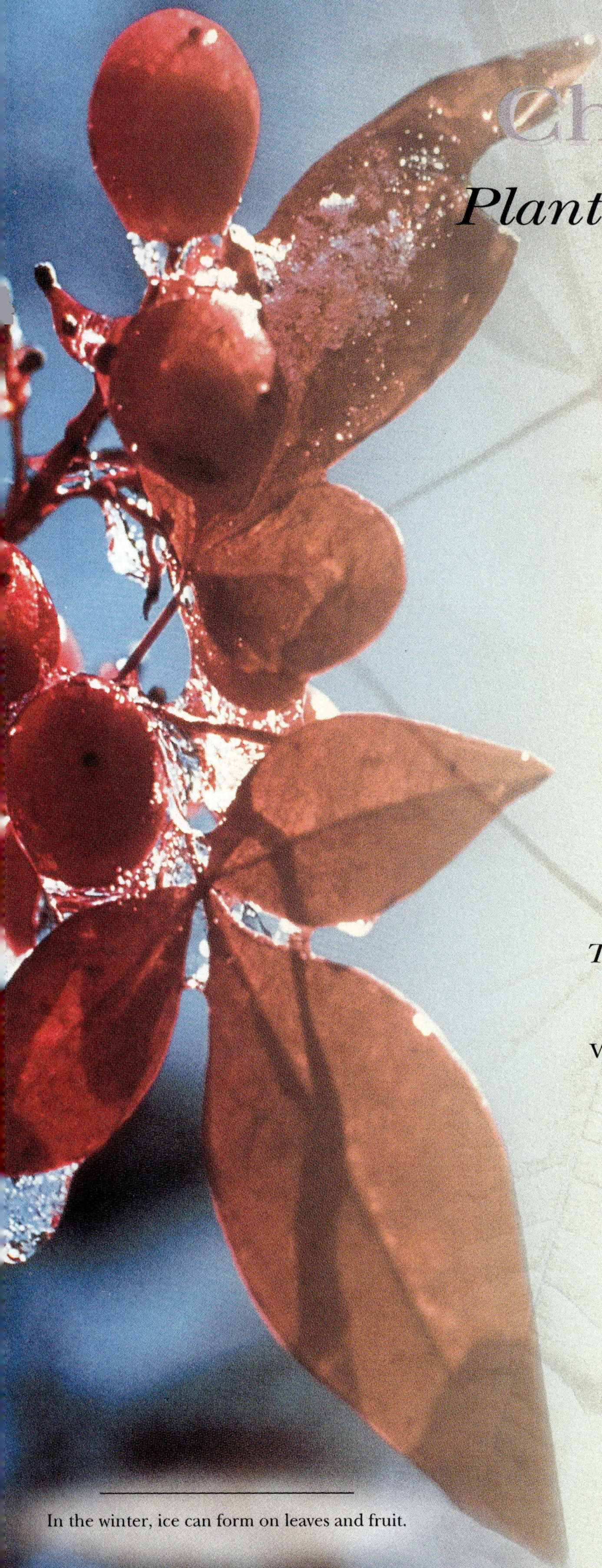

In the winter, ice can form on leaves and fruit.

Water and Nutrition

Importance of Water in Metabolism
Ionization, Dissociation, and pH
Essential Plant Nutrients

Soil

Soil Development
Soil Profile
Soil Texture

Water Movement

Kinetic Energy
Water Potential

The Soil-Plant-Air Continuum (SPAC)

Soil-Water Movement into the Roots
Water Movement Throughout the Plant Body
Transpiration
Cohesion-Adhesion-Transpiration Pull
Stomatal Regulation
Adaptations to Reduce Water Loss

Applications

Water
Fertilizer
Light
Soil
Compost
Potting
Pests

Hydroponics

This chapter presents the details of how a plant interacts with its physical surroundings—the soil and water. You will begin to understand why water becomes the ultimate limiting factor in plant distribution across the face of the earth.

The nature of soil is explained in terms of physical characteristics and how those attributes determine how much water can be held in reservoir. Then the soil/water reservoir is described in terms of how plants exploit the soil minerals and water through the roots, move them into and through the plant, and eventually lose most of the water out through the surface of the leaves in transpiration.

Of all the necessities for plant growth, water can be considered the most important. This may seem like an overly bold statement, since the absence of any one component essential for life will result in the organism's demise. In that sense, all essential components are equally important. Plants require water, carbon dioxide, oxygen, light, and certain minerals. Certainly there are a few environments in which light availability can be a problem (for example, forests with closed canopies or houseplants in dark rooms) (figure 12.1), and given soils vary in the presence and amounts of certain minerals. Consider the role of water in the activities of the other necessary factors.

WATER AND NUTRITION

The minerals required by plants are taken in from the soil and dissolved in water, and the CO_2 in the atmosphere is dissolved in water at the cell surface. Even the stomata, through which CO_2 and H_2O enter and leave the plant, are controlled by water availability to the cells "guarding" these openings. Photosynthesis, too, begins when the water molecule is split by light energy.

Importance of Water in Metabolism

Plants must have access to water all the time (figure 12.2). An aqueous environment is absolutely essential for all metabolic function; without it, life systems cease. Besides acting as a medium for dissolving solutes, water provides the **turgor pressure** for cells, much like the air pressure inside a balloon, and participates directly in several biochemical reactions.

Water is therefore critical to every plant function. For approximately 95% of all plant species this requirement is limited to freshwater. Marine plant species have an advantage, since about three-fourths of the earth is covered by oceans. Terrestrial and freshwater plants rely on a stable water cycle for their needs. Terrestrial plants, besides having to support their own weight, have another problem: since they are not surrounded by water, as aquatics are, they need a more complex method of obtaining and distributing water. Water uptake is further complicated by plant immobility. Water must come to the plant; the plant cannot go to the water.

If life as we know it exists on any other planet, it also requires water. Unmanned space exploration always includes investigation of whether there is or ever was any evidence of surface water or atmospheric water vapor.

Remember that water is unique among the various liquids on earth. Because of hydrogen bonding patterns—the continual breaking up and re-forming—water is a liquid; but the cohesive strength of those bonds is so great that water strongly resists the separation of its molecules from each other, and they can withstand the equivalent of 100 atmospheres of tension before breaking. Thus a capillary-sized column 100 m tall with a powerful suction applied at the top could easily pull a liquid column to the top. The actual height to which a column of water could be pulled is probably reduced because of impurities in most water that interfere with

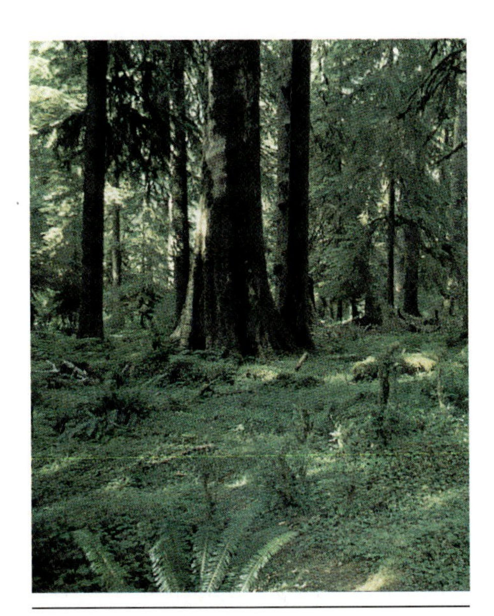

Figure 12.1
In some closed-canopy forests, water is not the limiting factor for growth, but low light intensity determines the kinds of understory species that can grow there.

Figure 12.2
A stable supply of water is absolutely critical to the growth and development of the rice plants in these paddies in Japan.

the perfect cohesive bonding of pure water. In addition, friction from the sides of the tube also affects the potential height reached. The final message, however, is the same: the great cohesive strength of water is an integral property in the movement of water through plants. The adherence of water molecules to other surfaces is also a function of hydrogen bonding. This property of water allows for capillarity, the process of climbing up the walls of a narrow-diameter tube. The resulting meniscus, or curved surface, of the water in such a tube is one of the first phenomena a student observes in a laboratory where accurate liquid measures are stressed. In plants, xylem vessels or tracheids exhibit capillary action.

Ionization, Dissociation, and pH

If minerals are to be taken up from the soil, they must first be dissolved in the soil water. When salts are dissolved in water, they **ionize,** which simply means that they dissociate into electrically charged particles. Not all salt molecules dissociate in water; some do so almost completely, others hardly at all. It is important to the plant because the root system must take up the nutrients in the ionized form.

One of the most important contributing factors in dissociation is the relative acidity of the soil solution. The acidity of any solution is determined by the number of charged hydrogen atoms or protons present. The term **pH** refers to the **hydrogen potential** and is measured on a scale from 1 to 14, describing the relative concentration of protons. The more protons present in soil solution, the greater the acidity. A pH value of 7.0 is **neutral;** pure water has a pH of 7.0. Values lower than 7.0 are considered **acidic,** and those greater than 7.0 are **basic** or **alkaline.** The pH of the soil solution determines the solubility of various soil salts. It is important to remember that all of the essential soil nutrients do occur as salts; therefore fertilizers are salts that can be good or bad, depending on their chemical makeup and concentration. Even potassium nitrate, a perfectly good fertilizer, can be a "bad" salt if it is applied in too high a concentration.

TABLE 12.1

Chemical Elements Important to Plants

ELEMENT	CHEMICAL SYMBOL	FORM AVAILABLE TO PLANT	ATOMIC WEIGHT	RELATIVE ABUNDANCE* IN PLANT TISSUE
Macroelements				
Hydrogen	H	H_2O	1	60,000,000
Carbon	C	CO_2	12	35,000,000
Oxygen	O	O_2, CO_2, H_2O	16	30,000,000
Nitrogen	N	NO_3^-, NH_4^+	14	1,000,000
Potassium	K	K^+	39	250,000
Calcium	Ca	Ca^{++}	40	125,000
Magnesium	Mg	Mg^{++}	24	80,000
Phosphorus	P	$H_2PO_4^-$, $HPO_4^=$	31	60,000
Sulfur	S	$SO_4^=$	32	30,000
Microelements				
Chlorine	Cl	Cl^-	35	3000
Iron	Fe	Fe^{++}, Fe^{+++}	56	2000
Boron	B	BO_3^-, $B_4O_7^=$	11	2000
Manganese	Mn	Mn^{++}	55	1000
Zinc	Zn	Zn^{++}	65	300
Copper	Cu	Cu^+, Cu^{++}	64	100
Molybdenum	Mo	$MoO_4^=$	96	1

With reference to molybdenum, the essential element required in smallest quantity.

Essential Plant Nutrients

Nutrition is just as important in plants as it is in animals. Cellular metabolism leading to the production of the organic molecules characteristic of all life requires only a few essential elements. In the surface layer of the earth there are some 92 chemical elements. In addition, scientists have synthesized several artificial elements in recent years. In spite of this chemical and geological diversity, only 16 elements are considered essential for all plants. Just as certain molecules are needed in greater amounts than others, so too are elements needed in greater or lesser quantities. Strictly on an arbitrary and approximate scale, essential plant nutrients are divided into those required in greater concentration, the **macroelements,** and those required only in minute quantities, the **microelements.** The designation has nothing to do with the size of the element itself, but only the quantity required for normal growth and reproduction.

The macroelements may best be remembered by the mnemonic "C HOPKNS CaFe Mg" (read this as C. Hopkins Cafe, Mighty Good). Each letter or letters represents a single element. The elements required in greatest quantity for plant growth, then, are carbon, hydrogen, oxygen, phosphorus, potassium, nitrogen, sulfur, calcium, iron, and magnesium. In addition, the microelements, also sometimes called **trace elements,** include manganese, copper, zinc, boron, molybdenum, and chlorine. These elements are required for essentially all plants. Table 12.1 indicates that iron should be a trace element,

TABLE 12.2

Functions and Deficiency Symptoms of Some Essential Plant Nutrients

ELEMENT AND NORMAL CONCENTRATION IN HEALTHY TISSUES	FUNCTION	DEFICIENCY SYMPTOMS
Nitrogen (15,000 ppm)	Constituent of amino acids, proteins, coenzymes, nucleic acids, and chlorophyll	Chlorosis, especially in the older leaves; stunting (small leaves and shortened internodes)
Phosphorus (2000 ppm)	Sugar phosphates, nucleotides, nucleic acids, phospholipids, coenzymes	Stunted; dark green color; some plants develop anthocyanin (red) pigments, especially in stems; delayed maturity
Potassium (10,000 ppm)	Not part of any known organic molecule; coenzyme and osmoregulator	Dicot leaves develop chlorosis and necrotic lesions (dead spots) on older leaves; in monocots, cells at the tips and margins of the leaves die first, spreading to the base of the leaves; stems are weakened
Sulfur (1000 ppm)	Constituent of sulfur-containing amino acids, coenzyme A, and vitamins	Deficiencies are rare; general chlorosis noted first in the young leaves
Magnesium (2000 ppm)	Constituent of chlorophyll; coenzyme, especially in reactions involving ATP	Chlorosis between the veins (the veins themselves usually remain dark green)
Calcium (5000 ppm)	Constituent of cell wall, middle lamella, and membranes; cofactor for some enzymes; may detoxify large concentrations of heavy metals	Inhibition of bud development and death of root tips; severely deficient plants become necrotic and die quickly
Iron (100 ppm)	Cofactor in enzymes of chlorophyll synthesis and other enzymes; constituent of cytochromes and ferredoxin, where it functions in electron transport	Interveinal chlorosis similar to that in magnesium deficiency, but on younger leaves; stunting

but it is often considered with the macroelements. In addition, two or three other elements may be required for certain kinds of plants. Sodium, for example, is an essential element in animal nutrition and appears to be required for certain kinds of higher plants. Its exact function is not known.

Each of these elements, whether a macroelement or microelement, does have a specific function in metabolism. Most are found as part of an organic molecule. Thus, if the element is missing, the molecule cannot be made in the cells. The only macroelement not found in organic molecules is potassium, and its function is thought to be primarily as a **cofactor,** or chemical helper in enzymatic reactions, and as an **osmoregulator.** It is required in large quantities to act as a solute in stimulating the movement of water across a membrane.

Although all carbon and some oxygen comes from the air as CO_2, and hydrogen and oxygen come from water, all the other elements must be absorbed through the roots as ions. Essentially all water, too, comes via the same pathway. The "trick" in metabolism is to bring together all the essential nutrients in the right cell at the right time so that the myriad of enzymes packaged in specific organelles can join them together with various chemical bonds to form the molecules necessary for life.

Table 12.2 gives a very brief introduction to the specific functions of some of the elements and the symptoms of plants deficient in these nutrients. Unquestionably, the single element most often deficient from soils over the face of the earth is nitrogen. That nitrogen exists in the atmosphere as the most abundant gas and in a very stable form is no assurance that nitrogen is plentiful for plants. Very few organisms have the ability to convert nitrogen gas to a form that can be taken up by plants. Certain microorganisms can do it, but higher plants have no ability by themselves to incorporate nitrogen. Even after nitrogen is fixed in the soil in a usable form, it tends to be readily lost to **volatilization** or **leaching.** One need look only at a few different kinds of organic molecules to realize the importance of nitrogen. It is found in all amino acids and therefore all proteins, in chlorophyll, in nucleic acids, and in ATP, to name a few.

Unlike animals, plants do not require any organic molecules for nutrition. They make their own, and many of these are required as essential molecules for animals. Take, for example, the requirement for vitamins. Most vitamins are fairly complex organic molecules made by plants. Since humans can synthesize only a few of the essential vitamins, we depend on plants to provide them for nutrition and good health. Green vegetables do provide certain nutritional components that you cannot get any other way except through dietary supplement.

Figure 12.3
Improper irrigation management may allow water to flow out of the furrow, wasting a valuable resource and carrying soluble plant nutrients with it.

Figure 12.4
If salts accumulate in soil or irrigation water, some crops may not grow. Regions of low rainfall are particularly susceptible to this hazard.

The essential nutrients must be available in the soil solution in the right place and at the right time. Placement in the soil is critical, for if a root does not come within a few millimeters of an ion, no uptake can occur. There is very little movement in the soil itself, and the roots must grow to the region where the nutrient happens to be. Nutrients certainly do move downward in the soil, but not in response to any magical "pull" from plants. Most movement is strictly gravitational. This movement is influenced by many factors, but perhaps the most important is the amount of rainfall or irrigation. Excessive water in the soil will tend to leach its nutrients as it moves downward below the root zone. This is a problem in many parts of the world, but particularly in regions of high rainfall. Even improper irrigation management in regions of low rainfall can aggravate the situation (figure 12.3). Loss of nutrients through leaching is a valid concern to developing agriculture in the moist tropics, and nitrogen availability is the biggest problem.

Some nutrients are far more soluble in water than others, and the pH factor is extremely important in determining **solubility.** Plant nutrients taken up as ions tend to be most soluble in a slightly acidic solution, at a pH of about 6.5. In regions of high rainfall CO_2 in the atmosphere is dissolved readily in H_2O to form **carbonic acid** (H_2CO_3). Even though carbonic acid is relatively weak, it too ionizes to produce a proton (H^+) and a **bicarbonate ion** (HCO^-_3). The high concentration of H^+ reaches the soil as **acidic rainfall** and displaces many of the ions attached to soil particles. Some ions are held more tightly than others, but H^+ can do a good job of replacing almost everything. Such soils in high-rainfall regions are referred to as acid soils, and it is almost impossible to add enough fertilizer to grow a crop; leaching removes the desirable nutrients almost as quickly as they are added. Nutrient leaching is a major problem in high-rainfall, tropical regions.

Many regions of the world now suffer from an acid rainfall caused by industrial pollution in which sulfur dioxide (SO_2) combines with moisture in the atmosphere to form sulfuric acid (H_2SO_4). This strong acid comes in contact with the soil through rainfall, and in some heavily industrialized regions, entire forests and lakes have lost most of their biological activity. This has become a severe international problem along the northeast United States-Canadian border, and is likely to become a problem in many other regions of the world.

The problem in arid regions is quite different. The rainfall there is so scanty that few salts are ever leached. Consequently, over time the problem of salt accumulation can become enormous unless there is sufficient rainfall to produce some leaching. Many western states in the United States suffer from this problem (figure 12.4).

Diagnosing plant nutrient deficiency symptoms is not easy. One must learn a great deal about the plant in question before drawing too many conclusions about nutritional disorders. Sometimes nutrient deficiencies are mistaken for insect and disease problems, drought, heat stress, cold stress, and salt stress. After some experience it becomes easier to spot specific deficiencies in some plants. In other plants, deficiencies are exceedingly difficult to diagnose, and both plant and soil analysis may be required before the diagnosis is certain.

Information concerning nutrient deficiencies in specific plants is obtained by growing the plants in a medium of known chemical composition, preferably water. By withholding one chemical element at a time, the nutritional disorder can be determined within a few weeks (figure 12.5).

(a)

(b)

(c)

Figure 12.5
Some plant nutrient deficiencies are easily diagnosed when plants are grown in a complete nutrient solution with elimination of one nutrient at a time. Sunflower plants respond rapidly to a deficiency of nitrogen (*a*), phosphorus (*b*), and calcium (*c*). Nutrient-deficient plants are shown on the left; healthy plants with a complete nutrient solution are shown on the right.

✔ Concept Checks

1. Why are the properties of water so important as a chemical vital to life?
2. What are the macroelements and microelements important to all plant life, and what is their function?

SOIL

The earth is approximately 13,000 km in diameter, but only a very shallow outer layer of the land surface sustains life. At one point in the earth's history that outer layer was essentially solid rock formed by the gradual cooling of the **igneous** material that still comprises most of the mass of the earth. An approximately 80 km thick crust of cooled **granite** now overlays a less solid (semiviscous) layer of **basalt** that gradually becomes molten farther from the surface (figure 12.6).

Granite occurs only where there is land; the ocean basins are composed of basalt covered by whatever sediments have slowly accumulated on the ocean floor (figure 12.7). In addition to the igneous material, primarily granite formed by the cooling of the earth's crust, **sedimentary rock** is formed by **cementation** and **solidification** of sedimentary deposits weathered from granite. These deposits may include shale, sandstone, and limestone. Finally, a third type of rock can develop from either granite or sedimentary rocks subjected to extreme temperature and

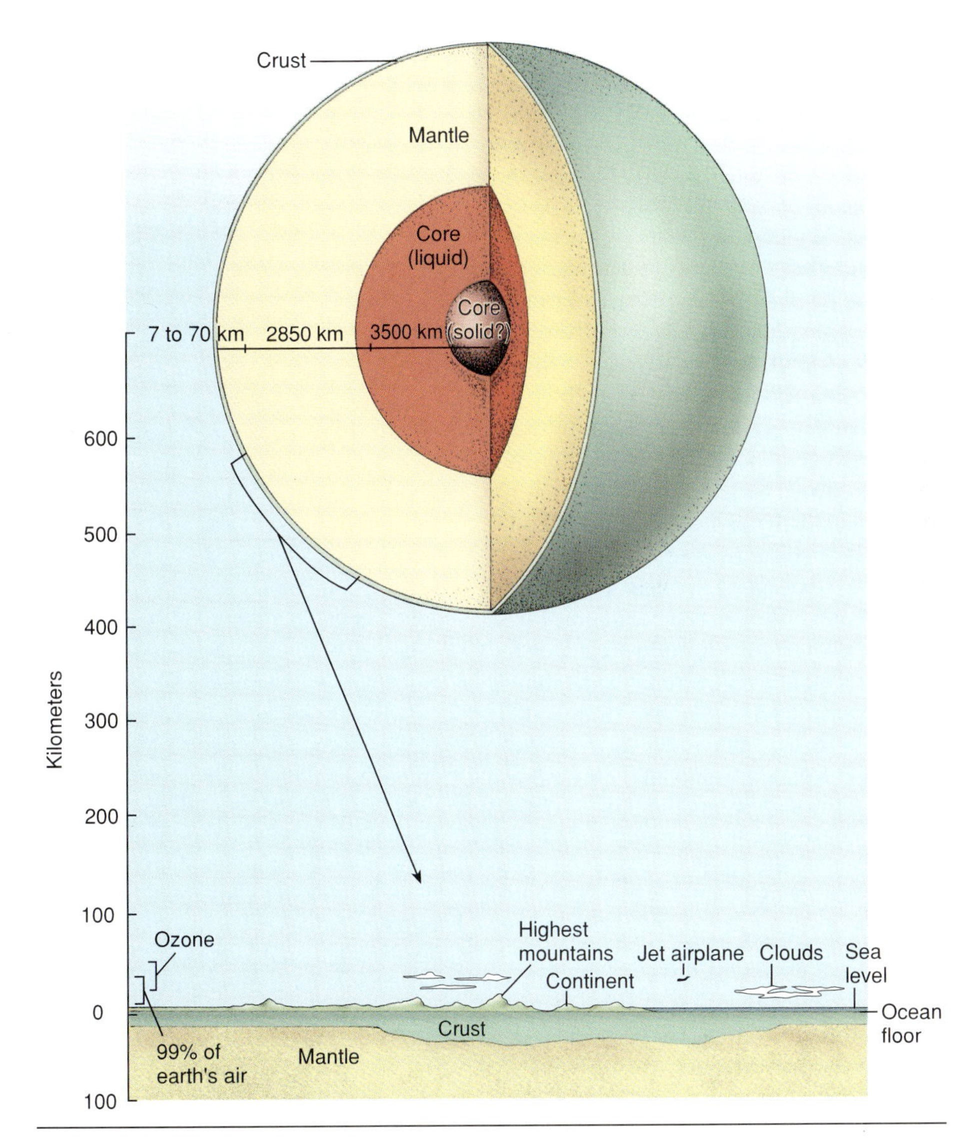

Figure 12.6
This model of the earth suggests that, below the relatively solid crust and mantle, there is a liquid core made of heavy metals. In the exact center there may be another core that is primarily solid and exceedingly dense. The scale of oceans, mountains, clouds, airplanes, and even satellites seems insignificant in comparison with the overall dimension of the earth.

pressure inside the earth. These **metamorphic rocks** are different structurally, although made from the same basic components. Quartzite may form from sandstone, slate may form from shale, and marble may form from limestone, for example.

Rocks are composed of many different minerals, some existing as single elements, such as iron, phosphorus, potassium, copper, sulfur, and magnesium; other minerals, such as quartz (SiO_2), are combinations of elements (Si, silicon; O_2, oxygen). Minerals found in rock are referred to as inorganic (not from living organisms).

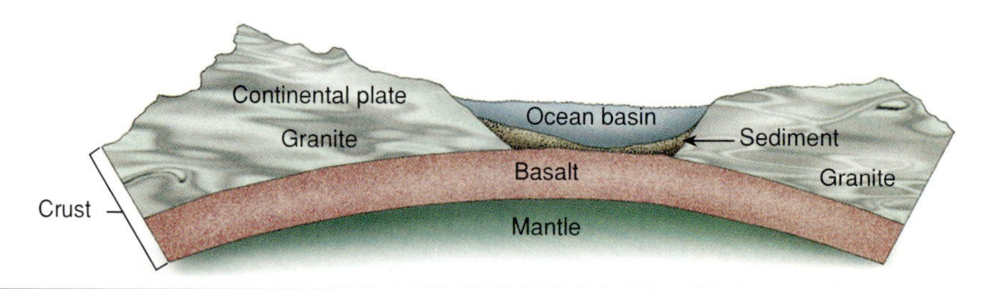

Figure 12.7

The earth's crust is thin in the ocean basins and thickest under the high parts of the continents. The continents are made of granite underlain by basalt, whereas the ocean floor is all basalt. Some sediment covers the basalt in the ocean.

Soil Development

Weathering of solid rock occurs through both physical breakage and chemical breakdown. Both freezing-thawing and heating-cooling cycles produce the expansion and contraction that crack and break solid rock into progressively smaller particles. Water runoff, root penetration, and wind also produce physical weathering of rock. Carbon dioxide and other gases combine with water to produce a weak acid that dissolves some minerals to chemically aid the weathering process. As these inorganic particles become smaller, more individual minerals become free and available.

The results of weathering are greatest near the surface, and soil formation is therefore limited to this relatively shallow zone. **Soil** is more than these weathered inorganic materials; it is a combination of **inorganic particles, organic materials,** and **pore space** occupied by water and air. In any case, these three basic components of soil play an important part in soil-plant-water relationships.

Soil scientists make clear distinction between soil and **dirt,** defining the latter as "misplaced soil." In other words, someone sitting on bare soil at a picnic will end with dirt on his shorts when he gets up.

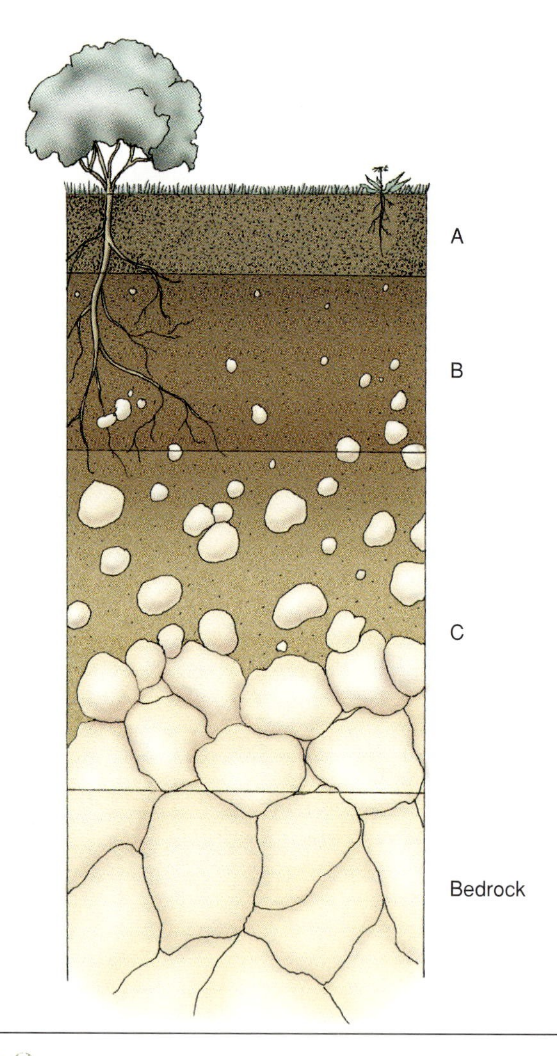

Figure 12.8

Soils become finer and more highly weathered toward the surface. The A horizon is readily distinguished by the dark color, caused by organic matter. The B horizon is less weathered but of such texture that water and nutrients are stored for uptake by deep root systems. The C horizon contains large particles and rocks more recently derived from the rocky, underlying parent material (bedrock).

Soil Profile

A vertical section of soil is called the **soil profile,** and it reveals more or less distinct horizontal layers—the soil **horizons** (figures 12.8 to 12.10). The uppermost layer contains the majority of the organic matter, the result of biological decomposition. Climate, the activity of animals, and even cultivation have a great effect on the development of this layer. This is the A horizon and is usually noticeably darker because of the presence of organic compounds. The A horizon varies in depth from a few centimeters to as much as a meter. Because most of the biological activity

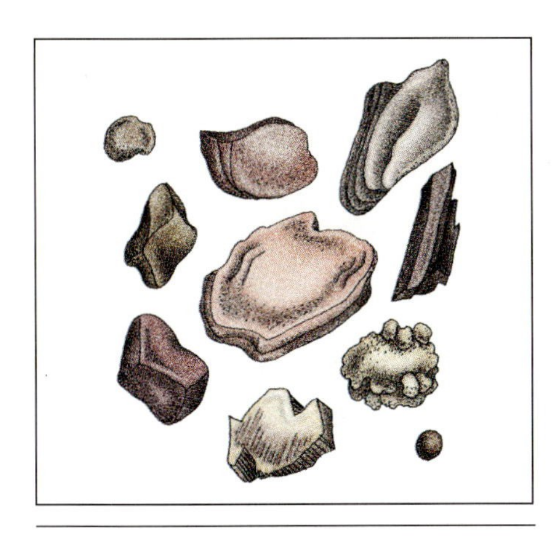

Figure 12.9
Although soil particles may appear quite uniform to the naked eye, viewed through the microscope they reveal many shapes and sizes. Some aggregates tend to stick together.

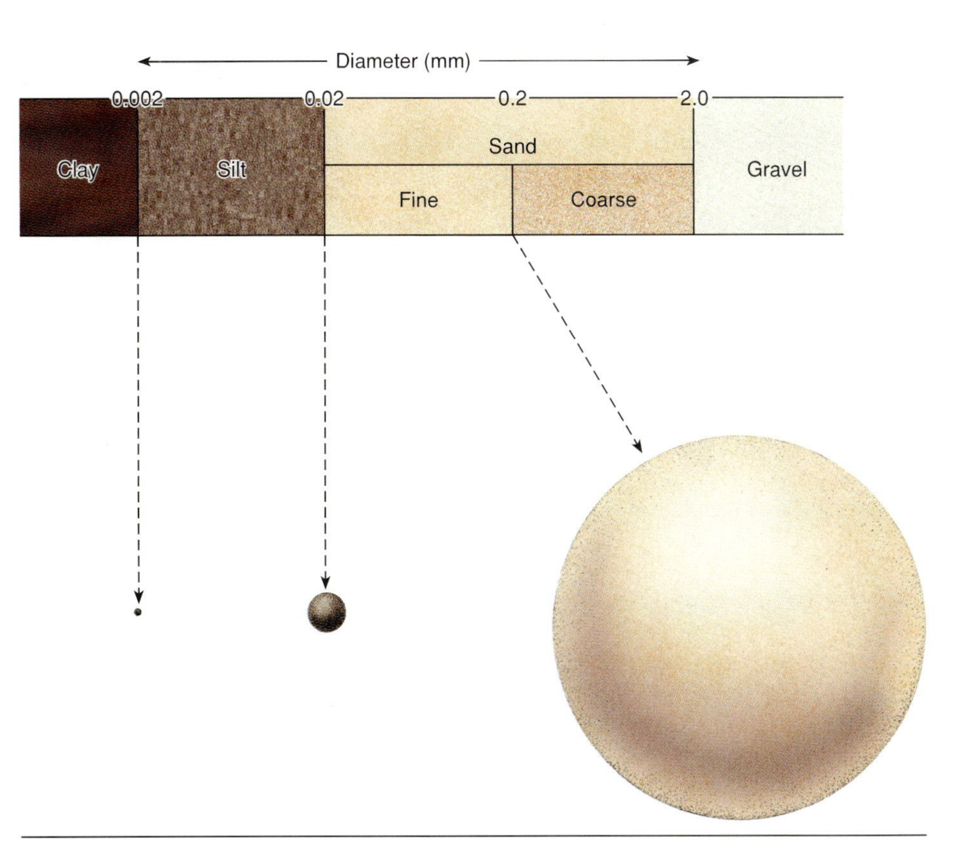

Figure 12.11
Soil particles vary tremendously in size. The classification into sand, silt, and clay is based on an arbitrary tenfold difference in size. Clay particles are small enough to behave as colloids and settle out of solution very slowly.

Figure 12.10
Some soils reveal a profile almost exactly like that described in figure 12.8.

occurs here, it is the richest horizon in terms of total nutrients and is often referred to as the **topsoil.** The maintenance of the A horizon is an important consideration in modern agriculture.

In addition to the organic materials present, the A horizon is composed of smaller inorganic particles. This is the result of the higher weathering activity near the surface. The **abiotic** (wind, water, temperature) and **biotic** (roots, decomposition, earthworm, and other organismal activity) components of weathering are more effective near the soil surface.

The B horizon is the next soil layer. It is characterized by having very little organic matter, being somewhat coarser in texture due to larger inorganic particles, and allowing deeply penetrating root systems. What organic matter is present is predominantly found near the top of the B horizon, close to the transition zone between A and B. The B horizon, therefore, is much lighter in color than the A horizon. Certain mineral components accumulate in greater amounts in the B horizon as well. Iron, calcium carbonate, gypsum, clays, and aluminum oxides are commonly found in this layer.

The C horizon is typified by very large inorganic particles (rock sized), essentially no organic material, and only occasional roots. This layer of the soil profile is actually a transition between the upper zones and the solid **bedrock** below. At this depth in the soil profile the effects of weathering are minimal. The bedrock is sometimes called **parent material** because it is the source of the inorganic component of the soil. It is often impervious to root and water penetration.

Soil Texture

How coarse or fine the texture of soil is depends on the size range of inorganic particles present. The particle sizes depend on the amount of weathering a given parent rock has undergone. The actual particle diameter determines whether it is generally classified as a **clay, silt, sand,** or **gravel** (figure 12.11).

The texture of a given soil is a direct measure of that soil's **porosity,**

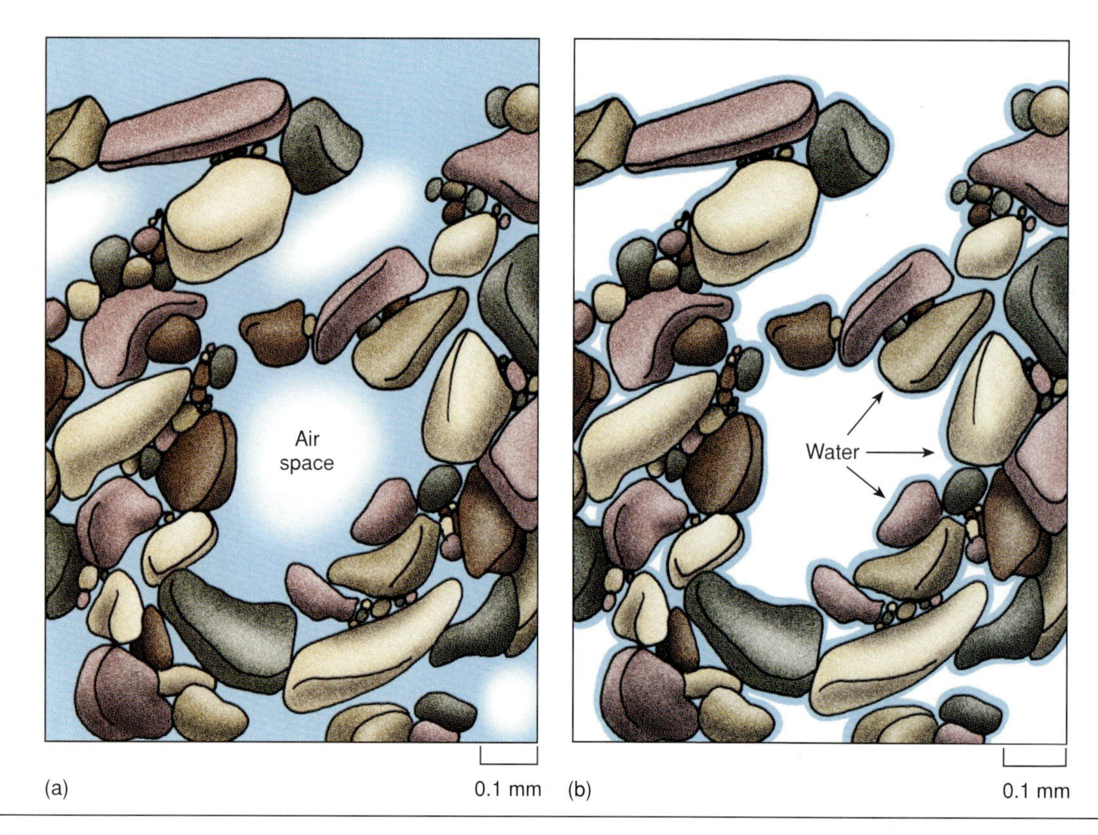

Figure 12.12

Water is held in the capillary pore spaces between soil particles. Large spaces between particles may be occupied by air spaces (*a*), and as the water is extracted from the soil by plant roots or evaporation, the layers of water become thinner and thinner (*b*).

or the space between particles (figure 12.12). That space contains either air or water, both essential to plant growth. Larger pores are termed **macropores,** and they drain water not held by capillarity. The smaller **micropores** hold water against gravity, so most of the water available to root systems is found there. Texture, then, dictates the water-holding capabilities of the soil; water that percolates downward out of the root zone is replaced by air.

Generally, the smaller the particle sizes, the closer they can pack together, thus lowering the porosity. It is important to remember that the adhesive property of water results in the surface of each soil particle being covered by water molecules. The greater the total surface area for a given group of particles, the greater the amount of water held in that type of soil. A 1 m cube of granite has 6 m^2 of surface area; breaking it in half results in the same volume of granite now having 8 m^2 of total surface area. If each of those halves are also broken in half, the total surface area exposed becomes 12 m^2. Each subsequent break results in smaller particles with increased total surface area for the same original 1 m^3 of rock. The process of weathering, which breaks down the mineral into smaller particles, increases the total surface area of a given volume of soil. The smaller the component particle sizes for a given volume of soil, then, the more water that can adhere to the surfaces. The surface area available to hold water ranges from less than 100 cm^2 per gram in coarse sand to over 1 million cm^2 per gram in clay.

The size of the clay particles largely determines the physical and chemical properties of mineral soils (figure 12.13). A pure clay soil, composed exclusively of clay-sized particles, therefore can contain a large quantity of water. In addition, water molecules adhering to soil particles are tightly bound, which means they are not pulled away by gravity or plant root systems. Clays, then, can hold a lot of water, but it is not readily available to plants; only the water molecules held in the small capillary pore spaces are easily removed.

Pure sand, on the other hand, has less total surface area for the same total soil volume, and therefore less moisture adhering to the particles. Because of the larger size and irregular shapes to the particles, however, sand does not pack as tightly together as clay soils, resulting in greater porosity. When water is added to a sandy soil, a great deal of it is readily removed by root systems and gravity, and it dries out rapidly. It should be noted that, except for the surface particles, few soils, even desert sands, completely dry out; there is always a layer of water molecules held tightly to the particles, but it is not available to plants.

Silt-sized particles provide water-holding and porosity capabilities between those of clay and sand. The best soil for plant growth is **loam.** Loams contain clay, silt, and sand-sized

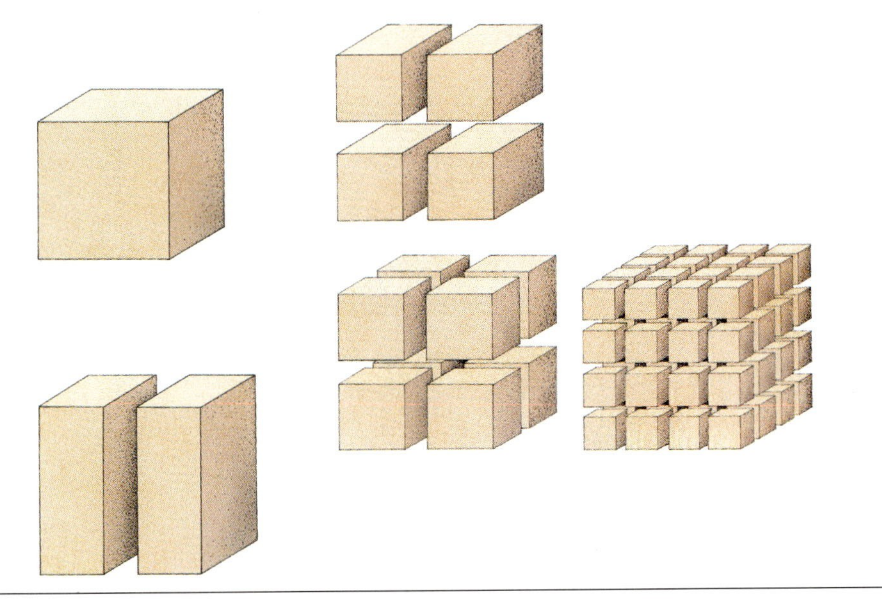

Figure 12.13
The surface area of any particle increases dramatically as the particle is split into smaller pieces. Tiny clay particles have a surface area per unit volume far greater than that of an equal volume of sand.

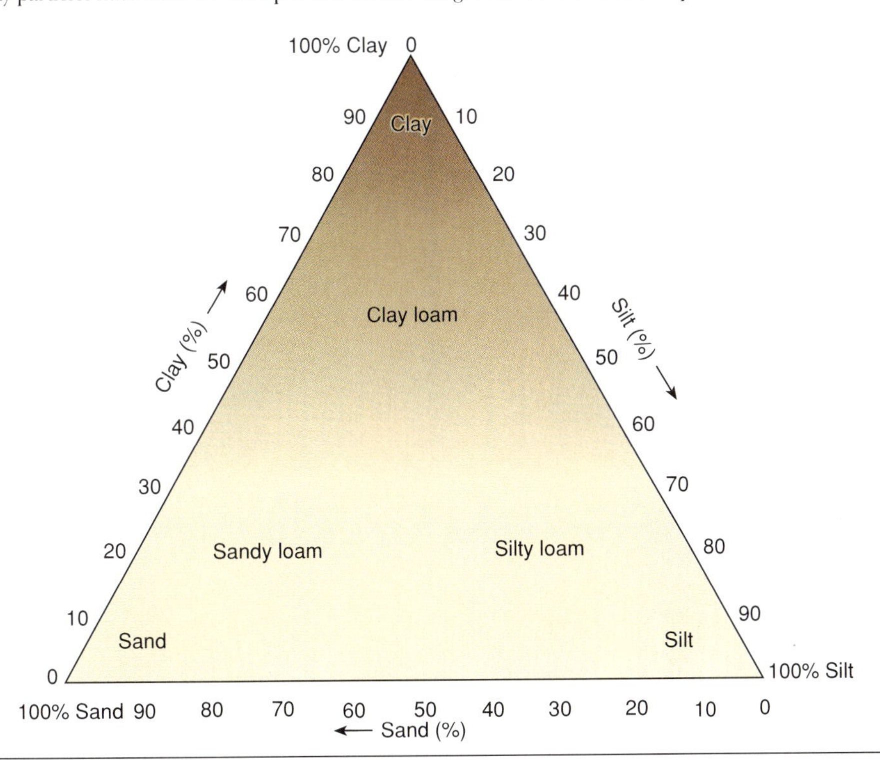

Figure 12.14
A triangle represents the three components of sand, silt, and clay that make up loams.

particles that provide a combination of water-holding capability (against gravity) and porosity. Plant root systems must have available water and air as well as sufficient looseness in the soil to allow root growth. Figure 12.14 gives the main categories of soils as determined by texture.

The upper limit of water available to plants is a soil's **field capacity.** This is determined by first oven drying a given volume of soil to remove all moisture. Next, water is applied to the surface of this soil much as a heavy, steady rain or irrigation might do. As the water percolates downward, it replaces air in the soil pores from top to bottom. The container holding the soil must have holes in the bottom to allow excess water to drain off. After the soil is completely saturated with water, application of water at the surface is stopped. There is a period of continued downward water movement out of the soil in response to gravitational pull. Once this downward flow has ceased, the soil is said to be at its field capacity. The soil is weighed at this point. The difference between dry weight and field capacity weight equals the amount of water held by the soil. At field capacity there is some air in the largest soil pores, but other pores are filled with water that is available to plants.

In nature a soil at field capacity loses its water through both surface evaporation and plant uptake. This occurs rapidly at first and then slows as the dry surface soil acts as a barrier to continued rapid surface evaporation. Eventually the plants lose water through daytime transpiration faster than they can take it up from the soil. For several days plants wilt earlier in the day but become turgid again overnight as their root systems continue to remove as much available water from the small soil pores as they can. Ultimately the plants are unable to rehydrate overnight and become permanently wilted. Although the plants are not yet dead, death will follow if additional water is not added to the soil. At this point the soil still contains a significant amount of water, but it is held in the smallest pores and on the particle surfaces so tightly that the plant's root system cannot remove it. The soil moisture content in such situations is termed the **permanent wilting point** (figure 12.15).

Different soils have different field capacities and wilting points. A large field capacity is no guarantee of an abundant supply of water for plants, nor is a low field capacity soil necessarily too dry for plant growth. Different plant species are adapted to different soils and use the available water in the most efficient way possible. Still, a loamy soil that possesses

> **✔ Concept Checks**
>
> 1. What makes up "soil"?
> 2. How does soil texture relate to water-holding capacity, and what soil texture is the most desirable for plant growth?

the best combination of porosity and water retention is generally the most productive soil for plants.

WATER MOVEMENT

Soil provides the reservoir for the water needed by plants. No single environmental factor affects the distribution of plants over the face of the earth as does water (figure 12.16). If you have the impression that we are preoccupied with water, your impression is well founded. We reemphasize water's importance in many places as a constant reminder of the critical role of water in living organisms. Since all life takes place in an aqueous medium, it is important to understand how water moves from one area to another.

Kinetic Energy

Kinetic energy results from the motion of all molecules that are in constant random movement. Even if you fill a bathtub with water and let it become perfectly still, each individual water molecule in the tub is still in constant motion. If several drops of dye are carefully added to one end of the tub, the color slowly spreads throughout the water. Initially the water at the end where the dye is added becomes intensely colored, and the other end remains clear. Gradually the color intensity at the application end decreases as the color spreads toward the other end. Ultimately the water in the tub is equally colored throughout, and all this takes place without stirring the water at all. What has happened is an example of the process of **diffusion.**

Diffusion

Diffusion is defined as a random movement of particles from a region of high concentration to a region of

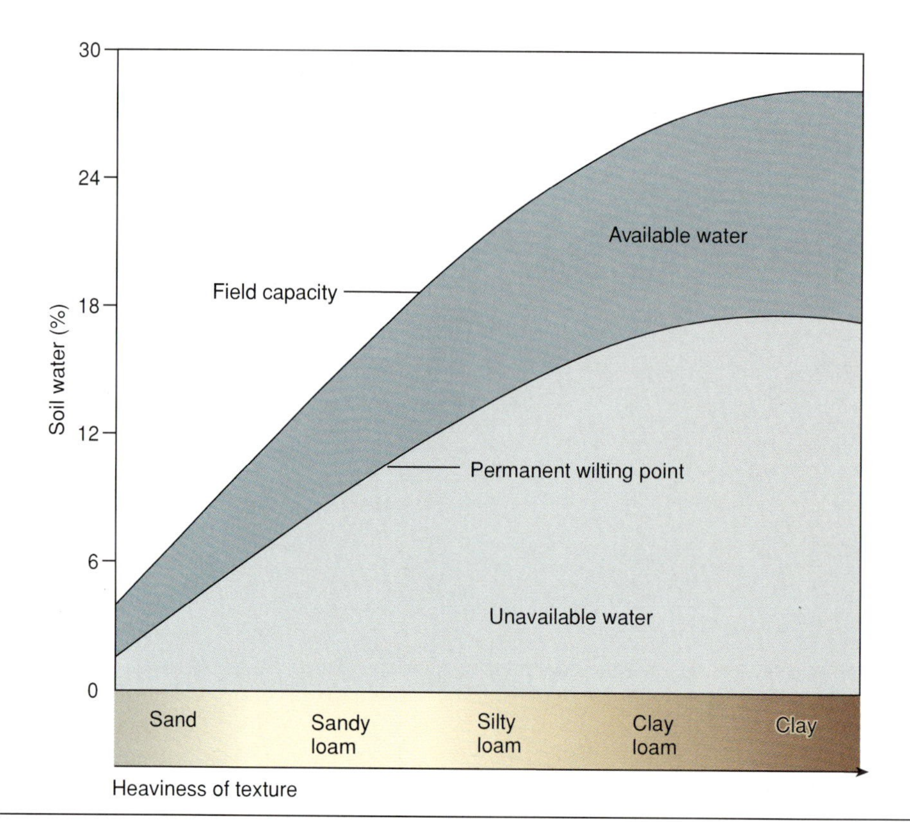

Figure 12.15
Water is available to plants only between field capacity and the permanent wilting point. The percentage of water available to the plant increases as the percentage of clay becomes greater.

Figure 12.16
Water availability dictates the distribution of plants. Both aquatic and land plants owe their positioning to the abundance and stability of a water supply as shown in the relative locations of aquatic plants.

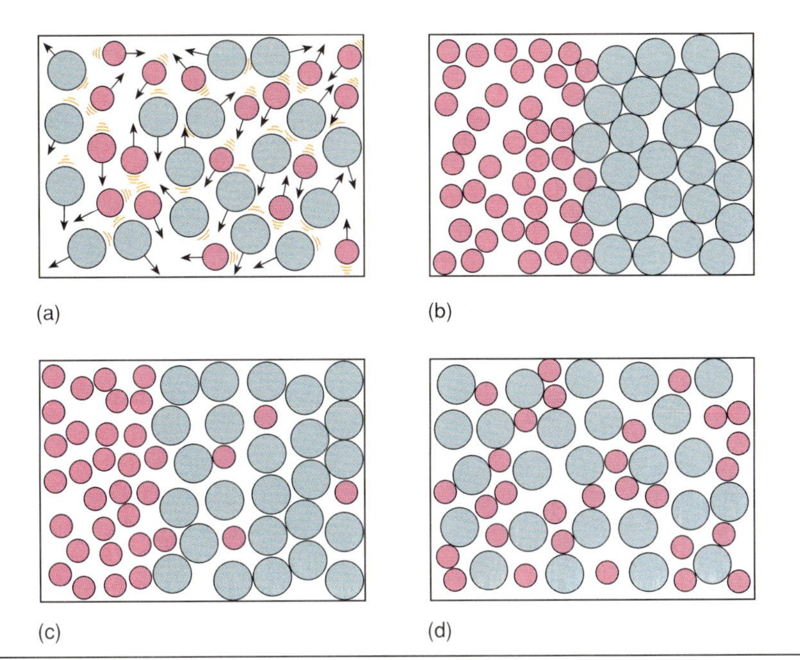

Figure 12.17

(*a*) Molecules are constantly in motion due to their kinetic energy. They move at random in straight lines until they collide with another molecule. (*b*) If large molecules are placed on one side of a container and small molecules on another, the colliding molecules at the interface gradually begin to intersperse (*c*) and finally come to equilibrium as the large and small molecules become equal in concentration at all locations (*d*).

lower concentration. Diffusion occurs because of the kinetic energy of matter, and it continues in the "downhill" direction until equilibrium is reached, that is, particle concentration is equal in all regions (figure 12.17).

Consider the example of the dye in the filled bathtub. If you examined any given individual dye molecule at the beginning of diffusion, you would not be able to detect that the net movement is from the application end toward the other end. Molecules move in all directions, and there is a net movement in one direction only because there are more dye molecules on one end. Thus a greater percentage of them are able to move toward the other end. Once equilibrium is reached, the water and dye molecules are equally distributed throughout the tub. Although they are still in constant **random motion,** there is no net directional movement due to a concentration gradient.

Diffusion occurs in living organisms as well as in bathtubs and in the air around us. But since living organisms are cellular in construction and since water and other molecules need to move from cell to cell within these systems, the process of diffusion allows for movement within living organisms.

Osmosis

Osmosis is a special kind of diffusion in which water molecules move across a **selectively permeable membrane** in response to a concentration gradient. In other words, water moves across a membrane from a region in which its molecules are greatly concentrated to a region in which its molecules are less concentrated (figure 12.18). Many students attempt to explain osmosis based on concentration of salt dissolved in the water, but to do so confuses the definition. All one needs to remember is that water molecules are most concentrated when there are no interfering or dissolved substances. As soon as some salt is put in pure water, the distance between the water molecules is increased and the water concentration is reduced.

If two adjacent cells have water concentrations of 90% and 92%, respectively, the 90% cell has 10% other molecules mixed in with the water, whereas the 92% cell has only 8% other molecules. Water moves from the region of higher concentration (92%) to the lower concentration (90%). Equilibrium is reached when both cells have the same water concentration.

If the concentration of solutes outside the cell is greater than the concentration on the inside, water may be pulled from the cell out into the surrounding solution. As it does so, the cell will begin to collapse. Even though the cell wall may impart rigidity, water will be lost from the vacuole and cytoplasm, causing the cell contents to collapse. This process is called **plasmolysis** and is, of course, disastrous for the cell. If water is quickly added to the external solution so that the solute concentration again becomes higher inside the cell, water will flow into the cell and bring about **deplasmolysis.** Plasmolysis occurs when salts in the soil solution, perhaps caused by overfertilization, become so concentrated that they create an inhibition to water absorption. Overfertilization in house or garden plants will first be indicated by severe wilting, and if water can be applied quickly enough to wash the salts from around the roots, the plant may revive.

In living systems, membranes are said to be selectively permeable because they allow some water-soluble materials to pass through while not allowing others. This natural selectivity enables cells to act as barriers to certain kinds of substances.

Water Potential

Water flows from one region to another because of a difference in potential energy. This potential energy is referred to as the **water potential.** Water melting from snow on a mountain, for example, has a great amount of potential energy, which it gives up in its travel down the mountainside. The **potential energy** might be used to turn a turbine to generate electricity as the water passes through, and the potential energy would be converted to a usable form of energy. Cell water, too, moves in response to gradients

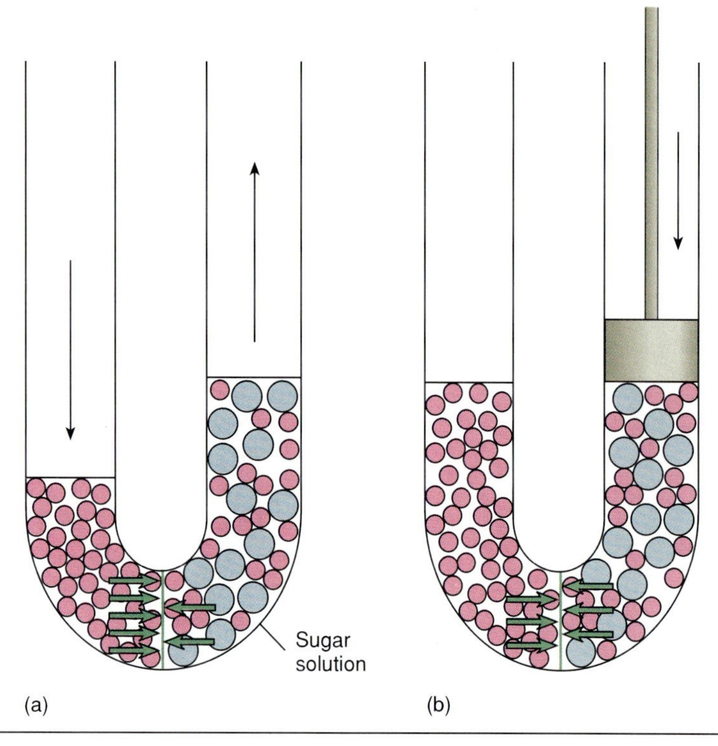

Figure 12.18

If a U-tube is filled with pure water on one side and a sugar solution on the other, water molecules bombard and move through the selectively permeable membrane from both sides. In this model, the sugar molecules are too large to cross the membrane. The pure water side has a greater concentration of water molecules, and therefore more water moves to the right than to the left (*a*). Water will rise in the right-hand tube until equilibrium exists. Even though an equilibrium in concentration can never exist in this model, pressure can be exerted on the right-hand tube to prevent the water from entering (*b*). The force exerted is equivalent to the osmotic pressure caused by the sugar particles in solution.

in water potential, and in living systems the water potential consists of the following three components.

1. The **solute potential** comes from the activity of particles dissolved in water.
2. The **matric potential** is developed by the interaction of water molecules adhering to a wettable surface. The force required to remove water from that surface is called the matric potential.
3. The **pressure potential** is the force created by a real pressure against a membrane. The turgor pressure caused by water inside the vacuole and cytoplasm pressing against the plasma membrane and cell wall develops a positive pressure potential.

The solute, matric, and pressure potentials combine to make up the water potential for a part of any system. These components can be measured, and they may be negative or positive. Adding the components together for one part of the system and comparing the value with the components for another part of the system results in a gradient. The water always moves to the *direction in which the water potential is most negative*. Water movement is strictly a physical process and it never requires metabolic energy.

THE SOIL-PLANT-AIR CONTINUUM (SPAC)

One can best conceptualize the movement of water in plants by considering **SPAC,** the **soil-plant-air continuum.** This is simply a consideration of water movement as a series of linked pulling forces that extend all the way from the soil, into the roots, into the xylem, up through the stem, out through the branches and petioles, into the blades of the leaf (figure 12.19), and finally out into the atmosphere. The transport cells of the xylem can be thought of as parts of a city water distribution system. As the main pipe moves water further toward the edge of the city, the size of the pipeline decreases in accordance with demand. In the leaf, vessel elements terminate in the mesophyll and supply water to a few living cells (figure 12.20). The water movement continues out through the stomata and finally into the atmosphere. From root to leaf, water moves as a liquid column, and there is no gaseous phase. From leaf to atmosphere, water changes to vapor. Water moves along this entire distribution system from the reservoir (the soil) along gradients of water potential, always moving to the most negative point. In almost every case that means from the soil, through the plant, and out into the atmosphere, providing the water needs of the plant along the way. The major driving force for this system is the **transpirational pull,** a tremendous wicklike force at the surface of the leaves that sucks the water into the atmosphere. It is no wonder that some 99% of all water taken up by plants ends up back in the atmosphere as a result of transpiration.

 *Concept Checks*

1. What is the difference between diffusion and osmosis?
2. What determines the direction of water movement in soil, plants, and the atmosphere?

Soil-Water Movement into the Roots

When there is precipitation and water percolates down into the soil pore spaces already occupied by the root system, water availability is no problem. The root system must be constantly growing and branching, exploiting newly available soil water to guarantee an adequate supply between periods of precipitation. It

Figure 12.19
In the soil-plant-air continuum, water moves from the soil, into the roots, into the stem, into the leaves, and finally into the atmosphere due to gradients in water potential.

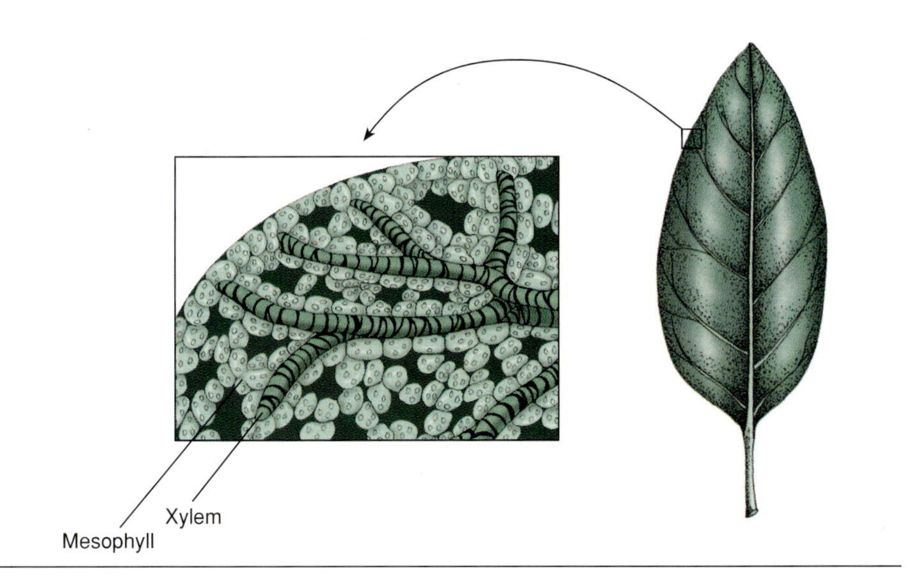

Figure 12.20
In the leaf, vessel elements terminate at the leaf margins, providing water to a few parenchymal cells.

is this new root growth, the youngest part of the root system, that is responsible for most water absorption. On this new growth the epidermal layer greatly increases its total surface area in the form of root hairs, which begin developing a few millimeters behind the root tip. As these young roots push their way between soil particles and come into contact with water in soil pore spaces, the water passes through the cell membranes of the epidermal cells by osmosis. Water also has no problem moving through cell walls, either at the root or internally.

Normally, soil water has fewer ions in solution than does the cytoplasm of root epidermal cells; thus cells have a more negative solute potential than does soil water. Added to the negative solute potential is the matric potential of the root external surface. Thus water is taken from the soil into the roots as a result of the more negative water potential at the roots.

From the epidermal cells, water and solutes move inward toward the center of the root, where the vascular tissues are located. Water moves from epidermal cells through the cortex of the root to the **endodermis.** Water movement is primarily along cell walls without having to pass through cell membranes, although some water does travel from cell to cell across membranes. Water movement along cell walls without passing through membranes is termed **apoplastic movement.** Once water reaches the endodermis, apoplastic water must pass through endodermal membranes because of a gasketlike band of **suberin** around each endodermal cell. Suberin is a waterproof waxy substance that seals off the route of water between cells. Figure 12.21 depicts the route of water to the endodermis where this waxy seal, the **Casparian strip,** is found.

In addition to diffusion, mineral ions are brought into the root epidermal cells by the expenditure of metabolic energy. These ions are said to be taken up by **active transport.** Once inside the epidermal cells, these ions can pass from cell to cell, through the cortex to the endodermis through plasmodesmata. This pathway is called **symplastic movement.** Whether water and mineral ions move apoplastically (between cells) or symplastically, everything entering the vascular tissues must pass through the cell membranes of the endodermis.

Once past the endodermis, water moves through the **pericycle,** a layer of cells from which the new lateral roots originate. Inside the pericycle, water enters the xylem and becomes a part of a continuous transportation network that terminates in the leaves.

Water Up the Drain

The question is often asked, "Why do I need to water the plants so often?" Or you may hear the comment: "It seems like the ground was just soaked from a big rain, but now it's dry again. We really need rain."

The soil is indeed a tremendous reservoir with great water-holding capacity, depending on texture and depth. If one considers all of that storage capacity, how can so much water be "lost"? Does it percolate out of the soil due to gravitational forces? Yes, some of it may do so. Quite a bit may go to runoff, especially during a high-intensity storm or if irrigation water is applied too rapidly.

But what of the bulk of water left in the root zone? Most of that water is "mined" by the plant's root system, which serves as the harvesting device to move water through the plant and out into the atmosphere. But can a plant really absorb *that* much water, and what happens when it travels through the main stem, branches, and finally the smallest twigs, to the petiole, and to the surface of the leaf blade? Surely not much water can get out of the leaves. It is true that a very small percentage of the total water absorbed by a plant is used to maintain pressure in all the cells, and it provides a medium for the biochemical reactions going on there. A few water molecules are also involved in the reactions themselves, but nearly all of the water simply moves through the plant.

One of nature's big surprises is the tremendous amount of water lost to the atmosphere through transpiration, the process of evaporation from a plant surface. Although most stems are covered with protective layers of cells and chemical barriers, the leaf blade is literally a sieve with stomata where pore size is controlled by guard cells. Environmental variables of light, temperature, wind, relative humidity, and other factors determine the size of the pore. In some cases, it may close completely, but even then, some water molecules manage to escape through the cuticle. The big loss, however, occurs when the stomates are open and a vapor pressure differential causes the water to be "sucked" out of the leaf.

One would think that there couldn't be enough of these pores to make much of a difference. After all, the stomates are microscopic, and they seem to be few and far between. Students in botany laboratories are often asked to calculate the number of stomates on the upper and lower side of a particular kind of leaf. Epidermal peels can be made rather easily for many kinds of leaves with a sharp razor blade, and then the peel can be observed through a microscope. Surprisingly, the lower surface of a cotton leaf, for example, may have more than 280,000 stomates. Some leaves have almost as many on the upper surface—others have almost none. The size of a water molecule is very, very small compared to the pore size of an open stomate. Thousands of water molecules can escape from such an opening at any given time. When one considers the number of water molecules, multiplied by the number of stomata per leaf, times the total number of leaves per plant (How many leaves might be on a large oak or maple tree?), one quickly begins to realize what happens to that large reservoir of water in the soil. A typical corn plant can lose two liters of water per day during the middle of summer. If the plant density is 100,000 plants per acre, it is not surprising that we are so dependent on predictable and frequent rainfall, or a carefully scheduled irrigation regime.

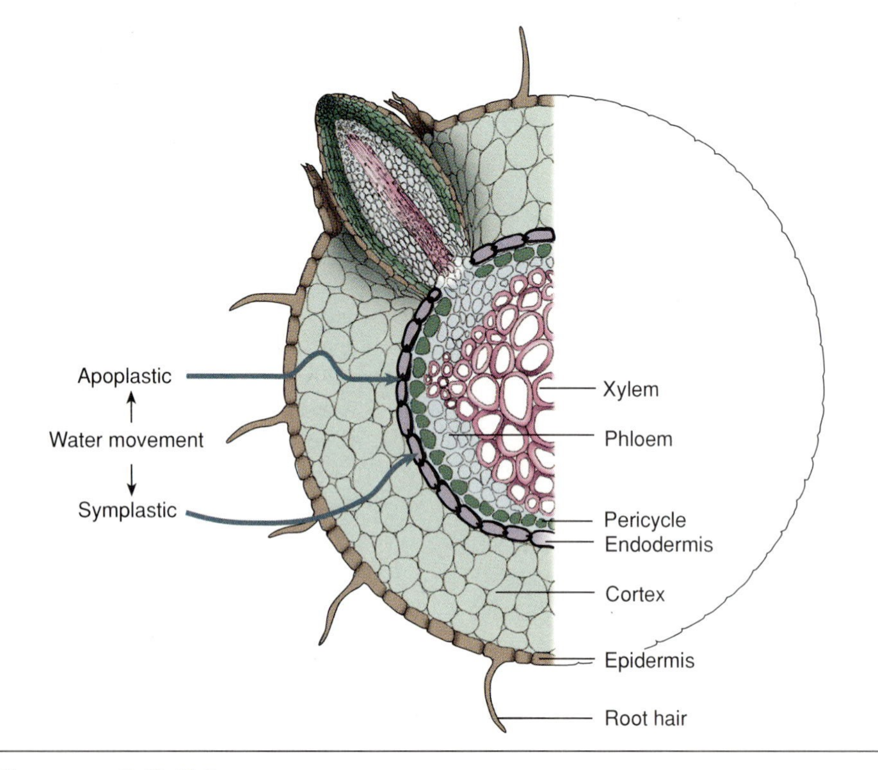

Figure 12.21
Apoplastic and symplastic water movement through a root.

Through this network, water moves from the roots to incredible heights without the use of a pump. The movement of water is triggered by transpiration at the leaves.

Even when there is no transpirational water loss, the roots continue to absorb water and to accumulate ions in the xylem. If water is not moving up into the stems and leaves, the ions accumulate in gradually increasing concentrations in the xylem, causing the solute potential of the xylem to gradually become more negative. This more negative water potential in the xylem causes water to move by osmosis into the xylem from surrounding cells, ultimately forcing water and soluble ions upward. This force produces **root pressure,** which in turn forces water to be exuded at leaf tips and margins through small pores of short plants.

The process of root pressure forcing water out of the leaves is termed **guttation** and only occurs when stomata

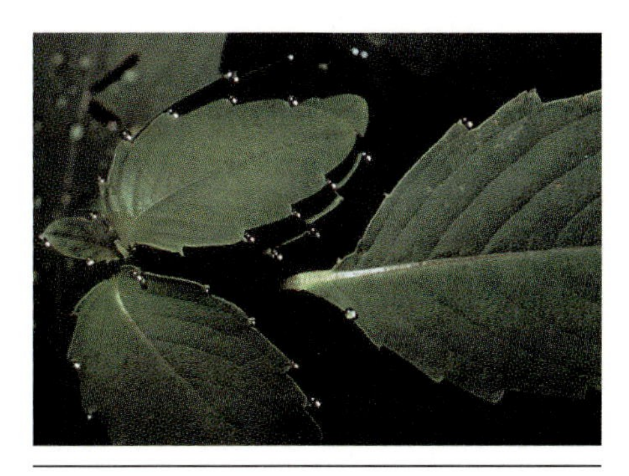

Figure 12.22
The process of guttation occurs in some leaves because root pressure forces xylem fluid out to the surface of the leaf. This condition occurs only during specific environmental conditions.

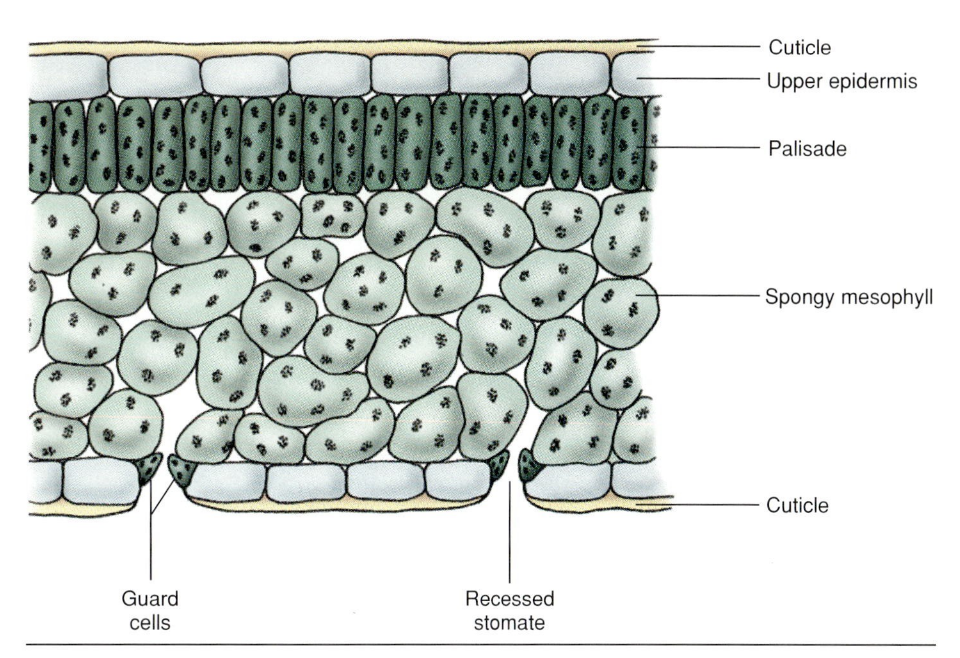

Figure 12.23
Cross section of a typical leaf, showing slightly recessed guard cells and a very thick layer of cuticle. CO_2 enters the leaf through the stomata, and water exits the leaf at the same place.

are closed and when water is readily available to the root system, usually overnight (figure 12.22). Although capable of forcing water up the xylem when all the stomata are closed, root pressure is a negligible component of water movement when they are open.

Water Movement Throughout the Plant Body

Xylem is the tissue through which water moves in plants. It is made up of elongated cells having secondary cell walls with holes in them. Water moves through the xylem via vessel elements and tracheids in angiosperms and in tracheids in gymnosperms. Both tracheids and vessel elements are nonliving at maturity, and both may contain pits in their cell walls. Vessel elements also have larger holes in their secondary end walls and occasionally in the side walls. These perforations are very important in allowing unimpeded water passage from one vessel element to the next.

Vessel elements are arranged end to end and form long, continuous strands throughout the plant. These strands are analogous to a long capillary-sized hollow tube because there are no cell membranes to retard water movement from cell to cell as there would be in living cells. Vessels move water more efficiently than do tracheids because the perforations in their end walls are larger than the tracheid pits, and because the cross-sectional area is much greater.

Transpiration

Unlike higher animals, which have circulatory systems that recycle fluids through the body, plants lose the vast majority of the water taken out of the soil. Transpiration is technically the evaporation of water from any portion of the plant body into the atmosphere. Actually, of the total water volume transpired by plants, so little of it is lost from plant surfaces other than the leaves that there is an understandable tendency to think of transpiration as exclusively a leaf-related phenomenon. In addition, the greatest bulk of water transpired by the leaves is lost through stomata. Although small, there are an amazing number of stomata on a leaf. For example, the lower surface of a typical cotton leaf averages 18,000 stomata per square centimeter.

Even though water does leave the plant and is lost in that sense, the term *lost* is misleading. In truth, the water that is transpired from stomata provides a critical service. The atmospheric CO_2 required for photosynthesis enters the leaf cells in an aqueous solution. The water at the surface of cells directly in contact with the atmosphere facilitates the intake of CO_2. This water also evaporates into the atmosphere (transpires) (figure 12.23).

Transpiration occurs whenever the stomata are open, and the cell surfaces inside the leaf are exposed to the atmosphere. A small amount of water is also lost directly through the cuticle and is termed **cuticular transpiration.** The reason water moves out of these cells is simple. The atmosphere normally has a lower water concentration than the cells that are in contact with it. Water passes through the cell membrane by osmosis and evaporates. These cells have a lower water concentration than adjacent interior cells that are not in contact with the atmosphere.

As water evaporates into the atmosphere, new water molecules are pulled up to take their place. Water moves along from cell to cell based on water potential gradients. In the xylem, the primary component of the water potential is the tension created by open stomata and transpiration pull. In the living mesophyll cells, the water potential is created primarily by the solutes and surfaces inside the living cells. Whether the water will remain in the leaf or dissipate through the stomata depends entirely on the overall water potential and the region where it is most negative. **Stomatal regulation** is a constant compromise for the plant. On one hand, the stomata should remain open so that CO_2 can enter and participate in sugar production. On the other hand, closing stomata conserves water and may allow the plant to adjust to a lack of water in the soil.

Cohesion-Adhesion-Transpiration Pull

Water moves from the roots to the leaves through a combination of the **cohesive** strength of water-to-water bonds, the **adhesive** attraction of water to vessel walls, and the transpiration at the leaves that pulls on the column of water below. It is the loss of water through transpiration that triggers the pulling tension on water in the xylem. As water evaporates through the open stomata, either from an open intercellular space in direct contact with the terminal xylem trace or from the outer surface of a cell, a negative water potential results, as compared with that of water in the xylem. Water in the xylem automatically moves out into the leaf in response to this negative water potential and pulls up water molecules from below in the xylem tract to replace it. The negative water-potential gradient produced in an actively transpiring leaf results from a combination of negative solute potential in the cells losing water, a negative matric potential (surface adhesion) on the cell wall and membrane surfaces, and especially a negative pressure potential caused by water being pulled out of the leaf into the atmosphere through evaporation.

Once the water starts moving from xylem cells toward open stomata because of this negative water potential, the phenomenal pull of the solid column of water in the xylem is transferred all the way down to the xylem originating in the roots. The great cohesive strength of water results in this column being pulled intact from bottom to top (root to leaf) without breaking.

There are a few instances in which the water transport system is reversed and water moves into the soil. In **fertilizer burn** the concentration of solutes in the soil is such a competing factor that the overall water potential is most negative in the soil. The atmosphere is pulling water one way, the soil is pulling water the other, and the direction with the most negative water potential wins.

Figure 12.24
Prosopis tamarugo is the only significant vegetation in the Atacama Desert. It provides forage for sheep that otherwise would not be able to survive.

The plant has the ability to modify water movement in various ways with a series of linked resistances, the most important of which is the guard cells on the leaf surface, which exert stomatal regulation. The ability to close stomata when placed under stress is like turning off a faucet. If the fertilizer concentration is too severe, so much water will be pulled from the plant that wilting will occur, and unless the water flow is reversed, the plant will die.

One interesting case of water-potential gradients occurs in the Atacama Desert of northern Chile. Rainfall in this desert is almost nonexistent, and no measurable precipitation may occur for a period of ten years. As one might suspect, vegetation is sparse, but the dominant feature of the landscape is a mesquite tree, *Prosopis tamarugo* (figure 12.24). This tree manages to survive because deep taproots mine water from underground aquifers. Few plants have such extreme adaptations to water stress.

Stomatal Regulation

The two **guard cells** bordering each stomatal opening determine whether the stoma is open or closed. The control is a physical response; the stomatal opening is produced by the adjacent guard-cell walls pulling apart when the cells are **turgid** and coming together when the guard cells are **flaccid** (figure 12.25). Current theory suggests that the movement of water in and out of guard cells is an osmotic response governed by potassium levels in the guard cells. If potassium concentrations in guard cells increase, a more negative water potential is created there and water moves osmotically into the guard cells, making them turgid. If the potassium levels drop, the water potential becomes more positive and water osmotically exits the guards cells, making them flaccid. A plant hormone, **abscisic acid,** is known to be somehow involved in the changing of potassium levels.

The guard cells are structurally unique. They have slightly thicker adjacent cell walls than outer cell walls. When the guard cells become turgid, the outer cell walls expand more than the inner ones, pulling the inner walls apart. As the cells lose turgidity, the space between the adjacent inner walls closes up as they come into contact again, preventing further transpiration and CO_2 uptake.

Because water loss and CO_2 uptake both occur when stomata are open, there is always a balancing act in progress. Plants must have CO_2 for photosynthesis, but on the other hand, only a given amount of water can be transpired and not replaced before the plant begins to wilt. The factors affecting this balance

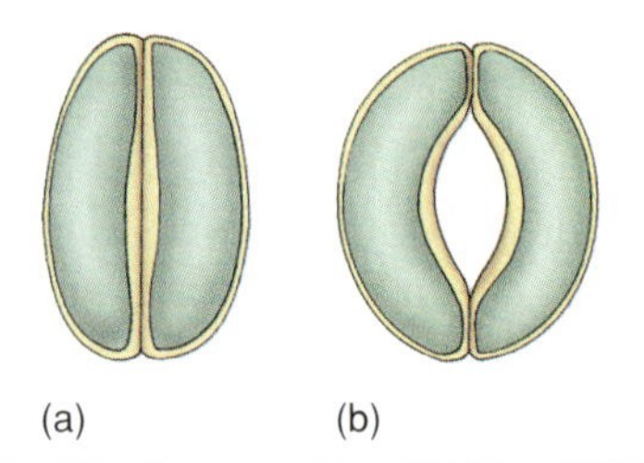

Figure 12.25
(*a*) The stomatal apparatus consists of two identical guard cells, the inner walls of which are slightly thickened. (*b*) Whenever turgor pressure increases inside these cells, they bend outward, exposing the stomatal pore.

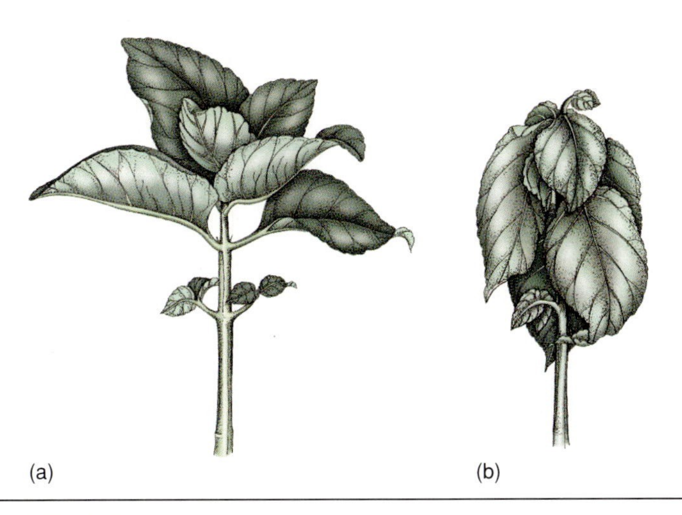

Figure 12.26
(*a*) When plenty of water is available, the leaves of *Coleus blumei* become turgid, exposing the blade to maximum sunlight. (*b*) As water availability decreases, the leaves become flaccid and the plant wilts.

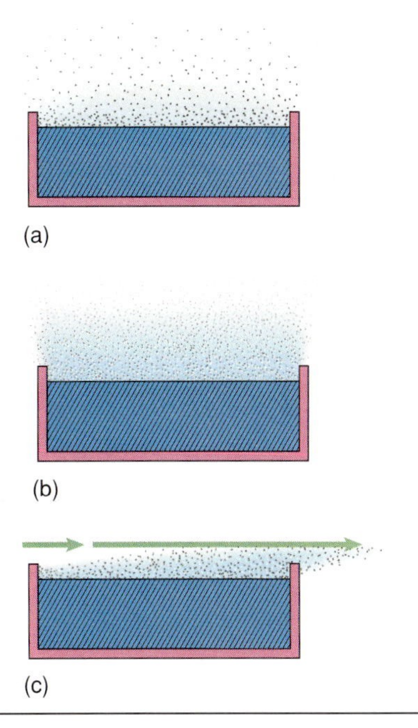

Figure 12.27
(*a*) A pan of water undisturbed by wind allows water molecules to escape from the surface, and a layer of high humidity occurs directly above the pan. (*b*) If the water is allowed to evaporate for some time and if the air is perfectly still, the layer of high humidity thickens. (*c*) If a wind blows across the pan, the water molecules are swept in the direction that the wind is blowing.

include both environmental and physiological controls.

Factors Influencing Transpiration

The single overriding control of stomatal activity is water availability. If there is not enough water available to the plant, all leaf cells, including guard cells, become flaccid as the leaf wilts (figure 12.26). Any time water availability fails to keep up with water loss and guard cells become flaccid, they close the stomata, preventing further transpiration until the roots can absorb additional water. Thus any factor that increases transpiration rate relative to water uptake rate will result in stomatal closing.

Temperature One of the most important regulatory factors is increased temperature. Above 30° to 35°C, stomata usually close automatically whether there is available water or not. At "normal" temperatures (10° to 25°C) an increase of 10°C results in a doubling in the rate at which water evaporates (transpires), thus affecting stomatal regulation by causing more rapid water loss than uptake. Many plants, especially those in arid and semiarid regions of the world, regularly close their stomata during the middle of the day when the temperatures are the highest. Such plants allow CO_2 uptake only in the morning and late afternoon, restricting their photosynthetic output.

Wind By itself, but especially in combination with increased temperatures, wind movement across leaf surfaces results in increased water loss by physically pulling away water vapor molecules from the open stomata (figure 12.27). Hot, windy days can impose a water stress on plants very quickly, resulting in early and sustained stomatal closure. On the other hand, wind may exert a cooling effect at the leaf surface and tend to offset the negative effects.

Humidity A third environmental variable affecting transpiration rates is the amount of water vapor in the atmosphere—**humidity.** Since evaporation occurs by water molecules physically escaping from the water-atmosphere interface, the net number that escape partially depends on how many water molecules are already in the atmosphere. Evaporation can be considered in terms of diffusion. When there are very few water molecules in the atmosphere (low humidity), the net flow of randomly moving water molecules is from leaf to atmosphere. Although the atmosphere will never have as many water molecules as a liquid surface, the closer the two densities become, the slower the net movement from water to air (evaporation). High humidity, then, slows the transpiration rate; low humidity increases it. A hot, windy, dry day therefore places a large stress on plants because of the combined effect on increasing the transpiration rate.

CO_2 Concentration High internal concentration of CO_2 will cause stomata to close, and low CO_2 concentrations cause stomata to open. Although these factors may be relatively unimportant on a normal day, if stomata have closed prematurely for other reasons, fixation of carbon into sugars may stimulate the reopening of stomata.

Diurnal Stomatal Closing

The vast majority of plants close their stomata at night even though their transpiration rates decrease significantly in the absence of the daytime heat. Since there is reduced photosynthetic activity at night because of the absence of the light energy needed to drive critical reactions, the need for CO_2 is less and continued stomatal activity is not required.

Conversely, there are plants that open their stomata at night and keep them closed all day. Cacti and other succulent plant groups use a different metabolic scheme to secure CO_2 for photosynthesis. The basis for this scheme is the conversion of CO_2 to certain organic acids at night, when water loss is lowest. During the day, when sunlight energy becomes available to drive photosynthesis, the CO_2 required is released by these organic acids. Photosynthesis can occur then with minimal water loss. This system is termed **crassulacean acid metabolism (CAM)** and will be discussed in more detail in chapter 13. It is worth noting that CAM plants are essentially all desert-dwelling plants, and the development of the CAM pathway is a very successful water-conserving adaptation for their hot, windy, dry environment.

Adaptations to Reduce Water Loss

The problem of balancing water loss via transpiration against the requirement for open stomata for CO_2 assimilation is most severe in hot, dry, and often windy environments. Plants that have successfully adapted to stresses have done so by the evolution of several morphological and physiological features. These adaptations are basically of two strategies: (1) reduction of water loss through open stomata and (2) different mechanisms for CO_2 utilization that avoid having stomata open as much.

Morphological Adaptations

The first group to adaptations are structural and have evolved in direct response to the primary environmental features affecting transpiration rates, temperature, wind, and humidity. One modification actually does not involve the stoma at all, but rather affects the remainder of the leaf surface—**cuticle thickness.** The thickness of the waxy layer covering all the epidermis of leaves varies from so thin as to be almost nonexistent to so thick that you can scrape the leaf surface with your fingernail and accumulate wax **(cutin)** underneath. Generally, the more arid the environment, the thicker the cuticle, which minimizes as much as possible water loss through and between the epidermal cells of the leaf. With this source of water loss essentially eliminated, the greatest amount of water lost from plants passes through open stomata. However, the evolution of a number of modifications has helped prevent excessive water loss in certain plants.

One of the most common modifications is the absence of stomata from the top surface of the leaf. The top surface is exposed directly to the sun, which increases the temperature significantly over that of the bottom surface. The higher the heat load, the faster transpiration occurs, so this modification lowers the total transpiration rate. The leaf surface is also very often highly **pubescent** in arid zone plants **(xerophytes).** The dense coverage of hairs is thought to add reflectance, protect the stomata against wind, and produce an increased humidity between the hairs and the leaf surface. The extra shading lowers the temperature while protecting against the wind, and the layer of increased humidity helps slow transpiration.

Another modification that increases humidity just outside the stomatal opening is the development of **sunken stomata** (figure 12.28). Instead of being flush with the epidermal cells, the guard cells are recessed. The result of a sunken stoma is protection from the wind and the formation of small pockets of high humidity immediately outside the stoma, which slows transpiration rates.

To provide shade and possibly create a zone of higher humidity, some leaves roll downward at the margins (figure 12.29). This creates essentially a recessed lower leaf surface that is partially protected from the wind and better shaded than flat lower leaf surfaces. The results are very similar to those in plants with sunken stomata, but for the entire lower leaf surface. Some plants use additional methods to avoid excessive transpiration. *Mimosa pudica* and other "sensitive" plants have leaflets that fold together in times of stress or disturbance, and other plants fold up their large leaves at night.

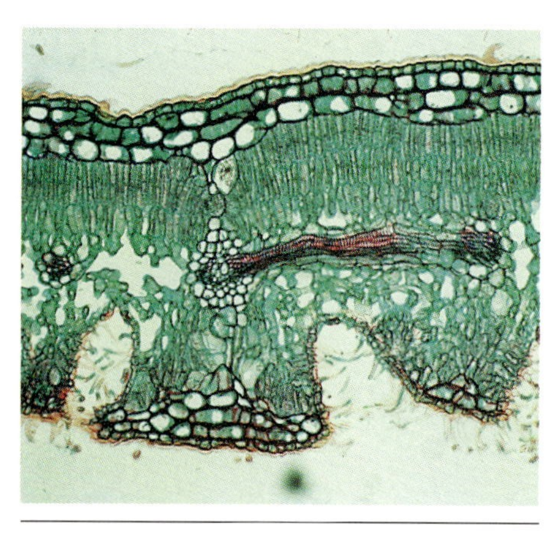

Figure 12.28
The leaves of *Nerium oleander* show the extreme modification of sunken stomata. These stomata on the lower side are deeply recessed in giant depressions filled with epidermal hairs, which add to the reduction in air movement. These modifications result in higher relative humidity around the stomata.

Figure 12.29
The leaves of some plants curve downward during periods of stress, slightly increasing the "pocket" of relative humidity. Others roll drastically, and the entire lower leaf surface becomes a barrier to air movement.

Grasses have modified upper epidermal cells, **bulliform cells,** which are enlarged for water storage (figure 12.30). In periods of water stress the bulliform cells lose their stored water,

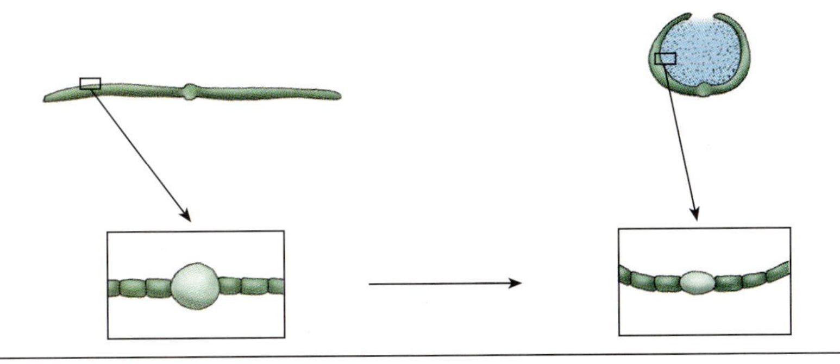

Figure 12.30
The upper leaf surfaces of some grasses have giant bulliform cells in the epidermis. Whenever these cells lose water, the leaf curls upward, creating a chamber of high relative humidity and thus reducing the rate of transpiration.

causing the grass leaf to roll up edge to edge, limiting further water loss from the upper epidermis.

Physiological Adaptations

Some plants have alternate pathways of carbon fixation and photosynthesis that provide for more efficient water use. Physiological changes have developed that allow such plants to survive in periods of low rainfall and in certain edaphic (soil) situations. The development of **succulent leaves** on plants that normally do not have them is yet another method of handling high water stress, this time by storing extra water.

APPLICATIONS

All of the aforementioned adaptations to survive water stress occur naturally and develop gradually over a long period. On a short-term basis, a plant not so modified cannot handle water stress well. Many houseplants have large leaves and are native to tropical regions, where abundant rainfall, high humidity, and moderate temperature have allowed plants to thrive without the development of any morphological or physiological water-conserving modifications. When such plants are introduced into a lower humidity and/or hotter climate as a houseplant, their inability to conserve water often results in minimal or no growth, dehydration, wilting, and death in extreme situations. Watering heavily does not solve the problem; such plants transpire water faster than their large surface area leaves are able to transport it in their new environment. As a result, these plants can spend most of their time with closed stomata and very little new growth.

The care of houseplants is not so difficult. The key to successfully growing houseplants is to understand the specific water, light, and fertilizing needs of each and to try to imitate as closely as possible the environmental conditions of their native habitat.

> ✔ *Concept Checks*
>
> 1. Why is the endodermis important in water movement into the plant? How does it work?
> 2. What is the significance of transpiration? How do stomates control transpiration rates?
> 3. What are some of the ways in which a plant may be able to cope with water stress?

Water

Houseplants vary in their water needs, but there are several general categories that apply to most common houseplants. Some must be kept well watered at all times; their soil must never be allowed to dry out. Very few plants thrive in a saturated soil, however. The key is to keep the soil moist but not flooded (figure 12.31).

Other plants require an alternation of thorough watering followed by a soil-drying period. This group of plants is most susceptible to overwatering. More houseplants are killed by overwatering than by not watering enough. Overwater can result in root rot, which is encouraged by insufficient root aeration. When a potted plant is watered, the entire root zone should be soaked. Shallow watering is the next most common mistake. When only enough water is applied to wet the top few centimeters of the potting soil, the plant looks well watered for a couple of days because the surface is damp. If the entire root zone does not receive water, however, the plant will ultimately die because the new growth areas of the roots will finally dehydrate. A good rule of thumb, or "rule of finger" in this case, is to stick your index

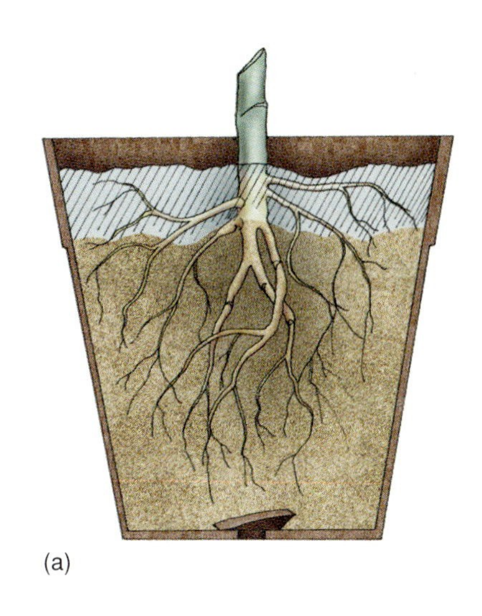

(a)

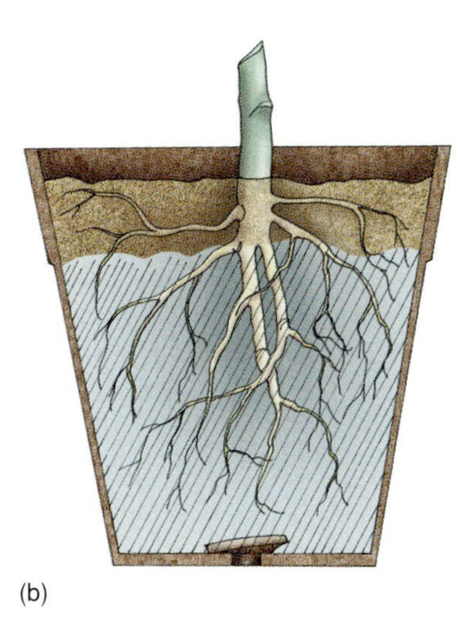

(b)

Figure 12.31
Proper watering is absolutely essential to growth and development. (*a*) The cross section depicts water that only wets the top of the pot. Most of the root system is still dry. (*b*) This pot has been watered thoroughly; the entire root system has adequate moisture, even though the surface dries within a few days. Overwatering can occur if more water is added too soon.

finger into the potting soil at least to the second knuckle to test soil moisture. If the soil is dry to that depth, the plant probably needs watering.

Cacti and other succulents form a third watering category. They should be thoroughly watered, but much less frequently than other plants. In addition, they must have a well-drained soil because they are especially susceptible to root rotting. In prolonged periods of cool or cloudy weather or in areas of high humidity, their watering schedule should be even less frequent.

Fertilizer

Each species has specific nutritional needs and should be fertilized accordingly. Unless a plant has an unusual requirement for given specific minerals, however, a general-purpose houseplant fertilizer will contain appropriate levels of macroelements and microelements when mixed in the proper concentrations. Do not fall into the trap of thinking that "if a little is good, a lot will be great" when it comes to fertilizer concentrations. As mentioned earlier, fertilizing with too high a concentration adds too many salts to the water in the soil, producing a more negative water potential in the soil than in the roots. Water is pulled from the roots in such a situation, producing fertilizer burn.

For safe concentrations and no loss of effectiveness, fertilize with half the recommended strength twice as often as is recommended. For many plants, in fact, such a fertilizing regimen actually produces better results. Table 12.2 summarizes some of the more common mineral deficiencies and their symptoms.

Light

Different plants also have different light requirements. Some must have full sunlight for at least half the day; others need only indirect or reflected light, and some require very low light intensities. Leaves that receive too high a light intensity will literally sunburn. The resulting bleached areas followed by dry brown splotches on the leaf surface are different from the gradually spreading brown leaf margin or apex that results from fertilizer burn or insufficient water.

Soil

Potting soil for houseplants can be purchased already mixed and packaged or can be mixed at home with equal results. Recalling soil composition will clarify what one needs to use in mixing potting soil: organic material, inorganic minerals, and porosity for air and water. For the vast majority of common houseplant species the following formula will produce a potting soil that provides these needs: equal parts of medium-coarse sand, peat moss, and local soil. The sand will provide porosity, the peat moss holds water well, and the native soil will add needed minerals. If local soils contain too much clay, are abnormally alkaline or acid, or are otherwise unsuitable, a rich compost should be substituted (see the next section). Local soils should be sterilized by oven heating at 350°F for one hour. To ensure proper packing in the pot, dampen the mixture to prewet the peat moss.

Because peat moss will ultimately compress and harden, houseplants need to be repotted every two to three years. A sure sign of peat moss packing is when a soil runs water all the way through the pot very quickly after watering. The water will not absorb because the peat moss fibers are too tightly packed together. Instead, water will often run down between the wall of the pot and the compacted block of soil. The replacement of fresh potting soil improves water availability, aeration, and organic and inorganic nutrient availability.

Cacti and other plants native to dry regions should have an adjustment made in their potting soil formula. At least 50% medium-coarse sand should be mixed with local soil and peat moss. Cacti do best when they have a high-porosity soil to allow water percolation and proper aeration. Many other plant groups are also better adapted to sandy rather than loamy soil. Care should be taken not to overwater cacti, even though they may be in an almost totally sandy soil. The top surface of sandy soil dries out rather dramatically in only one or two days after watering, even though the root zone has adequate moisture. Many people water again, based on the surface appearance. Cacti are very susceptible to root rot, even in a porous soil, if watered too often.

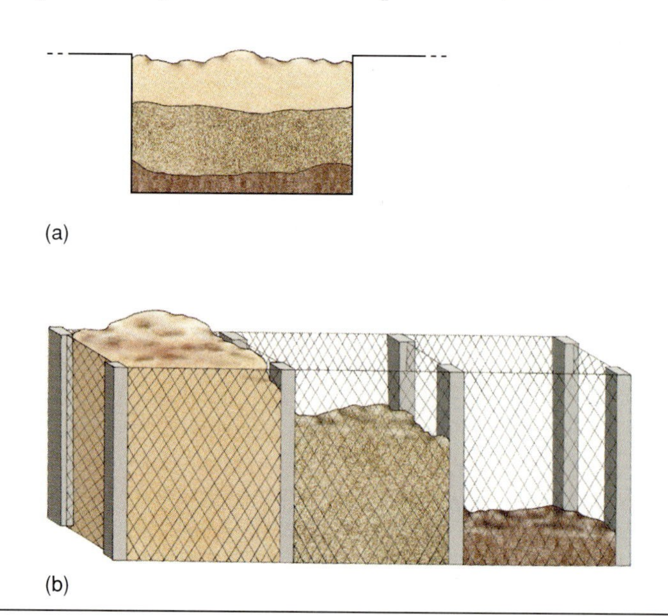

Figure 12.32

(*a*) An ideal compost pile consists of a "mature" compost on the bottom, a middle composting layer, and new organic material added at the top. (*b*) By constructing a three-partition pile, one section (*right*) can hold composted material ready for the garden. A middle section can hold material that has just been turned, and the first section (*left*) can hold new material.

Figure 12.33
A very simple compost pile can be made with a framework of loose boards and a mesh wire.

Compost

The addition of **compost** (figure 12.32) to a potting soil or garden enriches the soil with nutrients and texture. Compost is made by placing organic materials in a pit or small mesh-screened enclosure and providing an optimal environment for decomposition to occur (figure 12.33). Although animal tissues, woody plant parts greater in diameter than a pencil, and most Bermuda grass clippings are not to be used, essentially all other organic wastes can be composted. Leaves, garden prunings, kitchen vegetable waste, eggshells, coffee grounds, orange rinds, and an endless list of other plant parts can be placed in a compost. It is best to begin with an enclosure that provides a foot or two of depth. A shovelful of soil to provide the decomposing organisms and even a little fertilizer mixed in with the composting material will improve results. The top of the compost should be covered, and water should be added as necessary to keep the composting environment moist.

At the beginning of each year the compost should be turned. Remove the bottom layer, which should be ready to mix with potting or gardening soil. Periodic turning or loosening to improve aeration during the warm part of the year prevents compaction and the resulting suffocation of the organisms carrying out the decomposition. Regular addition of new materials to the top grades the layers of activity, with the bottom the most completely converted. Although composts are easier to develop in warmer and wetter climates, which also usually have longer growing seasons, with proper planning and care a compost can be successful anywhere.

Potting

When potting or repotting a houseplant, you must keep in mind several rules of green thumb. First, do not use a pot with a top diameter more than 5 cm wider than the root zone of the plant. Second, use a pot with a drain hole in the bottom, preferably a clay pot for most plants because it allows better soil aeration. Third, when roots are forced to grow in the shape of the pot, often growing through the drain hole, it is time to repot; the plant is probably becoming **root bound** (figure 12.34). Note that some plants need to be root bound to thrive. To repot those that do not, place one hand on top of the pot, turn the pot upside down, and strike the bottom of the pot firmly with the other hand until the root-soil mass slides out (figure 12.35). Without unnecessarily damaging the root structure, select an appropriately larger container and repot.

The new pot must first have the drain hole covered. A piece of broken clay pot (shard) that is larger than the drain hole is most commonly used, although rocks and other materials are acceptable. Cover the hole, but do not seal it; this prevents the new soil from washing out but allows water to drain freely. To hold the shard in place as well as provide a new growth area for the plant roots, add several centimeters of freshly mixed (damp) potting soil to the bottom of the new pot before inserting the plant. Finally, suspend the plant in the pot with one hand such that the stem will clear the top of the soil with 2 to 4 cm of space left between the soil and the top rim of the pot (figure 12.36). Add the soil and gently pack by tapping the pot on the tabletop; the soil should not be tightly hand packed. Thorough watering will help settle the new soil around the root system. An addition of soil to the surface may be necessary after several waterings have settled the soil and lowered the surface below the desired position.

(a)

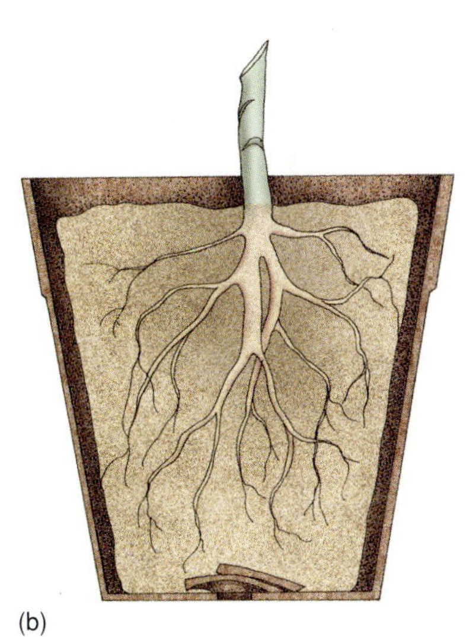

(b)

Figure 12.34
(*a*) This plant is root bound and needs to be repotted. The first indication of this condition is roots growing through the drainage hole in the bottom of the pot. This root system needs to be pruned, and the plant needs to be moved to a larger pot. (*b*) Repotting is also required when the soil compacts, pulling away from the sides of the pot.

Pests

The most dangerous houseplant pests are plant owners who do not know the needs of their plants. Overwatering, overfertilizing, underwatering, sunburning, and other problems of many houseplants are all a result of improper care. If these problems are eliminated, there are still several

Figure 12.35
In the repotting procedure a hand is placed over the top of the pot and around the plant, the pot is turned upside down, and a sharp blow to the bottom of the pot releases the entire soil ball.

Figure 12.36
Choice of a new pot should allow about 1 or 2 cm of new soil to be added around the original root ball. Place a shard in the bottom to cover the drainage hole, add soil mix to the bottom until the old soil ball is 2 to 4 cm below the rim of the pot (*a*), and add new soil around the root ball (*b*). Firm the soil by gently tapping the pot on a tabletop.

common insect pests that can cause problems. Houseplant gardening books should be consulted.

HYDROPONICS

Plants do not need soil per se, but only what soil provides: anchorage, support, water, air, nutrients. If all these needs could be met without the use of soil, the plants would thrive anyway. **Hydroponics** is the growing of plants in a liquid culture. Water is obviously available, and depending on the specific needs of given plants, air is available in moving water or aeration by bubblers, if necessary. A complete nutrient solution is added to the water that provides the correct concentration of all the macroelements and microelements. The recipe for this Hoagland's solution is presented in table 12.3. It can be used as a regular fertilizer solution for potted plants or for plants growing hydroponically. A number of common houseplants can be rooted in a glass of water, and some, such as ivy, can be maintained hydroponically.

Commercial hydroponic greenhouse owners suspend many plants with their roots in a trough that contains gently flowing water and the nutrient solution. Because these plants are in a greenhouse, anchorage against wind is not a problem; support, water, air, and nutrients are all provided, and soil is not necessary. This system removes the time-consuming soil preparation, potting and repotting, and the expense of the pots themselves. Hydroponically grown tomatoes are one of the most widely successful crops, but many others will be in large-scale hydroponic production in the near future. Although successful and able to produce crops out of season, hydroponic greenhouses are more expensive to operate than the traditional method of growing plants.

✔ *Concept Checks*

1. What is the "rule of finger" in watering houseplants?
2. How can fertilizers "burn" a plant root system?
3. What are the keys to proper potting of a houseplant?

TABLE 12.3

Modified Hoagland's Nutrient Medium

CHEMICAL	CONCENTRATION (MILLIMOLES/LITER)
$Ca(NO_3)_2$	5.0
KNO_3	5.0
$MgSO_4$	2.0
KH_2PO_4	1.0
NaFeEDTA	0.1
H_3BO_3	0.04
$MnCl{\cdot}4H_2O$	0.009
$ZnCl_2$	0.0008
$CuCl_2{\cdot}2H_2O$	0.0002
$NaMoO_4$	0.0001

Summary

1. Water is the most important factor in plant growth. It provides the solvent system for the distribution of nutrients and organic molecules, gives turgor to the cells, and participates in a few biochemical reactions. The special properties of water are due primarily to the hydrogen bonding between water molecules themselves and with other hydrophilic substances.
2. Soil develops from igneous, sedimentary, and metamorphic rock material in the earth's crust. Chemical and physical weathering break large mineral particles into smaller ones, giving rise to sand, silt, and clay. Combinations of these materials form loams, the soils in which plants grow. The highly weathered upper portions of the soil profile also contain organic matter and considerable air and water. The finer textured soils hold more water than do the coarse, sandy soils.
3. Diffusion is the physical process by which particles in random motion move from a region of high concentration to a region of lower concentration. Osmosis is a special case of diffusion in which water molecules move through a differentially permeable membrane in response to a concentration gradient.
4. The chemical potential of water is called the water potential, and the concept is used to explain how water moves from one part of a system to another. In the soil-plant-air continuum, gradients in water potential are developed by the components of solute, matric, and pressure potentials. They combine to determine water movement to the most negative water potential.
5. Transpiration is the evaporation of water from a leaf surface. It accounts for approximately 99% of the water absorbed from the soil. The control of transpiration is exerted by the cuticle and by stomatal regulation, the opening and closing of the pore between two guard cells. Various environmental factors influence stomatal regulation, and plants cope with water stress by various morphological and physiological adaptations.
6. An understanding of soil-plant-water relationships allows one to make intelligent decisions concerning the growth of plants—when to water, how to fertilize, how to mix soil properly, when to repot, and how to maintain healthy plants.
7. Hydroponics is the practice of growing plants in a defined nutrient solution without soil. It allows the grower to determine precisely the nutritional conditions of the plant, but the practice is primarily restricted to specialty crops and experimentation because of capital investment and labor costs.

Key Terms

abiotic 199
abscisic acid 208
acidic 194
acidic rainfall 196
active transport 205
adhesive 208
alkaline 194
apoplastic movement 205
basalt 197
basic 194
bedrock 199
bicarbonate ion 196
biotic 199
bulliform cells 210
CAM 210
carbonic acid 196
Casparian strip 205
cementation 197
clay 199
cofactor 195
cohesive 208
compost 213
cuticle thickness 210
cuticular transpiration 207
cutin 210
deplasmolysis 203
diffusion 202
dirt 198

Discussion Questions

1. How do plant nutrients get taken up by a plant?
2. Why is water so essential to plant growth?
3. Where does soil originate? What causes different textures?
4. What determines how much water can be stored in soil?
5. What causes water to move from the soil, into the plant, and out into the atmosphere?
6. How do plants survive in very dry regions?

Suggested Readings

Abou-Hadid, A. F., and A. R. Smith. 1993. *Symposium on soil and soilless media under protected cultivation.* International Society for Horticultural Science, Wageningen.

Alloway, B. J., ed. 1990. *Heavy metals in soils.* Halsted Press, New York.

Barton, L. L., and B. C. Hemming, eds. 1993. *Iron chelation in plants and soil microorganisms.* Academic Press, San Diego.

Chen, Y., and Y. Hadar. 1991. *Iron nutrition and interactions in plants.* Kluwer Academic Publishers, Boston.

El Bassam, N., M. Dambroth, and B. C. Loughman, eds. 1990. *Genetic aspects of plant mineral nutrition.* Kluwer Academic Publishers, Boston.

Glinski, J. 1990. *Soil physical conditions and plant roots.* CRC Press, Boca Raton.

Kabata-Pendias, A., and H. Pendias. 1992. *Trace elements in soils and plants.* CRC Press, Boca Raton.

Keister, D. L., and P. B. Cregan, eds. 1991. *The rhizosphere and plant growth.* Kluwer Academic Publishers, Boston.

Kolek, J., and V. Kozinka, eds. 1992. *Physiology of the plant root system.* Kluwer Academic Publishers, Boston.

Miller, R. W., and R. L. Donahue. 1990. *Soils: An introduction to soils and plant growth.* 6th ed. Prentice Hall, Englewood Cliffs.

Ming, D. W., and D. L. Henninger. 1989. *Lunar base agriculture: Soils for plant growth.* American Society of Agronomy, Madison.

Pillsbury, A. F. 1981. The salinity of rivers. *Sci. Am.* 245(1):54–65.

Reganold, J. P., R. I. Papendick, and J. F. Parr. 1990. Sustainable agriculture. 262(6):112–120. June 1990.

Rendig, V. V., and H. M. Taylor. 1989. *Principles of soil-plant interrelationships.* McGraw-Hill Publishing Co., New York.

Robson, A. D. 1989. *Soil acidity and plant growth.* Academic Press, San Diego.

Wright, R. J., V. C. Baligar, and R. P. Murrmann, eds. 1991. *Plant-soil interactions at low pH.* Kluwer Academic Publishers, Boston.

Chapter Thirteen

Energy Conversions

This chapter attempts to explain how the laws of thermodynamics are important to the botanical world as potential energy is converted into a usable form of energy so that biological "work" can be accomplished. In the process of photosynthesis, energy from sunlight is converted into chemical energy and stored in the bonds of various organic molecules. Once that energy has been captured, the atoms in these molecules can be rearranged, deleted, substituted, or supplemented so that other molecules can be constructed from the basic sugars formed in photosynthesis.

In respiration, these sugar molecules are broken down to release the energy needed in other biochemical reactions. The carbon "skeleton" of all metabolites are thus derived from photosynthesis, then changed and manipulated to produce all the molecules of life.

Banksia is one of the unique plants found in Australia.

Essentially all biological energy emanates from the sun. The center of our solar system is a fiery ball undergoing thermonuclear explosions at an incomprehensible intensity. Even though only a fraction of that energy reaches the surface of the earth, it is equivalent to 1 million Hiroshima-sized atomic bombs per day. Some of the light directed at the earth is screened out or reradiated back into space before it reaches the ground.

The thermonuclear explosions of the sun represent one type of energy that is converted into various other types of energy, notably heat and light. In addition, energy may exist as electrical energy, chemical energy, motion, sound, and molecular forces that hold together the atoms within molecules. Energy is really the ability to perform work. This is a relatively new concept, having come into the human vocabulary only during the past 200 years.

LAWS OF THERMODYNAMICS

Scientific questions are constantly being put to the test. Does the experiment work every time? What conditions modify the results? Can it be repeated by many investigators in many parts of the world? After such careful scrutiny, a number of natural laws have withstood the test of time and are now recognized as absolute. Two of these laws concern how energy flows through the solar system.

The first law of thermodynamics states: *Energy can neither be created nor destroyed, but it can be converted from one form to another.* It is possible, for example, to convert the energy locked in hydrogen in a hydrogen bomb to heat, light, short-wave destructive radiation, sound, and other forms of energy. Within the atoms a **potential energy** is held, capable of causing an incredible amount of "work" (and here the word is used loosely because many of us prefer not to equate widespread destruction with work). Just the same, the potential energy of the atoms and molecules is put to work. The energy itself is not lost but is converted from one form into several others.

Some forms of energy are used more efficiently in machines, but efficiency is nothing more than the percentage of energy that is converted into the human concept of work. The remainder is left in the original form or converted into some other type of energy that is not as desirable. Unlike water and nutrients, which cycle through ecosystems and are used over and over again, energy does not cycle; it changes, and eventually it ends up in a form that is of no tangible benefit to the system.

The second law of thermodynamics states: *When energy is converted, the potential energy of the final state will always be less than the potential energy of the initial state, provided that no energy enters or leaves the system.* This law helps us to understand that water at the top of a waterfall has potentially more energy than water in the pool below it (figure 13.1). Some of the energy is given up as the water spills over the face of the rocks. If a turbine happens to intercept the falling water, part of the potential energy may be turned into useful work. The system is never 100% efficient, and some of that energy is lost from the system as heat. Heat in itself can be an acceptable form of energy; if it is concentrated, it too may be used to generate steam and turn the same turbine. In many cases, however, the heat is so diffuse that it may not be practical to turn it into useful work.

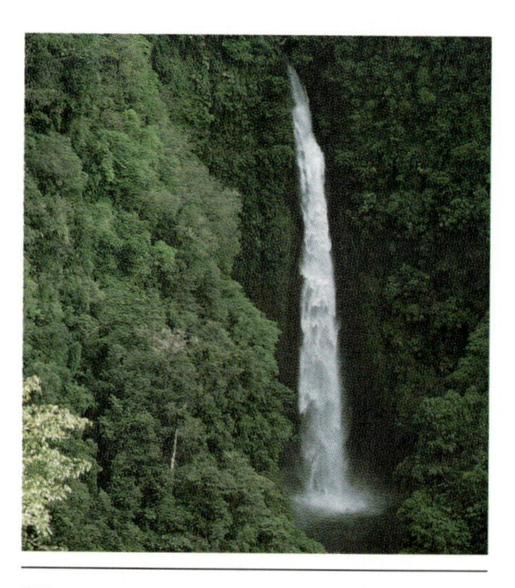

Figure 13.1
Water at the top of a waterfall is characterized by a potential energy that can be released in falling. If the water is used to drive a turbine, the potential energy may be turned into work in the form of electricity.

Consider all the electrical machinery used in factories. Engineers measure efficiency in terms of work produced per unit of electricity consumed. Electricity not converted to work eventually becomes heat, which is of no benefit to the factory. In fact, it means that the **ambient** or air temperature will increase, and the factories will have to be cooled to a greater extent than would be required if the machines were more efficient. The analogy holds in biological systems, although in warm-blooded animals the heat given off by the body processes of energy transfer becomes the body heat necessary to maintain a specific body temperature. In cold-blooded animals and in plants the heat is dissipated rapidly and totally lost from the organism. The energy was not destroyed but merely changed to a form that was of no use to the organism. What eventually happens to the heat? It warms the earth and is eventually dissipated into space. You have probably never thought of a tree or rose bush as warm, but in fact these plants produce body heat just as humans do, although at a much reduced rate. Lack of insulation in plants allows the heat to escape. It is possible to measure the interior temperature of large trees; it is several degrees higher than outside temperatures. But small plants seldom show a differential. In addition, some plant parts may produce a great amount of heat during reproduction, but that heat is quickly lost to the atmosphere.

Chemical reactions that occur spontaneously are called **exergonic** reactions (energy yielding), and the potential energy at the end of the reaction is less than the potential energy at the beginning of the reaction. An **endergonic** reaction is one in which energy must be added to the system for the reaction to occur. Endergonic reactions can never occur spontaneously.

One concept of energy concerns its organization or orderliness. Although difficult to define, **entropy** describes the degree of orderliness; potential energy has the greatest degree of orderliness. As the energy is converted and work proceeds, the matter becomes less structured and less orderly, and the randomness becomes greater.

The heat energy of the sun remains as heat, warming the earth and driving the hydrologic cycle and thus allowing life to exist. Light energy provides the potential energy for conversion in biological systems. Green plants begin the conversion process. In biochemical reactions, as in factories and power plants, no energy transfer is 100% efficient. Energy transfer in organisms that cannot use atmospheric oxygen, the anaerobic organisms, is relatively inefficient; organisms that do use atmospheric oxygen, aerobic organisms, transfer that energy with far greater efficiency.

OXIDATION-REDUCTION

Energy changes occur in chemical systems because of oxidation-reduction reactions. **Oxidation** may be defined as a *loss of electrons,* and **reduction** may be defined as a *gain of electrons.* In most biological systems the transfer of electrons is also accompanied by the transfer of protons. Thus, when energy transfer is described for biochemical systems, the changes in structure are usually written for the loss or gain of protons, rather than the electrons, although the electron transfer is implied. In this text and in others you will often see an electron-transfer molecule indicated as being oxidized (no special designation) or reduced (+2 H). For example, the primary electron acceptor molecule for photosynthesis is **nicotinamide adenine dinucleotide phosphate, (NADP).** Whenever you see NADP, the designation is for the oxidized form, and whenever you see $NADPH_2$, the designation is for the reduced form. The NADP has gained electrons to become reduced; however, this gain is indicated by the addition of hydrogen. Any oxidation of one molecule must be accompanied by a reduction of another molecule; one cannot occur without the other. Energy is transferred from one molecule to another, demonstrating the law of conservation of energy. Each step of photosynthesis that involves electron transfer must have a reduction accompanied by an oxidation, yet the net result produces an overall reduction of CO_2 to produce a sugar. By the same token, respiration involves oxidation-reduction that also occurs in distinct steps, but the overall result is to produce a net oxidation of sugar by losing electrons and protons. The primary electron acceptor in respiration is not NADP but **NAD;** the terminal phosphate is missing. In both photosynthesis and respiration **electron transfer** occurs in small steps with only slight changes in energy levels. This is made possible by an integrated chain of electron acceptors, the most important of which are a group of proteins called **cytochromes.** Cytochromes all have a similar structure, with an atom of iron at the center of the molecule. Since iron (Fe) can exist as Fe^{++} or Fe^{+++} it is possible to change the charge from one cytochrome to another simply by transferring the electrons on the iron atom. Cytochrome structure is very similar to the central portion of the **hemoglobin** molecule, which transports oxygen in red blood cells. The electrons are "carried" by these protein transfer agents in much the same way that the oxygen is carried by hemoglobin.

PHOTOSYNTHESIS

Little thought seems to have been given to photosynthesis and gases in the atmosphere prior to 1774, when the English chemist Joseph Priestley discovered oxygen and conducted experiments to show that animals cannot live without it. Later he was able to show that if one places a green plant in a closed container with an animal, the animal will live. He concluded that green plants growing in light were able to replenish the "poison air" (CO_2) by producing a substance called oxygen. A few years later a Dutch physician discovered that only green tissues produced oxygen, and then only in the light. When placed in the dark, plants consumed oxygen just as animals did. The pieces were all finally put together in 1804 when Nicholas Théodore de Saussure presented the formula for photosynthesis:

$$CO_2 + H_2O \xrightarrow{\text{Light}} CH_2O + O_2$$

This equation states that light energy is used to enable carbon dioxide and water to combine to form a sugar, giving off oxygen as a by-product. The light energy is thus stored as chemical energy in the sugar. Although the concept is simple, the exact mechanism by which photosynthesis occurs is complicated, and the process involves many steps.

For convenience and clarification the process is divided into **light** and **dark reactions.** The light reactions must have light energy to initiate them, but the dark reactions will go to completion without light. The dark reactions can take place in light or dark conditions but do not *require* darkness. Possibly they should be named the "light-independent reactions."

The light energy that arrives at the earth's surface has traveled the entire 150 million km in a matter of eight minutes. The light energy is contained in discrete bundles called **photons.** Some photons contain more energy than others, and the shorter the wavelength, the greater the energy per photon. Thus a photon of blue light (about 450 nm) has a great deal more energy than a photon of red light (about 650 nm). Wavelengths shorter than the visible spectrum, including ultraviolet light and x rays, contain considerably more energy than do the photons of the visible spectrum.

Pigments

Pigments are energy-capturing molecules that are specific for certain wavelengths. Although biological systems have only a few of these, plants do contain some significant radiation-capturing pigments.

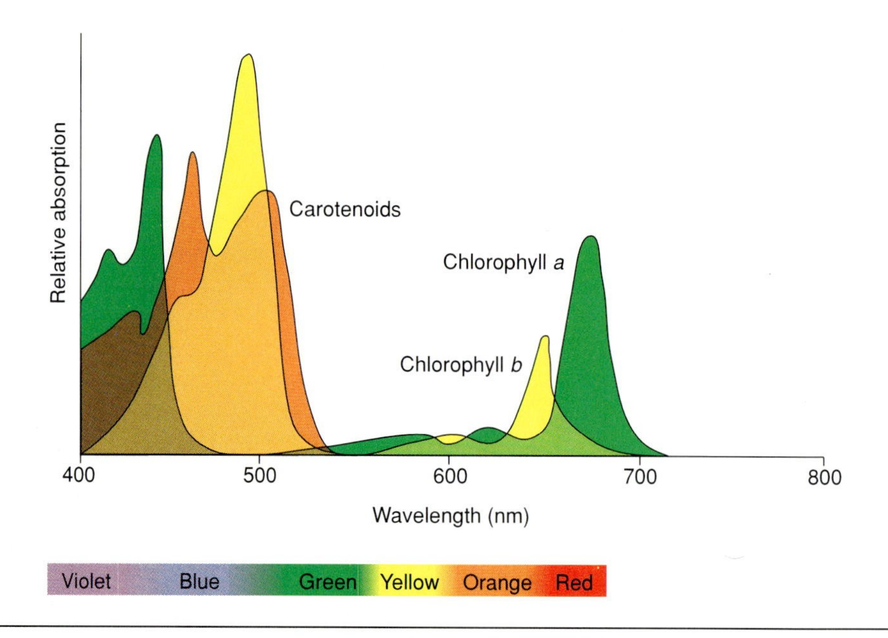

Figure 13.3
The absorption spectrum for chlorophyll *a*, chlorophyll *b*, and carotenoids. Note that the chlorophylls absorb in both the red and blue portions of the spectrum, whereas the carotenoids absorb only in the blue and blue-green portions.

Figure 13.2
The chlorophyll molecule consists of a magnesium atom core held in the center of a porphyrin ring. Attached to the ring is a long, lipid-soluble carbon chain. Chlorophyll *a* has a methyl (CH_3) group in the upper right-hand portion of the porphyrin ring. Chlorophyll *b* is exactly the same molecule, except that an aldehyde group (CHO) substitutes for the CH_3 group.

Chlorophyll

The basic green color of plants is caused by the pigment molecule **chlorophyll** (figure 13.2). Chlorophylls *a, b,* and *c* are slight chemical variations of a basic structure (according to species), but they perform essentially the same function. Most higher plants have varying ratios of chlorophyll *a* and chlorophyll *b,* with chlorophyll *a* always present in the higher concentration. Chlorophyll *c* replaces chlorophyll *b* in the brown algae and diatoms. Chlorophyll *a* is the primary pigment, and it occurs in all photosynthetic organisms except the photosynthetic bacteria. These "primitive" prokaryotes have a special kind of chlorophyll called bacteriochlorophyll.

Both chlorophyll *a* and chlorophyll *b* absorb the red and blue portions of the visible spectrum (figure 13.3). The midportion of the visible spectrum, yellow-green, is not absorbed but is reflected, and the human eye perceives it as green. Not all animals see the same colors. Dogs are relatively color-blind, honeybees see only in the shorter wavelengths of the visible spectrum and into the ultraviolet, and butterflies see only in the red or red-far portion. Such specificity has allowed coevolution of certain plants and animals.

The structure of the chlorophyll molecules allows for a **resonance,** or vibration of electrons, from one atom to another within the molecule. Electron movement, or **excitation,** is brought about by a photon of the correct wavelength (red or blue) striking the molecule. When the electron does become excited, the light energy becomes available to be (1) trapped and converted to chemical energy in the process of **photosynthesis,** (2) emitted at a longer wavelength with loss of energy as **fluorescence,** or (3) lost as heat. The first law of thermodynamics applies: light energy is converted into chemical energy in a complicated series of reactions called photosynthesis. In the intact leaf, fluorescence and heat loss are relatively unimportant. However, if chlorophyll is extracted by grinding leaves in an electric blender with acetone, the extract is poured into a test tube, and a bright incandescent light is directed at it, the extract will appear dark red rather than green. This is a demonstration of fluorescence, and it happens because the photosynthetic apparatus has been disrupted. The incandescent light continues to excite the chlorophyll molecule, but there is no electron acceptor to trap the energy. Therefore the light energy is emitted as light, but always at a longer and less

Capturing Photons

Whenever sunlight, a composite of colors of the rainbow visible to the human eye plus some other wavelengths that are invisible, strikes the plant leaf, a number of important things may occur. Sunlight comes in discrete bundles of energy, photons, which release that energy into various other forms of energy. Some of it may fail to hit the right molecules and simply pass through or bounce off the leaf. Other photons strike an important array of pigments, molecules capable of capturing that energy and using it for various purposes.

If the photon strikes a chlorophyll molecule (and there are several kinds depending upon the type of plant or other producer organism), it may set off an amazing chain of reactions that split a molecule of water and provide the energy to make complicated or simple molecules out of carbon dioxide. The carbon dioxide is said to be "reduced" as it takes part in the dark reactions of photosynthesis. Red and blue photons are particularly good for causing photosynthesis to occur. As plant pigments absorb the red and blue light, they reflect the yellow and green portion of the spectrum. Thus, our human eye perceives the reflected light, and we say that the plant leaf is "green." If you grow plants under green light and the only light the plant ever "sees" is green, then growth comes to a standstill. Don't forget that sunlight is virtually white light, but it is a painter's palette rich in all colors of the rainbow, including lots of red and blue.

There are also some other important pigments in plants. Carotenoids, for example, are red, orange, or yellow. Phycobilins, pigments restricted to cyanobacteria and red algae, are red and blue. They are called accessory or antenna pigments, and they transfer their light energy to a chlorophyll molecule so that photosynthesis can proceed.

Still another group of pigments are the **anthocyanins,** a group of red to purple molecules that are important in flower color, purple leaves and stems, etc. Anthocyanins are often associated with the onset of cold weather and the accumulation of red coloring in otherwise green stems.

So you see that pigments can come in almost any color, just as they do in a paint can. The color of a pigment represents only that portion of the visible spectrum that is *not* absorbed. Various organisms have different proportions of pigments to maximize their own advantage for survival.

You may wonder what causes trees or shrubs to change color in the fall. What is "fall color" anyway? With the change of seasons (shorter days, cooler nights, and a change in the quality of light), the chlorophyll pigments tend to become photooxidized, or bleached. As the green color begins to disappear, the protected pigments such as carotenoids begin to show through. They were there all the time, but the sheer numbers of chlorophyll molecules tended to mask them. Thus, bright yellow, orange, or red begins to show through. Some anthocyanins may add a purple touch. Not all species have the same pigments in the same proportions, so in a forest you may see many colors during the fall. You must look quickly, because the onset of very cold weather also causes a photooxidation of the accessory pigments, and in midwinter, all leaves are brown to straw colored.

energetic wavelength than originally absorbed. In the case of chlorophyll, the wavelength emission is in the red portion of the spectrum. For other pigment molecules, fluorescence may occur as green, yellow, blue, purple, pink, or other colors. Fluorescent paints that glow under black or ultraviolet lights function in exactly the same manner. Anytime fluorescence occurs, the additional energy originally held by the photon at the time it struck the pigment is lost as heat. In other words, part of the energy is retained in a photon with a lower energy level, and part of it is converted to heat. Again note that energy has not been created or destroyed, but merely converted to a different form.

Other Photosynthetic Pigments

In addition to chlorophyll, two accessory pigment systems perform in photosynthesis. These groups of molecules are the **carotenoids** and **phycobilins.** Carotenoids tend to be red, orange, and yellow, whereas phycobilins are red and blue. Both pigment systems function in electron capture and transfer. The phycobilins are restricted to **cyanobacteria** and the red algae. All these pigments can be excited by a photon of light of appropriate wavelength and then transfer that energy to a chlorophyll *a* molecule capable of continuing with the photosynthetic process. In other words, carotenoids and phycobilins act as antenna pigments, whereby they gather in the energy over a relatively large area and deliver that energy to a central molecule of chlorophyll *a*. Carotenoids also help protect the chlorophyll molecules from **photooxidation** or bleaching. When shade-adapted leaves are placed in bright sunlight, they often bleach because the chlorophyll loses structure and there are few accessory pigments. Accessory pigments help deter that bleaching. The accessory pigments, then, serve two important functions: (1) transfer of light energy to a chlorophyll *a* molecule and (2) protection of chlorophyll from photooxidation.

Concept Checks

1. What is potential energy, and why is the concept important in biological systems?
2. Why is it important for sunlight to encompass wavelengths of light from all parts of the visible spectrum?

Carotenoids are composed of carotene (orange) and xanthophyll (yellow) pigments. Carotenes serve several other important functions in nature unrelated to photosynthesis. **β-carotene,** for example, is the precursor for vitamin A and is exceedingly important in animal diets (figure 13.4). One of the standard measures of feed quality for livestock and

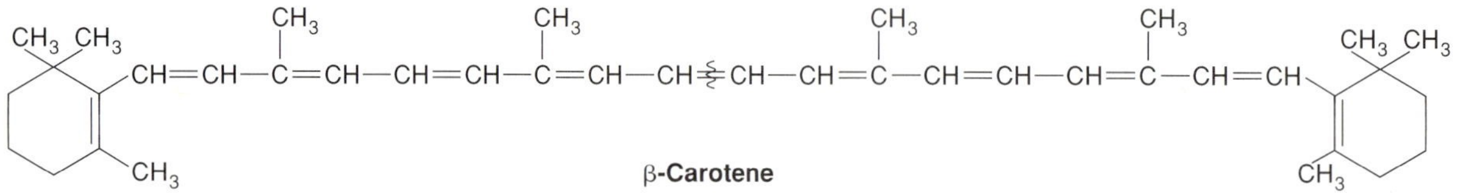

Figure 13.4

The β-carotene molecule is a long-chain hydrocarbon with a high degree of unsaturation (many double bonds). The entire chain consists of 40 carbon atoms.

wildlife is to measure the concentration of β-carotene. Carotene also gives egg yolks their characteristic yellow color and affects the pink pigmentation of flamingo feathers. The standard diet of flamingos includes algae rich in carotene. When dry food is used to feed flamingos in captivity, and when the carotene content is low, the pink pigmentation fails to develop, and the feathers are almost white.

The Chloroplast

In prokaryotic cells that carry on photosynthesis (photosynthetic bacteria and cyanobacteria) there are no chloroplasts. The chlorophyll molecules are attached to membranes in the cytoplasm, and the entire process takes place there. In all eukaryotic photosynthetic cells, however, the processes are neatly packaged into organelles, and the chloroplast functions in the conversion of light energy into chemical energy. The chlorophyll molecules are attached to membranes of the chloroplast.

The chloroplast is readily visible through a light microscope because of the green chlorophyll. The nucleus, which is larger, is rather transparent and visible only with special lighting and optical techniques. Chloroplasts are saucer shaped or flattened and elliptical and bound by a double membrane. They are approximately 5 µm in diameter, and they move freely in the cytoplasm. Cytoplasmic streaming, indicated by the movement of chloroplasts, tends to speed up considerably as light intensity increases and slow as light intensity decreases. The number of chloroplasts in a cell varies from about 20 to more than 100.

A thin section of a chloroplast can be viewed through the electron microscope to reveal an outer membrane and an inner membrane. A three-dimensional network of stacked **thylakoids** makes up the **grana.** Adjacent stacks of membranes are connected by strands that hold the entire system in place. The thylakoids support the enzymes associated with the light reactions. The matrix in between the grana is called the **stroma,** and the enzymes involved in the dark reactions are located there.

The Light Reactions

Light appears to saturate an area simply because the photons are so close together in time and space. In natural sunlight, photons of all energy levels (that is, all colors of the rainbow) strike the earth's surface at the same time and give the illusion of blended or relatively white light. Remember that the photosynthetic spectrum is only a tiny portion of the entire electromagnetic spectrum, and regardless of the energy levels of each photon, only those in the red or blue wavelengths are capable of exciting a chlorophyll molecule with energy levels correct for the photosynthetic process.

Whenever a red or blue photon strikes the chlorophyll molecule, it sets off a chain of reactions. Excitation raises an electron in the molecule to a higher energy level. The excitation is short-lived, and the electron tends to give up the energy in a matter of milliseconds. It is important, then, for the chlorophyll molecule to be located near the proteins that will allow the energy to be captured and converted to chemical energy. This is accomplished by an **electron transport system** that passes the electrons downhill as the energy is released. The electrons released by the chlorophyll molecule are replaced by those from a molecule of water.

There are two basic light reactions in photosynthesis: **photolysis,** or the so-called **Hill reaction,** and **photophosphorylation.**

Photolysis

In this overall reaction some of the energy gained from the excitation of the chlorophyll molecule is used to split a molecule of water (*photo,* light; *lysis,* to split). This is one of the few instances in which water actually enters into a biochemical reaction:

$$H_2O \xrightarrow{\text{Light}} 2\ H^+ \text{ and } \tfrac{1}{2}\ O_2 \uparrow$$

In this reaction, as in all chemical reactions, the beginning components on the left are called the substrates, and the resulting components on the right are called products. Oxygen is shown as half a molecule here because atmospheric oxygen is always O_2. In other words, this reaction must occur two times before a molecule of oxygen can be produced:

$$2\ H_2O \xrightarrow{\text{Light}} 4\ H^+ + O_2 \uparrow$$

Both products are exceedingly important in this case. The oxygen goes into the atmosphere as a gas (indicated by the arrow up), which becomes the source of oxygen for all aerobic organisms. If one extrapolates backward, oxygen can exist only if photosynthesis occurs. The ultimate conclusion is that aerobic organisms could not exist until photosynthesis occurred. Atmospheric oxygen has increased through geological history from none approximately

3.4 billion years ago to 21% of the air around us at the present time. This level of oxygen ensures that aerobic respiration can occur at a rate to produce sufficient energy compounds (primarily ATP) necessary for life.

Two of the H^+ produced in the splitting of water become the reducing source for driving the processes of photosynthesis. The hydrogen atoms reduce NADP to $NADPH_2$. The electrons are used to replace those given up by the chlorophyll molecule. Thus, electrons "flow" from one acceptor to another, and the overall process is called electron transport.

Photophosphorylation

The substrates that participate in biochemical reactions are called metabolites (a molecule involved in metabolism), and for many molecules to participate they must be phosphorylated. By examining a biochemical pathway, you can see that essentially all the intermediate metabolites are phosphorylated. It seems there is a great deal of phosphorus moving around in living cells. Where does it all come from, and how is the transfer made?

Recall that the "energy currency" for all living cells is ATP. This particular molecule functions so effectively in its role because the bonding of its (third) phosphate group creates a tremendous amount of stored energy—approximately 7000 calories/mole. (A mole is the molecular weight, in grams, of any substance.) Few other biological molecules can store that much energy. ATP is like a superbattery with exceptional storage capacity. In accordance with the law governing the conservation of energy, all of that stored energy has to come from somewhere. The energy stored in ATP, and essentially all usable energy, comes from the sun. It is stored directly in the ATP molecule by photophosphorylation—ATP formation using the light energy mediated by the chlorophyll molecule. This is the second general light reaction of photosynthesis. The reaction proceeds as follows (P_i is an unattached, inorganic phosphate):

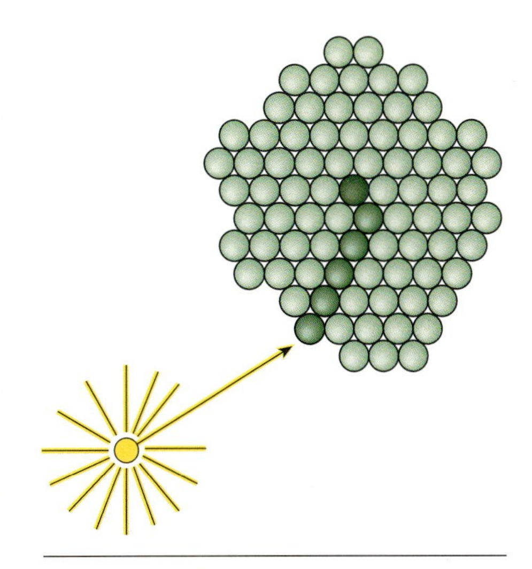

Figure 13.5
Each photosynthetic center consists of 300 to 400 pigment molecules, including a central molecule of chlorophyll *a*. The other antenna molecules may be chlorophylls or carotenoids, and a photon of light need strike only one of these molecules to set off a chain reaction that delivers energy to the central collector molecule. Energy is transferred from one molecule to another in approximately the shortest route to the center.

$$ADP + P_i \xrightarrow{Light} ATP$$

Since 7000 calories/mole of energy are stored in the terminal phosphate group of this molecule, that much energy (actually more because of the inefficiency of the system) had to be transferred from light energy coming from the sun.

To summarize, the two general light reactions of photosynthesis are:

Photolysis $H_2O \xrightarrow{Light} 2\ H^+ + \frac{1}{2}\ O_2\uparrow$

Photophosphorylation

$$ADP + P_i \xrightarrow{Light} ATP$$

The thylakoids of chloroplasts contain layers of proteins and various pigments, each photosynthetic unit being made up of approximately 300 to 400 pigment molecules (chlorophylls and carotenoids in higher plants) (figure 13.5). Photons of light may strike any of these accessory pigments to begin the chain reaction that eventually culminates in sugar formation. Only a single molecule of chlorophyll *a* is necessary to complete the photosynthetic process. Thus each photosynthetic system acts like a giant solar collector, and the efficiency of the system is greatly enhanced.

Current evidence indicates that there are two different kinds of photosystems: **photosystem I (PS I)** and **photosystem II (PS II).** In PS I the central chlorophyll *a* molecule absorbs most efficiently at a red wavelength of 700 nm and is usually called a P_{700} molecule. In PS II the central chlorophyll *a* molecule absorbs most efficiently at a red wavelength of 680 nm, hence the designation P_{680}. Even though both are chlorophyll *a,* the central molecules or reaction centers absorb at slightly different wavelengths.

The diagrammatic representation of electron flow in these light reactions is often referred to as the Z-scheme, in which the events of PS II occur before those of PS I (figure 13.6). Light entering PS II is absorbed directly (by the reactive center) or indirectly (by the antenna molecules), and the excited electron is transferred to an unidentified electron acceptor molecule. At the same time, photolysis of water occurs in which H_2O is split, giving its electrons to the electron-deficient P_{680} molecule. In the process, oxygen is given off into the atmosphere, and the protons become the reducing source to be used later in the series of reactions.

The excited electrons are accepted by an unidentified oxidized acceptor molecule, which becomes reduced. They are passed downhill, releasing energy through a series of protein molecules that comprise an electron transport chain. Cytochromes make up a portion of this chain. The energy is captured by ADP and P_i to form ATP through photophosphorylation. At that point another photon causes a similar excitation of the P_{700} reaction center as PS I, and again the excitation energy is transferred to electrons, raising them to an even higher energy level. Another unknown electron acceptor picks up the electrons and transports them through a similar electron transport scheme. This time

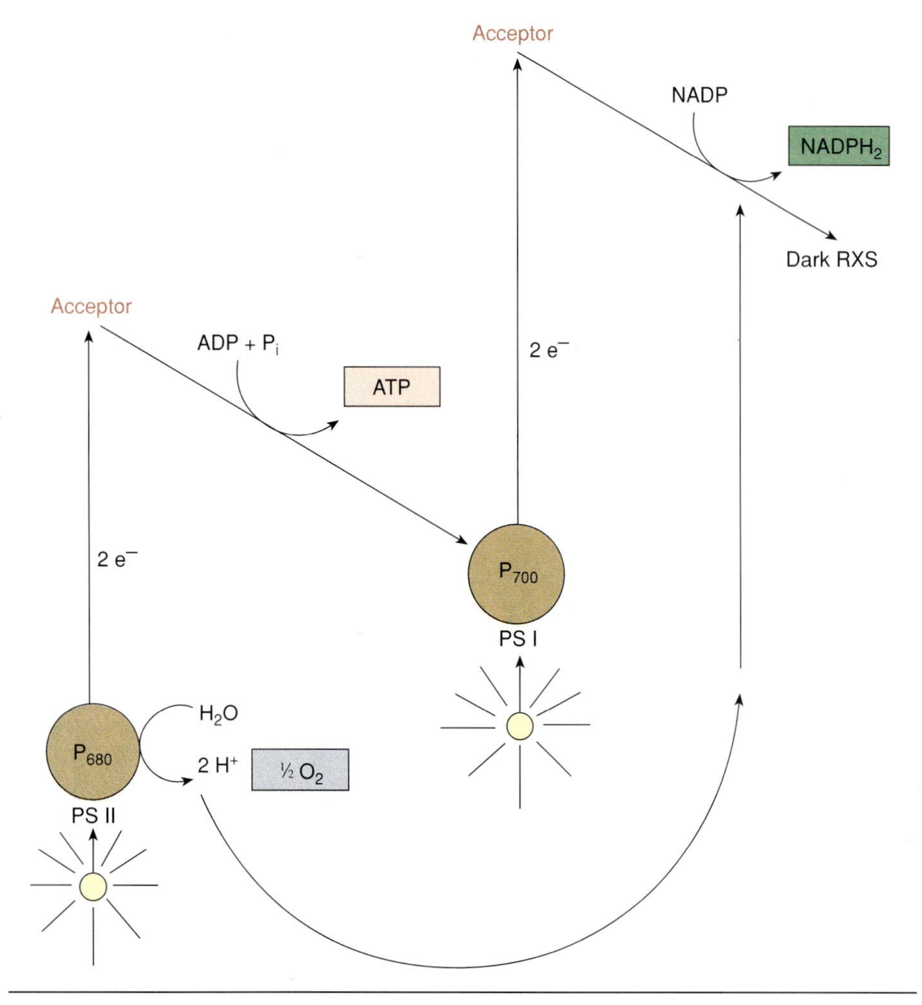

Figure 13.6
The Z-scheme of photosynthesis. The photoreactive center of PS II is activated by light to raise a pair of electrons to a higher energy level, where the electrons are captured by an acceptor. As the energy is passed downhill, ATP is formed. A second photon at PS I elevates the electrons to a higher level, and in the release of energy the electrons are delivered to NADP along with the 2 H^+ from water.

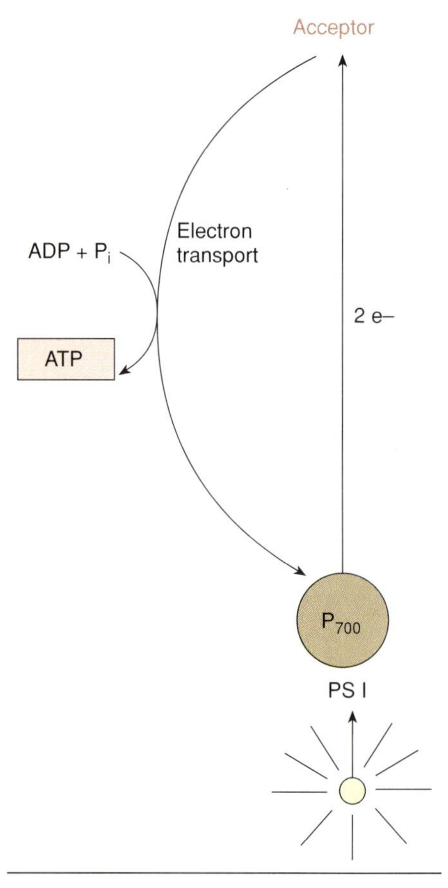

Figure 13.7
In cyclic photophosphorylation, high-energy electrons are passed directly to an electron transport chain that stores part of the energy as ATP. This process can occur without the benefit of PS II.

the terminal acceptor is NADP, and the electrons are delivered to that molecule. The protons to complete the reduction come from the molecule of H_2O that was split at the time of the initial PS II activation.

It is possible for PS I to work independently of PS II, causing electrons to be raised to a high energy state, received by the unknown acceptor, and then passed downhill by the electron transport scheme, forming ATP along the way through cyclic photophosphorylation (figure 13.7). Some prokaryotes and presumably all primitive photosynthetic organisms formed ATP this way, although no H_2O is split, no O_2 evolved, and no NADP reduced. Eukaryotic organisms can perform both cyclic and noncyclic photophosphorylation. Only in noncyclic photophosphorylation does the production of $NADPH_2$ occur, allowing CO_2 to be reduced to a carbohydrate.

The Dark Reactions

Even though photosynthesis is often considered a light-requiring process, the two processes just described are actually the only ones that require light. If, for example, one had the products of the light reactions, some chloroplasts with enzymes, and some carbon dioxide in the correct reaction mixture, the remaining aspects of photosynthesis could go on in a test tube in a dark room. In truth, the so-called dark reactions usually take place in the presence of light, but light is not required.

C_3 Metabolism: The Calvin Cycle

Little was known about photosynthesis until about 1950. An understanding of the way in which carbon dioxide is fixed into sugars came about primarily through the efforts of Melvin Calvin and his coworkers at the University of California. After the discovery of radioisotopes in conjunction with the World War II defense effort in the United States and elsewhere, certain isotopes became available for scientific purposes. The single isotope that has proven to be most effective and useful in biological research is carbon 14 (^{14}C), a radioactive form of the carbon atom. The stable isotope and the one found in greatest abundance in nature is ^{12}C, but ^{14}C is most useful in tracing the pathway of the carbon atom in metabolism. This isotope is exceedingly stable, having a half-life of more than 5000 years. This figure means that after a decay period of some 5000 years, one would still have half as much ^{14}C as in the beginning. This carbon isotope emits **beta rays** and is relatively safe in experimentation.

Calvin grew the green alga *Chlorella* in a complete nutrient medium in large round flasks and provided light from all directions. The efficiency of sugar production was quite good. The flask was equipped with a port near the top where radioactive carbon dioxide ($^{14}CO_2$) could be injected with a hypodermic needle. At the bottom a stopcock allowed the contents of the flask to be withdrawn rapidly and dropped into boiling methyl alcohol, where the cells would be killed instantly. Then sugars were extracted and spotted onto large sheets of filter paper for a testing process called paper chromatography. Certain solvent systems caused the paper to act as a wick; the chemical spot moved along with the solvent at a rate dependent on the physical and chemical properties of the substances in the spot. Some moved rapidly, others moved slowly, and by the time the solvent front reached the top of the paper, the chemicals had moved to various positions. The paper was dried, turned 90°, and subjected to another solvent system. Further separation occurred, and many spots appeared when the paper was sprayed with a suitable stain. These separate spots were then compared with known standards for identification. The same procedure forms the basis for many sophisticated biochemical techniques used extensively in laboratory research.

When Calvin injected $^{14}CO_2$ as a source of carbon and allowed the cells to carry on photosynthesis for a few minutes, he found that many spots had been labeled (become radioactive) with ^{14}C. This detection of radioactivity was made possible by placing an x-ray film over the paper chromatogram, allowing it to be exposed for a period of time, and then developing the x-ray film. Areas of radioactivity appeared as black spots on the negative. These spots were correlated with the positions of standard sugars and the specific spots identified. By gradually reducing the time of exposure to radioactivity, Calvin was able to work backward and gradually develop a pattern of labeling based on the rate of incorporation of various compounds.

This was not an easy task, and it finally became clear that the first labeled compound was the 3-carbon compound **3-phosphoglyceric acid (PGA).** Since CO_2 has only one carbon atom, and three carbons appeared in the product, the compound of incorporation would appear to be a 2-carbon fragment. However, after many months of searching, no 2-carbon compound was found, and the molecule eventually turned out to be a 5-carbon compound, **ribulose 1,5-bisphosphate (RUBP).** When CO_2 is combined with the 5-carbon molecule, it produces a very unstable 6-carbon intermediate, which immediately splits to two 3-carbon molecules, detected as PGA. Calvin went on to elucidate the entire pathway, a series of steps that involves a rearrangement of 3-, 4-, 5-, 6-, and 7-carbon sugars to produce a final product of **hexose** (6-carbon sugar) while regenerating the ribulose 1,5-bisphosphate (figure 13.8). For these efforts Calvin was awarded the Nobel Prize in Chemistry in 1961, and the process has come to be known as the C_3 pathway or the Calvin cycle.

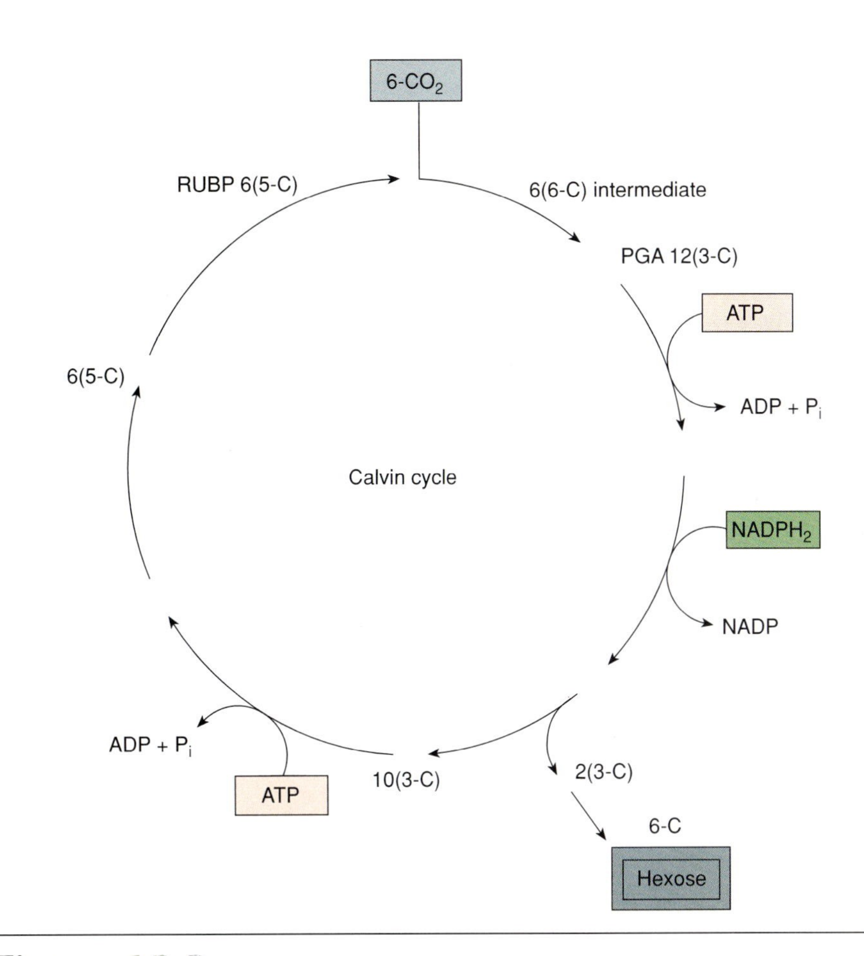

Figure 13.8

The Calvin cycle explains how CO_2 is incorporated into ribulose 1,5-bisphosphate to form a 6-carbon unstable intermediate. The first stable product is 3-phosphoglyceric acid (PGA). By incorporating six molecules of CO_2, the cycle generates a 6-carbon sugar while regenerating the RUBP.

As CO_2 enters the leaf, it is absorbed on the wet surfaces of mesophyll cells. The CO_2 crosses the plasma membrane, enters the cytoplasm, and crosses the chloroplast membrane, where it combines with the RUBP. The remainder of the cycle is merely a breaking down and restructuring of sugars so that hexose sugar is produced as a by-product of the cycle, and RUBP is regenerated. The $NADPH_2$ produced in the light reactions provides the reducing source (H_2) needed to convert CO_2 into a carbohydrate. The energy for this process is provided by the ATP from photophosphorylation.

The enzyme responsible for causing CO_2 to condense with RUBP is **RUBP carboxylase.** Not all enzymes are equally efficient in catalyzing their specific reaction, and this one is notoriously inefficient. RUBP carboxylase does not attach to CO_2 very well, and the only reason that carbon metabolism works as well in nature as it does is because large quantities of enzyme are synthesized by the plant. In some species this so-called fraction 1 protein makes up more than 50% of the entire leaf protein. As world food shortages become greater, more attention is being devoted to the use of leaf proteins. In most cases the bulk of that leaf protein will be the enzyme RUBP carboxylase.

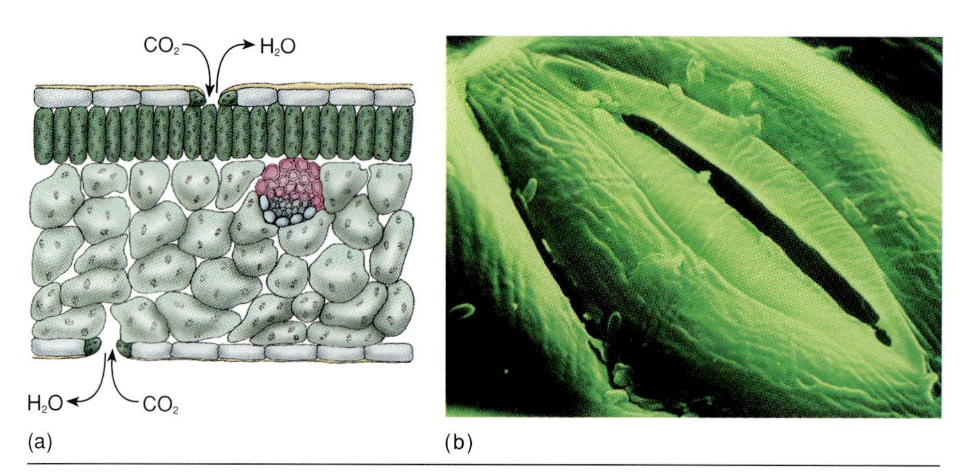

Figure 13.9

(*a*) Stomata on the leaf surface provide an opening for the uptake of CO_2 and the release of H_2O. (*b*) Scanning electron micrograph closeup of guard cells and stomatal opening.

The C_4 Photosynthetic Pathway

The Calvin cycle, which has also come to be known as the C_3 pathway because the first detectable product is a 3-carbon compound, is not the only means by which green plants fix carbon or incorporate it into sugars. In the early 1960s workers at the Hawaiian Sugar Planters Association found that if $^{14}CO_2$ was taken up by sugarcane leaves, the first detectable products were not 3-carbon molecules, but the 4-carbon organic acid **malate** and the 4-carbon amino acid **aspartate.** This system of carbon fixation is now known as the C_4 pathway of metabolism. It occurs in certain groups of plants and represents quite a departure from the C_3 pathway. In fact, C_4 metabolism still uses the Calvin cycle, but only in the later stages of fixation.

Anatomy of C_4 Plants The leaf anatomy of C_4 plants is quite different from that of typical C_3 plants. The vascular bundles of C_4 plants are surrounded by a group of cells called bundle sheath cells. The density of chloroplasts is much greater there than in the other spongy mesophyll cells. This is termed **Kranz anatomy,** and it occurs in almost all C_4 species.

Fixation in C_4 Plants In C_4 metabolism CO_2 enters through the stomata and is absorbed on the surface of the wet mesophyll cells (figure 13.9). Since CO_2 is readily soluble in water, it forms **carbonic acid** and immediately dissociates into H^+ (which makes the cell sap more acidic) and HCO_3^-, the bicarbonate ion. The bicarbonate ion actually enters into carbon fixation. Oxygen is released and may exit the stomata. The key to efficiency in C_4 metabolism is the initial enzyme that captures the bicarbonate, **PEP carboxylase.** This enzyme connects carbon to PEP (phosphoenolpyruvate), an important intermediate in respiration. PEP carboxylase is found almost exclusively in the spongy mesophyll cells, so carbon is fixed there as malate and aspartate. These two 4-carbon compounds are then transferred into the bundle sheath cells, where they are decarboxylated (carbon is dropped off). Then this carbon is fixed just as it is in normal Calvin cycle metabolism. The advantage here is that a very efficient carbon-capturing enzyme, PEP carboxylase, is present in mesophyll cells to collect great quantities of CO_2 and feed it into bundle sheath cells surrounding the vascular system. Once the sugars are finally made in the bundle sheath cells, they are readily loaded into the phloem, which is directly connected to the bundle sheath cells. The sugars can then be transported throughout the sieve tubes to parts of the plant where they are needed.

C_4 Efficiency It might seem that C_4 plants have evolved a cumbersome mechanism for absorbing CO_2 from the atmosphere. Why shouldn't they use the C_3 method, which is shorter and simpler? Many experiments and measurements in nature, however, have shown that, on the whole, C_4 plants perform better under conditions of high light intensity, high temperature, and low soil moisture than do C_3 plants. Many weed species are C_4 plants; they are more competitive than many crop plants for the reasons just mentioned. On the other hand, one must consider the environmental conditions for each particular situation before making broad statements about competitive ability. In a dark forest with adequate soil moisture, C_4 plants are not competitive at all. Under these conditions a typical C_3 weedy species would have every advantage. Evidence suggests that C_4 metabolism evolved in the dry tropics, where plants are often subjected to high light intensity, high temperatures, and drought.

The efficiency of C_4 plants arises from their ability to concentrate carbon even under conditions of low CO_2 in the atmosphere. They also have the ability to maintain respiration rates at a uniformly low rate, in both light and darkness. In C_3 plants, respiration rates increase in the light, a process termed **photorespiration,** and this increased use of photosynthate decreases the efficiency of carbon fixation. The first contact of CO_2 with an enzyme after entering the leaves is at the spongy parenchyma, where PEP carboxylase predominates. Once the initial fixation has occurred, it is possible to channel high concentrations of CO_2 into the bundle sheath cells, where the RUBP carboxylase predominates. Carbon fixation operates like a giant harvesting system in which "gatherers" at the outskirts feed the machinery at the center of the system. The physical relationships also place the bundle sheath cells adjacent to the phloem pipelines, which can readily load and transport the sugars. C_4 metabolism is like a special kind of partitioning in space: the cells with the correct enzymes are in exactly the right place in the leaf.

✔ Concept Checks

1. What are the basic light reactions in photosynthesis?
2. What is the source of carbon entering the Calvin cycle?
3. What is the difference between C_3 and C_4 plants?

TABLE 13.1

Typical C_3, C_4, and CAM Species

C_3 SPECIES	*Triticum vulgare* (wheat)
	Poa pratensis (Kentucky bluegrass)
	Gossypium hirsutum (cotton)
	Glycine max (soybean)
C_4 SPECIES	*Zea mays* (corn)
	Saccharum officinarum (sugarcane)
	Atriplex canescens (fourwing saltbush)
	Amaranthus retroflexus (pigweed)
CAM SPECIES	*Ananas sativus* (pineapple)
	Agave americana (century plant)
	Sedum prealtum
	Bryophyllum calycinum

The CAM Photosynthetic Pathway

A third type of carbon fixation has evolved for some plants, and the mechanism has many of the biochemical aspects of C_4 metabolism. Instead of partitioning metabolites in space (between mesophyll and bundle sheath cells), CAM plants use PEP carboxylase to fix carbon into malate, which is then stored until the next day, when the carbon is released and refixed in normal Calvin cycle metabolism. This kind of metabolism is often thought of as being partitioned in time rather than in space. Location of the cells is not so important, but the day/night time interval is very important.

Originally discovered in the Crassulaceae and subsequently named **crassulacean acid metabolism (CAM),** this system is now known to occur in many succulents and plants that grow in **xeric** (desert) environments. You might wonder why a plant would absorb CO_2 at night, fix it into malate, and then refix it into Calvin cycle sugars the next day. The answer appears to lie in the stomatal control of these plants. Unlike normal C_3 and C_4 plants, CAM plants open their stomata at night, when the relative humidity is higher and transpiration losses are lower. When sunlight reaches the plant, the stomata close, and new CO_2 cannot enter the plant. Although this slows the overall photosynthetic efficiency, it allows plants to live in more arid regions than would otherwise be possible. CAM plants manage to survive in the desert and in semiarid conditions, but the total biomass production is not as great as that of other plants. The only CAM plant with commercial importance is the pineapple (tables 13.1 and 13.2).

Ecological Aspects of Carbon Fixation

The unit of energy used to quantify metabolic demands and consumptions is the calorie. A **calorie** is the amount of heat (or heat equivalent) necessary to raise the temperature of 1 gm of water 1°C. The same unit of energy is used to measure the energy value of foods and feedstocks for combustion. The energy in any organic matter, such as a slice of bread, is determined by totally combusting the material in a **bomb calorimeter** (calorie meter) and measuring the calories of heat given off. The Calorie used by nutritionists is actually 1 kcal, or 1000 calories.

Each year the sun delivers 13×10^{23} calories to the surface of the earth.

TABLE 13.2

Some Photosynthetic Characteristics of Carbon Fixation

CHARACTERISTIC	C_3	C_4	CAM
Leaf anatomy	No distinct bundle sheath of photosynthetic cells	Well-organized bundle sheath rich in organelles	Usually no palisade cells; large vacuoles in mesophyll cells
Carboxylating enzyme	Ribulose bisphosphate carboxylase	PEP carboxylase, then ribulose bisphosphate carboxylase	Darkness: PEP carboxylase Light: mainly ribulose bisphosphate carboxylase
Transpiration ratio (gm H_2O/gm dry-weight increase)	450 to 950	250 to 350	50 to 55
Requirement for Na^+ as a micronutrient	No	Yes	Unknown
CO_2 compensation point (ppm CO_2)	30 to 70	0 to 10	0 to 5 in dark
Optimum temperature for photosynthesis	15° to 25°C	30° to 40°C	35°C
Dry matter production (tons/hectare/year)	22 ± 0.3	39 ± 17*	Low; highly variable

Data from Black, 1973 and Salisbury and Ross, 1978.
**Some C_4 plants are less efficient than C_3 plants, particularly at lower temperatures and lower light intensities.*

Knowing how long the sun shines at a given location for all seasons of the year, it is possible to calculate the number of calories per square meter per unit time. That value is the upper limit of productivity, since it is not possible to get more energy out of that square meter of the earth's surface than the sun put into it. Therefore we often state that the upper limit on crop yield is limited by the amount of light striking that surface. In fact, the actual upper limits of productivity are far less than the theoretical limits because of the inefficiency of the system. About one-third of the sunlight is reflected back into space, and much of it is absorbed by the earth and converted to heat, which becomes the driving force for operating the water cycle in nature. Green plants are able to convert only a small percentage of the total light available. Recall that many of the photons are not the correct wavelength to be absorbed by the chlorophylls or the accessory pigments. On an annual basis perhaps only 0.1% of the light energy that is capable of being converted to chemical energy is actually converted, and in many parts of the world the efficiency is considerably smaller. The opportunities are greatest where (1) the leaves are close together but not overlapping and (2) the leaves are exposed to light for the greater part of the year. Such conditions exist more often in the tropics, and the theoretical productivity is greatest in those regions.

The Leaf-Area Index

The leaf-area index (LAI) is defined as the ratio of total leaf area to a unit of ground area. It might seem that if leaf area were exactly equal to ground area (1 m^2 leaf area/1 m^2 ground area = LAI of 1.0), maximum efficiency could be achieved. Actually, fully grown plants often have LAIs that exceed 1.0 and in fact can reach values of 10 to 20. These plants grow where leaf canopy has many layers, and the bottom layer is very shady. The shaded leaves have little opportunity for photosynthesis.

Do shaded leaves benefit the plant? Part of the answer depends on the type of metabolism for that particular species. If the plant carries on C_4 metabolism, in which high light intensity is a major factor in photosynthetic efficiency, then a lower, shady leaf might be a liability or use more energy than it makes. On the other hand, the lower leaf of a C_3 species in a dense forest with little light might be perfectly adapted to low light intensity and carry on photosynthesis at a lowered but acceptable rate.

Respiration is constantly being balanced against photosynthesis in both nature and agriculture. The light intensity at which CO_2 being fixed via photosynthesis exactly equals the CO_2 being released by respiration is called the **compensation point,** and it has important ecological implications. A leaf just at the compensation point is contributing nothing to the buildup of metabolites or storage of materials in a plant. It follows that photosynthesis must exceed respiration by a considerable amount if plants are to accumulate biomass. In some cases senescent or shaded leaves carry on so little photosynthesis that the leaf falls below the compensation point and becomes a liability. When that happens, hormonal changes usually stimulate **abscission,** which helps the overall productivity budget for the plant.

The objective in agriculture is to provide as much productivity as

TABLE 13.3

Maximum Photosynthetic Rates of Major Plant Types under Natural Conditions

TYPE OF PLANT	EXAMPLE	MAXIMUM PHOTOSYNTHESIS ($mg\ CO_2/dm^2/hr$)
CAM	*Agave americana* (century plant)	1 to 4
Tropical, subtropical, and Mediterranean evergreen trees and shrubs; temperate zone evergreen conifers	*Pinus sylvestris* (Scotch pine)	5 to 15
Temperate zone deciduous trees and shrubs	*Fagus sylvatica* (European beech)	5 to 20
Temperate zone herbs and C_3 pathway crop plants	*Glycine max* (soybean)	15 to 30
Tropical grasses, dicots, and sedges with C_4 pathway	*Zea mays* (corn or maize)	35 to 70

Data from F. B. Salisbury and C. W. Ross, Eds., Plant Physiology, *2d ed., 1978. Wadsworth Publishing Co., Inc., Belmont CA.*

possible (table 13.3). Farmers attempt to achieve a photosynthesis/respiration ratio of about 40:1. Productivity is not always measured in total biomass, since only the selected biomass is important to the farmer. With wheat, for example, the selected biomass is only the reproductive structures; the roots, stems, leaves, and portions of the fruit are often discarded. With current emphasis on more comprehensive use of biomass, other parts of the plant have significance for an overall increase in productivity.

Farmers are always concerned about how patterns and density of plants maximize productivity. In temperate climates much of the caloric value from the sun is lost, because during certain seasons no leaves exist to carry on photosynthesis. In winter months fields may be bare. Even in early spring, when the seeds first germinate or when the buds on trees are just beginning to break, the LAI is so low that efficiency is poor. Later, leaves begin to expand, and new leaves are produced on many new shoots. The LAI may reach a value of 1.0 fairly early in the season, then rapidly reach a value of 6 or 8 by the time all leaves have been produced. Such values are considered acceptable for most crops and would imply that the spacing of the plants is correct. If the plants are placed too close together, they compete for light, water, and nutrients. If they are planted too far apart, much of the water and nutrient reservoir in the soil may not be used. Understanding crop ecology is important to the entire concept of productivity, and much is to be gained by knowing the nutritional and water requirements, pattern of rooting, type of carbon fixation, and other factors that contribute to yield.

Net Primary Productivity

Just how much biomass is produced by different types of vegetation? This is an important question both ecologically and agriculturally. Measurements have been made for various crop plants **(monoculture)** and for various types of natural ecosystems, including aquatic ones. At one time it was thought that the marine biome could be producing as much as 80% of the world's oxygen and therefore carrying on 80% of the world's photosynthesis. Those estimates have been revised downward, and ecologists currently believe that the oceans are responsible for about 33% of the total photosynthesis. Most of that is produced by the single-celled phytoplankton that live primarily in the surface waters. Depths below that level are too dark for effective photosynthesis by most organisms. The giant kelps carry on photosynthesis at a rapid rate, but their distribution is limited almost exclusively to coastlines.

Table 13.4 gives an example of net primary productivity in other ecosystems. Algal beds and reefs are exceptionally productive, as are tropical rain forests, swamps, and marshes. Notice that monocultures of agricultural production are not particularly productive, and because water is limited, desert biomes are notoriously poor in biomass production. In many cases ecosystems do not reach their potential productivity because of a number of catastrophic factors, including human activities. There can be no question that human activity, primarily in the past 100 years, has accelerated the loss of vegetation and greatly reduced the productivity of both terrestrial and aquatic ecosystems. In some grasslands of the central United States, overgrazing has led to the degradation of the landscape, elimination of grass species, and the invasion of weeds such as prickly pear cactus and broomweed. The "usable" productivity is almost eliminated, and the substituted species grow at a reduced rate or replacement vegetation fails to establish, leaving a bare soil.

TABLE 13.4

Estimates of Net Primary Productivity and Biomass for Various Ecosystems

ECOSYSTEM TYPE	AREA (10^6 km^2)	NET PRIMARY PRODUCTION PER UNIT AREA (gm/m^2/yr)	WORLD NET PRIMARY PRODUCTION (10^9 t/yr)	BIOMASS PER UNIT AREA (kg/m^2)	WORLD BIOMASS (10^9 t)
Terrestrial					
Tropical rain forest	17.0	2200	37.4	45	765
Tropical seasonal forest	7.5	1600	12.0	35	260
Temperate evergreen forest	5.0	1300	6.5	35	175
Temperate deciduous forest	7.0	1200	8.4	30	210
Boreal forest	12.0	800	9.6	20	240
Woodland and shrubland	8.5	700	6.0	6	50
Savanna	15.0	900	13.5	4	60
Temperate grassland	9.0	600	5.4	1.6	14
Tundra and alpine	8.0	140	1.1	0.6	5
Desert and semidesert scrub	18.0	90	1.6	0.7	13
Extreme desert, rock, sand, and ice	24.0	3	0.07	0.02	0.5
Cultivated land	14.0	650	9.1	1	14
Swamp and marsh	2.0	2000	4.0	15	30
Lake and stream	2.0	250	0.5	0.02	0.05
Total Continental	149		115		1837
Marine					
Open ocean	332.0	125	41.5	0.003	1.0
Upwelling zones	0.4	500	0.2	0.02	0.008
Continental shelf	26.6	360	9.6	0.01	0.27
Algal beds and reefs	0.6	2500	1.6	2	1.2
Estuaries	1.4	1500	2.1	1	1.4
Total Marine	361		55.0		3.9
Full Total	510		170		1841

Data from R. H. Whittaker, Communities and Ecosystems, *1970. Macmillan, London, England.*

Productivity, and how it is affected by improper human decisions, pollution, changes in climate, and other factors, is poorly understood. Many of the world's ecosystems are being and will continue to be upset or destroyed. A recent estimate places the loss of tropical rain forests alone at more than 500 km^2 per week, or more than 250,000 km^2 per year. Tropical rain forests currently produce about 22% of the world's net primary productivity (and oxygen). At the current rate of depletion, productivity loss on a world basis is about 1% every 3 years, and all tropical rain forests will be gone in approximately 65 years. Not only will this loss negatively affect oxygen and net primary productivity, but the tropical rain forests might also be a source for new crops and medicinal plants. Thus future gains for human welfare may never be realized unless their destruction is halted. Obviously, even those of us not living in the tropics have a moral obligation to become involved.

RESPIRATION

Respiration is the overall process by which the energy stored in carbohydrates is gradually released and transferred to ATP, the energy currency molecule. Unlike the burning of a bonfire, which results in combustion of the wood and the release of heat, the fuel of carbohydrates is "burned" at a slow and controlled rate so that the energy released can be partially captured and stored for later use. The beginning point for this orderly process is always the monosaccharide **glucose.** You will recall that glucose is the building block for **starch** and **cellulose.** Since cellulose is hard to digest, it makes an excellent structural material. However, digestion of starch is relatively easy and yields glucose. Glucose is also the chief transport carbohydrate in animals, but instead of storing it as starch, it is stored as a very similar macromolecule, **glycogen.** From this word is derived the term for the first phase of respiration—**glycolysis.** The processes of glycolysis and aerobic respiration are identical in plant and animal systems.

Aerobic respiration proceeds in three major phases: (1) glycolysis, (2) the Krebs cycle, and (3) electron transport. Any aerobic organism utilizes sugar in this manner, usually combusting the sugar molecule completely to form $CO_2 + H_2O$. Note that these products of respiration are the substrates of photosynthesis. In other words, respiration is an approximate reversal of photosynthesis:

$$CH_2O + O_2 \rightarrow CO_2 + H_2O + \text{Energy}$$

This time, instead of storing light energy as chemical energy, the energy already stored in the sugar molecule is released to form ATP. Also, remember that all reductions must be accompanied by an oxidation; whereas photosynthesis was a reduction process in which CO_2 was reduced by hydrogen produced in the photolysis of water, the burning of the sugar while utilizing oxygen is an overall oxidation process. All burning or combustion processes are oxidations. In the bonfire analogy the energy released by the wood escapes into the atmosphere as heat; in respiration the energy is slowly released and partially recaptured by ATP, although some is lost as heat. Again, the process is far from 100% efficient. Actually, the degree of efficiency varies according to a number of factors, which will be discussed later.

The Mitochondrion

The organelle containing the package of enzymes that function in respiration is the **mitochondrion.** Barely visible through the light microscope, mitochondria are round to rod shaped, 1 or 2 μm in diameter, and approximately the same size as a typical bacterial cell. The number of mitochondria varies in different tissues, depending on rates of respiration and need for ATP. Far more numerous than chloroplasts, they may number a thousand or more per cell. Although the origins of both the chloroplast and mitochondrion are subjects of debate, they both have double membranes and both contain DNA, which is responsible for part of their replication.

The internal membrane of the mitochondrion is involuted to provide a large surface area, much like a stack of shelves. These folds are called the **cristae,** and the total extent of the cristae is an indication of respiratory activity, although the number of mitochondria in a cell is probably a better indication. The matrix in between the cristae holds the enzymes responsible for the Krebs cycle, and the surface of the cristae holds the enzymes responsible for electron transport. Thus, even within a single organelle, the positioning of enzymes allows for a partitioning of functions. Mitochondria occur in all eukaryotic cells, both plant and animal. All living cells respire constantly, but the rate of respiration depends on many genetic and environmental factors.

Glycolysis

The first distinct series of reactions related to respiration—glycolysis—does not take place in the mitochondrion at all, but rather in the cytoplasm (figure 13.10). The enzymes and reactions function equally well under aerobic and anaerobic conditions. Glycolysis begins with glucose, which must first be phosphorylated by the transfer of ATP energy:

Glucose + ATP → Glucose 6-phosphate + ADP

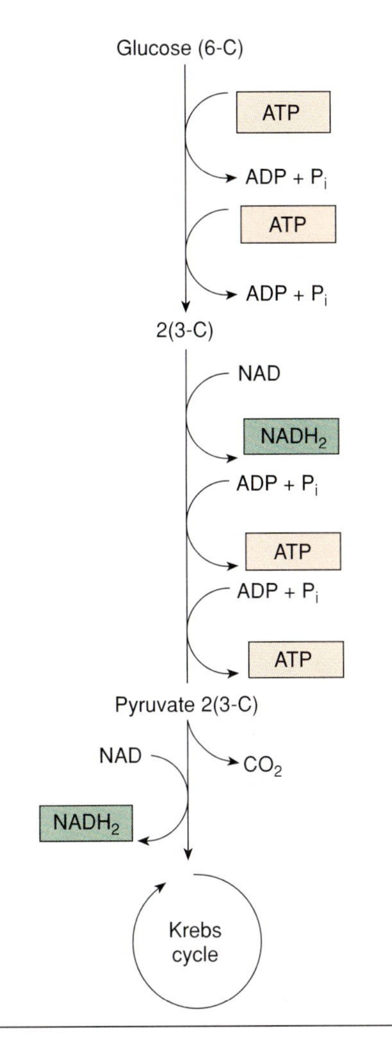

Figure 13.10
Glycolysis always begins with a molecule of phosphorylated glucose. In a series of reactions this 6-carbon sugar is broken down into two molecules of 3-carbon pyruvic acid (pyruvate). One carbon is lost from each molecule of pyruvate, and the remaining 2-carbon fragment enters the Krebs cycle.

Once "energized" by this phosphorylation process, glucose 6-phosphate undergoes conversion by splitting into 3-carbon sugars and finally, after partial oxidation, ending up as pyruvic acid. **Pyruvic acid** (also called **pyruvate**) becomes a key intermediate, which can function as a substrate under both aerobic and anaerobic conditions.

The Krebs Cycle

Pyruvate, which can readily penetrate the mitochondrial membrane, enters into the **Krebs cycle,** the second phase of aerobic respiration, named after the Nobel Prize winner Sir Hans Krebs (figure 13.11). This regenerating cycle is composed of a series of 4-, 5-, and 6-carbon organic acids. Prior to entering the cycle, the 3-carbon pyruvate is decarboxylated to give a 2-carbon acetate fragment. This fragment is activated by continuing with **coenzyme A** to become acetyl-coenzyme A. The carbon is lost as CO_2 and can be used as one measure of the rate of respiration. The acetate (2-carbon) fragment condenses with the 4-carbon oxaloacetic acid to form the 6-carbon citric acid. As the cycle proceeds, CO_2 is lost to produce a 5-carbon organic acid, which subsequently loses another CO_2 to become a 4-carbon organic acid. After a series of conversions and oxidations (loss of electrons and protons) the 4-carbon oxaloacetic acid is regenerated to combine with another acetate fragment, and the cycle begins anew. Since one carbon is lost from each pyruvate just before entering the cycle, and since two carbons are lost in one turn of the cycle, then two complete turns of the cycle are necessary to completely lose all carbons from a single molecule of glucose (6 carbons):

$$1 \text{ Glucose} \rightarrow 2 \text{ Pyruvate} \xrightarrow{\nearrow 2\, CO_2} \text{Krebs cycle} \rightarrow 4\, CO_2$$

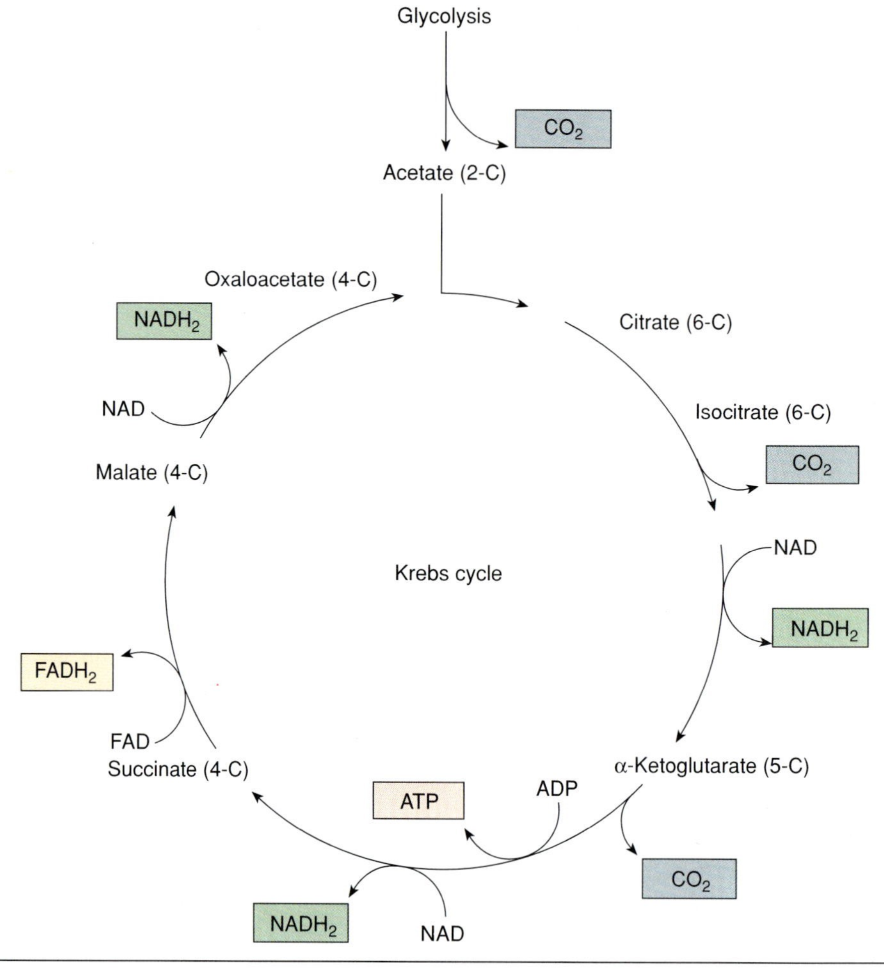

Figure 13.11
In the Krebs cycle, acetyl CoA condenses with oxaloacetic acid to form citric acid. This 6-carbon organic acid loses one CO_2, and later a second CO_2 is lost. Electron transport occurs at several locations during the oxidation process. Substrate level phosphorylation produces only 1 ATP, the FAD–electron transport generates 2 ATP, and the NAD–electron transport generates 3 ATP.

Electron Transport

Thus far there has been no discussion about how oxidation is coupled to electron acceptors so that the energy is captured in an orderly manner, or about H_2O as a product of the respiratory process. The third and final phase of aerobic respiration is electron transport (figure 13.12). Electron transport, also sometimes called **oxidative phosphorylation** (as distinguished from photophosphorylation) is the process by which electrons are passed from the oxidation of Krebs cycle organic acids to the electron acceptor NAD and subsequently to flavin mononucleotide (FMN), coenzyme Q, cytochrome b, cytochrome c, cytochrome a, and cytochrome a_3. In this chain reaction, energy is transferred from the electron transport chain during the coupling of P_i to ADP to form ATP:

$$ADP + P_i \xrightarrow{\text{Energy}} ATP$$

This is precisely the same reaction described as one of the light reactions of photosynthesis. In this case the energy is provided by the oxidation of sugar rather than by light energy.

If a pair of electrons is passed along the entire length of the electron transport chain, ATP is made in three separate places. Finally, after the electrons have lost most of their energy, they are transferred to molecular oxygen to produce H_2O:

$$2\, e^- + 2\, H^+ + \tfrac{1}{2}\, O_2 \rightarrow H_2O$$

Electron transport chains contain a group of cytochromes, the same iron-containing enzymes active in photosynthesis. These sites of electron transport exist anywhere electrons and protons become available during the oxidation process. Such opportunities occur at three sites of the Krebs cycle, and at one site just before entering the cycle. In addition, one pair of electrons

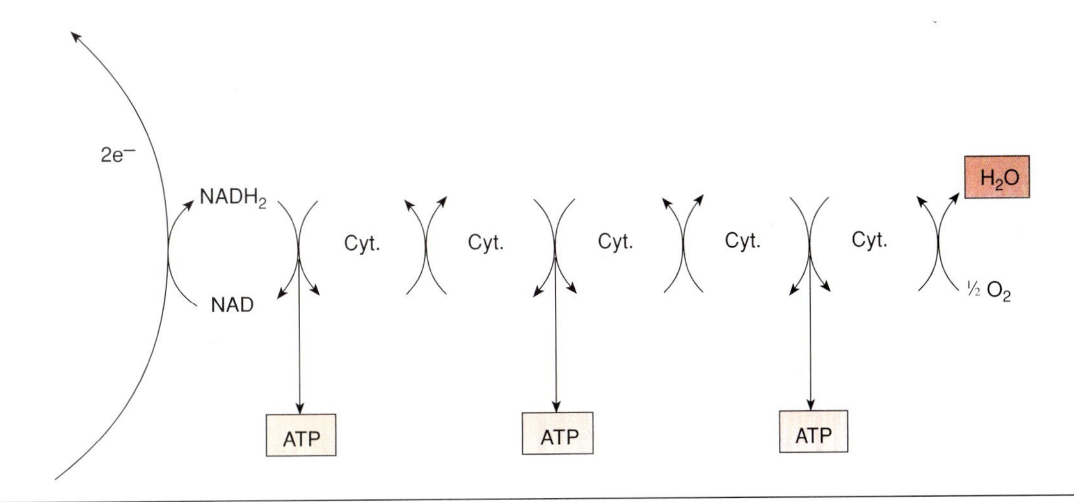

Figure 13.12
The electron transport chain begins as electrons and protons are given up by glycolysis or the Krebs cycle acids to NAD. The electrons are passed downhill through a series of proteins, including cytochromes, and the energy is converted to three molecules of ATP.

available in the Krebs cycle enters into the electron transport chain at a later position and makes only two molecules of ATP. Finally, one molecule of ATP is made within the cycle itself, and this method of making ATP is called substrate-level phosphorylation. The consequences of electron transport, then, are to synthesize ATP and deliver electrons and protons to oxygen, forming water. This is the only point at which oxygen is required by aerobic organisms. If it were not for the need of oxygen to combine with the terminal protons and electrons in the electron transport system, there would be no reason for us to breathe and distribute oxygen to all cells.

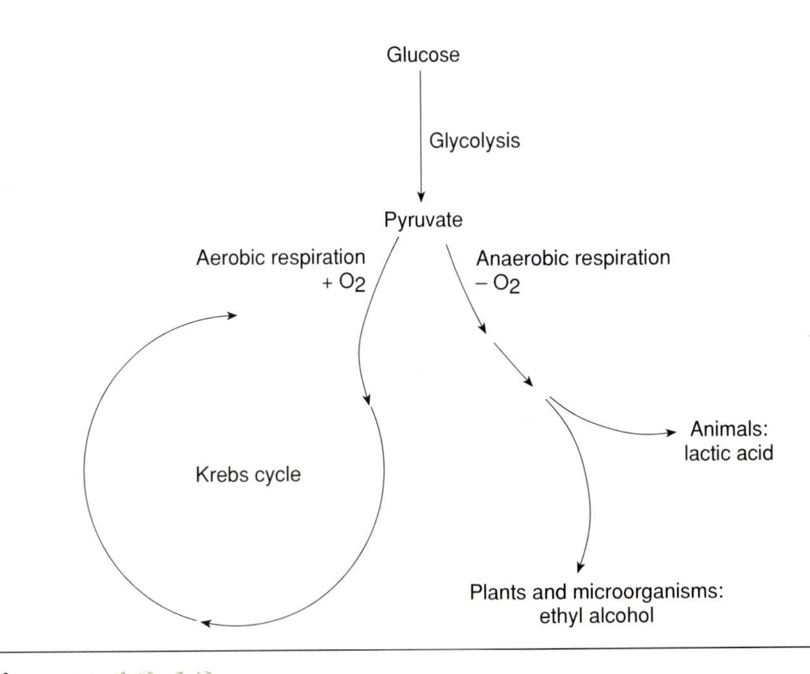

Figure 13.13
After glycolysis, pyruvate becomes the chemical compound that can be passed on to the Krebs cycle if oxygen is present, or used to produce lactic acid or ethyl alcohol if oxygen is absent.

Fermentation

Many microorganisms are able to carry on respiration without oxygen throughout their lives. Such anaerobic organisms contain a special group of enzymes that permit a partial oxidation of sugars to produce **ethanol.** This process is called **fermentation** and is to be distinguished from a similar process that occurs in animals, called **anaerobic respiration.** In anaerobic respiration the end product is **lactic acid** (figure 13.13). Even in aerobic systems such as our own, the lack of oxygenation in specific tissues can lead to the production of lactic acid. Whenever humans exercise excessively without having gradually built up muscle tone, the result is muscle soreness from lactic acid accumulation in those tissues. The red blood cells cannot transport oxygen to the muscle tissue quickly enough to supply all the oxygen needed in electron transport.

In fermentation, the natural process used in making beer and wine, alcohol is produced from many carbohydrate sources (figure 13.14). In any case, complex carbohydrates

✔ Concept Checks

1. How does the cell "burn" carbohydrates in respiration?
2. What is the primary function of the electron transport chain?

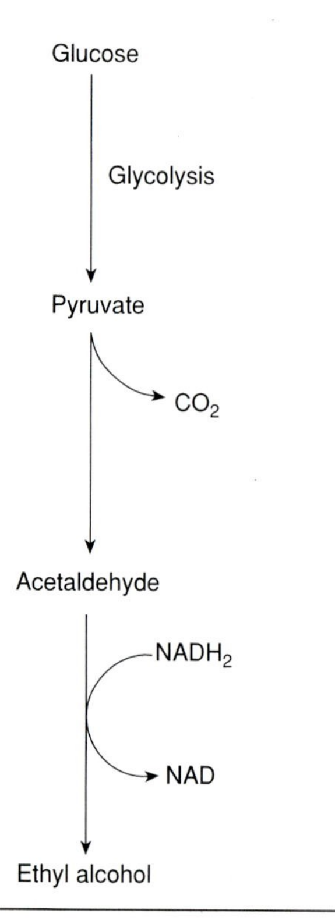

Figure 13.14
In alcoholic fermentation, glycolysis can occur under anaerobic conditions to produce pyruvate. Then decarboxylation (loss of CO_2) produces acetaldehyde, which is reduced to ethyl alcohol.

must be converted to glucose as the beginning substrate, just as in aerobic respiration. The process of glycolysis is exactly the same. (Recall that the enzymes of glycolysis perform equally well with or without oxygen.) Once pyruvate has been produced, however, the subsequent steps are unique, depending on whether conditions are aerobic or anaerobic. If no oxygen is present, pyruvate is decarboxylated, and the acetate fragment is oxidized directly to ethanol. This entire process goes on in the cytoplasm, and the mitochondria are not involved. Some microorganisms are able to switch metabolic pathways, depending on the presence or absence of oxygen. Such organisms, called facultative anaerobes, obviously have a great adaptive advantage. Some of these produce toxins and are exceedingly difficult to control.

Efficiency of Respiration

The total energy contained in a mole (1 gm molecular weight) of glucose is 686,000 calories. If this quantity of glucose, 180 gm, were burned completely in a bomb calorimeter, all of this energy would be released and the end products, as in complete aerobic respiration, would be CO_2 and H_2O. In aerobic respiration a fairly large portion of that energy is captured and "repackaged" as ATP. It is thus possible to calculate the efficiency of aerobic and anaerobic respiration by knowing the number of moles of ATP that are formed in the process.

During glycolysis two molecules of ATP are consumed in the initial stages of oxidation prior to the formation of two 3-carbon compounds. Then each of the two 3-carbon molecules undergoes oxidation, giving up electrons to NAD. This particular electron transport system is not complete, and each molecule of NAD releases only enough energy to make 2 ATP (a total of 4). Subsequently, each of the two 3-carbon molecules undergoes substrate-level phosphorylation on two different occasions. This gives an additional 4 ATP. At the end of glycolysis, pyruvate is formed, and the total ATP yield is –2 and +8, for a total of 6 ATP.

In the oxidation and decarboxylation of pyruvate, an additional pair of electrons is released to NAD, providing 3 ATP for each of the molecules, a total of 6 ATP. Subsequently, the Krebs cycle releases electrons and protons to NAD at three locations in the cycle, each forming 3 ATP, for a total of 9 ATP. In addition, 1 ATP is formed at the substrate level, and 2 ATP are formed by an oxidation that delivers electrons to FAD, forming only 2 ATP. (FAD is another electron acceptor that allows for only 2 ATPs to be made.) Since the Krebs cycle must go around twice for complete oxidation of the glucose, all of these values must be doubled for a total of 24 ATP (figure 13.15).

If each high-energy bond in ATP stores 7000 calories/mole, then $36 \times 7000 = 252{,}000$ calories. The remainder of the energy is converted to heat. The efficiency is thus 252,000/686,000, or almost 37%. Although seemingly low, this is actually a very good efficiency of energy conversion.

In anaerobic organisms the efficiency of conversion is much poorer. In typical lactic acid fermentation only two ATP molecules are produced, and the efficiency is 2%. In alcoholic fermentation the energy yield is only slightly higher. It thus appears that primitive anaerobic organisms were terribly inefficient in preserving the energy trapped by organic molecules. Only after the advent of aerobic respiration did the process reach a respectable level of efficiency.

The Substrate for Respiration

Once organic molecules have been formed, they function in biochemical reactions, as precursors for other metabolites, and in storage. Although one biochemical pathway may be involved in the synthesis of a compound, another may be involved in its breakdown. Large storage molecules are usually insoluble and therefore immobile. Some of these, such as cellulose, are used to make cell walls and become part of the permanent plant structure. Others, such as starch, are stored as insoluble macromolecules until needed. At that time, digestive enzymes begin to break down molecules into their smaller components. Each group of molecules has its own set of digestive enzymes:

$$\text{Starch} \xrightarrow{\text{amylases}} \text{Glucose}$$

$$\text{Protein} \xrightarrow{\text{proteases}} \text{Amino acids}$$

$$\text{Lipids} \xrightarrow{\text{lipases}} \text{Fatty acids}$$

Under most conditions a plant or animal uses a reserve organic molecule for respiratory energy only after it is first converted to glucose. In starch this is an easy one-step conversion. But

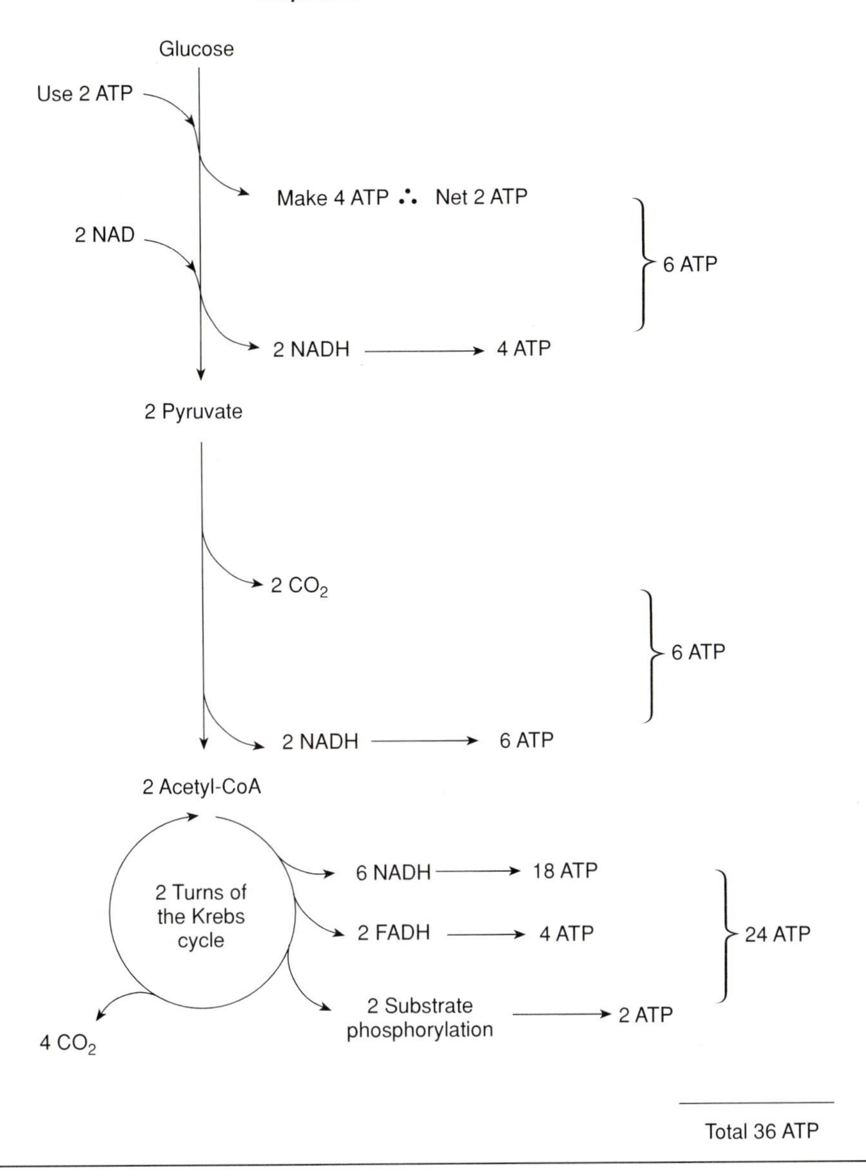

Figure 13.15
The energy balance sheet for respiration.

if the product is a lipid, lipases must break the lipid into fatty acids; then oxidation of the fatty acids produce 2-carbon acetate fragments that become available to the Krebs cycle.

Living organisms manage to store large quantities of energy in macromolecules. The most readily accessible and therefore the quickest energy comes from carbohydrates. Proteins are intermediate in terms of energy release, and lipids are broken down last. Consequently, the greatest amount of energy per unit weight is stored as lipids, and this is the reason that weight stored as fat is so difficult for dieters to remove.

The Implications of Metabolism

Organisms that carry on photosynthesis produce for themselves and for the heterotrophs all the beginning organic molecules of life. Starting with simple sugars, photosynthesis provides a "carbon skeleton" from which all organic molecules are synthesized (figure 13.16). It is easy to become so caught up in the details of photosynthesis and respiration that one forgets how they interact to perform all the feats of metabolism. Macromolecules are constantly being made from smaller molecules, and in a different part of the cell the macromolecules may be broken down into smaller molecules so that the carbon, hydrogen, and oxygen atoms may be rearranged to form an entirely different organic molecule. Some metabolic intermediates play key roles in providing large carbon pools from which other molecules are made. The Krebs cycle, for example, provides a 4-, 5-, and 6-carbon skeleton for the synthesis of many different organic molecules. One need only perform an amination reaction to add NH_2 (obtained from nitrogen in the soil) to one of these organic acids to form an amino acid. Purines and pyrimidines, lipids, hormones, sterols, pigments, alkaloids, starch, cellulose, proteins, nucleic acids, and many other compounds come directly from this backbone of photosynthesis and respiration.

Catalyzed by enzymes, these biochemical reactions are carried out with speed and perfection. Compartmentalization is the key to packaging enzymes in specific organelles so that one metabolic process is not hindered or canceled by another. You should begin to realize by now that the miracle of life must begin with the gene, for without the specificity of enzymes, biochemical reactions would not occur at a rate sufficient to sustain life. The living cell must surely be the most complicated system of operation that one could ever comprehend. The most sophisticated computer program should do so well.

✔ Concept Checks

1. How efficient is respiration?
2. From where does the "carbon skeleton" of all metabolism come?

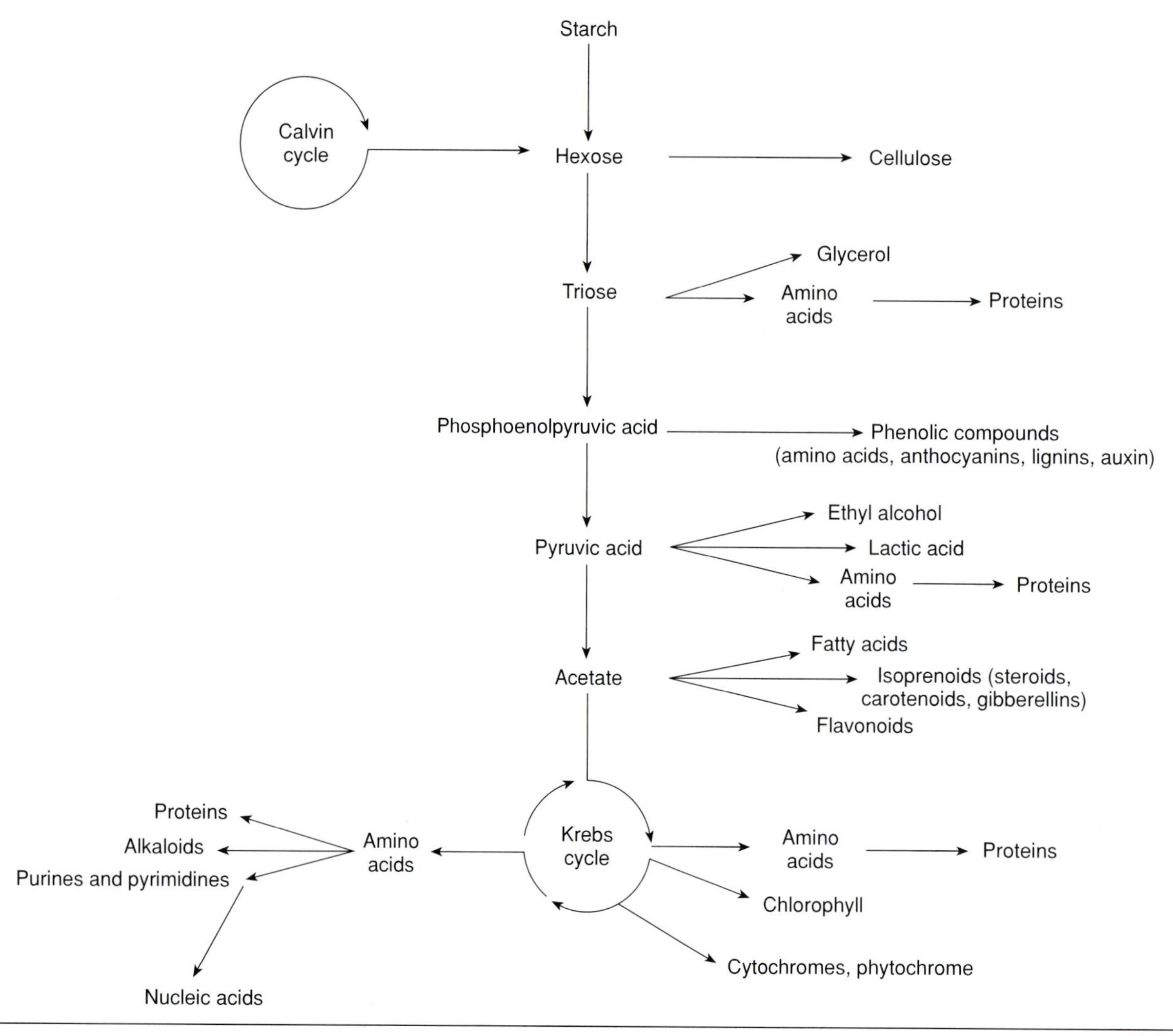

Figure 13.16

Photosynthesis and respiration combine in metabolism to produce all of the molecules important to life. The carbon skeleton of sugars, originally from atmospheric CO_2, are used in respiration to produce all organic molecules.

Summary

1. The laws of thermodynamics give important insight into the behavior of matter and its conversion into energy. The first law concerns the conservation of energy, and the second law allows one to make predictions about the probability that a reaction will occur. Energy is used throughout the living and nonliving world to perform work of various kinds.
2. Oxidation-reduction reactions in biological systems allow electrons and their accompanying protons to move from one molecule to another; every reduction must be accompanied by an oxidation. These reactions are all important in the energy conversion process of photosynthesis, in which CO_2 from the atmosphere is reduced to a carbohydrate through light energy.
3. A portion of that energy becomes stored in chemical bonds in the new organic molecule. Once the sugars have been made, the atoms may be rearranged, substituted, deleted, or supplemented to form new chemical compounds important to the cell and organism. Some of these stored sugars are converted to glucose and pass through aerobic respiration to provide the energy for the cell.
4. Photosynthesis consists of two basic light reactions, photolysis and photophosphorylation, followed by the dark reactions of the Calvin cycle. Some plants are especially adapted to high temperature, high light intensity, and water stress because the enzyme system for capturing CO_2 is different, and the processes are separated in time or space. Although such

plants do still use the Calvin cycle of normal C_3 plants, their additional enzymes cause them to be classified as C_4 or CAM plants.

5. In its simplest form respiration is the reversal of the chemical reactions for photosynthesis. Photosynthesis is an overall reduction process; respiration is an overall oxidation process.
6. The basic processes involved in aerobic respiration are glycolysis, the Krebs cycle, and electron transport. Glycolysis occurs in the cytoplasm under either aerobic or anaerobic conditions, but aerobic conditions must exist for further processing of the pyruvate in the mitochondrion. If oxygen is present, pyruvate is shuttled through the Krebs cycle, and the end products are CO_2 and H_2O. Much of the energy is stored as ATP. If oxygen is not present, anaerobic reactions lead to the formation of lactic acid in animals and ethyl alcohol in plants and microorganisms.

Key Terms

abscission 228
ambient 218
anaerobic respiration 233
anthocyanins 221
aspartate 226
β-carotene 221
beta rays 225
bomb calorimeter 227
calorie 227
CAM 227
carbonic acid 226
carotenoids 221
cellulose 231
chlorophyll 220
coenzyme A 232
compensation point 228
cristae 231
cyanobacteria 221
cytochromes 219
dark reactions 219
electron transfer 219
electron transport system 222
endergonic 218
entropy 219
ethanol 233
excitation 220
exergonic 218
fermentation 233
fluorescence 220
glucose 231
glycogen 231
glycolysis 231
grana 222
hemoglobin 219
hexose 225
Hill reaction 222
Kranz anatomy 226
Krebs cycle 231
lactic acid 233
light reactions 219
malate 226
mitochondrion 231
monoculture 229
NAD 219
NADP 219
oxidation 219
oxidative phosphorylation 232
PEP carboxylase 226
PGA 225
photolysis 222
photons 219
photooxidation 221
photophosphorylation 222
photorespiration 227
photosynthesis 220
photosystem I and II 223
phycobilins 221
potential energy 218
pyruvic acid 231
reduction 219
resonance 220
respiration 230
RUBP 225
RUBP carboxylase 226
starch 231
stroma 222
thylakoids 222
xeric 227

Discussion Questions

1. In chemical terms, what are oxidation-reduction reactions?
2. Explain the difference between the light and dark reactions of photosynthesis.
3. Why are plant pigments important to life?
4. Explain the process of fluorescence of a chlorophyll molecule.
5. What would be the adaptive advantage of CAM metabolism?
6. What is the ultimate limiting factor in food production?
7. Why is respiration a "necessary evil"?
8. What is fermentation?

Suggested Readings

Baker, N. R., and H. Thomas. 1992. *Crop photosynthesis: Spatial and temporal determinants.* Elsevier Science Publishers, New York.

Baker, N. R., and M. P. Percival, eds. *Herbicides.* Elsevier Science Publishers, New York.

Barber, J. 1992. *The photosystems: Structure, function, and molecular biology.* Elsevier Science Publishers, New York.

Bazzaz, F. A., and E. D. Jajer. 1992. Plant life in a CO_2-rich world. *Sci. Am.* 266(1):68–74. January 1992.

Bogorad, L., and I. K. Vasil. 1991. *The photosynthetic apparatus: Molecular biology and operation.* Academic Press, San Diego.

Coleman, G., and W. J. Coleman. 1990. How plants make oxygen. *Sci. Am.* 262(2)50–67. February 1990.

Douce, R., and D. A. Day. 1985. *Higher plant cell respiration.* Springer-Verlag, New York.

Galston, A. W. 1994. *Life processes of plants.* Freeman, New York.

Gregory, R. P. F. 1989. *Biochemistry of photosynthesis.* J. Wiley & Sons, New York.

Hinkle, P. C., and R. E. McCarty. 1978. How cells make ATP. *Sci. Am.* 229(3):104–123.

Kay, J., and P. D. J. Weitzman. 1987. *Krebs' citric acid cycle: Half a century and still turning.* British Biochemical Society, London.

Krebs, H. A. 1973. Recollections of Sir Hans Krebs. (Sound recording.) Harper & Row, New York.

Lawlor, D. W. 1993. *Photosynthesis: Molecular, physiological, and environmental processes.* J. Wiley & Sons, New York.

Schulze, E. D., and M. M. Caldwell, eds. 1994. *Ecophysiology of photosynthesis.* Springer-Verlag, New York.

Tobin, A. K. 1992. *Plant organelles: Compartmentation of metabolism in photosynthetic cells.* Cambridge University Press, New York.

Walker, D. 1992. *Energy, plants and man.* 2nd ed. Oxygraphics, Brighton.

Youvan, D. C., and B. L. Marrs. 1987. Molecular mechanisms of photosynthesis. *Sci. Am.* 256(6):42–49. June 1987.

Chapter Fourteen

The Control of Growth and Development

Thompson seedless grapes require gibberellic acid to fully ripen for commercial use.

This chapter looks at how plants grow and develop in response to both genetic and environmental factors. Consideration is given to temperature, water, both essential and nonessential salts, light, and the influence of humans on plant growth and development. Attention is given to the manner in which environmental cycles (diurnal and seasonal) affect these processes, including the environmental cues that cause and break dormancy.

Plant hormones and their interaction are studied in an effort to explain how growth promotion and retardation are under the control of the interaction of a series of chemical messengers. Finally, plant movements are discussed in terms of overall growth and development.

Plant growth, development, and reproduction depend on the interaction of the environment with the organism's genetic makeup, or **genotype.** Thus, environment and genotype combine to determine what an organism looks like—the **phenotype.** Even if a farmer plants the best quality seed available, the crop will be a failure if sunshine, water, and nutrients are not available in the right amounts at the right time. A light frost or a heat wave could spell disaster. In nature, too, the phenotype responds to the biotic and the abiotic environments. Soil factors, precipitation, and sunlight must be adequate for the plant to complete its growth cycle during the right season. Consider the consequences for a seed that happens to germinate at the beginning of winter. If that particular species were a warm-season annual, the seed would be killed by the frost. Another plant might survive the frost, lie under snow during the winter, and complete its life cycle the following spring and summer. Why does one survive and the other not?

The answer lies in the genotype. The winter annual must have genes that code for protection from cold. The warm-season annual does not have such genes. Scientists are now beginning to learn a great deal about how the expression of specific genes is modified by the environment. Even though experiments can be designed to look at specific gene-environment interactions, the phenotype represents the totality of thousands or even hundreds of thousands of genes, which have been influenced by the environment. Phenotype must be viewed as a summation of gene-gene interaction and gene-environment interaction.

You may have wondered why two seedlings derived from the same parent plant could grow at different rates, flower at different times, and assume shapes that are quite different. There are two basic reasons: **variability** and **environment.** Sexual reproduction leads to variability; for example, you do not look exactly like your brother or sister. Even identical twins possess slight differences. The environment also affects genetic makeup. It is certainly possible to identify a single gene as one that gives rise to the synthesis of a particular protein, which may represent an enzyme responsible for catalyzing a particular biochemical reaction. However, the timing of that gene expression, the amount of enzyme produced, and many other factors depend on the environment. Some genes are expressed during periods of light and others in the dark; some genes are turned on only when it rains; others are activated with the onset of autumn. All these environmental factors, acting in concert, bring about the gene expression, which is manifested as the phenotype.

THE PRINCIPLES OF GROWTH AND DEVELOPMENT

The terms **growth** and **development** are often used imprecisely. Sometimes the two are even used interchangeably, which is certainly incorrect. Growth refers to an increase in size or volume of a cell, tissue, organ, or organism. It happens because cell division is accompanied by an increase in cell size. One can even refer to a growth in populations, that is, an increase in numbers. There are some problems with the definition. A germinating seed imbibes water and therefore increases in volume; should this uptake of water alone be considered growth of the seed? Probably not, although the uptake of water is also accompanied by many metabolic changes. Some involve the digestion of storage macromolecules; others involve the synthesis of totally new molecules. For clarification, the definition for growth is usually modified to state *an irreversible increase in size or volume.* Thus permanent increases in dry matter are interpreted as growth.

Development is a summation of all the activities leading to changes in a cell, tissue, organ, or organism. The life cycle of an organism, for example, is a progressive change from a single fertilized egg, the zygote, through embryonic changes, and finally the changes that accompany full maturity of the organism. Growth, then, is a part of development. Development is an orderly sequence of events dictated by a precise set of genes turning on and off at every stage. One might compare the life cycle of an organism with a piano keyboard (the chromosomes). Music is produced only when the correct keys (the genes) are struck at the correct time. Some chords make sense if a group of keys is struck in unison, and the harmonic effect is pleasing to the ear. The correct sequence of genes has been turned on at the right time. Even a "good" chord must occur at the right time or it fails to integrate the musical composition.

Another important concept is **differentiation**—the chemical and physical changes associated with the developmental process. A meristematic cell in the vascular cambium undergoes certain chemical changes that eventually direct the cell to become a vessel element. In a short time that cell changes its metabolism, shape, and wall thickness, and finally the protoplast dies. Thus, while developing, that cell certainly underwent growth and differentiation. The two parts of the system constitute the developmental process.

Developmental biology is the study of how organisms, their cells, and their tissues achieve a final predictable form and function. It embodies genetics, physiology, biochemistry, biophysics, and many other disciplines, but gene regulation is the key in development.

LIMITATIONS TO GROWTH AND DEVELOPMENT (STRESSES)

Cells, tissues, organs, and organisms do achieve their predictable form and function as the result of environment exerting its influence on the genome, the entire complement of genes for the species. Many factors combine to give the overall environmental influence, such as temperature, light, and moisture. Even if a particular gene has been turned on, it may not be expressed, or the expression may be highly modified if the environment fails to meet certain limits of tolerance. Those environmental factors outside the normal range of tolerance for a given species are usually thought of as stress factors. Plants are constantly subjected to **environmental insults,** stresses from many sources that tend to limit growth and reproductive capacity.

The dry dormant seed lying in the soil has little resemblance to the organism that will spring from the embryo if the proper conditions are met. Seeds carry with them a "backpack" of stored molecules to nurture them in the early stages of germination. The early hours are precarious ones, and even the slightest desiccation at the wrong time may cause death. Although a few kinds of seeds will germinate on the soil surface (some actually require light and must lie on the surface), most seeds require the insulating and protective qualities of the warm, moist soil for germination. If stresses are not imposed and if the seed has imbibed water and oxygen, the machinery of germination is set in motion.

Macromolecules of starch, proteins, and oils are broken down, enzymes already present begin to function and new ones are rapidly synthesized, and metabolism gears up quickly. Respiration rates, for example, increase dramatically within a matter of hours, and the embryo begins to undergo mitosis and cell enlargement rapidly. Some cells of the embryo grow faster than others; the radicle is almost always the first organ to emerge. Why the radicle grows faster than the plumule is unclear. It is part of the entire developmental sequence that causes an organism to turn genes on and off at different times during the life cycle. Whether the genes actually get turned on and off is, again, determined by stress. The great unanswered questions in biology today are concerned with development: how embryos develop, what causes certain cells and tissues to grow more rapidly than others, what causes the onset of reproduction, and what causes aging. The mystery is gradually being unraveled, but the interactions are complex and the answers elusive. We do know that growth and development occur only within certain physiological limits imposed by the environment.

Figure 14.1
Temperature has a major effect on growth rate in plants. These cucumber plants have been grown for 14 days under controlled-environment conditions at five day/night temperatures, beginning at 75°/65°F and increasing at 5-degree intervals up to 95°/85°F.

Temperature

Fluctuations in temperature occur **diurnally** and **seasonally,** and in some parts of the world the fluctuations are much more pronounced than in others. As a general rule, greater seasonal variations occur as one moves toward the poles. Temperature extremes may be modified by factors such as altitude, clouds, relative humidity, ocean currents, and high- and low-pressure areas. Temperature affects each *biochemical* reaction in a particular way (figure 14.1). Generally, as the temperature increases within **physiological limits**—the tolerance range for that species—the rate of the reaction doubles for each 10°C rise in temperature. There are obvious limits; most biochemical reactions cease at approximately 43°C because the proteins are denatured and lose function at that point. This means that although the primary structure (amino acid sequence) may remain intact, the secondary, tertiary, and quaternary structures may collapse under severe heat conditions. The enzyme then ceases all activity. Some enzymes can resist changes under intense heat. Consider, for example, the cyanobacteria that manage to grow and reproduce in hot springs at or near the boiling point of water. Although there are exceptions, lower temperatures do not usually denature enzymes. Cold damage to plants may occur because the rate of the biochemical reactions is so slow that metabolic function is impossible. When temperatures fall below freezing, ice crystals may form and actually pierce membranes and cell walls. Although scientists do not fully understand how perennial plants are able to tolerate extremely low temperatures and survive, some believe the increasing viscosity of the cytoplasm of the cells to be a major factor in preventing ice crystal formation. Plant tissues also tend to dehydrate during the winter, and they rehydrate in the spring just as growth begins. Great gains in physiology and biochemistry will have been made when it is understood how genes of one organism allow survival under extreme heat and cold while the genes of another organism fail to be regulated, and the organism dies. Genetic engineering holds the potential of allowing stress-tolerant genes to be moved from one organism to another. One can visualize banana or pineapple being grown in cool temperate climates, and sugar beets being grown in the tropics.

Water

Although it certainly is true that many modern cities import water through pipelines or canals for many miles, human migration and settlement has been historically tied to locales where freshwater is found. Although water has been discussed in various contexts in this book, it still deserves special attention as a factor in controlling plant growth. As a medium for the support of all chemical reactions in living tissues, water dominates and controls the rates and combinations of reactions within living cells. **Water stress** in plant growth presents a major challenge for modern agriculture as growing human populations demand greater productivity. The stress problem is one of degree; each additional increment of water may increase **productivity** of plants a few percentage points, but the energy costs of water supply determine the amount of supplemental irrigation. Does one get enough in return to justify the additional costs?

In natural ecosystems the water-stress problem is quite different. Here survival is the key. Productivity may be exceedingly low in some desert ecosystems, but if the species can survive and manage to reproduce, even occasionally, then there is a good chance for relative success under those environmental conditions, provided that some outside force does not intervene. Plant geneticists and physiologists are particularly interested in understanding the mechanisms underlying **drought tolerance.** Perhaps, in the not-too-distant future, genes from drought-tolerant plants will be transferred to plants with less tolerance so that the "engineered" plants can be grown under a broader spectrum of environmental conditions.

Salts

One of the most serious stresses in the environment is imposed by salts, both naturally occurring and artificially applied. Plant owners sometimes fertilize too heavily. Even though a little bit is good, a lot is not necessarily better. Fertilizer burn actually comes about because water is "pulled" out of the plant and back into the soil; the plant cannot take up water fast enough to compensate for losses from the leaf surfaces.

Much salt damage is far more subtle than are the effects of overfertilization. Damaging salts, often not part of the essential nutrients, are found in most irrigation water. Such salts contribute to the overall problem of water quality. So-called freshwater used for irrigation contains some salts; even bottled drinking water contains some salts to keep the water from tasting "flat." Deionized and distilled water, on the other hand, has most of the salts removed. In nature, both underground water and surface water acquire dissolved salts, and these are transported along with the irrigation water. When the water is applied, much of it is taken up by plants, leaving the soil and leaf surface through the process of **evapotranspiration.** As the water molecules escape into the atmosphere, they leave the salts in the soil or on the leaves, where they accumulate unless removed by runoff or leaching below the root zone. Salt accumulation is generally not a problem in zones where adequate rainfall occurs; but in arid and semiarid regions, comprising approximately one-third of the entire land surface of the earth, salt accumulation is a major problem. Rainfall is not great enough to leach salts out of the root zone. These are the same regions to which water is transported for irrigation. The Imperial Valley of California is a good example of prime farmland plagued by salts. The ecological problem is enormous and worsening progressively.

Light

For any novice plant grower the importance of light becomes apparent very quickly. One of the first questions likely to be asked of a nursery salesperson is whether a new houseplant needs a sunny window or shade. Too many new "botanists" decide that the plant is not getting enough light and immediately transfer it from the dark corner to the hottest south window. The change in light might be too drastic for the plant, and bleached or dead leaves might be the consequence.

The process of photosynthesis involves the conversion of light energy into a stored chemical energy as sugars or other organic molecules. So light is absolutely essential for sustained plant growth, but the amount and kind can vary greatly. Plant physiologists are concerned with three factors of light: **duration, intensity,** and **quality.**

The duration is important in photosynthesis because the conversion process is an accumulative one. For all practical purposes, a plant can accumulate twice as much sugar in 10 hours of sunlight as it can in 5 hours. But there are limits, and the accumulation process may slow during very long light periods. The season and distance from the equator dictate the amount of natural sunlight to fall on any given spot, and it is quite predictable for any given date.

In a desirable climate at the equator every plant has 12 hours of sunlight each day in which to accumulate photosynthetic products; season has little if any effect on the amount of light received. But as one proceeds toward the north and south poles, the effect of this **photoperiod** becomes greater and greater. At the north pole during summer, daylight is almost constant, but during the winter there the days are quite short. Few plants are photosynthesizing at that time of the year. The seasons are exactly reversed at the south pole, and days are longest there when they are shortest at the north pole.

As one might expect, this finely synchronized system has important implications for reproduction. Like animals, plants reproduce during advantageous seasons. To be in the process of flowering at the time of the first blizzard and the onset of winter could result in death for a particular plant and even extinction for the species. Through the process of natural selection, species have ensured that the onset of reproduction will be triggered at an appropriate time of the year. The photoperiod system is a foolproof way to accomplish this. Its predictability is absolute. If the plant species continues to grow in the same place, the length of day is a certain, and thus ideal, cue. Most plants have developed a mechanism to signal and change the developmental process from strictly vegetative growth to reproductive growth and still allow time for the reproductive process to be completed before environmental conditions become unfavorable for growth. Although this environmental cue does not depend on the accumulation of a given amount of photosynthate, storage capacity is certainly important in providing energy for the reproductive process. Instead, the trigger itself is represented by a protein pigment called **phytochrome.** More will be said about this remarkable chemical in the latter part of this chapter.

Light intensity is important in the growth and development of all plants. Some plants are adapted to a shady, woodland environment, whereas others survive only under full sunlight. Light intensity is mediated not only by the shading of other leaves, but also by local weather conditions, including clouds. Sugarcane in Hawaii, for example, is not as productive as agriculturists would like because too many clouds surround the islands. As a C_4 plant (chapter 13), sugarcane is most productive under conditions of full sunlight. Many houseplants fail to survive because of low light intensity, whereas others die from excessively bright locations within the room.

Light quality is important in plant growth and development primarily because of the absorption spectrum of the chlorophyll and carotenoid molecules. Visible light without red or blue components will undermine plant growth and possibly inhibit flowering. Most sunshine contains a complete wavelength distribution, but artificial lights tend to be deficient in certain portions of visible light. Incandescent lamps produce light primarily in the red and far-red portions of the spectrum, whereas fluorescent lamps produce light primarily in the blue region. High-quality environmental-control chambers where plant experiments are conducted are equipped with a mixture of both incandescent and fluorescent lamps to simulate natural sunlight. With the correct mix of lamps, the quality can be duplicated rather well, but it is exceedingly difficult to get the intensity of natural sunlight. A special type of fluorescent lamp supplies more red light and therefore a more balanced spectrum than other fluorescent lamps.

✔ *Concept Checks*

1. What factors affect the phenotype of an individual organism?
2. What is meant by development with respect to botany?
3. What is the difference in water stress as applied to plant survival vs. plant productivity?
4. What is meant by fertilizer burn?
5. Why do some plants grow under lower light intensities than others?

Human Intervention

Each species has its own set of optimum conditions for growth and reproduction. By domesticating certain plants and animals, humans have learned about the environment best suited for those species. On the other hand, we really do not know the subtle requirements for optimum growth and reproduction for many species, even the domesticated ones. Each genome responds to the environment in a specific way. Change one or a few genes, and the tolerances may change. One plant may be able to tolerate conditions near the south window very well (many of the succulent desert plants), but others do not have the right combination of genes to tolerate that level of heat and light intensity. Put the cactus in a dark corner and it will not thrive at all, yet a philodendron may grow very well there.

For some plants one of the greatest stress factors is the human one. We may starve, overwater, underwater, overfertilize, underfertilize, overheat, freeze, desiccate, pollute, or crowd the plant to death. It certainly is not necessary for every person who ever looks at a plant to have a degree in agronomy, horticulture, or botany, but common sense concerning some of the principles of growth and development of plants can go a long way in making life greener, more pleasing and rewarding.

CYCLES

It is certainly not difficult to identify diurnal and seasonal cycles in our own lives. Work schedules, eating habits, body functions, and even psyches (many of us are cantankerous early in the morning) are affected. A modern ailment is jet lag, a rather uncomfortable physical/emotional state brought about by passing through several time zones in a short period of time. Seasonal cycles, too, cause us to have changes of moods, and we eat and dress differently.

Plants, too, are carefully attuned to these **biological clocks,** the most important of which is the 24-hour clock. Perhaps you've never really thought about why our days aren't 16 hours long, or maybe 32, or even 50. Would it make any difference to you if we had 8 hours of light and 8 hours of darkness year-round? Because of the physical design of our solar system, the earth rotates on its axis once every 24 hours to give us alternating exposure to the sun; at the same time, our planet follows an elliptical orbit around the sun, which is completed once each year. Because of the shape of that orbit, together with the tilt of the earth on its own axis, seasons are dictated for each precise location on the earth. Some other planet in some other solar system might have an entirely different biological clock. Ours is unique to this planet, and living organisms have evolved systems for responding to the days and seasons.

The system certainly is not perfectly predictable; we cannot foretell what the exact temperature and relative humidity will be on any particular day a year from now, but we have learned to expect some limits. The plants and animals that thrive at any particular locality do so because their ancestors have survived a rigorous screening over many generations. Those which do not have the correct genes to allow for survival under those conditions are eliminated.

It should be apparent that the day/night cycle allows a period of energy accumulation during the daylight hours; at sunset photosynthesis ceases. Daylight allows a plant to "recharge its batteries" and perhaps produce a little extra energy for storage. That's what agriculture is all about. Too much night, or even too much cloudy weather, could greatly reduce productivity. But is night actually necessary? Could plants photosynthesize in constant light? Experiments in controlled environment chambers with constant light have shown that some species seem to perform perfectly in constant light, but others do not. There is no rational explanation for this phenomenon.

So how do plants use their biological clocks to maximum advantage? The seasonal cue is rather predictable in some localities, but not so predictable in others. Frost kill date and long-term meteorological averages can vary over several weeks. Thus seasonal cues such as changes in precipitation and temperature are imprecise at best, but the photoperiod cue is absolute. Not all external control mechanisms are cued by daylength, but it certainly is a major factor.

The onset of autumn, when days become shorter, nights become cooler, and other climatic changes occur, provides a particular message for gene control. A specific daylength and temperature interaction cues the synthesis of abscisic acid in leaf petioles that causes leaf fall. After leaf fall, dormant buds are protected from winter injury. The following spring, daylength increases, nights become warmer, and the cue to the DNA calls for transcribing genes and synthesizing proteins responsible for renewed growth. Soon the buds break, cell division begins in the shoot apex, and new leaves are produced as stem internode elongation occurs.

How a plant perceives changes in daylength apparently has nothing to do with photosynthesis, and therefore chlorophyll is not involved. On the other hand, it is a light cue, so a pigment—a molecule that becomes activated or excited—must be responsible. The first clue about its identity came from the U.S. Department of Agriculture scientists at Beltsville, Maryland, in the 1920s. W. W. Garner and H. A. Allard were studying tobacco, which in Maryland normally flowers in late summer. One plant in their tobacco field failed to flower and continued growing until frost time. Intrigued by the plant, the scientists took cuttings of it, which they grew in a greenhouse. Various experiments with fertilizers, irrigation, temperature, and other factors failed to induce flowering. Finally, during December, the plant flowered. Although much shorter than field-grown plants, these plants had somehow delayed flowering until winter. Seeds taken from the greenhouse plants were planted the following season and again failed to flower until winter. It was clear that the factor responsible for the initiation of flowering was the length of day, all other factors being equal. The plants would simply not flower unless the length of day was shorter than *some critical number of hours.* Garner and Allard called this phenomenon photoperiodism and went on to work with other plants. Some species would flower only when the length of day was longer than some critical value, and that critical value was different for every species. Photoperiodism is an organism's response to a change in the proportion of light and dark in a 24-hour day.

These scientists were able to categorize plants as **long-day, short-day,** and **day-neutral** types (figure 14.2). Long-day plants flower in the summer, short-day plants flower in early spring or fall, and day-neutral plants will flower under a variety of light conditions. The absolute length of the light period is not the important factor, but whether it is longer or shorter than some particular interval. Consider the common cocklebur (*Xanthium strumarium*), which will flower anytime the length of the light period is less than 16 hours. Many varieties of chrysanthemum are also induced to flower anytime the daylength is about 14 hours or less. Both plants are considered to be short-day plants because they flower when the daylength is *less than some critical value,* although that critical value is different for the two species. Wheat, on the other hand, also flowers at a daylength of about 14 hours or longer. Thus wheat is by definition a long-day plant, as are many of the other cereal crops. A day-neutral plant is one in which flowering occurs in response to factors other than length of day.

Only in very recent years have scientists discovered that photoperiodism is at least partially controlled by a protein pigment called phytochrome. Unlike the other plant pigments, phytochrome is a very large protein molecule capable of existing in two different forms, referred to as P_r (phytochrome red) and P_{fr} (phytochrome far-red). The P_r form can be made to change into the P_{fr} form if a red light of 660 nm wavelength is shone on the pigment in solution. This conversion occurs rapidly in a matter of seconds, or at most, minutes. Once converted to P_{fr} form, it can be reversed to the P_r form if a far-red light of 730 nm wavelength is shone on the solution.

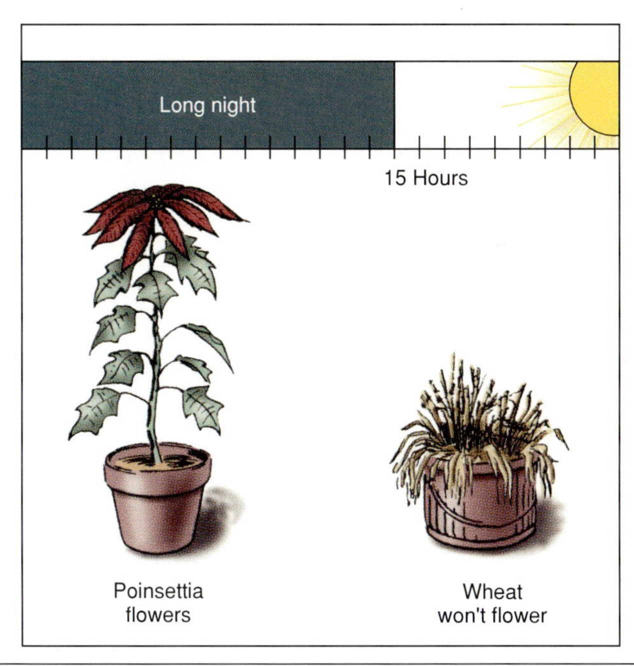

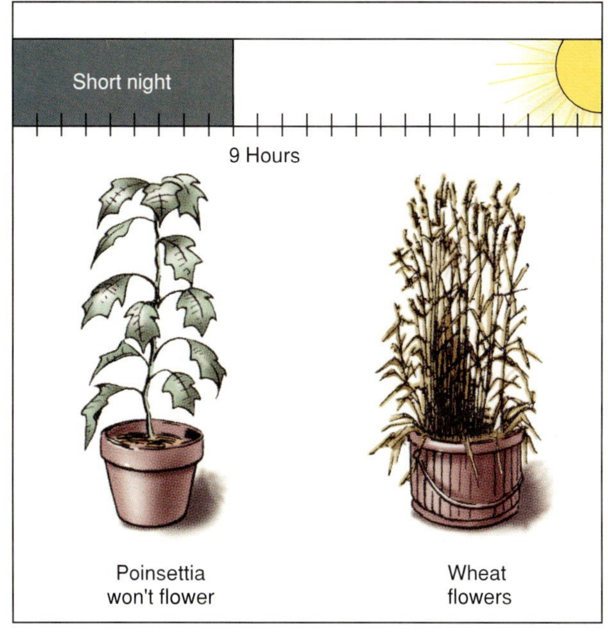

Figure 14.2

Some plants, such as poinsettia, require short days and long nights for flowering. Others, such as wheat, require long days and short nights. Times at the top of the figure refer to length of the dark period.

U.S. Department of Agriculture.

This reversal can go on indefinitely, and the pigment form at the end depends on the last light used to excite the molecule. The system can be depicted graphically:

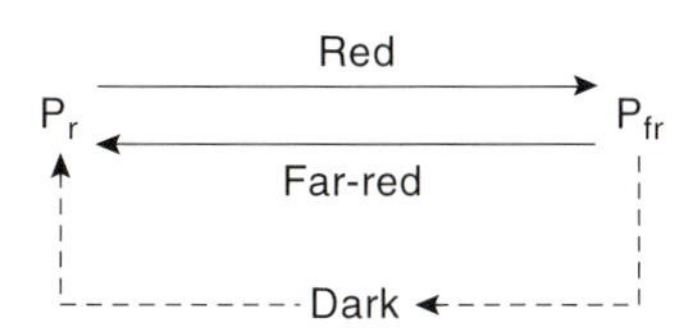

Note that once the pigment is in the P_{fr} form, it can gradually revert to the P_r form in darkness, even if no far-red light is present. This change is very slow, and the length of the night is a factor in determining the ratio of P_r to P_{fr}. In nature, the pigment is primarily in the P_{fr} form at the end of the day, because there is more red light in sunshine than far-red light.

This knowledge of photoperiodism has allowed plant scientists to take advantage of seasonal flowering and modify the flowering process to suit the needs of the grower. When it is known how long it takes to bring chrysanthemum flowers into production after the initiation of the floral stimulus, the length of the day can be controlled by pulling black cloth over the plants to shorten the day, or adding incandescent lights to lengthen the day, and the date for peak flowering can be precisely determined. This timing is particularly critical for specialized sale, not only in chrysanthemums, but also in Easter lilies and Christmas poinsettias.

Further experimentation has shown that the time-measuring device is not controlled only by the phytochrome. It has not been possible to draw conclusions about the ratios of P_r to P_{fr} and explain the phenomenon of long- and short-day plants. Although phytochrome is surely involved in the process, the full explanation is far more complicated and is in the realm of future experimentation.

> ✔ ***Concept Checks***
>
> 1. What is a biological clock?
> 2. How do plants perceive the length of a day?

PLANT HORMONES

Scientists are learning more about chemical messengers called **hormones,** which are at least indirectly responsible for much of the control of growth and development. Although it is difficult to demonstrate that a specific hormone can turn a particular gene on or off, the circumstantial evidence is great that hormones are intimately involved in the regulation process. In animal systems some hormones are **gene activators;** others clearly are not. It is tempting to suggest that some plant hormones are also gene activators, but there is no direct experimental evidence to prove the point. Hormones may be both stimulators and inhibitors, accelerating growth in one tissue and inhibiting in another. Sometimes the same hormone can perform both functions, depending on its concentration.

The regulation of growth and development by all the internal **(endogenous)** and external **(exogenous) factors** is partially due to these chemical messengers. By definition, a hormone is *an organic molecule synthesized in very small quantities in one part*

Plants "Reaching" for the Sun

In 1881, Charles Darwin and his son, Francis, reported on experiments performed with grass and oat **coleoptiles** (shoots) in *The Power of Movement in Plants.* They first described the phenomenon of **phototropism,** the bending of plants toward a unidirectional light source. The Darwins placed molded lead caps on the tiny shoot apex and noted that the shoot tip did not bend; when the lead cap was removed, the plant responded toward the light. Their notes recorded that "when seedlings are freely exposed to a lateral light some influence is transmitted from the upper to the lower part, causing the latter to bend." Essentially nothing was done about the Darwins' observations during the latter part of the nineteenth century and the early part of the twentieth century.

Then in 1926 the Dutch plant physiologist Frits W. Went, working in his father's laboratory at the University of Utrecht, performed a remarkable set of experiments. Went found that the tip of the growing shoot could be cut off and placed on a tiny block of **agar** (the gelatinlike material made from certain marine algae that maintains liquid in a semisolid state at room temperature). After about an hour the coleoptile was removed and the agar block placed on one side of another decapitated oat coleoptile. In a short time the upper part of the shoot began bending away from the side with the agar block. Went concluded that from the first coleoptile tip some chemical substance was diffusing into the agar block. This substance later diffused out of the block and into the second decapitated oat coleoptile, and moved only down the side of the tissue cylinder directly below it, causing the cells to elongate more rapidly than those on the opposite side of the cylinder. This differential in the rate of growth on opposite sides of the cylinder of tissue caused the entire cylinder to bend. This experiment was particularly significant because it proved for the first time that the stimulus described by Darwin was a chemical one, rather than physical or electrical.

of an organism and transported to another part of the organism, where the molecule exerts some profound physiological effect. You might think the definition sounds suspiciously like that for an enzyme, but enzymes are synthesized within the same cell in which they operate. Hormones may be transported over great distances before they reach a target cell or cells. The definition for plant and animal hormones is the same. Animal hormones tend to be very specific; in the human body, for example, literally hundreds of them have been identified. Plant hormones, on the other hand, tend to be very general, and each kind of hormone may perform many different functions.

Currently, there are five general classes of known plant hormones: **auxins, gibberellins, cytokinins, abscisic acid,** and **ethylene.** These molecules tend to be rather small, much smaller than the giant proteins that act as enzymes. Most of the plant hormones have molecular weights in the range of 200 to 300, which allow them to move through tissues with relative ease. Some animal proteins do act as hormones, but none are known in plants. The simplest kind of hormone is ethylene: $CH_2{=}CH_2$. Even though it is a very simple gas, it exerts a tremendous physiological influence on plants.

Many chemicals not found naturally in plants exhibit growth-regulating properties similar to those of the hormones. Since they are not synthesized in the plant, they do not meet the true criteria for a plant hormone and therefore are called plant growth regulators. By definition, all hormones are plant growth regulators, but not all plant growth regulators are hormones.

Auxin

Although most hormone research has been carried out since 1940, the history of interest in hormones is much older. The earliest recorded observations leading to the discovery of plant hormones are those of Charles Darwin. His and other researchers' experiments demonstrated that a chemical substance causes cells to grow faster in its presence (see "Plants 'Reaching' for the Sun"). This substance moves away from light, resulting in greater accumulations on the "dark" side of a growing stem tip, which in turn results in that side growing faster than the sunny side of the shoot. The growing tip bends toward the light (figures 14.3 to 14.5).

This chemical substance was named auxin after the Greek *auxein,* meaning "to increase." Although we now refer to any substance that will cause a similar bending response an auxin, the only naturally occurring one is **indoleacetic acid (IAA)** (figure 14.6). Many chemicals have auxinlike properties, and some, such as 2,4-dichlorophenoxyacetic acid (2,4-D), are used as **herbicides** or weed killers (figure 14.7). A low concentration of

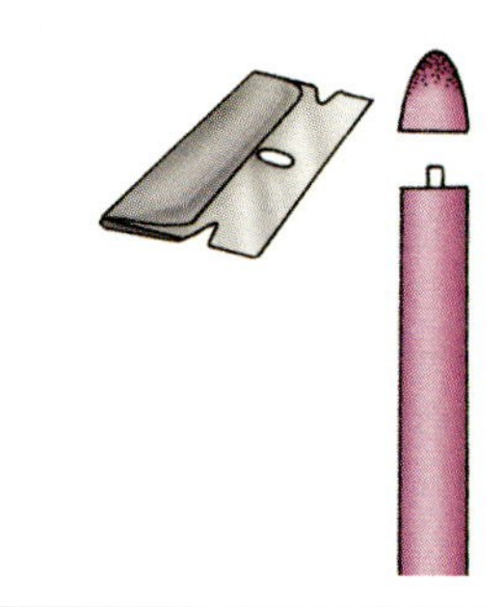

Figure 14.3
Auxin is produced in the tip of the *Avena* (oat) coleoptile, or growing tip. If the apex is cut off, no auxin is available for the remainder of the shoot.

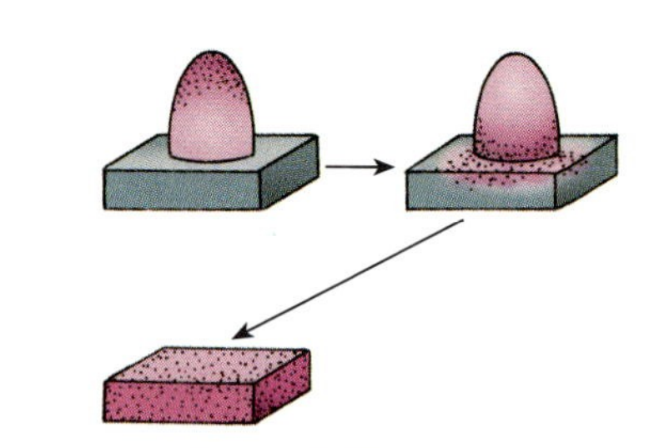

Figure 14.4
If the apex is placed on a tiny cube of agar, the auxin will diffuse into the agar within one hour.

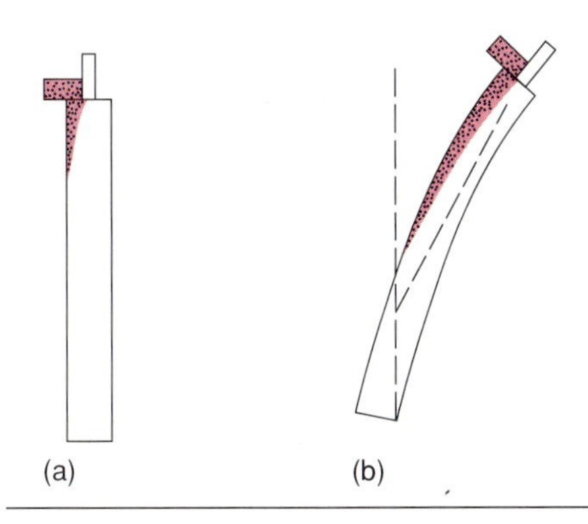

Figure 14.5
If the agar block is placed on one side of a decapitated coleoptile, the agar diffuses into the tissue directly below it, but not into the cells on the other side (*a*). After a short time the stem will bend away from the side of application because the cells receiving the auxin elongate more rapidly (*b*). The degree of curvature is proportional to the quantity of auxin in the agar block. This quantitative technique is called the *Avena* coleoptile bioassay.

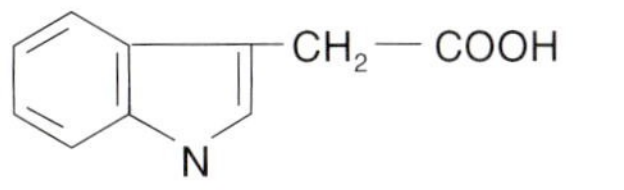

CH$_2$—COOH
N
Indoleacetic acid (IAA)

Figure 14.6
The chemical structure for the naturally occurring auxin: indoleacetic acid (IAA).

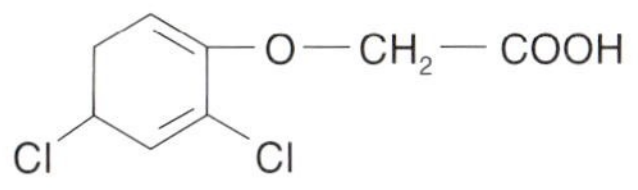

O—CH$_2$—COOH
Cl Cl
2,4-Dichlorophenoxyacetic acid (2,4-D)

Figure 14.7
The chemical structure for the herbicide 2,4-dichlorophenoxyacetic acid (2,4-D).

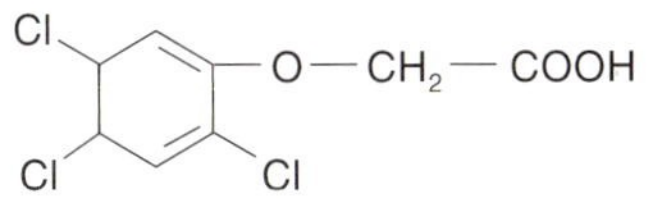

Cl
O—CH$_2$—COOH
Cl Cl
2,4,5-Trichlorophenoxyacetic acid (2,4,5-T)

Figure 14.8
The chemical structure for the powerful herbicide 2,4,5-trichlorophenoxyacetic acid (2,4,5-T). This weed killer, together with 2,4-D, was used to defoliate the jungle vegetation during the Vietnam War and is called by the code name Agent Orange.

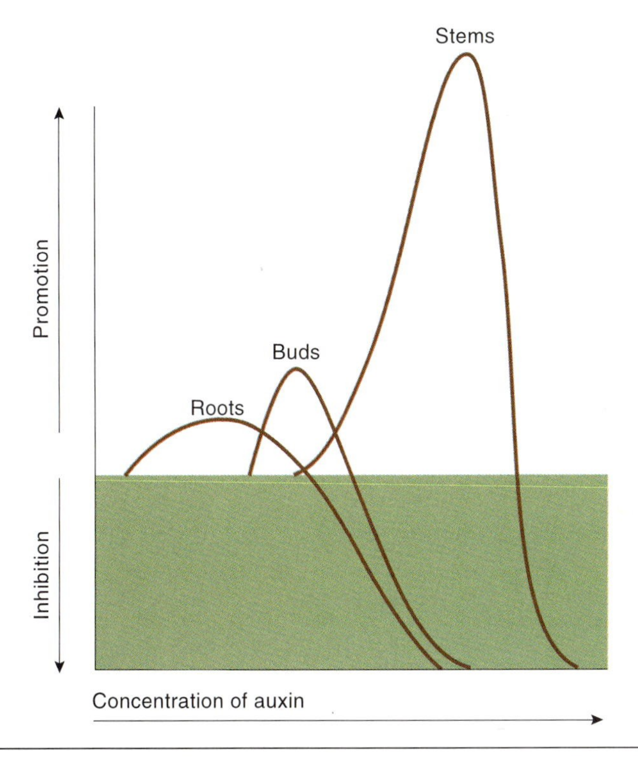

Figure 14.9
Tissue sensitivity is a major factor in the responses observed following the application of hormones. Auxin exhibits qualities of both growth promotion and growth inhibition, depending on concentration. Roots are most sensitive to auxin, buds are intermediate in sensitivity, and stems are relatively insensitive to high concentrations of auxin.

2,4-D is far more active than the same concentration of IAA in many plants, and small amounts can cause excessive respiration, excessive cell expansion, and finally tissue death. A mixture of 2,4-D and the closely related 2,4,5-trichlorophenoxyacetic acid (2,4,5-T) is the familiar **Agent Orange** used as a defoliant in the Vietnam War (figure 14.8).

IAA is synthesized from the common amino acid tryptophan; it is found as a by-product of both fungal and animal metabolism, and a particularly rich source is a pregnant horse's urine. However, these organisms neither use nor respond to it. Although no hormones are found in truly large concentrations, auxin seems to be concentrated at the site of synthesis in embryos, apical meristems, and in young leaves and fruit—all actively growing tissues. Auxin has directional or polar movement, from the growing tips of the plant toward the base of the plant; hardly any moves in the other direction. Movement generally occurs through parenchymal cells rather than through vascular tissues.

As with other hormones that exert tremendous influences at low concentrations, it is difficult to predict a specific response for auxin. It behaves differently as the conditions change; many dicots are more sensitive than monocots, and the root is more sensitive than the shoot (figure 14.9). From time to time and place to place, tissue sensitivity changes, which complicates predicting exactly what kind of response one might expect. It is quite apparent that very high concentrations of auxin are toxic; hence the use of 2,4-D and 2,4,5-T as weed killers. If extremely low concentrations of 2,4-D are used, it can be a very good synthetic auxin source for stimulation of growth.

Auxin Effects

Unquestionably, one of the most important actions of auxin is Darwin's observed effect of cell elongation.

Occurring as it does in meristematic tissues, auxin is apparently responsible for the rapid growth and elongation of tissues, and in the shoot apex it diffuses downward and causes the stem to elongate. Young, tender stems are most readily affected, and their cells elongate rapidly. If the shoot apex is removed from the plant, apical auxin supply is eliminated and the shoot stops elongating. If auxin is added to the cut surface, growth resumes. It appears that intact plants already have an optimum amount of auxin being produced in the shoot tips, and extra application generally fails to produce a response. The "auxin threshold" is apparently saturated under most growth conditions.

The same concentration of auxin that causes shoot elongation will cause an inhibition of root growth. This differential is probably due to a difference in tissue sensitivity. Although leaf growth does not appear to be directly controlled by auxin production, leaves do contain auxin. If for some reason, including seasonal triggers, the level of auxin in the leaf falls, there is a good chance that an **abscission layer** will form at the base of the petiole. Enzymes form, break down the cell walls, and weaken the petiole. The leaf eventually falls from the plant. Of course, this is normal for deciduous trees in the fall, but severe stresses during the growing season can also cause premature leaf fall. Such stresses may be induced by drought, flooding, high and low temperatures, light stress, and a number of other factors. Abscission zones also form at the base of the peduncle and may lead to premature fruit shedding.

One of the important commercial uses of auxin is to stimulate the production of **adventitious roots** on stem or leaf cuttings (figures 14.10 and 14.11). Again, depending on type of tissue, age, and a number of other factors, various concentrations of auxin may be needed to stimulate lateral roots. In commercial horticultural practice various synthetic auxins are used not only to speed the rate of root production, but also to increase the total number of roots produced.

Auxin is also important in the growth of fruit. Developing seeds are a rich source of auxin, and that source is responsible for the growth of the fruit surrounding the seed. If ovules fail to be fertilized or if for some reason the embryo aborts, the fruit will normally fail to develop. One of the best examples is the strawberry. The achenes are produced on the outer surface of the receptacle, and each one is responsible for a region of growth necessary to produce a normal accessory fruit. If the achenes on one side of the receptacle are removed when the fruit is small, that side of the fruit fails to develop, and an abnormal strawberry is produced. This is apparently why strawberry shape is so varied (figure 14.12). It is possible to substitute auxins for achenes and cause a normal fruit to develop. Such fruit

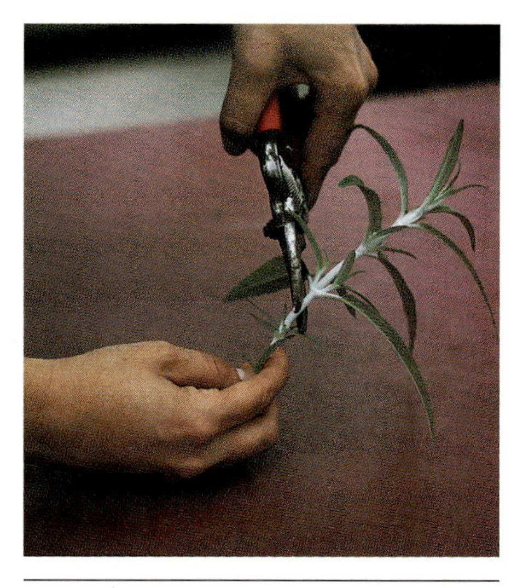

Figure 14.10

Vegetative cuttings are made by removing a section of stem tip 8 to 10 cm long, removing the lower leaves, and making a clean, sharp oblique cut at the base of the stem. Then the base of the stem may be treated with an auxin prior to insertion in the propagation medium.

Figure 14.11

Many woody cuttings, and some herbaceous ones, respond dramatically to an auxin treatment at the time the cutting is taken from the mother plant. Auxin not only causes more rapid rooting, but also increases the number of roots produced, ensuring the establishment of a healthier new plant. Cutting on left has been treated with high levels of auxin, in the middle with low levels of auxin, and on the right, no auxin was applied.

Figure 14.12

In normal strawberry growth and development, seeds are a rich source of auxin. They occur throughout the surface of the receptacle, and each seed is responsible for growth in its own region. A small fruit (*a*) with normal seed distribution will eventually lead to a normal fruit (*b*). If the seeds are carefully removed from one side of a developing fruit (*c*), growth on that side of the fruit will be inhibited, leading to a deformed mature fruit (*d*). If the seeds are removed from one side of the fruit early in development, but auxin paste is spread on the region where the seeds are removed (*e*), the fruit will develop normally, but without any seeds on the affected side (*f*).

Figure 14.13

Apical dominance is primarily controlled by the production of auxin at the shoot apex of a seedling. (*a*) As long as the apex is intact, growth proceeds normally from the tip, even though lateral buds are present at each node. (*b*) If the terminal bud is removed, lateral buds nearest the tip begin to grow and produce a lateral shoot. Buds near the base of the plant remain inhibited by the new auxin source produced by the new growing tips. (*c*) If the terminal bud is removed and an auxin paste is applied immediately, growth of the lateral buds is inhibited just as if the terminal bud were in place.

that develop without fertilization are called **parthenocarpic fruit.** Greenhouse tomatoes and cucumbers are often produced in this manner.

Auxin is also responsible for the phenomenon called **apical dominance,** which influences the shape of a plant (figures 14.13 to 14.15). If the shoot apex is removed, the level of auxin drops in the main stem, and the adjacent lateral buds (at the next lower node) are released from inhibition and begin to grow. Buds at successively lower nodes may also begin growing, but the influence is greatest at the buds nearest the shoot apex. If auxin is applied to these buds, they fail to grow, proving that they are inhibited by auxin moving **basipetally** (downward) from the shoot apex. Plants in nature that have one central leader, such as trees with a typical conical shape, are said to have very strong apical dominance. An oak tree, on the other hand, with much branching and no central leader, is said to have weak apical dominance. The ultimate in apical dominance is exhibited by columnar palm trees, which have a single meristem located at the shoot apex. Such trees die if the shoot apex is ever removed (figure 14.16).

Mode of Action

Obviously a plant hormone with so many different functions (only a few have been described here) must influence regulation of several different systems. Certainly one of the most

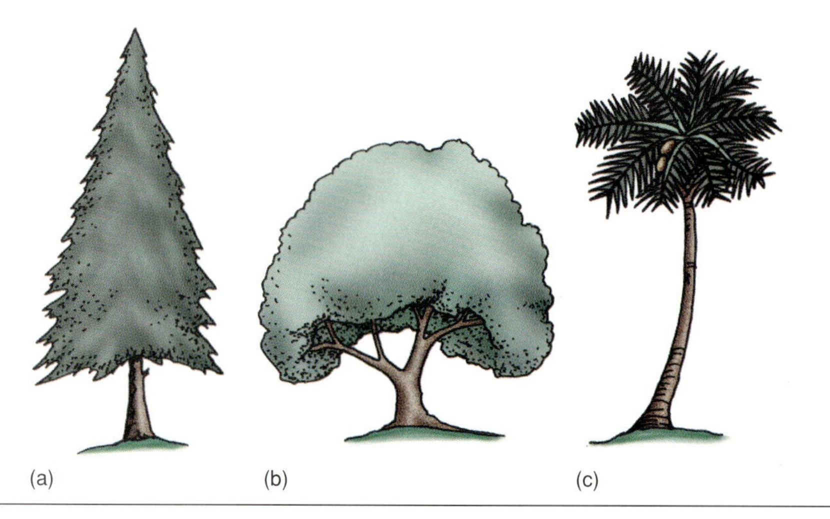

Figure 14.14
Apical dominance controls the ultimate shape of a plant. (*a*) Trees with strong apical dominance assume a conical, upright shape. (*b*) Trees with weak apical dominance have multiple branching and rounded shapes. (*c*) The ultimate in strong apical dominance is exhibited by columnar palm trees.

Figure 14.15
Lodgepole pine trees exhibit strong apical dominance with a central leader.

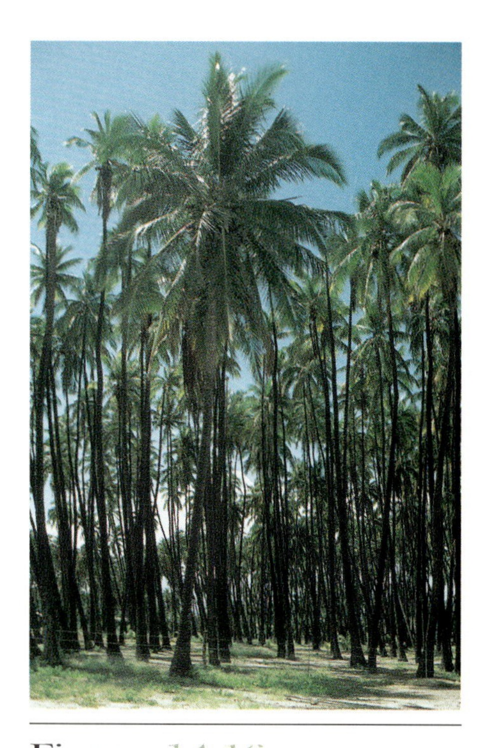

Figure 14.16
Coconut grove, Molokai, Hawaii. Columnar palm trees have no lateral buds, and therefore the apical bud controls the growth of the entire tree. If the apex is removed, the tree will die.

important is cell elongation. Detailed studies by many physiologists and biochemists have shown that auxin increases the **plasticity** or "stretchability" of the cell wall. Since living cells normally have turgor pressure that allows the plasma membrane to push against the cell wall, if the wall becomes loosened, then the cell can stretch until the wall again becomes a significant barrier. As long as the wall is rigid, no expansion can occur. As cells become older, they fail to respond to auxin as do young cells, and the elongation finally becomes impossible. Some rigidity is characteristic of older, mature tissues and is the foundation upon which structural wood is built.

✔ *Concept Checks*

1. What causes a shoot tip to bend toward a unidirectional light source?
2. What are the main effects of auxin on plant growth?

Gibberellins

At approximately the same time that Went was performing his classic experiments with auxin, Japanese scientists were investigating a fungal disease that caused excessive stem elongation, stem weakening, and finally stem collapse and death of rice plants. The disease was known locally as the *bakanae,* or crazy-seedling disease of rice. The organism was finally identified as *Gibberella fujikuroi,* and

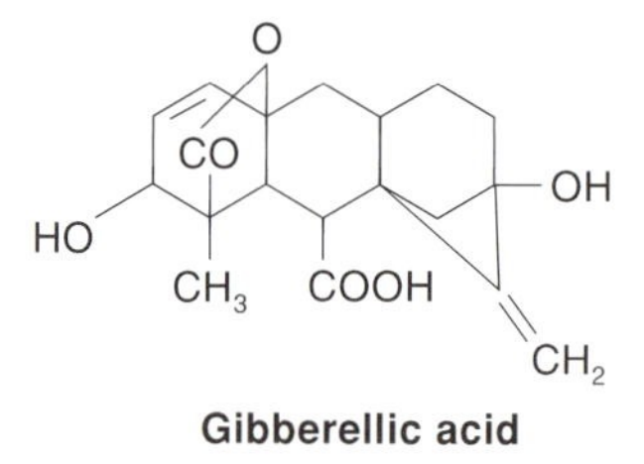

Figure 14.17
The chemical structure of the most commonly found gibberellin, gibberellic acid (GA_3).

the active growth compound causing *bakanae* was named **gibberellic acid** (figure 14.17). Because scientific communication was so poor between the Orient and the Western World at the time, the information was not "discovered" by Westerners until British workers reported in 1955 that minute quantities of gibberellic acid would cause genetically dwarfed pea plants to grow to a normal size. Within a few years dozens of laboratories were experimenting with gibberellins. In both dwarf pea and dwarf corn the amount of gibberellin applied was directly proportional to the amount of increased growth up to that for a normal plant. The obvious but incorrect conclusion was that the normal pea and corn plants contain plenty of endogenous (synthesized within the plant body) gibberellin, whereas dwarf mutants somehow

Figure 14.18
Cyclamen responding to Gibberellic acid treatment. Plant on left is the control (no treatment); plant on right shows elongated stems.

Figure 14.19
(*a*) Some long-day plants will not flower under short-day conditions unless treated with gibberellins.
(*b*) The hormone induces flowering just as if the plant were growing under long days.

failed to synthesize the hormone. But dwarf plants have even more gibberellin than do normal plants. Recent experiments suggest that light represses growth by causing an insensitivity of the tissue to endogenous gibberellin, perhaps through a gibberellin inhibitor in the dwarf plants. Dwarf pea and corn are the standard plants for gibberellin **bioassay,** a method of quantitatively determining activity of a substance using a living organism. (Both straight growth and curvature of oat coleoptiles are used as bioassays for auxin.)

As one might expect from the original rice disease, gibberellic acid proved to be a growth promoter. There are now more than 50 slightly different compounds, all closely related chemically, classified as gibberellins. Less than half occur naturally, and most have no known biological activity. Gibberellin has been extracted from seeds, young shoot and root tips, and young leaves. Its concentration diminishes greatly in older tissues. In the 1950s botanists found that some plants that flowered only under long days could be induced to flower under short days if given gibberellic acid (figures 14.18 and 14.19). This exciting news led to the incorrect conclusion that gibberellin was unequivocally identified as the flowering hormone. It does cause flowering in some species, but not in others.

A good example of how a hormone might act as a chemical messenger is the effect of gibberellin on the germination of barley seed. In the normal germination process, following imbibition of water, various enzymes begin to break down macromolecules. Barley seeds have a large starchy endosperm that must be broken down into sugars to provide energy for the germinating seed. This process is triggered by the synthesis of the enzyme α-amylase, which causes starch to be converted into glucose. Some early experiments showed that this conversion process was speeded up dramatically with the addition of gibberellic acid, and later studies showed that the gibberellin caused the production of the enzyme in the aleurone layer, a group of cells surrounding the endosperm. Even if exogenous gibberellin is not applied, the process eventually occurs; the embryo itself produces the gibberellin, which causes the synthesis of the enzyme, eventually breaking down the starch. If the seed is split in half so that the embryo is removed, no α-amylase is produced. However, if gibberellin is applied to the same seed half without the embryo, α-amylase is produced. It is still difficult to prove that the gibberellin is the direct trigger that turns the gene on and off, but the circumstantial evidence is most convincing.

The brewing industry uses gibberellin to speed up the conversion of starch to sugar in barley seeds. In beer brewing, malting barley is used as the sugar source for making alcohol. The malting process is rather slow if seeds are allowed to germinate normally, but if gibberellin is

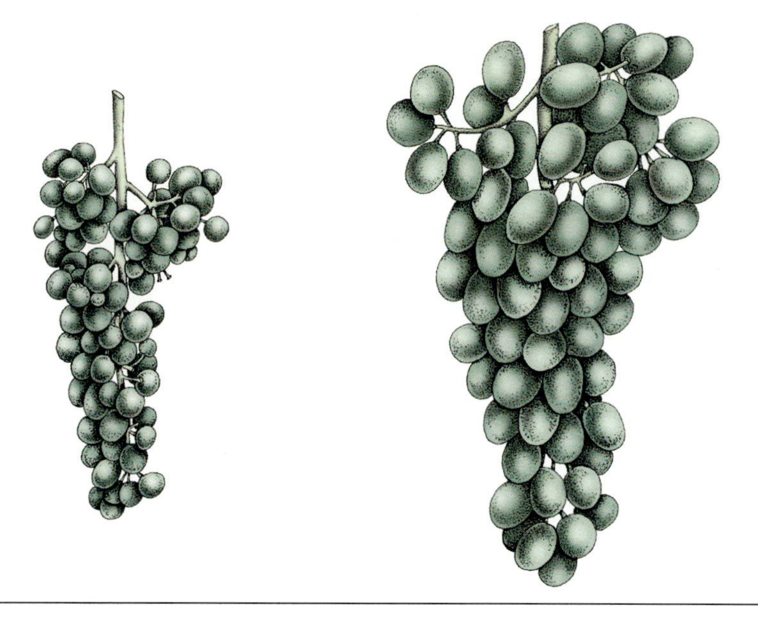

Figure 14.20
Cluster size in Thompson seedless grapes can be improved dramatically with a single spray of gibberellic acid. Most commercial grapes of this variety are treated in this manner.

Kinetin

Figure 14.21
The chemical structure of a naturally occurring cytokinin, kinetin.

Abscisic acid

Figure 14.22
The chemical structure of abscisic acid.

added to the germinating seeds, the conversion to sugar is greatly expedited and the entire process becomes far more efficient. Increased efficiency translates into greater income for the brewer.

Gibberellin has also been used commercially to increase the fruit size and cluster uniformity in Thompson seedless grapes. The embryo in each of these grapes fails to provide the optimum amount of gibberellin, and if the clusters are sprayed with a dilute concentration of gibberellin very early in fruit development, the ultimate size may be doubled or tripled, and the quality of the product is not usually changed (figure 14.20).

Cytokinins

In the 1940s, workers at various laboratories began studying plant tissue culture as a means of isolating specific tissues or organs, removing them from external influences, and then studying nutritional and growth factors one at a time. Although tissues taken from various plants, including tobacco and carrot, grew for a while in culture, the cells eventually stopped dividing. Many culture media were tried, but liquid coconut milk proved to have some factor that immediately caused the cells to divide. After years of attempts at isolation, the factor was finally identified as a derivative of the purine adenine, the same substance found in nucleic acids. The compound was named kinetin (figure 14.21), and now the entire class of compounds isolated from plants that cause rapid cell division are called cytokinins (*cyto,* cell; *kinin,* division). Since cytokinins are so effective in promoting cell division, it was thought that this compound or a similar one might be responsible for the uncontrolled growth of cancer cells. During the 1960s a great deal of research effort was devoted to trying to associate cytokinin with tumor development in both plant and animal cells, but no direct connection was ever made. Instead, cytokinin appears to be a naturally occurring compound responsible for the control of the cell division process in meristematic regions. Cytokinins also seem to slow the turnover of proteins in plant tissues, which retards the aging, or senescence, process. For example, if cytokinin is sprayed on lettuce leaves, the storage life is greatly enhanced, and the product can be shelved for long periods. This allows supermarkets to distribute fresh vegetables more efficiently and ultimately buffers price fluctuations in a highly volatile market.

Abscisic Acid

Just as auxins, gibberellins, and cytokinins are generally considered to be promoters of growth, there are a group of compounds considered inhibitors of growth. They reduce the rate of growth or induce dormancy at critical times, such as the onset of fall and winter. The primary molecule that controls these inhibitory processes is abscisic acid (figure 14.22). Originally identified in dormant buds of ash trees and potatoes, it was later found to be the same molecule that caused the abscission of leaves, flowers, and fruit. When days become shorter in the fall and temperatures begin to decrease, levels of abscisic acid gradually rise in the abscission zones of petioles. Leaves fall, and the level of abscisic acid in the dormant buds remains very high throughout the winter. With the onset of spring, levels of abscisic acid

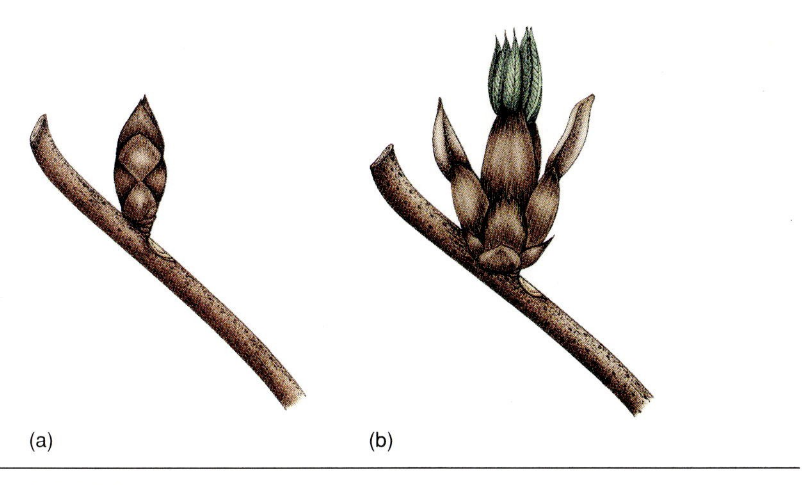

Figure 14.23
(*a*) Dormant, woody buds are high in concentrations of abscisic acid. (*b*) At the end of winter the level of abscisic acid drops, and the bud begins to develop, revealing a new shoot.

decrease at the same time that levels of auxin, gibberellin, and cytokinin begin to rise. Buds are activated and stems begin to grow and develop (figure 14.23).

Whenever a water stress occurs in leaves, abscisic acid levels rise in the guard cells, which triggers the movement of K^+ out into adjacent cells, making the solute less concentrated and causing water to move out of those cells. When the pressure decreases in the guard cells, the stomates close, and transpiration is greatly reduced. Thus it appears that abscisic acid is intricately involved in plant water stress, and scientists are attempting to understand how that information might be put to practical use in reducing transpiration or the effects of the stress itself.

Ethylene

A recently "discovered" plant hormone is ethylene, although this simple molecule has long been recognized as an important gas (figure 14.24). For many years greenhouses were heated with open space heaters. Occasionally an attendant would arrive at work after a cold night to find that all the leaves had fallen off the plants. Various toxic gases were suspected, but many years went by before ethylene was recognized as a major component of exhaust gases, and the association was made between leaf abscission and ethylene. Even more powerful than abscisic acid, this gas can trigger unusual growth and development responses even in trace amounts. The ripening of fruit involves a number of chemical and anatomical changes, and in the process of ripening, storage molecules change to soluble sugars, as in bananas. This is part of the respiratory process and is accompanied by the release of large quantities of ethylene. As bananas ripen naturally, the dark flecks that appear on the peel are pockets of concentrated ethylene. One can increase that concentration and speed the ripening process by enclosing the fruit in a plastic bag so that all of the ethylene given off by the ripening fruit is concentrated and not allowed to diffuse. Green stalks of bananas are ripened in this manner at the supermarket, or they are placed in a small gassing room and treated with bottled ethylene gas.

The study of postharvest physiology is a major segment of the field of plant science and horticulture. Scientists study the environmental and biochemical factors that influence the storage and ripening process of fruit and vegetables. Some fruit, such as banana and avocado, produce large amounts of ethylene and ripen very suddenly, and this ripening is accompanied by a rapid rise in the

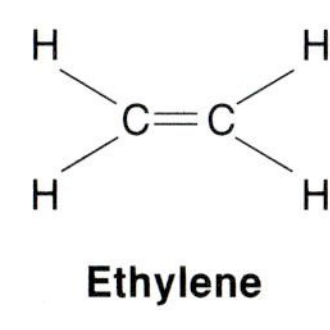

Figure 14.24
The chemical structure of ethylene.

rate of respiration of that fruit tissue. This sudden increase in respiration is called the climacteric rise and at one time was the subject of many laboratory studies.

The storage of apples has received particular attention, because apples are a major crop and subject to rapid deterioration under improper storage conditions. Postharvest physiologists have learned that if apples are stored in a controlled atmosphere of temperatures just above freezing, high relative humidity, low oxygen, and high carbon dioxide levels, storage life can be increased dramatically. It is now possible to store apples year round with essentially no reduction in quality. This finding has obvious implications for availability of fresh fruit for the consumer and the stabilization of prices.

Although the effects of ethylene are well documented, some of the effects are so interrelated with auxin that it is difficult to determine which is causing the effect. In addition to its effects on fruit ripening, ethylene causes leaves to abscise, chlorophyll to bleach, flower pigments to fade prematurely, and leaf petioles to grow more rapidly on the upper side and therefore curve downward. The petiole's response to ethylene is called epinasty and should not be confused with the phototropism attributed to auxin.

✔ Concept Checks

1. How do gibberellins and cytokinins differ in their mode of action in controlling plant growth?
2. What is the most important applied use of ethylene?

Figure 14.25
These two chrysanthemum plants demonstrate the effectiveness of a growth retardant, Phosfon D, thought to be an antigibberellin. Minute quantities of the chemical are added to the soil, which results in shortened internodes and produces a more compact, salable plant (*right*). The plant on the left has received no growth retardant.

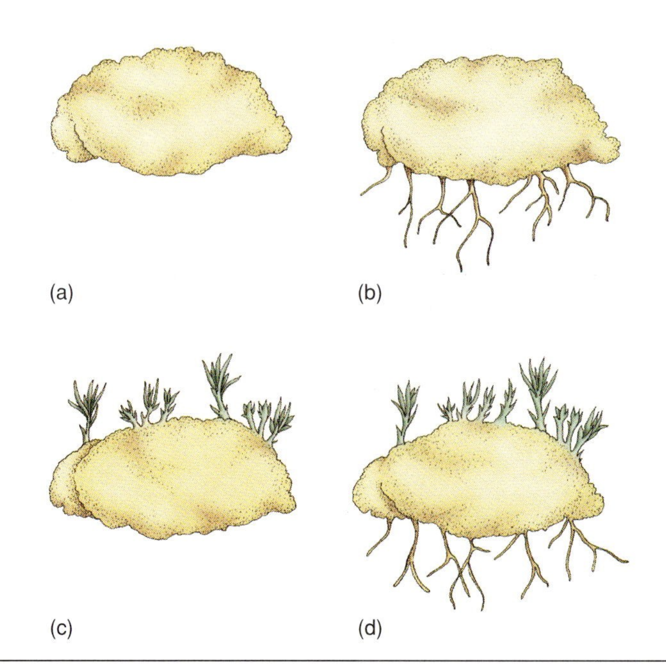

Figure 14.26
Hormone interactions are important in the growth and development of all tissues. In sterile tissue culture many species of plants produce an undifferentiated callus tissue (*a*). By manipulating the concentrations of auxin and cytokinin, the callus can be induced to form roots (*b*), shoots (*c*), or both to eventually regenerate complete plants (*d*).

Growth Retardants

Many commercial growth regulators are advertised as growth retardants and are sold for the purpose of reducing pruning or mowing costs, decreasing stem length in flowering plants such as azaleas and poinsettias, and decreasing biomass production to reduce fire hazards. It appears that several of these compounds act by blocking the synthesis of gibberellin. Others may act as antiauxins, and in fact several possibilities exist for counteracting the effects of hormones. Common growth retardants include maleic hydrazide, CCC, and Phosfon D (figure 14.25).

Hormone Interactions

It is probably apparent by now that hormone action is rather complicated. Change the concentration, and the promoter may become an inhibitor. Both auxin and gibberellin act on cell elongation; both ethylene and abscisic acid cause abscission. It is important to remember that hormones never act alone in any tissue, and the bottom line of growth and development depends on the relative concentrations of all the hormones acting in concert to produce the final product. Tissue culture experiments have revealed a great deal about the interaction of auxin, gibberellin, and cytokinin (figure 14.26). If the ratio is changed slightly, one may get only **callus** (undifferentiated tissue); change the ratio again, and shoots may be produced; change it again and only roots may be produced; and if the ratio is just right, both roots and shoots may be produced, eventually leading to a complete new plant from clonal tissue. Many exciting results have been obtained in understanding how the chemical messengers do their work, but many additional experiments will need to be conducted before we understand all of the ramifications of hormone interaction.

The Flowering Hormone

Thousands of hours of research have gone into the search for the flowering hormone. Various laboratories have isolated an extract that will cause a vegetative shoot apex to become reproductive and produce flowers, but so far no one has ever chemically identified the compound. The floral stimulus is obviously transported from leaves to the shoot apex. For example, it is possible to take a short-day plant such as cocklebur (*Xanthium*) and place it under long-day conditions so that the plant remains entirely vegetative. If a single leaf is enclosed by a dark box so that it receives only short-day light, the plant will flower. In addition, if a cocklebur plant is growing under long-day conditions and if a leaf from a flowering short-day plant is grafted onto the vegetative plant, the chemical stimulus is transferred to the apex and the noninduced plant will flower. This elusive chemical stimulus has been given the name **florigen,** although nothing is known about its true chemical nature or structure. It seems particularly ironic that, in an age of modern miracles and molecular biology, such a compound could not be isolated, but scientists are coming more and more to the conclusion that florigen may be a combination of substances, perhaps including gibberellin and other hormones.

Figure 14.27
(*a*) If a shoot apex is removed from the plant and placed on an agar block, the auxin diffuses evenly throughout the block, provided that light is being received uniformly from all directions. (*b*) If the apex is placed on two separate blocks and if a unidirectional light is beamed from the left side, far more auxin accumulates in the right-hand block than in the left-hand block, proving that in the phototropic response, auxin migrates to the dark side of the tissue.

Figure 14.28
In this field of sunflowers, all the flower heads point in the same direction. These flowers track the sun from sunrise to sunset.

PLANT MOVEMENTS

Phototropism

You have already learned that some plant tissues and organs are capable of modifying their rate of growth so that there is a change in direction. Darwin's experiments with the oat coleoptile showed that cells on the dark side of the stem elongate more rapidly than those on the light side. Went later demonstrated that the auxin levels are higher in the cells that elongate more rapidly, and even if the rate of cell division is unchanged, the cylinder of tissue will curve toward the light. An explanation of this phenomenon is that light causes a migration of auxin molecules from the light side to the dark side of the stem (figure 14.27). It has been shown that seedlings kept in constant darkness elongate much more rapidly than seedlings grown in light (from the time of emergence from the soil). Such plants are said to be **etiolated,** and they have tiny underdeveloped leaves and weak, pale internodes. If stem tissues are analyzed, the etiolated (dark-grown and spindly) seedlings contain much more auxin. Etiolation in the dark helps the seedling to force its way through the soil before expending all its energy. Plants in shaded conditions tend to elongate faster than bright light-grown plants, which allows them to compete more effectively for the light.

Plant leaves and flowers also move a great deal, often exactly tracking the sun. Sunflowers, for instance, track the sun from early morning to sunset, and leaves of many other plants do the same. This allows the plant to readjust constantly so that only perpendicular rays from the sun hit the leaf surface, and therefore the photosynthetic efficiency is increased (figure 14.28).

Geotropism

The earth's gravitational pull may cause growth responses. Roots grow down into the soil (positive geotropism), and shoots grow upward from the ground (negative geotropism). All these tissues start out from the same embryo, originally consisting of only one cell. As soon as the first division of the fertilized egg occurs, a polarity is established, and it is possible to determine which direction is heads and which is tails, that is, shoot and root. Tissues in the same organism react positively at one end and negatively at the other, and one theory suggests a differential tissue sensitivity to certain hormones, perhaps auxin. Gravity may cause a higher concentration of auxin on the lower side of the tissue than on the upper side. Thus it is possible for a given auxin concentration to accumulate on the lower side of the plant embryo and stimulate those cells to elongate faster. That tissue would curve upward and become the shoot. At the other end of the organism the same concentration of auxin might be inhibitory to the cells and slow the rate of elongation. Thus cells on the upper surface would grow more rapidly, and the tissue would curve downward and become the root.

Thigmotropism

This word derives from the Greek *thigma* (touch) and refers to directional growth caused by plants touching a solid object. This phenomenon is often exhibited by climbing plants such as English ivy, which send out aerial roots when a shoot comes in contact with a wall. The roots attach the plant to the surface so that it may continue to climb upward. Hormones are undoubtedly involved in the contact response and subsequent differentiation of tissues.

Nastic Movements

Nastic movements are those made in response to some stimulus but are not oriented relative to the direction of the stimulus. Some of these movements are reversible and do not involve true growth. Others are true growth responses, as in the epinastic responses described for ethylene, and the changes are relatively permanent. Other nastic responses include thermonasty, as in the opening and closing response of tulips due to fluctuations in temperature; nyctinasty, the so-called sleep movements of leaves due to changes in turgor of certain cells located at the base of the petiole; and seismonasty, a response to shaking or some other mechanical disturbance. The sensitive plant (*Mimosa pudica*) and the Venus fly trap (*Dionaea muscipula*) both display leaf movements not related to differential growth (the response is much too rapid). They are due instead to turgor movements. Large, turgid cells act as a hinge for these organs. When the trigger is released, an electrical impulse is transmitted instantaneously to these storage cells, membrane properties are changed, and the water pressure is released. After a period of time the water pressure is regained and the trap is reset.

Time-lapse photography of a growing plant will convince the doubter that plants do indeed move a great deal. Most of these movements appear to be in response to day and night cycles, but other influences may cause the movement. Some of the movements seem to have little meaning, but others are related to specific adaptations that may aid that particular species in survival.

Concept Checks

1. Why do scientists think there is a flowering hormone?
2. Why do plants grown in the dark tend to be tall and spindly?
3. Why do roots grow downward?
4. What are examples of plant movements in response to other stimuli?

Summary

1. The genetic composition of a plant—the genotype—is modified by the environment to produce a phenotype, the morphological features that characterize a plant. The genotype specifies the potential characteristics of the organism, but whether they are in fact expressed depends on the environment.
2. Growth is defined as an irreversible increase in size or volume. Development is a summation of all the activities leading to changes in a cell, tissue, organ, or organism. Development is an orderly sequence of events dictated by a precise set of genes turning on and off at specified times. Differentiation comprises the chemical and physical changes associated with the development process.
3. Developmental biology is the study of how organisms, their cells, their tissues, and their organs achieve a final predictable form and function.
4. Growth and development are often limited by environmental extremes that suppress genetic function. Temperature, water, salt, light, and other factors combine to limit gene expression by influencing the turning on and turning off of specific genes.
5. Plants are carefully tuned to biological clocks, the most important being the 24-hour and the day/night cycles. Seasonal changes provide environmental signals to make certain changes in growth and development. Reproductive cycles are usually attuned to these environmental cues, and the flowering process in many plants is due to photoperiodism. The primary receptor in this system is a protein-pigment that senses a change in the daylength.
6. The primary plant hormones are auxin, gibberellic acid, cytokinin, abscisic acid, and ethylene. They act as stimulators and inhibitors of growth by directly or indirectly influencing gene regulation.
7. Plants respond to a unidirectional light source by compensatory growth, causing the cells on the dark side to grow more rapidly than those on the light side. The curving response occurs because of change in growth rates of localized cells, but the change in growth rate is brought about by a migration of auxin from the light side to the dark side. This bending response to light is called phototropism, and other tropisms include a response to gravity (geotropism) and a response to touch (thigmotropism).
8. Other plant movements may be temporary and occur in response to a change in membrane properties so that the turgor in certain target cells changes rapidly.

Key Terms

abscisic acid 246
abscission layer 248
adventitious roots 248
agar 246
Agent Orange 247
apical dominance 249
auxins 246
basipetal 249
bioassay 251
biological clock 243
callus 254
coleoptile 246
cytokinins 246
day-neutral plant 244
development 240
developmental biology 240
differentiation 240
diurnal 241
drought tolerant 242
endogenous factors 245
environment 240
environmental insults 241
ethylene 246
etiolated 255
evapotranspiration 242
exogenous factors 245
florigen 254
gene activators 245
genotype 240
gibberellins 246
growth 240
herbicide 246
hormone 245
indoleacetic acid (IAA) 246
long-day plant 244
parthenocarpic fruit 249
phenotype 240
photoperiod 242
phototropism 246
physiological limits 241
phytochrome 243
plasticity 250
productivity 242
seasonal 241
short-day plant 244
variability 240
water stress 242

Discussion Questions

1. What is the importance of the genotype in determining what an organism looks like?
2. Why should there be variability in two organisms that are derived from the same parents?
3. What is developmental biology?
4. What effect does temperature have on plant growth?
5. Why is light so important in plant growth?
6. What is a biological clock?
7. What are the major plant hormones and what is their principle effect on plant growth and development?
8. What are tropisms and how do they influence plant growth?

Suggested Readings

Gresshoff, P. H., ed. 1992. *Plant biotechnology and development.* CRC Press, Boca Raton.

Hoad, G. V., ed. 1987. *Hormone action in plant development.* Butterworths, Boston.

Jenkins, G. I., and W. Schuch, eds. 1991. *Molecular biology of plant development.* Company of Biologists Ltd., Cambridge.

Lyndon, R. F. 1990. *Plant development: The cellular basis.* Unwin Hyman, Boston.

Moore, T. C. 1989. *Biochemistry and physiology of plant hormones.* 2nd ed. Springer-Verlag, New York.

Pech, J. C., A. Latche, and C. Balague, eds. 1993. *Cellular and molecular aspects of the plant hormone ethylene.* Kluwer Academic Publishers, Boston.

Purohit, S. S., ed. 1987. *Hormonal regulation of plant growth and development.* Kluwer Academic Publishers, Boston.

Rivier, L., and A. Crozier, eds. 1987. *Principles and practice of plant hormone analysis.* Academic Press, Orlando.

Roberts, L. W. 1988. *Vascular differentiation and plant growth regulators.* Springer-Verlag, New York.

Sachs, T. 1991. *Pattern formation in plant tissues.* Cambridge University Press, New York.

Salisbury, F. B., and C. W. Ross. 1985. *Plant physiology.* 3rd ed. Wadsworth Publishing Co., Inc., Belmont.

Part Five

Evolution and Diversity

The evolution of reproductive structures in this chickweed flower ensures successful cross-pollination and fertilization.

Chapter Fifteen

Meiosis, Sexual Reproduction and Inheritance

Meiosis

Inheritance

A general understanding of plant structure and ecological interactions gained in earlier chapters sets the stage for examining the importance of genetic variability and the inheritance of traits. The existence of variability means that some individuals are better suited to a given environment than are others. Thus, as environments gradually change (and they do), each individual responds. If the environmental change does not suit some organisms, there might be others that find success within the new conditions. Because of the essential role of plants in all biological systems, variability in plants is necessary for the success of all species. How, where, and when sexual reproduction takes place are crucial questions in understanding both variability itself and its long-term role as the key to change. Understanding how change occurs permits a clear appreciation of past, present, and future diversity among organisms and the mechanisms by which that diversity continues to exist. This chapter addresses the process of sexual reproduction and the inheritance of genetically controlled traits.

Colorful flowers of a *Heliconia* in Brazil.

You do not need to be a scientist to notice the incredible array of living things that exist on earth. Not only are there millions of different kinds of organisms, but within each kind no two individuals are exactly alike.

Prokaryotic organisms, and some eukaryotic ones, reproduce **asexually.** Usually a single-celled individual splits into two genetically identical individuals. At best there can be slight genetic differences between the two offspring as a result of mutations. Most eukaryotic organisms, on the other hand, produce offspring by **sexual reproduction** as well as asexually. Sexual reproduction ensures that all individuals are at least slightly different genetically. Identical twins are an exception, but even here, single gene-coding mistakes or mutations can result in minor differences.

Without the resulting **genetic variability** among individuals that sexual reproduction produces, you and your friends would look alike, and you would not be able to distinguish your parents from your neighbors, or your dog from any other. All other individuals of each kind of organism would look and function the same. There would be no cultivated food crops because there never would have been a larger tomato, a juicier grapefruit, or higher-yielding wheat inflorescence to select as the start of a better crop plant.

MEIOSIS

The key to sexual reproduction is a process of nuclear and cellular division different from that of mitosis (chapter 11). The critical feature of **meiosis** is the production of daughter cells having only one set of chromosomes, half the number of the parent cell that began the process. **Diploid** cells have two sets (2n) of chromosomes; half that number, one set (n), is termed the **haploid** number. The fusion of two haploid cells during **fertilization** completes the sequence of events in the reproductive cycle, producing a genetically unique organism.

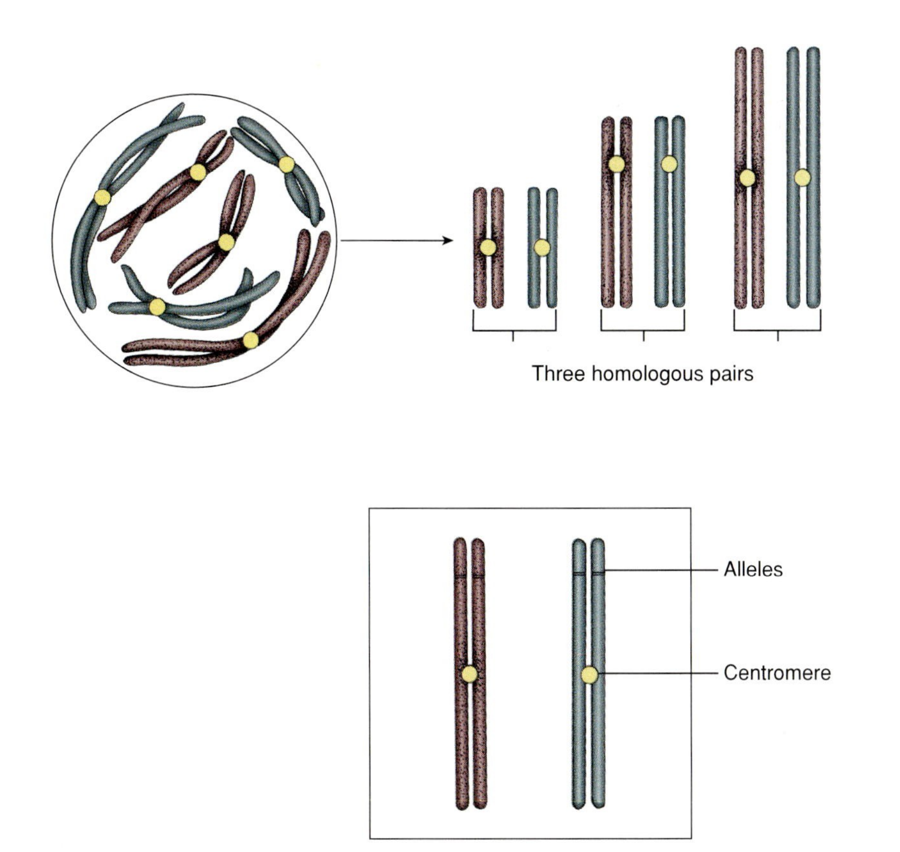

Figure 15.1
Three homologous chromosome pairs. Homologues are the same length, have their centromeres in the same position, and contain genetic messages for the same set of traits. Each of these chromosomes is composed of two genetically identical chromatids. Genes for the same trait are called alleles. The bands on the boxed chromosomes represent alleles replicated on the two chromatids at the same locus (site) of each chromosome.

Most sexually reproducing higher organisms are diploid, which means that all the cells making up the body of the organism, that is, **somatic** cells, contain a nucleus with two sets of homologous, or similar, chromosomes. These **homologues** are similar in size, position of their centromere, and, most important, in their genetic composition. Each of the two chromosomes of a homologous pair contains genes for the same group of traits, and these genes occur in the same sequence from one end of the homologue to the other. It is important to remember that homologues contain genetic instructions for the same traits but *not necessarily the same instructions.* Thus the expression of each trait is controlled by two genetic messages, one gene from each of the two homologous chromosomes. These two genes are called an allele pair or **alleles.** Alleles are always in the same position (locus) on the two homologous chromosomes. All of the functions of each diploid somatic cell, therefore, are genetically encoded twice, once per haploid set of chromosomes. A set of chromosomes is composed of one chromosome from each different homologous pair, and it does not matter which of the two homologues is present (figure 15.1).

The following analogy may help you visualize homologous chromosomes. Hold your hands together so that your palms are facing each other and your fingers are touching and pointing in the same direction. Your two thumbs and each of your four fingers are paired, with similar digits across from each other. This is analogous to five homologous chromosome pairs in that each member of each pair is the same size and length and they look similar. The matching joint creases in your fingers are analogous to the loci for the two genes for the same traits—alleles. Meiosis is a process that mostly occurs in specific tissues of sexually reproducing organisms. In angiosperms the flower is the reproductive structure, but meiosis occurs in specific cells in the anthers and ovules. In each of these tissues a different sequence of events takes place. Like mitosis, the process of meiosis is a continuous series of activities, but it has been subdivided into phases for convenience of discussion (figure 15.2). Unlike mitosis, meiosis consists of *two sequential nuclear divisions,* not just one. The first of these, meiosis I, is termed the **reduction division** because it is during this sequence of events that the haploid condition is produced from a diploid parent cell. The second meiotic division, meiosis II, is similar to mitosis in its activities. For convenience, meiosis is discussed using the subdivisions of meiosis I and II with each further subdivided into the phases of nuclear division presented for mitosis.

The importance of a reduction division preceding the fusion of two

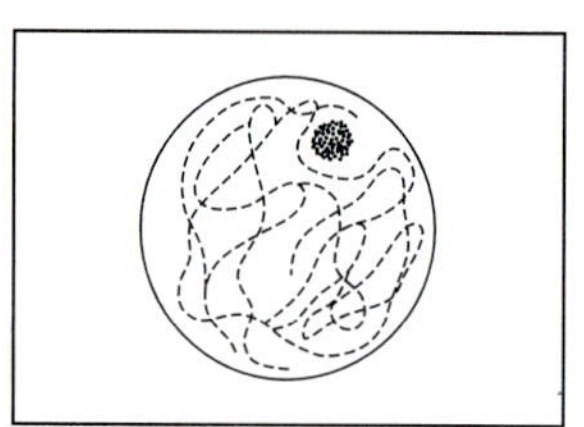
(a) Interphase

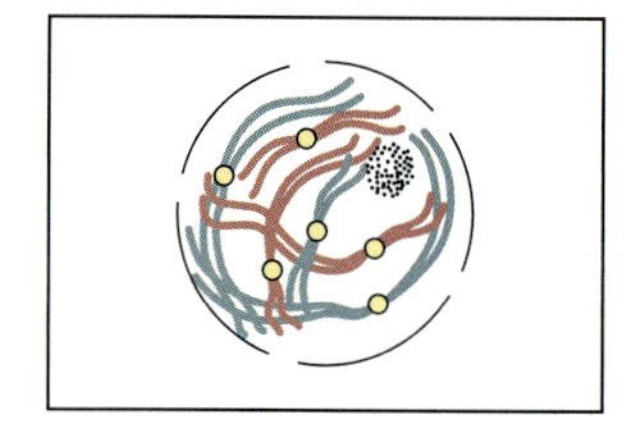
(b) Early prophase I

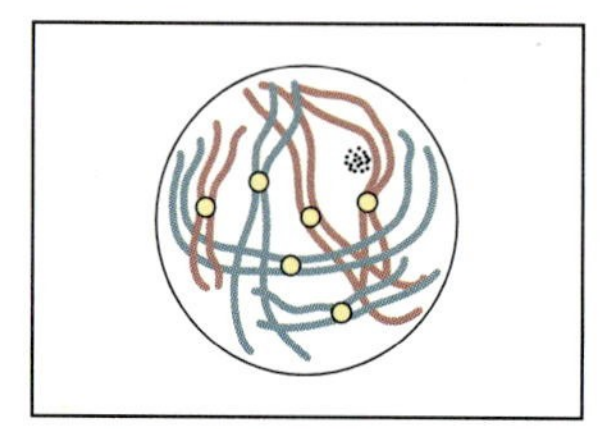
(c) Mid-prophase I

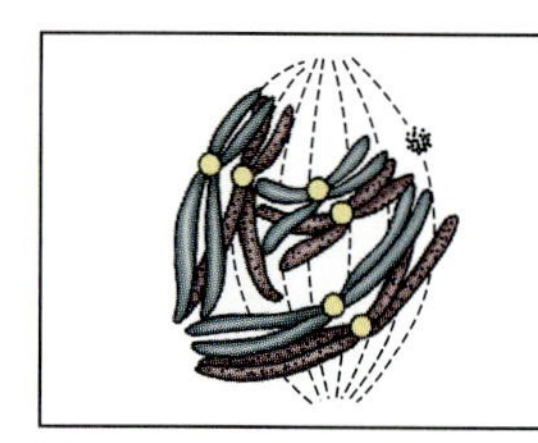
(d) Late prophase I

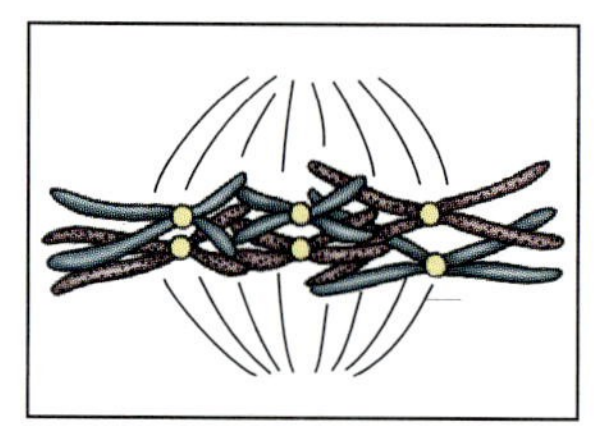
(e) Metaphase I

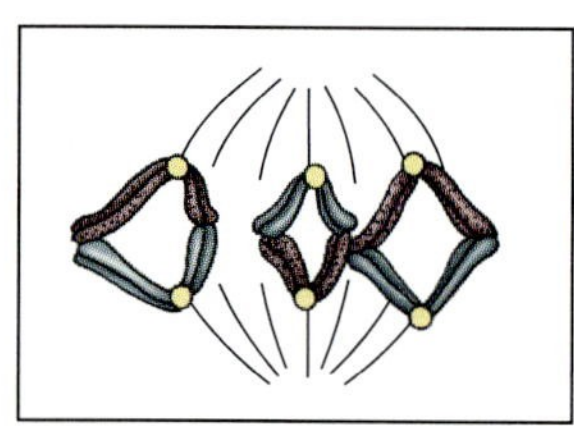
(f) Early anaphase I

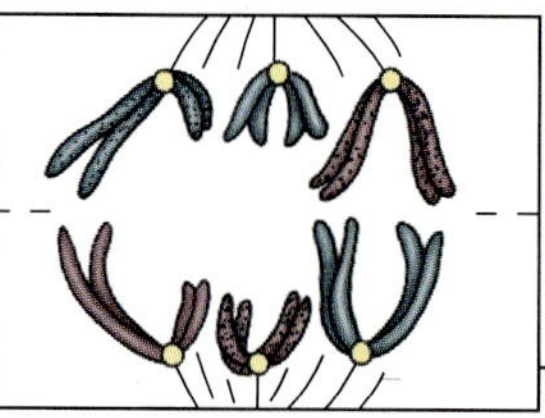
(g) Late anaphase I

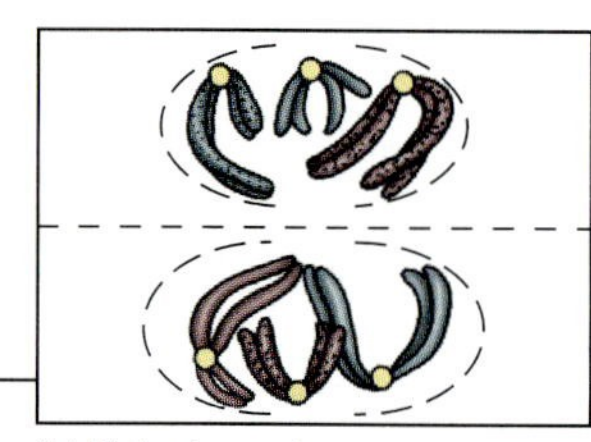
(h) Telophase I

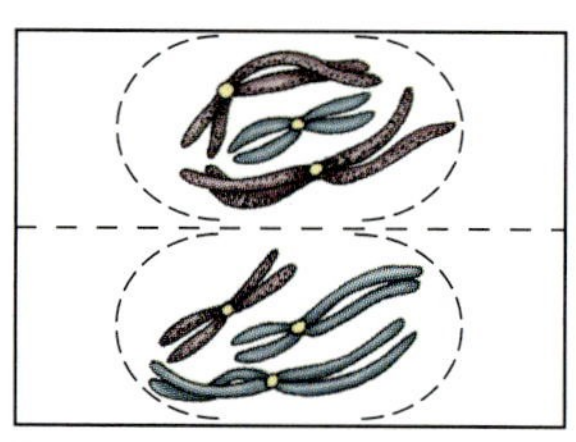
(i) Prophase II

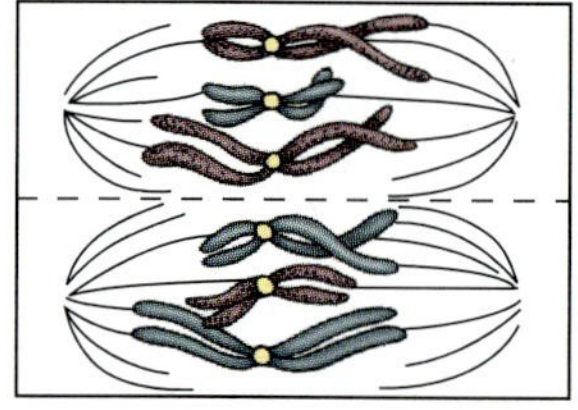
(j) Metaphase II

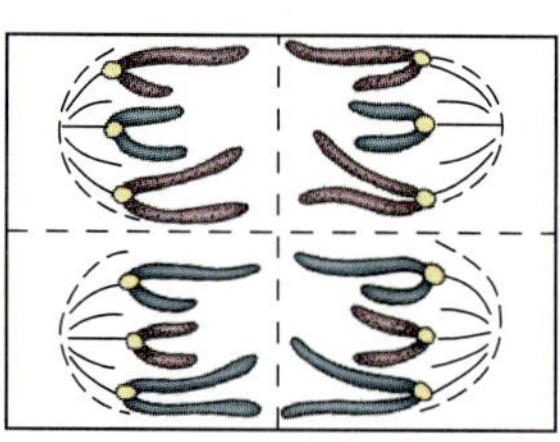
(k) Anaphase II

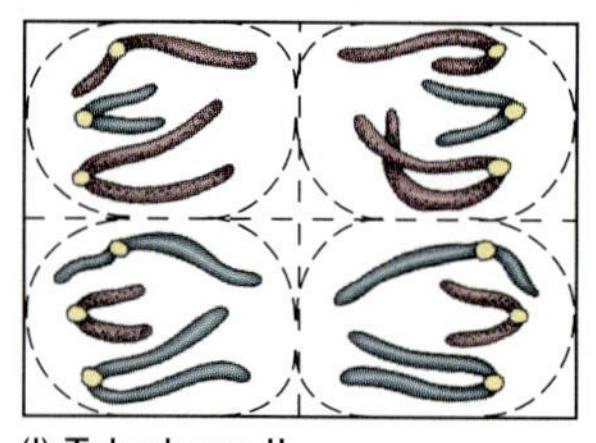
(l) Telophase II

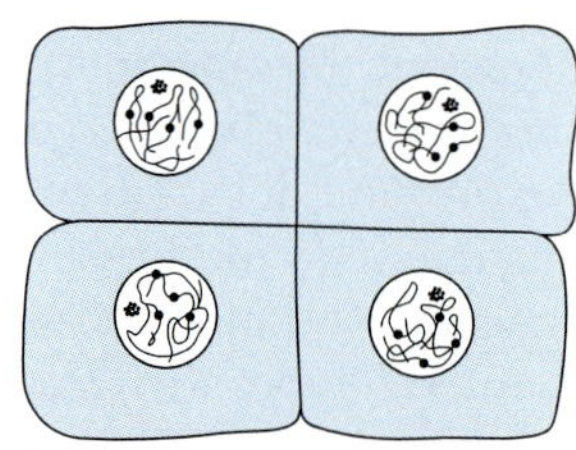
(m) Late telophase II

Figure 15.2

Stages of meiosis, from a single diploid cell having six chromosomes through representative stages of the two meiotic divisions.
(*a*) Interphase. (*b*) Early prophase I. The chromosomes are not readily visible, and the nuclear envelope is just beginning to disappear.
(*c*) Mid-prophase I. The six chromosomes actually should be visible. (*d*) Late prophase I. Each chromosome can be seen to contain two chromatids, and they are condensed.
(*e*) Metaphase I. The bivalents attach to spindle fibers at the equatorial plane of the cell.
(*f*) Early anaphase I. The centromeres of the two homologues of each bivalent separate, forming a doughnut shape. (*g*) Late anaphase I. The homologues separate and move apart.
(*h*) Telophase I. Three chromosomes are at opposite ends of the original cell. The reduction from a diploid number (6) to a haploid (3) chromosome condition is complete.
(*i*) Prophase II. Each set of the three chromosomes prepares for the second meiotic division, which will separate the chromatids.
(*j*) Metaphase II. The centromeres attach to the spindle fibers at the equatorial plane of each haploid cell. (*k*) Anaphase II. The chromatids separate and move apart. (*l*) and (*m*) Telophase II. Four haploid nuclei begin to form, resulting in four genetically different haploid cells.

TABLE 15.1

Selected Plant Chromosome Numbers

COMMON NAME	SCIENTIFIC NAME	DIPLOID CHROMOSOME NUMBER
Adder's tongue fern	*Ophioglossum reticulatum* (largest known)	1260
	Haplopappus gracilis (smallest known)	4
Dandelion	*Taraxacum officinale*	24
Lettuce	*Lactuca sativa*	18
Annual sunflower	*Helianthus annuus*	34
Watermelon	*Citrullus vulgaris*	22
Cotton	*Gossypium hirsutum*	52
Morning glory	*Convolvulus arvensis*	50
Wheat	*Triticum aestivum*	42
Green beans	*Phaseolus vulgaris*	22
Soybeans	*Glycine max*	40
Potato	*Solanum tuberosum*	48
Douglas fir	*Pseudotsuga menziesii*	26
Redwood	*Sequoia sempervirens*	66
Ponderosa pine	*Pinus ponderosa*	24
American elm	*Ulmus americana*	56

gametes is apparent if you consider what would happen without it. If a diploid organism produced gametes through mitosis, each would be diploid, just like all of the somatic cells. Human gametes (egg and sperm) each would have 46 chromosomes, the diploid number for all somatic cells. Fusion of the two gametes would result in a single-celled **zygote** having 92 chromosomes in its nucleus. As the zygote develops mitotically into a mature multicellular organism, each individual cell in that organism would have 92 chromosomes. Two such individuals, each producing gametes without a reduction division, would have an offspring with 184 chromosomes in each cell. By the fifth generation the offspring would have 1472 chromosomes in every cell of every individual. Obviously this doubling of chromosome number each generation could not continue for very long before there would be an unsolvable space problem. In addition, problems with multiple genetic messages would occur.

Most organisms within a given species, therefore, are constant in their chromosome number. As already stated, all humans have a diploid number of 46 chromosomes (23 pairs). By comparison, the diploid number for plants ranges from a low of 4 to a high of 1260. A sampling of plant chromosome numbers is presented in table 15.1.

Meiosis I

As in mitosis, the cellular and nuclear activities of **interphase** prepare the cell for division. These activities include DNA replication to produce chromosomes composed of two genetically identical chromatids united at the centromere. **Prophase I** is underway when the threadlike chromosomes first become visible. Next the nuclear membrane starts to break down, and the spindle fibers begin forming. Later the nucleolus becomes gradually less distinct.

As the chromosomes coil, becoming more condensed and visible, the homologous pairs physically come together. This process of **synapsis** does not occur in mitosis. Synapsis produces an exact pairing of the two homologues with the same genetic regions side by side for the full length of the two chromosomes. The pairing is evident as soon as the chromosomes become clearly visible. The paired homologues, each composed of two chromatids, are called a **bivalent** (*bi,* two; *valent,* combined, associated). Thus each bivalent contains four chromatids. During bivalent formation the nuclear membrane and nucleolus continue to disappear, the spindle forms, and the paired chromosomes become fully condensed and begin moving toward the center of the cell. Prophase I ends at this point.

Metaphase I begins as the chromosomes arrive at the equatorial plane of the cell and their centromeres attach to spindle fibers. Typically, the bivalents are doughnut shaped because their centromeric regions push away from one another during full chromosome condensation. As each bivalent lines up on the spindle, the two centromeres attach to fibers on opposite sides of the equatorial plane. For example, in a plant having a diploid chromosome number of 6, there are three bivalents attached to the spindle at this point in meiosis. Each of these three bivalents contains a chromosome from each of the two parents that formed the reproducing plant. Which member of the homologous pair lines up on which side of the equatorial plane is a random process.

The next event during meiosis results in the actual reduction in the number of chromosomes each daughter nucleus receives. The centromeres of each of the two homologous chromosomes paired during the first part of meiosis I are pulled apart and move toward opposite poles of the cell. This occurs in **anaphase I.** Each chromosome is still composed of two chromatids. Half the total number of chromosomes go in one direction and half in the other. It is important to remember that *in each resulting group there is one chromosome from each bivalent.* Thus each pole has a complete set of chromosomes with one genetic message for each trait. In a plant where 2n = 6 represents the diploid condition, n = 3 represents the haploid condition that results from meiosis I.

As the separated homologues reach opposite ends of the parent cell, several events occur at once—in what is considered **telophase I.** During telophase I the uncoiling of the chromosomes, the reformation of the nuclear membrane, the reappearance of nucleoli, and the complete disappearance of spindle fibers occur. Here it is important to point out that in different organisms these events vary in their level of completion. In some organisms telophase I events never really occur before the chromosomes begin the sequence of activities leading to the second meiotic division. In others, telophase I proceeds almost to completion prior to the start of the meiosis II division. Cytokinesis can also occur completely, partially, or not at all, depending on the organism. In instances when telophase I events do take place to completion, there can be a period termed interkinesis, but it differs from interphase in that there is no DNA replication.

Meiosis II

Each of the two daughter nuclei resulting from meiosis I is haploid, having one set of chromosomes composed of two chromatids each. The events that these two haploid daughter nuclei go through are essentially the same as those in mitosis. In prophase II the chromosomes become tightly recoiled and are visible as having two chromatids each. The nuclear membrane and the nucleolus gradually disappear, and spindle fibers begin forming. Once the spindle is formed and the chromosomes are fully condensed, the chromosomes move to the equatorial plane between the ends of the spindle. This is metaphase II. Each chromosome acts independently, with centromeres attaching to spindle fibers at the equatorial plane. The centromeres then divide at the beginning of anaphase II, and the two chromatids separate and move toward opposite poles. As these newly formed chromosomes reach the opposite poles, telophase II events begin. These are the same as those in telophase I. At the end of telophase II there are four resulting daughter nuclei, each one containing a single complete set of chromosomes and each one being genetically different from the others.

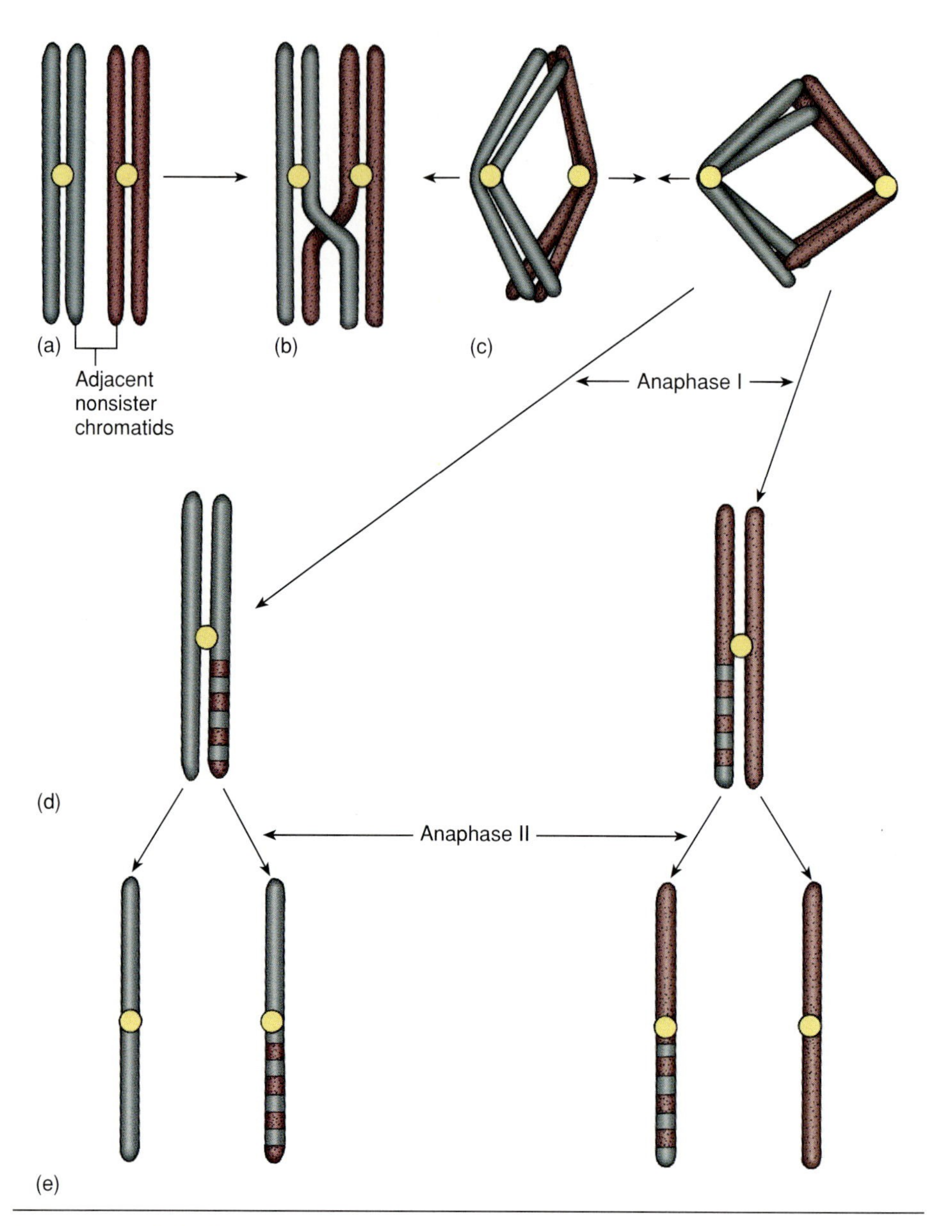

Figure 15.3

Crossing over. Segments from nonsister chromatids within a bivalent (*a*), physically cross over (*b*), *and* exchange genetic material from one homologue to the other (*c*). Each genetically different chromatid of a homologous chromosome (*d*), will separate during anaphase II of meiosis, resulting in four haploid nuclei (*e*), each having a chromosome with a genetic composition different from each original homologous pair.

Crossing Over

It would appear that of the four haploid daughter nuclei produced by meiosis, there should be two pairs that contain genetically identical members. In other words, each of the two daughter nuclei resulting from meiosis I will be genetically different because of the random assortment of homologues, but each chromosome in those two daughter cells should have identical chromatids. These chromatids then separate in meiosis II. But the chromatids are not identical, specifically because of the process of **crossing over,** an exchange of genetic material during the early part of meiosis I (figure 15.3).

This exchange is between adjacent chromatids on the two homologous chromosomes of a bivalent. As homologues undergo synapsis during early prophase I, each chromosome is composed of two genetically identical, or sister, chromatids. During the bivalent condition, adjacent nonsister chromatid arms often physically cross over each other and associate for a portion of their length. As the chromosomes become more condensed and visible, this configuration can be seen as a crossover of chromatid segments and is termed a **chiasma.** There is an average of one chiasma per bivalent, although not every bivalent necessarily has a chiasma form; others form more than one.

As bivalents reach their most condensed state, the homologues begin to repel one another and physically separate, except at the crossovers. The crossed-over region moves toward the ends of the chromatid arms exchanging genetic information along those chromatid arms. This results in an exchange of genetic material between two of the four chromatids in a bivalent. The exchange produces two chromosomes, each composed of two genetically different chromatids.

When the chromatids separate during meiosis II, therefore, the four resulting nuclei all are genetically different from each other. This genetic variability, combined with the mixing of genetic information that occurs as a result of random assortment and separation of homologues during meiosis I, provides the potential for a phenomenal number of original genetic combinations in gametes. To this genetic diversity add the random possibilities of which two genetically different gametes will fuse during fertilization, and it becomes clear why no two sexually reproducing individuals are ever alike.

Mutations

Although the replication of genetic information normally occurs flawlessly, coding mistakes, or **mutations,** do occasionally occur. Many mutations are not lethal to the organism, and some are thought to have absolutely no effect at all; however, many others produce changes in the appearance or function of an organism. Mutations are another source of variability in sexually reproducing organisms.

Between random assortment of homologues, chiasmata, mutations, and chance combinations of gametes during the process of sexual reproduction, genetic variability is ensured. Variability is essential for the success of organisms as their environments gradually change over many generations.

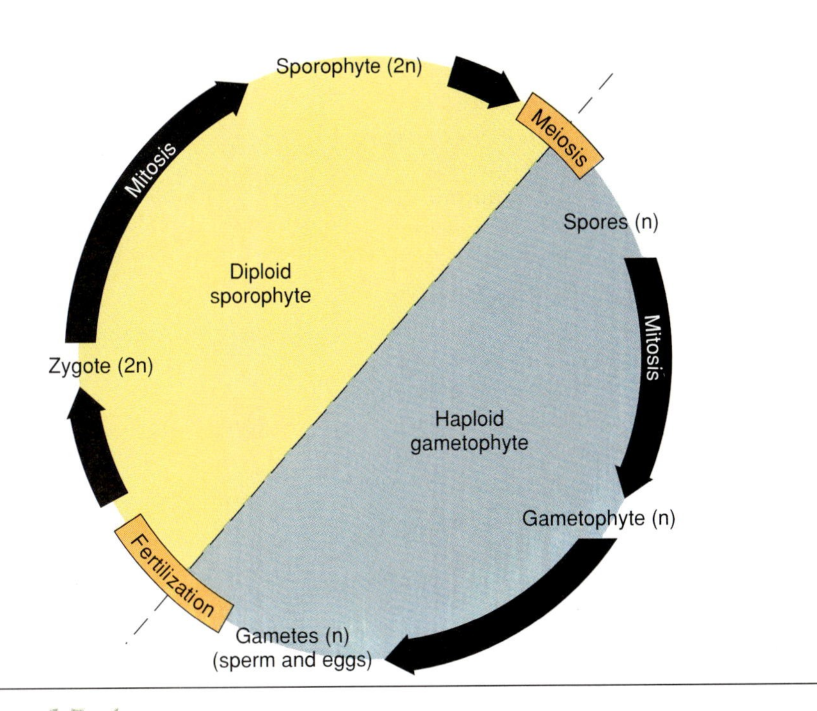

Figure 15.4
Alternation of generations: the alternation between the sporophyte and gametophyte stages in the life cycle of all plants.

> **✔ Concept Checks**
>
> 1. What is the significance of meiosis I, the "reduction division"?
> 2. In what ways are homologous chromosomes similar?
> 3. How does crossing over help increase genetic diversity?
> 4. How do mutations help increase genetic diversity?

Alternation of Generations

The actual process of meiosis is essentially the same in all sexually reproducing organisms. What the resulting haploid cells (nuclei) are called and what they do next differ. In animals, the haploid cells normally differentiate to become gametes, which fuse during fertilization into a single-celled diploid zygote. That zygote can then begin dividing mitotically to form a multicellular organism that grows, matures, and reinitiates the entire cycle. In plants, however, a different sequence of events usually occurs. Plants have two generations, sporophyte and gametophyte. Both gametophyte and sporophyte are multicellular vegetative plants. The **sporophyte generation** produces spores, each of which develops into a **gametophyte,** which in turn produces gametes. The gametes fuse to form a sporophyte, and the alternation between these phases continues in successive generations. This is called the **alternation of generations** in plants (figure 15.4).

The sporophytes are diploid, producing haploid spores by meiosis. These haploid spores divide mitotically to produce the multicellular, haploid gametophytes. Specific cells produced by the haploid gametophytes are gametes, which may fuse with other gametes to form a diploid zygote. The zygote divides mitotically to form a two-celled embryo, which continues dividing mitotically to ultimately form the diploid sporophyte stage of the life cycle.

The gametophyte (haploid) stage or generation is the dominant form in the lower (nonvascular) plants, the sporophyte being less evident and smaller and often dependent on the gametophyte. With increasing complexity from less complex (lower) to more complex (higher) vascular plant forms, this relationship gradually changes until in the angiosperms (the most complex plants) the sporophyte is the dominant, long-lived, independent stage, and the gametophytes are small, short-lived, dependent groups of specialized cells that differentiate within the flowers of the sporophyte in a connected sequence of events. They occur in two separate tissues of the flower, one in the anthers and the other in the ovules. In the anthers, microsporogenesis (the making of small spores) is followed by microgametogenesis (the making of small gametes). In the ovules, megasporogenesis is followed by megagametogenesis.

In the developing anthers of a flower there are four areas of sporogenous (fertile) cells, one in each of the four sacs comprising the anther. These cells initiate the developmental sequence to produce pollen grains. The tapetum is a layer of nutritive tissue that provides nourishment for the sporogenous cells as they divide. The tapetum, as well as the sporogenous cells, is diploid. The cells in the sporogenous areas that undergo meiosis are called microspore mother cells, or microsporocytes. Each diploid microspore mother cell produces four haploid microspores. In monocots, cytokinesis normally occurs as each of the two meiotic divisions occurs, whereas in dicots the four haploid nuclei are produced first, and then cytokinesis forms all four microspores simultaneously at the end of meiosis.

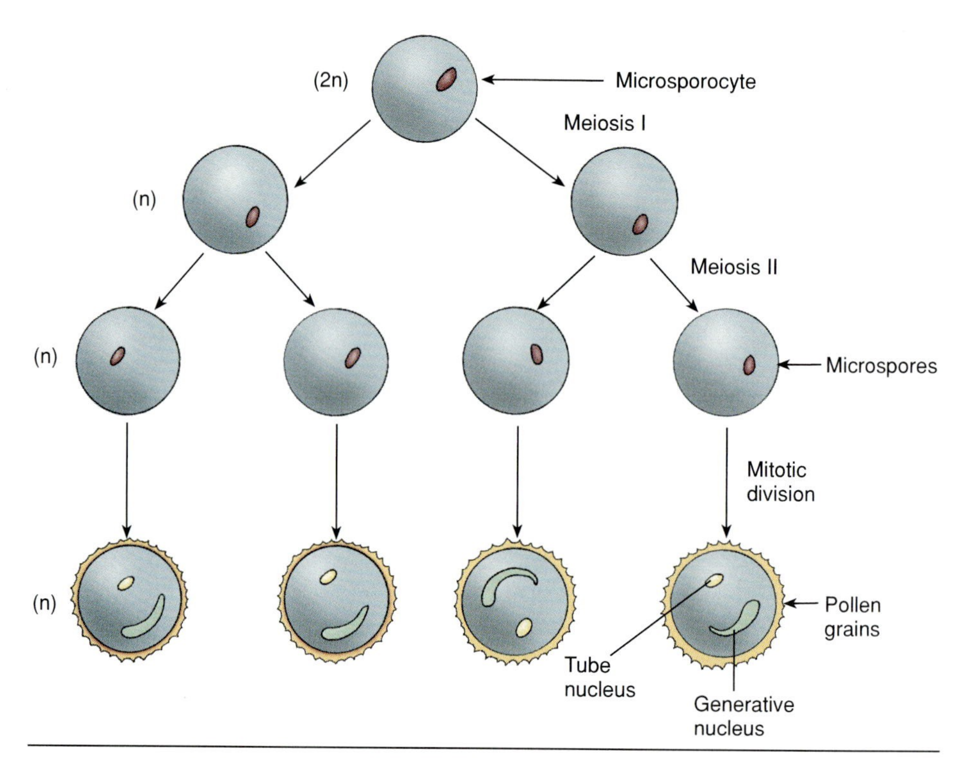

Figure 15.5

Microsporogenesis and microgametogenesis. The end product of these activities is a pollen grain having either a binucleated condition (tube and generative nuclei) or a trinucleated condition (tube and two sperm nuclei). Shown here is the binucleated pollen with the elongated generative nucleus prior to its mitotic division to produce two sperm nuclei.

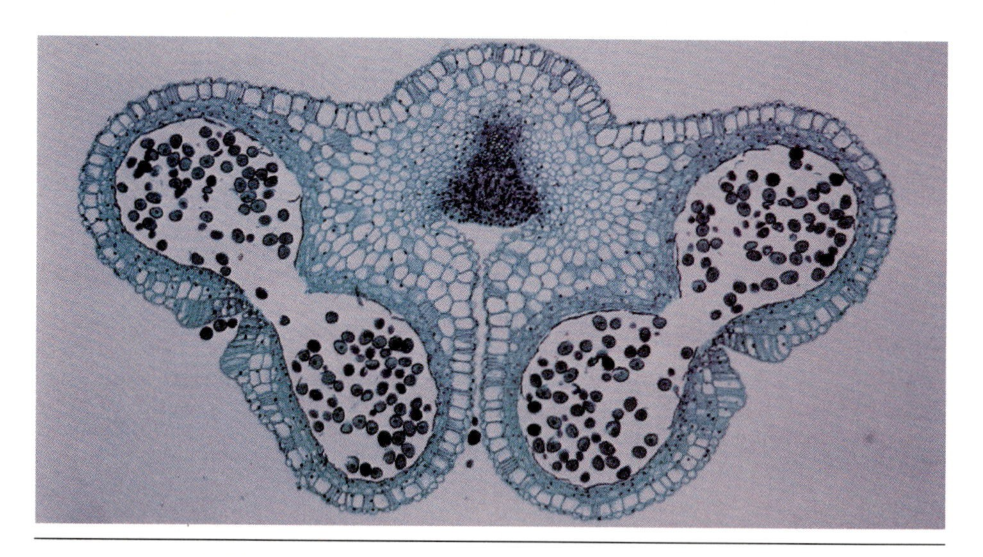

Figure 15.6

Developing pollen grains within lily anther sacs.

Each of the four haploid microspores enlarges and develops into a pollen grain (figures 15.5 and 15.6). The outer coat of a pollen grain, the exine, is a tough layer that makes pollen one of the most persistent plant parts over the entire geologic fossil record of higher plants. The inside layer of the pollen grain the intine, is not as resistant.

In angiosperms, the pollen grain is the gametophyte stage, the gamete-producing organism in the alternation of generations. The haploid

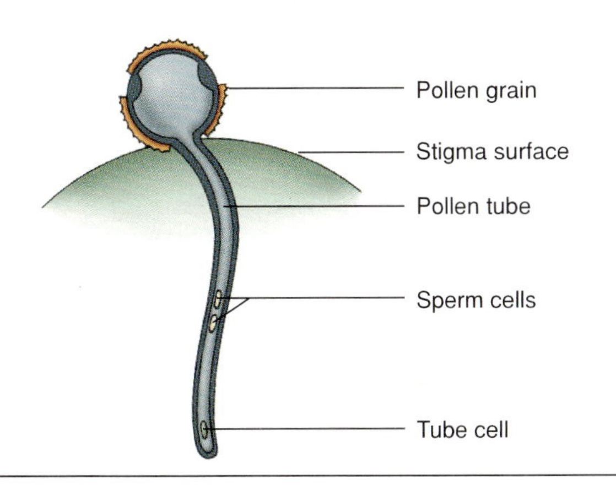

Figure 15.7
After pollination the pollen grain germinates, and a pollen tube forms intercellularly down the style. The tube contains the tube cell and two sperm cells.

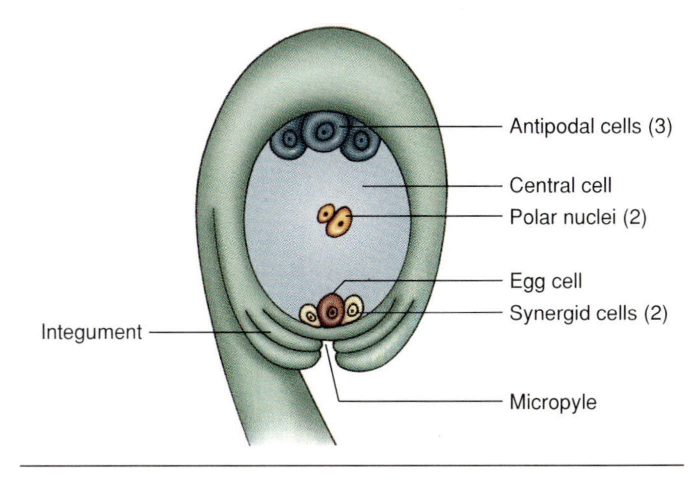

Figure 15.8
The ovule with seven cells having eight nuclei. The three antipodal cells are on the end of the embryo sac away from the micropyle; the two nuclei of the central cell are in the center; and the egg cell, flanked by the two synergid cells, is at the micropyle end.

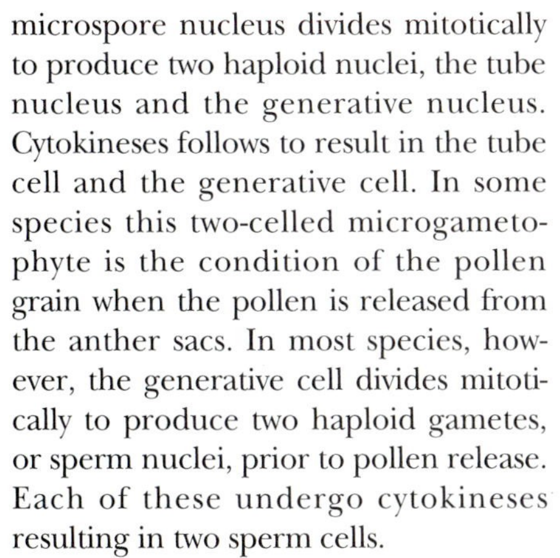

microspore nucleus divides mitotically to produce two haploid nuclei, the tube nucleus and the generative nucleus. Cytokineses follows to result in the tube cell and the generative cell. In some species this two-celled microgametophyte is the condition of the pollen grain when the pollen is released from the anther sacs. In most species, however, the generative cell divides mitotically to produce two haploid gametes, or sperm nuclei, prior to pollen release. Each of these undergo cytokineses resulting in two sperm cells.

If the pollen is released in the two-celled condition, the generative cell divides mitotically sometime during or soon after pollination to produce the two nonmotile sperm cells, since both must be present for double fertilization to occur. The tube cell is so called because after the pollen grain lands on a receptive stigma, the tube cell directs pollen germination through one of the pores in the exine and forms a pollen tube down through the tissue of the style (figure 15.7). The tube cell directs the production of an enzyme that dissolves intercellular tissue of the style and forms a small tube through which the tube cell and two sperm cells travel to the ovary.

The ovary of a flower may produce from one to hundreds of ovules, depending on the species. Each ovule is connected to the placental tissue of the ovary by its stalklike funiculus. The ovule develops one or two outer layers, the **integuments,** which grow upward from the base of the nucellus toward its apex to enclose the nucellus tissue within. There is a small opening, the **micropyle,** at the end where the integuments come together. Through this opening the pollen tube containing the sperm cells enters the ovule prior to fertilization.

In the nucellus tissue there is a single diploid megasporocyte that undergoes meiosis and cytokinesis to produce four haploid megaspores. At the conclusion of meiosis these four megaspores are usually aligned in a chain. Normally the three cells nearest the micropyle degenerate, leaving only one functional megaspore to develop into the megagametophyte.

This single haploid megaspore undergoes three consecutive mitotic divisions to produce two, then four, and finally eight haploid nuclei within a single cell in the ovule. This cell is the **megagametophyte.** The eight nuclei are normally arranged with four at the micropyle end and four at the opposite end (figure 15.8). Subsequently one nucleus from each group of four migrates to the middle of the megagametophyte. These are the two **polar nuclei.** Cell walls are synthesized around the remaining three cells at the micropylar end, resulting in a middle

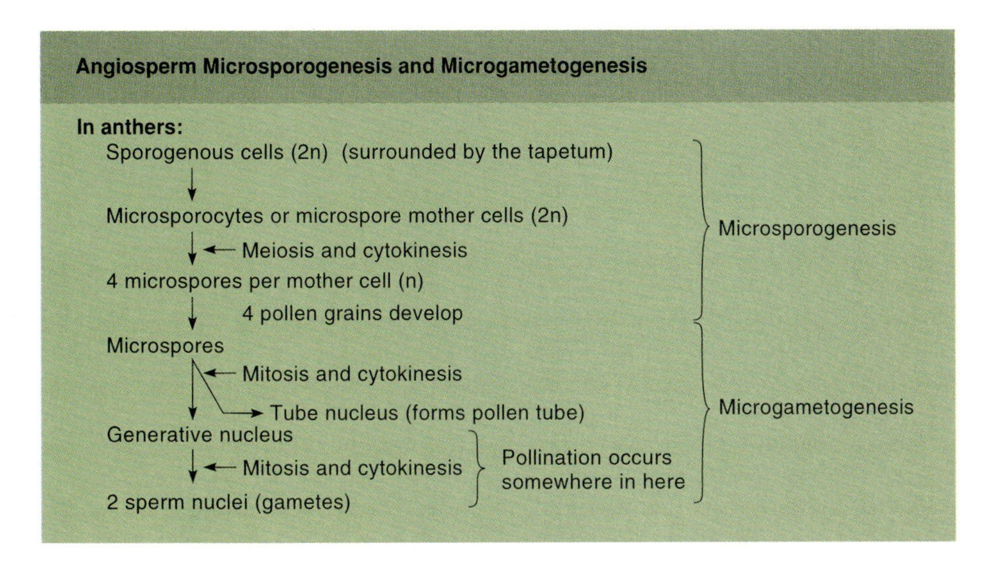

Angiosperm Microsporogenesis and Microgametogenesis

In anthers:

Sporogenous cells (2n) (surrounded by the tapetum)
↓
Microsporocytes or microspore mother cells (2n)
↓ ← Meiosis and cytokinesis
4 microspores per mother cell (n)
↓ 4 pollen grains develop
} Microsporogenesis

Microspores
↓ ← Mitosis and cytokinesis → Tube nucleus (forms pollen tube)
Generative nucleus
↓ ← Mitosis and cytokinesis
2 sperm nuclei (gametes)
} Pollination occurs somewhere in here
} Microgametogenesis

egg cell flanked by two **synergid cells**. The three nuclei at the other end of the megagametophyte also form cell walls and are called the **antipodal cells.** Both polar nuclei remain in the center of the original cell, becoming a binucleated **central cell.** This eight-nucleate, seven-cell condition is the mature megagametophyte, a sexual plant having a single egg cell, the gamete. Each ovule in a mature ovary contains a megagametophyte just as each pollen grain contains a two- (or three-) celled microgametophyte.

TABLE 15.2

Traits of Pea Plants Studied by Mendel

TRAIT	DOMINANT EXPRESSION	RECESSIVE EXPRESSION
Seed color	Yellow	Green
Seed coat	Round	Wrinkled
Flower color	Red	White
Pod shape	Inflated	Constricted
Stem length	Long	Short
Flower position	Axial	Terminal
Pod color	Green	Yellow

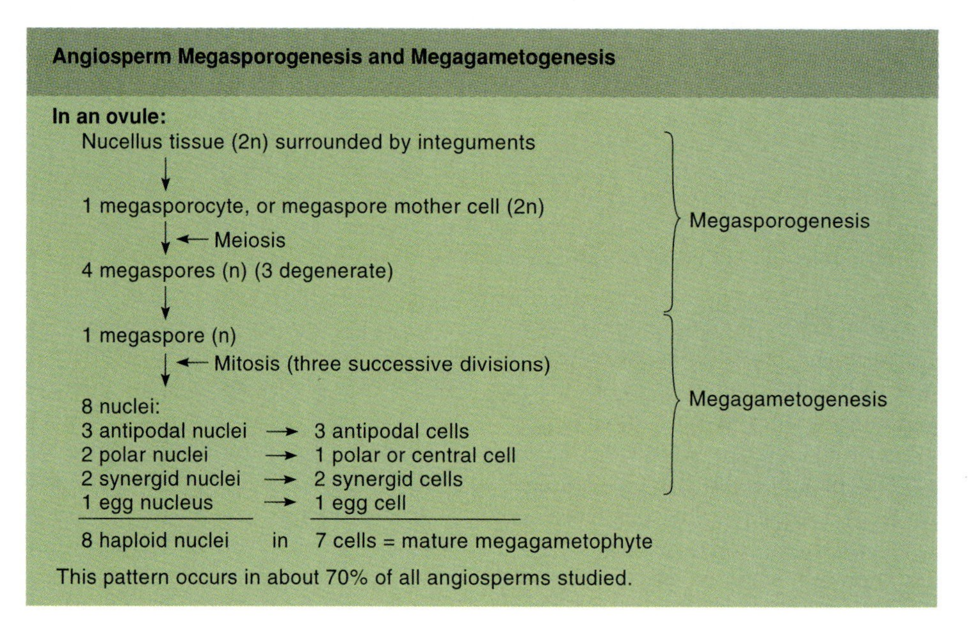

Fertilization

The culmination of sexual reproduction in angiosperms is a unique form of fertilization occurring in the ovule. The pollen tube normally enters the ovule through the micropyle opening. Once adjacent to the egg, it forms a pore in its wall to allow the two sperm cells to exit. Fusion between the haploid egg and one sperm nucleus occurs, and a diploid zygote cell is formed. This is true sexual fertilization. Angiosperms are said to have **double fertilization** because the other sperm nucleus fuses with the two polar nuclei to form a triploid (three sets of chromosomes) nucleus of the primary endosperm cell. This double fertilization is one feature that makes angiosperms unique among plant groups.

The zygote forms the embryo of the seed through repeated mitotic division, whereas the triploid **endosperm** tissue of the seed forms mitotically from the triploid primary endosperm cell. The seed coat develops from the ovule's integuments. The new individual becomes relatively dormant in its seed, ready to be dispersed, germinate, and grow to maturity to initiate the reproductive cycle again. It is worth repeating that the most important consequence of sexual reproduction is *variability.*

✔ Concept Checks

1. How does meiosis in the alternation of generations in plants differ from the role of meiosis in animals?
2. What are the gametophytes in angiosperms?
3. Describe double fertilization in plants.

INHERITANCE

As covered earlier in this chapter, sexually reproducing diploid organisms have two genetic expressions for each trait, one provided by each parent. These two alleles are located at the same position, or locus, on homologous chromosomes. The study of genetics entails an understanding of how these alleles are passed from one generation to the next and how their messages control the appearance of an organism. The appearance of an organism is termed its **phenotype;** the genetic information controlling it is the **genotype.** Only the phenotype can be seen; thus the study of inheritance stems largely from studying phenotypic ratios of the traits that make up the organism.

The first such extensive study in which the results were carefully recorded and published was in 1866 by an Austrian monk named **Gregor Mendel.** Mendel studied a number of traits of the common pea plant (*Pisum sativum*), each of which had two different expressions. Between 1856 and 1868 he made hundreds of carefully recorded crosses and observations about the inheritance of these traits through several generations. Mendel made these observations and drew sound conclusions about the control of inheritance without knowing about chromosomes and meiosis; these were first described later in the century. Table 15.2 lists the seven traits with which Mendel worked.

TABLE 15.3

Mendel's Experimental Crosses: Traits and Results

PARENTAL TRAITS	F_1 RESULTS	F_2 RESULTS	F_2 RATIOS
Yellow × green seeds	100% yellow	6022 yellow; 2001 green	3.01:1
Round and wrinkled seed	100% round	5474 round; 1850 wrinkled	2.96:1
Red × white flowers	100% red	705 red; 224 white	3.15:1
Inflated × constricted pods	100% inflated	882 inflated; 299 constricted	2.95:1
Long × short stems	100% long	787 long; 277 short	2.84:1
Axial × terminal flowers	100% axial	651 axial; 207 terminal	3.14:1
Green × yellow pods	100% green	428 green; 152 yellow	2.82:1

Mendel's Experiments

"Pure parents" are plants that produce offspring bearing only one expression for a given trait. When Mendel crossed plants that were pure for one phenotype of a given trait (for example, yellow seeds) with plants pure for the other expression of that trait (for example, green seeds) the results were probably a bit unexpected. It seems logical that the offspring from such a cross would have some plants with green and some with yellow seeds. But, as table 15.3 shows, the F_1 or first filial (*filial,* offspring) generation for every such cross had only one of the two possible expressions appear. These results were consistent regardless of which parent variety provided the pollen and which provided the ovule. Mendel termed the expression of the trait that appeared in the F_1 plants the **dominant** character; the other he called the **recessive** character. Thus yellow is dominant and green is recessive for pea color. This means that if the hereditary factors for both yellow and green are present, the expression of the trait will be yellow. The term *gene* was not proposed until much later, but Mendel's *factors* are equivalent to genes. Note that the seed characters are expressions of the genotype of the parent plants, not that of the embryo (next generation plant) contained therein.

When Mendel let the F_1 plants self-pollinate, he noticed that in the next generation, the F_2 generation, approximately one-fourth of the plants had green seeds. As can be seen from these results, the hereditary factor for the recessive expression of the trait reappeared in the F_2 generation after being absent in the F_1 generation. So the recessive factor can be carried through a generation unchanged even when it is not expressed. This phenomenon held true for each of the other six character combinations studied; the recessive expression was totally absent from all F_1 plants but reappeared in the F_2 generation when F_1 plants were allowed to self-pollinate. Most important, these recessive characters appeared in the F_2 generation in approximately a 25% frequency in every case.

Mendel concluded from these results that each plant has two hereditary factors for each trait studied. When gametes are formed, these two factors segregate, and only one factor of each pair is included in each gamete. The formation of a new individual following fertilization has two of these factors coming together, one from each parent, to direct the expression of the trait in question. Mendel's **principle of segregation** explains how a recessive character can be hidden, but not lost, from one generation to the next.

By combining what is now known about genes, meiosis, and the separation of homologous chromosomes with Mendel's experimental crossing results, an explanation is possible of how these traits are inherited and controlled. Here letters represent the two possible alleles of a given trait, the capital letter for the dominant allele and the lowercase of the same letter for the recessive. The mathematical possibilities can be traced for the experimental crosses to show what happens to these genes. *Y* represents the allele (hereditary factor of Mendel) for yellow, *y* the allele for green. A pure parent producing yellow seed would have two alleles for yellow, one on each of a homologous pair of chromosomes and at the same loci on those chromosomes. The genotype for this yellow-seed parent would be represented by *YY*. Similarly, a pure parent for green seed would have a *yy* genotype. During meiosis in each of these two parents the two alleles for seed color segregate into different gametes. So 50% of the gametes produced are expected to contain one of the *Y* alleles and the other 50% the other *Y* allele. In the *yy,* or pure green-seed parent, the two alleles also segregate: 50% of the gametes produced contain one of the *y* alleles, 50% the other. In a yellow (*YY*) × green (*yy*) cross, therefore, all offspring produced during fertilization will have a *Y* allele from one parent and a *y* allele from the other. All the F_1 plants thus produced would have a *Yy* genotype and a yellow phenotype. Although only the phenotype can be seen, the genotype of the F_1 is known because the genotypes of the pure parents were known.

This cross also shows how the yellow allele, *Y,* has complete dominance over the allele for green seed color—*y*. When there is one of each of these alleles, the phenotype is always yellow. A Punnett square visually demonstrates how the alleles for a given trait from two parents segregate. Fifty percent of the gametes receive one of them, and 50% receive the other. Joining each possible gamete type of one parent with each possible gamete type of the other parent is done by

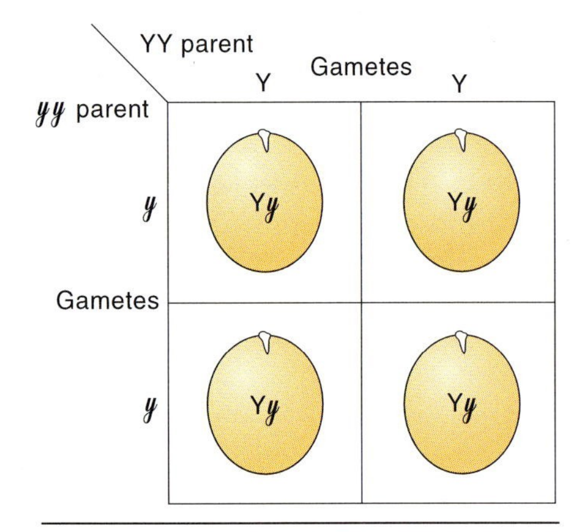

Figure 15.9
Punnett square for $YY \times yy$ cross. Each of the possible F_1 offspring types has a heterozygous *Yy* genotype.

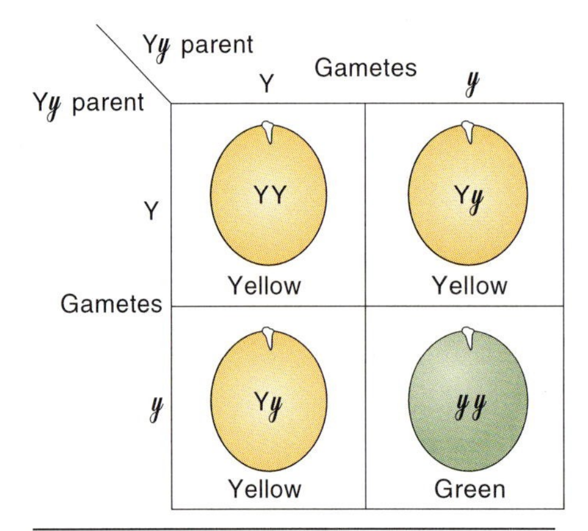

Figure 15.10
Punnett square for $Yy \times Yy$ cross of F_1 plant interbreeding. One of four (25%) of the possible F_2 offspring is *YY* (homozygous dominant); two of four (50%) of the possible F_2 offspring are *Yy* (heterozygous); one of four (25%) of the possible F_2 offspring is *yy* (homozygous recessive). Phenotypically, three of four (75%) are yellow (*YY* and *Yy* genotypes will be yellow), and one of four (25%) is green with a homozygous recessive genotype of *yy*.

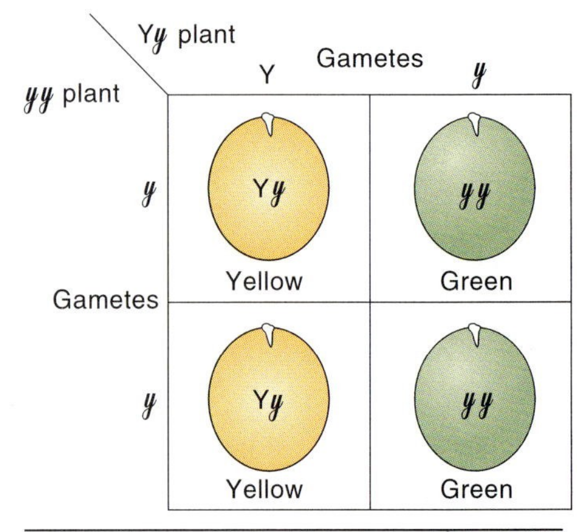

Figure 15.11
A backcross of an F_1 plant, *Yy*, with the homozygous recessive parent, *yy*. The genotypic and phenotypic ratios are 50:50.

connecting each of the two letters across the top of the square with each of the letters down the side of the square. In figure 15.9 it can be seen that each union of gametes produces a *Yy* genotype. Four out of four genotypes are *Yy*, and this is termed a **heterozygous** genotype (*hetero*, different). Both *YY* and *yy* genotypes are called **homozygous** (*homo*, same); *YY* is homozygous dominant and *yy* homozygous recessive. Only the homozygous recessive genotype, *yy*, will produce green-seeded pea plants, because this is the only one of the three possible genotypes that does not include a dominant allele for yellow.

When Mendel allowed the F_1 plants to self-pollinate, he allowed plants all having the heterozygous genotype, *Yy*, to interbreed. Each F_1 plant produces 50% *Y* and 50% *y* gametes because of the gene segregation during meiosis. The resulting F_2 genotype frequencies are 1*YY*:2*Yy*:1*yy* (figure 15.10). These genotypes produce a 3 yellow/1 green phenotypic ratio because *YY* and *Yy* will yield the same phenotype—yellow seed. This 3:1 ratio was approximated by all of Mendel's crosses (table 15.3). Of course, a Punnett square establishes the *theoretical* mathematical frequencies for each possibility, and the larger the number of crosses, the closer the observed ratio should approximate the expected ratio. Mendel's results are considered to be amazingly close to the expected, possible frequencies. Remember, Punnett squares *reflect ratios, not the actual number of offspring.*

Backcrosses

When an offspring is crossed with one of the two parental types that gave rise to it, a **backcross** has occurred. For instance, if one of the F_1 plants having a *Yy* genotype was backcrossed to the homozygous recessive parent with green seeds (*yy*) what would be the genotypic and phenotypic ratios produced? Figure 15.11 is a Punnett square setting up this backcross. The resulting progeny should have a 50:50 ratio of *Yy* and *yy* genotypes. Since all *Yy*s will be yellow and all *yy*s will be green, this will result in a 50:50, yellow/green phenotypic ratio. The backcross between an F_2 (*Yy*) with the homozygous dominant parent (*YY*) will yield a 50:50, *YY*:*Yy* genotypic ratio but 100% yellow seed. Backcrosses between one of the F_2 offspring and one of its parents (all *Yy* because they are F_1 plants) is also possible; however, there is a problem with visually interpreting the results because there are two possible genotypes, *YY* and *Yy*, for yellow seed. All that can be seen is yellow seeds; the genotype cannot be seen directly. In this situation there is a way to determine the genotype for a plant with a dominant phenotype (in this case yellow seed); a **testcross** is performed.

Testcrosses

When the phenotype for a given trait is the dominant expression of that trait, the genotype could be either homozygous dominant or heterozygous. For the yellow- versus green-seeded pea plants just discussed, yellow could be either *YY* or *Yy*. To determine which genotype a plant possesses, a testcross is performed. Since the yellow-seeded plant has the "unknown" genotype, a cross must be made with a plant with a known genotype. This makes possible offspring with different phenotypes, depending on which of the two dominant genotypes the unknown plant possesses. A plant with the homozygous recessive genotype, a green-seeded plant, is used for the known genotype.

Thus a testcross would be a green-seeded plant (*yy*) × a yellow-seeded plant (either *YY* or *Yy*) (figure 15.12). Establish two Punnett squares and work through the two possible crosses: *yy* × *YY* and *yy* × *Yy*. You should find that the first cross of homozygous recessive × homozygous dominant

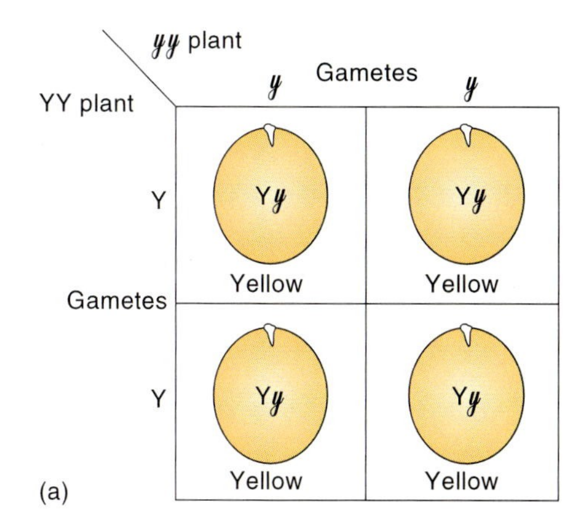

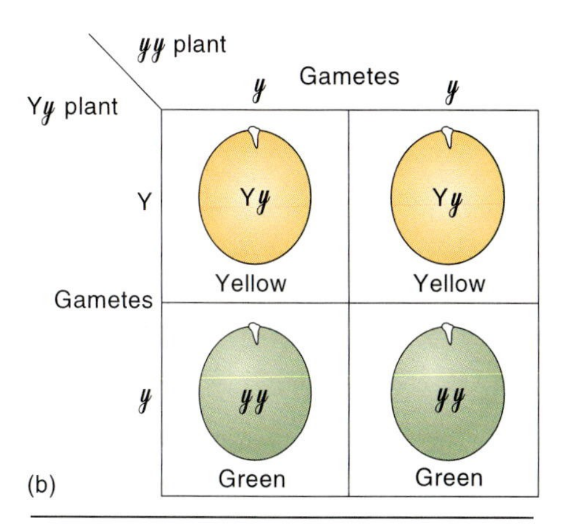

Figure 15.12

Testcross possibilities. (*a*) A *YY* × *yy* cross will result in 100% heterozygous (*Yy*), yellow offspring. (*b*) A *Yy* × *yy* cross will result in 50% heterozygous (*Yy*) and 50% homozygous recessive (*yy*) offspring and therefore a 50:50 ratio of yellow to green seeds, respectively.

would result in 100% yellow-seeded offspring. The second cross of homozygous recessive × heterozygous would result in a phenotypic ratio of 50:50, yellow- and green-seeded offspring. Depending on the results of the testcross, therefore, the genotype of the yellow-seeded plant can be determined. Such a cross between plants with dominant and recessive phenotypes will work for any trait that is controlled by complete dominance, as were the seven pea-plant traits studied by Mendel.

Incomplete Dominance

In all seven of the pea-plant traits studied by Mendel, one allele always showed complete dominance over the other allele for that trait. For other traits, however, this is not always the case; the expression of many traits is controlled by the alleles blending their contributions. Such genetic controls are said to be examples of **incomplete dominance.** In Mendel's pea plants, red flowers were completely dominant over white. A heterozygous genotype, therefore, would have a red flower. The genotype frequencies for incomplete dominance crosses are based on the same principle of segregation; only the phenotypic expression of heterozygous individuals will differ. In a plant with incomplete dominance for flower color, such as snapdragon, heterozygous plants would be pink. Two pink-flowered plants cross-pollinating would produce 25% red-, 50% pink-, and 25% white-flowered offspring. The 1:2:1 genotypic ratio of homozygous dominant/heterozygous/homozygous recessive is directly reflected by the phenotype in plants with incomplete dominance.

✔ Concept Checks

1. Define homozygous and heterozygous genotypes in a Mendelian complete dominance situation.
2. Explain how the principle of segregation protects a recessive trait from being lost to the gene pool.
3. What is a backcross and a testcross?

Dihybrid Inheritance

To this point we have discussed the inheritance of only one trait at a time, or monohybrid crosses. To consider only a single trait is convenient and serves to demonstrate the basic mechanisms of Mendelian genetics, but it is not a realistic approach. A **dihybrid cross** involves inheritance patterns for two traits considered simultaneously, a trihybrid cross involves three simultaneous traits, and so on. There are two different situations possible when more than one trait is being studied. The alleles controlling one trait can be on a different chromosome pair from the alleles controlling the other trait, or both allele pairs can be on the same chromosome pair. In the latter case the two traits are said to be **linked.** We discuss linkage later in this chapter, but first let us consider dihybrid inheritance of traits located on different chromosome pairs (unlinked traits).

If we were to select two of the morphological traits studied by Mendel, we would study their inheritance patterns together much as we did with a single trait. If a plant that is pure, or homozygous, for round yellow seed is crossed with a plant that has wrinkled green seed, we would have a cross similar to the single trait of seed color examined earlier in this chapter. Each parent plant would be homozygous—one homozygous dominant for both traits, the other homozygous recessive for both traits. In addition to the letters *Y* and *y* used earlier for seed color, the letters *R* for round and *r* for wrinkled will also be used. The homozygous dominant parent, then, would have a *YY/RR* genotype with the two capital *Y*s representing the two alleles for yellow and the two *R*s representing two alleles for round, *located on different pairs of homologous chromosomes.* The other parent will be homozygous recessive for each trait and would have a genotype of *yy/rr.* Homologous chromosomes line up and separate during meiosis I independently of other pairs. This is called the **principle of independent assortment** and is very important in the production of genetic variability.

The more homologous pairs there are, the greater the number of possible alignments. The formula for determining the possible independent alignments is 2^n, where n = the number of bivalents involved. Using the formula for figuring the possible combinations in a dihybrid cross, we can see that there are 2^2 or 4 possible combinations of alleles going to different gametes. Three bivalents (2n = 6) gives 2^3 or 8 possible alignments. The eight possibilities when using a second color for one each of the homologues are seen in figure 15.13. Even though each gamete has the same genetic composition for this cross, it is important that the Punnett square be set up using all possible combinations. A plant having 2n = 12 chromosomes would have 2^6 or 64 possible alignments. Humans, then, would have 2^{23} or 8,388,608 different possible configurations! Of course, only one of the total number of alignments that could occur actually happens in each parent cell undergoing meiosis.

As can be seen in figure 15.14, all the F_1 plants resulting from this cross are genotypically heterozygous for both traits and all have yellow and round seeds, since both yellow and round expressions are completely dominant over their recessive counterparts. To this point the dihybrid cross is essentially identical to a sample monohybrid cross, except with two traits. The next step, however—crossing F_1s among themselves—is a different situation. The principle of independent assortment applies in determination of the possible genetic combinations of this cross. The four genetically different gamete types that each parent can produce in equal numbers is depicted in figure 15.15. The genotypic combinations possible when these four gamete types fuse are also enumerated in this Punnett square. The resulting phenotypes are in a 9:3:3:1 ratio of yellow-round/yellow-wrinkled/green-round/green-wrinkled (figure 15.15). Again, these are *ratios of expected phenotypes;* the total of 16 does not represent the number of F_2 offspring produced. The actual data that Mendel obtained for this cross are presented in table 15.4.

As complicated as this dihybrid cross was, imagine such a cross with characters displaying incomplete dominance or a trihybrid cross (three nonlinked traits). The use of a Punnett square becomes a bit unwieldy beyond dihybrid crosses, and even trying to represent visually the possible phenotypes in an incomplete dominance dihybrid cross is more confusing than clarifying. The point is that when the study of inheritance includes the more realistic situations involving multiple traits or expressions other than complete dominance, the complexity of such a study is much greater.

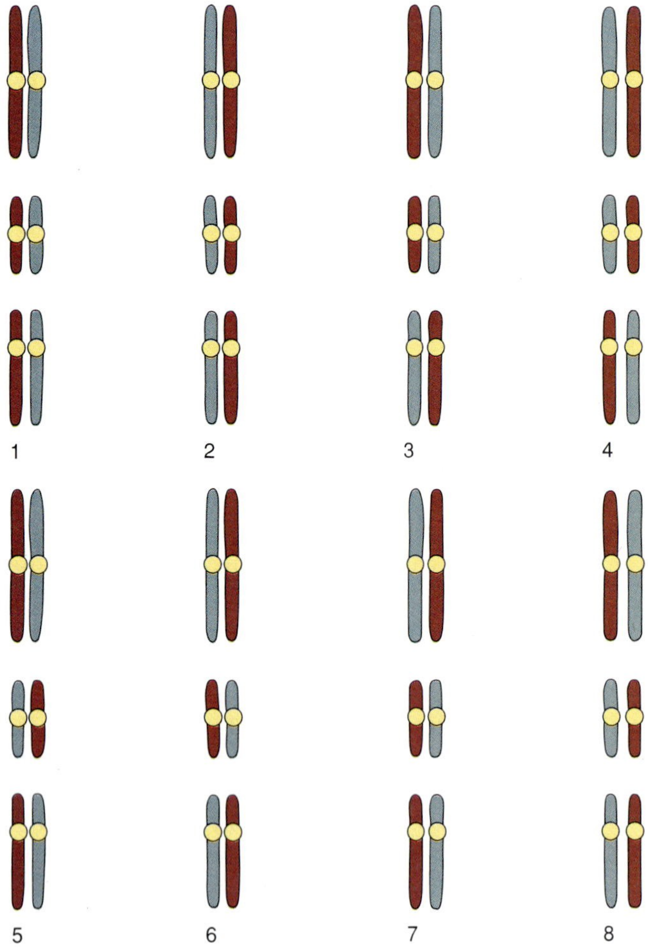

1 2 3 4

5 6 7 8

Figure 15.13

Random assortment of homologous chromosomes results from random alignment at metaphase I of meiosis. Note that for simplicity the chromosomes are represented diagrammatically here as a single unit, not as composed of two chromatids each—as they actually are. For three bivalents there are eight possible alignments.

Linkage

It should be obvious that many genes occur on a single chromosome, and therefore at telophase I they must segregate as a group. Such genes or alleles are said to be linked. Linkage serves to stabilize the genome, allowing coordinated genes to continue acting together. If all genes sorted independently, there would be a significantly greater chance for unsuccessful recombinations to occur, resulting in poorly adapted individuals. Thus linkage is important in the balance between genetic stability within a generation and the production of new recombinations and variability. Two linked traits will not produce the typical 9:3:3:1 ratio in the F_2 generation of a dihybrid cross (as previously discussed), because the genes are not subject to independent assortment during meiosis. Until the mechanism of crossing over was elucidated by a geneticist working on fruit flies (*Drosophila*), such varied crossing results were difficult to understand. Now it is easy to see how the linked traits can become "unlinked" by a crossover. The farther apart physically

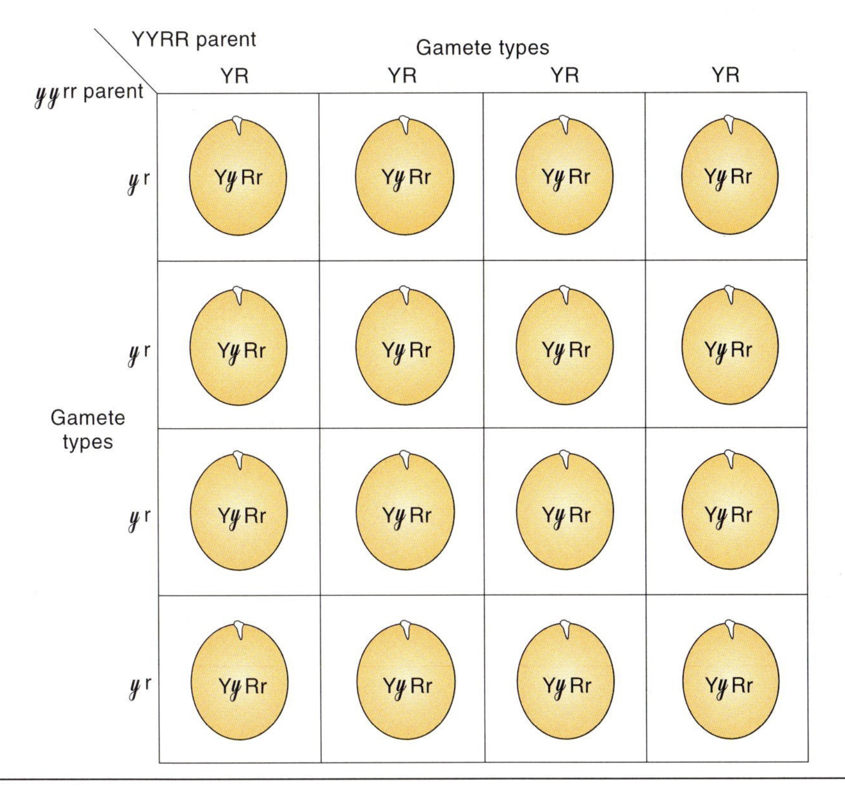

Figure 15.14

Dihybrid cross of two pure parents, one homozygous dominant for two traits (*YYRR*), the other homozygous recessive for those two traits (*yyrr*). All F_1s are heterozygous for both traits (*YyRr*), and all have the dominant phenotype of yellow and round seeds.

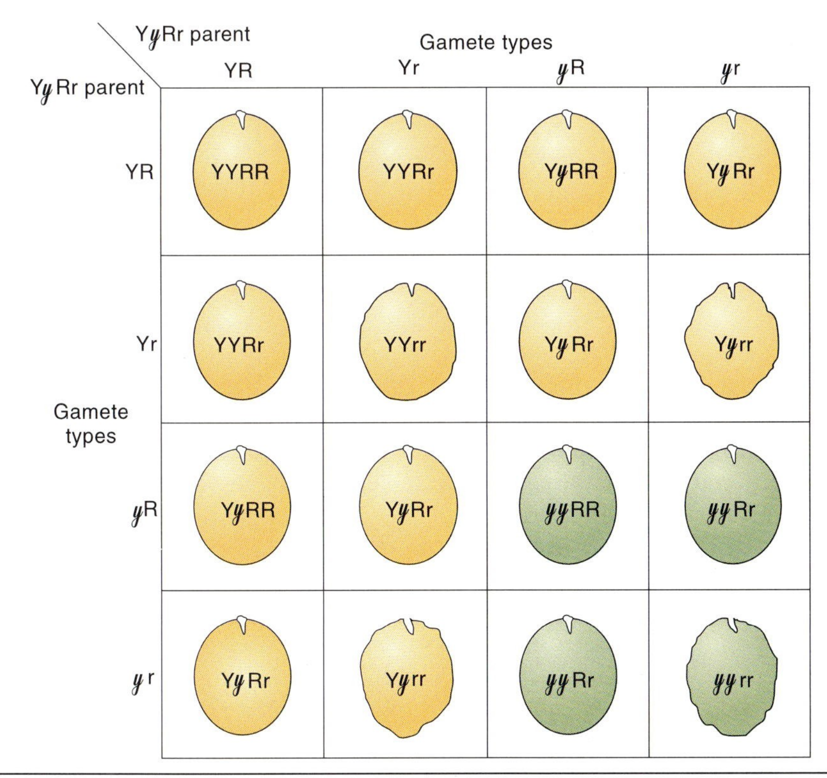

Figure 15.15

Dihybrid cross of two plants (F_1s from the cross in figure 15.14) that are heterozygous for both traits. The resulting 9:3:3:1 ratio in the F_2 generation is depicted visually with the four possible phenotypes in the corresponding squares of the Punnett square.

such genes are, the greater the likelihood that they will cross over; the closer together, the stronger the linkage. However, genes that are very far apart on a long chromosome arm may actually have a higher linkage frequency than others closer together, because two crossovers on the same arm can relink them.

Chromosome mapping is the process of assigning unit distances between genes based on their frequency of recombination due to crossovers. The assumption must be made that the probability of chromatid breakage is essentially equal for the entire length of the chromosome. If that is true, then the likelihood of a break between any two genes increases as the number of units between them increases. The frequency of crossing over between any two linked genes, therefore, is directly proportional to the distance they are apart; that is, crossover frequency is proportional to relative map unit distances. One unit on a chromosome map is the length in which a crossover is expected to occur 1% of the time. If two linked genes exhibit recombination in 15% of the offspring, they are said to be 15 map units apart. This is not an absolute measure that can be used to give actual distances between genes; it is only a relative measure of distance.

Gene Interactions

Up to this point inheritance has been discussed in terms of a single pair of alleles controlling one trait. For purposes of demonstrating the mechanism of genetic control and inheritance of traits, a one-gene/one-trait model is easy to understand. However, probably very few traits are exclusively controlled by a single allele pair. It is very likely that most genes do not act alone; rather, interactions of nonallelic pairs of genes control the expression of the phenotype. It is truly remarkable that Mendel was lucky enough to have chosen those seven traits in the garden pea that exhibit complete dominance, no

TABLE 15.4

Mendel's Dihybrid Cross F_2 Results

TRAITS	NUMBER OF PLANTS*
Yellow-round	315
Yellow-wrinkled	110
Green-round	108
Green-wrinkled	32

**315:110:108:32 is approximately a 9:3:3:1 ratio.*

linkage problems, and apparently single allelic control. Apparently they are single-gene traits, because it is very possible that other gene interactions also are involved in the control or expression of these traits, even though there is no overt incidence for such action.

Epistasis

One of the most common examples of gene interactions is one gene having a masking effect on the expression of another, nonallelic gene. It is worth noting that this is not a case of dominance, because epistatic genes influence nonallelic genes. Dominance is the effect of one allele over the other in an allele pair.

The actual effect of the epistatic genes often varies, but the general control of such genes is predictable within any genetic system studied.

Pleiotropy

Not only do genes interact to control the expression of a single trait, but many genes affect more than one trait. Single genes that can affect the expression of several traits are said to be pleiotropic. Because of the nature of all gene action—the production of proteins and resulting enzymes—it is probable that most genes have some influence on other traits during the development of the organism. Thus not only does more than one allele pair often influence the phenotypic expression of a single trait, but also single genes can be involved in the control of many different characteristics.

Multigenic Inheritance

Most of the genetic control of phenotypic characteristics discussed thus far have been discrete or discontinuous. There is no blending; rather, the phenotype represents one expression or the other. Flowers are red or white, seeds are yellow or green, and seed coats are round or wrinkled. Even considering the effects of incomplete dominance, epistasis, and pleiotropy as modifiers of genetic control, the possible phenotypes change only in degree or numbers of discrete possibilities. For a trait to vary continuously, there must be **multigenic** or **polygenic** control.

Multigenic control is one of combined additive effects of two or more allelic pairs of genes. Height, color, and shape of plants and plant parts are typical examples of traits with continuous variation of expression. Fruit or inflorescence size in many plants, and human skin, hair, and eye color also vary in a continuous fashion. The more allele pairs involved in the multiple gene control of a trait, the more continuous is the blending of possible phenotypes. Of course, it is unrealistic to assume that all allele pairs affecting the same trait additively do so with equal influence. Environmental changes and modifier genes also can come to bear on final phenotypic expression. In spite of how little is really known about multigenic inheritance, we can assume that many traits are controlled in this manner, and considerable research effort has been directed at better understanding the phenomenon of multiple gene control of heredity.

Molecular Genetics

Since Mendel's pioneering work, the revelation of DNA structure must be ranked as the most significant discovery in the area of genetics. Recently reported discoveries hold the promise of even greater significance. Since the mid-1970s the area of molecular genetics has seen incredible breakthroughs in understanding gene action. Researchers study inheritance at the nucleotide level. They have learned to cut, splice, and insert genes into organisms and then study the effects. Most of this **genetic engineering** has been with bacteria and viruses, but some molecular geneticists feel that it is only a matter of time before human hereditary diseases will be cured through the use of engineered genes (figure 15.16).

Thus far engineered genes have been used in bacteria to produce small amounts of insulin, interferon, human growth hormone, and a vaccine against a cattle disease called scours. The genes that have been painstakingly produced for these and other experimental uses are often termed **recombinant DNA.** Basically, any genetic material that is produced synthetically or in a different organism and then introduced into the test organism is called recombinant DNA. The two areas given greatest attention by molecular geneticists are the production of new medicines and the improvement of domesticated plants and animals.

Recombinant DNA-modified plants have been known since 1983, and by 1987 genetically engineered crop species were commercially available for field trials. By the end of 1992, over 40 species of DNA-modified food and fiber crops had been described, and several hundreds of field trials are currently ongoing in over 20 countries around the world. It has been estimated that the world market for genetic engineering in agriculture alone could easily be $50 billion to $100 billion by the year 2000. In Canada, researchers have genetically engineered oilseed rape (*Brassica napus* in the mustard family), by inserting a gene that chelates metals. The goal is to grow such plants

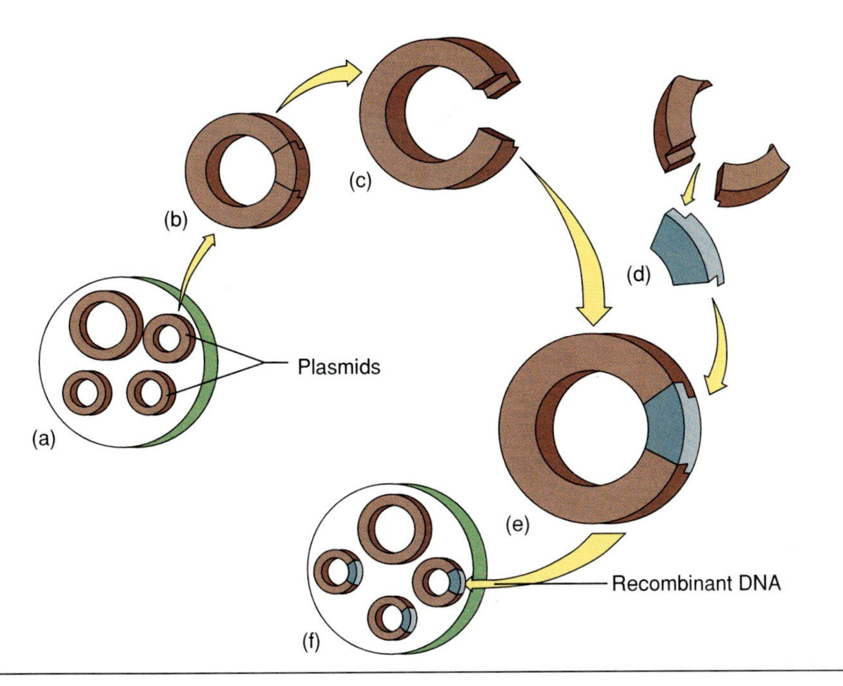

Figure 15.16

Gene splicing. A bacterium containing small rings of genetic material called plasmids (*a*) is broken open; this allows isolation of the plasmids (*b*). The plasmids are treated with a restriction enzyme, which cuts the DNA at specific sites, causing the plasmid to open (*c*). With the same restriction enzyme in use, a specific gene is snipped out of the DNA of a different organism, such as a virus (*d*). The second gene is inserted into the plasmid, where it fits exactly (*e*). This new recombinant DNA is then reinserted into the bacterium (*f*), where it is replicated when the bacterium divides.

in mine spoil areas to trap metals and clean up polluted soils and water that drains through such areas. In addition, once trapped, these metals can be disposed of safely or possibly even reclaimed by burning the plant. Gold and platinum, for example, might be "mined" from these genetically transformed plants.

Another commercially valuable example involves the first genetic transformation of a grape variety. Using a bacterium as the delivery system, researchers have successfully inserted into the chardonnay grape cultivar the gene for a protein in the coat of the Grapevine Fanleaf Virus, a lethal viral pest that is a threat to infect about 10% of the world's vinyards. In order to infect the plant's genome, the virus has to shed its protein coat; the genetically engineered grape plants produce the protein, keeping the virus "dressed" with its protein coat—thus rendering it noninfective. So far, the chardonnay cultivar is the only grape varietal able to maintain this genetic change from generation to generation.

Although the potential for introducing new genes into food crops and livestock to enhance their productivity is great, these applications will probably be slower to develop than in the area of new medicines. Pharmaceutical companies are investing heavily in this area of science and have already dominated the workforce of scientists trained in genetic engineering.

The "Gene Machine"

The production of a synthesized gene once took four to eight months of tedious biochemical manipulations. Since the advent of **gene synthesizers** in 1982 a functional gene segment of one's choosing can be automatically produced in a single day. By simply typing in the desired genetic sequence on the gene machine's keyboard, geneticists can splice the appropriate nucleotides automatically, and a synthetic gene fragment is ready to be introduced into the DNA of an experimental organism. Existing sequences can be duplicated or modified, or researchers can even design totally new sequences.

By using this jointly with a protein sequencer, a machine that can read out the exact sequence of amino acids in a protein, researchers can now carry out sophisticated experiments rapidly. Even tiny amounts of a useful enzyme (protein) can be analyzed by the protein sequencer in only a few hours. Which amino acids are in the protein, and in what order, can next be programmed into the gene synthesizer. This machine will produce a genetic fragment to be inserted into bacteria to direct the production of large quantities of the desired protein.

Many medically and industrially important substances are proteins. The availability of this new technology will allow researchers to work much more efficiently and carefully toward ethically acceptable uses for recombinant DNA in genetic engineering applications. Most of the past gene splicing has consisted of taking existing genes out of organisms capable of a desired function, such as drought tolerance, and inserting those genes, or copies of them, into the organism in which such capabilities would be desirable. Now it is possible to try modifications of existing genes and even completely new genes in such efforts.

There is no doubt that few fields of biology have ever moved forward as quickly or generated as much interest as the area of molecular biology. Especially intriguing are the areas of genetic control and the potential for manipulating that control.

✔ *Concept Checks*

1. What is incomplete dominance?
2. How does the principle of independent assortment in a dihybrid cross ensure genetic variability?
3. What are linked genes?

The Germplasm Market

Are the plants of one country the property of that country, like oil or metal ores, or are they common property of the global community? More specifically, is the wild germplasm of a major food-crop species an asset that the country possessing it can market to the rest of the world, like gold ore in its unrefined state? This question revolves around the fact that approximately six areas of the world, containing nearly all of the underdeveloped nations, are also the source for the genetic base for over 95% of the world's total food-crop production. Although the United States is one of the world's leading agricultural producers, we can claim no major food-crop species as native. Our crop scientists and plant breeders have developed from wild genomes highly productive, elite hybrid crop varieties—which are often unaffordable to the less developed countries from which the raw genetic material originally came.

Despite many advances in biotechnology, such as the gene machine and protein sequencers, natural genetic diversity is impossible to replicate; crop scientists must rely on the irreplaceable sources of natural genetic variation, the wild plants found mostly in tropical, underdeveloped countries. These "wild genes" are not only the potential source for the development of new hybrid crops. They are the backup for the highly refined super crops that are potentially susceptible to a "monoculture crash" due to a disease or new pest for which their genetic protection has been removed by the very breeding process that created their high-yield productivity. But who controls these wild genes? If they are the common domain of all countries then shouldn't the hybrid crops developed from them also be the common property of all countries? Not so, claim the seed companies who spend hundreds of thousands of dollars and many years to develop a new "elite" crop variety. Why is one resource free while a derivative of that resource is unaffordable to the country from which the "raw material" was originally "mined"?

This debate rages through the United Nations' Food and Agriculture Organization (FAO). Seed banking is becoming standard practice. The Consultative Group on International Agricultural Research (CGIAR) supports over a dozen centers scattered throughout the world that provide a seed-bank network for exchanging and developing agricultural information worldwide. But the ultimate question remains unanswered—who owns the natural resource of raw genetic potential found in the wild plants of the world, and as we lose them, does the "price" of those species that are protected increase? If you discovered a new species found only on your land and that species was found to contain a gene that produces a compound to cure cancer, could you patent that gene? If the gene increased crop productivity to feed millions of people, could you sell the rights to it? Food for thought!

Summary

1. Sexual reproduction ensures variability among offspring. The process of meiosis is the key to producing that variability. Meiosis produces haploid cells from a diploid cell. The fusion (fertilization) of two haploid cells produces a new, single-celled zygote with a unique combination of genes. In angiosperms, meiosis occurs in the anthers and ovules of the flower.
2. Meiosis is two sequential nuclear divisions: the first is a reduction division producing nuclei with only a single set of chromosomes (haploid); the second is like mitosis, further producing genetically different cells. The sequence of nuclear and cellular events is similar to that of mitosis in both meiosis I and II except that in the former, homologous chromosomes pair up and then separate.
3. Crossing over, the formation of chiasmata, results in additional genetic recombinations, thus increasing variability. Mutations may also result in genetic variability.
4. All plants have an alternation of generations between a haploid spore-producing (sporophyte) stage and a gamete-producing (gametophyte) stage. In angiosperms, both stages occur within the tissues of the flower. In the anthers, microsporogenesis is followed by microgametogenesis. In the ovules, megasporogenesis and then megagametogenesis occur. The resulting gametes fuse in a unique double fertilization to produce a new sporophyte stage in the form of a diploid zygote that will develop into the embryo of the seed, complete with a triploid nutritive tissue.
5. All sexually reproducing, diploid organisms have two genetic expressions (alleles) for each trait. The study of inheritance patterns in peas by Gregor Mendel helped elucidate the ideas of gene control of phenotypic expression in offspring. Traits with complete dominance can be homozygous dominant, homozygous recessive, or heterozygous dominant genotypes. Experimental crosses between pure parents producing F_1s, and then between F_1 plants producing F_2 offspring, can demonstrate the patterns of gene control.
6. Backcrosses between offspring and parent plants, especially a testcross between a dominant plant and a homozygous recessive plant, enable researchers to study gene control. Some traits have incomplete dominance, allowing for a blending of genetic control and intermediate phenotypic traits.
7. Dihybrid crosses involve two traits. When the alleles controlling those traits are on the same chromosome, they are said to be linked. Other gene interactions include the modifying effects of epistasis and pleiotropy and the complex control of multigene influences.
8. Molecular genetics is an area of science that studies the modification of genetic control. Genetic engineering has allowed the artificial production of recombinant DNA. Technological advancements in the form of gene synthesizers (gene machines) and protein sequencers are enabling scientists to study gene function much more rapidly than ever before.

Key Terms

alleles 261
alternation of generations 265
anaphase I 263
asexual reproduction 261
backcross 270
bivalent 263
chiasma 265
crossing over 264
dihybrid cross 271
diploid 261
dominant 269
double fertilization 268
endosperm 268
fertilization 261
gametophyte 265
gene synthesizers 275
genetic engineering 274
genetic variability 261
genotype 268
haploid 261
heterozygous 270
homologues 261
homozygous 270
incomplete dominance 271
integuments 267
interphase 263
linked genes 271
megagametophyte 267
meiosis 261
metaphase I 263
micropyle 267
multigenic 274
mutations 265
phenotype 268
polygenic 274
principle of independent assortment 271
principle of segregation 269
prophase I 263
recessive 269
recombinant DNA 274
reduction division 262
sexual reproduction 261
somatic 261
sporophyte generation 265
synapsis 263
telophase I 264
testcross 270
zygote 263

Discussion Questions

1. What would be the long-term advantages and disadvantages of being able to clone (reproduce asexually) any living organism?
2. What would be the consequences if we could "see" an organism's genotype as easily as we can the phenotype?
3. Why is it important for recessive genes to be maintained through Mendel's principle of segregation?
4. What is the significance of genetic engineering especially in relation to introducing modifier genes?

Suggested Readings

Adams, R. P., ed. 1994. *Conservation of plant genes II: Utilization of ancient and modern DNA.* Missouri Botanical Garden, St. Louis.

Bowler, P. J. 1989. *The Mendelian revolution: The emergence of hereditarian concepts in modern science and society.* Johns Hopkins University Press, Baltimore.

Committee on Managing Global Genetic Resources: Agricultural Imperatives. 1991. The U.S. National Plant Germplasm System. National Academy Press, Washington, D.C.

Corcos, A. F., and F. V. Monaghan. 1992. *Gregor Mendel's experiments on plant hybrids: A guided study.* Rutgers University Press, New Brunswick.

Hawkes, J. G. 1991. *Genetic conservation of world crop plants.* Academic Press, San Diego.

Holden, J., J. Peacock, and T. Williams. 1993. *Genes, crops, and the environment.* Cambridge University Press, New York.

John, B. 1990. *Meiosis.* Cambridge University Press, New York.

Juma, C. 1989. *The gene hunters: Biotechnology and the scramble for seeds.* Princeton University Press, Princeton.

Mendel, G. 1963 (1865). *Experiments in plant hybridisation.* Harvard University Press, Cambridge.

Paabo, S. 1993. Ancient DNA. *Sci. Am.* 269(5):86–92. November 1993.

Rothwell, N. V. 1993. *Understanding genetics: A molecular approach.* Wiley-Liss, New York.

Tice, R. R., and A. Hollaender, eds. *Sister chromatid exchanges: 25 years of experimental research.* Plenum Press, New York.

Chapter Sixteen

Evolution and Taxonomy

Using the information in the preceding chapter as a base for understanding the sources of genotypic and phenotypic variation, we can explore the concepts of organismal and populational change. The recognition of natural genetic units and the mechanisms involved in changes in the composition of those units are central to understanding the concepts of evolution and taxonomy. For many groups of organisms, their biogeography and how continental drift may have affected their distribution are also important.

These small wild orchids belong to one of the largest and most highly evolved plant families.

The concepts of natural selection, adaptation, and evolutionary change have been introduced in earlier chapters. It would be very difficult, if not totally impossible, to discuss ecological balance, development of characters, structure, function, genetics, or almost any other subject in biology without including natural selection, adaptation, and evolution. However, prior to 1859, when Charles Darwin published *On the Origin of Species,* the generally accepted view was that the biological world was static, or unchanging. The earth itself was thought to be only about 6000 years old; all organisms were thought to have been created at the same time and not to have changed since then.

VARIABILITY AND NATURAL SELECTION

A number of important discoveries in the physical sciences and mathematics were made by the ancient Egyptians, Babylonians, and Greeks. Aristotle, along with a few of his students, produced classic studies on plants and animals during the period of 400–300 B.C. Religious and philosophical dogma, however, inhibited much original thought and questioning in the biological sciences until the middle of the nineteenth century. Then, in addition to Darwin's theories, the cell theory was developed, and Pasteur's work disproving spontaneous generation of life helped awaken biological researchers. The modern era of research in biology, however, arose primarily in direct response to Charles Darwin's "theory of evolution by natural selection." No single event had ever caused such an explosion of biological observation, experimentation, and critical thought; nor has anything since been so instrumental in producing such a reaction. And yet, Darwin gave no early indication that he would have an intellectual impact on even a small segment of the world.

Charles Darwin

Born in 1809 to a wealthy English family, Darwin was a sickly, spoiled, and unmotivated youth whose early academic career was undistinguished. After Darwin graduated from high school, his father sent him to Edinburgh University to study medicine. Because of his family's wealth, Charles saw no need to make any effort to do well, but he did find some areas of the natural sciences interesting. A year of medical school proved to be as much as he could endure, and in 1826, at the age of 17, Darwin entered Cambridge University to study theology. His three years at Cambridge were no better academically than any of his other educational experiences, but while there he did become close friends with a botany professor named John Henslow. Through Henslow, Darwin was invited to be the ship's naturalist for the H.M.S. *Beagle.* In 1831, at the age of 22, Charles Darwin embarked on a 5-year voyage around the world. His observations on that voyage provided him with many of the ideas that ultimately resulted in his theory of evolution by natural selection. These ideas were primarily developed over a 20-year period between 1840, when he published *Zoology of the Voyage of the Beagle,* and 1859, when *On the Origin of Species* was printed. Even his book stating his natural history observations on the *Beagle* voyage included comments that revealed a preliminary questioning of the theory of special creation.

Biological Variation on the Galápagos Islands

The Galápagos Islands are found near the equator some 950 km west of the coast of Ecuador (figure 16.1). As the ship's naturalist, Charles Darwin and the other passengers on the *Beagle* spent five weeks visiting these islands, and many of the observations made by Darwin while he was there were central to his theories. The key to his concepts is the phenomenon of **variation**—genetically controlled differences in individual or populational phenotypes. Darwin was impressed by two aspects of the variations he saw on the islands. First, although many of the plants and animals of the Galápagos were similar to those he had seen on the South American mainland,

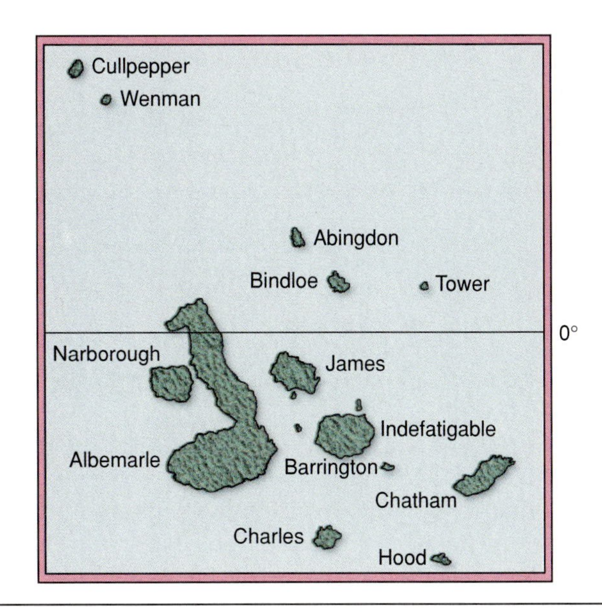

Figure 16.1
The Galápagos Islands are located approximately 950 km west of the coast of Ecuador.

they were also distinctly different. Second, and more important, each island contained clearly different forms of these organisms.

The giant tortoises, for which the islands were named, were clearly different from one island to the next. Even sailors who visited the islands to take on freshwater, fruit, and live tortoises (to be used later for fresh meat) were able to name the island from which each tortoise came.

The most common large tree on the islands was in the genus *Scalesia* of the sunflower family. In other parts of the world this genus is always represented by herbaceous or small shrubby species. Although each island contained a form that was morphologically distinct from those on the other islands, they were obviously closely related to each other, since they were the only trees in the genus. Darwin also noted races of finches, lizards, and other plants and animals with different morphologies on each island.

It is important to point out that within any given island there is variability among individuals of each species. This variability is within the range typical for all sexually reproducing organisms, but it does not approach the variation present between islands. In other words, although morphological differences do exist between individual tortoises on any given island, all these individuals are more similar to each other than to the tortoises on any other island (figure 16.2).

These observations raised several questions in Darwin's mind. If groups of organisms were all specially created independently, then why did the tortoises of the Galápagos Islands more closely resemble the tortoises of South America than those of some other more distant part of the world? Why were the tortoises not all specially created identically and unchangeably at the same time on all the islands?

Darwin's Theory

These and other observations made by Darwin on the voyage of the *Beagle* formed the basis for his theory of evolution by natural selection. The theory has two central components. First, Darwin could not accept the idea that living organisms are "specially created" or that they are unchanging, appearing today exactly as their ancestors have always appeared. He proposed, rather, that recent organisms are a product of change, having gradually descended from ancestral types that once looked quite different. Darwin theorized that the tortoises, sunflower trees, finches, lizards, and other unique plants and animals of the Galápagos Islands were similar to forms on the mainland of South America because they were descended from those organisms. The differences between each island group resulted from the inhabitants of each island having been isolated from each other and changing slowly and separately in response to the conditions on their own island. Today, as a result, they are quite distinct in appearance even though obviously more similar to each other than to their mainland relatives.

Figure 16.2
The giant tortoises that inhabit the Galápagos Islands have recognizably different appearances on each island.

The second part of Darwin's theory took years to develop. Although he recognized that changes had occurred, the mechanism that brought about those changes was not immediately evident to him. It was not until much later that he conceived of the process of **natural selection.** A combination of (1) his observations of nature, (2) his interest in plant and animal domestication and breeding, and (3) a work written by Thomas Malthus led to the development of his theory of natural selection. Malthus's work, *An Essay on the Principle of Population,* warned of possible future human population explosion and the dangers of an unnaturally large population size which would result in competition for limited food resources.

Darwin noticed that the natural habitats of closely related, but distinct, individuals were usually slightly different. Food and water availability, space, nesting sites, and other organisms competing for the same essentials were habitat differences Darwin noted. He was also aware of the method by which plant breeders produced new domesticated forms. The variability among wild or undomesticated organisms makes available to the breeder certain individuals with larger fruit or a more colorful flower than others of the same type. Whatever character a breeder selects to enhance, only those individuals having that trait are used to produce the next generation. For example, if increased flower size is the desired trait, only the plants having large flowers are cross-pollinated. The resulting offspring would normally display a range of flower sizes, but the average would be larger than in the previous generation. From this generation, again, those individuals having the largest flowers are selected for breeding; the others are not used. This agricultural selection demonstrated to Darwin how changes could be produced; but he also realized that natural systems would not work the same way as the **artificial selection** practiced in the domestication of plants and animals.

It was the Malthus paper on population size that provided Darwin with the basis of how selection works in natural populations. Plants and animals maintain reasonably stable population sizes in nature. For the energy flow to remain in balance, the different trophic levels must maintain proper proportion. This means that each type of organism has certain natural limits to its population size. For example, if a rabbit population has 100 individuals composed of 50 mating pairs, for the population of 100 to remain stable, each pair would need to exactly replace themselves. If there were any fewer than an average of two offspring per pair, the population would decline; with any more than two offspring per pair, the population would increase. In reality, essentially all wild organisms produce more than an average of two offspring per pair.

But not all of the progeny produced by any given species can survive. If each of the many offspring produced had an equal capability of surviving, there would be little impetus for change. In artificial selection, plant and animal breeders decide which trait is to be selected and thus what changes will occur. In nature, some individuals are better adapted to specific environmental situations. Some are genetically defective and unable to survive; others are "normal" but differ in size, coloration, growth rate, and other phenotypic characters. A commonly used brief definition for natural selection is "the survival of the fittest." Darwin did not define his theory in this way, and in fact it is a poor definition because it requires too much interpretation. Strictly speaking, the survival of the "fittest" *implies* the survival of only one individual; it *means* the survival of the more fit phenotypes. More importantly, this definition refers only to survival. A more accurate and only slightly longer definition is "the survival and failure to survive of individuals competing for similar resources." In addition, population geneticists use the term *fit* to refer to reproductive success. Those individuals producing the greatest number of viable offspring are said to be the most fit. Of course, only those individuals that are well adapted have the opportunity to produce offspring; therefore adaptation, survival, and reproductive fitness are all important components of natural selection.

Since Darwin's original work, thousands of scientists and nonscientists have worked to understand the theory of evolution by natural selection. Although still technically a theory, there are enough natural examples, experimental results, and artificial simulations of the selection process to consider it just that—*a process*. We now have a much better understanding of how natural selection works. Following are the basic criteria that must be met for natural selection to occur:

1. *Individuals vary;* no two individuals are exactly alike, and this *variability is heritable.*
2. *More individuals are produced each generation than can possibly survive.*
3. *Competition:* some individuals have characteristics that give them an advantage over other individuals in the struggle to survive. There is a struggle because there is a *limit to resources.*
4. Natural selection acts on the *individual phenotype,* and thus selects individuals which, in turn, change the populational gene pool.
5. Individuals that are selected must be able to produce *viable, fertile offspring* to pass on the successful genes within the population.
6. Long periods of *time* must be available, since some changes occur slowly.

It is this last requirement that needed the support of the geologists of Darwin's time. If the earth was only slightly older than 6000 years, as proponents of the special creation theory believed, natural selection would not have sufficient time to produce much change. Geologists had evidence then that the earth was more than 2 billion years old, and since then the accumulation of geological, fossil, and radioisotope dating evidence has established the age of the earth as between 4.5 and 5 billion years. The comprehension of such a time span is difficult, as it is for any concept of great magnitude, such as how many stars are in the universe. Without at least some awareness of the geological time scale, however, the significance of how much evolutionary change has been produced by natural selection since life on earth began is not appreciable.

Lamarckian Evolution

Charles Darwin was not the first scientist to propose a theory of evolution; Jean Baptiste Lamarck (1744–1829) proposed that all existing species are descended from other species. The core of his theory was the inheritance of acquired characteristics. According to Lamarck's theory, those traits acquired by an individual could be passed on to its offspring. For example, today's long-necked giraffe evolved from ancestors who increased their neck length by stretching up to feed on higher tree branches (figure 16.3). The increased neck length was passed on to the next generation, where it was increased more by further stretching, and so on.

Darwin's theory would propose that ancestral giraffes had a variety of neck lengths. The long-necked individuals survived a period of food shortage because they were able to reach the upper branches. The short-necked giraffes did not survive this selective pressure, and thus the genes they carried for shorter neck lengths were removed from the breeding population. All surviving giraffes were better adapted to a period of shortage, and their genes were maintained in future populations.

We know now that individuals do not inherit acquired characteristics. A weightlifter will not produce offspring with increased strength; they too will have to develop increased strength through the same process as their parent.

Figure 16.3

Lamarckian evolution by inheritance of acquired characteristics proposed that the giraffe gradually developed longer necks from one generation to the next by stretching to reach higher branches. In the successive generations represented in (*a*) to (*c*), the giraffes have longer and longer necks according to the Lamarckian theory, and this trait is passed on to the offspring. Darwinian evolution holds that the natural morphological variability within a population, here represented by different neck lengths (*d*), allows for some individuals to be better adapted in times of shortages. (*e*) The giraffe with the shorter neck is unable to reach the upper branches and will perish. (*f*) In subsequent generations, only the genes for longer neck length have been passed on.

Alfred Russell Wallace

An often overlooked name in the world of evolutionary theory is that of another British naturalist, Alfred Russell Wallace. About 20 years after his return from the 5-year voyage on the *Beagle,* Darwin still had not completed the task of putting his ideas on paper. A methodical and careful man, Darwin continued accumulating data to support the theory that had actually been formulated within the first 6 years after his return from the voyage. In 1858 Wallace sent Darwin a short manuscript that concisely paralleled Darwin's theory point for point. Being an honorable person, Darwin turned the paper over to Charles Lyell, a highly respected geologist, and Joseph Hooker, an esteemed botanist. It was resolved that Darwin and Wallace would jointly present their writings to the Linnaean Society. This was decided because a number of his colleagues knew that Darwin had developed his ideas many years before Wallace had done so (Wallace himself even acknowledged this), because Darwin's name preceded Wallace's at the Linnaean Society meetings, and because Darwin published his 490-page *On the Origin of Species* only 16 months after those meetings. Charles Darwin was a household name, whereas Wallace was little known outside of evolutionary biology disciplines.

Rates of Evolutionary Change

Because of the impact that Darwin's theory of evolution had on both the scientific and religious communities, it has been constantly debated, scrutinized, tested, and reevaluated. The most rigorous and challenging critics have been other evolutionary biologists, and this is as it should be. The objective and rational approach taken by scientists in their challenges and criticisms of existing theories is the very basis on which scientific investigation is built; they are doing nothing that Darwin would not do were he alive today. It is amazing that after over 100 years of constant research and evaluation the theory of evolution is more firmly established than ever before.

Not surprisingly, however, some of the data continue to be subject to interpretation in relation to the *specific components* of evolutionary theory. There is no question in the minds of these scientists that evolution occurs; there are some questions concerning how fast it occurs. Darwin felt that most evolutionary change was slow and gradual, responding to the forces of natural selection over long periods of time. Some of today's evolutionary biologists think that most major changes are relatively sudden,

(a)

(b)

Figure 16.4
Archaeopteryx, one of the most significant fossils ever found. (*a*) Although birdlike in many skeletal features and with feather imprints in one especially well-preserved specimen, *Archaeopteryx* also has the reptilian features of a long bony tail and clawed fingers, making it a true "missing link" between two major groups of animals. (*b*) Photo of *Archaeopteryx* fossil imprint.

followed by long periods without change. The basis for this interpretation, called **punctuated equilibria,** is the animal fossil record. Paleontologists have known for decades that there are huge gaps in the fossil records of closely related species, and that in spite of a few discoveries of missing links there is little fossil evidence supporting slow, gradual "Darwinian" evolution. Staunch supporters of **gradualism,** as it has been called, point out that such an incredibly small percentage of past organisms are represented in the fossil record that no conclusions could be based on it, in either direction.

One of the most significant missing link discoveries was that of *Archaeopteryx,* a Jurassic fossil thought to be a small dinosaur until an exceptionally well-preserved skeleton was found with the imprint of feathers around it (figure 16.4). This find has clearly pointed to a connection between dinosaurs and modern birds. Even with such discoveries, the preponderance of unfilled gaps indicated that at least some sudden evolutionary events occurred over a few generations, and probably in isolated areas, such that the likelihood of finding intermediate fossil forms would be essentially nil. An earlier term for rare, sudden change was **quantum evolution.**

The principal component of the punctuated equilibrium argument is not rapid evolutionary change, but the long periods of **stasis,** or no change. Proponents of this theory point to the fossil remains which show that for millions of years a particular species stayed the same. What little variation is present is relegated to the normal range for any species, living or extinct. Other scientists contend that fossil evidence for stasis is misleading. Fossilized bones or other hard parts such as teeth and shells may remain essentially unchanged for long periods of time while considerable evolutionary change is taking place in the soft parts of these animals.

Still other evolutionists take the stand that, because there are so few environments in which animal remains can be preserved, it is unlikely that intermediate forms will ever be found. The gradual changes that occur do so in direct response to a changing environment, very probably in a restricted area. Such change is not likely to be recorded in the fossil record until the successful new species expands its range, increasing in total numbers and into habitats in which some individuals might become fossilized.

And then there is the perception of what is "rapid" and what is "slow." These terms are relative to how many years and how many generations are involved in producing the evolutionary changes observed in the fossil record. Some evolutionary biologists believe that under strong natural-selection pressures a new species could evolve in as few as 50 to 100 generations. Although this would be considered rapid evolution, it still occurs in response to the same broad criteria that Darwin outlined in his theory.

The paleontologists most supportive of the punctuated equilibrium theory consider 100,000 years a brief period when compared with the millions of years of stasis that follow this "rapid" evolutionary event. Depending on the species in question, 100,000 years is a long time, allowing for gradual evolutionary changes. Even in species that are considered long-lived, this period of time would allow for several thousand generations. A great deal of evolutionary change can occur in that many generations if certain environmental pressures exist.

Although there is no final answer to these debates (there have never been any absolutes concerning evolutionary change), the academic exchange of ideas continues. The general framework of Darwin's theory of evolution by natural selection has been repeatedly tested, and it has held up. Specific questions concerning rates of evolution are unresolved, but most evolutionary biologists feel that both gradual change and more rapid events have taken place. The new areas of evolutionary research will continue to feed data into these theories, and as new discoveries expand understanding, the existing theories will be modified to accommodate this knowledge.

EVOLUTION

In general, with variation among individuals acting as the raw material, the process of natural selection is now almost universally accepted by scientists and nonscientists alike. As with any other area of science, however, constant questioning, testing, challenging, observing, and objective evaluating must continue to further refine and understand its impact on the biological world.

There are very few "laws" of nature that can be repeatedly tested and supported. The law of gravity and the laws of thermodynamics have achieved such status, but most natural processes are theories with varying amounts of supporting evidence. **Evolution** has been one of the most extensively studied of such theories since Darwin first proposed his ideas of change by means of natural selection. Because long periods of time are thought to be required for a new group (species) to evolve, the physical process of evolution cannot be observed and substantiated.

To "prove" evolution would require a documented study of natural selection producing change over a long enough period of time that the resulting organisms could be said to be a new species distinct from the original forms. It is this gap between knowing that the process works and demonstrating that it works that keeps evolution a theory. In spite of this lack of absolute documentation, however, scientists have accumulated so much evidence supporting the theory of evolution that biological work is much clearer when studied from an evolutionary standpoint.

What Is a Species?

Since evolutionary biologists place a great deal of emphasis on the existence of individual species of organisms, the concept of **species** is important. It is difficult to formulate a single, comprehensive definition for all living organisms. Generally, a species can be defined as *a group of similar individuals that are capable of successfully reproducing among themselves.* Many biologists would add that, to be a distinct species, individuals of one species are not able to interbreed successfully with individuals of a different species. This **reproductive isolation** is the key to the **biological species concept,** as it is commonly known.

Although this concept works well in separating most animal groups (especially the higher animals), it is not as consistent in the plant kingdom because plants have a much higher frequency of natural hybridization between plant species than occurs between animal species. **Hybridization** is the production of offspring by members of different species. The use of the term *hybrid* in domesticated plant and animal breeding is different from its use in nature. A new "hybrid" rose cultivar (cultivated strain or variety) is not usually the offspring of two different species, but rather a new phenotype of a cultivated species. A poodle and a cocker spaniel, therefore, do not produce a true "hybrid" as used in this context. One of the few examples of *true* hybridization among animals is the production of a mule by a male donkey and a female horse. Since mules are reproductively sterile, horses and donkeys are technically reproductively isolated and will remain distinctly different species.

A much larger number of plant species hybridize in nature, making strict acceptance of the biological species concept difficult. The ability for some plant species to combine genetic information in hybrid offspring produces new and often more adaptive combinations of morphological and physiological traits. It is also this propensity for hybridization that has stimulated plant breeders to produce new hybrid crop strains artificially. Many examples of wild species are known to hybridize, including two California columbines, *Aquilegia formosa* and *A. pubescens,* several species of *Baptesia* in the coastal area of Texas, members of the genus *Iris* in Louisiana, and three species of *Tragopogon* in the western United States.

Examples of cultivated (artificial) hybrids are almost as numerous as the number of plants in domestication: corn, wheat, oats, tomatoes, squash, roses, tobacco, and many more. One of the most interesting hybrid grains is actually a result of crossing members of two different genera. *Triticale* results from hybridizing *Triticum*

✔ *Concept Checks*

1. What questions were raised in Charles Darwin's mind by the variation he observed on the Galápagos Islands?
2. What were the keys to Darwin developing the process of natural selection?
3. What are the basic criteria that must exist for the process of natural selection to work?
4. How do the rates of evolution differ in gradualism vs. punctuated equilibria?

(wheat) and *Secale* (rye). This new cereal has great promise for areas that are drier and colder than optimum for either of its parents. Although first produced in the United States, thus far there has been little produced as a commercial crop. The Soviet Union, on the other hand, has initiated extensive *Triticale* farming.

Thus, although reproductive isolation may be a useful guideline in distinguishing between many plant species, it is not absolute in separating others because of the existence of natural hybridization often accompanied by some low percentage of fertility in the hybrid offspring. Species that produce hybrid offspring are usually thought to be genetically similar. Recognizing one group of organisms as one species and another group as a different species simply indicates genetic differences at a level sufficient such that all individuals of one group are more similar genetically to each other than they are to individuals of the other group.

Individuals, and thus groups of individuals, are constantly changing through the action of natural selection, and this change is usually thought to be a slow process. Therefore, genetic divergence over time could easily be visualized in some instances as not yet having occurred to a sufficient degree to have two recognizably distinct genetic units (species). Other groups, conversely, may have diverged enough to have two easily recognizable and genetically distinct species. The grouping of similar individuals into their own species, then, is an attempt to indicate genetic separation from other closely related species.

Speciation and Speciating Mechanisms

Speciation is the process of evolutionary change producing a group of individuals that have genetically diverged sufficiently to be recognized as a new species (figure 16.5). For a group of individuals to diverge from others in their species, there must be an accumulation of morphological, anatomical, and physiological features that are

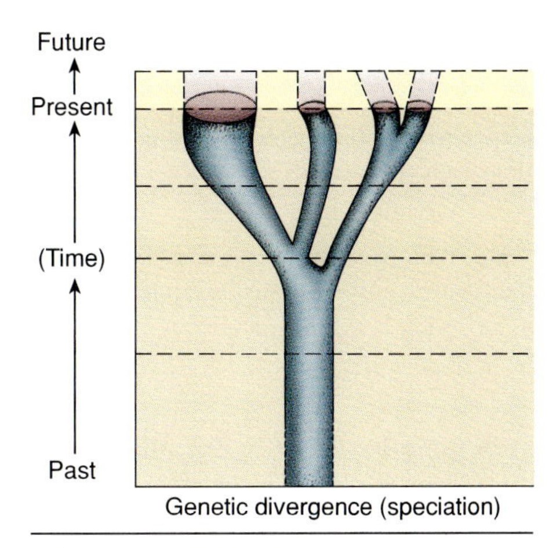

Figure 16.5

Speciation. Genetic divergence can result in one species giving rise to several more-or-less distinct genetic units. The two branches of the tree on the far right represent recent genetic divergence that has produced two groups very closely related to each other. These two may not yet have evolved enough to deserve recognition as distinct species. The vertical dotted lines represent what might occur in the future.

not like those of individuals from which they are evolving. As was pointed out in chapter 15, sexual reproduction and mutation constantly produce variability among individuals, so the raw material for speciation is provided. The intermixing of genetic information between individuals in a given population, however, discourages divergence and isolation from the population. For speciation to occur, a group of individuals must become removed from the genetic exchange in the parent population, thus allowing the accumulation of unique gene combinations and resulting in genetic divergence from the individuals of the parent population. The key to speciation, then, is **genetic isolation.**

For a long time evolutionary biologists acknowledged speciation only in populations isolated by geographical separation. Geographically isolated populations are termed **allopatric.** Individuals of a geographically separated population can accumulate genetic changes, making them different from the parent population. The speed of genetic divergence partially depends on how different the selection pressures are for the separated population. Ultimately, however, geographical isolation can result in the development of a new species.

Groups of individuals having geographically overlapping distributions are said to be **sympatric.** It is now understood that mechanisms exist which allow for sympatric speciation to occur. The key factor is, again, isolation. Minor changes in flower shape, size, color, or time of opening can develop as a result of very little genetic change. Any such flower modification can result in reproductive isolation by the flower being less attractive to the original pollinator, who then fails to visit that flower. Many flower species are pollinated because a given insect species is attracted to a specific flower shape, color, size, or fragrance. Any change that results in the pollinator not recognizing some individual flowers, will effectively remove those flowers from the interbreeding group. Species having sympatric distributions, therefore, can still have genetic isolation—the key for allowing individuals to diverge into a new species.

In addition to isolation due to pollinator failure, there are other sympatric speciation mechanisms in plants. The role of hybridization is important to the production of genetic divergence by providing a large number of new genetic recombinations in the hybrid offspring. The more genetically dissimilar the two parent species are, the more distinct their hybrid offspring will be. In addition, hybrids are often better suited to the habitat in which they find themselves than are either of their parent species. These hybrids often have very low fertility, however, and are unable to produce viable gametes. Hybrids, then, may be unique genetically but unable to pass on any advantageous traits. With their unique genetic composition and successful adaptation to their habitat, however, a new species could emerge in a reasonably short period if they were able to reproduce successfully.

Polyploidy

When an organism has two sets of homologous chromosomes, it is a diploid; when it has more than two complete sets, it is a **polyploid** (*poly,* many; *ploid,* sets). The most common type of polyploid is a plant with four sets of chromosomes, a **tetraploid.** Plants have a rather high incidence of polyploidy; 47% of all angiosperms are estimated to be polyploids.

The role of polyploidy in higher plant evolution is primarily one of increasing fertility in hybrids. Most hybrids are sterile because the parental chromosomes are too dissimilar to form homologous pairs in meiosis. This lack of bivalent formation produces sterile gametes. Doubling the chromosome number of the hybrid results in the exact duplication of every chromosome present in the original diploid hybrid. This type of polyploidy is termed alloploidy or amphiploidy. The exact chromosome duplicates act as homologous pairs, and the fertility of the hybrid is greatly enhanced. The more chromosomally dissimilar the parent species are, the greater the chances for sterile hybrid offspring. Such hybrids will also have a greater range of genetic variability from which better-adapted individuals might result. The more sterile the hybrid in the diploid condition, the more fertile the allopolyploid will be. With this increased fertility, the chances of hybridization resulting in a new species is much greater than without polyploidy.

Doubling of the chromosome complement within a single species is termed autoploidy, and it does little to increase speciation potential in plants. In fact, true autoploids actually have an increased incidence of sterility compared with their diploid progenitors. Sterility in itself is nonadaptive; that, combined with the fact that autoploids have no new genetic combinations, makes autoploidy a nonspeciating mechanism.

✔ Concept Checks

1. What is the key to defining a biological species?
2. What is hybridization?
3. What is the difference between allopatric and sympatric speciation?
4. What is polyploidy?

POPULATIONAL DIVERSITY

Evolutionary biologists study the genetic variability of individuals and of populations. The cumulative genetic variability of all individuals in a population is the **gene pool.** Population geneticists study the changes in frequencies of different alleles in the gene pool over time. Although the genotype of an individual does not change except through mutations, the genetic composition of a population does change as those mutations result in either death of individuals or new advantageous recombinations in offspring. All the alleles carried by the individuals within the breeding population determine the genetic makeup of the gene pool, just as all alleles carried by an individual that is not successful and is thus no longer part of the breeding population are removed from the gene pool.

Changes in a population's genetic makeup can also occur when individuals migrate from one population to another. They remove alleles from one population and add them to the other, changing the relative genetic frequencies of both populations. This **gene flow** between populations allows for increased populational variability and new diversity for the process of natural selection to act upon.

Plants not only display a range of morphological variability as a group, but individuals also can vary in direct response to environmental changes. A plant can display different growth forms in a single growing season if some significant change occurs in its habitat. For example, if a plant growing in open, hot, dry conditions is shaded and watered, it can respond by changing from a low-growing small-leafed form to a taller plant with larger leaves. The total range of variability possible by a single individual is often called **morphological plasticity.** Neither the short, small-leafed nor the taller, luxuriant growth form of the preceding example is genetically locked in. Both forms are within the range of morphological plasticity for the species.

In some cases, individuals of a given species occur throughout a large geographical range that displays a continuous change in climatic features. A south to north geographical range, for example, could show a gradual change from taller individuals in the south to increasingly shorter individuals farther north. Such gradual change is termed a clinal gradation in morphology.

The lack of significant populational isolation results in the formation of a **cline.** Although gene flow is possible throughout the cline, individuals are not likely to demonstrate the total morphological plasticity of the entire species. For example, northern individuals may not grow as tall as the southernmost individuals in the cline if transplanted to the south. There are not many species having true clinal gradations in morphological expression. Populations of a species often become geographically isolated from each other when spread over a broad range of habitats.

Several such populations having discontinuous distributions (disjunct populations) can develop a suite of traits unique to each population and genetically locked in to the individuals of that population. Populations genetically adapted to their specific habitats have been termed **ecotypes** within a given species. The individuals of one ecotype may look significantly different from individuals of another ecotype and will respond differently if transplanted into each other's habitat. Even with such differences, the individuals of one ecotype are fully fertile with the individuals of the other ecotype within the species.

A classic set of studies conducted by Jens Clausen, David Keck, and William Heisy demonstrated genetic adaptation of isolated populations to different habitats (figure 16.6). These researchers examined the morphological variation seen in a number of different plant species native to California and other western states. Of

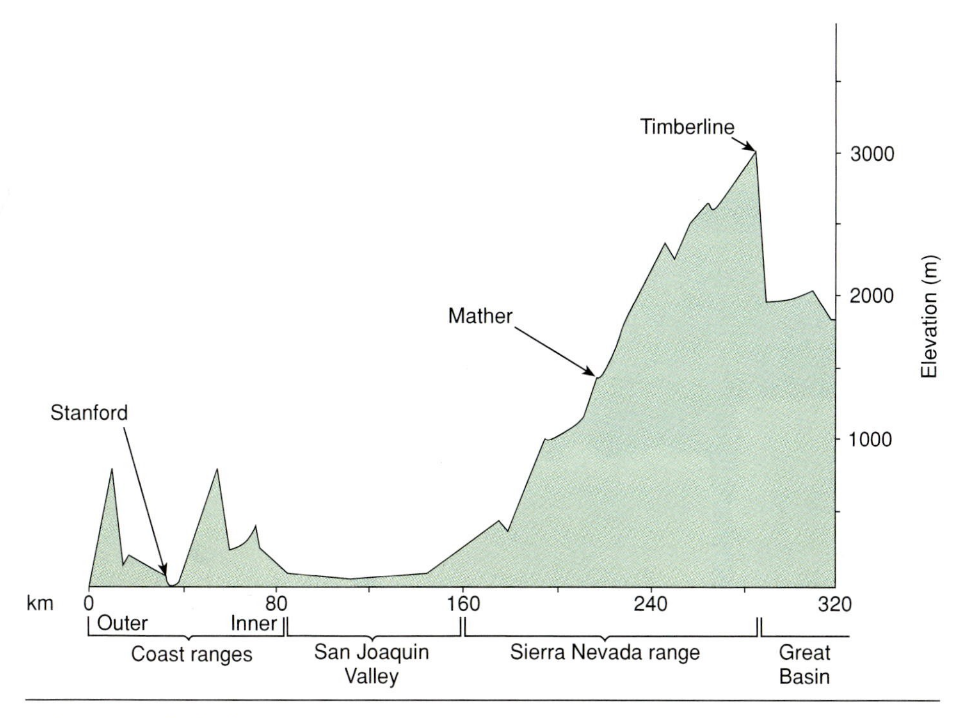

Figure 16.6
The three field transplant stations at Stanford (30 m elevation), Mather (1400 m elevation), and Timberline (3050 m elevation) used by Clausen, Keck, and Heisey during their ecotype research.

particular interest was a member of the rose family, *Potentilla glandulosa.* In addition to normal sexual reproduction, this species propagates asexually from runners. This feature made their studies of environmentally induced variation more meaningful, because they could use genetically identical cloned material for comparison.

Potentilla glandulosa is found from sea level to above 3000 m in elevation in isolated populations near and in the Sierra Nevada. Three field stations were established to allow for comparisons of plants found naturally at different elevations when transplanted into each other's habitat. The species has three main morphologically distinct growth forms. The coastal ecotype occurs near sea level, a Sierran foothill ecotype is found from 900 to 2100 m elevation, and an alpine ecotype grows in areas from about 2400 to 3500 m in elevation. These three ecotypes display morphological variation within each population and between the populations in their normal distribution, but they are distinct from each other in several features. In addition to morphological differences, each of these three ecotypes flowers at significantly different times of the year in response to the different lengths of their growing seasons. The coastal ecotype normally flowers in mid-April, the foothills variety first flowers in late May to early June, and the Alpine ecotype does not flower until late August.

The three field transplant stations were the Stanford station at 30 m, the Mather station at about 1400 m, and the Timberline station at 3050 m elevation. When either the coastal or foothills ecotypes were transplanted to Timberline, they nearly always failed to survive. When the Alpine ecotype was transplanted to either Mather or Stanford, however, it would not only survive but also grow larger than in its native habitat.

These researchers found that each different ecotype responded differently when transplanted to the three field stations, and the response of a given ecotype was consistent. Their data demonstrate that isolated populations of a species found in different habitats will produce genetically adapted ecotypes. Such genetic divergence is not enough to call each a different species, but it does demonstrate a speciation mechanism in action. Not all ecotypes are genetically different enough to assign an official name; others are. The *Potentilla glandulosa* ecotypes have, in fact, been recognized as distinct subspecies, a taxonomic rank indicating a given level of observable genetic differences below the level of species.

TAXONOMY

Taxonomy, the science of classifying and naming organisms, is probably the oldest of all scientific disciplines. Prehistoric people needed to be able to communicate with each other about what was edible and what was not, what could be used for a tool, and what made a good shelter for the night. Their world was considerably more limited in scope, and the numbers of different things needing a name was proportionately smaller than today. With increasing complexity in human society, and as scientists learn more and more about the world around us, the number of items needing names has increased manyfold.

Biologists have to deal with a mind-boggling number of organisms, and they need to do so with accuracy. The naming of organisms cannot be done haphazardly or differ from one country to the next or from one language to the next. Strict international rules governing the naming of plants have existed since the eighteenth century. The most basic of these rules involves the use of a **binomial** (two name) system of nomenclature (naming) in which all organisms have a scientific name consisting of a **genus** and a species. All humans, for example, are *Homo sapiens,* all giant redwoods are *Sequoia sempervirens,* all dandelions are *Taraxacum officinale,* and so on. The genus is capitalized, and the specific epithet is lowercase (occasionally, one will find the specific epithet incorrectly capitalized when named after a person). In addition, no two

Uncovering the Silversword Story—Sherwin Carlquist*

Events that happened even yesterday are often difficult to reconstruct. However, circumstances have permitted several of us to reconstruct with remarkable precision an event that happened about five million years ago. This event was the dispersal of the seeds of a tarweed on the surface of a bird that flew from California to Hawaii over at least 2700 miles of open ocean. The establishment of this plant in Hawaii led to an amazing, explosive evolution of diverse descendants—the silversword alliance. It is my opinion that this group of plants easily qualifies as the most amazing evolutionary happening known to island biologists.

When I first visited Hawaii as a beginning graduate student in 1953, the Hawaiian silverswords (*Argyroxiphium*) and their allies, *Dubautia* and *Wilkesia,* which were also restricted to the island, were not thought to be related at all closely to the tarweeds, a group of annuals or small shrubs found in California and nearby regions (figure 1). When I visited Haleakala caldera on Maui in

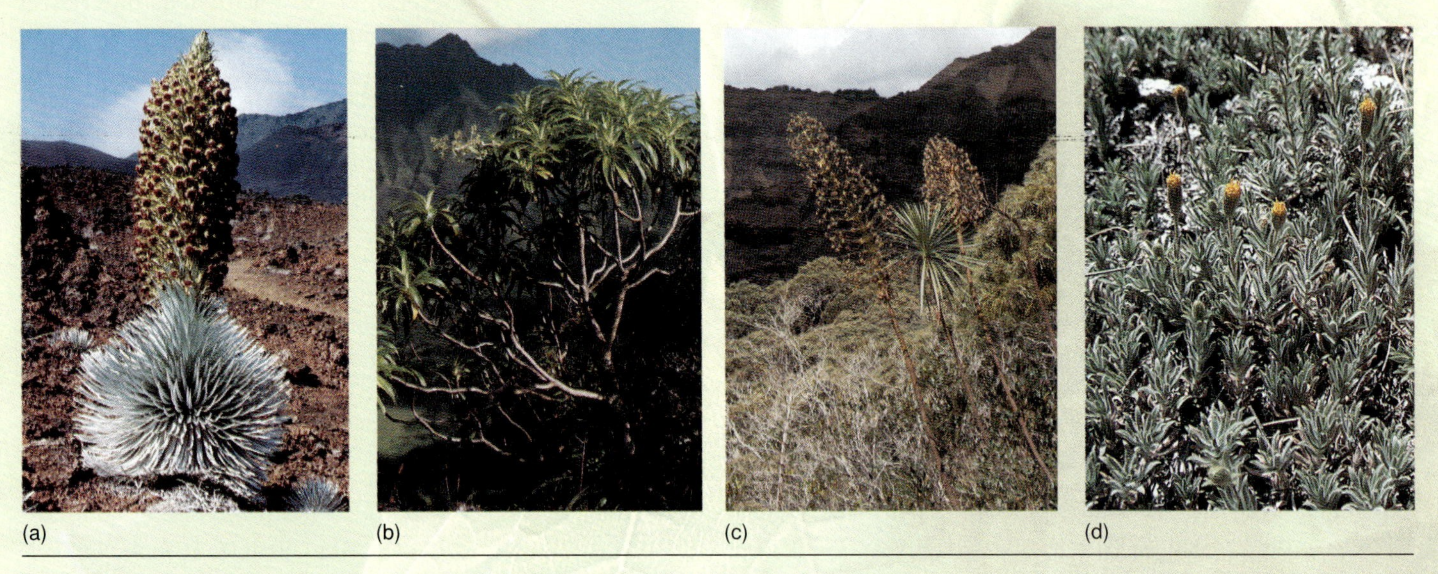

Figure 1

(*a*) Silversword (*Argyroxiphium,*) (*b*) *Dubautia,* (*c*) *Wilkesia,* (*d*) Tarweed (*Raillardiopsis*).

**A specialist on island biology and an internationally renowned plant anatomist, Dr. Carlquist was the recipient of the 1991 Allerton Medal for excellence in tropical botany sponsored by the National Tropical Botanic Garden. His substantial contributions to these fields include over 180 research articles and papers and 7 texts, including* Island Life, Island Biology, Hawaii: A Natural History, *and* Ecological Strategies of Xylem Evolution, *which are often used by undergraduate and graduate students.*

plant species can have the same name. Some genera have only a single species contained within it; others have two or more species. For example, the genus *Homo* has only one species, *H. sapiens,* and the plant genus *Ginkgo* has only the single species *G. biloba. Quercus,* the oak genus, contains many species, including *Quercus suber* (cork oak), *Q. virginiana* (live oak), *Q. havardii* (shin oak), *Q. alba* (white oak), and *Q. macrocarpa* (bur oak), among many others. All these oak species have more in common with each other than with any member of a different genus, such as *Fagus,* the beech genus, or *Castanea,* the chestnut genus. Grouping individuals into a given species and grouping related species into genera are not done out of convenience, but to represent natural genetic entities and evolutionary relationships.

The second basic rule of naming organisms is that the scientific binomial is in Latin. Since Latin is a foreign language, scientific binomials are always italicized or underlined. It is true that many groups of organisms have common or non-Latinized names, but their use by the scientific community is not desirable because of the need for accuracy. Biologists rely on accurate worldwide information exchange. Some species have more than one common name in accepted use in different regions. *Ulmus rubra,* for example, is known as both red elm and slippery elm, depending on location. Another problem with use of common names is the application of the base name to members of different genera. Thus, the she-oak in Australia is in the genus *Casuarina,* and poison oak in the United States is a member of the genus *Rhus,* neither of which is a true oak in the genus *Quercus.* (*Rhus* also includes poison ivy, which is not an ivy).

1963 and saw the Haleakala silversword in flower, I caught the same scent I remembered when I had seen tarweeds in California. Could they be related—and could the techniques I had learned in plant anatomy demonstrate such a relationship?

Despite the advice of a well-known botanist that there was nothing to be learned studying tarweeds under a microscope, I spent several years collecting members of the Hawaiian silversword alliance, the California tarweeds, and some other plants the silverswords had been claimed to resemble. When these plants had been converted into thin sections that could be analyzed with a microscope, not one but several exciting stories emerged. The Hawaiian silversword alliance was the only group in the sunflower family at all closely related to the California tarweeds. The ways tarweeds have of surviving hot California summers, flowering just before the weather begins to cool in fall, were evident in their peculiar leaf structures. Moreover, they had passed on these and other unusual structures to the Hawaiian silverswords and their relatives. The microscope revealed intricate ways in which the Hawaiian plants had evolved to suit diverse Hawaiian habitats—*Dubautia* and shifted from small shrubs to trees and even vines; *Wilkesia,* a strange unbranched tree, had originated on Kauai, and *Argyroxiphium* had produced species like the silversword with its shiny leaves suited to the extremely dry alpine climate of Haleakala Crater, but had also gone on to produce plants adapted to sopping mountain bogs. Studying the anatomy of these plants made me sure that one plant—*Raillardiopsis,* a tarweed from California, belonged to the tarweed-silversword group although nobody had thought before that it was at all closely related to the silverswords.

To explain this, I hypothesized that on the outer surface of some bird, tarweed seeds had been able to travel across the unbroken ocean distance that separates California from Hawaii resulting in the magnificent assemblage we call the silversword alliance. I didn't realize that few biologists believed the dramatic events I had hypothesized were likely—perhaps they regarded evidence from plant anatomy as not very convincing. However, Dr. Gerald D. Carr, studying chromosomes of these plants at the University of Hawaii; Dr. Bruce G. Baldwin, working with their DNA at the University of California, Davis; and other workers attracted by the striking plants, provided key evidence that have been accepted as overwhelmingly convincing. These workers even crossed a Californian tarweed—the one nobody but me had believed was a tarweed, *Raillardiopsis*—with a Hawaiian *Dubautia,* and successfully grew seeds of the hybrids, demonstrating a close genetic relationship. Computerized analysis (cladistics) of the results of various workers showed exciting details of the story. They showed *Raillardiopsis,* or something very closely related to it, was the genus that had likely given rise to the Hawaiian silversword alliance. A migratory bird such as a Pacific golden plover must have flown with seeds of that plant from the Californian coast ranges to the island of Kauai about five million years ago.

Beginning with a likely arrival on a high Kauai mountain similar in climate to the Californian coast ranges, the silverswords evolved so rapidly and thoroughly to suit various Hawaiian localities that they look like representatives of unrelated families—*Wilkesia* looks more like a yucca than a tarweed, for example. We who have worked on the tarweed-silversword story like to think that there is no more amazing or dramatic story to be found among plants or animals of islands.

It is often asked why scientific names are in Latin. Why not English, German, Russian, French, or Swahili? At the time that attempts were being made to standardize nomenclatural practices, Latin was the language of the scholar and had been for centuries. Equally important is that today Latin is a "dead" language, which means it is not an actively used language of any nation or people. Thus it is politically neutral and does not change through usage, both important considerations in a world where nationalistic jealousies can arise so quickly. Finally, Latin is descriptive. The species epithet actually originated as a word that would aid in identifying the species by describing one of its more salient features. *Quercus macrocarpa* has large fruit (*macro,* large; *carpa,* fruit), and *Q. alba* has whitish gray lower leaf surfaces (*alba,* white). The use of a single diagnostic species name originated with the famous biologist, Carolus Linnaeus (1707–1778). Linnaeus is called the Father of Taxonomy because he popularized the use of the binomial system of naming plants and animals. In fact, 1753, the date of his monumental publication *Species Plantarum,* is the date recognized worldwide as the beginning of modern plant taxonomy.

Above the level of genus, classification into groups does not become part of the scientific name of the organism (table 16.1). Similar genera are grouped into **families,** whose names normally end in *-aceae.* There are eight plant families having both a name ending in -aceae and an alternative name that is also valid. Because of the more common historical usage, throughout this text we have chosen to include the older name in parentheses following the family name having the -aceae ending. The family names ending in -aceae are formed by using an important genus within the family for the

TABLE 16.1

Plant Nomenclature

KINGDOM	Plant	Plant	Plant	Plant	Plant	Plant
DIVISION	Angiosperm	Angiosperm	Angiosperm	Angiosperm	Angiosperm	Angiosperm
CLASS	Dicotyledon	Dicotyledon	Dicotyledon	Dicotyledon	Dicotyledon	Monocotyledon
ORDER	Fagales	Fagales	Fagales	Fagales	Urticales	Liliales
FAMILY	Fagaceae	Fagaceae	Fagaceae	Betulaceae	Ulmaceae	Liliaceae
GENUS	*Quercus*	*Quercus*	*Fagus*	*Betula*	*Ulmus*	*Allium*
SPECIES	*virginiana*	*macrocarpa*	*grandifolia*	*nigra*	*rubra*	*sativum*
COMMON NAME	Live oak	Bur oak	Beechnut tree	River birch	Red elm	Garlic

**To demonstrate the relatedness indicated by grouping within the plant taxonomic hierarchy, this table lists two closely related species of oaks, a species in a closely related genus* (Fagus) *in the same family, a closely related family (Betulaceae) to the Fagaceae, and so on.*

root, for example, the genera *Aster* for Asteraceae and *Lamium* for Lamiaceae.

Old Name	**New Name**
Leguminosae	Fabaceae
Compositae	Asteraceae
Cruciferae	Brassicaceae
Gramineae	Poaceae
Labiatae	Lamiaceae
Umbelliferae	Apiaceae
Palmae	Arecaceae
Guttiferae	Clusiaceae

Rosaceae refers to the rose family, which includes a large number of genera that share several morphological similarities. Even though most plant families are genetically natural units, grouping above this level is not always as accurate in reflecting genetic relationships. As will be seen in the next chapter, we use the five-kingdom system of classification in this text.

Within the plant kingdom the most recognizable groups are the vascular plants, which include the ferns, gymnosperms, and angiosperms. The most recently evolved, most successful, and most widely distributed of all the plant groups are the flowering plants or angiosperms. It is for this reason that they have received the greatest emphasis and coverage throughout this text.

✔ *Concept Checks*

1. What is morphological plasticity?
2. What are ecotypes?
3. What is a binomial?
4. What taxonomic groups usually end in -aceae?

BIOGEOGRAPHY AND CONTINENTAL DRIFT

Biogeography

Biogeography is the study of where organisms occur, the ecological constraints controlling the extent of their occurrence, and how they came to be where they are now. Taxonomists and other evolutionary biologists are especially interested in understanding not only what species exist and how they are currently related to each other, but also the evolutionary history of these groups. The establishment of the **phylogeny,** or evolutionary history, for species, genera, or even families of plants is more difficult than for most animal groups because of the relative absence of a plant fossil record. Unlike many animals, which have bones, teeth, or exoskeletons (external supportive structures as in insects or marine crustaceans), plants have few parts that resist decomposition and allow for fossilization.

Scientists have developed phylogenetic "trees" for many groups by using the fossil record that does exist in the form of petrified wood, fossilized leaf, flower, and seed imprints, and the resistant pollen grains, and by drawing heavily on comparisons among existing groups of plants. It has become increasingly clear that the current distributions of plant groups cannot be explained exclusively in terms of the existing world geography and its current ecological habitats.

Both the current biogeography and the phylogeny of plant and animal species must be studied in a historical perspective. The earth itself and the living organisms on it are, and have been, constantly changing. The fluctuations in how many and which species occur in a given habitat are greatly influenced by both existing and past conditions. Past climates, ecological interactions, available habitats, and the time scale during which these past conditions existed are the bases for what, where, and why given species exist today.

Combining what is known about present conditions and existing groups of organisms with evidence from the fossil record and what is known about past climates and geological conditions is the only way scientists can hope to ultimately unscramble the sequences of events leading to the current flora and fauna of the earth. Until relatively recently such reconstructions have been much more difficult because it was virtually impossible to reconcile existing worldwide plant and animal distributions using the long-standing belief in a rigid and unchanging earth.

Continental Drift

Today it is almost humorous to think that anyone could have ever believed that the world is flat. Five hundred years ago, however, only a few radicals thought otherwise. Until Columbus and other explorers sailed their tiny ships great distances past where the edge was thought to be and ultimately sailed around the entire earth, there was little evidence to support a spherical shape. Once it became clear that the earth was not flat, some visual observations that had previously been confusing suddenly made sense (such as the curved horizon). A similar sequence has taken place relative to the long-held belief that the continents have been essentially where they now are since their formation.

Although the concept of continents moving has existed for over 100 years, it was either ignored or ridiculed until fairly recently. A few staunch believers published their arguments between 1915 and the 1940s, but it was not until 1956 that serious converts revived active support of the concept based on new geological evidence. Finally, in the 1960s enough evidence became available to explain how something as massive as an entire continent, even a small one, could be moved. Today the idea of drifting continents is in the same category as a spherical earth. There is simply too much scientific evidence to believe any longer in a static and unchanging earth.

Continental Outlines

Even as far back as the earliest mapping of the shores of the Atlantic, observations were made concerning the similarity of the opposing coastlines of Africa and South America. It was not until 1858 that the concept was first publicly proposed that the continents were once fitted together into a single supercontinent and have subsequently moved apart to their present positions. Earliest attempts to fit the outlines of the continents together were only moderately successful because the shoreline was used, not the true edge of the continents. This was because the underwater continental shelf was not well mapped. The world's continents can be fitted into a single supercontinent, **Pangaea** (figure 16.7). Although the fit is not perfect, it is easy to see why even the earliest proponents of the existence of Pangaea felt strongly that the continental outlines fit too well to write it off to mere coincidence.

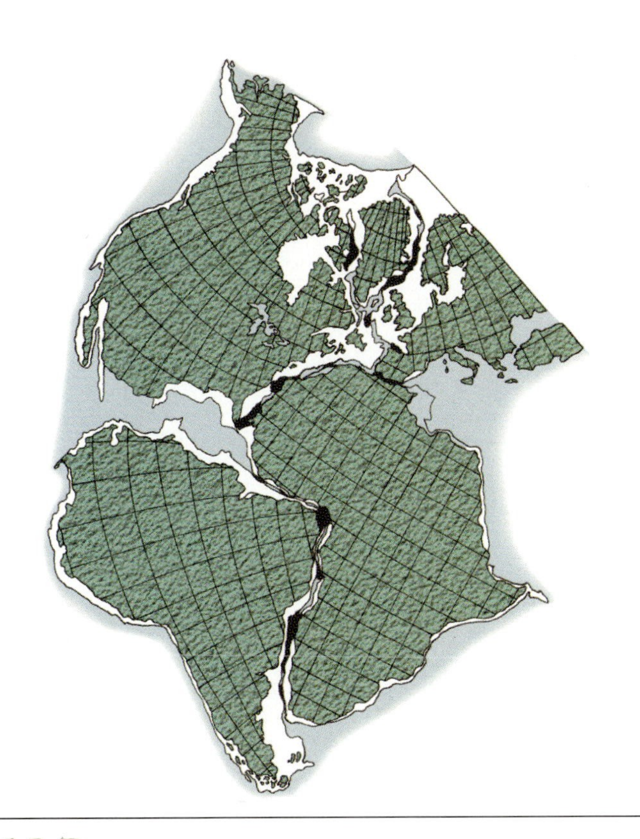

Figure 16.7
Within the outline of the continental shelves the continents fit together amazingly well, certainly much better than within the outline of the shoreline (hatched area).

The breakup of Pangaea began during the Triassic period of the Mesozoic era some 200 million years ago. By the end of the Triassic, 180 million years ago, a northern group of continents called **Laurasia** had moved away from the southern mass, **Gondwana.** In addition, Gondwana had begun to break up, with India splitting away and Africa–South America rotating away from Antarctica–Australia.

By the end of the Jurassic period, 135 million years ago, India had moved even farther from Antarctica, and the North Atlantic Ocean was well established by North America separating from Europe–North Africa. At about this time India began to pass over a hot spot of volcanic activity, and the South Atlantic first started forming between Africa and South America. During the next 70 million years, until the end of the Cretaceous period 65 million years ago, continental movement continued with India moving steadily toward Asia, Madagascar splitting away from eastern Africa, and South America and Africa becoming well separated. It is interesting to note that 65 million years ago, North America and Eurasia were still connected through what would become Greenland, and South America, Antarctica, and Australia were still essentially all connected to each other (figure 16.8).

In the past 65 million years, India collided with Asia, forming the Himalayan Mountains; Australia and

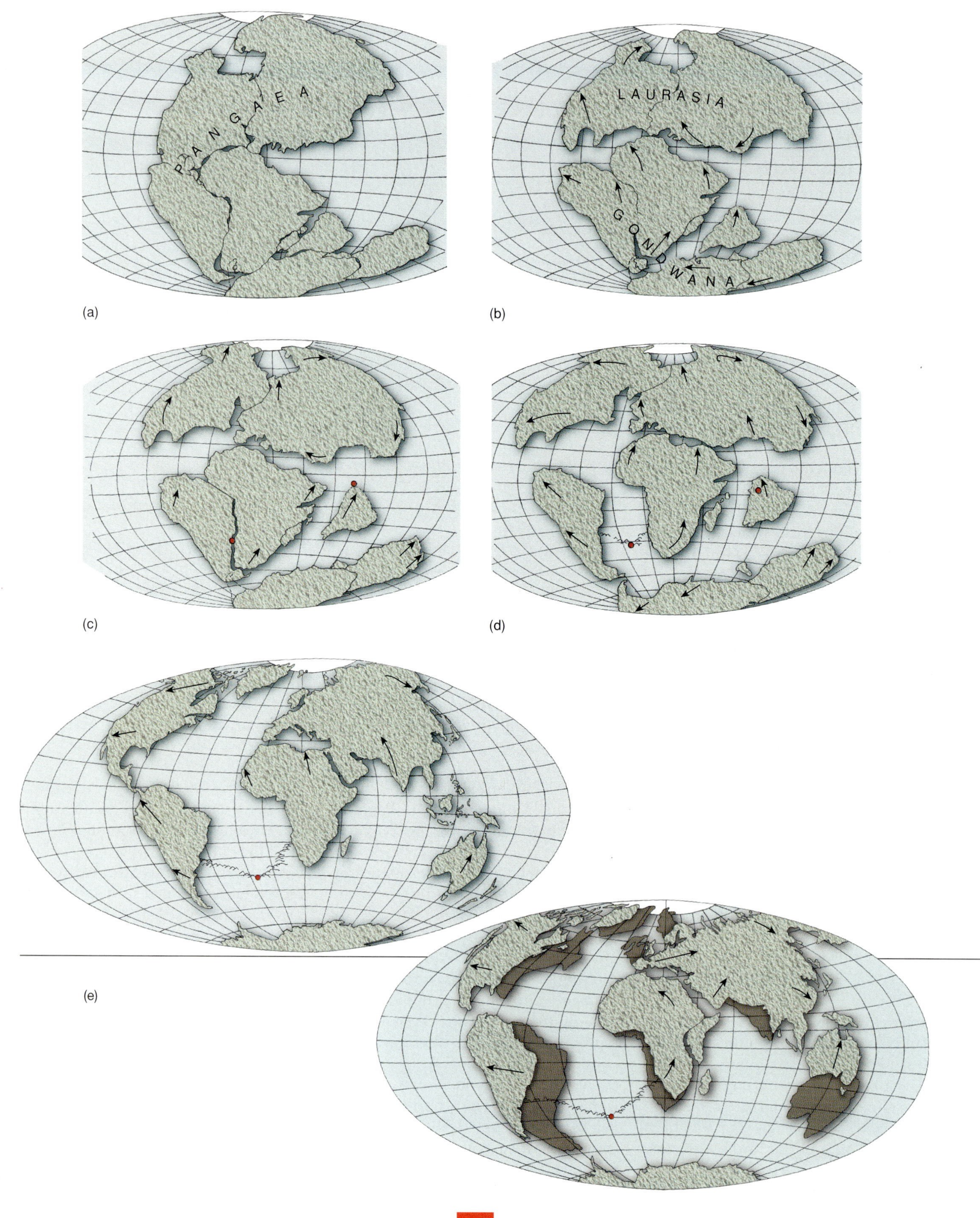
PANGAEA
(a)
LAURASIA
GONDWANA
(b)
(c)
(d)
(e)

South America split away from Antarctica; Greenland, North America, and northern Europe all separated; and North and South America became united. Based on current directions and rates of movement, eminent geologists have postulated what will likely be the positions of the continents 50 million years from now. Among the more interesting changes are the separation of North and South America, the movement of Australia to tropical latitudes adjacent to Southeast Asia, and the separation and northern movement of Baja, California, and a slice of coastal California west of the San Andreas fault, which includes Los Angeles. It is estimated that in 10 million years Los Angeles will be due west of San Francisco.

Plate Tectonics

Lines of evidence supporting the existence of a supercontinent include remarkable matching of geological formations and the alignment of continental paleomagnetic lines to true north-south when the continents are positioned together. Each continental mass has a magnetic north that varies from true north (the pole) according to the rotation the moving continents have gone through since their alignment as part of Pangaea. Realignment of the continents improves the agreement of magnetic and polar north. The combination of such evidence with the fit of the continental edges supports the idea of the existence of Pangaea and subsequent movement of continents, but it does not explain how a continent moves.

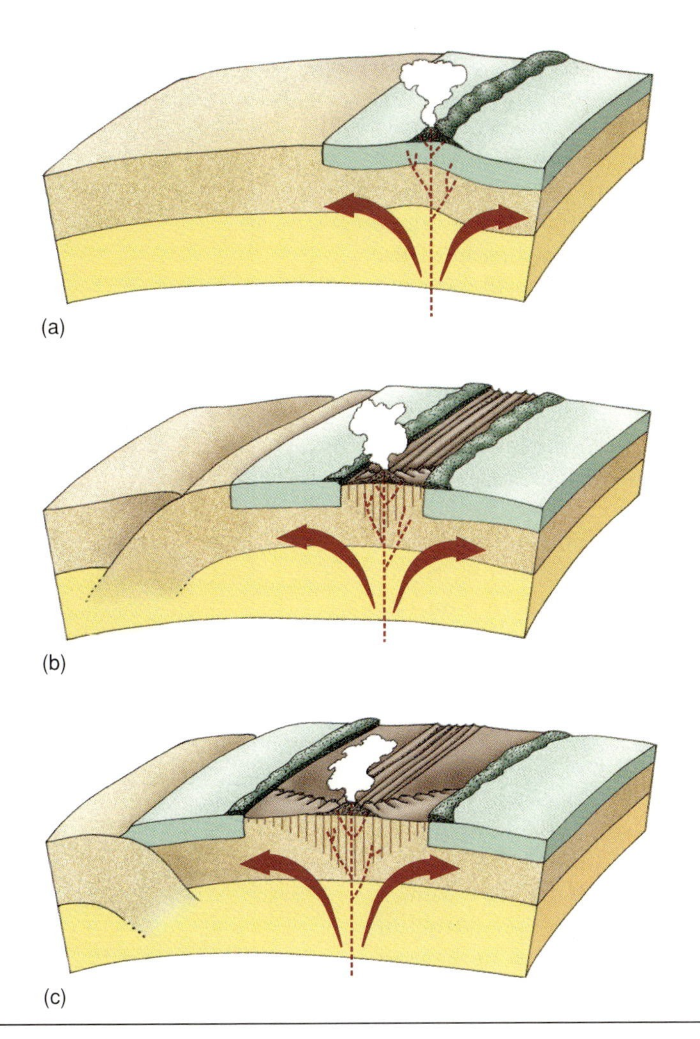

Figure 16.9

Convection currents in the molten core of the earth can break a continental plate and push the parts away from each other. The South American and African plates were formed in this way (*a*), with a north-south median ridge forming on the ocean floor where they were originally together (*b*), and lateral ridges forming from a hot spot in the median ridge (*c*).

Figure 16.8

The single supercontinent Pangaea (*a*) began breaking up about 200 million years ago. By approximately 180 million years ago the northern Laurasia and southern Gondwana continents appeared (*b*). The positions of the continents by 135 million years ago (*c*) and 65 million years ago (*d*) resulted from the movement of each continental plate (*arrows.*) In the past 65 million years the continents have reached their current positions (*e*). It is predicted that 50 million years from now the continents will be in the positions depicted in (*f*).

Plate tectonics is the study of continents as distinct granite plates "floating" on a more dense, semiviscous layer of basalt. Ocean floors are basaltic in composition, and they also move. The driving force producing continental and ocean plate movement comes from the molten core of the earth as it swirls slowly and powerfully beneath the mantle and crust. The movement of these **convection currents** produces a powerful lateral pressure on continental plates. The force of these convection currents moves the continents slowly across the earth's surface (figure 16.9).

Several things can happen as moving continent and ocean plates collide with or move away from each other. Two continents (or one that has become split) moving away from each other leave behind a new ocean floor as basaltic materials cool and are deposited by the convection currents. For example, Africa and South America have been pushed apart, depositing a median ridge where the two convection currents push toward the same part of the crust. Also formed is an ever-increasing Atlantic Ocean floor between these two continents. The first part of the Atlantic Ocean floor formed, then, would be that part adjacent to Africa and South America when they had just begun separating. The most recently formed ocean floor would be nearest the median ridge.

The depth of nonbasaltic sediment deposits should be greatest where the ocean floor has been in existence the longest, near the continents. Not only the depth of sediments, but also the age of some fossilized organisms equidistant on each side of the median ridge should be the same if seafloor formation and spreading has been as just described. Extensive sampling has shown this to be true; sediments are deepest and oldest near the continents and youngest and most shallow near the median ridge.

The median ridge of the Atlantic Ocean runs north and south equidistant between the continents, North America and South America on one side and Europe and Africa on the other. In the South Atlantic, there exists a chain of underwater mountains called the lateral ridge. A thermal volcanic area, or **hot spot,** that remains stationary relative to a plate movement in the crust forms volcanic "mountains" as the continents move apart. Each such mountain moves with the newly formed ocean floor toward either Africa or South America. The oldest underwater mountains are closest to the continents because they were formed by the volcanic hot spot when the two continents were still close together. The youngest mountains, then, are nearest the median ridge where they were formed.

There are other stationary hot spots under the earth's crust but none located in a median position between two separating continents. There is one in the Indian Ocean over which the Indian continental plate passed as it moved north, and a chain of volcanic mountains running north to south through India were formed by the periodic volcanic eruptions. Another hot spot is found under the Pacific Ocean, and it has produced a long chain of mostly underwater mountains extending all the way to the Aleutian Trench. Several Pacific Islands are part of this chain, including Midway and all the islands of the Hawaiian chain. These were formed as the Pacific Ocean floor moved north and west across this volcanic site (figure 16.10).

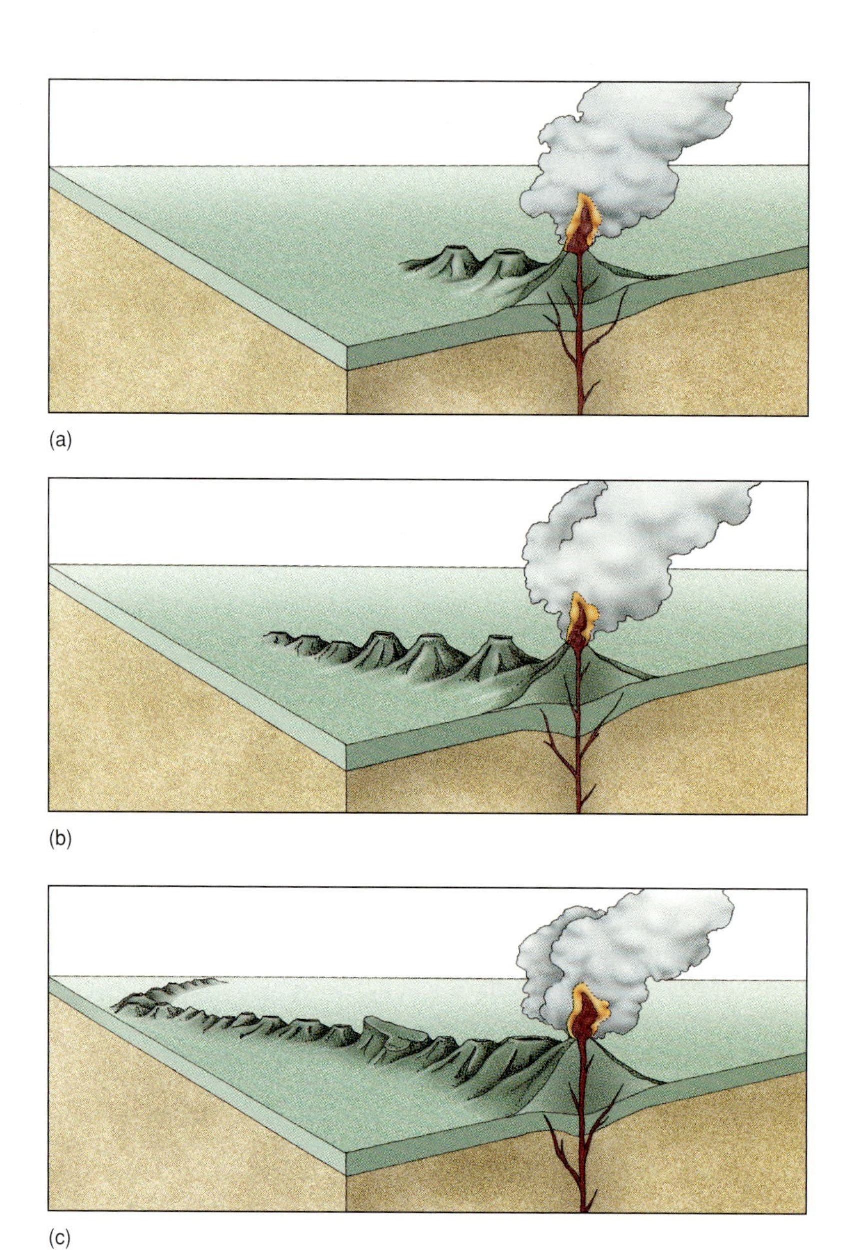

Figure 16.10
The Pacific Ocean plate has moved west across a thermal volcanic area or "hot spot," forming a chain of underwater mountains and volcanic islands. The Hawaiian Islands are the most recent of this volcanic chain. (*a*) Seventy million years ago. (*b*) Forty million years ago. (*c*) Ten million years ago.

The concurrence of volcanic eruptions and earthquake activity with the edges of continental and ocean plates and with median ridges adds yet another line of evidence supporting plate tectonics and continental movement. The combination of continental outlines, paleomagnetic line convergence, median-ridge formation, ocean-floor sediment depth, hot-spot activity, earthquakes and volcanic eruptions, geological formations (figure 16.11), mountain building, and other structural crust modifications from plate interactions clearly demonstrates that continents are moving and have been for some 200 million years.

Biological Implications of Continental Drift

Before theories of continental drift were formulated, the occurrence of closely related organisms in widely separated continents defied explanation. Plant and animal species that do not swim, float, or fly would not likely have

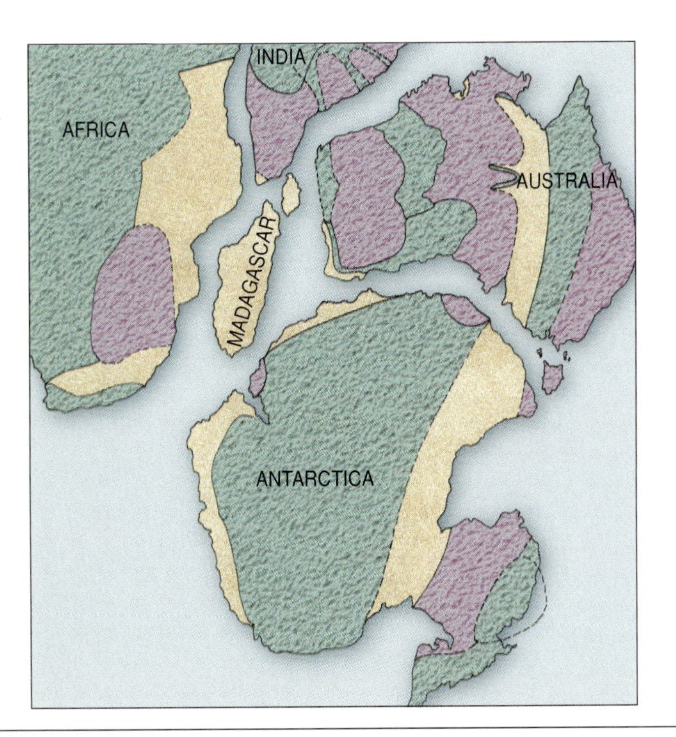

Figure 16.11
Identical geological formations on different continents match up very well when those continents are repositioned as they would have been in Gondwana.

close relatives across an ocean that has been there since the earth was formed. Certainly there are organisms that do occur worldwide because of their dispersal mechanisms, and coconut palms are one of the most noteworthy. Coconuts have a coarse, fibrous husk that provides protection from salt water and buoyancy for the seed within as they float from beach to beach all over the tropical regions of the world.

Other species, however, do not have seed-dispersal mechanisms that would account for the great distances that separate them today. Two closely related species of sycamore trees (*Platanus*) occur in eastern North America and eastern Europe. They are so closely related that, when experimentally crossed, they proved still to be interfertile. Such fertility suggests either a common ancestor giving rise to these two species or one giving rise to the other followed by some genetic divergence. In either case, that angiosperms were very common and diverse in the fossil record well before North America and Europe physically separated during the past 65 million years offers an explanation for this situation. A common ancestor could have easily existed throughout central Laurasia and given rise to two different species after east-west separation of North America and Europe.

The silver beech *Nothofagus menziesii* now occurs in New Zealand, but it is a relic of a group of *Nothofagus* that is found in the fossil records of southern South America, Australia, and New Zealand. These trees existed during a period when South America, Antarctica and Australia–New Zealand were all in contact and when the climate in Antarctica was significantly warmer than today. Undoubtedly *Nothofagus* and many other plant and animal species (such as the marsupials) had such a distribution prior to the breakup of these three continents, which began some 60 million years ago. There are, in fact, many more similarities between the flora and fauna of southern South America and Australia than between Australia and Asia, even though the latter are in much closer geographical association today. This is because Asia has a predominantly Laurasian biota, whereas that in Australia is from Gondwana ancestral stock. Because the separation of these two supercontinents occurred prior to or at approximately the same time as the evolution of the angiosperms, the subsequent worldwide diversity of flowering plants has occurred since the Laurasia–Gondwana separation.

✔ Concept Checks

1. When was Pangaea intact?
2. What causes continents to move?
3. What does a "hot spot" cause in the ocean? What does it cause on the land?
4. How does continental drift help explain the distribution of *Nothofagus*?

Summary

1. As the naturalist on the H.M.S. *Beagle,* Charles Darwin observed many examples of natural variation between individuals, populations, and species. These observations, especially on the Galápagos Islands, were the basis for his theory of evolution by natural selection.
2. The process of natural selection involves several basic criteria: individual variation; larger numbers of offspring are produced than can survive; competition for resources among individuals; selection acts on individual phenotypes; selected individuals must be able to produce viable fertile offspring; and long periods of time must be available for this process to produce significant change.
3. Lamarckian evolution proposed the inheritance of acquired characteristics, whereas Darwin's theory of evolutionary change suggested the continuance of traits that surviving individuals possessed. Alfred Wallace also produced a theory of evolution

paralleling Darwin's but developed it many years after Darwin returned with his ideas from the voyage of the *Beagle.*

4. Darwinian evolution has been referred to as gradualism because he proposed a slow rate of change. Punctuated equilibria is a theory of rapid evolutionary change followed by long periods of stasis, or no change. This latter theory has also been called quantum evolution.
5. The process of natural selection is well documented, but evolution is still a theory because of the time needed for sufficient change to produce a new species and the inability to observe these changes directly.
6. The concept of what is a species is complicated in the plant world because of natural hybridization. The mechanism that produces evolutionary change resulting in a new species is called speciation. Both allopatric and sympatric speciation depend on genetic isolation. Polyploidy is one of the important isolating mechanisms.
7. Populational diversity can include morphological plasticity or genetically programmed ecotypic diversity. A thorough study on ecotypic diversity was done by three scientists in California on *Potentilla glandulosa.*
8. Taxonomy is the science of classifying and naming organisms. Using a binomial system (genus and species) of Latin scientific names, scientists can communicate more accurately about the organisms on which they work than if common names were used. Genera are grouped by relationship into families, the names of which normally end in -aceae.
9. Biogeography is the study of distributions of species and the ecological considerations controlling them. The establishment of the evolutionary history or phylogeny of a group is a goal of taxonomists, who use the fossil record, biogeography, and other pertinent information.
10. Continental drift helps clarify worldwide distributional patterns of some plants and animals not otherwise explained by normal distributional mechanisms. Plate tectonics explains how continents have moved from the single supercontinent Pangaea through the two continents of Laurasia and Gondwana to their present positions.

Key Terms

allopatric 285
artificial selection 280
binomial 287
biogeography 290
biological species concept 284
cline 286
continental drift 291
convection currents 293
ecotypes 286
evolution 284
families 289
gene flow 286
gene pool 286
genetic isolation 285
genus 287
Gondwana 291
gradualism 283
hot spots 294
hybridization 284
Laurasia 291
morphological plasticity 286
natural selection 280
Pangaea 291
phylogeny 290
plate tectonics 293
polyploidy 286
punctuated equilibria 283
quantum evolution 283
reproductive isolation 284
speciation 285
species 284
stasis 283
sympatric 285
taxonomy 287
tetraploid 286
variation 279

Discussion Questions

1. Why did Darwin's Theory of Evolution by Natural Selection cause such an impact on the world?
2. What did Thomas Malthus write about that was important to Darwin's theory?
3. Why do plants not fit as cleanly as animals within the biological species concept?
4. What can we expect to happen to the positions of existing continents in the next 50 million years?

Suggested Readings

Bowler, P. J. 1993. *Darwinism.* Maxwell Macmillan International, New York.

Grant, V. 1981. *Plant speciation.* Columbia University Press, New York.

Rensberger, B. 1982. Evolution since Darwin. *Science* 82 3(3):40–45.

Robson, N. K. P., ed. 1991. *Introduction to the principles of plant taxonomy.* 2nd ed. Cambridge University Press, New York.

Somit, A., and S. A. Peterson, eds. 1992. *The dynamics of evolution: The punctuated equilibrium debate in the natural and social sciences.* Cornell University Press, Ithaca.

Stace, C. A. 1989. *Plant taxonomy and biosystematics.* 2nd ed. Routledge, Chapman and Hall, New York.

Stuessy, T. F. 1990. *Plant taxonomy: The systematic evaluation of comparative data.* Columbia University Press, New York.

Sullivan, W. 1991. *Continents in motion: The new earth debate.* American Institute of Physics, New York.

Wesson, R. G. 1993. *Beyond natural selection.* MIT Press, Cambridge.

Williams, G. C. 1992. *Natural selection: Domains, levels, and challenges.* Oxford University Press, New York.

Wilson, E. O. 1989. Threats to biodiversity. *Sci. Am.* 261(3):108–116. September 1989.

Chapter Seventeen

Life's Origins and Prokaryotes

Evolution provides for a high degree of structural complexity among organisms.

The origin of life and its evolution have been linked to the chemical and physical changes of the earth. This chapter will introduce theories concerning the origin of life and geological time. A five-kingdom concept for the classification of organisms is presented along with the structural, beneficial, pathogenic, and ecological aspects of prokaryotes and viruses.

The number of living species in the world is estimated to be approximately 10 million. Some 500,000 of these are plant species. Taxonomists—scientists who classify organisms—are constantly discovering new species, changing the lists, and deciding that old species should be combined or divided. All this shuffling may seem unnecessary to the layperson, but accurate classification is exceedingly important in industry, medicine, agriculture, biology, and even in landscaping. In spite of all this naming and renaming, there are still probably hundreds of thousands of species that have never been classified nor seen by humans; we know that they exist, however, because new species are continually discovered when biologists sample new niches or study more intensely previously known niches.

Close observation of organisms reveals that some are quite similar, whereas others are very different. Some are complex, and others are relatively simple. Some have only one or a few cells; others have billions of cells. Where did all these different kinds of organisms come from? Have they always been here? How and when did life begin?

A STAR IS BORN

Before there can be life, there must be a source of energy, and the energy for our planet comes primarily from the sun. The absolute dimensions of space and time are incomprehensible to most of us. Where did the earth and the sun come from? Where did our entire solar system come from? How long has it been here? And finally, when and how did life begin? These questions have occupied the minds of many scholars and laypersons for centuries. You have already been exposed to the scientific method and have some appreciation for the difficulty in explaining theories, establishing hypotheses, and then establishing laws. Absolute proof of the age and origin of the solar system is probably impossible, but recent advances in radioisotope dating of minerals and organic materials has led to some startling generalizations that are now accepted by most of the scientific world. Recent results from lunar and planetary missions, as well as the study of moon rocks and meteorites have strengthened the hypotheses in many cases, and each bit of evidence helps reinforce the efforts.

The Big Bang

The theory with the greatest credence concerning the origin of our universe is the big bang theory. This concept suggests that all of the universe's 10 billion galaxies came into being about 20 billion years ago. The colossal event required only 180 seconds of "creative chaos," which sent everything flying through space on a journey that led to the establishment of all planetary systems, including our own solar system. The big bang is actually an abstraction or a sort of model. Even though the proposed dates and times may be uncertain, the theory portrays a scheme that attests to the vastness of space and suggests how billions of stars originated. Our sun is a star whose birth was no doubt unspectacular among the births of other stars. Stars form when a molecular cloud (hydrogen, helium, and dust) collapses under the influence of its own gravity. As the cloud contracts, it breaks into smaller volumes of gas, each destined to become a star by collapsing on itself, much like a giant balloon losing air and becoming smaller. One of these balls of gas became our sun, at first with a slight spin, which gradually increased as the gas ball flattened out. Gas gathered at the center, and the new star-to-be was still without light of its own. As the collapse continued, the pressure of the star at its center increased, compressing and heating the interior until thermonuclear reactions began, causing the new star to burn. The big bang led to many pockets of gas and dust in space, some of them becoming stars with their own planetary systems.

Formation of the Earth

After compression and formation of the sun, cosmic dust and gases were left over and held in flattened disks around the sun by gravitational forces. Eventually, smaller centers of condensation began to form, and each of these became the planets of our solar system. Scientists believe that the formation of the sun and the origin of the planets of our solar system occurred between 4.5 and 5 billion years ago.

As the earth condensed, the heavier elements were drawn toward the molten center while release of radioactive material kept the interior very hot (the earth is still very hot at a depth of 50 km). For example, nickel and iron were drawn toward the core, whereas the lighter elements such as hydrogen and helium formed the primitive atmosphere. The size of a planet is critical in retaining an atmosphere, and the evidence suggests that, unlike the big planets Jupiter and Saturn, Earth was too small to retain the gases and they escaped into space. Thus the primitive earth was at first a giant rock without any atmosphere. Later, continual compression and heating within caused the buildup of enormous pressures, causing volcanoes to erupt on the surface, which spewed hydrogen (H_2), ammonia (NH_3), water (H_2O), and methane (CH_4) into the atmosphere. As the volcanoes erupted with gaseous explosions not unlike those occurring today, a second atmosphere was formed. It is interesting that all these primitive gases were inorganic except methane, and evidence suggests that methane did indeed exist long before any organisms.

In its final formative stages the earth had a molten core of nickel and iron covered by a very thick layer of silicates of iron and magnesium. Finally, the outer crust was composed of the very light silicates that make up the current surface of the planet. This layer is only 8 to 65 km thick. It is, of course, from this layer that radioisotope dating has revealed the age of the earth. The oldest known rocks are about 4 billion years old.

DATING THE EARTH

Even if you have never been on a thrilling raft trip, encountering the white water and rapids, you have probably observed this spectacular form of recreation on television or in the movies. Exerting a tremendous force, the moving water carries a great deal of mineral and organic matter. Giant boulders, pebbles, tiny grains of sand, clay, branches of trees, seeds, leaves, and giant logs can be carried along. In times of flash floods even entire buildings, wildlife, domestic animals, and humans are swept away. These events may have been the early stages of fossil formation (figure 17.1). When the stream or river slows as it meets a pond, lake, or ocean, the bits and pieces being carried by the water begin to settle out according to weight; the heavier particles are deposited first, the silt and clay sediments later. Such sized deposits lead to striations and the formation of deltas; the deltaic deposits reveal layers of varying thickness and components. Finally, the sediments may be converted into rock, the sand becoming sandstone, and the mud becoming shale. Such rocks are called sedimentary rock, and they reveal a detailed record of geological history. Plant materials that were carried along with the mineral particles may have been incorporated with the sediments and later found within the rock, provided that they did not decay before fossilization. The rock formation process involves compaction, cementation, and removal of water.

There is reason to believe that sedimentation has been going on since the crust of the earth first formed. No single sedimentary profile exists to depict the entire history of the earth, but segments of the record occur throughout the world, and rocks of the same age, as determined by isotope dating, may be found on all landmasses.

Scientists who study ancient life are called **paleontologists,** and those who study ancient plants are paleobotanists. These scientists use the fossil record to explain the evolution of groups of organisms and to put the relative age of these organisms in proper perspective. A **fossil** is any trace of a former living organism, and fossils may include all or part of an intact specimen, often found embedded in amber (the resin of ancient trees), in tar pits, or in bogs and other places where decay has been prevented (figures 17.2 and 17.3). Sometimes fossils are molds, casts, or impressions in which the shape of the original organism is retained, but the organism itself has decomposed. These impressions are often left if the plant material is rapidly covered with fine sediments or volcanic ash.

Another kind of fossil is a compression, formed when carbonized plant material is still present in the original shape but is greatly compressed and reduced in size because of exceptional pressure. In petrifications, as in petrified wood, the plant parts are infiltrated or replaced by mineral substances such that the structure is preserved but the fossil is actually rock. Petrifications result in the preservation of minute detail.

A final form of fossil evidence is the accumulation of products once associated with the living organism. Coal and oil, for example, represent incomplete decay products of plants and are part of the fossil record.

Fossil evidence from many localities throughout the world and at various depths has allowed paleontologists to reconstruct the fossil record, and from this record biologists have developed a detailed scenario of life at various times throughout the history of the earth (figure 17.4).

The sedimentation record is obviously restricted to surface features

Figure 17.1
Sedimentation carried by normal runoff, floods, and stream flow contributes to the debris that may become fossilized in sedimentary layers.

Figure 17.2
Because sedimentation occurs quickly, usually as the result of floods, volcanic eruptions, and similar catastrophic events, portions of plants and animals can be embedded and protected from decay. These large dinosaur bones were found in such a sedimentary layer.

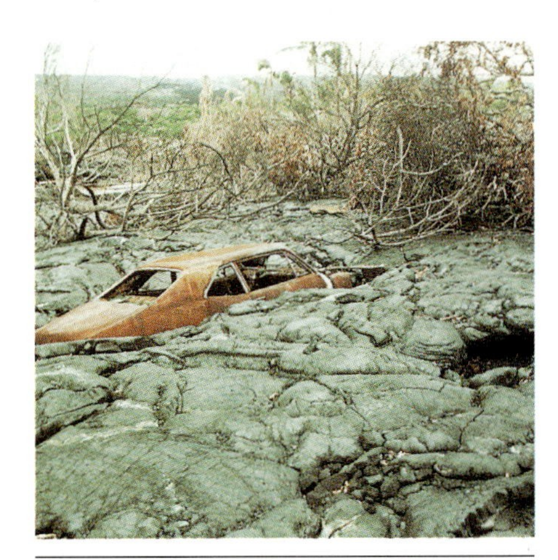

Figure 17.3
Fossil formation is quite likely from volcanic eruptions, lava flows, and ash deposits that cover an area quickly before decomposition can occur. This car was covered by lava flow and ash after this Hawaiian eruption in Maunu Ulu. Many plants and animals are preserved in this sedimentation.

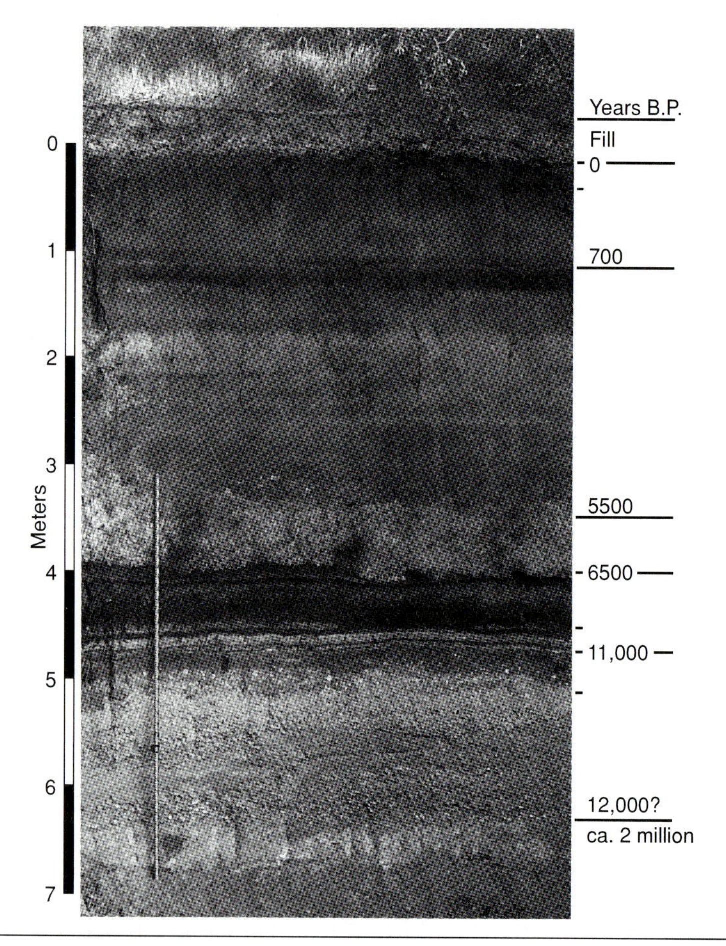

Figure 17.4

A sedimentary profile depicts the life history of an area and can sometimes be used with exceptional accuracy for dating archaeological and anthropological events for that area. This sedimentary profile from the Lubbock Lake National and State Landmark at Lubbock, Texas, provides an excellent account of early human occupation of the site dating back approximately 12,000 years. The 12,000-year level provides good evidence that early humans (Clovis man) were hunters with a kill and butchering area located at the archaeological dig. Researchers have subdivided the layers according to past events to categorize the artifacts found there.

characterized by low-lying areas, such as river basins, lake beds, valleys, and canyons. Hills and mountaintops would be poor places to search for the fossil record. There is considerable evidence, for example, that angiosperms evolved in the highlands at a date much earlier than is actually indicated by the fossil record. One must also be careful to consider the effects of upheaval or sinking in modifying the sedimentary layers. Some fossil evidence may, as a result, appear in the "wrong" place. The dating of rocks and organic material is accomplished with radioactive isotopes.

Isotopes

You have already seen that the number of protons, the atomic number, determines the nature of any chemical element. The number of neutrons, however, may be the same as the number of protons or different. Atomic forms of the same chemical element that differ only in the number of neutrons are called isotopes. Ordinary hydrogen, with an atomic weight of 1, has a single proton, a single electron, and no neutron. A second isotope, called deuterium (^2H), has one proton, one electron, and one neutron. A third isotope, called tritium (^3H), is extremely rare and contains one proton, one electron, and two neutrons. All three of these isotopes are indeed hydrogen, but their atomic weight differs. The chemical behavior of all three isotopes is essentially the same, since the chemical nature of the element is determined by the electrons. The difference in atomic weight can be detected with a mass spectrophotometer, a modern laboratory instrument that is used in the identification of compounds.

Most of the common elements have naturally occurring isotopes, and some of these are quite stable; others are radioactive. In radioisotopes the nucleus is unstable and emits energy of various wavelengths as the nucleus gradually changes to a more stable form. The energy can be detected and measured by a Geiger counter, and it can also be detected on x-ray film. The energy of radioactive decay can be used in radiation therapy in medicine, and it can be used to trace the path of specific "tagged" molecules in living organisms or in biochemical reactions. Isotopes can also be used for the radioisotope dating of rocks and organic materials found in the fossil record.

Radioactive decay is the emission of a specific kind of energy at a fixed rate for any given isotope. The rate of decay is measured in terms of half-life, the time required for half of the radiation to be emitted. Half-lives vary tremendously, and isotopes with very short half-lives are used for short-term experiments. If an isotope has a half-life of one year, at the end of that time only half the amount of isotope remains. At the end of the second year one-fourth remains, and at the end of three years only one-eighth remains. The relative proportion of stable isotope versus radioisotope in a given sample is a good indication of the age of the sample.

Radioactive carbon 14 (six protons, eight neutrons) can be used for dating carbon-containing rocks and organic matter between the ages of about 100 and 50,000 years. The ratio of carbon 14 (^{14}C) to carbon 12 (^{12}C) is constant

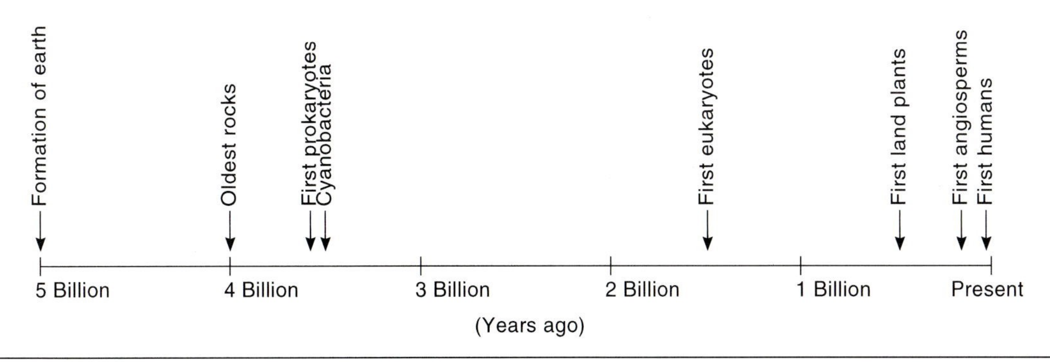

Figure 17.5
Radioisotope dating may be used to establish important events during the formation and aging of the earth. Arrows indicate specific dates of events as determined by many independent investigators. Paleontologists are confident about the approximate dates for these events.

on earth and in living organisms. Carbon 14 emits β-rays at a given rate such that the half-life is 5730 years. Thus, at the end of 11,460 years, only one-fourth of the radioisotope remains. After approximately 50,000 years so little radioactivity remains that the technique is no longer accurate. For materials older than 50,000 years, other isotopes must be used in the dating process (figure 17.5).

When an organism dies, it no longer obtains new ^{14}C by photosynthesis or the products of photosynthesis, yet the amount of ^{14}C in the atmosphere remains constant (^{14}C comes from the bombardment of nitrogen with cosmic rays). The ^{14}C in the dead organism decays at a known rate, and the ratio of ^{14}C to ^{12}C can be used to determine the age.

The radioactive isotope for uranium (^{238}U) has a half-life of 4.5 billion years. Therefore any sedimentary rock that contains ^{238}U may be used to date those rocks. This isotope decays through several intermediates into lead (^{206}Pb), and at the end of 4.5 billion years one would have only half as much ^{238}U as in the beginning. The ratio of ^{238}U to ^{206}Pb has been used to date the earliest rocks at nearly 4 billion years. Other radioisotopes used in the study of fossil rocks are potassium (^{40}K), with a half-life of 1.3 billion years; rubidium (^{87}Rb), with a half-life of 48 billion years; thorium (^{232}Th), with a half-life of 14 billion years; and a different uranium (^{235}U), with a half-life of 710 million years.

Geologic Time

The evidence for life on earth as early as 3.5 billion years ago comes from the experiments of a British engineer named William Smith (1769–1839). "Strata" Smith recognized that each of the different layers of rock, or strata, had its own characteristic fossils and that the lower the layer, the less its fossils resembled modern forms. The shallower the strata, the more closely the fossils resembled living forms. He found vertebrates in the shallowest layers and invertebrates in the lower levels. So the fossil record for plants and other organisms also is revealed in the profile of sedimentary rock. The most primitive plants occur near the bottom.

Geologists have compared rocks from many parts of the world, used radioisotope dating, and finally reconstructed a geological time scale (table 17.1) based on the evidence from that record. This system of naming rock units evolved slowly, and the names were often taken from the locality where the sedimentation was first discovered. **Cambrian** comes from the medieval Latin name for Wales. **Ordovician** and **Silurian,** derived from the names of ancient Welsh tribes, are used to describe the rocks in that part of the world. **Devonian** is used to describe the fossil-bearing strata near Devonshire. **Permian** comes from the Perm province of the former Soviet Union. The **Carboniferous** period comprises the **Pennsylvanian** and **Mississippian,** in which many of the world's fossil fuels were produced; they were named from rocks found in those states. Geologists and paleontologists now agree that geological time can be divided into four eras: **Cenozoic, Mesozoic, Paleozoic,** and **Precambrian.** Each of these eras is further divided into periods, and the most recent Cenozoic era is further subdivided into epochs. The boundaries between any two eras, periods, and epochs are somewhat arbitrary, but the time frame is accepted by most scientists. Current geological history is described as being in the Cenozoic era, the Quaternary period, and the Recent epoch.

CHEMICAL EVOLUTION

There was no free oxygen in the primitive atmosphere. As water vapor spewed into the atmosphere from the earth and other water vapor condensed from outer space, the seas were filled. Gases from the atmosphere

✔ Concept Checks

1. Discuss the big bang theory on the origin of our universe. How was earth formed?
2. What is the age of planet Earth? How can its age be determined?
3. Reconstruct the geological time scale, first listing eras, then periods, then epochs. During what epoch did the earliest prokaryotes appear? The earliest vascular plants?

TABLE 17.1

Geological Time: Characteristics of Organisms and Physical Features of the Landscape

ERA	PERIOD	EPOCH	BEGAN (MILLIONS OF YEARS AGO)	PHYSICAL AND BIOLOGICAL CHARACTERISTICS
Cenozoic	Quaternary	Recent Pleistocene	2.5	Rise of civilizations; fluctuating climate, including Ice Age; uplift of Sierra Nevadas; extinction of many large mammals, including woolly mammoth; increasing aridity and widespread deserts; first true humans (*Homo*) at beginning of the period
	Tertiary	Pliocene	7	First hominids; cooler climate with mountain formation and widespread extinction; uplift of Panama to join North and South America; herbaceous plants become abundant in diverse climates; large carnivores appear
		Miocene	26	Grasslands increase and forests decrease as climate moderates; grazing animals and apes are prominent
		Oligocene	38	Lands low, but uplift of Alps and Himalayas; volcanoes in Rocky Mountains; many flowering plants evolve; apes appear; expansion of scrub forests
		Eocene	54	Mild to tropical; India collides with Asia; Antarctica separates from Australia; grasslands expand; horses, camels, giant birds; western North America covered with shallow lakes
		Paleocene	65	Mild and cool; shallow seas covering continents recede; first primitive primates and carnivores
Mesozoic	Cretaceous		136	Reptiles abundant; angiosperms appear and spread rapidly; last of the dinosaurs; uplift of Rockies forms a rain shadow; Africa and South America separate
	Jurassic		190	Mild climate; large areas of Europe covered by seas; cycads and ferns abundant; many dinosaurs, small mammals, flying reptiles; birds first appear
	Triassic		225	Continents still joined and mountainous; large areas are arid; volcanoes in eastern North America; first primitive mammals; first dinosaurs; ferns and gymnosperm forests are extensive

were brought down in the rain, minerals from the surrounding crust leached into the water, and the blazing sun powered a hot, churning primordial soup which was also fed by violent thunderstorms with severe lightning. This hot brew was bombarded with high-intensity ultraviolet radiation from the sun, since there was no filtering by ozone.

The Heterotrophic Theory

Recent experiments involving chemical evolution under laboratory conditions have shown that when the same gases are used in the approximate concentrations believed to have been present at the time the earth was cooling, and when an electrical discharge is placed through the gaseous mixture, the simple organic molecules common to life begin to appear in minutes or hours. Sugars, amino acids, and other molecules self-assemble; the statistical chance of their doing so depends on the number of molecular collisions of the primitive gases. Some of these molecules begin to take on more complex structures, such as chains of amino acids (polypeptides) and nucleotides necessary for nucleic acids. Thus, over a

TABLE 17.1

Geological Time: Characteristics of Organisms and Physical Features of the Landscape—cont'd.

ERA	PERIOD	EPOCH	BEGAN (MILLIONS OF YEARS AGO)	PHYSICAL AND BIOLOGICAL CHARACTERISTICS
Paleozoic	Permian		280	Uplift of the Appalachians; most of the seas drained from North America; earliest cycads, gingkos, and conifers; first reptiles
	Pennsylvanian Mississippian	Carboniferous	325 345	Mountain formation in eastern North America, Colorado, and Texas; extensive forests of primitive vascular plants, including ferns and lycopods, gymnosperms; abundant insects, first reptiles, many sharks; climate moist and warm; many fungi; many amphibians
	Devonian		385	Europe mountainous and arid; volcanoes in eastern North America; western North America low and covered by seas; evolution of vascular plants; bryophytes appear; algae and fungi; amphibians and bony fish, many mollusks
	Silurian		430	Earliest vascular land plants; mild climate; continents generally flat but mountains building in Europe; continents generally flooded; fish with jaws; arthropods invade land; coral reefs
	Ordovician		500	Mild climate; landmasses low and covered by shallow seas; limestone deposits; first fungi; some microscopic plants; shell-forming sea animals
	Cambrian		570	Mild climate; landmass covered by seas; first fishes; many invertebrates, trilobites
Precambrian			4600	Diverse climates; formation of earth's crust; cooling of surface layers; earliest prokaryotes; some autotrophs by middle of era; eukaryotes and multicellularity by close of era

long period of time, some of these molecules appear to have become self-replicating, a prerequisite for cell formation and the perpetuation of life. Groups of molecules aggregated into droplets much like oil in water. The first forms of life no doubt occurred in the sea and fed on these same molecules from which they were made. Such organisms are said to be **heterotrophic.** These organisms existed only in the ancient seas, since the ultraviolet radiation from the sun was too intense out of the water. Too great an exposure to the mutagenic ultraviolet radiation would have killed them. Surely no bacteria could have survived on the land. Today, short-wave ultraviolet radiation is used to sterilize hospital equipment and barber's tools.

As the number of cells began to increase, the number of organic food molecules began to decrease, and the entire living system appears to have come very close to starving itself to death. Organisms that did survive gradually developed the ability to obtain their energy from the sun, rather than having to depend on the dwindling supply of organic molecules in the sea. This theory of the origin of life has gained wide acceptance in recent years and is referred to as the heterotrophic theory.

Autotrophs

Studies of **stromatolites** in Western Australia have confirmed estimates of the age of life on earth by dating several different organisms found in these peculiar rocks. Photosynthetic organisms that closely resemble the structure of modern filamentous cyanobacteria have been discovered. They have been dated to be at least 3.5 billion years old, which means that nonphotosynthetic bacteria would have to be even older if the heterotrophic theory is valid. Most scientists now believe that bacteria are indeed older than 3.5 billion years.

As the supply of organic molecules began to run out in the oceans, competition among organisms, even as now, became fierce. Only those that could make the best use of limited resources were the ones most likely to survive. Over hundreds of thousands of years, cells evolved that were able to make their own energy-rich molecules out of simple inorganic substances. These new **autotrophs** were the keys to survival, and without their evolution, life on earth would have ceased. This ability to utilize inorganic molecules to synthesize organic molecules is thought to have come about approximately 3.2 billion years ago. The photosynthetic bacteria were able to utilize light energy to obtain a reducing source (hydrogen) from a substance other than water, usually hydrogen sulfide (H_2S). The reaction, much like the Hill reaction in photosynthesis, produces no oxygen, and therefore these photosynthetic bacteria failed to contribute oxygen to the atmosphere.

It was not until approximately 3 billion years ago that the cyanobacteria evolved a scheme for using chlorophyll *a* and developed the mechanism for splitting a molecule of water, thus producing H^+ as a reducing source and releasing oxygen into the atmosphere. This oxygen accumulation was not substantial until approximately 1 billion years ago. By that time there were enough photosynthetic organisms on earth to begin to make an impact, and the concentration of oxygen in our atmosphere has steadily increased since that time to the present. Approximately 21% of the atmosphere is now oxygen, and it is relatively stable so long as the rate of photosynthesis on a worldwide basis does not decrease.

Only in the past 400 million years has the level of oxygen become great enough to produce a substantial quantity of ozone (O_3), a natural by-product of oxygen. As an efficient absorber of ultraviolet radiation, ozone dramatically reduced the amount of harmful ultraviolet radiation arriving from the sun. Consequently, photosynthetic organisms were able to make the transition from ocean to land.

DIVERSITY OF LIFE: A FIVE-KINGDOM SCHEME

You might think that any nonbiologist could certainly tell the difference between plants and animals. After all, there is no mistaking an elephant from a rose bush; yet the differences become less discernible in the simple organisms, particularly the single-celled ones. The debate still rages about whether such organisms should all be called plants or animals or put into a different category entirely. After completing a course in botany, and perhaps one in zoology or microbiology, you will be impressed with the similarities that pervade the biological world. Recall, for example, the process of aerobic respiration and how the biochemical reactions of glycolysis, the Krebs cycle, and electron transport are similar in all aerobic organisms. The same is true of the cellular organelles and of the processes of mitosis and meiosis. Although slight differences exist here and there, a common thread binds the basic processes in all life.

Prior to the explosion of information in the life sciences, it seemed that a general classification of the living world would have to start with two kingdoms: plants and animals. With the advent of the microscope, biochemistry, and molecular biology, it became apparent that some organisms failed to fit neatly into either category. Plants, for example, are generally expected to carry on photosynthesis, yet fungi, all of which are heterotrophic, have traditionally been studied within botany. Some organisms, such as *Euglena,* a single-celled microscopic organism, have the ability to act as autotrophs in sunlight but grow and reproduce normally if fed a carbon source and kept in the dark. What should we do with these organisms that confound our plant and animal scheme?

There are several approaches to observing living organisms, and they may be classified in various ways. Certainly not all biologists, but a rapidly growing number of them, have established a five-kingdom scheme for classifying organisms (figure 17.6). The purpose of this section is to introduce the kingdoms and their broad characteristics. You will no doubt be able to see the limitations of any organizational scheme as soon as you understand the mechanics of the organization.

Kingdom Monera

The only living members of the kingdom Monera are the **bacteria, archaea,** and **cyanobacteria** (figure 17.7). The cyanobacteria were formerly called blue-green algae, but because of confusion with classification of the true algae, the term cyanobacteria is more descriptive. Monerans are the simplest of all organisms, and as **prokaryotes** they have no nuclear envelope. The genetic material is in the form of one long, circular molecule of DNA that is not combined with protein, as it is in higher organisms. Cytokinesis in these organisms is not accompanied by mitosis but is accomplished by transverse fission. The cytoplasm does contain ribosomes that function in protein synthesis, but membrane-bound organelles such as chloroplasts and mitochondria are absent. The cytoplasm is enclosed by a plasma membrane and a cell wall. The composition of that wall is not cellulose, but a complex polysaccharide-protein, peptidoglycan, unlike that found in any other type of cell.

Kingdom Protista

The protists include eukaryotic heterotrophs and autotrophs that are unicellular, colonial, or filamentous. Some protists are large multicellular forms. This group is by nature extremely diverse with poorly defined boundaries and is considered a catchall by many scientists. Some of these organisms are animal-like, some are plantlike, and still others are fungal (figure 17.8). The unifying theme is that they are all eukaryotic and produce spores and gametes within the boundary of a single cell (i.e., they have unicellular gametangia

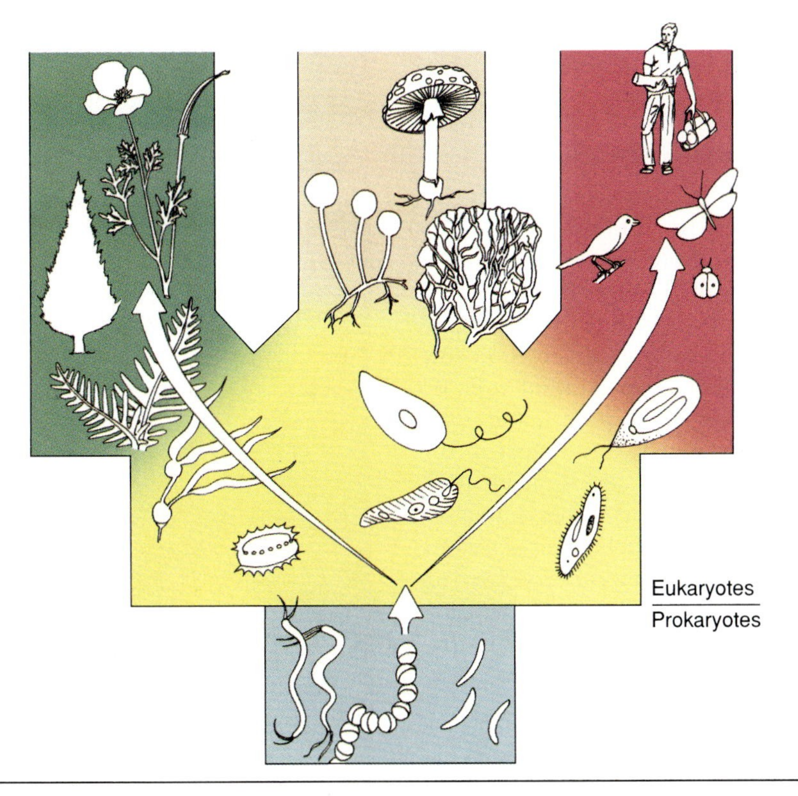

Figure 17.6
Simplified illustration of the five-kingdom system of classification. The oldest organisms are the prokaryotic monerans, from which the eukaryotic protists evolved. Plants, fungi, and animals arose from the protists.

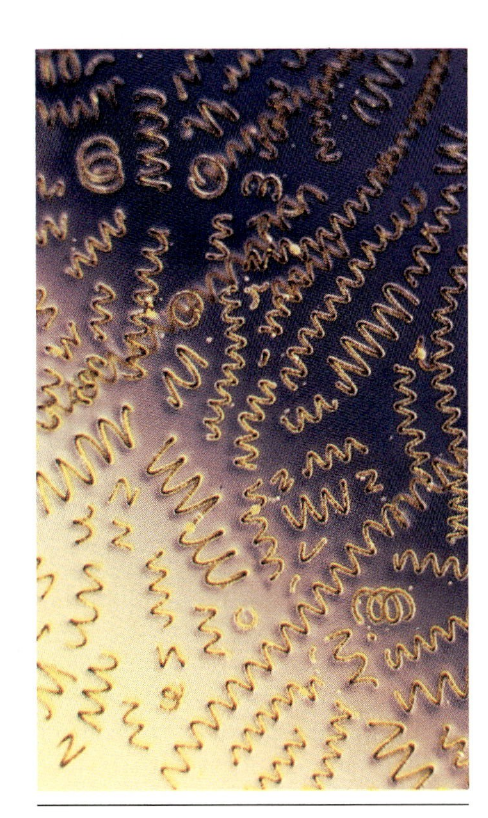

Figure 17.7
Anabaena. Cyanobacteria are representative of the kingdom Monera, comprising all the prokaryotes (×150).

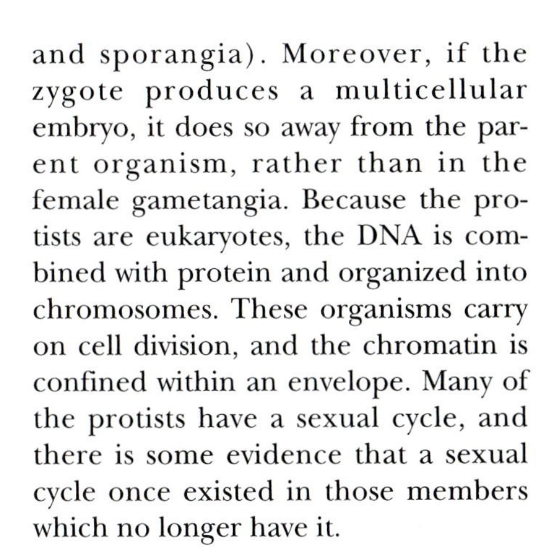

and sporangia). Moreover, if the zygote produces a multicellular embryo, it does so away from the parent organism, rather than in the female gametangia. Because the protists are eukaryotes, the DNA is combined with protein and organized into chromosomes. These organisms carry on cell division, and the chromatin is confined within an envelope. Many of the protists have a sexual cycle, and there is some evidence that a sexual cycle once existed in those members which no longer have it.

Kingdom Fungi

The fungi have been a taxonomic problem for many years. Although often taught with botany, they have many characteristics that distinguish them from both plants and animals (figure 17.9). They are all heterotrophs, and although some cellulose may be found in certain species, the cell wall is composed of **chitin,** a polysaccharide that is never found in

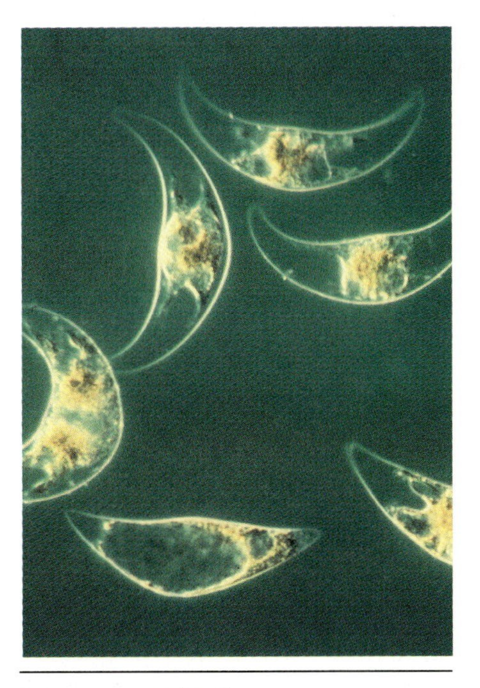

Figure 17.8
Procystis. The kingdom Protista is characterized by unicellular, filamentous, colonial, and multicellular eukaryotic organisms. These dinoflagellates are typical of the protists and are bioluminescent.

Figure 17.9
This poisonous mushroom, *Amanita verna* (Death Angel), is a representative of the kingdom Fungi.

plants. Chitin is, however, the principal component of the exoskeleton of insects and crustaceans. Most fungi reproduce both sexually and asexually. Fungi typically obtain nutrients by secreting digestive enzymes termed **exoenzymes** into the food source and absorbing the smaller organic molecules that are released. They live in soil, water, or other media that contain organic substances. As parasites and saprophytes, they act in consort with bacteria as the world's decomposers.

Kingdom Plantae

The plant kingdom includes autotrophic, eukaryotic, multicellular organisms (figure 17.10). The common ancestor for plants is probably a fairly complex embryo-forming and retaining green alga that invaded land some 450 to 500 million years ago. Terrestrial plants typically have a waxy cuticle that protects all aboveground parts from desiccation. Most of them also have stomata that allow for gas exchange. In the early land plants, stomata were poorly developed and in some cases nonexistent. Sex cells are surrounded by protective layers, which also prevent desiccation, and all plants have a reproductive cycle that involves an **alternation of generations.** In this cycle, the fertilized egg and the multicellular embryo which develops from it are retained and nurtured in the female sex organ.

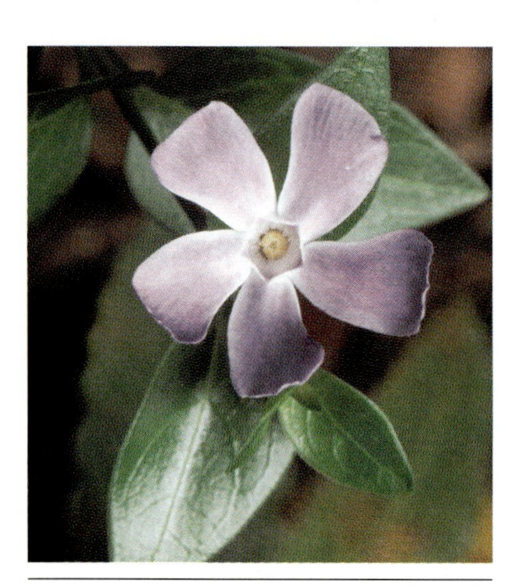

Figure 17.10
The plant kingdom includes eukaryotic, multicellular photosynthetic organisms. Flowers of the periwinkle, *Vinca major.*

Kingdom Animalia

Like the fungi, animals are multicellular, eukaryotic, heterotrophs that depend on photosynthetic organisms for their nourishment. Food is transported as glucose, but it is stored as glycogen, a complex macromolecule made from glucose. Animal cells do not have walls, and both cell and tissue movement is much easier than in the rigid-walled plant cells. Reproduction is usually sexual. The higher animals—arthropods and vertebrates—are the most complex organisms, having many specialized tissues, including sensory and neuromotor mechanisms not found in other organisms (figure 17.11).

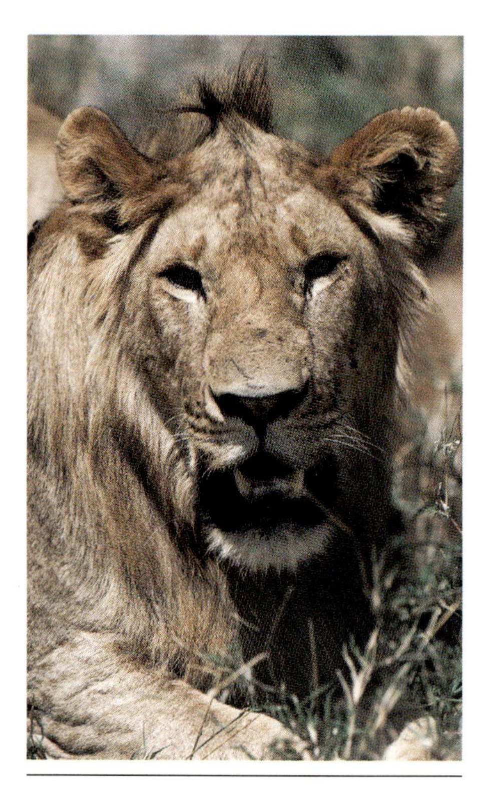

Figure 17.11
The animal kingdom includes the eukaryotic, multicellular heterotrophs such as this lion.

VIRUSES, VIROIDS, AND PRIONS

In any discussion of the primitive organisms, viruses must be considered. Virologists and microbiologists agree that these particles are *noncellular* and are therefore generally not included in the five-kingdom classification scheme. Virologists do not refer to viruses as living or dead, but rather as active (infectious) or inactive (noninfectious). They contain either DNA or RNA (never both) and are categorized as either DNA viruses or RNA viruses (table 17.2).

The nucleic acid core is surrounded by a coat of protein termed the **capsid,** and this organization constitutes the virus. In some viruses, such as the HIV (human immunodeficiency virus), responsible for Acquired Immunodeficiency Syndrome (AIDS), the capsid is surrounded by an envelope (figure 17.12). Although it would be easy to describe an evolutionary scenario in which a simplified group of molecules—the nucleic acids and proteins—gradually gave rise to prokaryotes, such does not seem to be the case. All viruses are parasitic, that is, they must have a living host cell to reproduce. Therefore it does not seem likely that viruses could have been the first organisms on earth. The morphology of viruses is variable, ranging from round (polyhedral symmetry) to rod (helical symmetry), to complex (helical and polyhedral) shapes, the subparticles forming a crystalline arrangement. Adenovirus, for example, one of the many viruses that cause the common cold in humans, is an icosahedron, or 20-sided figure, made up of equilateral triangles of protein subunits termed capsomeres. The total number of capsomeres is 252. The tobacco mosaic virus is a rod-shaped RNA. The protein coat is made up of 2200 protomeres, protein molecules, which are identical, each with 158 amino acids (figure 17.13).

Viruses are generally smaller than bacteria (figure 17.14), some as small as 20 nm (picornaviruses) and others nearly as large as a bacterial cell (poxviruses). Most viruses pass through filters designed to prevent the passage of bacteria, and other means must be found for protection from infection and disease.

Viruses change the function of a host cell by substituting their own DNA or RNA for that of the host. Viral diseases in humans include the common cold, influenza, measles, and polio. There are many viral diseases of plants that are often apparent from mosaic or white patches on leaves, reduced rates of growth, and in extreme cases death (table 17.2). Virologists now feel that virus-free plants are rare, and many crop yields are reduced because of undiagnosed viral infection.

Bacteria themselves have their own set of viruses known as **bacteriophages** (figures 17.15 and 17.16). Whenever a bacteriophage (also known as phage) infects a bacterial cell, six fiberlike protein structures attach to the cell wall, the baseplate pins of the phage are pulled close to the wall, and digestive enzymes dissolve the bacterial wall to allow penetration of the cell. The DNA, stored in the nucleocapsid of the virus, is injected into the bacterial cell. Within a matter of seconds the DNA from the phage takes control of the cell's machinery and metabolism, directing the synthesis of additional viral DNA and its assembly into the new phage particles exactly like the original. Within half an hour the cell lyses, or disintegrates, and the new viral particles, fully infectious, are ready to attack additional bacterial cells.

The evidence is now convincing that viruses cause some cancers and other uncontrolled growths in many plants and animals. One does not usually detect the virus in cancer cells, however, because once the viral nucleic acid enters into a cell, it becomes incorporated into the genetic material of the host cell. Such cells are said to contain a provirus, and the trigger for their malignancy could be caused by any number of conditions, including tars and nicotine, environmental pollution with heavy metals or complex synthetic molecules, and short-wave radiation. Cells also contain oncogenes, or cancer-causing genes, that can be "turned on" by viruses or other environmental factors named above.

Even more puzzling than viruses is a group of "organisms" called **viroids,** which are responsible for a number of destructive diseases of cultivated plants, but may also occur and cause disease in animals. Unlike the viruses, viroids are short, naked strands of RNA, usually no more than 300 to 400 nucleotides long. There is no protein coat, and they appear to have enough genetic information to code for only a single protein. Viroids are known to be the causative agents of death in seed potatoes, citrus and avocado groves, chrysanthemum, and coconut

TABLE 17.2

Classification of Viruses and Viroids

CATEGORY	EXAMPLES OF DISEASES PRODUCED (VIROID ABBREVIATION)	
DNA Viruses		
Adenoviruses	Acute to mild respiratory infections	
Caulimovirus	Strawberry vein banding; figwort mosaic; dahlia mosaic; cauliflower mosaic	
Hepadnavirus	Acute and chronic hepatitis	
Herpes viruses	Fever blisters; chicken pox; jaundice; certain eye, skin, and genital infections; implicated in infectious mononucleosis	
Papovaviruses	Warts in humans	
Poxviruses	Smallpox; cowpox; formation of fibromas (nodules or benign tumors) on hands and skin	
RNA Viruses		
Closterovirus	Chlorotic leaf spot	
Comoviruses	Bean curly dwarf; bean-pod mottle; radish and squash mosaic	
Cucumovuirs	Cucumber mosaic; peanut stunt	
Enteroviruses	Diarrhea; polio in humans	
Influenza viruses	Influenzas	
Necrovirus	Cucumber necrosis; tobacco necrosis	
Picornavirus	Infectious hepatitis, croup	
Retroviruses	AIDS, certain tumors (sarcomas); leukemia	
Rhabdoviruses	Rabies, barley yellow striate mosaic; beet leaf curl	
Rhinoviruses	Common colds	
Tobamovirus	Tobacco mosaic virus (TMV); tomato mosaic	
Tymovirus	Turnip yellow mosaic	
Viroids		
	Potato spindle tuber viroid	(PSTV)
	Citrus exocortis viroid	(CEV)
	Chrysanthemum stunt viroid	(CSV)
	Chrysanthemum chlorotic mottle viroid	(ChCMV)
	Coconut cadang-cadang viroid	(CCCV)
	Cucumber pale fruit viroid	(CPFV)
	Hop stunt viroid	(HSV)
	Avocado sunblotch viroid	(ASBV)
	Tomato apical stunt viroid	(TASV)

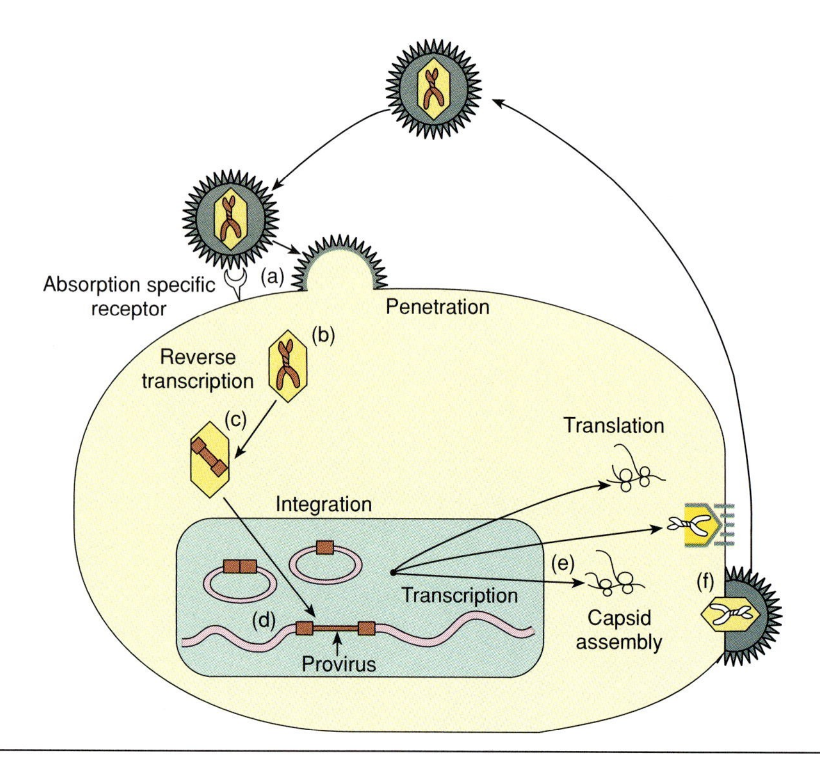

Figure 17.12

Replication of the human immunodeficiency virus. (*a*) Attachment of virus to a specific cell-surface receptor. (*b*) Penetration of nucleic acid-containing capsid into host cell (note that viral envelope incorporates with the membrane of host cell). (*c*) RNA of virus is converted to a double-stranded DNA through process of reverse transcription. (*d*) DNA transit to host-cell nucleus and integration into host-cell DNA to form the provirus. (*e*) Using the host cell's nuclear and cytoplasmic machinery, the provirus acts as a template for manufacturing of new viral RNA (transcription) which become encapsulated by newly manufactured protein. (*f*) Assembled viruses depart host cell and in the process new envelopes are formed.

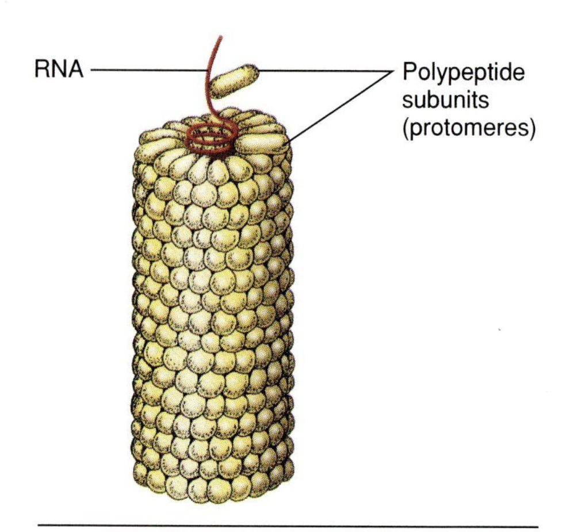

Figure 17.13

Although viruses come in various shapes and sizes, they all consist of a protein coat surrounding a core of nucleic acid (either DNA or RNA, but never both). The core of this tobacco mosaic virus is RNA surrounded by 2200 identical protein molecules, each containing 158 amino acids.

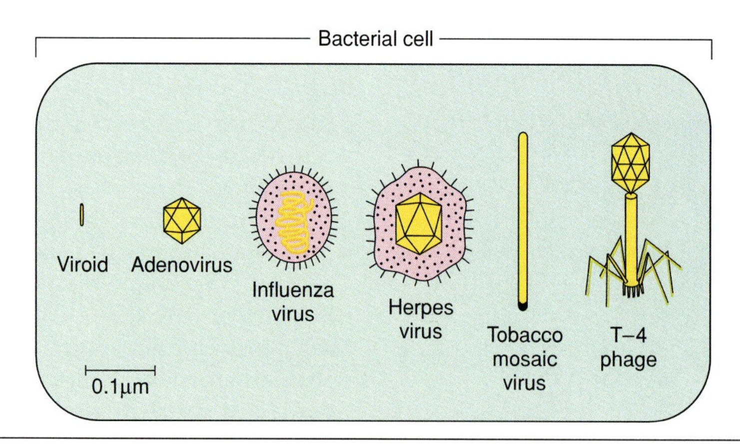

Figure 17.14

The relative size of viroids and viruses to the bacterial cell *E. coli*.

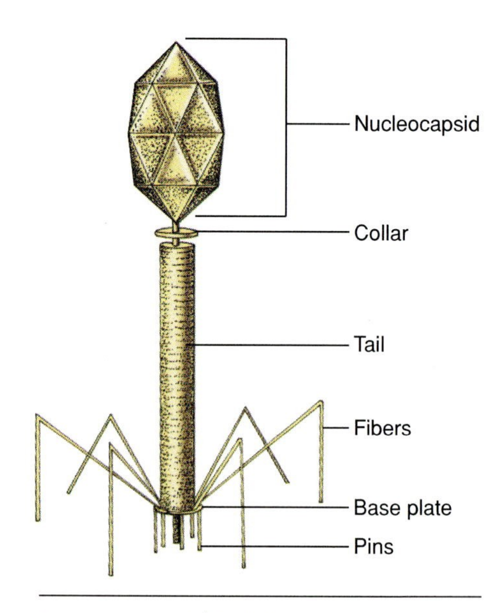

Figure 17.15

A virus that attacks bacterial cells is called a bacteriophage or phage. Its space-age construction, consisting of fibrous "feet," a main body, and "head," provides maximum efficiency in injecting the nucleic acid into its host bacterial cell.

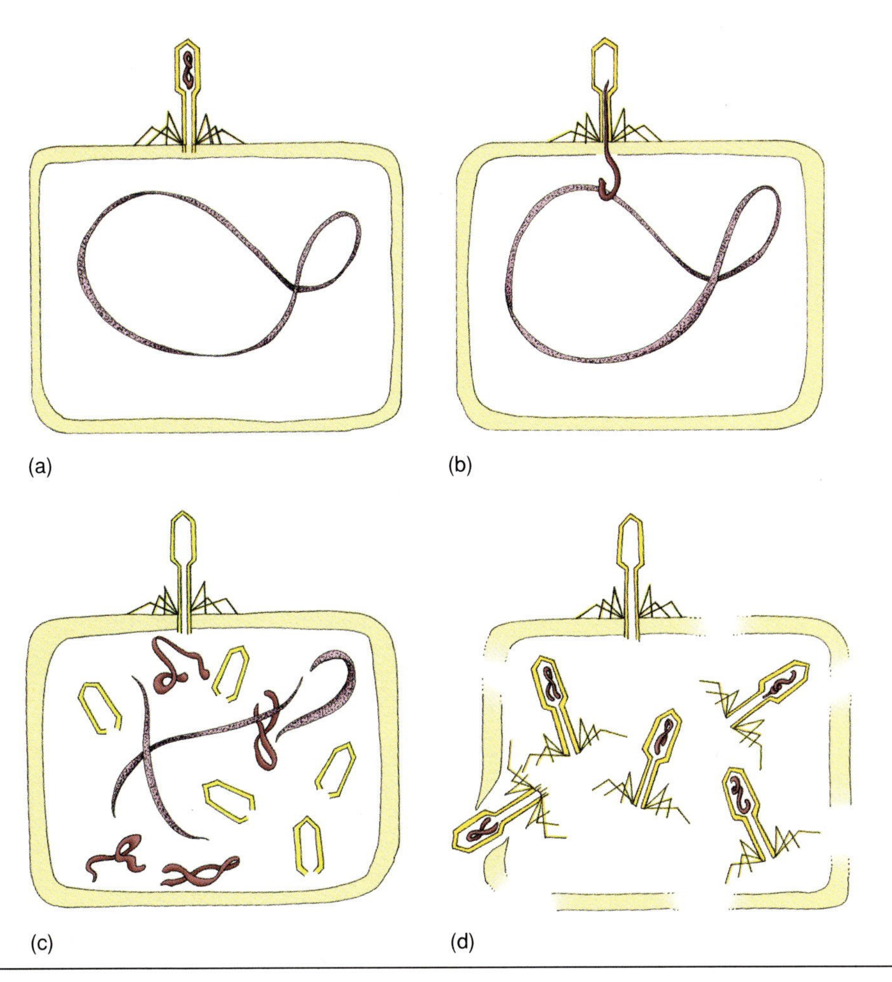

Figure 17.16
Whenever a bacteriophage attaches to the bacterial cell wall (*a*), the six fibers attach to the cell wall, pulling the baseplate to the cell, and the nucleic acid contained in the head is injected into the cell (*b*), much like the action of a syringe. A nuclease enzyme hydrolyzes the bacterial chromosome. The viral nucleic acid takes over the reproductive functions, forming new phages (*c*). Within 30 minutes the cell lyses, and several hundred newly formed phages emerge, fully capable of infecting new bacterial cells (*d*).

palms. More recently, prions, small proteins that do not seem to be associated with any nucleic acid, yet are capable of expressing infectivity, especially neurological diseases such as scrapie and kuru, have been identified. Little is known about these particles, and their discovery further complicates the entire scheme of biological organization and has far-reaching implications for human and veterinary medicine as well as for molecular biology.

> **✔ Concept Checks**
>
> 1. Discuss the heterotrophic theory for the origin of life.
> 2. Construct a diagram placing the five kingdoms into an evolutionary tree arrangement.
> 3. What is a virus? A viroid? A prion? How do they fit into the five-kingdom scheme of classification?

THE PROKARYOTES

Bacteria

Bacteria are not only the smallest and oldest group of organisms, they are also the most abundant. Many of them are approximately 1 μm in diameter (about the size of a mitochondrion), although some may be 60 μm long. With a cell regeneration time as short as 10 to 12 minutes in some species, the world would literally be covered up with bacteria in a short time if it were not for reproductive constraints imposed by nutrient availability, temperature, moisture, space, and other factors. To be convinced of the magnitude of the reproductive potential, calculate the number of cells produced at the end of 2 hours, beginning with a single cell and allowing division to occur every 10 minutes. In that short period there would be 4096 cells. Imagine the consequences of uncontrolled reproduction for 24 hours, or a year!

Even though bacteria are usually thought of as being **pathogens,** or disease-producing organisms, few of them actually cause disease in humans. Instead, they represent a remarkable force in the balance of nature. Together with the fungi, they function as decomposers of organic matter, returning the released nutrients to the soil. Bacteria are also important in many industrial functions and in the processing of food and medicines, and as models in genetic engineering and recombinant DNA technology.

Bacteria are single-celled organisms. For such organisms, size, shape, and function become more important than external features such as appendage shape and size, reproductive structures, and other features. A number of criteria are used to classify bacteria, and this classification is necessary in medicine, food production, and industrial processes in which pure strains are essential.

Classification

There are three basic shapes of bacteria, and these morphological differences are used as the beginning basis of classification (figure 17.17).

1. **coccus** (pl. cocci)—spherical
2. **bacillus** (pl. bacilli)—oblong or rod shaped
3. **spirillum** (pl. spirilla)—corkscrew shaped

Various aggregations of these basic cell shapes also are used in classification. For example, cocci may occur in pairs and are termed diplococci, in chains (as in *Streptococcus*), or

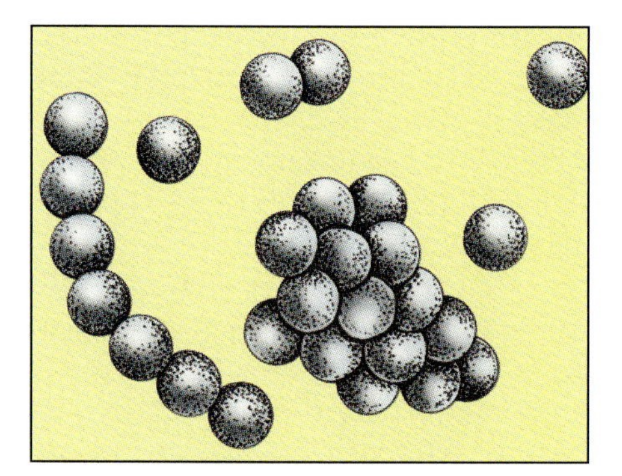

(a)

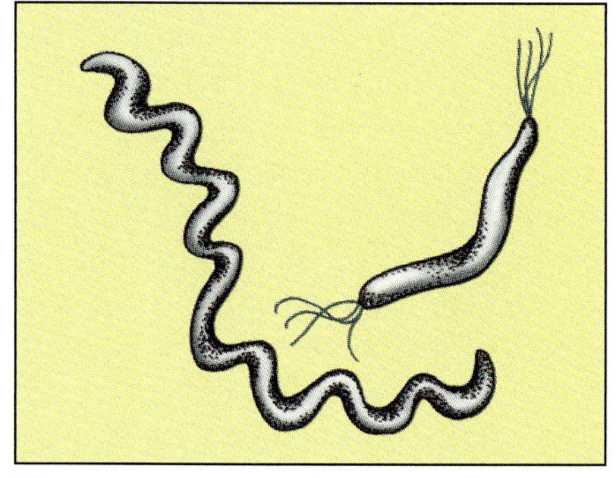

(b)

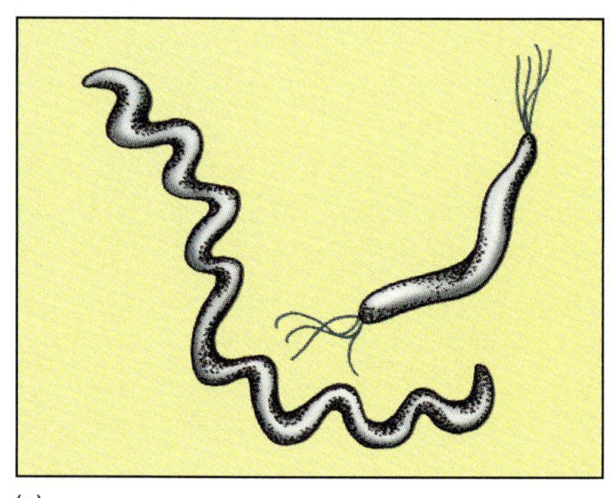

(c)

Figure 17.17

The classification of bacteria is based primarily on shape. (*a*) Spherical cells are referred to as cocci, doubles are termed diplococci, chains are streptococci, and clusters are staphylococci. (*b*) Elongated or rod-shaped cells are termed bacilli, and they may occur singly or in chains. (*c*) Spirilla are corkscrew-shaped cells, and they almost always occur as individuals. Often they have flagella on one or both ends or even across the entire cell wall.

in grapelike clusters (as in *Staphylococcus*). Bacilli often occur in chains. Spirilla usually occur singly. Some bacteria are motile, able to move about in aqueous environments by using simple flagella, long whiplike projections. Many bacteria also possess pili, very short whiskerlike projections that have special antigenic properties or may serve as attachment sites for viruses.

The nature of the cell wall and plasma membrane is such that all bacterial cells can be readily stained with crystal violet and iodine. Cells that retain the stain following destaining are called gram-positive, whereas those which do not retain the stain are called gram-negative. This quick Gram-staining procedure is another means of initial separation of species and classification.

Another diagnostic technique takes advantage of the oxygen requirements of some bacteria. In a defined medium in the presence of oxygen **aerobic** organisms grow rapidly; **anaerobic** organisms grow only when oxygen is removed from the medium; and facultative anaerobes grow anaerobically but can switch to aerobic metabolism and reproduce with or without atmospheric oxygen.

When bacteria are placed on particular kinds of nutrient media, they can grow rapidly. Within a day or two, millions of them may produce a particular size, shape, color, and texture of a **colony.** These colony characteristics can be used to help identify an organism.

The defined medium itself may also be used as a diagnostic tool. Some bacteria cannot live without specific nutrients in the medium. For example, eliminating nitrogen from a defined medium may facilitate the separation of similar organisms.

The following description of a particular organism demonstrates how the characteristics just described are used to classify organisms:

Escherichia coli: a bacillus commonly found in the human digestive tract; gram-negative; facultative anaerobe.

Cell Structure

Bacterial cells are always enclosed by a cell wall of a polysaccharide-protein complex quite different from the cellulosic wall of a higher plant. The plasma membrane has typical unit membrane characteristics, and bacterial ribosomes, slightly smaller than eukaryotic ribosomes, carry out the same function of protein synthesis. Recall that prokaryotic cells (bacteria and cyanobacteria) have no chromosomes and no nucleus. Cell replication is ensured by a single, naked circular thread of DNA. Many bacteria also have plasmids small, autonomously replicating, circular segments of DNA. At the time of division the DNA molecule is replicated, and the cell simply divides into equal halves by a process called fission, or pinching in half (figure 17.18). Although bacterial cells do not have sexual reproduction per se, sometimes the mixing of two closely related species will lead to an intermediate form through the processes of conjugation, transformation, and transduction. Although not called sexual reproduction, the intermediate or hybrid form does represent a sharing of DNA. This genetic recombination is relatively unimportant as a source of variability in bacteria when compared with mutation, which, combined with a high rate of reproduction, leads to much variability.

Bacterial Ecology

Because reproductive rates are so rapid in bacteria, large numbers can be produced in a short time. Perhaps no other group of organisms is so responsive and sensitive to environmental factors that influence growth and reproduction. The number of species of bacteria is truly enormous, and at least one of those species seems to fit into some niche or available space in almost every place on the planet. From the frozen depths of Antarctica to the hottest geothermal pools, certain species manage to survive. Those which tolerate temperatures near boiling are called thermophilic, and those which tolerate high concentrations of salt (as in the Great Salt Lake) are called halophilic bacteria. In spite of exceptional diversity, any given species is very sensitive to slight changes in its particular nutrition, pH, osmotic concentration, temperature, light, and moisture. Thus populations are kept under control by conditions

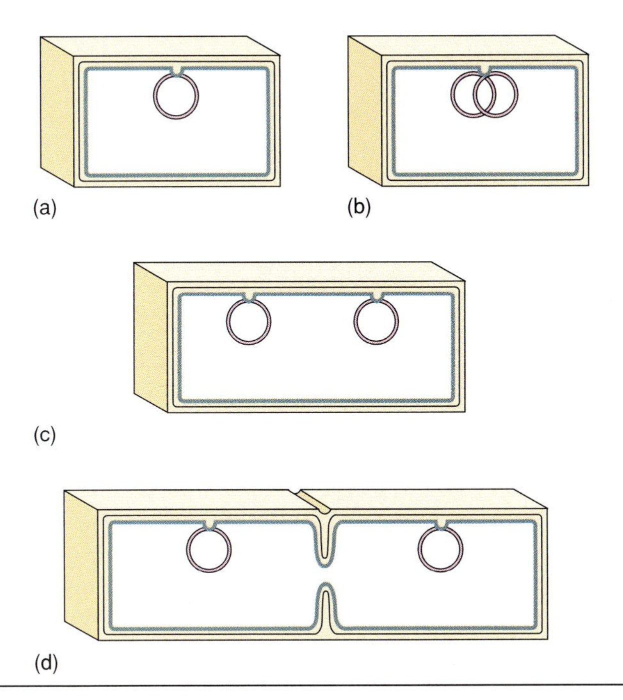

Figure 17.18
In the process of division of the bacterial cell, the circular thread of DNA is attached to the plasma membrane (*a*) and undergoes replication (*b*). The cell elongates and the two strands of DNA separate (*c*). Finally, the new wall is laid down between the nuclear threads (*d*). The new cells are pinched apart and separated completely. Cyanobacteria also divide by this fission process.

that fluctuate constantly in nature. Certain bacteria in soil, for example, may reach very low populations during a drought. However, when the rains come, the population level may reach hundreds of thousands per gram of soil within two or three days. Many species fail to die out completely during drought because of spore formation—a dormant form resistant to environmental stresses.

Beneficial Bacteria

Bacteria act as significant biological agents for human, ecological, and economic purposes. In addition to the role of microorganisms in fermentation of sugars into ethyl alcohol, certain species of bacteria are capable of carrying out other types of fermentation that lead to the production of compounds such as butyric acid and propionic acid.

Bacteria are also very important in many industrial processes. They are used to produce certain chemicals that would be either more expensive or impossible to synthesize in a chemical laboratory. These products include vinegar (acetic acid), butyl alcohol, acetone, lactic acid, and a number of vitamins. Bacteria are often used to decompose the pectins that inhibit the breakdown of fibers in flax and hemp. Both the curing of tobacco and the tanning of leather involve bacterial preparations.

In addition to functioning as decomposers in nature and therefore being able to exist on essentially any organic substrate, bacteria perform many useful functions. In **nitrogen fixation** the bacterium *Rhizobium* infects the roots of leguminous plants, causes a nodule to form, and proceeds to utilize nitrogen from the atmosphere, fixing it into ammonia that can be utilized by itself and by the plant. This must undoubtedly be one of the most beneficial processes in nature, since it allows a large number of higher plants to compete at a level that would be impossible without the supplemental nitrogen. Scientists are only now beginning to appreciate fully the importance of biological nitrogen fixation in nature. Another group of bacteria, the actinomycetes, are capable of fixing nitrogen in many other plant families. Thus far it appears that *Rhizobium* confines its activities strictly to the legume family. Genetic engineering may allow scientists to implant the genes responsible for this remarkable symbiosis into other plants that cannot now fix nitrogen.

Bacteria also have a role in the production of many foods, including yogurt and cheeses. The flavor and aroma characteristic of many kinds of cheese is imparted by a specific strain of a bacterium producing a particular chemical in the curd. Swiss cheese, for example, derives its flavor and odor from a strain of bacteria that produces propionic acid. Corn and hay silage for cattle feed are preserved by acids that result from bacterial action.

In recent years the manufacture of antibiotics has revolutionized the pharmaceutical industry, and this has come about because it is economically feasible to produce antibiotics by growing pure strains of bacteria. These antibiotics were selected for production because they are inhibitory to the growth and reproduction of pathogenic bacteria. Most of the antibiotic drugs being used today are derived from the actinomycetes, including streptomycin, actinomycin, and tetracyclines. Penicillin is an exception and is not derived from this group; it is produced by the fungus *Penicillium.*

The future of industrial microbiology looks very bright. Every day new uses are being discovered for products made by bacteria. One of the most exciting recent discoveries concerns the possibility of using bacteria as biological control agents against other types of organisms. One such organism, *Bacillus thuringiensis,* has proven effective as a selective agent that attacks certain kinds of chewing insects. The bacterial suspension is sprayed on leaves. When the insect eats the leaf, the rod-shaped bacteria are ingested and secrete enzymes that dissolve the gut of the insect. After about 24 hours the insect stops feeding, and death occurs within a few days. This organism is specific against only certain orders of insects and is not pathogenic to humans.

Pathogenic Bacteria

The mention of bacteria to most people conjures up visions of deadly diseases, and indeed a few of them deserve that reputation. Some bacteria are also pathogenic to other animals and to plants, although the largest number of plant diseases are caused by fungi. Human bacterial diseases include strep throat, bacterial pneumonia, leprosy, scarlet fever, whooping cough, meningitis, syphilis, gonorrhea, cholera, bubonic plague, Lyme disease, tuberculosis, typhoid fever, and boils. In addition, Rocky Mountain spotted fever and typhus fever are caused by rickettsias (very small and specialized bacteria) transmitted by ticks, mites, and fleas.

Chemoautotrophic Bacteria

One interesting group of bacteria does not obtain energy from sunlight or organic molecules but from the oxidation of nitrogen, sulfur, iron, or gaseous hydrogen. One subgroup, the methanogenic (methane-producing) bacteria, obtains energy from CO_2 and H_2. These organisms have cell walls quite different from those of other bacteria, and their metabolism appears to be fundamentally different. These organisms are strict anaerobes, and very recent investigations have suggested that they may have evolved more than 3 billion years ago, when the anaerobic atmosphere was rich in CO_2 and H_2. Some microbiologists have suggested that **Archaea,** also termed **Archaebacteria,** composed not only of methanogenic bacteria, but thermogenic species, are so fundamentally different from the other prokaryotes (e.g., in the base sequences of their rRNAs) that they should be placed in a separate, sixth kingdom. These organisms are confined to the deep-sea trenches, where oxygen levels are extremely low and temperatures are far above boiling.

Cyanobacteria

Structure

Formerly called **blue-green algae,** cyanobacteria constitute the remainder of the kingdom Monera (prokaryotes). Their structure is quite similar to that of the true bacteria, except they all contain chlorophyll. These are single-celled organisms shaped like rods, disks, or spheres. Sometimes a group of cells embedded in a gelatinous sheath appear to be a multicellular organism (figure 17.19). In most cases, however, they function as single-celled organisms. They tend to be somewhat larger than bacterial cells. Although these organisms do contain chlorophyll *a,* it is not localized within chloroplasts as it is in the eukaryotes. Instead, the pigment molecules are arranged in flattened, membrane-bound vesicles or in pigmented lamellae (figure 17.20). Cyanobacteria, of course, contain no nuclei and no large vacuoles characteristic of plants. Likewise, they have no mitochondria, endoplasmic reticulum, or Golgi apparatus. They do, however, have ribosomes for protein synthesis and store both proteins and carbohydrates in granules. None have flagella, and any motion they may exhibit, such as the gliding motion exhibited in *Oscillatoria* is still unexplained.

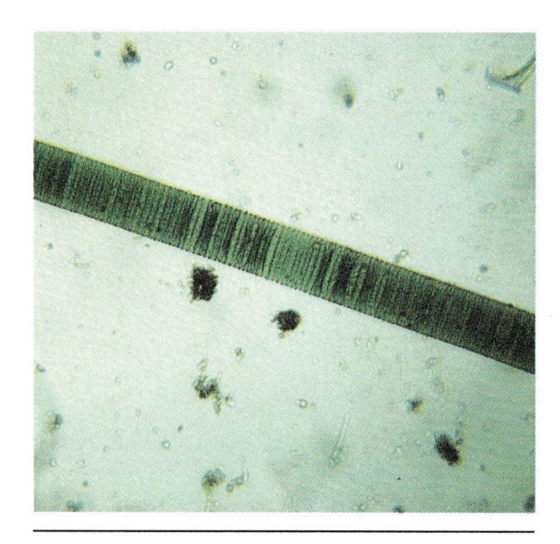

Figure 17.19
The cyanobacterium *Oscillatoria princeps* is typical of this group of organisms. Groups of platelike cells are held together to form a long filament, which oscillates slowly.

In addition to having chlorophyll *a* and certain carotenoids, cyanobacteria are characterized by the pigments **phycocyanin** (blue) and/or **phycoerythrin** (red). The phycocyanin plus chlorophyll makes the cyanobacteria blue-green. However, not all blue-green algae are actually blue-green. Depending on the various concentrations and mixtures of pigments, organisms may be black, brown, yellow, red, bright green, or other colors. The Red Sea gets its name from a particular species of cyanobacterium, *Trichodesmium,* rich in phycoerythrin.

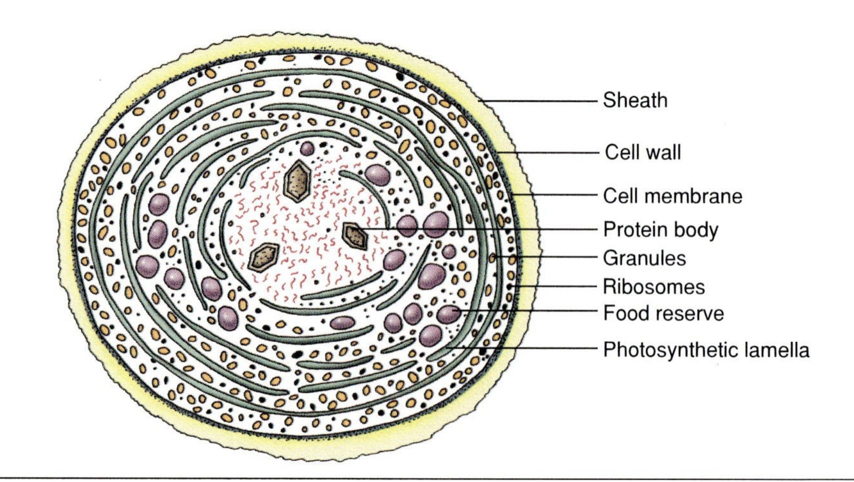

Figure 17.20
Prokaryotic organization of a cyanobacterium cell illustrating the extensive photosynthetic lamellae. The DNA is located in the central region of the cell.

Cyanobacteria Ecology and Economics

In recent years the ecological importance of cyanobacteria has been realized. Many species are excellent nitrogen fixers, taking gaseous nitrogen from the atmosphere and fixing it into ammonia. This anaerobic respiratory process is accomplished by an enzyme called nitrogenase. Even though the organisms grow under aerobic conditions, nitrogen fixation occurs in certain specialized cells called **heterocysts** (figure 17.21), whose particularly thick walls exclude oxygen. It is also interesting to note that these very specialized cells have chlorophyll and carry on only the photosystem I portion of the light reaction. Therefore no oxygen is produced that would inhibit the enzyme, a remarkable adaptation that allows an organism to exist in waters or soils where nitrogen is deficient.

Cyanobacteria are far more ubiquitous than originally thought. They seem to occur everywhere and are particularly good indicators of polluted waters, where they often impart a fishy smell. A number of species of the genus *Trichodesmium* can produce "blooms" in freshwater and give off a toxin harmful to fish and mammals. They grow in nitrogen-deficient tropical soils, in cracks in desert rocks, under the snow, and in rice paddies. It is believed that the continuous cultivation of paddies for thousands of years has been possible only because of the nitrogen input from cyanobacteria. Some species grow in hot springs, where temperatures approach boiling, and others grow in equally inhospitable places. They are also sometimes found as the photosynthetic organism in lichens.

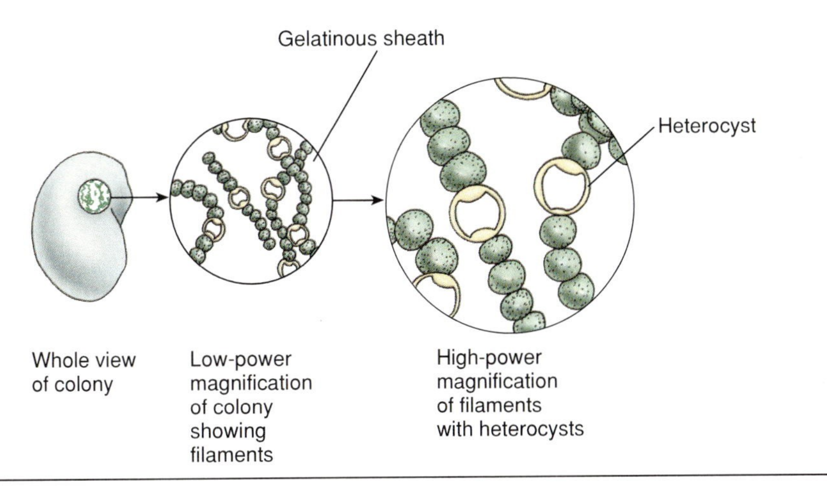

Figure 17.21
Nostoc, a cyanobacterium, exists as a colony of filaments ensheathed by a common gelatinous mass. The heterocysts are the site of nitrogen fixation.

✔ Concept Checks

1. Describe the three basic morphological forms of bacteria. Distinguish between gram-negative and gram-positive bacteria; aerobic and anaerobic bacteria.
2. Distinguish between thermophilic and halophilic bacteria. Discuss several ecological and economic attributes of bacteria.
3. Construct an evolutionary tree, showing the Archaebacteria as a sixth kingdom.
4. How do cyanobacteria differ from bacteria? Describe the economic and ecological importance of the cyanobacteria.

Summary

1. Of approximately 10,000,000 species in the world, almost 500,000 are plant species. The tremendous diversity of life is manifested in similarities and differences that allow organisms to be grouped and classified.
2. The big bang theory suggests that our solar system arose about 5 billion years ago within a universe that was created by a colossal explosion that occurred about 20 billion years ago. Our own planet formed from cosmic dust and gases left over after the formation of the sun.
3. The primitive earth was a giant rock without atmosphere, but heating and compression from within caused the evolution of an atmosphere composed of hydrogen, ammonia, water, and methane. This primordial soup is believed to be the precursor of the first organic molecules that eventually led to the beginning of life.
4. The earth is dated by studying sedimentary layers that have fossilized organisms within their profile. Fossil evidence from many localities throughout the world and at various depths has allowed paleontologists to reconstruct the fossil record, and from this record biologists have developed a detailed scenario concerning life at various times throughout the history of the earth.
5. The heterotrophic theory of the origin of life suggests that the first organisms were heterotrophs, and only after they began to run out of organic molecules that had been

self-assembled did the selection pressure favor the evolution of autotrophs. Approximately 3.5 billion years ago the cyanobacteria evolved a scheme for utilizing chlorophyll *a* and splitting a molecule of water for both a reducing source and molecular oxygen.

6. The five-kingdom scheme divides all organisms into Monera, Protista, Fungi, Plantae, and Animalia. Among the lower organisms there is a great deal of overlap, and it is particularly difficult to separate photosynthetic protists from the lower plants. Some scientists have proposed that the Archaebacteria should represent a sixth kingdom.
7. Since viruses require a host cell to reproduce, there is little evidence to suggest that they were the most primitive organisms. Consisting of only a nucleic acid covered by a protein coat, viruses are seemingly at the borderline of the living and nonliving.
8. The prokaryotes include both bacteria and cyanobacteria. They are not only the most primitive organisms, but also the most abundant throughout the world. Bacteria are classified by shape, size, oxygen requirements, Gram staining, and a number of nutritional requirements. Some of them reproduce every 10 to 12 minutes. Together with the fungi, bacteria are the primary decomposers in the biosphere. Although some bacteria are pathogenic, many others are important to humans as medicinals and in the production of food and industrial products.

Key Terms

aerobic 310
alternation of generations 306
anaerobic 310
Archaea (Archaebacteria) 312
autotrophs 304
bacillus 309
bacteria 304
bacteriophages 307
blue-green algae 312
Cambrian 301
capsid 306
Carboniferous 301
Cenozoic 301
chitin 305
coccus 309
colony 310
cyanobacteria 304
Devonian 301
exoenzymes 306
fossil 299
heterocysts 313
heterotrophic 303
Mesozoic 301
Mississippian 301
nitrogen fixation 311
Ordovician 301
paleontologists 299
Paleozoic 301
pathogens 309
Pennsylvanian 301
Permian 301
phycocyanin 312
phycoerythrin 312
Precambrian 301
prokaryotes 304
Silurian 301
spirillum 309
stromatolites 303
viroids 307

Discussion Questions

1. Why are viruses not considered to be living organisms? Outline the infectious replication cycle of a virus.
2. What are some of the important viral diseases of plants and animals? Of viroids? Of prions?
3. List the five kingdoms. What group might represent a sixth kingdom? For each, cite whether the basic cell type is prokaryotic or eukaryotic; whether it is composed primarily of multicellular organisms or not; and the modes by which the organisms primarily derive their nutrition.

Suggested Readings

Adams, D. P. 1991. *The greatest good to the greatest number: Penicillin rationing on the American home front, 1940–1945.* P. Lang, New York.

Balows, A., ed. 1992. *The prokaryotes: A handbook on the biology of bacteria: Ecophysiology, isolation, identification, applications.* 2nd ed. Springer-Verlag, New York.

Buncel, E., and J. R. Jones, eds. 1987. *Isotopes in the physical and biomedical sciences.* Elsevier Science Publishers, New York.

Carmichael, W. W. 1994. The toxins of cyanobacteria. *Sci. Am.* 270(1):78–86.

Clewell, D. B. 1993. *Bacterial conjugation.* Plenum Press, New York.

Diener, T. O. 1981. Viroids. *Sci. Am.* 244(1):66–73.

Hardy, K. G., ed. 1993. *Plasmids: A practical approach.* 2nd ed. IRL Press, New York.

Hartman, H., and K. Matsuno, eds. 1992. *The origin and evolution of the cell.* World Scientific Publishing Co., River Edge.

Hawking, S. W. 1993. *Hawking on the big bang and black holes.* World Scientific, Singapore.

Lerner, E. J. 1991. *The big bang never happened: A startling refutation of the dominant theory of the origin of the universe.* Times Books/Random House, New York.

Macfarlane, G. 1985. *Alexander Fleming, the man and the myth.* Oxford University Press, New York.

Maramorosch, K., and J. J. McKelvery, Jr., eds. 1985. *Subviral pathogens of plants and animals: Viroids and prions.* Academic Press, Orlando.

Margulis, L. 1988. *Five kingdoms: An illustrated guide to the phyla of life on earth.* 2nd ed. W. H. Freeman, New York.

Matthews, R. E. F. 1992. *Fundamentals of plant virology.* Academic Press, San Diego.

Moberg, C. L., and Z. A. Cohn, eds. 1990. *Launching the antibiotic era: Personal accounts of the discovery and use of the first antibiotics.* Rockefeller University Press, New York.

Mohan, S., C. Dow, and J. A. Cole, eds. 1992. *Prokaryotic structure and function: A new perspective.* Cambridge University Press, New York.

Rose, A. H. 1981. The microbiological production of food and drink. *Sci. Am.* 245(3):126–139.

Schopf, W. J. 1993. Microfossils of the early archean apex chert: New evidence of the antiquity of life. *Science.* 260:640–646.

Watanabe, M. E. 1994. Hot-vent microbes: Looking backward in evolution for future uses. *The Scientist,* May, 30:14–15.

Wilson, E. O. 1992. *The diversity of life.* Belknap Press, Cambridge.

Wilson, M. A., and J. W. Davies, eds. 1992. *Genetic engineering with plant viruses.* CRC Press, Boca Raton.

Woese, C. R. 1981. Archebacteria. *Sci. Am.* 244(6):98–125.

Chapter Eighteen

Diversity: Nonvascular Eukaryotes

Our planet's biological diversity—the variety and variability within and among living organisms and the ecological complexes in which they occur—should be considered our greatest treasure. It is valuable as a source of intellectual and scientific knowledge, recreation, and aesthetic pleasure. This chapter develops a sense of appreciation for organisms assigned to the kingdoms Protista and Fungi, noting their influences on our daily lives. The chapter concludes with a description of bryophytes, organisms belonging to the plant kingdom.

Diversity of living organisms includes the complex reproductive structures of *Eucalyptus*.

This chapter describes the diversity that exists among so-called lower, or nonvascular, eukaryotic organisms. Botanists often refer to those organisms that have a well-developed xylem and phloem system as vascular plants and those that lack such tissue systems as nonvascular organisms. Although it is true that most lower organisms are also short or small in stature, this is not always the case. This chapter introduces some lower organisms that are indeed large.

Even though plants represent a significant number of all known species, other organisms in entirely different kingdoms are often considered within the study of botany. This grouping is partly traditional and partly functional, since many of the organisms are autotrophs or otherwise more closely related to plants than to animals. The intention in this section is not to be exhaustive in looking at every organism within a group and considering its life cycle, but merely to demonstrate the diversity of life forms. We begin with the photosynthetic protists and work our way along the evolutionary ladder, considering each group of organisms, how they are distinguished from all the others, and how they influence human life. The extinct species, although important in evolution and paleontology, are considered only briefly in this text. The prokaryotes were covered in chapter 17; the following groups of organisms will be discussed here: **photosynthetic protists** (algae), **heterotrophic protists (water molds** and **slime molds), fungi,** and **bryophytes.**

THE EVOLUTION OF EUKARYOTES

The first cells, bacteria and cyanobacteria, lacked most cellular organelles in their cytoplasm, but they had other structures that performed similar roles. Folded membranes were the site of enzymes essential for cellular function, and some folds extended into the interior of the cell. Even the photosynthetic pigments were attached to these membranes, although some had discrete spherical bodies termed chromatophores, which carried the pigments and enzymes important in photosynthesis. Today prokaryotes perform metabolic tasks in the same manner and with the same structure.

The first eukaryotic organisms were probably haploid and asexual. There is no direct evidence for the evolution of diploidy, but many organisms of the fossil record of 700 million years ago were diploid. It is probable that diploidy arose when two haploid cells fused to form a zygote. The establishment of meiosis may have occurred at approximately the same time, so that the zygote could return to the haploid condition. This significant event established sexual reproduction and led to new opportunities in genetic variability. Perhaps by accident, some diploid cells divided by mitosis rather than meiosis, giving rise to organisms that consisted of more than one (diploid) cell.

The evolution of compartmentation and the "division of labor" in eukaryotic cells is subject to a great deal of conjecture, but one theory suggests that organelles arose as a result of **endosymbiosis,** the incorporation of prokaryotic cells into the cytoplasm of a eukaryote, a theory with overwhelming evidence.

Although both photosynthetic bacteria and cyanobacteria arose long before the first eukaryotic cells, the cyanobacteria seem to be the most likely progenitors of chloroplasts. They have the same basic chlorophyll *a* pigment, and they both use similar pathways of noncyclic photosynthetic phosphorylation.

Similarly, mitochondria are thought to have arisen from aerobic bacteria; they possess the same pathway of electron transport, and the internal membrane structure may be considered analogous to the cristae of the mitochondria. The outer membrane of the mitochondrion is similar to the endoplasmic reticulum of the host cell and presumably was derived from it. Such an incorporation would be to the mutual advantage of both cells and therefore would be considered true symbiosis.

Both chloroplasts and mitochondria contain DNA and ribosomes similar to that present in prokaryotic cells (further evidence for the endosymbiotic theory), divide, grow, and develop at least partially autonomously—that is, partially on their own—without total direction from the nucleus. In fact, genes from both the nucleus and the organelle direct the function of the organelle. It would appear that, during the evolutionary process, the eukaryotic nucleus incorporated some of the metabolic functions of the organelles into its own mechanism, so that the cell division processes, both mitosis and cytokinesis, are synchronized fully.

PHOTOSYNTHETIC PROTISTS (ALGAE)

The photosynthetic protists (algae) are an assemblage of diverse mostly autotrophic, eukaryotic organisms. They are very different in terms of habitat, habit, nutritional requirements, and reproduction. Although the protists are predominantly unicellular, there are several large, multicellular representatives. It is important to remember that each of the three multicellular kingdoms, Plantae, Animalia, and Fungi, originated from a different ancestral group among the Protista and that multicellularity has arisen independently in each of the eukaryotic kingdoms. In addition to many heterotrophic protists not described here, such as protozoans and sponges, the groups of interest to botanists include the **Chrysophyta, Pyrrhophyta, Euglenophyta, Chlorophyta, Phaeophyta,** and **Rhodophyta** (table 18.1).

Classification

This diverse group of organisms comprises about 23,000 described species, which are grouped into six divisions by pigment forms, composition of the cell wall, and stored carbohydrates (table 18.1), collectively termed **algae** (singular **alga**). They vary from microscopic organisms to the giant kelps that can attain lengths of 30 or 40 m.

TABLE 18.1

Characteristics of Photosynthetic Protists (Algae)

DIVISION (COMMON NAME)	PHOTOSYNTHETIC PIGMENTS	CELL-WALL COMPOSITION	STORED CARBOHYDRATES	HABITAT AND SEXUALITY
Chrysophyta (diatoms, golden brown algae, and yellow-green algae)	Chlorophylls *a* and *c*, carotenoids, including fucoxanthin	Silicon dioxide (SiO_2) and pectic compounds	Leucosin	Both marine and freshwater
Pyrrhophyta (dinoflagellates)	Chlorophylls *a* and *c*, carotenoids	Cellulose	Starch	Both marine and freshwater, sexual reproduction by cell division
Euglenophyta (euglenoids)	Chlorophylls *a* and *b*, carotenoids	No cell wall (pellicle)	Paramylon	Mostly freshwater, no sexual reproduction
Chlorophyta (green algae)	Chlorophylls *a* and *b*, carotenoids	Cellulose, hemicellulose, and pectins	Starch	Mostly freshwater but some terrestrial and marine; very diverse organisms, including some single celled
Phaeophyta (brown algae)	Chlorophylls *a* and *c*, carotenoids, mostly fucoxanthin	Cellulose and alginic acids	Laminarian or mannitol	Temperate marine
Rhodophyta (red algae)	Chlorophylls *a* and *d*, carotenoids, phycobilins	Cellulose and pectins; coralline algae impregnated with calcium carbonate	Floridean starch	Mostly tropical marine but a few freshwater species; oogamous; sexuality lacking in some members

The single-celled algae make up the phytoplankton, the producer organisms of the marine and freshwater biomes. Confined largely to the open sea, they photosynthesize only in the upper layers of water, where the light intensity is still great enough to carry out the photosynthetic light reactions. Most organisms fail to photosynthesize at greater depths, although some of the algae grow at depths of 200 m in clear, tropical waters. The macroalgae, including the giant kelps, are largely restricted to the intertidal zone because they must secure themselves to the bottom of the ocean to survive. One macroalga that manages to do otherwise is Sargassum, a large, free-floating or anchored, brown alga that inhabits the Sargasso Sea, a vast area of the Atlantic Ocean between the West Indies and North Africa.

It is important to realize that none of the algae have a true vascular system, and therefore they are classified with the nonvascular plants. Some of the large multicellular types do have phloemlike conducting systems, but the cells are not comparable to the phloem and xylem of vascular plants. Although some taxonomists argue that single-celled algae should be included with the kingdom Protista, and multicellular algae with the kingdom Plantae, for purposes of consolidation all algae are included in the kingdom Protista.

Chrysophyta

Chrysophyta includes the diatoms, golden brown algae, and yellow-green algae, which derive their names from the various pigments stored within their bodies. The thousands of species of diatoms are both botanically interesting and economically important. Diatoms come in a myriad of shapes and sizes, each species exhibiting a characteristic cell-wall pattern as aesthetically pleasing and diverse as snowflakes (figure 18.1). Diatoms are often used by artists and designers as patterns for cloth, jewelry, and paintings. The cell wall becomes impregnated with pectic substances and silicon dioxide (SiO_2), the same substance from which glass is made, and these single-celled organisms literally live in glass houses. They produce millions of cells in both the ocean and in freshwater environments, such as stock tanks. When the cells die, they fall to the bottom of the ocean, lake, or tank, but the glass walls remain. Millions of these microscopic, single-celled broken remnants create a sediment that can become very thick. In pure layers it is referred to as **diatomaceous earth.** In some parts of the world once covered by the ocean, diatomaceous earth occurs in 300 m thick deposits. These broken bits of glass walls are excellent filters for water treatment and are used extensively in the manufacture of fine polishes, ranging from silver polish to toothpaste. Diatomaceous earth is also used as an additive in cement and explosives and as insulation in fireproof safes.

Although we emphasize the importance of dead diatoms, the billions and billions of living diatoms in aquatic biomes carry on an extensive amount of photosynthesis. A single diatom contributes little to carbon

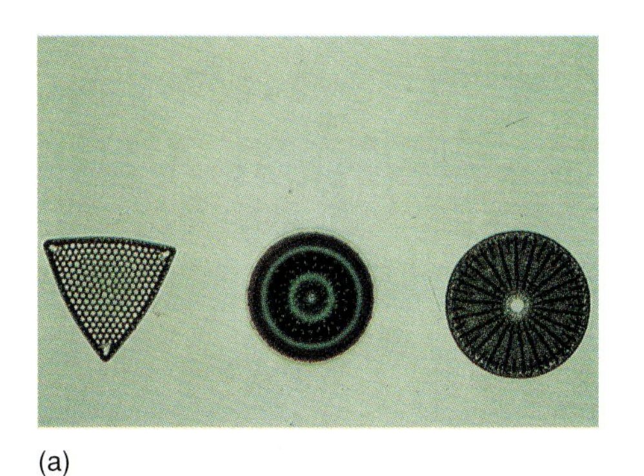
(a)

(b)

Figure 18.1
(*a*) Diatoms come in a myriad of shapes, sizes, and patterns. These three organisms represent some of the diversity. (*b*) The intricate pattern of sculpturing of the cell wall and symmetry is revealed by scanning electron microscopy.

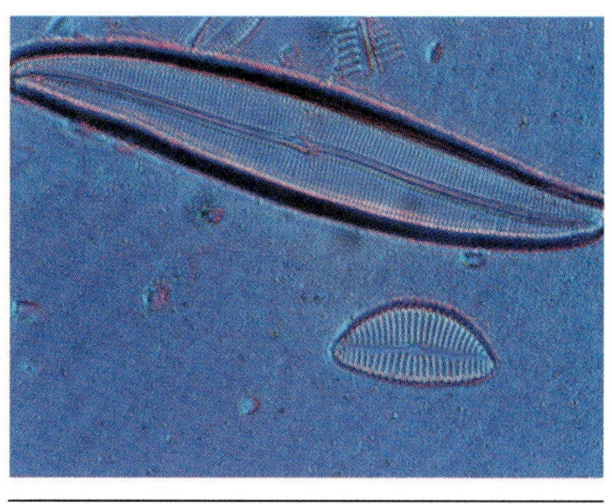

Figure 18.2
These diatoms (*top*) *Gyrosigma* and (*bottom*) *Cymbella* are representative of the phytoplankton so important in aquatic food webs.

fixation and oxygen evolution, but on a worldwide basis they are an important part of the phytoplankton in aquatic food chains (figure 18.2). Although they can trap light effectively only in the top few meters of open ocean, the vast area of the oceans (three-fourths of the earth's surface) ensures that diatoms, golden brown algae, and yellow-green algae contribute significantly to the world's carbon fixation and gas exchange.

Pyrrhophyta

The Pyrrhophyta, also called dinoflagellates, are an ecological factor in localized areas where water temperature and nutritional conditions encourage their rapid reproduction to enormous population densities in a short time. They occur in both freshwater and in the oceans, and they are major contributors to total primary production (figure 18.3). Some species produce vast numbers of individuals, sometimes reaching millions per liter of seawater, sufficient to color the water reddish brown, a condition termed the red tide along many coastlines, including those of California, Florida, and Texas. These single-celled algae rarely reproduce sexually; they store photosynthate as starch and often give off a neurotoxin that is ingested by fish and crustaceans. In fish, the neurotoxin causes blood vessels in the gills to constrict, resulting in suffocation and large quantities of dead fish washing onto the beach. If the accumulation of toxin in fish or shellfish is high enough it may cause illness or even death in humans who eat the contaminated organisms. Some beaches have posted warning signs to indicate times of the year when it is not safe to eat local seafood. The toxins also can cause skin reactions in swimmers and mucosal irritation from airborne particles.

These organisms also exhibit an interesting phenomenon called bioluminescence, the production of light through the use of ATP. Whenever seawater containing high populations of dinoflagellates is disrupted, as in the churning effect of the surf as it encounters the beach, the energy is released and the entire wave may appear to glow. Someone running on wet sand containing these algae may cause sufficient disturbance to make the footprints glow. If broken pieces of kelp are washed up on the shore, swinging the plant in the air may provide sufficient release of ATP to cause bioluminescence, and even shaking a glass jar of seawater can produce the effect.

Figure 18.3
The dinoflagellates are bizarre organisms with heavily sculptured cell walls consisting of a number of plates covered by the cell membrane. *Gonyaulax* is one of the organisms that causes the red tide.

Euglenophyta

The autotrophic euglenoids have long been a favorite research subject for biologists. In many respects they behave like a typical plant—they possess chlorophyll, and in light they carry on photosynthesis as would any single-celled green plant. However, if fed a carbon source such as sugar and kept in the dark, they continue to grow and reproduce like a typical heterotroph. Some species are unpigmented and strict heterotrophs; others are incapable of synthesizing vitamin C and certain B vitamins. Euglenoids do not have a cell wall, but a protein layer called a pellicle is located just inside the plasma membrane. Thus *Euglena* (figure 18.4) and related species possess both plant and animal characteristics. They are motile, possessing a single whiplike flagellum, and respond to light by swimming toward the source or turning away if the intensity becomes too great. Although not important economically, these organisms continue to be a favorite

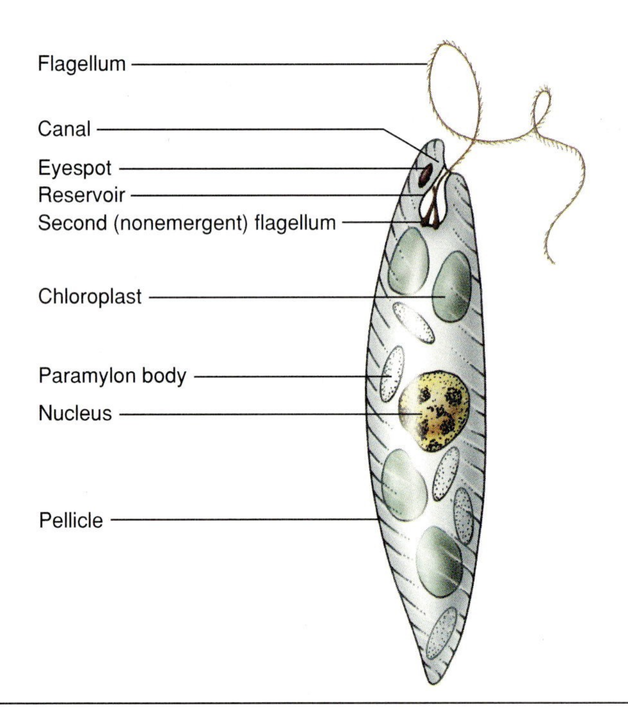

Figure 18.4
Possessing both photosynthetic and animal-like characteristics, *Euglena* is common in ponds, puddles, and farm tanks.

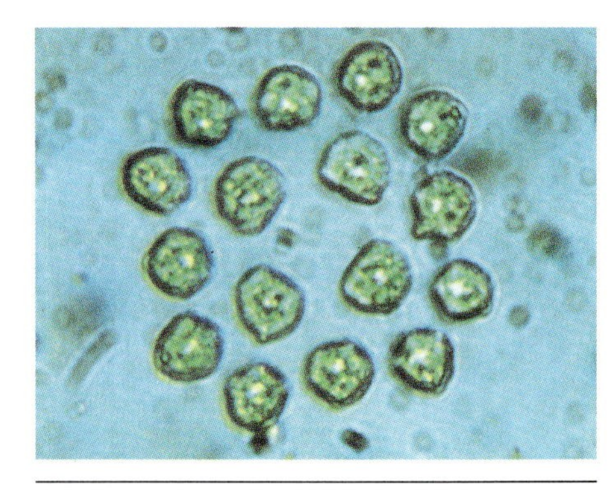

Figure 18.5
The motile colonial green algae include the genus *Gonium,* with 4, 8, 16, or 32 cells, depending on species. This 16-celled *Gonium pectorale* is typical.

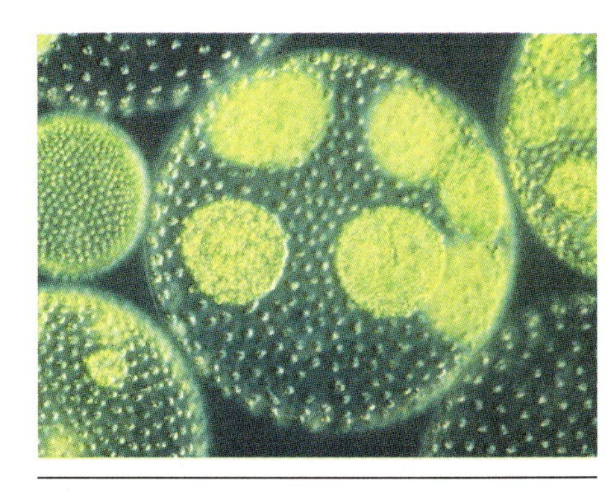

Figure 18.6
The highest degree of specialization in the colonial algae occurs in *Volvox,* with five hundred to many thousands of cells held together in a hollow sphere. The green clusters in the center are new colonies that have already multiplied and will soon break free to begin their own colonies.

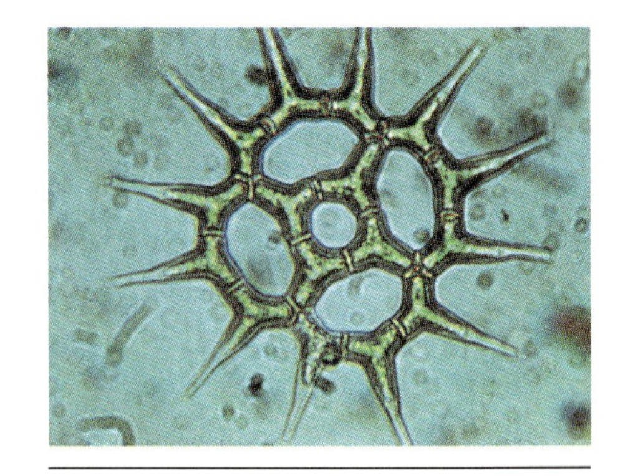

Figure 18.7
Pediastrum duplex is a colonial green alga that grows on the bottom of quiet pools and lakes. Its colonial organization is artistically interesting, and it can be readily grown in laboratory culture.

experimental subject because they may act as both an autotroph (plant) and heterotroph (animal).

Chlorophyta (Green Algae)

The green algae are the most diverse group of algae, including some 7000 species that vary from unicellular to large multicellular forms more than 8 m long. For the most part they are aquatic and occupy freshwater, although some occur in the oceans and on land, including the unicellular ones that are symbiotic with lichens. The photosynthetic products are stored as starch, and the cell wall is composed of cellulose and hemicellulose, as in the higher plants. Lines of evolution among the green algae have been studied in depth, beginning with single-celled species and advancing to species with 4, 8, 16, 32, and larger numbers of cells and hollow balls of thousands of cells, such as *Volvox.* These studies allow botanists to better understand the relationships between colonial organisms (those that form colonies) and multicellular organisms as well as understand the relationship between adjacent cells of the same organism (figures 18.5 to 18.7).

Reproduction in green algae, as in most algae, may occur asexually by (1) cell division, (2) fragmentation, or (3) zoospore formation. Fragmentation is simply a breaking apart of the plant body, often by the mechanical action of waves or the separation into two parts when a fish or other animal takes a bite out of the center. Zoospores are flagellated motile spores formed within an algal cell without sexual reproduction. Sexual reproduction may occur by **isogamy** or **oogamy.** In isogamy the gametes are of equal size and indistinguishable. In oogamy the nonmotile female gamete is much larger than the motile male gamete (sperm). One particularly common and interesting green alga is *Spirogyra,* which is **filamentous** and possesses one or more spiral, ribbonlike chloroplasts (figure 18.8). Sexual reproduction in this organism is via a type of isogamy known as conjugation, in which posi-

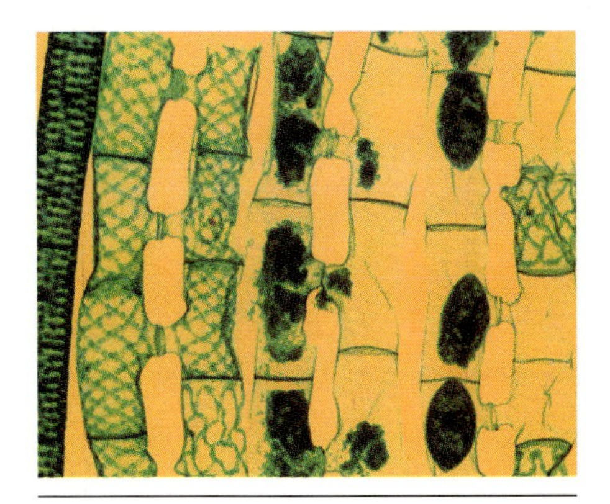

Figure 18.8

In the filamentous green alga *Spirogyra* the chloroplast consists of one or more ribbonlike organelles that coil the full length of the cell (*left*). Sexual reproduction (*middle*) is by conjugation, a process in which two mating strains of filamentous, haploid cells align and form a conjugation tube between adjoining cells. When the passageway is complete, the entire contents of one cell pour into the other cell, giving rise to a diploid zygote. Following fertilization (*right*) rounded zygospores are formed with heavy cell walls. These dormant spores are resistant to environmental stresses and lie on the bottom of a lake or stream until spring turnover brings them to the surface. At that time light and warm temperatures bring about spore activation, accompanied by meiosis and formation of a new haploid filament that floats on the surface of the water.

tive and negative mating strains line up side by side, and adjacent cell walls expand to form a conjugation tube. The entire protoplast of one cell simply migrates into the other cell, with the protoplasts fusing to form a diploid zygote. This zygote develops a thick resistant coat and is termed a zygospore. In freshwater lakes the zygospore overwinters until the following spring, when the spore is brought to the surface. Germination and meiosis occur then, and the new haploid filament is formed.

The tiny, single-celled *Chlorella* has been used in many experimental studies, including those of Melvin Calvin in the discovery of the dark reactions of photosynthesis. More recently this organism has been studied as a potential high-protein food source, particularly by the Japanese. In culture, *Chlorella* may produce up to 50% protein and all the essential human amino acids. Under controlled cultivation, the net production of *Chlorella* exceeds that of most terrestrial crops (12.3 g organic matter/m^2/day). At some time in the future we may find that "algae cakes" are a regular part of the human diet, and we may eat them much as we would crackers or cookies.

Several lines of evidence suggest that the progenitor of land plants was a multicellular green alga. Although the species cannot be positively identified, it was probably a branched plantlike body, that retained and protected the embryo and that somehow managed to survive when washed up on the shore. Shared characteristics of these aquatic green algae and the derived land plants include chlorophylls *a* and *b,* carotenoids, photosynthate accumulation as starch, cell walls composed of cellulose, and similarities in nuclear structure and the pattern of cell-plate formation in cytokinesis.

The **charophytes** or stoneworts, often classified with the green algae, are so distinct that they are sometimes classified separately. Some species are characterized by heavily calcified cell walls, and they grow in fresh or brackish waters. The rigidity of the cells causes them to be particularly well preserved and represented in the fossil record. The charophytes are the only green algae that form multicellular sex organs; motile sperm are produced in **antheridia,** and eggs are borne in **oogonia.** In one charophyte, *Coleochaete,* placental cells are formed that supply nutrients to the retained embryo. Since this feature is unique among algae, many evolutionary biologists have postulated that the primitive land plants may have originated from a *Coleochaete*-like green alga.

The evolutionary history of the green algae is particularly interesting, because individual species range from those that are truly unicellular, to those with filaments (one plane of division), to those made up of sheets (two planes of division), to complex three-dimensional organisms similar to the higher plants.

Figure 18.9

The sargassum weed is one type of brown algae that may not attach to the ocean floor, thereby forming large floating masses.

Phaeophyta (Brown Algae)

This group of macroalgae is almost entirely marine and is represented by the seaweeds or giant kelps of temperate regions (cool waters). There are only about 1500 species of brown algae, but they totally dominate the rocky shorelines in many parts of the temperate world. Except for the sargassum weed (figure 18.9), these giant plants are confined to regions where they can attach to the ocean floor with rootlike holdfasts. Their internal structure is fairly complex, and some taxa (e.g., *Macrocystis* and *Neurocystis*) have sieve tubes comparable to the phloem in vascular plants. If the coastline drops off quickly, the kelps may not extend very far out into the ocean. However, if the ocean floor slopes gently, these giant seaweeds may extend 5 or 10 km from the coastline and attach to the seafloor at a depth of 20 to 40 m (figure 18.10). Scuba divers may view a spectacular underwater forest inhabited by many fish and other marine animals. Sportfishing is often confined to the kelp beds because the concentration of fish is greater there, since they find abundant food and protection from predators.

Figure 18.10
The brown algae include the giant kelps of cool, temperate waters. This one was photographed near the Falkland Islands, off the coast of Argentina.

Figure 18.11
The red coralline algae calcify and become part of a reef. A sea urchin rests on this rock, surrounded by a bright red sponge and darker red coralline algae. Photographed at Saba Bank in the Caribbean.

In addition to chlorophyll *a,* the brown algae also have chlorophyll *c* and various carotenoids, including fucoxanthin, which gives the plants their characteristic dark brown color. The cell wall is made of cellulose, but on the outside is a layer of mucilage, primarily alginic acid or algin. For this substance the brown algae are harvested; its uses include stabilizers for ice cream, chocolate milk, and other dairy products. It is also used as an emulsifier, particularly in paints and cosmetics, and sometimes algin is used as a coating for paper. In addition, certain species of brown algae are consumed by humans, particularly in the Orient.

Rhodophyta (Red Algae)

This group of some 4000 species is largely confined to warm, tropical waters, although a few species are freshwater and are sometimes found in temperate regions. These algae have pigmented groups of cells containing phycobilosomes, which serve as light-gathering antennae for chlorophyll *a*. The red algae owe their characteristic color to the presence of accessory, water-soluble proteinaceous pigments, termed phycobilins. Phycobilins, consisting of red-colored phycoerythrin and blue-colored phycocyanin, typically mask the green pigments, chlorophyll *a* and *d* so that many red algae appear violet or brownish in color. Phycobilins increase the effectiveness of light absorption. At a depth of 200 m practically all light has been filtered out by the water. In this dark undersea world, faint rays of orange to red light are captured by these pigments. The energy is transferred to the chlorophyll *a,* and photosynthesis proceeds without competition from any other photosynthetic organisms, except photosynthetic bacteria. Most red algae attach onto rocks or other algae, although some are free floating and a few more are unicellular or colonial. The cell walls contain mucilages with a galactose base. This gives the red algae their slick, leathery feel. The multicellular plant body generally consists of densely packed filaments of cells, and growth is from the tips of these filaments. Some of these species found in tropical waters may become encrusted with calcium carbonate, and these coralline algae in some cases become the principal component of coral reefs. Although one generally thinks of reefs as being formed from encrusted marine animals, the red algae play a significant role in many locations (figure 18.11).

Most of the seaweeds that constitute a significant part of the human diet in the Orient are red algae. For example, the dark seaweed wrapped around the raw fish of sushi is typically a red alga. In addition, the mucilaginous cell wall is extracted for **agar** and **carrageenan** for many cosmetic, pharmaceutical, and industrial uses. The commercial and experimental culturing of bacteria and fungi is made possible by growing the organisms on media using agar as a solidifying agent. Gelatin capsules are also made from agar.

✔ Concept Checks

1. What is the concept of endosymbiosis?
2. List six divisions of eukaryotic algae. What features distinguish the divisions?
3. Describe the ecological value and commercial uses of algae.

HETEROTROPHIC PROTISTA AND FUNGI

Fungi are placed in a kingdom by themselves because they exhibit characteristics quite different from those of plants, animals, and the photosynthetic protists. In spite of a tremendous diversity in identified species, all are heterotrophic, being either parasitic or saprophytic at some time during their life history. In addition the vegetative body of fungus is haploid. The Oomycota (water molds) and slime molds, traditionally included within the kingdom Fungi, are now interpreted by scientists as having no direct evolutionary relationship with the fungal kingdom. Because the terminology used to describe these heterotrophic protists is the same used for the fungi, we have elected to discuss these taxa together.

Fungal Growth

Fungal cells tend to grow rapidly in long threads, termed **hyphae.** Hyphae may be **septate,** (that is, divided into compartments by cross walls), or **nonseptate.** These threads branch abundantly, giving rise to a tangled mass, a **mycelium.** In some, this hyphal mat becomes organized into a dense reproductive fruiting body, as in mushrooms. In others, the fruiting bodies are microscopic. The fungi are exceptionally fast growing, extending at the tip of the hyphae at an incredibly rapid rate. Fungi have the greatest surface-to-volume ratio of all organisms, which provides advantages and disadvantages in exposure to the environment. As long as the organism is growing in an area with adequate moisture and humidity, the large surface area allows for rapid absorption of water and nutrients. On the other hand, if the fungus finds itself in an inhospitable environment, this large surface area may be detrimental to the organism as a result of excessive heat, desiccation, or damaging light.

Fungi occur everywhere—in water, in soil, and in any other place where organic substrates are found. Since most fungi do not possess a motile phase, they spread into new environments primarily by the dissemination of spores through the air or water and by growth of the hyphae. Most fungi obtain food by secreting enzymes, termed exoenzymes, that digest (decompose) the food source, breaking it down into smaller molecules that can be readily absorbed through the hyphal wall and membrane. Because this degradation is extracellular, some fungi are capable of degrading many pollutants including cyanides, polycyclic aromatics, pesticides, such as DDT and chlordane, and munitions, such as TNT.

In 1992 a team of American and Canadian biologists reported the discovery of a single fungus extending across 37 acres in a northern Michigan oak-maple forest. Billed as the world's largest fungus, this organism is composed of a vast array of underground hyphae, representing over 1500 years of continued growth.

Beneficial and Pathogenic Heterotrophic Protista and Fungi

Fungi are responsible for many plant diseases, yet they are not all detrimental to other organisms (including humans). Certainly the **yeasts** are of major economic importance in the production of bread, beer, and wine. Other fungi are grown by pharmaceutical companies to produce drugs and antibiotics such as cyclosporine and **penicillin,** and still others produce the characteristic flavoring in certain cheeses.

Perhaps the most dramatic example of a heterotrophic protist as a plant pathogen is the organism that causes potato blight, *Phytophthora infestans.* During one summer week in 1846, nearly all the potatoes in Ireland were destroyed. The great famine that subsequently developed there and in other parts of Europe between 1845 and 1847 changed the course of world history and significantly influenced the ethnic population of the United States. More than 1.5 million people died of starvation in Ireland, and another 1.5 million emigrated to the United States. Within ten years the population of Ireland decreased from 8 million to 4 million. This great tragedy occurred because essentially all of the potato plants were genetically identical and lacked any resistance to the disease. The dangers of a narrow genetic base for any crop, evident from this experience and similar ones described later, should be enough to convince the world of the importance of conserving a broad genetic base. Crops selected and bred for uniformity often lose resistance to disease, insects, and environmental stresses. If the wild relatives of domesticated crops become extinct, the world may have lost the last opportunity to save the crops themselves when such problems arise.

Later in the nineteenth century another related heterotrophic protist, *Plasmopora viticola,* which causes downy mildew of grapes, almost destroyed the entire wine industry of France and southern Europe. In 1970 a fungal blight threatened the entire corn crop of the United States. Fortunately, because of a lack of favorable environmental conditions for the fungus and because of control measures rapidly instigated, epidemics failed to develop in succeeding years. Plant pathologists are constantly watching for potential new diseases that might destroy important agricultural and forest crops.

Generally, fungal disease in humans is not a major problem. Ringworm of the body and **athlete's foot** may be caused by one or more fungal species. A number of other skin disorders are caused by fungi, and although not usually serious, they are difficult to treat and cause chronic problems.

Fungal Reproduction

Fungi reproduce both sexually and asexually. Occasionally asexual reproduction occurs by fragmentation of the hyphae, but most reproduction is through the spreading of asexual spores by wind or water. The bright colors and powdery textures often exhibited on the surface of a fungus are due to spores. In some, the spores are produced in a large spore case called a **sporangium.** The common bread mold *Rhizopus* is an example (figure 18.12). Sporangia appear as tiny black dots on the bread, and they

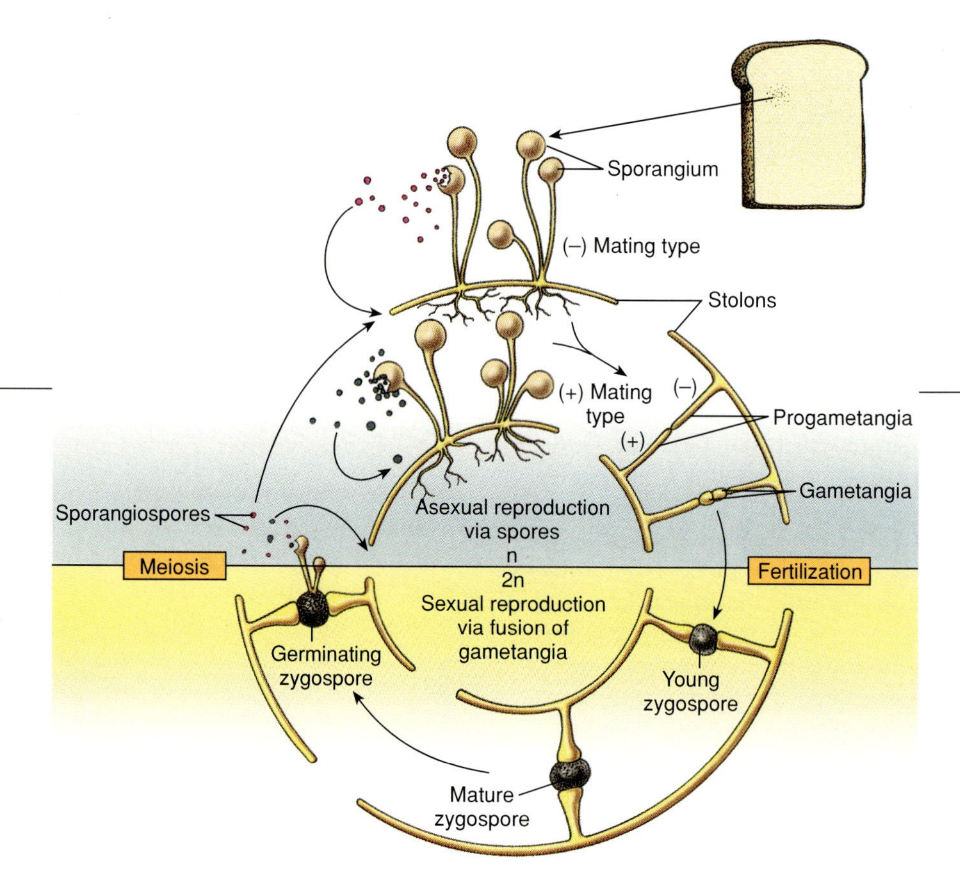

Figure 18.12
Life cycle of the black bread mold, *Rhizopus stolonifer*. At the upper right, positive (+) and negative (–) mating strains come together and touch. As they do so, the tips increase in size and they rise above the substrate. These progametangia are soon separated from the remainder of the hyphae by a new cell wall, and the adjacent enlarged cells are called gametangia. After the wall between the gametangia dissolves, the nuclei fuse to form a diploid zygote, now called a zygosporangium with a single zygospore. As the zygospore germinates, it undergoes meiosis and sends up a new hypha, which gives rise to asexual spores capable of germinating and starting new hyphae. Two mating strains of hyphae may come together to renew the sexual process.

TABLE 18.2

Fungi and Heterotrophic Protista

KINGDOM	DIVISION (NUMBER OF SPECIES)	HABITAT	MYCELIUM
Protista	Oomycota (500)	Some aquatic, some terrestrial	Nonseptate, growth not extensive
Fungi	Zygomycota (600)	Mostly terrestrial	Nonseptate, growth not extensive
Fungi	Ascomycota (30,000)	Mostly terrestrial	Septate, extensive in growth
Fungi	Basidiomycota (25,000)	Mostly terrestrial	Septate, extensive in growth
Fungi	Deuteromycota (25,000)	Mostly terrestrial	Septate, extensive in growth

occur so close together that the entire bread surface may appear dark gray or black. Once spores have been released from the sporangium, they may be carried for great distances, and they occur almost everywhere. Every household has thousands of bread mold and other spores that find their way into the kitchen where, under the proper conditions of moisture and temperature, they can germinate on bread, vegetables, fruit, or other organic matter to form a new mycelium.

Sexual reproduction in *Rhizopus* is initiated by the growing together of two mating strains. These strains may form swollen hyphae, the progametangia, which fuse at the point of contact. **Gametangia** are formed, and nuclei from both strains fuse to form a zygote containing many nuclei. The zygospore may be very resistant to environmental stresses and survive a long period of dormancy. Under suitable conditions the zygospore germinates, the nuclei undergo meiosis, and within the germ sporangium **sporangiospores** are formed and dispersed that can germinate to produce a new haploid mycelium. The details of reproduction vary for different groups of fungi, and these differences aid in classification.

Classification

The classification of a number of lower organisms, including fungi and heterotrophic protista, is subject to a great deal of interpretation and personal prejudice. Distinct morphological characteristics that would put related groups of organisms in definitive groups are difficult to determine; one character may group several kinds of organisms; yet another character may separate them. For the purposes of simplicity, we have chosen to include four distinct groups of fungi commonly termed the **Zygomycetes,** the **Ascomycetes,** the **Basidiomycetes,** and the **Deuteromycetes** (table 18.2).

The characteristics of sexual reproduction usually dictate that these groups are indeed different. Students should be aware that most mycologists would classify the **Oomycetes,** traditionally included in the fungal kingdom, as heterotrophic protists, along with another group of organisms called the chytrids. Like the Oomycetes, the latter have motile zoospores powered by flagella, and in that respect they are different from the other true fungi.

We have also included the **lichens** with the kingdom Fungi. Since the lichens are symbiotic organisms composed of a fungus and a green alga or cyanobacterium, they might be classified in any of the three groups. Lichens are, however, always named according to the fungal member of the association. Therefore it is appropriate that they be included with the kingdom Fungi. The slime molds are divided into two distinct groups, the cellular slime molds and the plasmodial slime molds. These organisms

SEXUAL AND SPORE STAGE	ASEXUAL SPORE STAGE	MOTILE CELLS	REPRESENTATIVE GENERA
Thick-walled resting stage called OOSPORE, not in a fruiting body	Motile spores (ZOOSPORES) produced inside a sporangium	Present	*Saprolegnia* (water mold) *Phytophthora* (late blight of potato)
Thick-walled resting stage called ZYGOSPORE, not in a fruiting body	Nonmotile spores (APLANOSPORES) produced inside a sporangium	None	*Rhizopus* (bread mold)
ASCOSPORES (usually eight) borne inside a sac (ascus), usually in a fruiting body; many lack the sexual stage	Nonmotile spores formed on the tips of specialized filaments (CONIDIOPHORES); some lack an asexual stage	None	*Morchella* (morels) *Saccharomyces* (yeasts)
BASIDIOSPORES (usually four) borne on the outside of a clublike cell (basidium), usually in a fruiting body	Nonmotile spores (CONIDIA) formed on the tips of specialized filaments (CONIDIOPHORES); many lack an asexual stage	None	*Agaricus* (common mushroom) *Amanita* (fly agaric)
None	Nonmotile spores (CONIDIA) formed on the tips of specialized filaments (CONIDIOPHORES)	None	*Penicillium* (some spp) *Aspergillus* (some spp)

have been particularly difficult to classify, but are best included with the heterotrophic protists.

Oomycota

Members of this division of less than 600 species are aquatic for part of their life cycle. They live in water and soil, and others are parasites of plants and animals (figure 18.13). The organism that causes potato blight is one example of a plant parasite that first becomes established when rain or irrigation water splashes its spores on the underside of plant leaves. The Oomycetes have motile asexual spores, called zoospores. Unlike fungi, their cell walls are composed not of chitin, but of other polysaccharides, including cellulose. The hyphal body is typically without cross walls (nonseptated). Such organisms, possessing a large mass of cytoplasm with many nuclei, are called **coenocytes.** In contrast the fungi possess chitinous cell walls, have cross walls (septa) that divide the hyphae into cells containing one to many nuclei, and are termed septate.

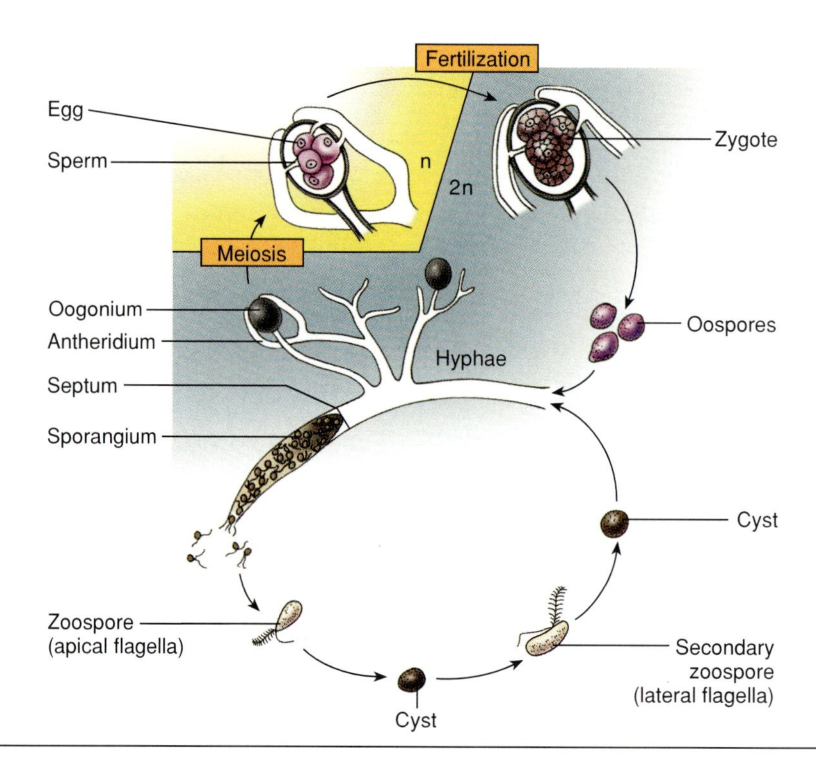

Figure 18.13

The life cycle of *Saprolegnia,* an Oomycete. Reproduction is largely asexual from the diploid mycelium. The tip of a hypha differentiates into a sporangium, which releases zoospores with flagella. Each zoospore changes to an encysted zoospore, followed by a new zoospore, followed by yet another encysted zoospore, which germinates to produce a new hypha. In sexual reproduction antheridia and oogonia are produced on the mycelium, and meiosis occurs there. Sperm from the antheridium fertilize the eggs within the oogonium to produce a zygote. This thick-walled oospore germinates to produce a new mycelium.

Zygomycota (Conjugation Fungi)

These organisms are also coenocytic, but the asexual spores and gametes are not motile. They live in the soil, obtaining energy from decaying plant or animal matter. They also include parasites of plants and animals. This division includes approximately 600 species, and the name is derived from their production of zygospores, a highly resistant diploid cell formed at the time of conjugation, when two gametangia unite, just prior to fertilization. Whereas the water molds reproduce by oogamy, the union of a large nonmotile egg with a small motile or nonmotile male gamete, the Zygomycetes reproduce by isogamy, the union of gametes of equal size. Isogamy is characteristic of sexual reproduction in many lower organisms, and oogamy is typical of organisms more advanced evolutionarily, including humans. The black bread mold (figure 18.12) described earlier is a typical Zygomycete.

Ascomycota (Sac Fungi)

This group of organisms comprises some 30,000 species, many of which are economically and ecologically important. Food spoilage is usually caused by blue-green and brown molds belonging to the Ascomycetes. The powdery mildews of grapes and other fruit belong to this group, as do the highly destructive organisms that cause Dutch elm disease [*Ophiostoma* (*Ceratocystis*) *ulmi*] and chestnut blight [*Cryphonectria* (*Endothia*) *parasitica*]. Perhaps the most economically important group of organisms in this class are the yeasts, which produce the carbon dioxide that causes bread to rise. They also produce ethyl alcohol by fermentation, which occurs in beer and wine production. The strains used in wine production can grow under anaerobic conditions until the alcohol content of their environment reaches 12% to 14%. These organisms bring about their own demise by alcohol toxification, and therefore wines are limited to this alcohol concentration. Distilled spirits with higher alcohol content must have water removed to achieve higher concentrations.

The morels and truffles, highly prized by gourmets, are Ascomycetes (figure 18.14). Defying widespread commercial culture, the truffles grow in a symbiotic relationship with certain oak trees, the fruiting body occurring below the ground. The structure must be rooted out by highly trained pigs or dogs, bred specifically for that purpose.

Rye plants used to make bread flour are occasionally parasitized by a sac fungus, *Claviceps purpurea* (figure 18.15). This fungus produces ergot bodies, containing many alkaloids and the hallucinogen lysergic acid. (The closely related lysergic acid diethylamide, LSD, is produced synthetically and is far more potent.) During the Middle Ages as many as 40,000 people are known to have died from ergot poisoning. Even today, ergot occasionally contaminates rye flour and causes illness or death.

The distinguishing characteristic of this division of fungi is the ascus, a sac containing sexual spores, termed **ascospores.** Asexual reproduction usually occurs by specialized structures called conidiophores, which produce **conidia.** There are no motile spores in the sac fungi.

In sexual reproduction the ascus (sac) that bears ascospores is usually produced from a very dense compact mass of hyphae, or the fruiting body (figure 18.16). These fruiting bodies are often a means of identification; they may be large and cup shaped, enclosed, or flask shaped with a long slender neck. The fruiting bodies assume many colors and sizes, depending on the species. Four or eight ascospores are produced in asci, usually on the inner surface of the ascocarp. Each ascospore can produce a new hypha.

Reproduction in yeast is unique among the Ascomycetes. As unicellular organisms, they reproduce by budding or pinching off from the mother cell (figure 18.17). This asexual reproduction is the general rule, although sexual reproduction does occur. The single organism *Saccharomyces cerevisiae* is used in both alcoholic fermentation and bread making, although many different strains are selected for specific flavors.

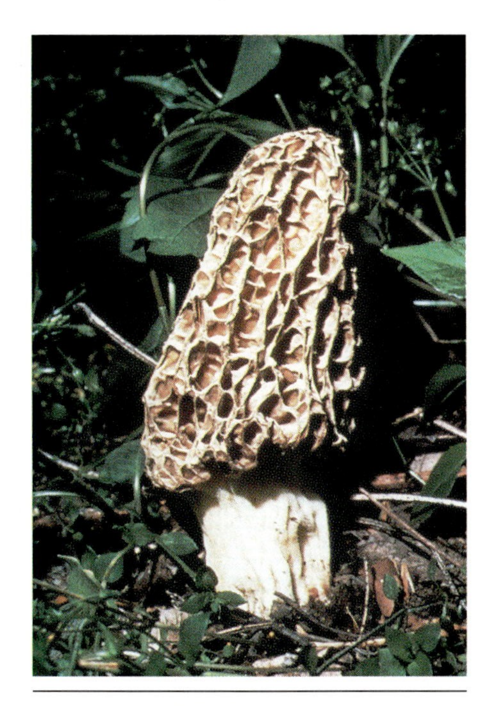

Figure 18.14
The morels belong to the genus *Morchella* and have a stalked, spongelike fruiting body or apothecium. They are all edible, but the season extends for only about one month in the spring, and they are difficult to grow commercially.

Figure 18.15
The fungus *Claviceps purpurea* parasitizes the kernels of rye and other grains, producing lysergic acid, which leads to the condition known as ergot poisoning. If infected flour is made into bread and then eaten, hallucinations, serious illness, and death can result.

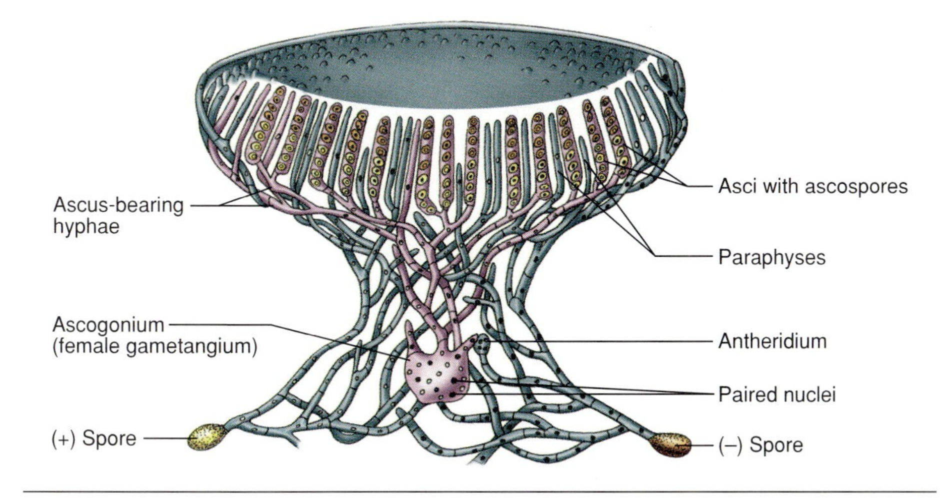

Figure 18.16
Diagram depicting ascocarp development. This cup-shaped ascocarp, termed an apothecium, contains asci in which ascospores form.

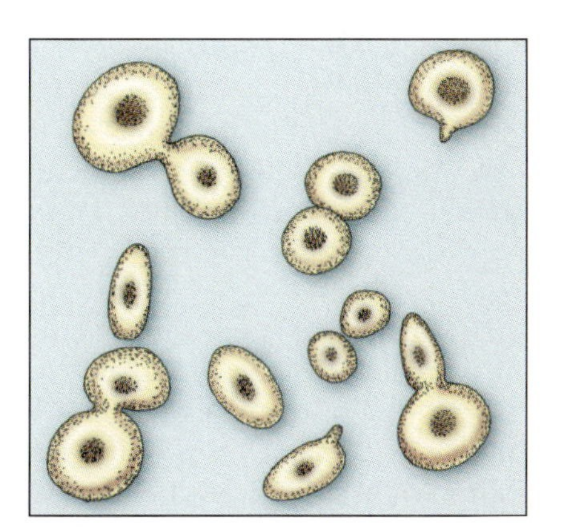

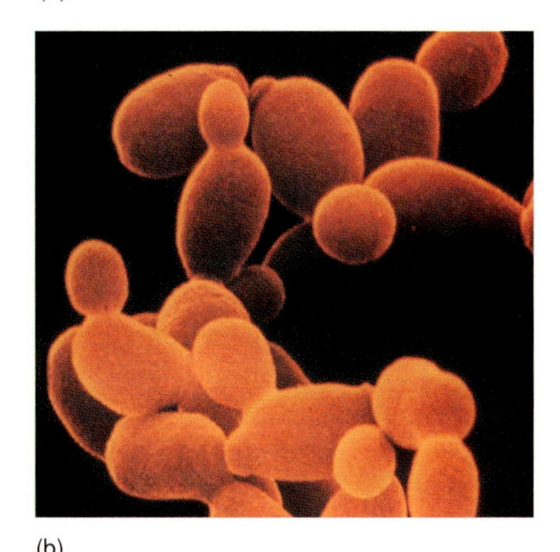

Figure 18.17
(*a*) and (*b*) Yeasts generally reproduce by an asexual process called budding, in which the mother cell pinches off a portion of itself to begin a new cell.

Deuteromycota

Among the fungi there is a large group of organisms in which no sexual reproduction is known to occur. This may be because the sexual phase has not been described or because the sexual phase has been lost during the process of evolution. Such organisms are termed imperfect or Fungi Imperfecti (Deuteromycetes). When the perfect or sexual phase in the life cycle is discovered, in most instances **mycologists** (scientists who study fungi) find that the organisms belong to the Ascomycetes. There are perhaps 25,000 described species in the Deuteromycota. Some of these molds are of particular economic importance. The unique flavor of highly prized cheeses such as Roquefort and Camembert is produced by particular strains of *Penicillium. Aspergillus* (figure 18.18) is used to ferment soy sauce in the Orient, and another species of this genus is important in the initial stages of brewing sake, although the actual fermentation is accomplished by the same yeast used to make beer, *Saccharomyces cerevisiae.*

In 1928 Sir Alexander Fleming noticed that *Penicillium* mold had contaminated a nutrient agar plate of *Staphylococcus,* a bacterium, and had totally inhibited the growth of the bacteria. This observation led to the concept of antibiosis and the development of antibiotics, substances produced by one microorganism that in minute quantities injure or inhibit the growth of another microorganism. The later purification of this substance in *Penicillium* led to the discovery of penicillin, which has been responsible for saving millions of lives during the last 50 years. This drug is useful in treating pneumonia, rheumatic fever, diphtheria, scarlet fever, syphilis, and gonorrhea.

Other imperfect fungi include the organism (commonly termed dermatophytes) that causes ringworm and athlete's foot. These disorders are difficult to treat. Still other imperfect fungi spores are sometimes inhaled, causing respiratory tract infections or allergic responses.

Basidiomycota (Club Fungi)

This large group of some 25,000 described species includes the smuts, rusts, mushrooms, stinkhorns, puffballs, and shelf fungi. The spores produced by these fungi, following meiosis, are called **basidiospores** and are borne on a club-shaped structure called a **basidium.** The mycelium is always septate, and there are two subclasses, one that includes the mushroom/puffball group and the other the organisms that produce rusts and smuts. The *Agaricus bisporus* mushroom is a significant commercial crop and is now grown all over the world. Originally cultivated only in caves, they are now grown in all sorts of structures where light, temperature, and relative humidity can be controlled. Many wild mushrooms are edible, but others are quite toxic, including the beautiful but deadly *Amanita phalloides.* Unless you are completely certain about the species identification, leave the mushroom collecting to the professionals! In the mushroom the fruiting body, called a **basidiocarp,** is formed from a dense mycelium, and basidia and basidiospores are produced from the gills that hang down on the underside of the mushroom cap. One cap may produce millions of spores, each capable of germinating and starting a new hyphal network in the soil. Few of these spores, of course, ever do produce new mushrooms in nature because they fail to encounter the proper environmental conditions for completing the life cycle (figures 18.19 to 18.23).

Lichens

Usually studied with the fungi, most lichens are Ascomycetes that grow in a symbiotic relationship with green

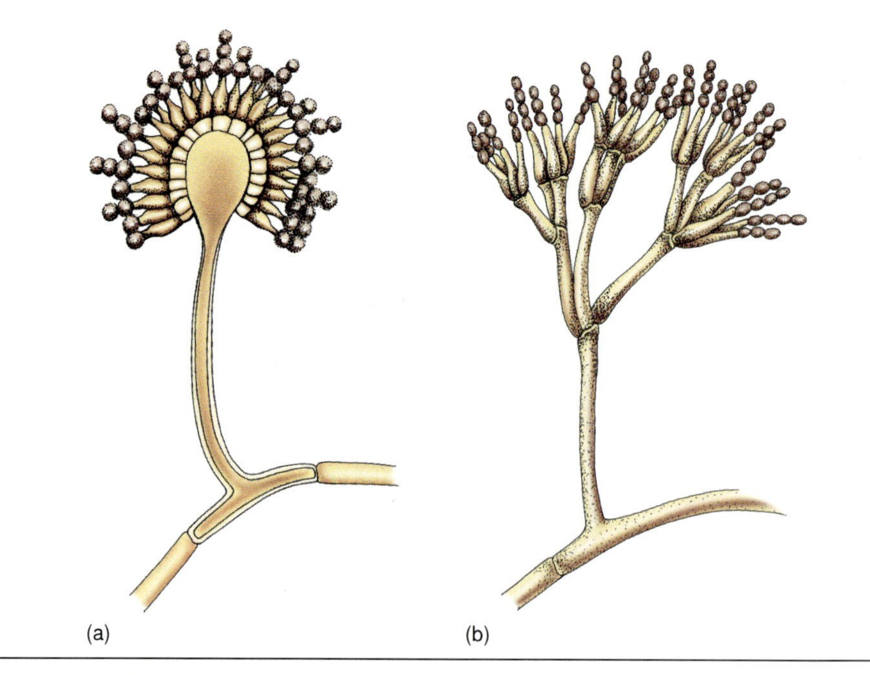

Figure 18.18

Some genera of the imperfect fungi are important both as decomposers and as producers of chemicals for food and pharmaceuticals. (*a*) The conidia of *Aspergillus* are tightly clumped on the spore-bearing stalk, called the conidiophore. (*b*) In *Penicillium* the conidia are borne on loosely branched stalks of the conidiophore.

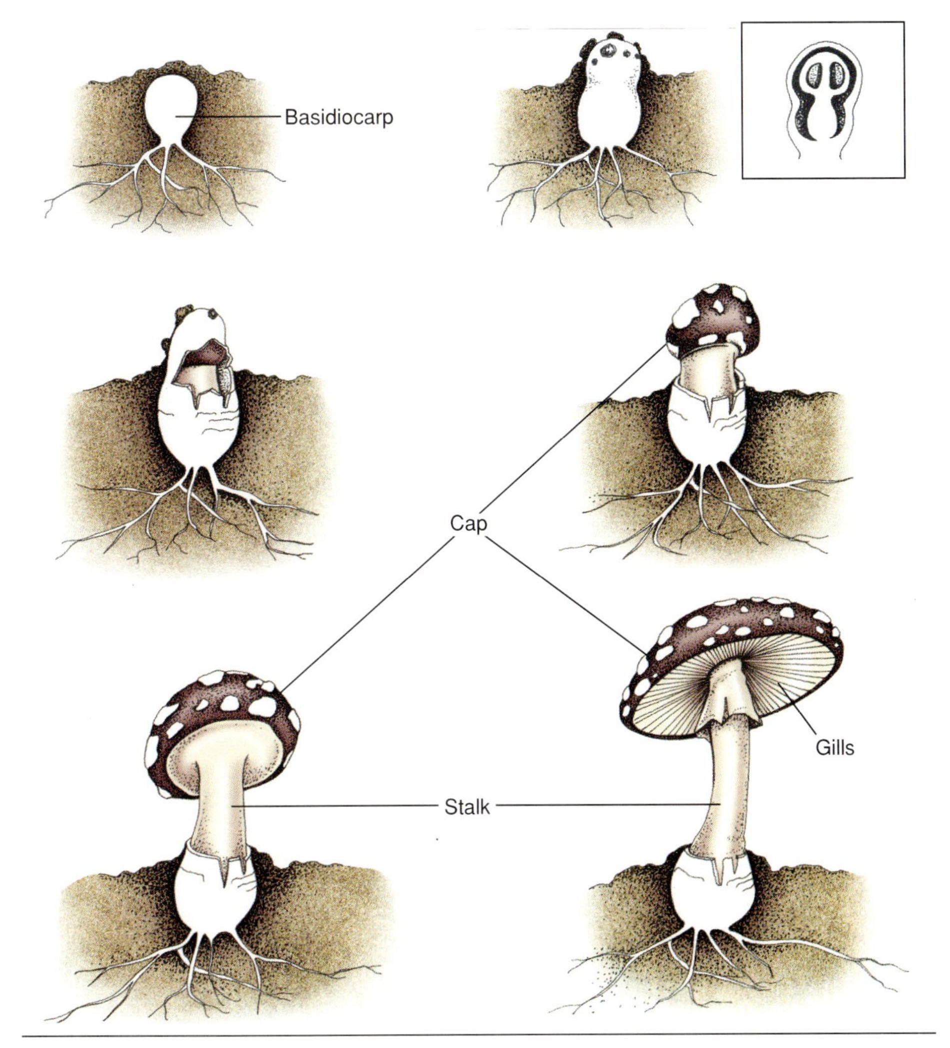

Figure 18.19

Mushrooms are produced from a mass of hyphae that forms a spherical structure underground, the basidiocarp. As it enlarges, the top portion breaks open to reveal a cap. Eventually the cap expands until the membrane underneath breaks, revealing the gills or basidium from which the spores are produced. A single mushroom can produce many thousands of spores.

algae or cyanobacteria (figure 18.24). There are approximately 25,000 species, which manage to survive in the most extreme environments, often on bare rock. Such organisms are called colonizers or pioneer organisms, and they represent the first step in developing a biotic environment where there was only an abiotic one, as on a lava flow. They grow in the driest deserts and the coldest, windy mountaintops throughout the world. Lichens come in a myriad of colors—black, white, and all colors and intensities of green, orange, yellow, and red. Lichens range in size from almost invisible to large, leafy forms that cover vast regions of the Arctic to a depth of 10 to 15 cm and are known as reindeer moss (figure 18.25). (Note that this organism is indeed a lichen and not a true moss.) Of the lichens, only this plant has economic significance for humans as a source of feed for reindeer, but ecologically all lichens play a major role. It is important to note that the fungi in this group always occur with the mutual photosynthetic species, whereas the associated algae or cyanobacteria may occur independently. As heterotrophs, the fungi must obtain their energy from organic molecules produced by the photosynthetic algae or cyanobacteria. As might be expected, lichens growing on a bare rock surface do so very slowly, photosynthesis occurring only when body moisture levels are high. When desiccation occurs, the lichens become dormant; they are able to survive extremes in light intensity, temperature, and drying. When it rains, they imbibe water rapidly and again assume rapid rates of photosynthesis and respiration.

Lichens are capable of absorbing nutrients from their substrate, but in most cases they rely on rain and wind-blown nutritional deposits for a source of food. Consequently, they are exceedingly susceptible to environmental pollution, and lichen growth (or lack of it) is often used as a measure of pollution in industrialized areas. There are three general types of lichens: crustose, or crustlike; fruticose, or shrublike; and foliose, or leaflike.

Slime Molds

Slime molds hold a fascination for biologists, not for economic reasons but because they behave as if they were part plant and part animal. They are all heterotrophic, lack a cell wall for most of their life cycle, and display movement. Taxonomists classify these organisms with the heterotrophic Protista. They are divided into two groups, the noncellular or plasmodial slime molds and the cellular slime molds. Species of the first group possess a very thin, streaming mass of protoplasm that varies from microscopic to a square meter or more in area. The naked protoplast is

Figure 18.20

The fruiting bodies of mushrooms come in many colors, shapes, and sizes. (*a*) *Hygrophorus* sp., Arkansas. (*b*) *Marasmius* sp., rain forest in Tingo Maria, Peru. (*c*) *Pholiota* sp., Arkansas. (*d*) *Marasmius siccus*, Missouri. (*e*) *Amanita caesarea*, Arkansas. (*f*) *Polyporus* sp., Manaus, Amazon. (*g*) *Cantharellus cinnabarinus*, Maryland. (*h*) *Amanita* sp., Arkansas. (*i*) *Flammulina velutipes*, Missouri. (*j*) *Pleurotus serotinus*, Iowa. (*k*) *Pleurotus ulmarius*, Iowa. (*l*) *Oudemansiella radicata*, Iowa. (*m*) *Amanita flavoconia*, Iowa. (*n*) *Pleurotus ostreatus*, Iowa. (*o*) *Dictyophora* sp., Cameroons.

Figure 18.21
The gills of a common mushroom as seen on the underside of the cap (*a*). The basidiospores are produced from the margins of the gills (*b*).

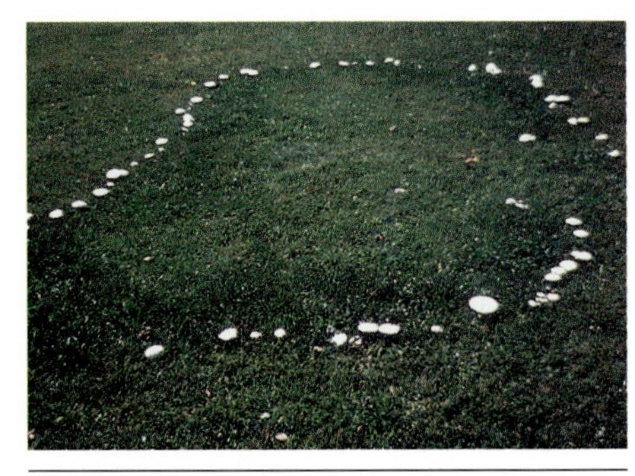

Figure 18.22
Some fungal mycelia grow through the soil from the center, gradually dying out in the middle and producing new fruiting bodies at the outer edge. Such conditions lead to a "fairy ring" common in turf grasses following rain.

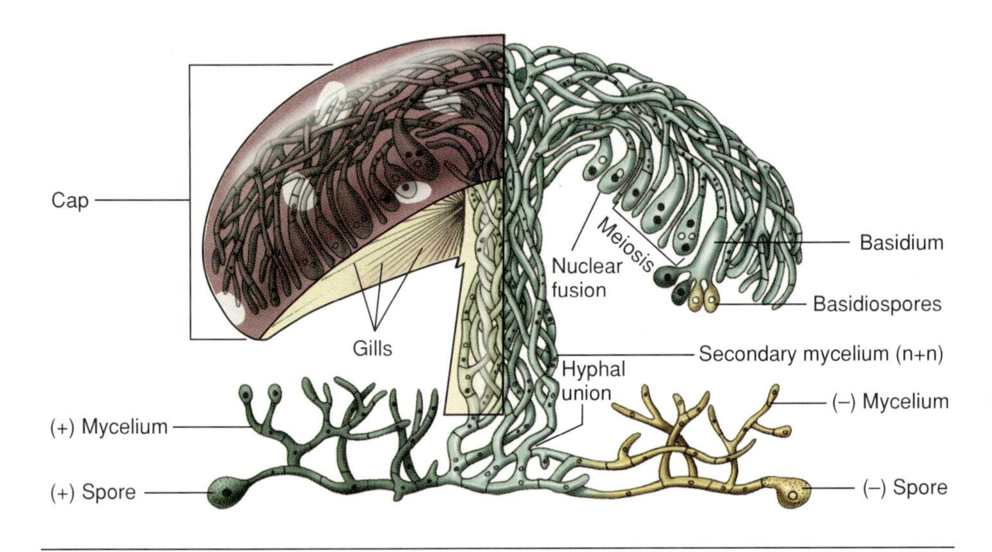

Figure 18.23
Diagram depicting how a basidiocarp develops. Following fusion of + and – hyphae, dikaryon hyphae are formed, which grow in concert to form a basidiocarp. In the basidiocarp illustrated, the basidia are attached to gills on the underside of the cap. Four basidiospores are produced from each basidium.

termed a plasmodium; it has no cell wall but may contain thousands of nuclei, all of which divide synchronously. As the plasmodium travels, it ingests bacteria, yeasts, fungal spores, and decayed plant and animal matter. Occasionally such large plasmodia are found in back yards or forests; the startled discoverer may think the "Blob" has arrived from outer space. When conditions are unfavorable for further growth (lack of water or food), the plasmodium will mound up in various places to produce sporangia, containing spores. These are enclosed by a cell wall and become resistant to drying. Later each spore may germinate and produce myxamoebae or zoospores and eventually a new plasmodium (figure 18.26) via a sexual process.

The cellular slime molds also spend part of their life cycle as an amoeboid mass, but the individual cells that make up the mass retain their cellular character and aggregate together. In response to a chemical trigger, the cells rapidly migrate toward a center and produce a stalk with many spores. Each spore is capable of producing a new amoeboid cell.

✔ Concept Checks

1. Why are fungi classified into a separate kingdom from the algae?
2. Discuss both beneficial and harmful aspects of fungi.
3. What characteristics separate the four divisions of fungi? What characteristics separate fungi from Oomycota?
4. What are the biological components of lichens? Discuss their importance in ecological succession.

BRYOPHYTES

From many lines of evidence it appears that the bryophytes, consisting of **liverworts,** hornworts, and the true **mosses,** and the vascular plants have evolved from some ancient

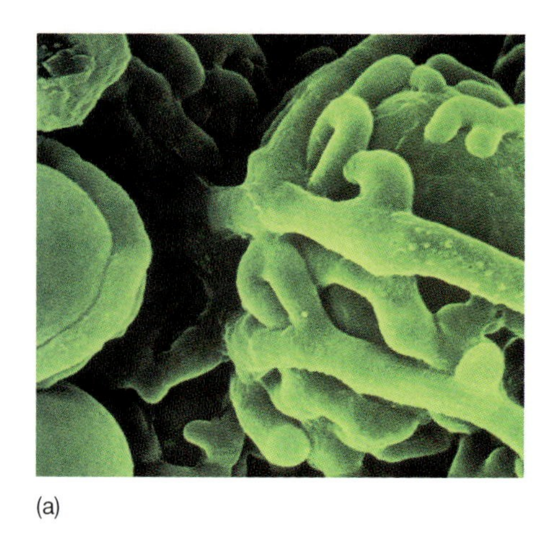

(a)

(b)

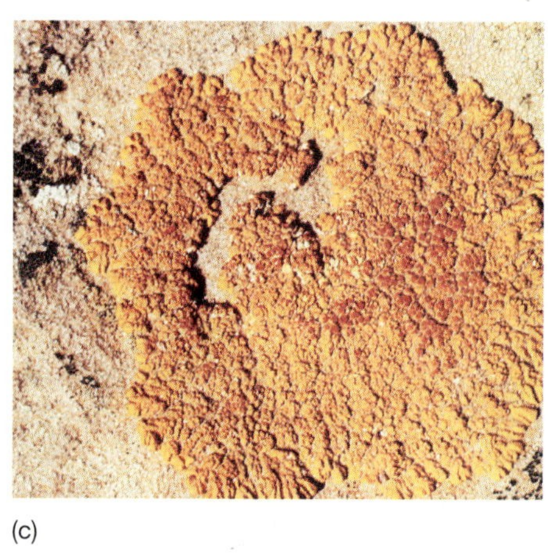

(c)

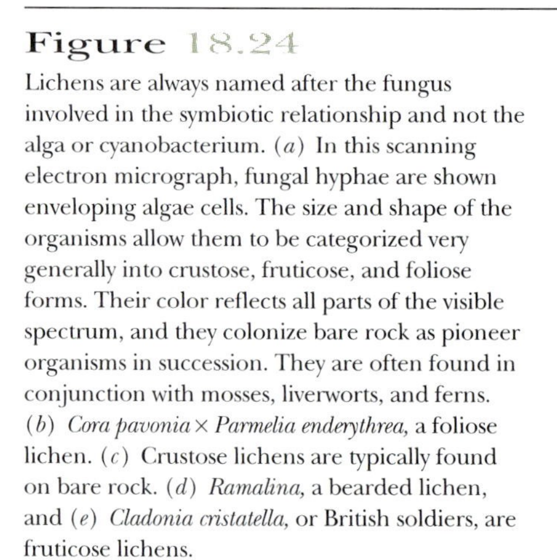

Figure 18.24

Lichens are always named after the fungus involved in the symbiotic relationship and not the alga or cyanobacterium. (*a*) In this scanning electron micrograph, fungal hyphae are shown enveloping algae cells. The size and shape of the organisms allow them to be categorized very generally into crustose, fruticose, and foliose forms. Their color reflects all parts of the visible spectrum, and they colonize bare rock as pioneer organisms in succession. They are often found in conjunction with mosses, liverworts, and ferns. (*b*) *Cora pavonia* × *Parmelia enderythrea,* a foliose lichen. (*c*) Crustose lichens are typically found on bare rock. (*d*) *Ramalina,* a bearded lichen, and (*e*) *Cladonia cristatella,* or British soldiers, are fruticose lichens.

(d)

(e)

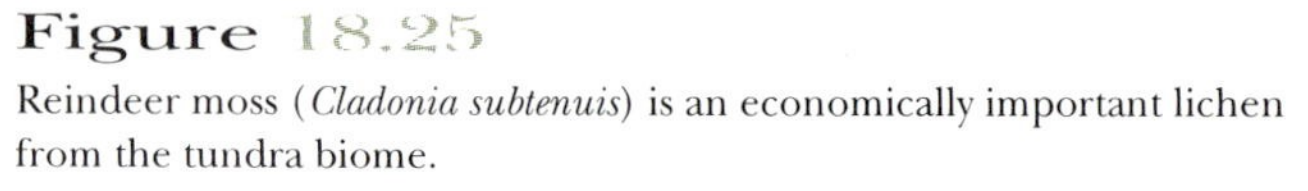

Figure 18.25

Reindeer moss (*Cladonia subtenuis*) is an economically important lichen from the tundra biome.

Figure 18.26

Fruiting bodies of the plasmodial slime mold *Stemonitis.*

group of green algae. Pigmentation, mode of cell division, construction of the cell wall, the presence of lignin, chloroplast structure, and storage of starch are common threads for all these groups. It is probable, then, that the bryophytes and vascular plants evolved as two distinct branches of the evolutionary tree after the multicellular green algae successfully inhabited land. The bryophytes first appear in the fossil record of about 350 million years ago, and there is little evidence that fossil species are very different from those living today. In the transition to land, desiccation was a major problem in survival. Likewise, movement onto land is correlated with the retention of the zygote within the female gametangium, or sex-cell-producing structure, where it develops into an embryo. In algae the multicellular embryo, if present, is an independent entity and has minimal protection from the environment; in the ocean there would have been no selection for such a feature.

In the vascular plants, evolution on land is also correlated with the development of the cuticle, the waxy layer that protects all of the aboveground parts of a plant from desiccation. Gas exchange takes place via the stomata, tiny pores in the surface of leaves that are regulated by both internal and external factors, which allow some control over gas exchange, including water loss. The bryophytes have failed to develop either an effective cuticle or effective stomata, although a thin cuticle may be present.

All bryophytes and vascular plants reproduce by oogamy. They also display **alternation of generations,** a reproductive cycle in which a haploid phase, the gametophyte, produces gametes (the sex cells; sperms and eggs), which fuse to form a zygote, which in turn gives rise to a diploid sporophyte. Spores produced by meiosis in the sporophyte give rise to new haploid gametophytes. In the bryophytes, sperm are produced in a multicellular gametangium termed an **antheridium;** the multicellular egg-producing structure is termed the **archegonium** and is typically vase shaped consisting of an extension termed the neck, and a swollen basal region called the venter in which the egg is protected.

In the process of land survival and adaptation, the bryophytes have failed to occupy the great diversity of habitats characteristic of the vascular plants. Because the bryophytes lack xylem and phloem for the long-distance, rapid transport of water and nutrients, lack an effective cuticle to minimize desiccation, and because sperm need water to swim, they are relegated to moist, shady habitats not so very different from those of their aquatic progenitors. Some mosses, however, are found in shady cracks in niches of dry habitats, and some form patches on cracks in hot, exposed rocks. A few even grow above timberline. Like the lichens, mosses and liverworts are particularly susceptible to pollution, and their numbers decrease rapidly with the onset of synthetic pollutants. Bryophytes are seldom greater than 20 cm in height, and therefore they are hardly a dominant feature of the landscape. They do, however, fill an important ecological role in reducing soil erosion along stream banks and on other slopes.

It is interesting to note that of more than 24,000 identified species, a very large plant group, no bryophytes are marine. They never have true xylem or phloem, and the alternation of generations is characterized by a dominant gametophyte, unlike the vascular plants, in which the gametophyte is very much reduced. The sporophyte is embedded in and dependent on the gametophyte. Although the bryophytes have no true stems, leaves, or roots, the gametophyte is often "leafy." The bryophytes also include a small group of less than 100 species known as the hornworts, but they are not considered here.

Liverworts

The tiny liverworts are generally less conspicuous than mosses. Although most species possess "leafy" gametophytes, the name of the group is derived from the shape of the flat thalloid plant body of some species, roughly shaped like a liver. At one time they were thought to be useful in the treatment of liver ailments. These liver-shaped structures grow much like a green ribbon on the surface of moist soil, branching dichotomously to give two ribbons, which may branch further. The original body gradually dies away, and many plants may be derived asexually from a single gametophyte. One of the more common genera, *Marchantia,* grows in this manner and is a common specimen for a terrarium (figure 18.27). In addition to the fragmentation of the ribbonlike body, which gives rise to new plants asexually, *Marchantia* also produces a specialized cuplike structure of the gametophyte, called a gemmae cup, from which gemmae are produced. Swelling mucilage and raindrops dislodge the gemmae and carry them to new substrates, where they may produce a genetically identical gametophyte.

Sexual reproduction occurs in specialized umbrella-like structures termed gametophores, which arise from the gametophyte. The male and female sex organs are borne on separate gametophores termed the **antheridiophore** and **archegoniophore** respectively (figure 18.28). Following fertilization a sporophyte develops, composed of foot, seta, and **capsule.** The nuclei of the sporocytes within the capsule undergo meiosis, and spores are formed. Following the rupture of the capsule the spores are released, particularly through the twisting and turning of hygroscopic **elaters,** and in favorable habitats they germinate to produce new gametophytes.

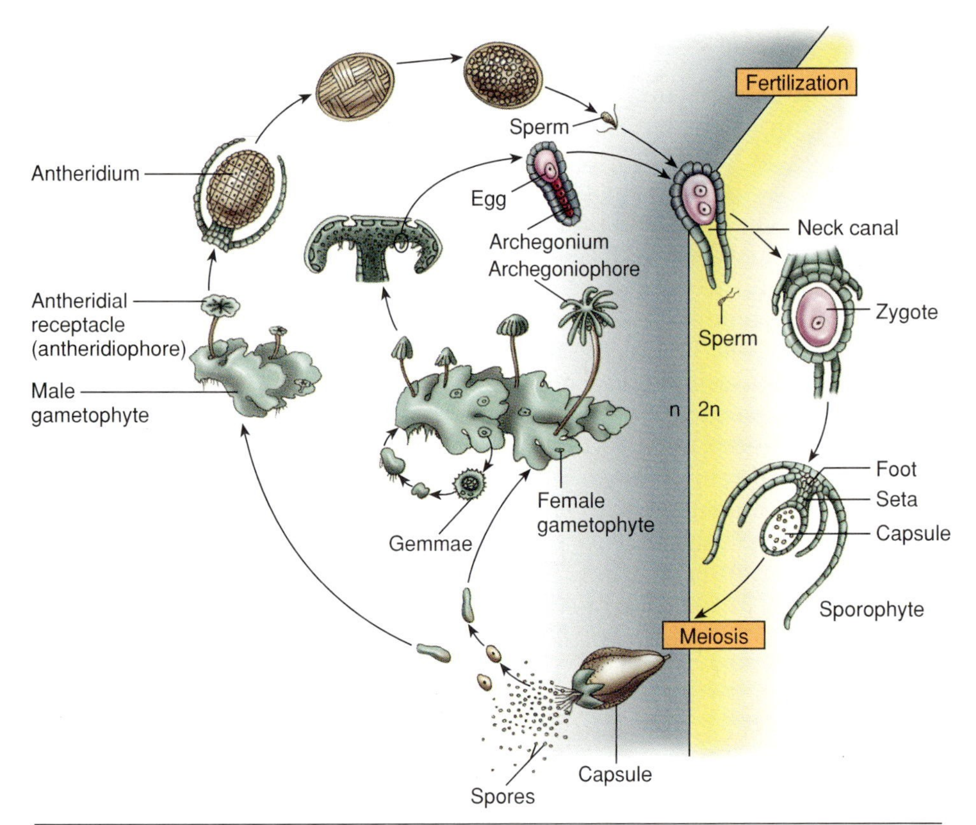

Figure 18.27
The life cycle of *Marchantia.* Spores produced in the capsule germinate to produce independent and separate male and female gametophytes. The antheridial receptacles produce antheridia, which in turn produce flagellated sperm. The archegonial receptacles produce archegonia, each containing an egg. Fertilization is achieved by the motile sperm being "splashed" to the archegonial receptacle, then swimming into the neck canal of an archegonium and fusing with the egg. The sporophyte develops within the archegonium to produce a capsule with spores.

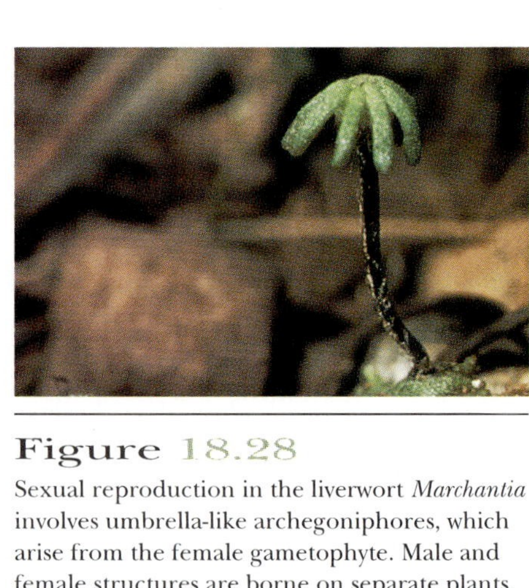

Figure 18.28
Sexual reproduction in the liverwort *Marchantia* involves umbrella-like archegoniphores, which arise from the female gametophyte. Male and female structures are borne on separate plants.

Figure 18.29
The true mosses produce conspicuous gametophytes with the sporophyte embedded in the gametophyte. The capsule is borne on a long, slender filament (*Bryum argenteum*).

Mosses

Like the liverworts, the conspicuous plant body in mosses is the gametophyte stage, which is usually upright and leafy (figures 18.29 and 18.30). Mosses are often so tiny that the individual body is almost microscopic. With high-power magnification a population of moss plants may resemble a very dense forest. These, too, are popular as terrarium plants, and they spread rapidly under ideal conditions of humidity, light, and temperature. In most mosses the sporophyte is embedded within the gametophyte and depends on it for nutrition. The sporophyte grows upright, terminating in a spore-bearing capsule (figure 18.31). Sporocytes in the capsule undergo

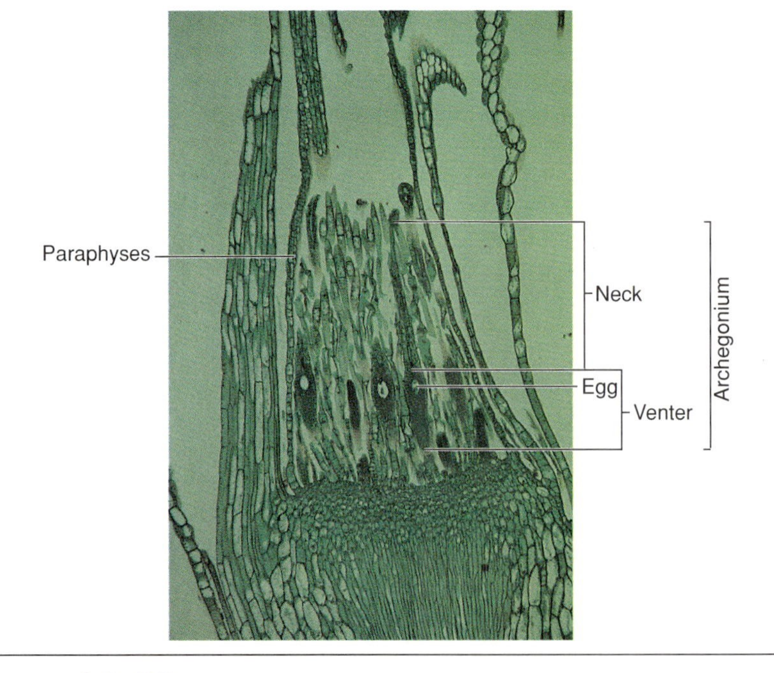

Figure 18.30
Gametangia of *Mnium,* a true moss. Archegonial head with archegonia surrounded by sterile paraphyses.

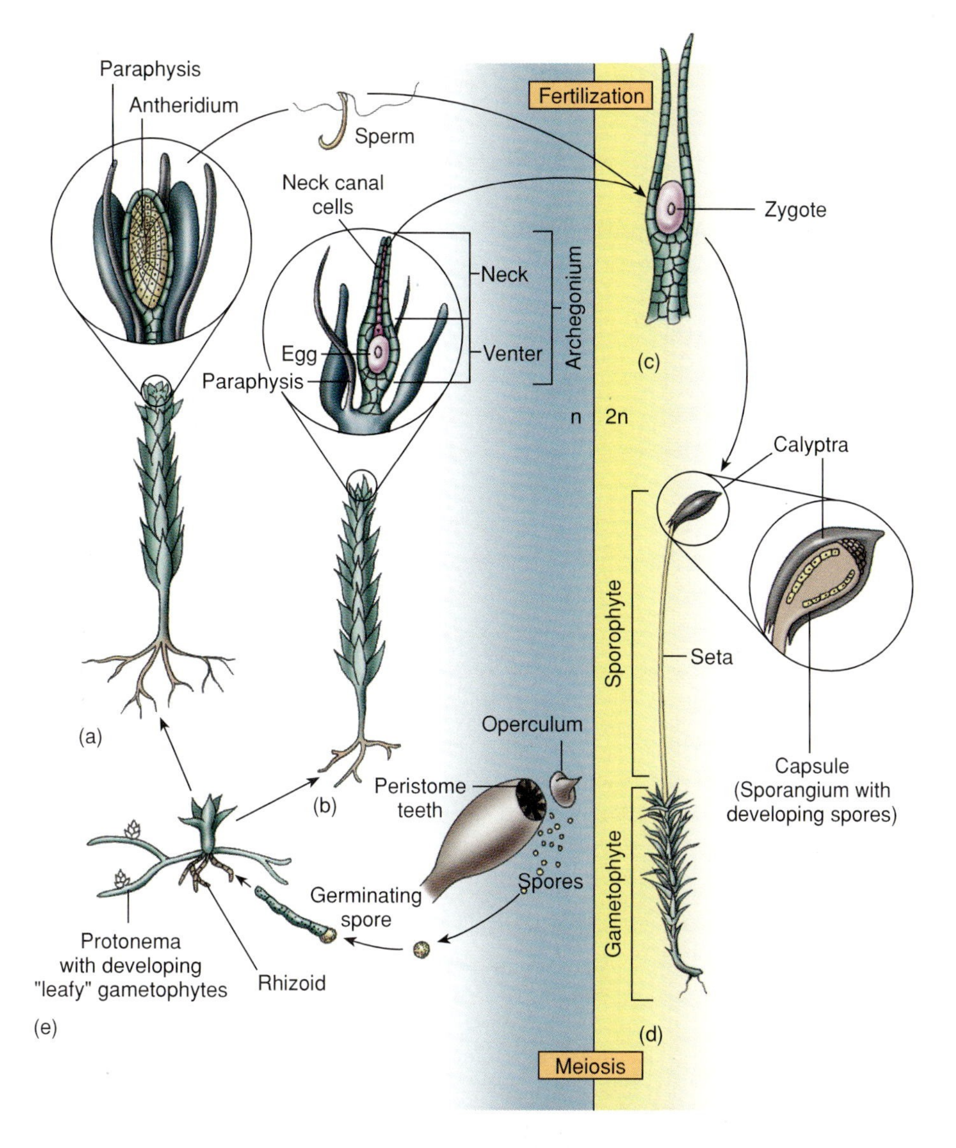

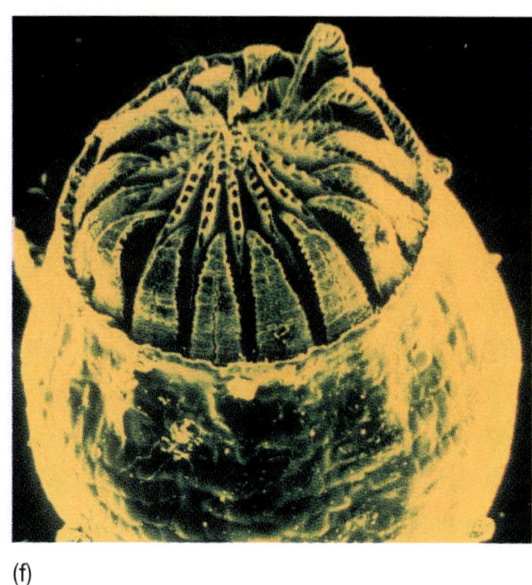

Figure 18.31

Reproduction in mosses. (*a*) Male gametophyte plant showing antheridia and sterile hairs (paraphyses). (*b*) Female gametophyte plant with archegonia and paraphyses; note neck and venter regions of archegonium and neck canal cells and egg in longitudinal section. Flagellated sperm are liberated and dispersed by droplets of water to the female gametophyte, where the sperm enter the neck canal and fuse with the egg to form the diploid zygote (*c*). (*d*) Capsule of maturing sporophyte containing spores, which are dispersed with the aid of peristome teeth. (*e*) Germinating spore forms an alga-like branching system of cells termed the protonema. (*f*) Scanning electron micrograph of a moss spore capsule revealing peristome teeth.

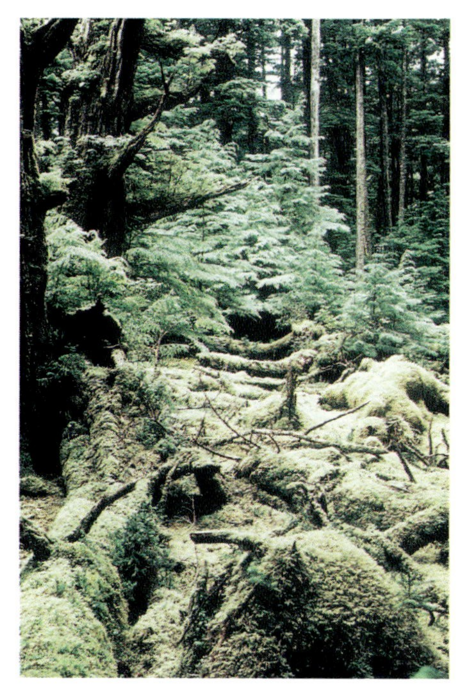

Figure 18.32

Mosses sometimes cover fallen logs and the forest floor.

meiosis to produce spores. On maturation of the capsule microscopic spores are cast into the air. If a spore lands in a suitable environment, a new **protonema** gametophyte will develop (figure 18.32).

The group of mosses with the greatest economic impact are the **peat mosses,** some 300 species belonging to the genus *Sphagnum.* The gametophyte is leafy with many dead cells between rows of photosynthesizing cells. These large, dead cells are the site of water storage, and a sphagnum moss can hold many times its own weight in water. In the colder, temperate regions of the world, peat bogs fill a very important ecological role. Peat bogs form from lakes that gradually fill in with sphagnum moss; the dead but undecayed plants fall to the bottom. Gradually the bog fills in completely, and the peat moss contributes to its own acidic environment. By the time the surface becomes dry, the pH of the lake may be as low as 3.0. In some peat bogs this high acidity effectively eliminates the competition, except for blueberry (*Vaccinium*) species, which thrive in the rich acidic medium. Gradually the blueberries are replaced by larch trees, a deciduous gymnosperm, which form the climax vegetation in peat bogs. Such bogs remain spongy, and it is possible to stand on one side of a bog, jump up and down, and cause the large larch trees on the other side of the bog to sway. Commercial peat moss is harvested by removing the trees, cutting giant squares of the moss, compressing and drying it, and packaging it for shipment around the world. The best quality peat comes from Canada, Germany, and the Scandinavian countries, although peat bogs are found in many countries with cool climates, including the northern United States.

✔ Concept Checks

1. From what group of plants are the bryophytes thought to have evolved? What evidence supports this theory?
2. Describe both the sporophyte and gametophyte plant of either a moss or liverwort. What is meant by alternation of generations? What two major phenomena result in the alternation of generations?
3. Describe the ecological role of *Sphagnum*. Can you think of some commercial applications of *Sphagnum?*

Summary

1. For both traditional and functional reasons many organisms that are not assigned to the plant kingdom are considered within the study of botany. This text considers, in addition to the prokaryotes, the photosynthetic protists (algae) and the fungi.
2. The evolution of diploidy dates back to at least 700 million years ago. This significant event established sexual reproduction and led to new and increased variability among organisms.
3. The theory of endosymbiosis suggests that some organelles arose in eukaryotic cells through the incorporation of prokaryotic cells into a eukaryote. Chloroplasts could have arisen by incorporating a cyanobacterium, and mitochondria could have arisen through the incorporation of an aerobic bacterial cell.
4. The photosynthetic protists (algae) include Chrysophyta, Pyrrhophyta, Euglenophyta, Chlorophyta, Phaeophyta, and Rhodophyta. Most occur in both marine and freshwater, and they contribute significantly to primary productivity and oxygen supply. The diatoms (Chrysophyta) are particularly important in producing diatomaceous earth, a product of great industrial importance. The macroalgae are a source of many food and industrial products.
5. The fungi are all heterotrophic and characterized by a chitinous cell wall (in most groups). They have high surface-to-volume ratios, which allow them to perform well as parasites and saprophytes. The major groups are the Zygomycetes, Ascomycetes, Basidiomycetes, and Deuteromycetes. Fungi are the major sources of plant diseases and contribute to disease in many other organisms, including humans. They also, however, perform many beneficial functions, including gas and alcohol production important in the baking and brewing industries.
6. Lichens are organisms named after the fungal symbiont but function with a green alga or cyanobacterium as a colonizer organism. The Oomycota (water molds), plasmodial and cellular slime molds are classified in this text with the fungi, but are often classified with the heterotrophic Protista.
7. The bryophytes, the liverworts and mosses, are nonvascular land plants that apparently evolved shortly after a green algal ancestor made its way from ocean to land. They are interesting plants botanically but have little economic significance, except for the sphagnum mosses.

Key Terms

agar 322
alga 317
algae 317
alternation of generations 333
antheridia 321
antheridiophore 333
antheridium 333
archegoniophore 333
archegonium 333
Ascomycetes 325
ascospores 327
athlete's foot 323
basidiocarp 328
Basidiomycetes 325
basidiospores 328
basidium 328
bryophytes 317
capsule 333
carrageenan 322
charophytes 321
Chlorophyta 317
Chrysophyta 317
coenocyte 326
conidia 327
Deuteromycetes 325
diatomaceous earth 318
elaters 333
endosymbiosis 317
Euglenophyta 317
filamentous 320
fungi 317
gametangia 325

hyphae 323
isogamy 320
lichens 325
liverworts 331
mosses 331
mycelium 323
mycologist 328
nonseptate 323
oogamy 320
oogonia 321
Oomycetes 325
peat moss 335
penicillin 323
Phaeophyta 317
photosynthetic protists 317
protonema 335
Pyrrhophyta 317
Rhodophyta 317
septate 323
slime molds 317
sporangiospores 325
sporangium 323
yeasts 323
Zygomycetes 325

Discussion Questions

1. Identify several organisms that are classified in the kingdom Protista; the kingdom Fungi; the division Bryophyta of the plant kingdom. What are the major characteristics that are used to separate the protists from the fungi?
2. Construct a life history using the following terms placed in proper sequence: spores, gametophyte, sporangium, meiosis, sporophyte, fertilization, gametangia, egg, sperm.
3. Describe the development of a basidiocarp from the time before and after hyphae unite until the release of basidiospores.
4. Discuss the ecological importance of photosynthetic protista, heterotrophic protista, and fungi.
5. Discuss how algae and fungi are economically important.

Suggested Readings

Ahmadjian, V. 1993. *The lichen symbiosis.* Wiley, New York.

Bates, J. W., and A. M. Farmer. 1992. *Bryophytes and lichens in a changing environment.* Oxford University Press, New York.

Bold, H. C., and M. C. Wynne. 1985. *Introduction to the algae: Structure and reproduction.* 2nd ed. Prentice-Hall, Englewood Cliffs.

Bold, H. C., C. J. Alexopoulos, and T. Delevoryas. 1987. *Morphology of plants and fungi.* 5th ed. Harper & Row, New York.

Chang, S. T., J. A. Buswell, and P. G. Miles. 1993. *Genetics and breeding of edible mushrooms.* Gordon and Breach, Philadelphia.

Crumm, H. A. 1988. *A focus on peatlands and peat mosses.* University of Michigan Press, Ann Arbor.

Farr, D. F., ed. 1989. *Fungi on plants and plant products in the United States.* American Phytopathological Society Press, St. Paul.

Luning, K. 1990. *Seaweeds: Their environment, biogeography, and ecophysiology.* Wiley, New York.

McQueen, C. B. 1990. *Field guide to the peat mosses of boreal North America.* University Press of New England, Hanover.

Newhouse, J. R. 1990. Chestnut blight. *Sci. Am.* 263(1):106–111.

Okaichi, T., D. M. Anderson, and T. Nemoto. 1989. *Red tides: Biology, environmental science, and toxicology.* Elsevier Science Publishers, New York.

Phillips, R. 1991. *Mushrooms of North America.* Little, Brown, Boston.

Saltarelli, C. G. 1989. *Candida albicans: The pathogenic fungus.* Hemisphere Publishing Corp., New York.

Sandgren, C. D., ed. 1988. *Growth and reproductive strategies of freshwater phytoplankton.* Cambridge University Press, New York.

Shubert, L. E., ed. 1984. *Algae as ecological indicators.* Academic Press, Orlando.

Chapter Nineteen

Vascular Plants

Movement to Land

Adaptations of Land Plants
Evolution and Distribution of Vascular Plants
Extinct Vascular Plants

Today's Seedless Vascular Plants

Psilophyta
Lycophyta
Sphenophyta
Pterophyta: The Ferns

The Seed Plants

Gymnosperms
Angiosperms

This chapter begins with a discussion of early plant life in their "new" land environment. Once on land, diversity of plants accelerated, and in a matter of 300 million years a vast assemblage of different forms evolved. This chapter introduces this diversity of land plants and the major evolutionary changes that have taken place in the plant kingdom.

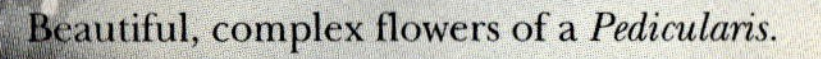

Beautiful, complex flowers of a *Pedicularis*.

One of the most significant, yet unsolved problems in the history of plant life on earth was the migration from the oceans and subsequent occupation of the land by descendants of green algal ancestors. There is no fossil evidence to suggest that life on land existed prior to 450 million years ago. Various marine organisms must have repeatedly washed up on the shore, but the rocky, barren environment was inhospitable to them. This period also coincided with the receding of the oceans and the exposure of algae to the land. The seawater offered protection from desiccation, and the sun's destructive ultraviolet radiation. The seawater, furthermore, acted as a buffer so that temperature fluctuations were moderate and allowed organisms to absorb nutrients directly from the minerals dissolved in it. Marine organisms did and still do cope with the osmotic problems of a saline environment, but otherwise growth and development in the ocean was without many of the problems that were to be encountered on land.

MOVEMENT TO LAND

Although some oxygen was apparently produced by electrical discharges that split water molecules, it was not until cyanobacteria began photosynthesizing and thus obtaining their reducing source (H^+) from the photolysis of water that oxygen was put into the atmosphere in appreciable quantities. The oxygen molecule, O_2, automatically forms a small amount of ozone, O_3, which has the ability to absorb ultraviolet radiation. Although the proportion of ozone is relatively small compared with the concentration of oxygen, it is a significant factor in the earth's atmosphere. The fossil record indicates that cyanobacteria began photosynthesis about 3 billion years ago, but evidence suggests that significant oxygen accumulation did not occur until about 1 billion years ago. It was not until approximately 450 million years ago that a sufficient concentration of ozone finally began to absorb enough harmful radiation that organisms could survive on the land, exposed directly to the sun's rays.

Single-celled marine organisms existed as the sole life forms for a long time. Then true multicellularity evolved, first as a single-dimensional filament, then as a two-dimensional sheet of cells, and finally as three-dimensional organisms with considerable differences in the external and internal environment of the organism as a whole. Colonial organisms such as the green alga *Volvox* developed modified internal concentrations of nutrients and gases. As multicellular marine plants became larger and more complex, specialized cells formed to accommodate and alleviate problems of "communication" between cells and tissues. Survival is more likely for organisms that evolve a system of coping with environmental changes.

There is no evidence that any prokaryotes or eukaryotes survived on land until a multicellular green alga established itself on land in the **Silurian** period, about 430 million years ago. Evidence suggests that it may have evolved symbiotically with a fungus, perhaps like the modern lichens. Probably thousands of algae were washed onto land before successful establishment was achieved. At best, the terrestrial habitat was precarious, but surviving offspring had improved structure/function relationships, and the landscape finally had its first occupants.

Adaptations of Land Plants

The fossil record suggests that early land plants were not significantly different from their marine progenitors. Some of the simplest plants are the bryophytes, a group that is commonly considered an evolutionary dead end. While the bryophytes followed one line of evolution, another group developed a level of specialization never before achieved on land. Elongated, thin-walled cells had already developed, and mutation of those cells gradually led to conducting cells that eventually evolved into xylem and phloem. With a relatively inefficient cuticle and system of gas exchange, plants could not grow very tall, and the surface area of aboveground parts had to remain small. Such plants were branched and had almost no leaves. As xylem and phloem became more efficient, water and nutrients moved faster, leaves expanded, and the area available for capturing light energy increased; thus more photosynthate could accumulate, creating greater biomass. Those plants that developed an efficient transport system for water and another for the transport of organic solutes are called **vascular plants.**

The vascular plants and bryophytes diverged long ago, shortly after the progenitor green alga made the transition from ocean to land. Since all land plants have a well-developed alternation of generations, it is believed that the green algal ancestor must also have had this characteristic.

The rate of evolution was rapid after the movement to land. Extreme environments lead to a more expeditious selection of variants than does a stable environment; therefore fluctuations in temperature, changes in light conditions, nutritional scarcity, changing relative humidities, and unpredictable water availability promoted swift plant evolution on land. One of the critical features of survival on land is the protection of sex cells; organisms successful in making the transition must have gametes with an outer layer of protective cells. The development of the antheridium (male multicellular sex organ) and the archegonium (female multicellular sex organ) were important advancements in the success of land plants.

Early in the movement to land, organisms developed elevated, multicellular sporangia well-protected by walls that prevented desiccation. Spores produced by these structures could be dispersed readily by wind. At the same time, plants were beginning to develop a cuticle to keep the plant body from drying. Efficient conducting systems evolved from primitive transport cells hampered by friction to more streamlined cells with little or no frictional loss. Gradually from the alternation of generations viewpoint, the sporophyte evolved an increasingly dominant role; the gametophyte gradually reduced in size and eventually became dependent on the sporophyte.

There is considerable diversity among plant groups in their protection from drying. The bryophytes are very poorly protected, with little or no cuticle and little stomatal development, whereas the more advanced vascular plants have very thick cuticles and excellent stomatal control, which regulates water loss efficiently. Such vascular plants are able to fill environmental niches with little competition from the less complex vascular organisms.

During the latter stages of the Paleozoic era the climate was stable and mild, similar to the climate of modern subtropical and tropical regions. Although mountains were forming in what is now the eastern United States, Texas, and Colorado, large areas of the United States were flat, and the far western portion of the continent was still covered by the sea. In most places the saline waters were not deep; some were vast coal-producing swamps. These very stable, warm, humid regions gave rise to lush plants with a great deal of biomass. This era, known as the Carboniferous period, produced most of the vegetation that decomposed and formed vast deposits of coal, oil, and natural gas.

Evolution and Distribution of Vascular Plants

The vascular plants include those with and without seeds. A great deal has been said in older botanical literature about the **seed ferns,** a group of extinct plants. The fossil record clearly shows that they were not ferns at all, but primitive **gymnosperms** and were therefore seed bearing (figure 19.1). We will examine three extinct and four living divisions of non-seed-producing plants. A comparison of vascular plants is provided in table 19.1 and their evolutionary relationships are illustrated in figure 19.34. The evolution of vascular plant form may be explained in part by the **telome theory.**

TABLE 19.1

Comparative Morphology of Extant Vascular Plants

MORPHOLOGICAL FEATURE	PSILOPHYTA	LYCOPHYTA	LYCOPHYTA
Gametophyte	Underground cylindrical, associated with fungus; antheridia and archegonia embedded in thallus; independent of sporophyte	Underground, associated with fungus; antheridia and archegonia embedded in thallus; independent of sporophyte	Antheridia and archegonia on separate gametophytes; each independent of sporophyte
Fertilization	Motile sperm	Motile sperm	Motile sperm
Embryo development	Attached to gametophyte	Attached to gametophyte	Within megagametophyte
Form	Herbaceous	Herbaceous	Herbaceous
Morphology of sporophyte	Branching stem; enations; mycorrhizal rhizomes	Prostrate branching stem with upright branches, microphylls, sporophylls, and roots	Prostrate branching stem with upright branches, microphylls, sporophylls, and roots
Vascular system	Simple	Simple	Simple, but with vessels
Secondary growth	None	None	None
Reproductive structure	Synangium, no strobilus	Sporophylls, sometimes aggregated into strobilus	Strobilius composed of sporophylls
Spores	Homospory	Homospory	Heterospory (megaspores, microspores)
Representative genera/groups	*Psilotum*	*Lycopodium*	*Selaginella, Isoetes*

Figure 19.1

Medullosa. The seed ferns are not really ferns at all, but primitive gymnosperms. Their fossilized remains are found frequently in sediments from the Carboniferous period. Their appearance resembles that of ferns. They were the size of shrubs or small trees.

Extinct Vascular Plants

All three divisions of extinct vascular plants shared several traits. They were leafless, dichotomously branched, bore sporangia either at the tips of their branches or laterally on the stem, lacked roots, and had a stature of a few centimeters to less than one meter. **Rhyniophyta** are the oldest group, dating back more than 400 million years into the Silurian period (figure 19.2). The presence of well-developed xylem in *Rhynia* distinguishes it from similar nonvascular plants. These simple, small vascular swamp plants were the ancestors of another ancient group, **Trimerophytophyta,** which are first found in the fossil record of the early Devonian period, some 375 million years ago. Trimerophytophyta (figure 19.3) are thought to be the progenitors of both the ferns and the progymnosperms. A third group, **Zosterophyllophyta,** dates to early **Devonian** (figure 19.4). This group is believed to have had an origin from Rhyniophyta and to have separately given rise to **Lycophyta,** a group

SPHENOPHYTA	PTEROPHYTA	GYMNOSPERM	ANGIOSPERM
Green thallus; gametangia embedded in thallus; independent of sporophyte	Heart-shaped thallus; independent of sporophyte	Reduced archegonium embedded in the female gametophyte (within ovule); pollen grain and tube are male gametophyte	Female gametophyte within ovule; pollen grain and tube are male gametophyte
Motile sperm	Motile sperm	Male gametic cells in pollen tube (conifers); motile sperm in pollen tube in a few genera	Male gametic cells in pollen tube
Attached to gametophyte	Attached to gametophyte	Within a seed	Within a seed
Herbaceous	Herbaceous or woody	Woody shrubs or trees	Herbaceous, woody shrubs or trees
Rhizome producing upright, jointed branches; small, scalelike leaves, roots	Stem-rhizome or upright; fronds and roots	Stems, leaves, and roots	Stems, leaves, and roots
Simple but with vessels	Vascular strands, but vessels only in a few genera; sieve cells, but no companion cells	Core of xylem surrounded by phloem; vessels rare; sieve cells without companion cells; pith in stem but not in roots	Core of xylem surrounded by phloem; sieve-tube members with companion cells; pith in stems, but not in dicot roots; pith in some monocot roots
None	None	Cambium in all species	Cambium in perennial species, absent in annuals
Strobilus with sporangiophores	No strobilus, sporangia aggregated into sori	Pollen and ovulate strobili	Flowers with stamens and carpels
Homospory	Mostly homospory	Heterospory	Heterospory
Equisetum	*Pteridium*	*Pinus, Cycas, Ginkgo*	*Magnolia,* composites, orchids

Evolution of the Sporophyte Axis

One of the most striking features of vascular plants lies in their variety of form. Surprisingly, this diversity is believed to have been derived from a common architectural form. The telome theory, developed by the noted German botanist Walter Zimmermann in 1930, proposes that the ultimate branches of the primitive, dichotomous rhyniophyte plant, termed telomes, became modified by producing leaves, various axis patterns, and reproductive structures. The change from a rhyniophyte plant results from the tendency of branches to grow unequally. According to this theory, some of the ultimate telomes were overtopped (figure 1) by larger branches that continued as the main branch. Initially the smaller, side branches were three-dimensional branching units. The development of a branching system in one plane, a process termed planation, and the coalescence of apical meristems giving the appearance of parenchymatous growth among the axes is thought to have resulted in webbing, to produce a leaflike branch. The veins in such a leaflike structure represent the branch axes. These processes of the telome theory together with reduction, the loss or suppression of growth, and recurvation, the bending of the sporangial stalks, can be used to explain the origin of the synangia in *Psilotum,* sporophylls in club mosses, the sporangiophores in horsetails, frond and sori in ferns, and the origin of the ovule. The telome theory has less application when attempting to explain the origin of gymnosperm and angiosperm reproductive features.

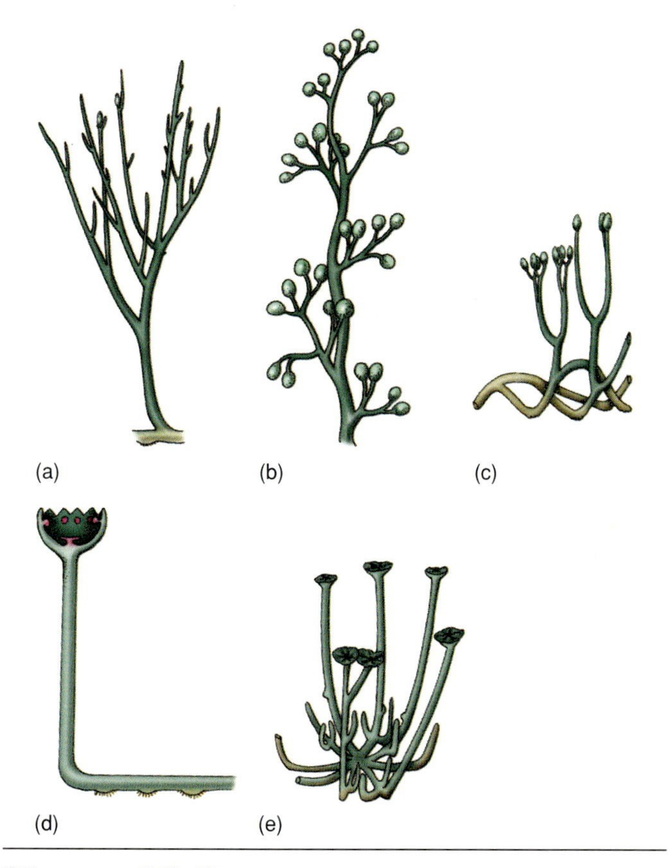

Figure 19.2

Rhyniophyta comprises the oldest known division of vascular plants, dating to the Silurian period of 400 million years ago. In the genus *Rhynia* (*a*) the shoot is leafless and dichotomously branched. *Cooksonia* (*b*) is the oldest known vascular plant. *Aglaophyton* (*c*), formerly called *Rhynia major,* lacks definable annular or helical tracheids and is now considered to be nonvascular. It is considered to be intermediate between the Rhyniophyta and Bryophyta. *Lyonophyton* (*d*) and *Sciadophyton* (*e*), two recently described fossils from the same geological deposits as *Rhynia,* represent gametophytes.

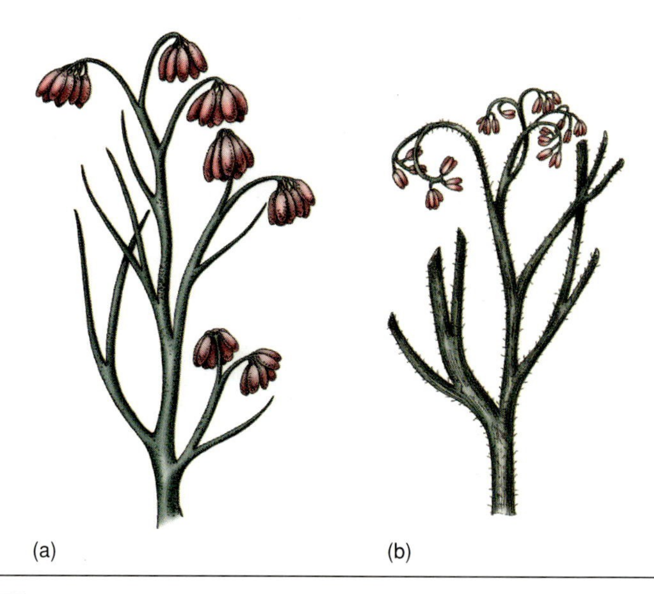

Figure 19.3

Trimerophytophyta specimens are believed to have evolved directly from Rhyniophyta and are found in the fossil record of 360 million years ago. Slightly more advanced, they were leafless but branched more profusely than Rhyniophyta. Some branches terminated in a group of elongated sporangia, as seen in both *Trimerophyton* (*a*) and *Psilophyton* (*b*).

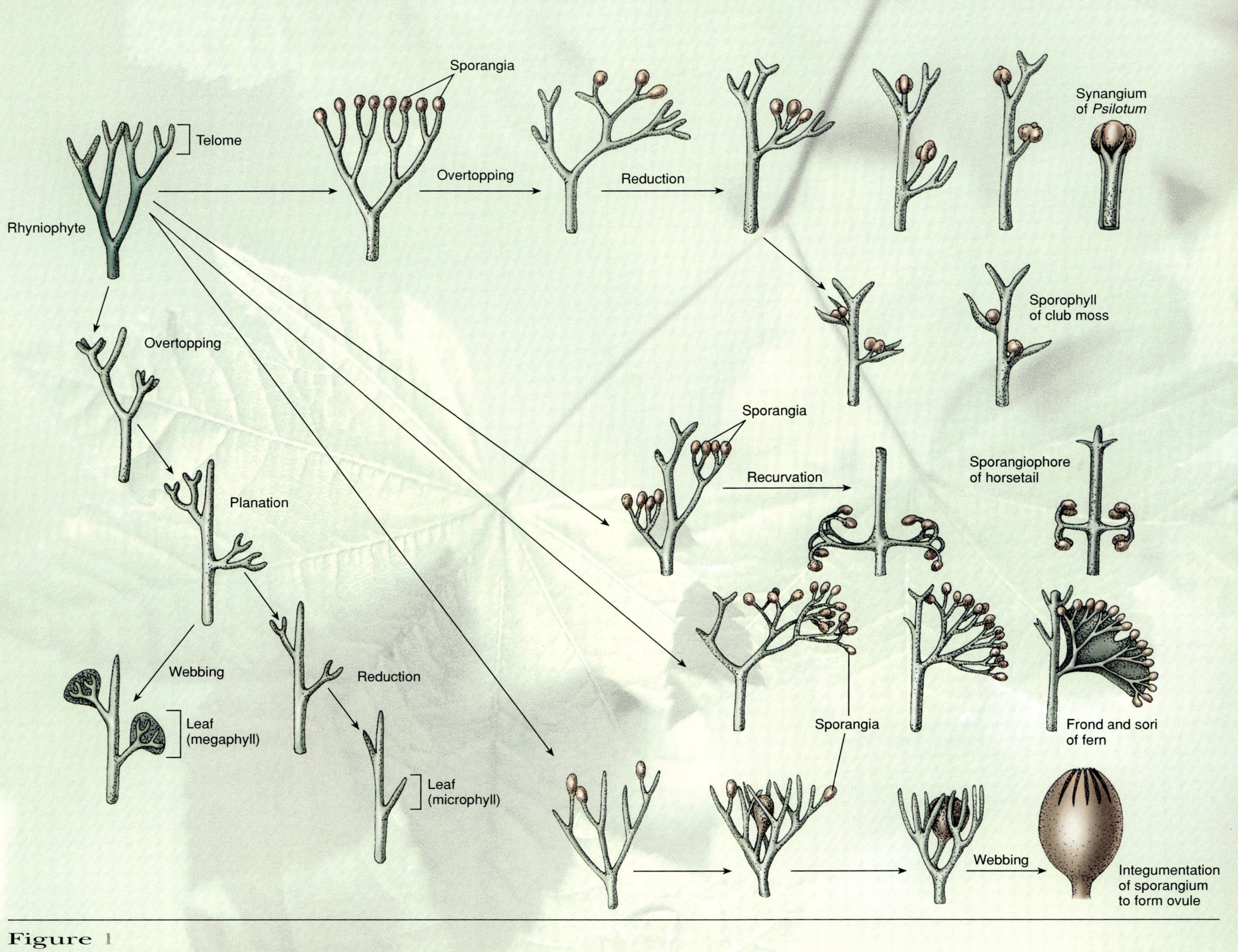

Figure 1
Diagrammatic representation of the evolution of leaves and reproductive structures according to the telome theory.

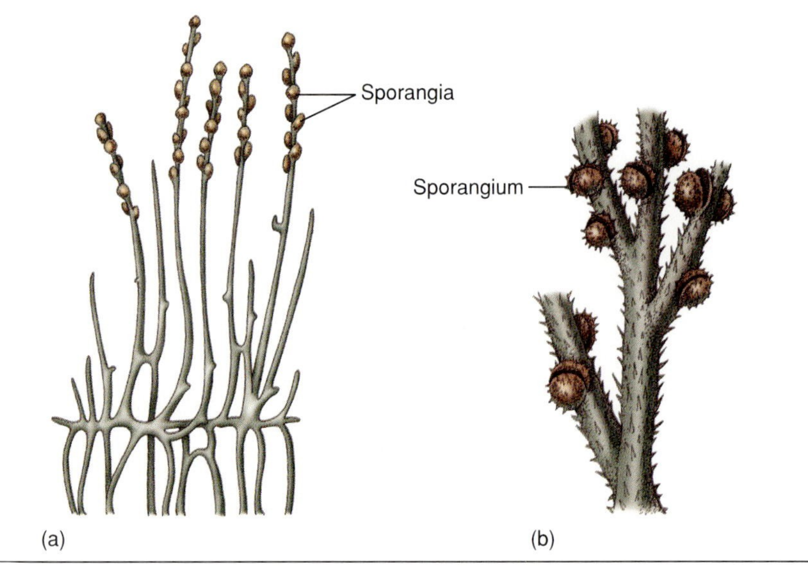

Figure 19.4

Zosterophyllophyta were leafless and dichotomously branched, but frequently lateral branches further differentiated into one axis that grew upward and another downward, as in *Zosterophyllum* (*a*). The sporangia were borne laterally on short stalks, as seen on *Sawdonia* (*b*).

Figure 19.5

The lycopod trees, like this *Lepidodendron* (*a*) and *Sigillaria* (*b*), grew to heights of 25 m and dominated much of the Coal Age. Their preservation in the fossil record is excellent; the shoot system branched successively, terminating in tufts of long leaves. The roots were shallow, and evidence indicates that they were easily blown over in the coal swamps.

represented today by a herbaceous line but which included tree lycopods during the Carboniferous period (figure 19.5). These were among the most common forest plants in the coal-forming process of that period.

TODAY'S SEEDLESS VASCULAR PLANTS

There are four major divisions of living seedless vascular plants, but only one is still a significant component of today's flora. The other three are only briefly mentioned.

Psilophyta

The division, **Psilophyta,** contains only two living genera, *Psilotum* and *Tmesipteris*. *Psilotum* is leafless and lacks roots. *Tmesipteris* is an epiphyte found only in Australia, New Zealand, and nearby South Pacific Islands. *Psilotum* is **homosporous,** producing spores that are of the same size and shape. The sporophyte is dichotomously branched. The stem produces small, nonvascularized, toothlike outgrowths referred to as enations. Three fused sporangia occur, grouped into a complex termed a **synangium.** The synangia are positioned on the end of short branches subtended by two enations (figure 19.6). Some botanists feel that this division should be included with the primitive ferns. Others maintain Psilophyta as a separate group, with its origins in the Devonian as a branch of Trimerophytophyta or possibly even Rhyniophyta.

Lycophyta

Lycophyta is a division most notably represented by the genera *Lycopodium* (club mosses), *Selaginella* (including the resurrection plant), and *Isoetes* (quillwort). *Lycopodium* includes some 200 homosporous species, including tropical epiphytes, temperate forest-floor species, and even some arctic members (figure 19.7). The lycopods produce sporangia, which are produced on the

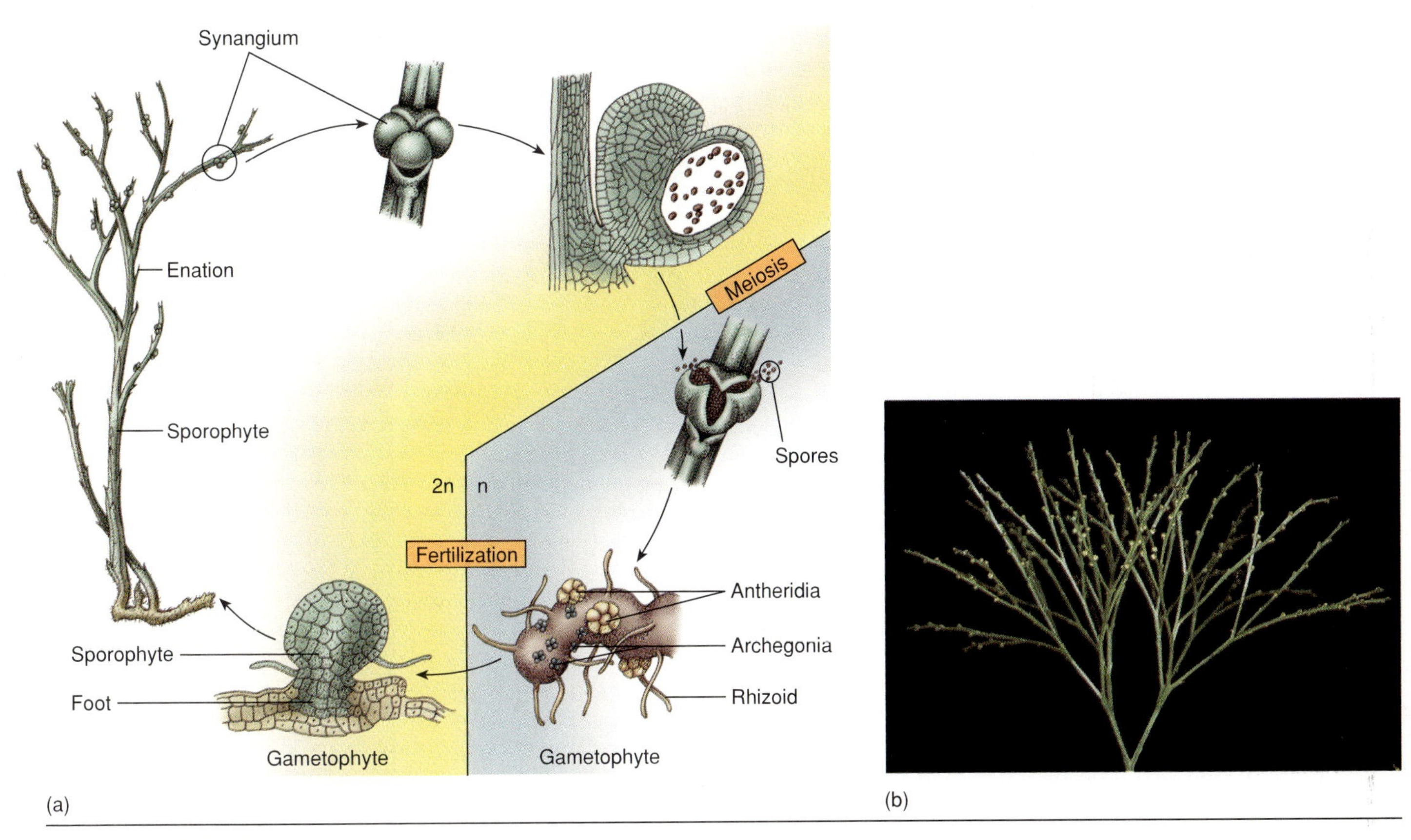

Figure 19.6

The *Psilotum* life cycle (*a*). The sporophyte bears short, lateral branches that produce three-fused sporangia and spores. On germination the spores give rise to bisexual radially symmetrical gametophytes, which resemble the sporophyte rhizome. The subterranean gametophyte is symbiotic with a mycorrhizal fungus. The flagellated sperm require water to swim to the archegonia for the fertilization process. The young sporophyte is nurtured by the gametophyte but eventually becomes detached and independent. (*b*) The sporophyte of *Psilotum* depicting the characteristic dichotomous branching and absence of leaves.

upper surface of modified leaves termed **sporophylls.** The sporophylls may be grouped together at the tip of the shoot to form a **strobilus** (cone), or may occur interspersed with microphylls, small, non-sporangia-bearing leaves. *Selaginella* is an even larger genus with over 600 species. This genus is characterized by heterospory, the production of spores of two sizes, each with a different function (figure 19.8). In the evolution of land plants the occurrence of heterospory represents an important evolutionary step toward the seed habit. Although relatively few of these species are found in desert environments, those which are have developed a notable xeric adaptation. In periods of water stress the small, rock-dwelling tufts close inward and look like a dry, gray-brown, dead clump. When watered, they quickly open to reveal a green, healthy bird's-nest-shaped mat—their sudden "resurrection" from apparent death and hence their common name. *Isoetes* is an aquatic, heterosporous lycopod having a sporophyte stage with an underground corm that bears long, quill-shaped leaves on its upper surface and roots on the bottom.

Sphenophyta

Sphenophyta is a division known to have been present in the Devonian period, including tree species present during the Carboniferous period approximately 300 million years ago. Today only a single genus remains. *Equisetum,* commonly known as horsetail or scouring rush, is often found in moist habitats and is characterized by jointed stems that are often unbranched. The stems contain large amounts of silicon in their cell walls, thus giving them an abrasive quality. Pioneers used a handful of crumpled stems to scour their pots and pans (figures 19.9 and 19.10).

Sphenophyta are known from fossil deposits in the middle Devonian period and are thought to have evolved from ancestral lines in Trimerophytophyta. Most successful

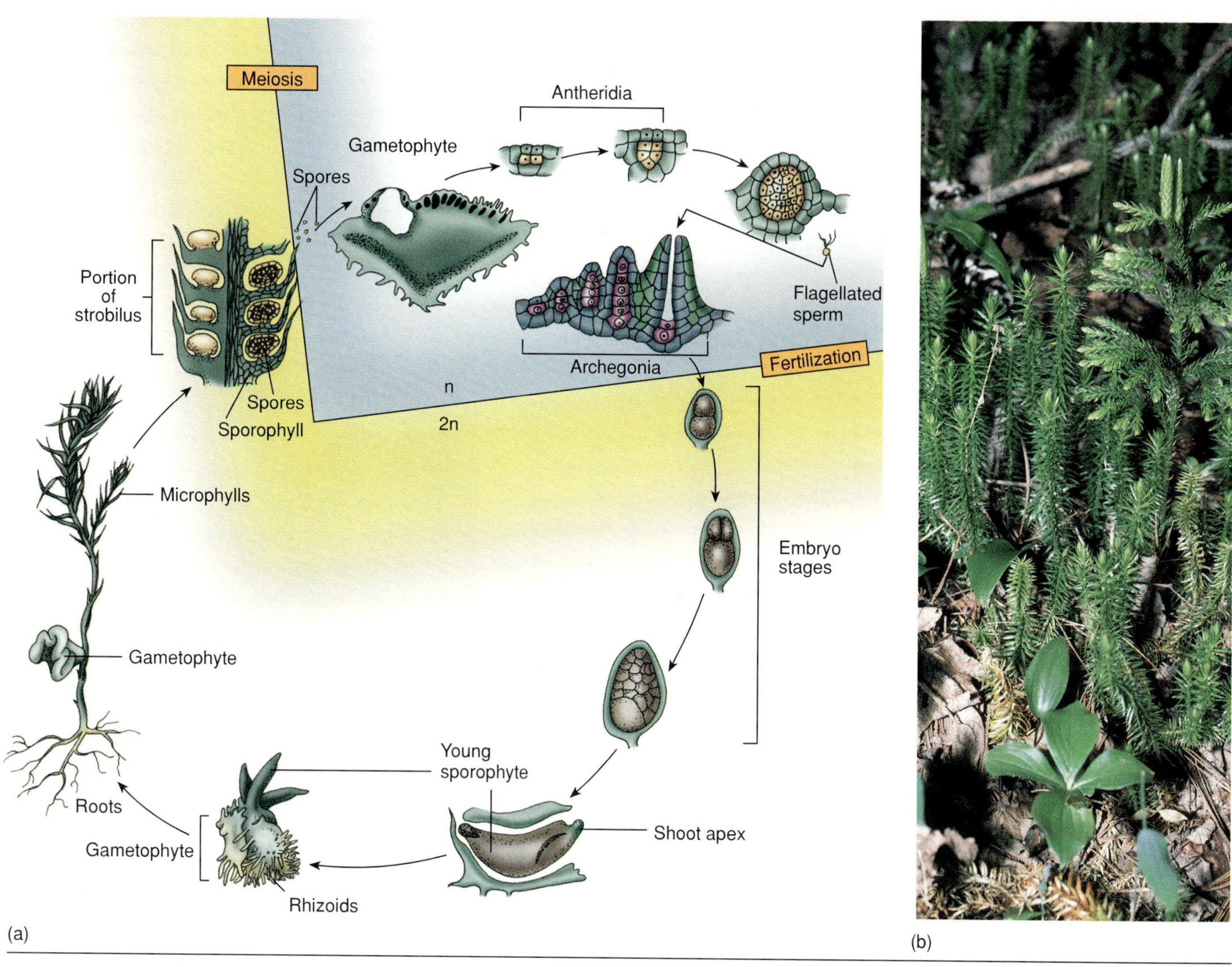

Figure 19.7

The *Lycopodium* life cycle (*a*). Each spore is capable of forming a gametophyte. On germination a gametophyte is produced (often underground and in association with a mycorrhizal fungus), which produces both antheridia and archegonia on the same thallus. Following fertilization in the base of the archegonium via flagellated sperm swimming in water, the zygote begins development. The mature sporophyte may produce a strobilus with sporangia, which produce the spores. The spores all look alike (homosporous), but are genetically different as a result of their meiotic production. (*b*) The sporophyte of two species of *Lycopodium*.

in the Carboniferous period, Sphenophyta even included tree-sized species.

Pterophyta: The Ferns

Evolution and Distribution

The fourth division, **Pterophyta,** or ferns, is the largest and most evident group (figures 19.11 and 19.12). There are approximately 12,000 living species, and they are cosmopolitan (found throughout the world). The so-called Age of Ferns in the fossil record dates from the Carboniferous period, a time in which they were the dominant vegetation. Water was plentiful, and the climate was warm, humid, and unchanging. In many respects the world was like a giant greenhouse with ideal growing conditions for many plants. Some ferns grew to 8 m tall and had broad trunk bases with many aerial roots as props.

Most extant ferns are tropical, but some are found in temperate regions, including the mountainous regions of the United States; a few are adapted to aridity and inhabit deserts. Living species vary in size from the tiny aquatic *Azolla microphylla* (figure 19.13) with **fronds** or fern leaves less than 1 cm long, to the giant tree ferns, some of which may reach almost 25 m in height and 30 cm or more in diameter. All of this stem tissue is primary

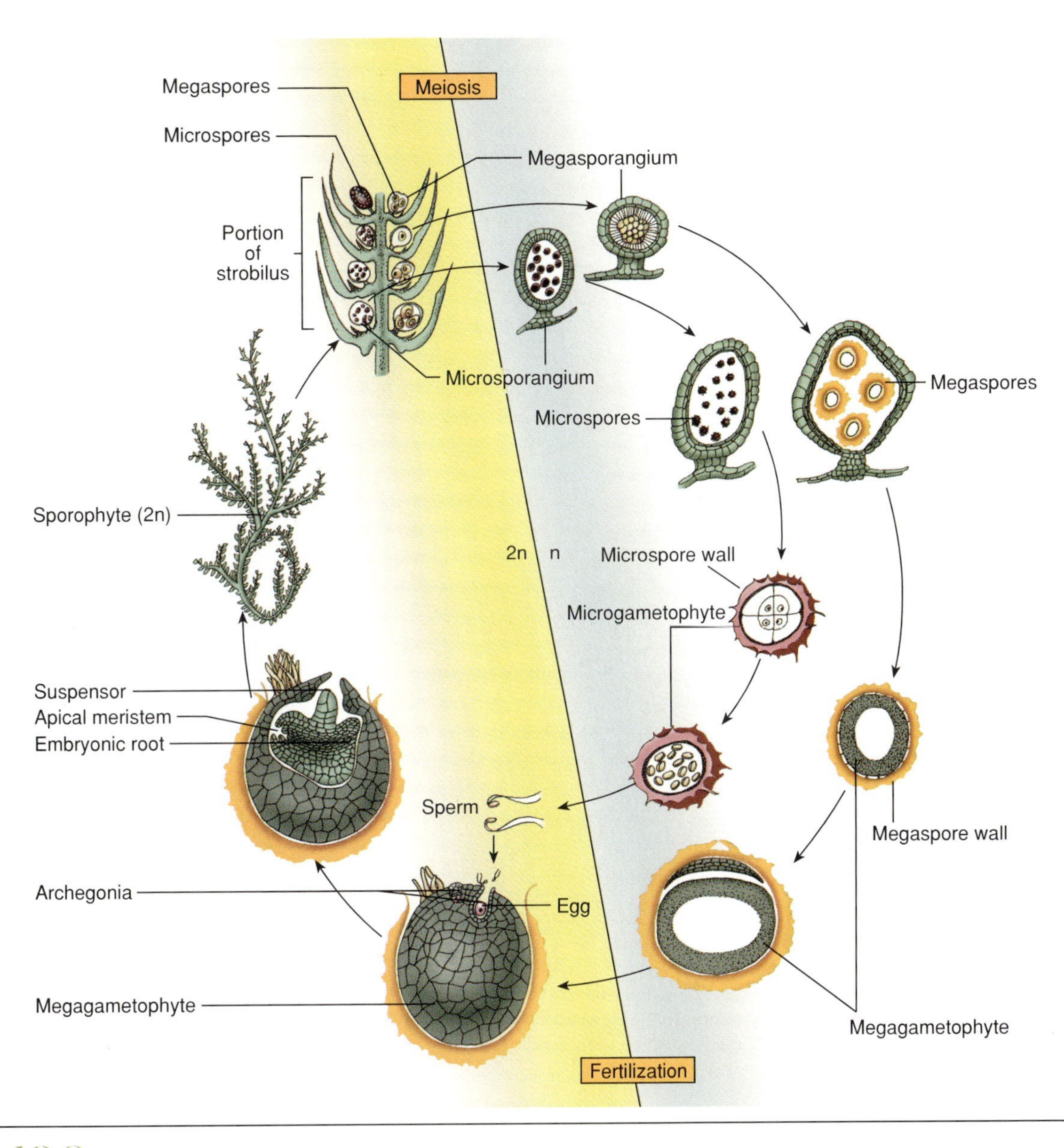

Figure 19.8

The *Selaginella* life cycle. The sporophyte produces two kinds of spores (heterospory) following meiosis. Each gives rise to morphologically distinct, endosporic gametophytes. The megasporangia produce megaspores. Megaspores, in turn, house the megagametophyte on which develop the archegonia. Microspores encase the microgametophyte, which produces flagellated sperm. Following fertilization the young sporophyte is nurtured by the megagametophyte during early development.

Figure 19.9

The vascular plant *Equisetum,* or horsetail, is usually found in marshy habitats. The stem is the dominant organ; tiny leaves are present for a short period of time but become dry and scalelike.

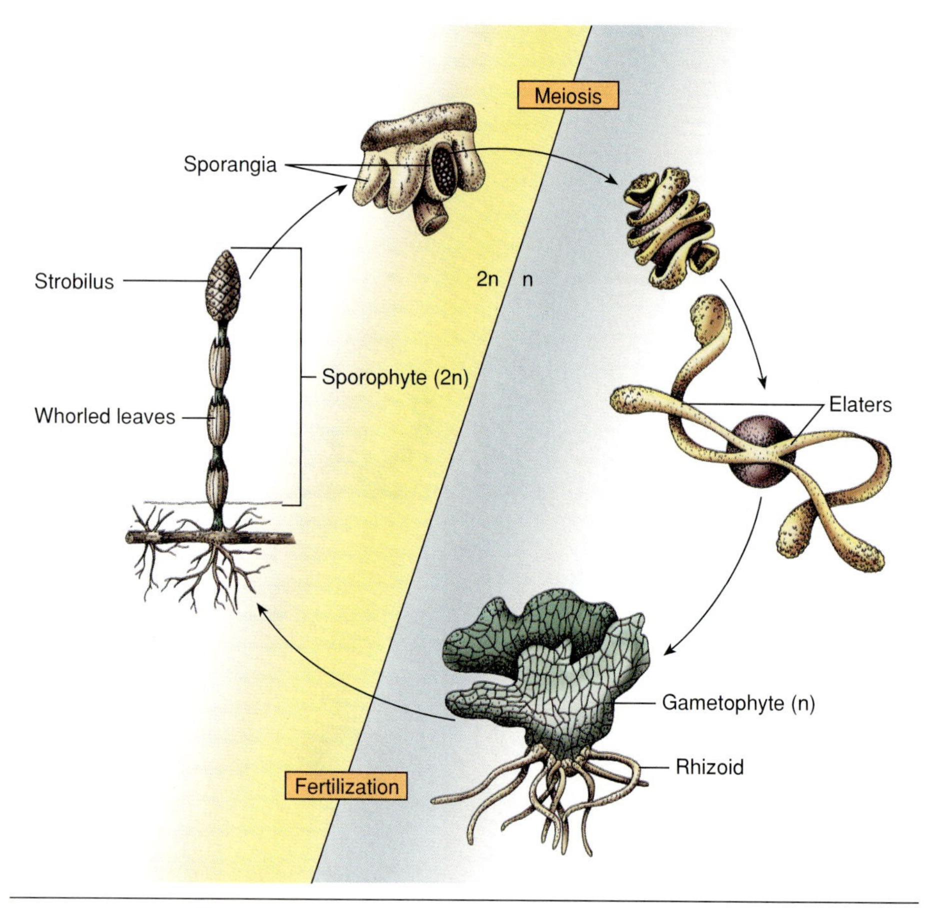

Figure 19.10

The *Equisetum* life cycle. The sporophyte bears a single type of spore (homosporous) along the margins of small sporangiophores clustered into strobili at the apex of the stem. When the spores are mature, the sporangia split and are released into the wind, their dispersal aided by winglike adaptations of the spore wall termed elaters. Following germination pinhead-sized green, free-living gametophytes are produced, either bisexual or strictly male. Following maturity of the antheridia and archegonia the sperm swim through water to fertilize the egg, which then develops into the sporophyte.

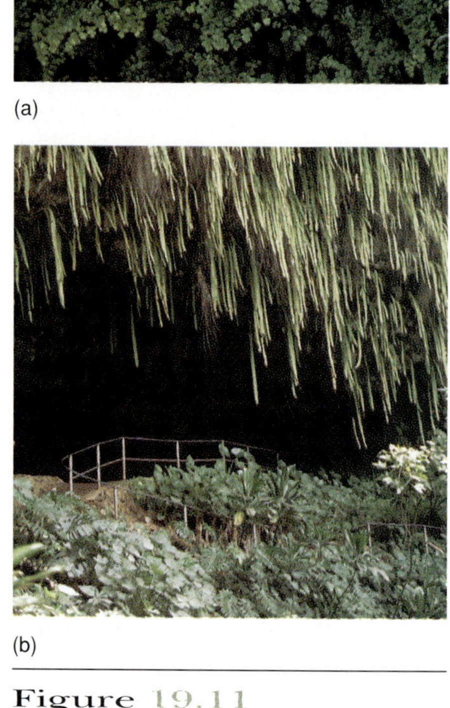

Figure 19.11

Thick populations of ferns are common in moist areas. (*a*) The maidenhair fern, *Adiantum,* in McKittrick Canyon, Guadalupe National Park, Texas. (*b*) Sword ferns at the Wailua River State Park Fern Grotto, Kauai.

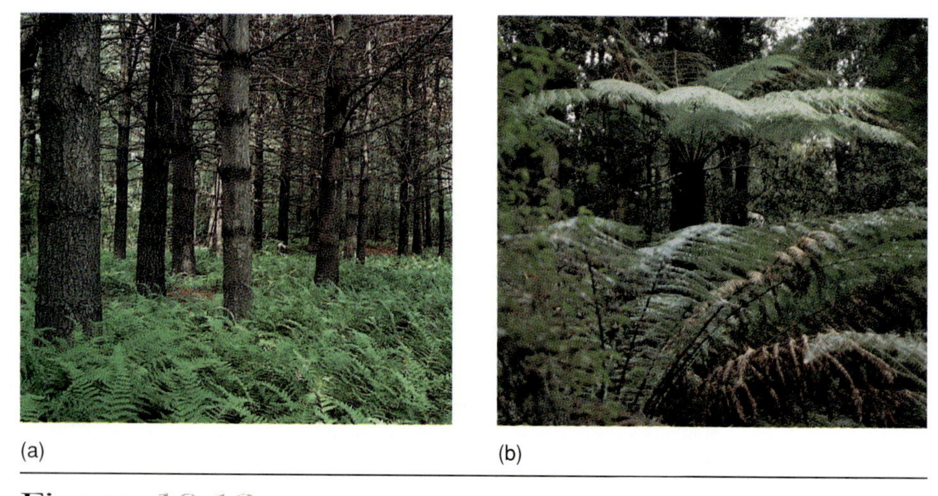

Figure 19.12

(*a*) Woodland ferns are typical of the ground cover in temperate forests. (*b*) Tree ferns reaching five meters in height may be found in some rain forests, such as in the Blue Mountains of eastern Australia.

Figure 19.13

Ferns are sometimes inconspicuous and deceiving, as the tiny *Azolla microphylla.*

Figure 19.14
Fern leaves or fronds unroll from base to tip. In the coiled condition they are referred to as a crozier or fiddlehead.

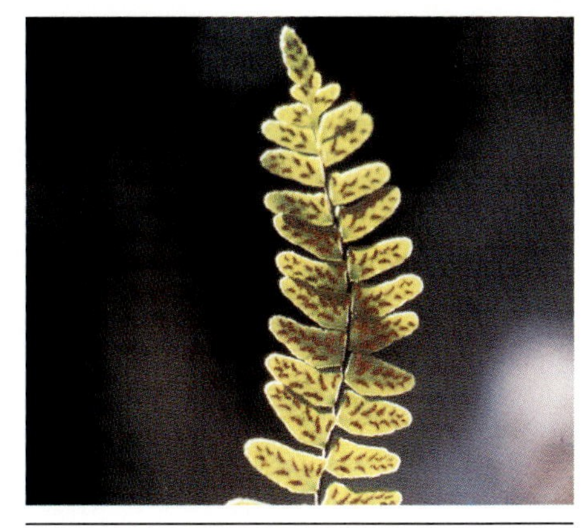

Figure 19.15
Fertile fronds of ferns produce sori, aggregations of sporangia on the underside of the fronds.

in origin, and only two genera of ferns (*Botrychium* and *Ophioglossum*) have a vascular cambium.

Fern fronds come in myriad shapes and sizes, and many are highly dissected. Unlike the seed plants, ferns produce leaves from a coiled position. The frond expands by unrolling from base to tip with new growth produced at the apex. Such new leaves are often covered with surface hairs. The coiled frond is called a **fiddlehead** because of its resemblance to the neck and head of a violin (figure 19.14).

Reproduction

In almost every case ferns are homosporous, that is, produce only a single kind of spore. These spores are borne on the underside of fertile fronds, as opposed to the non-spore-bearing sterile fronds (figure 19.15). Sporocytes in the fronds undergo meiosis. The spores are produced in **sporangia,** and many sporangia are grouped together in a region to make up a **sorus.** The sori (figure 19.16) are arranged on the underside of the leaf in varying patterns, depending on the species, and are often mistaken for insect eggs. Each sorus may be covered with a membranelike protective layer called the **indusium.** The indusium acts as a protective umbrella until the spores mature; then the membrane shrivels or breaks at the time of spore release. The sporangium is surrounded by a specialized layer of thick-walled cells, the **annulus,** except on one side where the cells are thin walled. These thin-walled cells, termed **lip cells** split, and the outer wall of the sporangium pulls back as if attached to a rubber band. When the relative humidity decreases, the annulus is drawn back, then released suddenly, and the spores are catapulted through the air for maximum dispersal. The microscopic, dustlike spores are readily carried by the wind. Millions of them can be produced by a single sporophyte or fern plant.

If the spore lands on a suitable substrate in an environment conducive to germination, the haploid spore will give rise to a chain of cells. After a few divisions in the single dimension, mitosis begins in two dimensions, giving rise to a free-living bisexual gametophyte—a thin, green, heart-shaped structure capable of producing both antheridia and archegonia on the underside of the gametophyte. This tiny, 1 or 2 cm structure lies flat on the ground and usually goes unnoticed by the untrained eye. Absorption of water and nutrients is achieved by **rhizoids,** which grow like root hairs on the underside of the haploid plant body.

The aquatic-dwelling ancestry of the ferns is revealed by a reproductive system in which motile sperm are produced within the antheridia, swim through water to the neck of the archegonium, attracted by chemicals released during the breakdown of the neck canal cells, and fertilize the egg to produce a new diploid zygote. After numerous divisions, the sporophyte embryo begins to take shape while still nurtured by the photosynthetic gametophyte. Finally, a tiny new frond emerges, roots are formed at the base, and a new fern sporophyte is ready to produce the familiar plant. Once the sporophyte has become autotrophic, the gametophyte withers and dies. The new sporophyte may grow for months or even years before producing new fertile fronds with spores, thus completing the life cycle (figure 19.17).

(a)

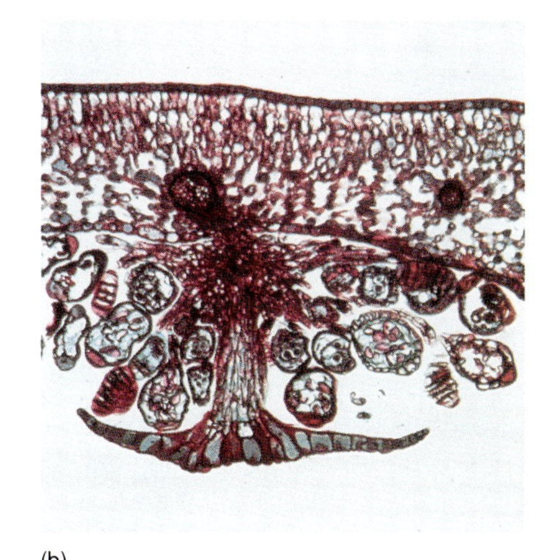

(b)

Figure 19.16
The sori, which contain sporangia with spores, occur in various patterns on the underside of the frond, depending on species. (*a*) *Asplenium dregeanum,* Ghana. (*b*) Cross section of fertile frond of *Cyrtomium,* showing a sorus with mature sporangia and spores and protective umbrella-like covering termed the indusium.

Ecology and Importance

One does not usually think of ferns as having great economic significance, but they are in such demand as ornamental plants that their economic value is great. Since no secondary growth is formed in ferns, the structure of the trunk is rather weak and unsuitable as timber. The fibrous tissue is often used, however, as a support for viny ornamentals such as *Philodendron cordatum.* The dead stem is also rather resistant to decomposition and is often used as a soil amendment where good drainage is essential. Ferns are easily grown in climates where the relative humidity and precipitation are sufficient to prevent desiccation. Tropical ferns are sensitive to low temperature, but woodland ferns become deciduous in the fall, overwintering as fleshy rhizomes that produce adventitious roots and new shoots the following spring.

The ecological niche occupied by ferns is generally moist, warm, and shady. The root system is effective in reducing soil erosion by stabilizing stream banks. The bird's-nest fern (*Asplenium nidus*) behaves like an epiphyte. It grows in the crotch of trees high above the tropical rain forest floor. Spores germinate anywhere organic debris accumulates between two branches, and magnificent specimens may have fronds that reach 2 or 3 m in length. This fern has adapted well to cultivation and is sold as a potted plant throughout the United States.

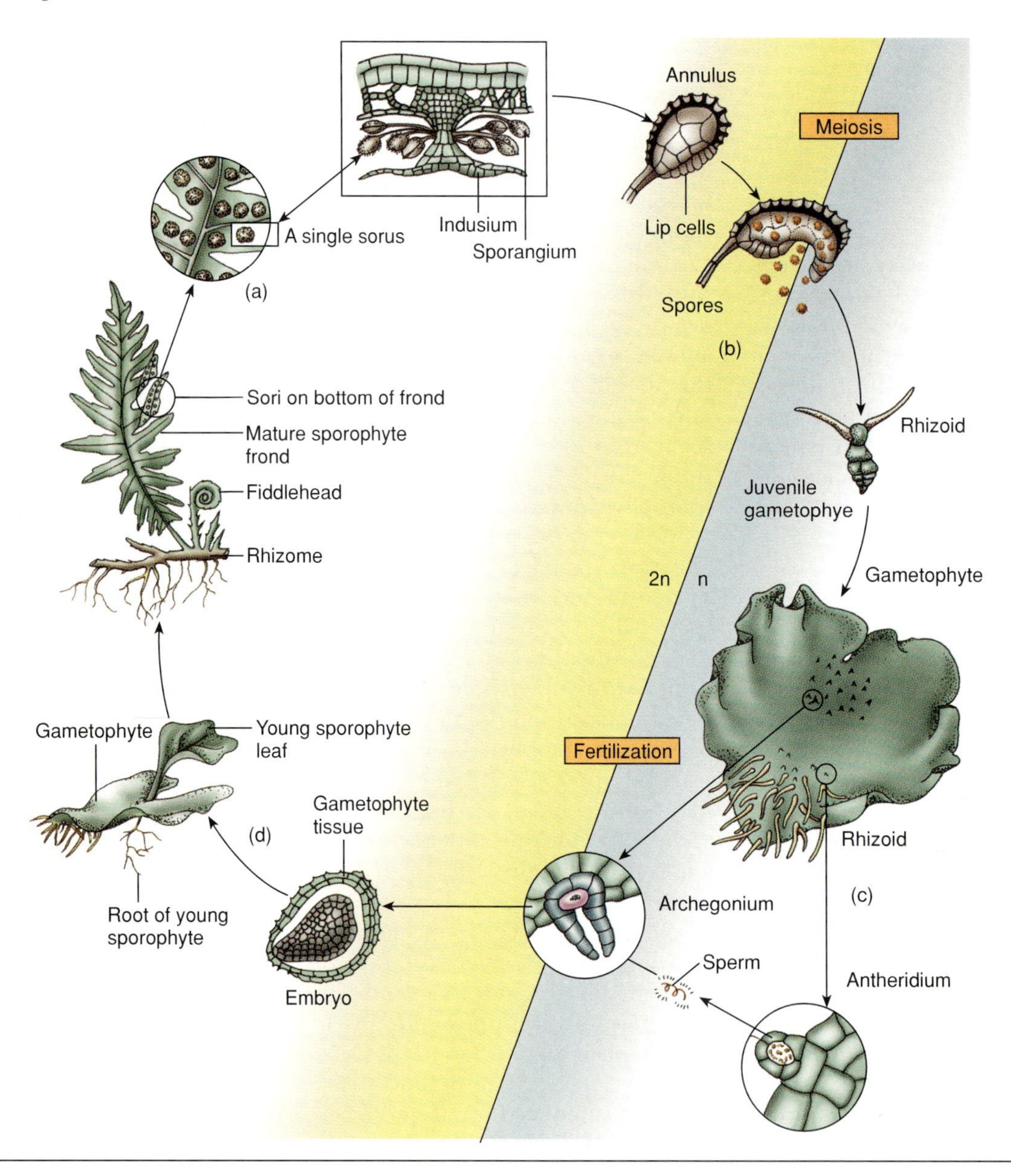

Figure 19.17

The fern life cycle. (*a*) Fertile fronds of the sporophyte produce spores within sporangia. (*b*) When mature, the spores are dispersed by the wind to a suitable substrate, where they germinate and eventually form a gametophyte. (*c*) On the underside of the gametophyte, flagellated sperm are released from the antheridia and swim through water to the neck of the archegonium, where they enter and fertilize the egg. (*d*) The new zygote undergoes mitosis to produce a new sporophyte, which emerges from the body of the gametophyte and eventually becomes nutritionally independent of it.

> **✔ Concept Checks**
>
> 1. Discuss the adaptations that the first land plants had to develop to survive.
> 2. Describe the three extinct divisions of early land plants. Cite a representative genus for each.
> 3. Determine which division of vascular plants is associated with the following terms: elaters; sori; synangium; fiddlehead; enation; indusium; sporangiophore; megaspores; antheridia. Describe how the flagellated sperm reach the egg in non-seed-producing plants.

THE SEED PLANTS

The oldest seeds have been found in rocks from the Devonian period, some 350 million years old. Today seed-producing plants thrive in a diverse range of climates.

Modern seeds develop from a mature **ovule,** the female gametophyte being embedded within a fleshy **nucellus.** The nucellus is enclosed by one or two **integuments,** each of which may be several cell-layers thick and that develop into the seed coat. The thickness and chemical composition of the integuments ultimately determine the nature of the seed coat and thus its protective quality. Some seed coats become exceedingly thick and highly impermeable to water and gases, which gives them a high survival capability.

In most modern seeds the embryo matures before dispersal, whereas ancient seeds apparently completed embryo development after dispersal. Survival ability would appear to be best with embryo development prior to dispersal, and such embryos might have a selective advantage. Even today some seeds have an after-ripening dormancy because the embryo is not fully developed. Newly harvested sugar beet seeds exhibit this characteristic.

Storage of reserve food within the seed is critical for survival. Within the endosperm, or cotyledons, various storage molecules (starch, protein, lipids) become available for energy during the germination process. As germination proceeds, respiration rates increase dramatically, and storage molecules are hydrolyzed to produce the glucose substrate for respiration. This food reserve is usually more than adequate, and sometimes stored reserves are available weeks or months after the germination process begins. Such storage capacity is vital to survival during long periods of dormancy, perhaps hundreds of years for some species.

This remarkable adaptation—the seed—once increased the chances of survival on land. Today it allows for continuation of the life cycle under conditions unfavorable for growth. Frozen soil, drought, extreme heat, and other environmental factors usually inhospitable to plant growth are endured in the seed stage; this resistant, reproductive structure simply goes dormant until the unfavorable conditions have passed.

There are five extant groups of seed plants: the **cycads,** the **ginkgos,** the **conifers,** the **gnetophytes,** and the **angiosperms.** The first four of these comprise the gymnosperms, or plants with naked seeds; the angiosperms, or flowering plants, have enclosed seeds.

Gymnosperms

The gymnosperms apparently evolved during the Paleozoic era from an intermediate group of plants that were more highly vascularized than the lower plants. This now-extinct group was the Trimerophytophyta. They had no leaves; the main axis formed lateral branches that divided several times. Some of the lateral branches terminated in sporangia, which produced a single type of spore. Trimerophytophyta eventually gave rise to **progymnosperms,** plants with fernlike foliage and woody stems (figure 19.18).

Figure 19.18
The progymnosperms exhibited characteristics intermediate between those of Trimerophytophyta and the gymnosperms. They reproduced by spores but produced secondary xylem, like the gymnosperms. Some of these reached a height of more than 20 m. *Archaeopteris* tree (*a*) and branch (*b*).

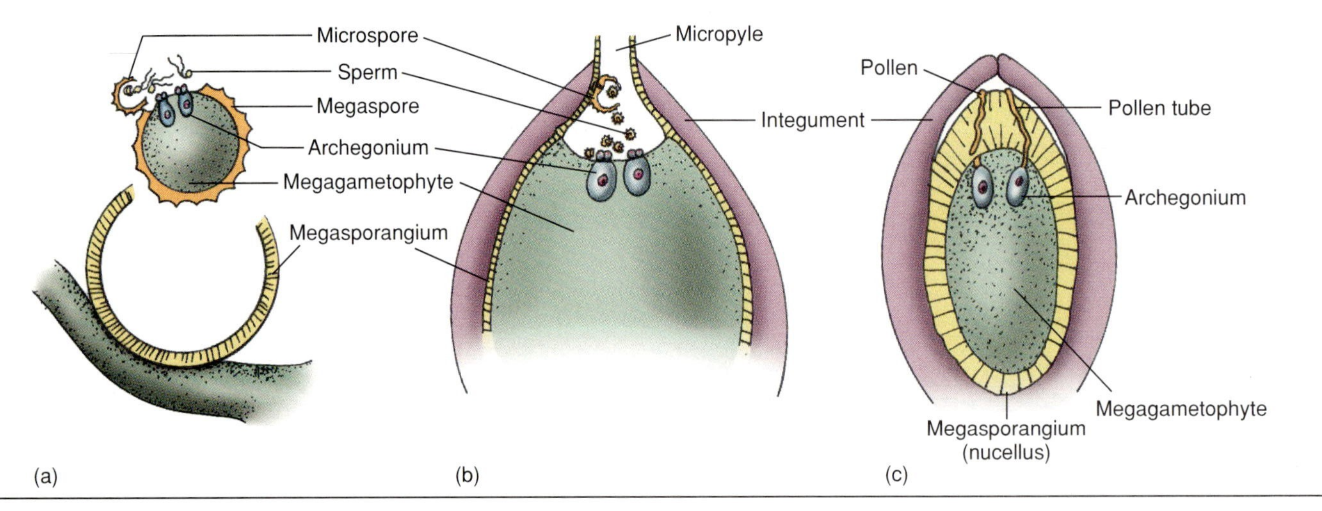

Figure 19.19
Evolution of the ovule and siphonogamy. (*a*) Fertilization process in *Selaginella*, a non-seed-bearing plant in which the sperm-bearing microspores land near the megaspore, which contains the archegonia. With the availability of water, the flagellated sperm swim to an archegonium and achieve fertilization. (*b*) Fertilization process similar to that occurring in *Ginkgo* and cycads. Multicellular microspores, now termed pollen, enter through the micropyle canal to release flagellated sperm near archegonia. Note that the megasporangium containing a single, functional megaspore, in which develops the megagametophyte, is now surrounded and protected by the integument. (*c*) Fertilization process as in conifers. Pollen tubes grow through the megasporangium wall, also termed the nucellus, to the archegonia and discharge nonflagellated sperm.

The movement to land and increasing aridity presented the problem of moving male gametes to the female gamete for fertilization. Whereas ferns and lower plants facilitate the reproductive process by having motile sperm swim through water to reach the egg, most large land plants failed to develop a mechanism for ensuring that water-based reproduction could continue. Only in the cycads and in *Ginkgo* do the relic motile sperm swim to the region of fertilization. Instead, airborne pollen evolved, and pollination had to be accomplished by wind or by animals. In all gymnosperms the immature male gametophyte, the pollen grain, is borne by the wind to the vicinity of the female gametophyte within an ovule. Following pollination the pollen grain germinates and produces a pollen tube. In the gnetophytes and conifers the pollen tube delivers the sperm directly to the archegonium, a process termed **siphonogamy** (figure 19.19).

The Conifers

Evolution and Distribution
Although the total number of conifer species is not large, they are the largest group of living gymnosperms, comprising about 50 genera and 550 species (figure 19.20). These plants have adapted remarkably to aridity by means of sunken stomata and thick cuticles, and the needlelike leaves have a much reduced surface area (figures 19.21 and 19.22). There is reason to believe that the evolution of conifers coincides with the tremendous selection pressure during the worldwide aridity of the Permian period. During the early Tertiary period some genera were more prominent than they are in the modern landscape. On all the northern continents vast regions were and still are covered with conifers.

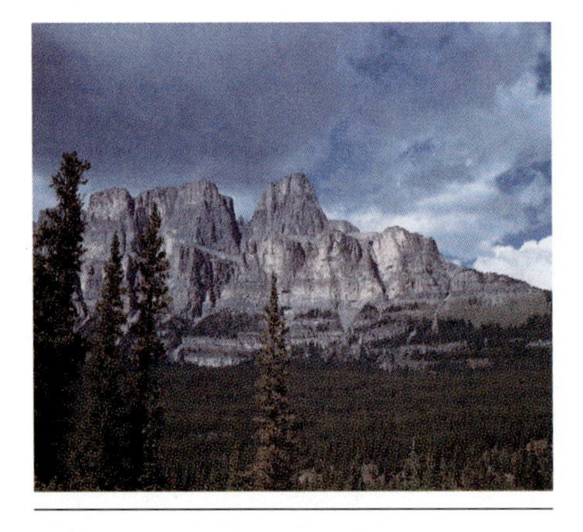

Figure 19.20
Conifers dominate much of the landscape in forests throughout the world.

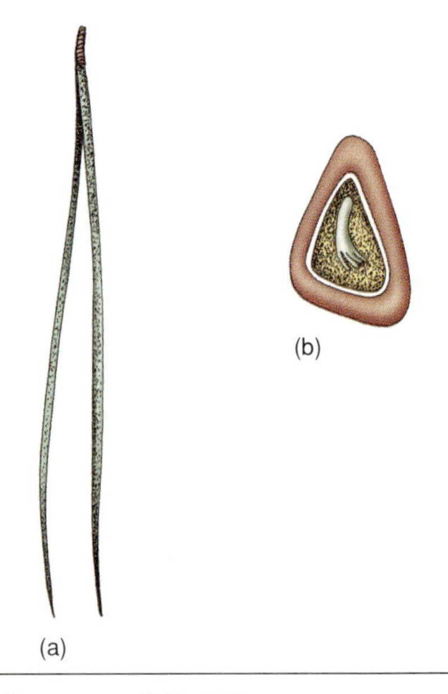

Figure 19.21
(*a*) Conifer leaves consist of needles born singly or in bundles called fascicles. (*b*) The seed consists of a multicotyledon embryo embedded in nutritional megagametophyte, haploid tissue.

The pines are unquestionably the most economically important conifers. There are about 90 species widely distributed throughout North America, Europe, and Asia.

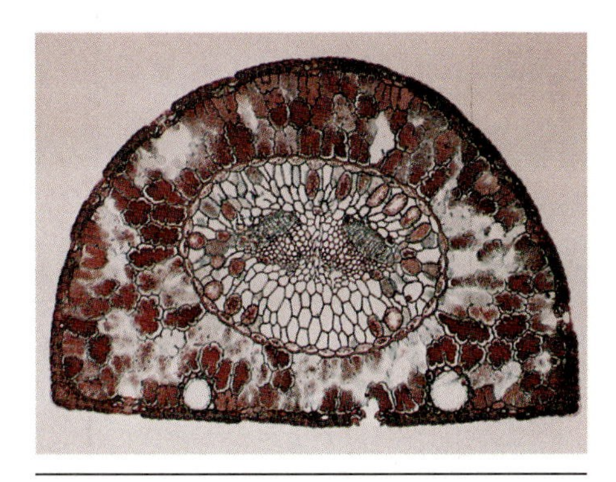

Figure 19.22
Cross section of a pine leaf. The needle is characterized by a very thick cuticle, sunken stomata, an endodermis, transfusion tissue, and resin ducts.

Figure 19.23
Aggregations of pollen cones together with large ovulate (seed) cones in a lodgepole pine, *Pinus contorta.*

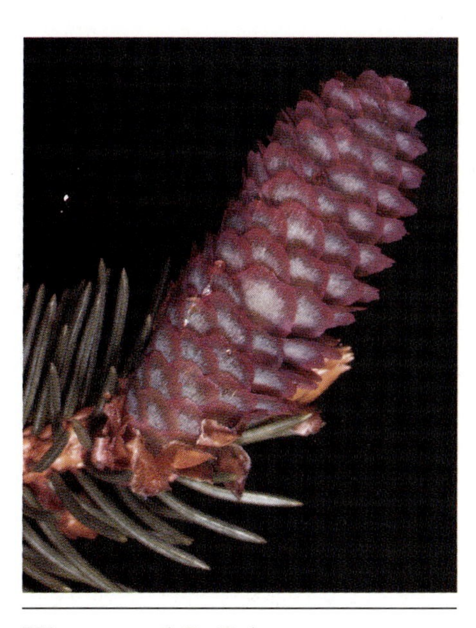

Figure 19.24
A pollen-receptive (ovulate) cone of Norway spruce (*Picea abies*). Pollen cones shed pollen when the ovulate cones are in this condition.

Reproduction The reproductive structure in conifers is a cone (strobilus). Pollen and seeds are produced in separate cones. Both types of cone are found on the same plant; the pollen cones occur on the lower branches, and the more conspicuous seed or ovulate cones are borne above (figure 19.23). It is not uncommon to find 1-, 2-, and even 3-year-old seed cones on the same tree in *Pinus,* although most conifers develop mature cones in one season. Compared with that of angiosperms, the reproductive cycle in conifers is very long (figures 19.24 to 19.27). The period between pollination and fertilization—when the pollen grains logged in the micropyle of the ovule begin to form pollen tubes and the time required for the tubes to grow to the archegonia—may be many months, whereas in angiosperms the period is a matter of hours or days.

Pollen cones are quite small, usually no more than 1 or 2 cm in length. Each of the spirally arranged **microsporophylls** (scales) contains two **microsporangia,** and each microsporangium contains many microsporocytes, which undergo meiosis to produce four haploid **microspores.** Each of these develops into a winged pollen grain with two **prothallial cells** (having no apparent function), a **generative cell,** and a **tube cell.** The mature pollen grains are shed by the pollen cones in great abundance in the spring, and sometimes coniferous forests are shrouded in yellow clouds of pollen, even covering the entire surface of lakes. The bladderlike air sacs increase the buoyancy of the pollen, and it can travel great distances before landing inside the scales of an ovulate cone.

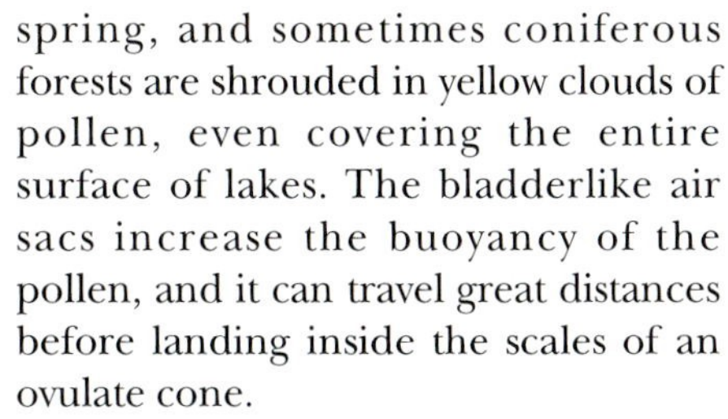

The ovulate cone is technically a modified branch with spirally arranged scales known as seed-scale complexes, or **ovuliferous scales.** Each scale bears two ovules on the upper surface. Each ovule consists of a nucellus surrounded by the integument, which eventually gives rise to the seed coat. A tiny opening at the base of the integument, the micropyle, is the site of penetration of the pollen tube. Within the megasporangium a megasporocyte undergoes meiosis, producing four megaspores. The three nearest the micropyle disintegrate, leaving a single megaspore to produce the megagametophyte containing the archegonia, each with an enclosed egg. Pollination occurs in the spring. The ovulate cones open the scales so that the pollen grains are aerodynamically drawn into the crevice. A drop of sticky fluid at the micropyle captures the pollen grain, and as the fluid evaporates, the pollen grain is pulled closer to the ovule. During this process the scales again close, providing additional protection from the environment. When the pollen grain comes in contact with the nucellus, it forms a pollen tube that grows slowly toward the megagametophyte. This process is exceedingly slow, sometimes requiring

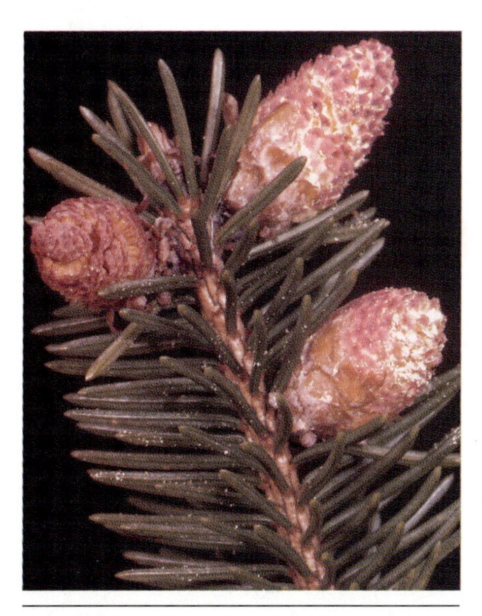

Figure 19.25
Pollen cones of Norway spruce (*Picea abies*). The pollen is just beginning to be released.

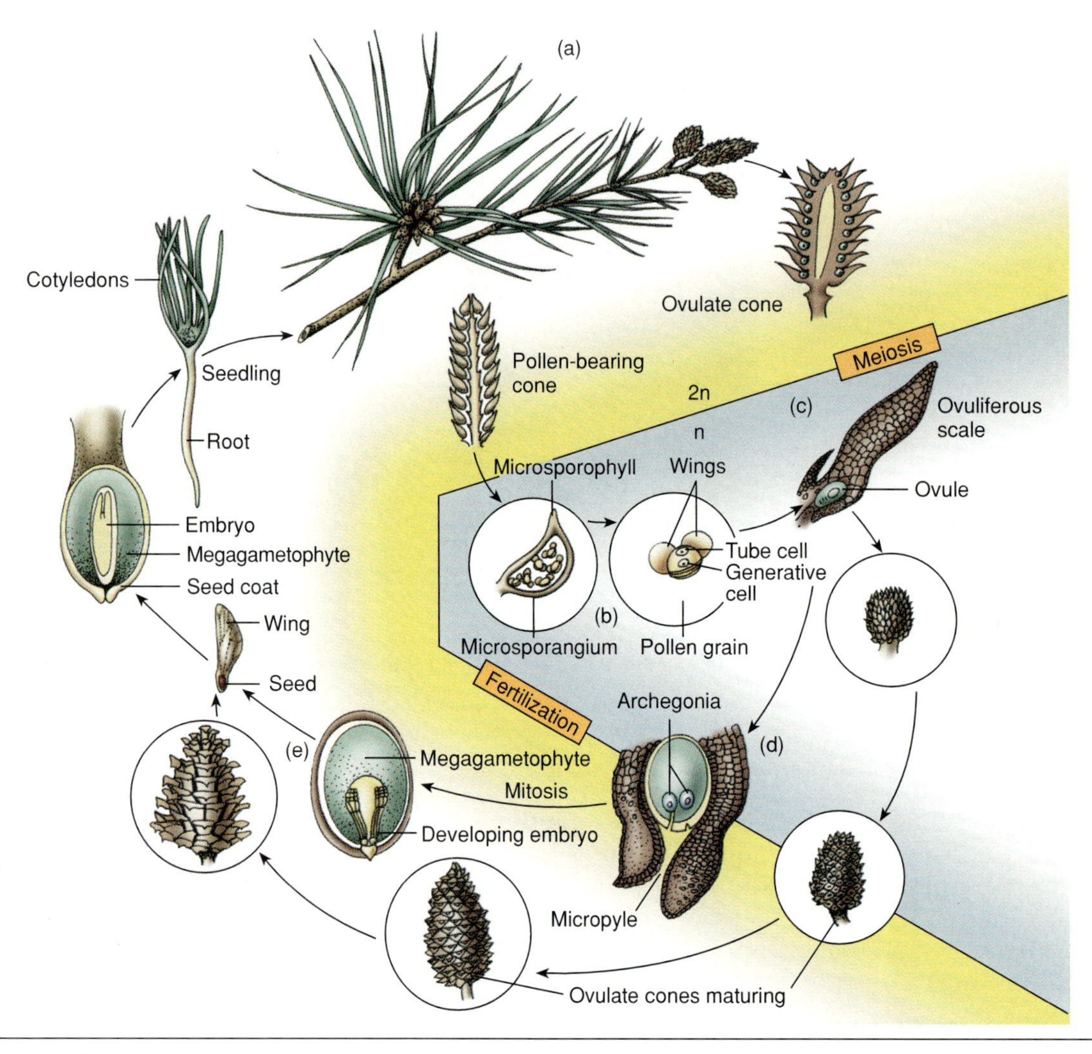

Figure 19.26

The pine life cycle. The pine seed germinates to produce a pine tree, which, at maturity, produces both pollen and ovulate cones (*a*). Pollen (*b*) is blown inside the scales of the female cones (*c*), where the long process of pollen-tube growth and fertilization leads to the production of a zygote (*d*). Following mitosis an embryo is developed to produce new seeds on each scale (*e*).

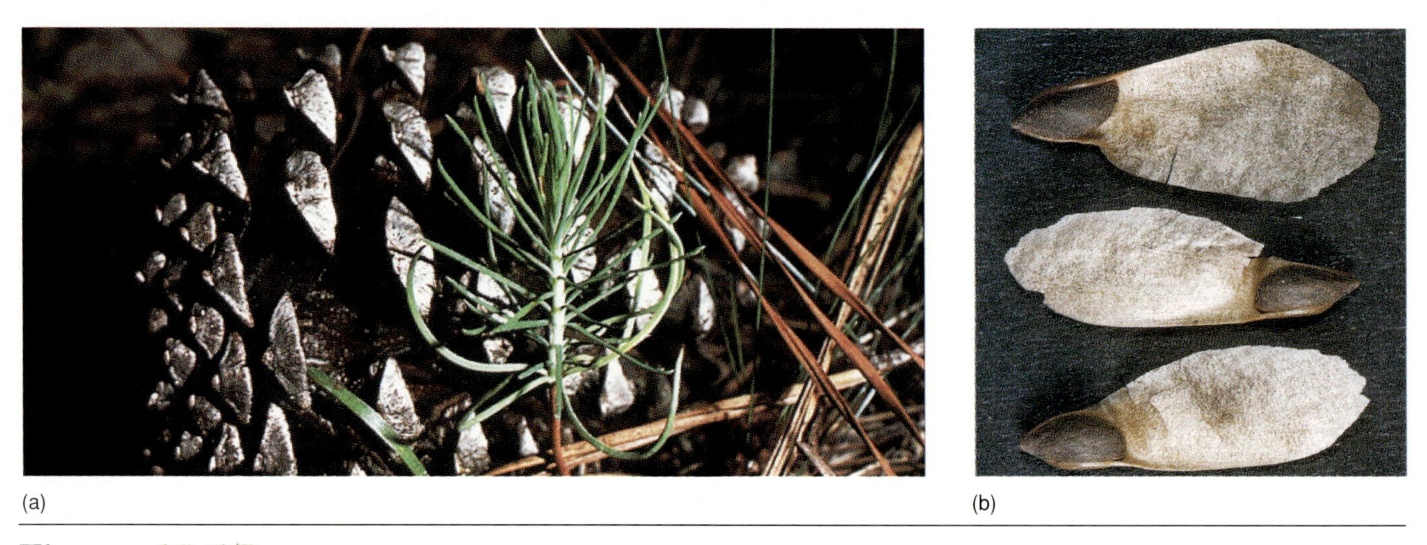

Figure 19.27

The ovulate cone (*a*) opens to expel winged seeds (*b*) from each scale. Some fall to a suitable substrate and produce new seedlings.

as long as 6 months for initiation and an additional 6 months for complete development. Fertilization is not accomplished until approximately 15 months after pollination.

In conifers, two nonflagellated sperm cells are eventually produced, but one disintegrates and the other fertilizes the egg. Even though more than one egg may be fertilized and more than one embryo may start to develop—a process known as polyembryony—all but one generally abort, and a single embryo develops to maturity within each seed. The mature cone completes development during the fall of the second year following pollination. The scales of the cone open, releasing the winged seeds that can be carried a considerable distance by the wind. The mature conifer embryo consists of a root/shoot axis with several, usually eight, cotyledons or seed leaves. The environmental conditions essential for germination are similar to those for the angiosperms.

Ecology and Economic Importance

The conifers' adaptation to aridity, particularly true of the pines, has allowed colonization in many parts of the world. Vast areas are covered by coniferous forests, particularly the temperate regions of the northern hemisphere. Secondary growth is rapid and extensive; thus the conifers form the base for the lumbering industry throughout the world. Although conifer wood is usually considered to be softwood and angiosperm wood is considered to be hardwood, the distinction is far from absolute. Many conifers produce relatively hard wood. The largest and oldest living organisms in the world are conifers (figure 19.28).

Ginkgo

One of the broad-leaved, deciduous street trees of many cities throughout the world is the maidenhair tree, or *Ginkgo biloba.* Few nonbotanists recognize this tree as a gymnosperm, yet its morphological features, including naked seed, place it within this group (figures 19.29 and 19.30). The unusual leaves are fan shaped and sometimes deeply lobed. *Ginkgo* is the only living species of a group of organisms that extended into the late Paleozoic era and were widespread in the Mesozoic era. *Ginkgo* is very limited in the wild, growing naturally along the northwestern border of Chekiang and southeastern Anhwei in China. The species has spread rapidly during recent years because of its excellent qualities as a street tree for metropolitan areas and its particular resistance to air pollution. Tokyo's famous cherry trees, many of which have succumbed because of industrial pollution, have been replaced with *Ginkgo.* Ginkgos are **dioecious** trees; that is, pollen and ovulate structures are borne on different trees. The ovules are borne in pairs on the end of short stalks with seeds produced in the fall. Fertilization is very much delayed and may not occur until after the ovules have been shed from the tree. The seeds produce butyric acid, which has an offensive odor, and consequently pollen trees are usually propagated for the ornamental trade. The species is readily propagated vegetatively, and therefore it is easy to perpetuate only the pollen-producing plants.

Figure 19.28
One of the ancient gymnosperms that has been saved from extinction is the dawn redwood, *Metasequoia glyptostroboides.* Although it was first described in 1941 as a fossil *Sequoia*-like genus from the Pliocene, living specimens were found to be growing in the People's Republic of China in 1945. Since that time this deciduous tree has been distributed worldwide and has proven to be adaptable to many climatic conditions.

Figure 19.29
Ginkgo biloba is exceptionally resistant to air pollution and makes an excellent street tree. This large specimen is growing in the Missouri Botanical Gardens, St. Louis.

Figure 19.30
A branch of *Ginkgo biloba,* the maidenhair tree. Both leaves and seeds would, at first glance, suggest that the tree is not a gymnosperm, but in fact the seeds are borne naked, and the fossil record places this single species as a primitive gymnosperm.

Cycads

These are palmlike gymnosperms, very different from either the conifers or *Ginkgo* and native to subtropical and tropical regions.

In the Mesozoic era, during the reign of the dinosaurs, cycads were a dominant vegetation. There are approximately 10 genera with 100 species living at the present time. The only species native to the United States is *Zamia pumila,* which grows in southern Florida. Cycads produce a trunk with some secondary tissue, and the leaves are usually clustered at the apex to produce a palmlike effect, which makes these plants useful in landscaping. Pollen and seed cones are borne on separate plants, and the cone can be very large in some species (figure 19.31).

Gnetophytes

This unusual group of plants consists of three genera with some 70 species. Like the angiosperms, this single group of gymnosperms conducts water via vessels. *Ephedra,* which inhabits arid regions, is a shrubby gymnosperm with extensive branching and scalelike leaves (figure 19.32). It is the only genus of this division native to the United States. One species, *Ephedra antisyphilitica,* the so-called Mormon's tea, was brewed by the early settlers as a substitute for true tea, and the liquid was thought to have medicinal properties. *Ephedra* also produces ephedrine, an important bronchodilator.

Gnetum is a group of trees and climbing vines found throughout tropical rain forests. They have thick, leathery leaves and are easily mistaken for typical dicots. *Welwitschia* is one of nature's strangest products. It grows only in the desert regions of Southwestern Africa. The aerial parts consist of a woody, saucer-shaped disk that produces only two elongated, straplike leaves. On old plants the leaves are torn and tattered, the splits giving the impression of more than two leaves (figure 19.33).

(a)

(b)

Figure 19.31

Cycad plants are dioecious. (*a*) Pollen cones are quite large and produce an abundance of pollen. (*b*) Seeds develop either in cones or on exposed appendages as shown in this Australian cycad, *Cycas basaltica.*

✔ Concept Checks

1. What is meant by the term gymnosperm? List four groups of extant gymnosperms with a generic example of each.
2. Define the following terms: ovule, seed, integument, prothallial cell, polyembryogeny, siphonogamy.
3. What economic value do gymnosperms have?

(a)

(b)

(c)

Figure 19.32

The genus *Ephedra* is confined to arid and semi–arid regions. It is a profusely branched shrub, with small scalelike leaves. The species shown (*a*) is dioecious with the pollen-producing strobili (*b*) on different plants from the ovulate stobili (*c*). The integument of each ovule forms an extension termed the micropylar tube, which aids in capturing pollen.

Angiosperms

Evolution and Distribution

Unquestionably, today's dominant worldwide vegetation consists of the flowering plants. Besides their agricultural importance, their applications are abundant in commerce and industry. Most of the plant products for consumer use come from the angiosperms. This giant group of plants, which includes some 170,000 species of dicots (grouped into about 250 families) and 65,000 species of monocots (grouped into about 60 families) is characterized by a diversity not seen in any of the other plant groups (see figures 19.35 to 19.42).

Figure 19.33
Welwitschia mirabilis growing in the Namib Desert of Southwest Africa. Strobili are produced around the periphery of a central bowl. Only two leaves are produced, but they become split into a series of twisted and buckled segments.

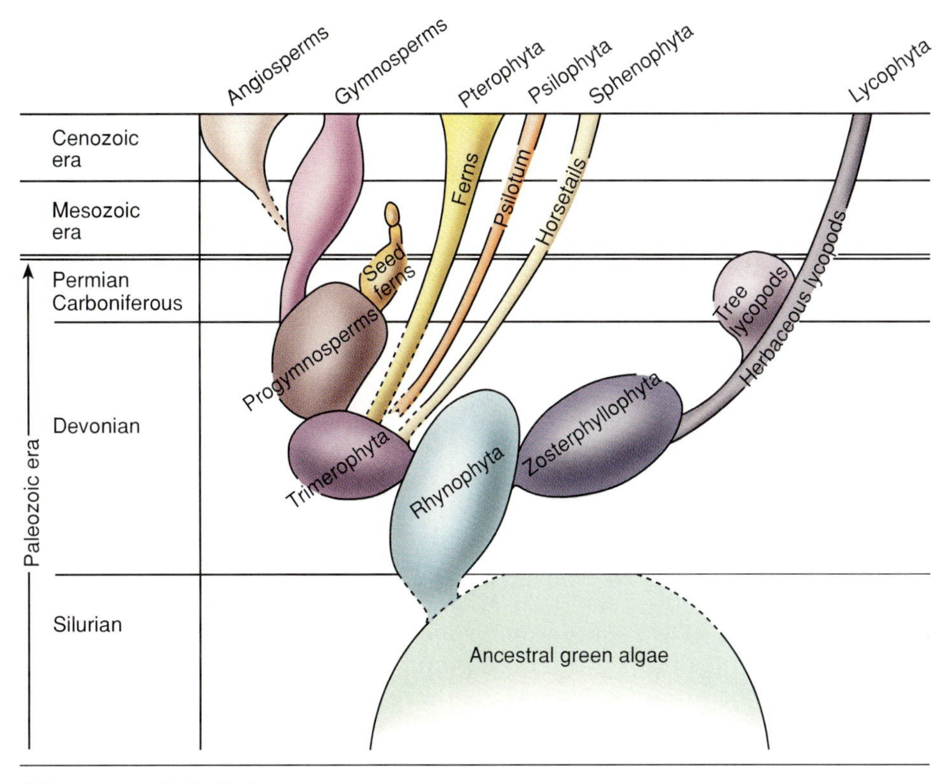

Figure 19.34
This phylogenetic chart demonstrates the rise and fall of various kinds of vascular plants. It is interesting that some evolutionary lines have been very successful, whereas others have not. Gaps occur in some records, indicated by the dotted lines. The angiosperms appear to have arisen from the gymnosperms sometime during the last 200 million years; the exact date is not certain. The angiosperms are particularly successful under the present climatic conditions.

Figure 19.35
Among the supposedly primitive angiosperms is *Franklinia alatamaha,* named after Benjamin Franklin. The species was found on a small plot of land in Georgia in 1765 but was extinct in nature by the early 1800s.

Figure 19.36
(*a*) One of the smallest angiosperms is the tiny aquatic duckweed, *Lemna minor* that floats on the surface of ponds and lakes and is readily cultivated in aquaria. (*b*) The tiny roots are not anchored but absorb water and nutrients directly from the surface layer of the pond.

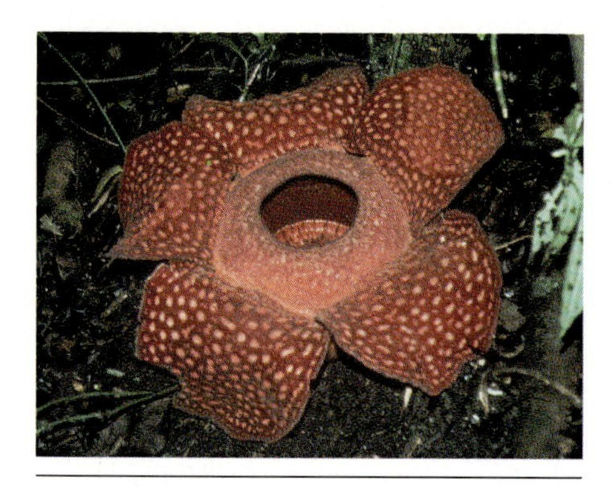

Figure 19.37
Flowers have evolved in a myriad of shapes and sizes. The largest flower known to the scientific world is *Rafflesia arnoldi* from Sumatra. Near extinction, this magnificent flower is approximately 1 m in diameter, and the petals are approximately 3 cm thick.

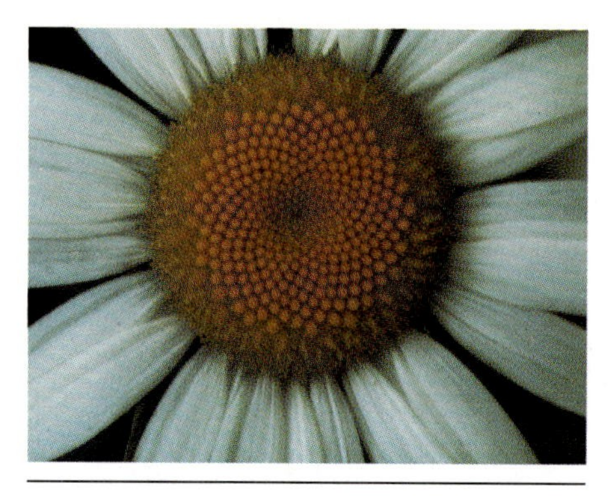

Figure 19.38
The daisy (Asteraceae) typifies flower appearance to the nonbotanist. It actually is an aggregation of many flowers.

From the sedimentary strata of the fossil record, one can determine layer upon layer of ferns and primitive gymnosperms, the lush vegetation of the Carboniferous forests about 200 million years ago. Then suddenly, angiosperms appear in sediments from about 127 million years ago. Paleobotanists have generally concluded that angiosperms did indeed arise at that time, but there is also a theory that angiosperms evolved considerably earlier than the fossil record indicates—perhaps as long as 175 to 200 million years ago. The argument suggests that conditions that lead to selection pressure (changing and diverse climates) existed in the tropical highlands, not in the valleys and lowlands where the sedimentation would be expected to occur. It would be possible for a group of organisms to have evolved at such high elevations and have no fossil record for many millions of years.

Known distribution patterns and the fossil record suggest that angiosperms evolved from now extinct broad-leafed gymnosperms (figure 19.34), in the hills and uplands of Gondwana, which began breaking apart about the same time the angiosperms appeared. The separation of South America and Africa was not complete in the tropical regions until about 90 million years ago. Presumably the early angiosperms radiated into Laurasia in the region of what is now North Africa, the Iberian Peninsula, and the Middle East. Even though the Indian subcontinent was moving rapidly northward at this time and eventually collided with Asia, there is little evidence of major angiosperm radiation crossing this link: the Indian subcontinent had undergone such major climatic changes during its move northward that most species became extinct.

The fossil record also suggests that by 75 million years ago many flowering plant families were well established, and some of these are common in today's angiosperm flora, including birch (*Betula*), alder (*Alnus*), oak (*Quercus*), elm (*Ulmus*), sycamore (*Platanus*), basswood (*Tilia*), chestnut (*Castanea*), maple (*Acer*), beech (*Fagus*), sweet gum (*Liquidambar*), hickory (*Carya*), and *Magnolia*.

A number of modern angiosperms have returned to an aquatic habitat where the environmental stresses are not serious. Whether the plant is submerged or floating, water stress in an aquatic environment is seldom a factor in growth and reproduction. Light and gas exchange can be compromised, but species that have managed to survive did so through specific anatomical and morphological adaptations, such as through the formation of extensive internal air spaces or canals for gaseous transport or storage. In *Lemna minor* (duckweed), for example, the tiny leaves float on the surface of the water, and the roots lie just below the surface (figure 19.36).

Angiosperm evolution, as in other organisms, is characterized by the development of new species as world climates change. Unquestionably, one of the major selective forces is the lack of a constant water supply, and many species have managed to survive because variants within the species were capable of coping with aridity. Because lack of water is such a compelling selective factor, proportionately fewer organisms manage to compete well in arid ecosystems. Species that do survive are often very different morphologically from their ancestors.

Evolution of the Flower The shoot apex of flowering plants is a remarkable structure: it sometimes produces leaves and sometimes produces flowers, depending on a number of complex internal and external factors. A vegetative apex that produces leaves is said to be **indeterminate,** that is, new leaves arise whenever climatic conditions are suitable for growth and development. Even in periods of dormancy the vegetative apex remains intact, and primordial leaves wait for the onset of favorable conditions. This same shoot apex, cued by certain environmental or hormonal factors, suddenly stops producing leaves and begins producing reproductive organs. The terminal apex generally ceases growing after producing a flower, fruit, and seeds; thus it becomes a **determinate** organ. The conversion from indeterminate to determinate status is characteristic of angiosperms. Perhaps no single event of plant evolution has had such a significant impact as has the evolution of the flower. It is important to remember that floral parts have evolved from leaflike organs, and indeed the flower is really nothing more than a modified shoot apex. Much of what we know about the evolution of flowers is based

Figure 19.39

No group of plants displays as much diversity as the angiosperms. (*a*) *Passiflora* sp. from tropical and subtropical regions. (*b*) Cup of flame bromeliad (*Neoregelia carolinea*). (*c*) *Tillandsia cyanea* bromeliad, Ecuador. (*d*) *Tillandsia* sp. epiphyte, Brazil. (*e*) *Nymphoides indica,* Sri Lanka. (*f*) Honeyflower, (*Protea cymaroides*), South Africa. (*g*) Horticultural variety of *Hibiscus* sp. from tropical and subtropical regions. (*h*) *Lantana,* a creeping shrub or vine in tropical regions. (*i*) *Banksia attenuata,* Australia. (*j*) California poppy (*Eschscholzia californica*). (*k*) The succulent *Huernia zebrina,* South Africa. (*l*) *Spathophyllum* sp. (*m*) *Nymphaea,* the pink opal water lily. (*n*) Tropical torch ginger (*Alpinia malaccensis*). (*o*) Staminate flowers of *Alnus* sp. (*p*) Indian pipe (*Monotropa uniflora*) is a chlorophyll-deficient angiosperm that forms a mycorrhizal relationship with a fungus associated with another plant. The fungus acts as a bridge, supplying nutrients to the Indian pipe.

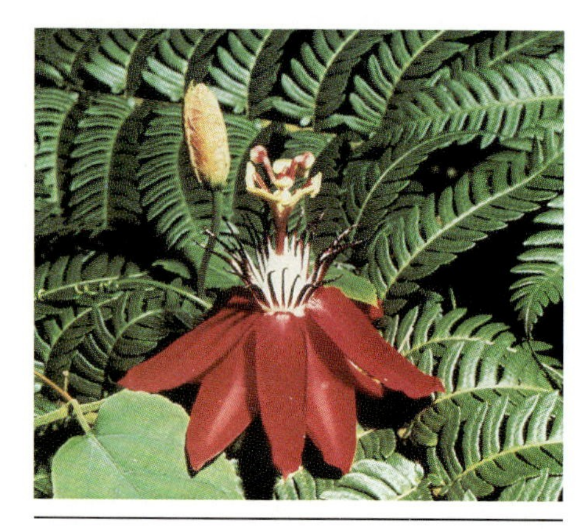

Figure 19.40
One of the more spectacular red flowers is *Passiflora coccinea* from Brazil.

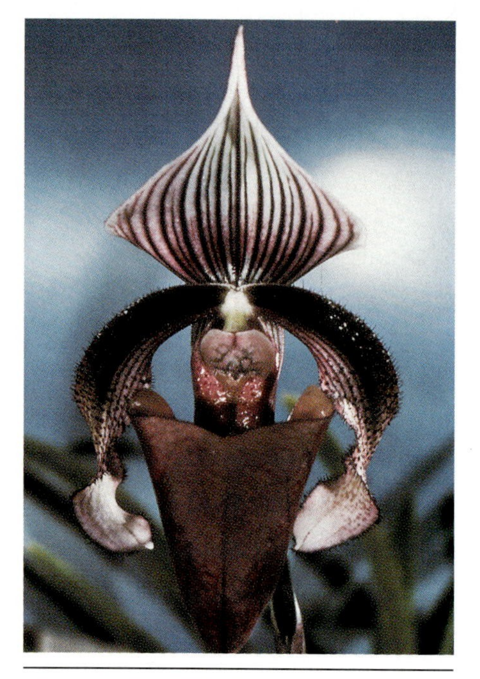

Figure 19.41
The most specialized monocot family is Orchidaceae. This *Paphiopedilum curtisii* is from Sumatra.

Figure 19.42
Variability is great even within a family, as indicated by this orchid (*Bulbophyllum medusae*) from Malaya and Borneo.

on morphological comparisons of modern forms, because flowers are generally too delicate to be preserved in the fossil record. During the past decade, however, a number of fossil angiosperms from the early Cretaceous have been discovered. As more discoveries are made the history of floral evolution will be better understood.

Recall that a flower is a determinate shoot bearing various leaflike appendages. Studies of early fossil flowers indicate that the carpel is derived from a folded blade without a well-defined stigmatic surface to which pollen grains can adhere. Hairs on the margin of the blade probably acted as pollen traps. Gradually, with the evolution of carpels, the stigma became more specialized and relocated near the top of the structure. In several early angiosperms the ovules were arranged in rows near the edges of the inner surface. As evolution proceeded, the number of ovules decreased. Likewise, these primitive flowers contained a number of separate carpels within the ovary, which have been fused and/or reduced in number with increasing specialization.

Stamens, too, have evolved from leaflike organs, although most modern stamens bear them little resemblance. The extant flowering plant, *Magnolia,* produces flat, broad stamens, similar to the stamens observed in several early angiosperm fossils. It is thought that ancestral stamens were nothing more than leaf blades with elongated sporangia near the center of the blade. As specialization took place, the blade narrowed to produce what is now the filament, and the sporangium was left at the tip of the modified leaf. Increased specialization has led to fused stamens. In some cases they are fused to each other to form columns; in other cases they are fused to the corolla. Some very advanced species have sterile stamens that have become modified to produce nectar, although typical nectaries are not derived from stamens.

Most sepals are green, photosynthetic, and leaflike. They have apparently been derived directly from leaves with little modification. Petals are occasionally derived from sepals, but generally they are specialized from stamens and have become broadened, pigmented, and otherwise modified for pollinator attraction. Petals, like stamens, generally have a single vascular strand, whereas sepals, like leaves, have three or more vascular connections. In the more advanced families the petals are fused into a tube, and the stamens are often fused to that tube. Likewise, the sepals may fuse into a tube in the more recently evolved groups.

Some primitive flowers were composed of many carpels, stamens, petals, and sepals (all distinctly separated), and all structures were spirally arranged on the tip of the stem. A modern flower that still retains those characteristics but is probably much larger, is the *Magnolia.* By comparing *Magnolia* with a more advanced species such as an orchid, one can make some generalizations about the trends in floral evolution.

1. Flowers have evolved from many indefinite free (unfused) parts to a few definite fused parts.
2. Starting with four types of floral organs: (1) gynoecium; (2) androecium; (3) corolla; (4) calyx; the types of floral organs have been reduced from four in the primitive flower to three, two, or one in specialized flowers.
3. The primitive position of the ovary was superior; the advanced position is inferior.
4. The radial symmetry of primitive flowers has given way to bilateral symmetry in the more advanced flowers.

The most specialized dicot family is the sunflower family (Asteraceae), and the most specialized monocot family is the orchid family (Orchidaceae). The Asteraceae has the largest number of dicot species.

Evolution of the Fruit A fruit is a mature ovary. In some cases it retains floral parts. Depending on the arrangement of the carpels, fruit can be simple, multiple, or aggregate (figures 19.43 to 19.46). Modification of floral parts has, in some cases, led to fruit that include the ovary but are largely composed of other organs. In the apple, for example, the floral tube enlarges, growing around the carpels to become the fleshy portion of the fruit. The ovary itself becomes the apple core (and is of secondary importance as a food source).

Many fruit have developed hard walls that protect the seeds inside. Such a protective mechanism helps to

(a) (b) (c) (d) (e) (f) (g) (h) (i)

Figure 19.43

Fruit shapes, sizes, and colors are as varied as those of flowers. (*a*) Tomato. (*b*) Pear. (*c*) Lemon. (*d*) Grapefruit. (*e*) Coffee (*Coffea arabica*). (*f*) Breadfruit. (*g*) Cacao. (*h*) Fruit of the lipstick tree (*Bixa orellana*) from Colombia, where the Indians use the juice for red dye. (*i*) Palm fruit of *Caryota mitis*, Indonesia.

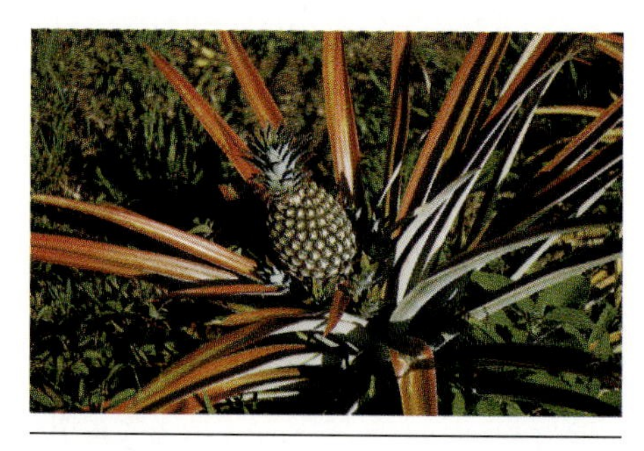

Figure 19.44
Pineapple is one of the more important multiple fruit.

Figure 19.45
The fleshy part of the apple is actually the floral tube, consisting of the perianth and androecial parts fused together and swollen, which grows around the ovary.

Figure 19.46
The fruit wall, as well as the seed coat, may become sclerified and very hard.

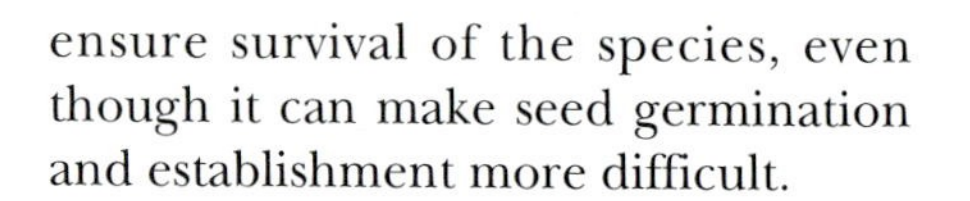

ensure survival of the species, even though it can make seed germination and establishment more difficult.

Coevolution

The evolutionary process is normally described in terms of individuals and populations of a given species changing in adaptive (or nonadaptive) ways. The variability in sexually reproducing individuals allows natural selection to function in several ways. The selective forces, however, are usually described in general terms of the organism's "environment," which all too often brings to mind only abiotic and climatic features. The complete environment acting on organisms also includes interactions with other living organisms. The term coevolution refers to two or more groups of organisms evolving in parallel and interdependently. Neither group "causes" the other to change; both groups develop new types independently. However, whether a new type succeeds is often a direct function of the interrelationships between the groups. Coevolution, then, refers to successful adaptations only.

Insect-Flower Coevolution One of the most striking examples of coevolution is that of flowers and insects. Although today there are some 235,000 species of flowering plants and well over 700,000 species of insects, 200 million years ago angiosperms were just beginning to evolve. Within the Insecta, most orders [Coleotera (beetles), Diptera (flies), Hymenoptera (wasps, ants, and bees)] already existed by the Triassic, before the ascendancy of the angiosperms. By the mid Cretaceous, however, a dramatic increase appeared in the number of plant and insect taxa below the familial level, attesting to the high success of plant-insect relationships. It is one of the most successful examples of coevolution known to biologists (figures 19.47 to 19.51).

This interdependence revolves around the selective advantage it provides both groups. Flowers are more efficiently cross-pollinated by insect visitors, and the insects are provided a reliable food source. Since gymnosperms are wind pollinated today as they were when angiosperms evolved, undoubtedly insect pollination is one of the advantages that allowed flowering plants to become more successful than gymnosperms.

The earliest visitations were probably by beetles, one of the most abundant insects during the lower Cretaceous, feeding randomly on soft plant tissues, pollination droplets of the gymnosperm ovules, and sap or resin exuded from leaves and stems. During their random foraging, the beetles accidentally carried pollen from plant to plant. If primitive flowers, formed by broad-leafed gymnosperms, accidentally produced a more nutritious tissue or fluid that increased beetle visitations, pollen sticking to their bodies would be carried from flower to flower. This would have increased the effectiveness of cross-pollination over wind pollination, and increased cross-pollination frequencies resulted in more ovules developing into seed and greater total genetic variability in successive generations of offspring.

As you will recall, most flowers are self-incompatible—they cannot be fertilized by their own gametes. Thus new genetic material from a different plant of the same species can constantly be introduced. Cross-pollination ensures maximum genetic variability possible through sexual reproduction.

As the numbers of surviving offspring and variability increased, flowers having even greater insect attraction developed, producing even more regular and frequent visitation and cross-pollination. For the insects, more plants with more flowers meant greater amounts of nutritional material and a reliable food source. These flowers were numerous enough to become primary or even exclusive food sources for the insects. This interdependence has occurred in a spiraling way: more and different insect types, resulted in increased cross-pollination efficiency, which resulted in increased flower diversity from which insects could feed, and so on. Unlike the chicken and egg quandary (which came first?) flowers and insects diversified, multiplied, and succeeded inparallel—they coevolved.

It is probable that several features unique to angiosperms evolved as a direct response to insect pollination. Beetles are generally not very dainty or specific feeders; thus an ovule is as desirable a food as other floral tissues.

(a)

(b)

(c)

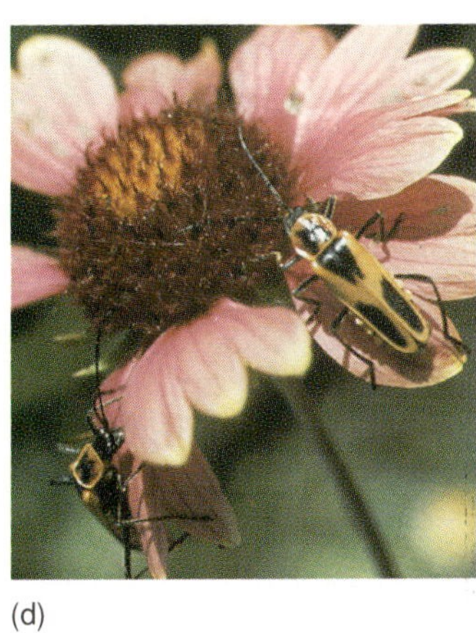
(d)

Figure 19.47
A variety of insects visit flowers to feed on nectar or pollen, effectively cross-pollinating those flowers. Bees are among the most specific and numerous visitors. (*a*) *Rudbeckia*. (*b*) *Gaillardia* (where they are also using the flower head as a mating site). (*c*) *Cirsium*. (*d*) Soldier beetles are common visitors on many Asteraceae species but do not necessarily function as pollinators.

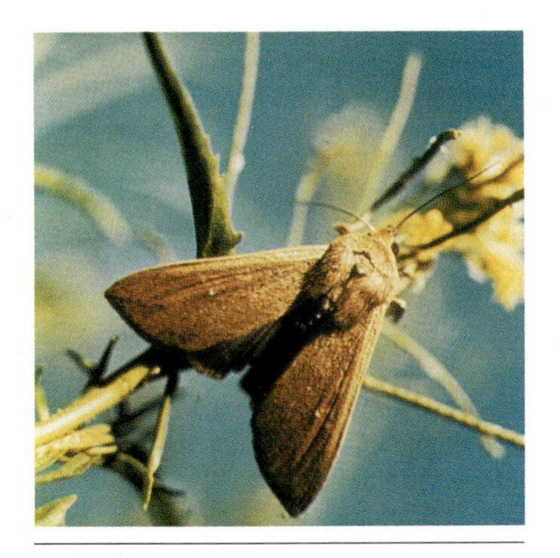

Figure 19.48
Moths normally act as nocturnal pollinators, but some species also visit flowers during the day.

Figure 19.49
A sunbird extracting nectar from a bottle brush.

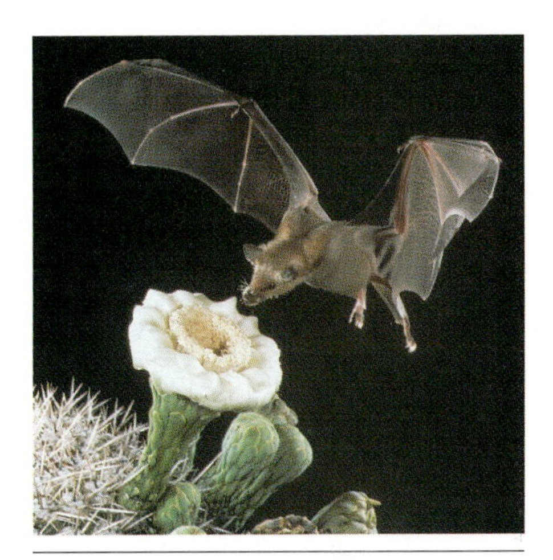

Figure 19.50
Certain species of bats are effective pollinators of night-blooming plants.

Early flower types that lacked sufficient surrounding protective tissue for their ovules did not survive. Today all flowering plants have protected ovules. Subsequent development of entire ovaries surrounded by protective tissue has resulted in inferior ovary position.

As insect groups other than beetles evolved, even greater flower variability and insect-flower specificity evolved. From approximately 50 million years ago on, the fossil record indicates continual diversification among insect groups such as the bees, butterflies, and moths. Correspondingly, flower size, shape, color, and organizational complexity also have undergone remarkable change. The efficiency of bisexual flowers, in which an insect is able to pick up and deliver pollen in the same visit, exemplifies continuing flower evolution. Groupings of flowers into inflorescences have also expedited pollination by insect visitors. The development of petals from stamens resulted in an additional floral structure having a phenomenal number of different shapes and specific modifications.

The large number of flower and insect types that exist today is a result of this process of coevolution. Visitation specificity ranges from general, as in the primordial relationships, to specific one-to-one relationships. It is

Figure 19.51
Monarch butterfly larvae feed on the leaves of milkweeds, accumulating toxic compounds. This example of the milkweed (*Asclepias*) shows two butterflies, neither of which is a monarch.

Hot Flowers

Although the production of heat to maintain body temperature above surrounding air temperature is a characteristic feature of higher animals (birds and mammals), it is uncommon among plants. Six plant families are currently known to produce thermogenic (heat producing) flowers and inflorescences. These include the Annonaceae (sweetsop and soursop), Araceae (aroids), Aristolochiaceae (aristolochias and asarums), Cyclanthaceae (Panama hat plant), Nymphaeaceae (water lilies), and Cycadaceae (cycads). In flowering plants the biological significance of heat production is well understood. The flower has the biological function of pollination, fertilization, and subsequent seed-set. This process of reproduction perpetuates the species, providing genetic variation in the offspring and population. Increased floral temperatures facilitate volatilization and the rapid diffusion of aromatic compounds. The odor of the flower, which becomes more apparent due to thermogenesis, attracts pollinators and increases the chance of pollination and subsequent fertilization and seed-set.

To understand the physiology of this dramatic process of heat production one must recall the basic structure of the plant cell, which contains numerous mitochondria. These organelles, commonly nicknamed "the powerhouses of the cell," are the site of cellular respiration where foods (sugars and lipids) are chemically broken apart (oxidized), releasing energy that is used to form the high-energy phosphate bonds of adenosine triphosphate, or ATP. In thermogenic plants an "alternate pathway," termed the cyanide-insensitive or resistant pathway, is present. In this pathway, rather than producing ATP, energy is released in the form of heat. Recent research on thermogenic plants has identified acetylsalicylic acid (aspirin) as the chemical trigger initiating heat production.

Investigations on heat production in the water lilies, *Victoria*, the giant amazon lily, and *Nelumbo*, the Oriental and American lotus, have revealed that thermogenesis occurs in specialized parts of the flower (see figure 1). These parts are rich in starch, and the thermogenic process results in temperature increases of between 6°C to 12°C above ambient air temperature in the flowers. In other plants, such as the voodoo lily (*Sauromatum*), increases to 40°C above the surrounding air temperature have been recorded.

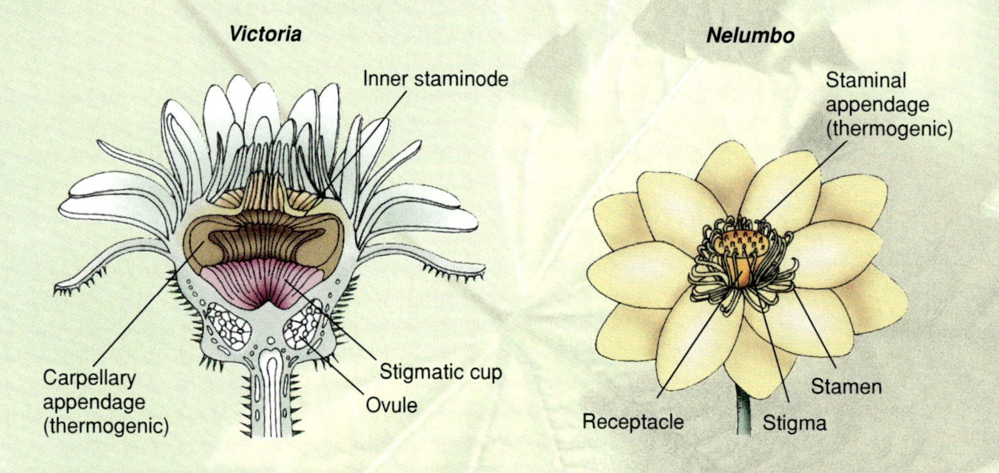

Figure 1
Two thermogenic taxa illustrating the site of heat production. In *Victoria*, thermogenesis occurs in modified floral parts termed carpellary appendages. In *Nelumbo*, heat production occurs in modified appendages at the distal end of the stamens.

important to point out that, although the general trend in pollinator-flower coevolution is toward greater specificity, the resulting advantages are there only as long as both partners are successful in all other ways as well. If either member of a highly interdependent association fails, the other will fail (unless it is able to survive by previously unused mechanisms).

Dispersal into new areas is one of the critical requirements for plant success that is very closely tied to pollinator range expansion. In a sense, pollinators are the only legs plants have. Even if seed is dispersed into new areas, unless those plants are able to sexually reproduce, they are ultimately doomed to failure in that habitat. Interestingly, the most successful colonizers are weedy plants found in new, disturbed, and changing habitats; these plants have unspecialized pollination systems.

Pollinator Specificity The importance of pollinators for sexual reproduction in most flowering plants cannot be overemphasized. Because plants lack mobility, cross-pollination and the genetic variability it allows would be very limited without the activities of pollinators. The role of the "birds and the bees" is vital, since these two groups participate in some of the most reliable and species-specific pollinator-flower relationships. A brief and very generalized description of the most common pollinator organisms and the kind of flowers they visit will provide a better perspective of the preceding discussions.

Beetles Even though beetles were probably the earliest pollinators and their role in the successful evolution of angiosperms is important, today they are relatively overlooked as pollinators. Generally, beetles are large, awkward, and poorly adapted to flower pollination. Their mouth parts are adapted for chewing; thus they feed on edible tissues and pollen, seldom nectar. Most beetles that do visit flowers do not depend exclusively on flowers for their nourishment; rather, they feed primarily on other plant parts, dead animals, or animal dung. Their flower visits, therefore, are largely accidental, and as pollinators they are undependable. Certainly, not all beetle visitation is random; there are many examples of regular, predictable, and even highly specific visitations. There are even some beetles with mouth parts modified for nectar feeding and flowers that have developed attractants for beetle visitation.

Even highly modified flowers, such as some orchids, are pollinated by beetles, but the typical flowers visited by beetles are far less specialized. Usually large and solitary or small and grouped into a large inflorescence, these flowers are often dull white to greenish with strong fruity or decaying proteinlike odors. Beetles' sense of smell is much more acute than is their vision, and they are attracted by odors that simulate their common food sources—fruit, carrion, and dung.

The flowers are usually open, flat or bowl shaped, and shallow, with plentiful pollen and accessible sexual organs. In addition, the ovary is normally well protected, safe from the beetles' indiscriminate feeding habits and chewing mouth parts. Some well-known solitary flowers pollinated by beetles are species of the *Magnolia* genus, poppies, cactus, and lily. Inflorescences in the carrot family (umbels), dogwoods, and tropical members of the Fagaceae (beech family) are also beetle pollinated.

Bees Of all the flower-visiting–pollinating organisms, bees are definitely the most well adapted, specific, and numerous (figure 19.47). From small nonsocial bees to the larger social honeybees and bumblebees, these insects are experts in flower recognition, feeding, and pollination. Bees have hairy bodies, ideal for pollen transport; bumblebees are known to carry as many as 15,000 grains per individual. Bees are readily able to learn shapes, patterns, and colors, and they have mouthparts modified for nectar feeding and pollen collecting. In addition, social bees have a large food demand, providing for themselves and their brood. They also have a communication system to inform others in their colony of the exact location of a bountiful food source.

An interesting phenomenon of bee vision is their range of color perception. Bees have a visible spectrum that is shifted into the ultraviolet wavelengths but out of the red range. Therefore they are able to detect patterns produced on petals by ultraviolet-reflecting pigments but are essentially red color-blind. One of the groups of secondary plant compounds, the **flavonoids,** includes some pigments often found in petals that reflect ultraviolet patterns. Two related species, both with the same color (to us) yellow flower, have distinct colors to a bee visitor—hence differential visitation.

Bee and flower coevolution is specific and highly complex; as a result, some of the most unusual and highly modified flowers are bee pollinated. Typical bee-pollinated flowers are irregular, sturdy, fairly deep, and often have a "landing platform." Usually bright yellow or blue (but not red), bee-pollinated flowers commonly have colored nectar guidelines running from the outer edges of the top surface of lower petals down the tube of the flower. Many also have semiclosed flower throats, which help prevent nectar thieving by smaller insect visitors too weak to force their way past the closure. Many flowers have hairy areas near the stigma, which groom pollen off the bee's body to ensure pollination.

The flower tubes are often structurally designed so that only the appropriate bee species can get to the nectar—the reward for visiting that flower. In addition, the pathway to the nectar supply also requires that the visiting bee come in contact with the stigma, guaranteeing pollination. Pollen from the flower being visited is also deposited on the bee's legs, back, or underside, often by complicated modifications that break open anther sacs or force the departing bee to come into contact with the mature anthers.

One of the most specialized and highly evolved bee-flower interdependencies involves a wild orchid in the genus *Ophrys.* The flowers of a given species open at a time in the spring when males of the coevolved bee species emerge. The flowers are the same size and shape as the female of the bee species (which does not emerge until later). The male bees are instinctively attracted to the flower and attempt to copulate with it. Although notably unsuccessful in this effort, the thrashing around on the flower does result in pollen deposition on the male bee's body. Repeated attempts to copulate with these flowers guarantees cross-pollination. When the female bees finally emerge some time later in the spring, the males are very quick to guarantee the next generation of bees. Pseudocopulation, as this bee-flower activity is termed, occurs between several specific members of the *Ophrys* genus and given species of bees, wasps, and even some flies.

Common to many bee flowers is a postpollination change in the appearance of the flower, probably induced chemically by the development of pollen tubes. Flowers are known to have their visual attractant markers change after pollination. The potential bee visitors then do not recognize the flower and thus pass it by. Were these bees to visit such flowers anyway, they would find no nectar, and the desirability of future visits to flowers of that species would be diminished. Changes such as the fading of the nectar guidelines, wilting of the landing platform petal, general flower closure and withering, and color dullness are adaptations of some bee flowers that secure even greater visitation, pollination efficiency, and success.

Butterflies and Moths Even though butterflies and moths are closely related and similar morphologically, as pollinators they are quite different. Butterflies are active in the daytime and have good vision but a weak sense of smell. Moths are nocturnal and have a well-developed sense of smell. Butterflies light on the flower, whereas moths hover. Both suck the thin nectar through their long, thin, hollow tongues, and neither have to provide food for developing young.

Usually yellow, blue, or in some cases red, butterfly flowers have a long, thin floral tube with a sturdy outer flower structure for adequate landing. Butterfly flowers include many members of the sunflower family (Asteraceae), *Lantana,* and various trumpet-shaped blossoms.

Moths visit white, pale yellow, or pink flowers that are open at night and produce a strong, heavily sweet perfume (figure 19.48). These flowers also have deep tubes but with open or bent-back margins that allow the hovering moths to reach the nectar with their long tongue. Because hovering expends more energy, moths generally visit flowers that produce more nectar. The hawk moth is a competitor of hummingbirds for their food sources. Essentially as large as and remarkably similar in appearance to a hummingbird, this moth has a long proboscis for sucking up the thin sugary nectar of typical bird flowers. More commonly, however, moth flowers are larger than bird flowers and open only at night. *Yucca,* evening primrose (*Oenothera*), and a number of night-blooming cactus species are moth-pollinated flowers.

The *Yucca* is pollinated only by the four species of a single moth genus (*Tegeticula*), which in turn depends on the *Yucca* flower for its entire reproductive cycle. The female moth collects pollen from one flower and transports it to another. It pierces the ovary wall, lays eggs inside the ovary, and then deposits the mass of pollen on the stigma. When the moth eggs hatch, the larvae feed on the developing seeds within the ovary. Only about 20% of the developing *Yucca* seed is destroyed by the feeding larvae, which eat their way out of the ovary when they are mature so that they may pupate on the ground below the plant. The female *Yucca* moth ensures a food source for her larvae while pollinating the flower.

Birds Different kinds of birds act as pollinators by feeding on nectar or on insects within the flowers (figure 19.49). Sparrows are known to visit spring crocus and have been observed pollinating pear trees. Honey creepers pollinate the Hawaiian lobelias, and the sunbirds of Africa and Asia are known pollinators. The most well-known bird pollinators, however, are the hummingbirds of North America. These tiny animals expend incredible amounts of energy in flight, especially in hovering while feeding. They have keen eyesight, being most responsive to reds and some yellows, but do not have a well-developed sense of smell. Their long thin beak enables them to reach the abundant, thin, sweet nectar. Their feathers carry large amounts of pollen, picked up primarily on the front and top of their heads when they come into contact with stamens extruded beyond the floral tube.

Typical flowers pollinated by hummingbirds are red with a good supply of thin, sweet nectar found at the bottom of a slender floral tube. The lips of the flowers are usually curved back out of the way, and they are normally a solid color, lacking nectar guides. These flowers usually have little or no scent, which, in combination with their color and long slender tubes, usually makes them unnoticeable to most insect visitors. Sugar ants are a notable exception, but they provide very little competition to hummingbirds. Striking examples of bird flowers include *Erythrina, Aquilegia* (columbines), orchids, *Salvia, Mimulus,* and *Lobelia* species that have red flowers.

Flies, Ants, and Mosquitoes
Flies display greater variation in their methods and tendencies of pollination than any other insect group. Primitive flies parallel beetles in their lack of sophistication; highly specialized flies are comparable to bumblebees and hawk moths in complexity. Some South African flies have a 50 mm long proboscis and the ability to hover while nectar feeding. In spite of this range of variability, the most well-known and discussed fly pollinators are the carrion and dung flies. Because they lay their eggs in diseased or decaying animal flesh or on fresh dung, they are attracted by the putrid odors of decomposing protein. Flowers that are pollinated by these flies usually attract with strong odors but offer no reward to the visitor. The "carrion" plant *Stapelia* is the best known of the fly-pollinated flowering plants, carrying the attraction mimicry even a step further. *Stapelia* flowers vary in size depending on the species, but they are flat and open with the dull yellow color of decaying meat, complete with reddish streaks. Most notable, however, is the odor, which does not invite a second sniff!

Ants are exceptionally fond of sweets, from table sugar to flower nectar. They are also so small that they can raid a flower without touching anthers or stigma. In addition, their bodies are hard and not well adapted to pollen transport, and they are aggressive defenders of a new-found food source. As pollinators, therefore, ants are essentially noneffective, although there are a few isolated examples of larger ants providing nonspecialized or accidental pollination for some flowers.

Their aggressive defense of a food source often chases away other insects, including those that would actually effect pollination were they allowed to visit the blossom. Some flowering plants have actually evolved "ant guard" adaptations that deter ant visitations. One of the most successful is the ring of sticky glandular hairs on the stem immediately below the flower of *Viscoria vulgaris*. Stiff hairs projecting downward on the stem or outward in the throat of a corolla tube are other common ant guards that have evolved in some species.

Mosquitoes also are too small and ill designed for effective pollination; however, some flowers are, in fact, mosquito pollinated. Certain small and inconspicuous orchids are visited by both male and female mosquitoes, which feed on nectar rather than blood. The mouth parts of several mosquito species are actually modified to be better adapted for nectar feeding than blood sucking. The food demands of mosquitoes are very small, and even though they carry pollen with them as they visit from flower to flower, they seldom need to visit flowers beyond those on the same branch of one plant. As cross-pollinating agents, then, they are less effective than other insects that also visit many of the same flowers.

Bats Like birds, bats have several effective pollinating features (figure 19.50). They are large, have a rough (furry) surface for holding large amounts of pollen, and can move rapidly across large distances. Most bats are insect eaters, but a number of vegetarian bat species exist. Fruit-eating bats are found worldwide, and it is hypothesized that nectar and pollen feeding developed from such lines, as it did in bird species.

Pollinator bats are nocturnal, have an acute sense of smell over great distances, and, unlike most bats, have acute vision (although colorblind). Their sonar system is less developed, and therefore flying in densely vegetated areas is difficult. Many of these bats can also hover like hummingbirds and have long, slender noses and long tongues.

As these bats fly from flower to flower, they lap up the nectar, often eat parts of the flower, and transport large amounts of pollen in their head fur. Although some species ingest pollen only accidentally while feeding on nectar, others use long tongues to lick pollen off their heads, a major portion of their diet. It is now known that some bats actually depend exclusively on nectar and pollen for their food, and the flowers that these bats visit contain high amounts of protein in their pollen.

Although flower-visiting bats enjoy mostly tropical distribution, some migrate in the summer as far north as the southern United States and northern Mexico, feeding on *Agave* (century plant) and cactus flowers. Bat pollination is also known in northern Australia (on an introduced *Agave*) and in Asia as far north as the Philippines. In Africa, bat pollination does not extend north of the Sahara, and in the Pacific islands their distribution extends south to Fiji. Certain plant distributions can be explained by knowing about bat-pollinated flower types. For example, there is a Hawaiian banana plant (*Musa fehi*) that is adapted to bat pollination. However, it is thought to be introduced to the island, not native, because bats are not indigenous to Hawaii.

Typical bat-pollinated flowers open only at night and only for one night. They are often drab greenish to pink-purple and sometimes white or creamy. They emit a strong fruity or fermenting odor at night and have a very large quantity of both nectar and pollen held in large or numerous anthers. These flowers are generally large, sturdy, and solitary or in inflorescences positioned outside the foliage of the plant for increased pollinator accessibility.

Other Angiosperm-Animal Coevolution

In the evolution, diversification, and adaptation of the flowering plants other specific coevolution with various animal groups has taken place. Fruit- and seed-dispersal mechanisms and morphological adaptations to prevent herbivory are obvious and easily demonstrated examples. Less evident are chemical groups that are now thought to exist as a result of coevolutionary interactions. These were once regarded by plant biochemists as dead-end or nonessential secondary compounds. But many of these chemical groups are now thought to be involved in specific processes of plant protection, pollinator attraction, and feeding stimulation.

We have already mentioned certain flavonoid compounds—ultraviolet wavelength pigments in some flower petals that are visible to bee pollinators. Closely related chemically are the anthocyanins, which are visible wavelength flower pigments. The *Ophrys* flowers provoke pseudocopulation because of visible patterns produced by anthocyanin pigment.

Plant Palatability Plant palatability is significantly influenced by secondary compounds. Both insect and vertebrate animal groups, including humans, are attracted or repulsed by a range of secondary compounds that affect the sense of taste. Members of the mustard family, for example, contain compounds that deter many insects and people from feeding on them, whereas other insects feed only on these plants. The acrid taste and pungent odor of horseradish, cabbage, cauliflower, radish, watercress, mustard, brussels sprouts, and rutabaga are caused in all by the same class of chemical compounds. These undoubtedly coevolved with specific insect groups, protecting the plants from excessive herbivory by most insects while providing a constant food source for insects adapted to tolerate or even actually attracted to these compounds.

Alkaloids **Alkaloids** are a large group of loosely related, nitrogen-containing compounds, many of which are toxic. Toxicity, combined with an unappealing bitter taste, protects plants containing such compounds from being eaten. In the milkweed family (Asclepiadaceae), for example, alkaloid compounds and cardiac glycosides combine to act as severe toxins to most vertebrate animals. Foxglove (*Digitalis purpurea* in the Scrophulariaceae) also produces cardioactive glycosides that are used as a treatment for heart disease, but in higher doses they can induce a heart attack in vertebrate consumers. Certain insects, however, are unaffected by these compounds and preferentially feed as larvae or as adults on the tissues of such plants. In fact, the bodies of these insects contain high concentrations of alkaloid compounds. By being brightly colored and patterned, these insects are recognizable to potential predators, which are predominantly vertebrates. After one experience of acute digestive upset, vomiting, and diarrhea, the predator avoids further ingestion of such insects.

Monarch butterflies feed on milkweeds, and their distinctive visibility to birds acts as a protective device rather than as an attractant. So effective is this coloration and pattern that the viceroy butterfly has evolved a similar appearance to protect itself from predation, even though it does not feed on milkweed. The evolution of the viceroy butterfly is an example of mimicry, a process that has occurred repeatedly in both plant and animal groups (figure 19.51). For many plant species that contain toxic or unpalatable compounds there is one or more species that have evolved in a parallel manner, protected by their close resemblance (figure 19.52).

Many of the secondary compounds that have evolved in plants as protective adaptations or as attractants produce strong effects in humans. The opium poppy, the hemp plant (*Cannabis,* or marijuana), and peyote cactus are among these plants. All contain secondary compounds, mostly alkaloids, that occur in nature because of specific coevolutionary phenomena.

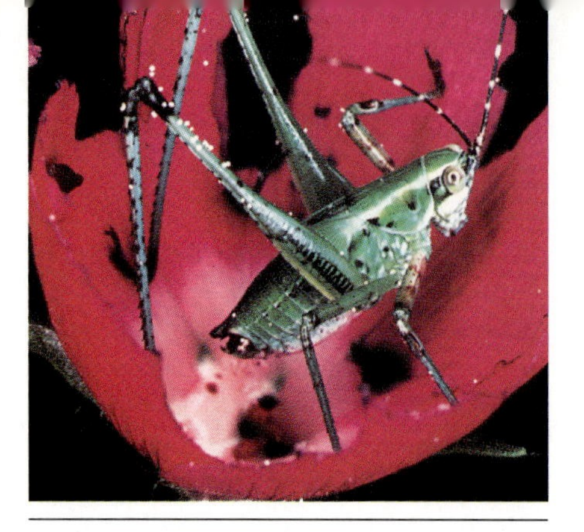

Figure 19.52
Not all plants have secondary compounds that prevent insect foraging. This grasshopper is causing a considerable amount of feeding damage on a winecup (*Callirhoe involucrata*) flower.

Biochemical Evolution As plants evolve into different forms and develop new strategies for adaptation, they may also be selected against if they do not also develop new modes of metabolism; in some cases the biochemistry actually changes. There is strong evidence that both plants and animals adapt to environmental stresses by synthesizing isozymes, or alternate forms of an enzyme, which may allow the plant to carry on photosynthesis, respiration, protein synthesis, or other roles under changed environmental conditions. Given plants appear to synthesize isozymes for water stress, salt stress, heat stress, and cold stress. Those individuals of a population that do so may survive in a changing environment, whereas the plants capable of synthesizing only the normal enzyme may fail.

Biochemical evolution also includes the alternative strategies for carbon fixation (photosynthesis). Most plants fix carbon via the C_3, or Calvin, cycle. Other plants that have evolved under conditions of high temperature, high light intensity, and water stress have developed an alternative method, the C_4 pathway. The C_4 phenomenon is widespread among monocot and dicot families, which have evolved in the hot and drier parts of the world. Such plants have managed to survive continental movements and changes in world climates, largely because they have been able to change certain aspects of their metabolism.

Ecology and Importance of Angiosperms

A unifying theme of this text is the applications of botany in a human-dominated world. Everywhere one looks, flowering plants influence our lives. Except for the coniferous forests, which provide timber for housing and other construction, angiosperms are the dominant plants for food, forage for animals, commercial fibers, industrial products, and medicinals. Even most of our landscaping materials are flowering plants. There is no accurate method to assess the absolute value of angiosperms in our lives. A price tag can be placed on food or other items of trade, but our very being is essentially and complexly influenced by higher plants. Societies that depend on firewood for survival are exquisitely aware of the importance of angiosperms.

Finally, the angiosperms have so totally invaded the terrestrial ecosystems that species are found almost everywhere. No other group of plants has been so successful in filling niches under the most extreme environmental conditions. Such plants compete most effectively for available light energy and contribute incalculably as producer organisms.

✔ Concept Checks

1. Discuss the following concepts relative to the evolution of angiosperms. What is the ancestral group? During what geological period did they first appear? What did the flower of the early angiosperm look like? What kinds of trends have taken place in the evolution of the flower?
2. What are the major pollinators in flowering plants? How does the structure of a beetle-pollinated flower differ from one that is bee pollinated?
3. Define coevolution and explain how flowers and insects have coevolved.

Summary

1. Approximately 450 million years ago primitive multicellular marine green algae were able to survive the transition onto land. They were able to survive the ultraviolet radiation of the sun because of an atmospheric ozone layer that had been accumulating since the first cyanobacteria began photosynthesizing some 3 billion years ago.
2. In addition to the bryophytes, primitive vascular plants began evolving. Their vascular system in addition to protected sex cells, spores, and cuticle development allowed them to succeed and diversify rapidly on land. Three extinct primitive vascular groups, the Rhyniophyta, the Trimerophytophyta, and the Zosterophyllophyta, gave rise to the four major divisions of living seedless vascular plants.
3. The Lycophyta once included tree lycopods but today is herbaceous and contains *Lycopodium, Selaginella,* and *Isoetes.* The Sphenophyta were also a common group during the Carboniferous but are represented today only by *Equisetum.* The Psilophyta now have only two genera, *Psilotum* and *Tmesipteris.* The fourth group, the Pterophyta, or ferns, contains almost 12,000 living species.
4. Ferns have fiddlehead coiled leaf fronds, and the reproductive structures, sori, occur on the underside of the fronds. The sporangia within each sorus produce spores that germinate and develop into heart-shaped gametophytes. Ferns are most important economically as ornamentals.
5. Seed plants are even better adapted to extreme climates than are spore-producing plants. The two seed-producing groups are the gymnosperms and the angiosperms.
6. Gymnosperms include the cycads, gnetophytes (including *Ephedra*), *Ginkgo,* and the conifers. By far the largest and most evident group are the conifers, which reproduce by having separate male and female cones. From wind pollination to fertilization takes about 15 months in pine, with an additional year before mature seeds are ready to be released by the female cone.
7. The angiosperms are today's dominant vegetation, comprising some 235,000 species. They are thought to have evolved from a broad-leafed gymnosperm between 125 and 190 million years ago. Flowering plants occupy a wide range of ecological habitats and vary in size from a few centimeters to 100 m tall.
8. Flower parts evolved from leaflike organs, and several primitive flowers commonly have many of each floral part. More highly derived flowers have reduced numbers of floral organs, an inferior ovary, fused parts, and bilateral symmetry. The Magnoliaceae are considered to be one of the most primitive angiosperm families; the Orchidaceae and the Asteraceae are among the most highly evolved families. Fruit have also evolved more complex types.
9. Flowering plants and insects have coevolved, providing both groups with survival advantages. Originally beetles acted as the primary pollinators; now wasps, bees, butterflies, moths, ants, mosquitoes, birds, and bats all act as pollinators for an incredible array of flowering plants.
10. Beetles are clumsy, general visitors. Bees and wasps are highly specialized visitors attracted by complex shapes, colors (except red), markings, and even ultraviolet patterns. Butterflies are active in the daytime, visiting yellow, blue, and red flowers with thin nectar and sturdy corollas. Moths are nocturnal and prefer stronger-smelling light-colored flowers. Hummingbirds are the most common bird pollinators, visiting red, tubular flowers with lots of thin nectar. Flies, ants, and mosquitoes effectively pollinate a variety of flower types, including some that attract flies by their rotting meat smell. Bats visit large, nocturnal flowers with a strong perfume smell, thin nectar, and copious pollen.
11. Secondary plant compounds are thought to protect plants from herbivores and insects. Some of these compounds have strong physiological effects on humans. No other plant group has more utility to human society than the angiosperms.

Key Terms

alkaloids 368
angiosperms 351
annulus 349
conifers 351
cycads 351
determinate 358
Devonian 341
dioecious 355
fiddlehead 349
flavonoid 365
fronds 346
generative cell 353
ginkgos 351
gnetophytes 351
gymnosperms 340
homosporous 344
indeterminate 358
indusium 349
integument 351
lip cells 349
Lycophyta 341
microsporangia 353
microspores 353
microsporophylls 353
nucellus 351
ovule 351
ovuliferous scale 353
progymnosperms 351
prothallial cells 353
Psilophyta 344
Pterophyta 346
rhizoids 349
Rhyniophyta 341
seed ferns 340
Silurian 339
siphonogamy 352
sorus 349
Sphenophyta 345
sporangia 349
sporophyll 345
strobilus 345
synangium 344
telome theory 340
Trimerophytophyta 341
tube cell 353
vascular plants 339
Zosterophyllophyta 341

Discussion Questions

1. Construct an evolutionary tree, beginning with the most primitive, vascular land plant, showing the relationships among the following groups: trimerophytophyta, progymnosperms, ferns, lycopods, and gymnosperms.
2. Diagram a generalized life history of a non-seed-producing, homosporic plant. Is each gametophyte produced from a spore genetically similar or different? Explain.
3. Discuss the evolutionary significance of the following: vascular tissue, heterospory, siphonogamy, the seed, self-incompatibility.

Suggested Readings

Allen, K., and D. Briggs, eds. 1990. *Evolution and the fossil record.* Smithsonian Institution Press, Washington, D.C.

Barth, F. G. 1991. *Insects and flowers: The biology of a partnership.* Princeton University Press, Princeton.

Blackmore, S., and S. H. Barnes. 1991. *Pollen and spores: Patterns of diversification.* Oxford University Press, New York.

Cronquist, A. 1993. *The evolution and classification of flowering plants.* 2nd ed. New York Botanical Garden, New York.

Endress, P. K. 1987. The early evolution of the angiosperm flower. *Tree* 2(10):300–304.

Gifford, E. M., and A. S. Foster. 1989. *Morphology and evolution of vascular plants.* W. H. Freeman and Co., New York.

Gleason, H. A. 1991. *Manual of vascular plants of northeastern United States and adjacent Canada.* New York Botanical Garden, New York.

Graham, Linda E. 1993. *Origin of land plants.* John Wiley & Sons, Inc. New York.

Heywood, V. H., ed. 1993. *Flowering plants of the world.* Oxford University Press, New York.

Johri, B. M. 1982. *Experimental embryology of vascular plants.* Springer-Verlag, New York.

Jones, D. L. 1987. *Encyclopaedia of ferns: An introduction to ferns, their structure, biology, economic importance, cultivation and propagation.* British Museum of Natural History, London.

Jones, D. L. 1993. *Cycads of the world.* Smithsonian Institution Press, Washington, D.C.

Marshall, C., and J. Grace. 1992. *Fruit and seed production: Aspects of development, environmental physiology, and ecology.* Cambridge University Press, Cambridge.

Ottaviano, E., ed. 1992. *Angiosperm pollen and ovules.* Springer-Verlag, New York.

Remy, W., P. Gensel, and H. Hass. 1993. The gametophyte generation of some early Devonian land plants. *International Journal of Plant Science* 154(1):35–58.

Rudall, P. 1992. *Anatomy of flowering plants: An introduction to structure and development.* 2nd ed. Cambridge University Press, Cambridge.

Rudwick, J. S. 1992. *Scenes from deep time: Early pictorial representations of the prehistoric world.* University of Chicago Press, Chicago.

Thomson, K. S. 1990. Benjamin Franklin's lost tree. *American Scientist* 78:203–206.

Vakhrameev, V. A. 1991. *Jurassic and Cretaceous floras and climates of the earth.* Cambridge University Press, Cambridge.

Part Six

Plants and Society

The flower of the morning glory untwists as it opens.

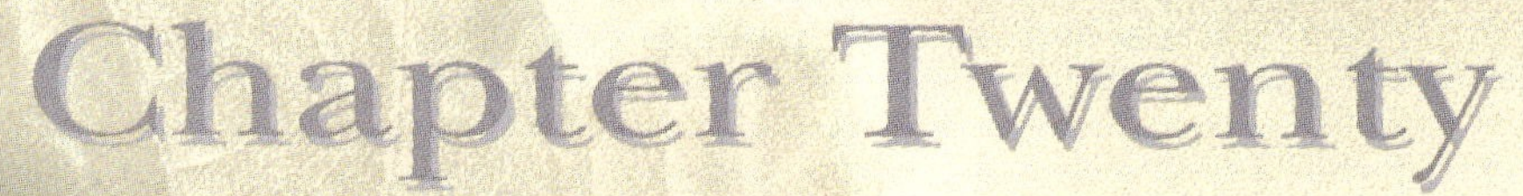

Chapter Twenty

The Roots of Culture and Modern Agriculture

Seeds of the cycad, *Zamia.*

This chapter takes a look at our historical roots—how humans evolved on this planet as hunters and gatherers, and how they successfully made the transition to organized agriculture. Even today, hunting and gathering tribes exist in several areas of the world. Their methods recreate a historical past rich in patterns of sharing, values, and sensitivity to fellow humans.

We look at the origins of agriculture, including theories on why it happened in various parts of the world at about the same time in history. Knowing the kinds of multipurpose crops that were grown by early cultures helps us understand the pressures for food and other products. Continued pressures for food have caused leaders throughout history to pursue new crops that might be moved to a new land. The breadfruit story of Mutiny on the Bounty *is an excellent example of the struggle for political and financial gain by moving crops to a new locale.*

The Industrial Revolution led to a new era in agriculture, which has grown and continued unabated. New biotechnologies seek to improve the quality and quantity of agricultural products while attempting to stay ahead of a burgeoning world population that seems not to recognize any limits.

Sometime during the past few million years, hominid creatures began walking upright. Because of their posture, we consider them the first human beings. It is believed, primarily from the anthropological studies conducted by Dr. Louis B. Leakey, his wife Mary, and their two sons, Richard and Charles, that the transition to upright stature occurred somewhere in Africa, since the fossil evidence comes from the Olduvai Gorge region in Tanzania. Like most other events in evolution, upright posture was probably achieved over a long period of time. Eventually it proved advantageous; the hands were available for manipulative functions instead of being used for four-legged movement. An upright posture also freed the mouth from grasping and carrying functions, which ultimately allowed for the development of speech.

Through the past several hundred thousand years these hominids have undergone tremendous **evolution,** including development of the ability to reason as a result of an increased brain size, changes in dexterity as the arms, hands, and fingers became more adapted, and modifications in skull features and dentition. Such changes are well documented by skeletal remains. However, for the period comprising the past 30,000 to 40,000 years (the advent of Cro-Magnon man) there is essentially no evidence of biological evolution. Certainly there has been remarkable cultural evolution, but no real changes in brain size, skeletal features, organs, or appendages. Even today human beings have vestigial organs, such as the tonsils and the appendix, that have no essential function. Perhaps they served some function in the past.

Population sizes increased in the Old World, and the fossil record indicates that early humans had spread widely by some 50,000 years ago. But what of the New World? When did humans arrive in the Americas, and how did they get here? The fossil record indicates that about 25,000 years ago there was a land bridge at the Bering Strait connecting North America with Asia. The earliest humans must have gradually migrated across this land, throughout North America, into Central America, and finally into South America. There is excellent evidence of early people in many sites in North America, including finds in the Southwest, which indicate that a so-called Clovis man, armed with carefully chipped stone weapons, roamed the plains about 12,000 to 14,000 years ago. Indian civilizations were well developed in Mexico and Central America some 5000 years ago, and the advanced Aztec and Mayan civilizations of more recent times are well known.

The advent of stationary human societies and the consequent development of civilization were possible only after the establishment of agriculture. Humans did not "put down their roots" and cease their nomadic wanderings until they learned to cultivate the land and grow crop plants for food and other needs on a 12-month basis. Thus agriculture provided release time; not everyone was needed to provide food, so other societal occupations developed.

HUNTING AND GATHERING

For over a million years humans obtained food by hunting wild animals and gathering plants. This was an endless activity throughout the longest period of human history, the Paleolithic or Old Stone Age, which is thought to have begun between 1 and 2 million years ago. During this time at least four separate glacial periods occurred; ice advanced over large areas of the northern hemisphere, later retreating during interglacial periods to occupy an area similar to the current ice caps. The most recent glacial period ended some 15,000 to 20,000 years ago. Currently we live in an interglacial warm period. **Hunting and gathering** societies existed throughout the Paleolithic period until as recently as 10,000 years ago in Europe and still exist in isolated groups in various parts of the world.

By necessity these Paleolithic societies were nomadic, constantly wandering in small family groups in search of game and edible plants. Their primary source of protein was meat; sugars and vitamins were provided by wild berries and other fruit; carbohydrate sources included various roots and grass seed rich in starch supplies; oils were provided by nuts. As the seasons produced changes in forage availability for the animals, causing them to move on, these nomadic people followed to hunt them and gather the fruit, roots, nuts, and other edible plant parts as they went. Modern hunters and gatherers often follow a complicated yearly cycle as the seasons change, moving to where food is available. This demands a thorough knowledge of the environment and resources. Such knowledge, developed over thousands of years, led to the domestication of plants and animals.

Twentieth Century Hunters and Gatherers

There are hunting and gathering cultures in Australia, South America, New Guinea, and Africa. At the present time about 5000 bushmen in the Kalahari Desert of southwest Africa still pursue the hunting-gathering, nomadic way of life. One group of approximately 1000 is the !Kung tribe, whose members live in the Nyae Nyae region (figure 20.1). Twenty-seven bands of this tribe continue the traditional existence, roaming through a region of about 10,000 square miles. Each band lives in a fairly well-defined territory and owns the wild plant foods and water resources, but not the land, within its territory. The primary food staples are the mongongo nut (*Ricinodendron rautaneii*) and morama bean (*Bauhinia esculenta*). Both plants produce a nutritious nut that contains a high concentration of protein, oil, and carbohydrate. Highly adapted to sandy soils and extensive droughts, both plants grow on sand dunes, usually at the top. The mongongo grows as a tree and the morama bean as a large vine. The site is inhospitable for most plant growth, and one might expect palatable plants to be collected to extinction. Instead, plants are protected and cared for with the same enthusiasm one might expect in the husbandry of domesticated animals.

Figure 20.1
The morama bean (*Bauhinia esculenta*) is a food staple of the !Kung tribesmen of the Kalahari Desert. The viny legume grows on the sand dunes and provides a high-protein bean as well as a starchy tuber, being dug here.

The !Kung know that their future depends on the survival of these important food resources, and they have learned to live with and respect the environment.

These hunter-gatherers live in small nomadic or seminomadic bands composed of loosely integrated families. They usually lack more organized kinship, such as clans and lineages. The family itself assumes the dominant organizational role, and there are no specialized economic, religious, or political groups. The groups are not closed societies; some people may leave when food is scarce and regroup for collective hunts, to share a water hole, or to participate in some important ceremony. Sometimes conflicts arise, and a member may be ostracized and have to seek a new band. Men do most of the hunting and women do most of the gathering. Hunter-gatherers have little control over the plants and animals on which they depend, and they have to adjust to seasonal and annual fluctuations in resources spread over wide areas. Bands migrate within their territories according to availability of food and water. The food resources are communal property, and all families have equal rights to them. Tribe members often have to protect their resources from the encroachment of strangers.

Plant resources are guarded by the family unit far more jealously than are the animal resources. Even though the morama bean dies back to the ground level each year, the women and children remember their location and return to the exact spot where the storage tuber will renew the plant's growth when the rains come. The group may have wandered over hundreds of miles since the last collection, but they manage to return at the correct time when a particular plant produces ripe seeds. They also seem to be acutely aware of the ecological principle that some seeds must be left to replenish the old plants that die.

The headman of the band determines its movements and decides when to visit each food resource area and whether the band should break up into smaller units. !Kung hunters do not confine themselves to territories, and game animals do not belong to anyone until they are shot. A hunting party usually consists of two to five men, usually relatives and friends; there is no formal leader, but the more experienced hunter usually makes the accepted suggestions concerning when and where to go. Hunters are the most interested in eland, kudu, gemsbok, wildebeest, hartebeest, and springbok. Hunting is done with poison arrows, and sometimes a dying animal must be tracked for several days. It is important to reach the dying animal before the predators arrive.

Distribution of meat follows an elaborate system not organized along family lines. The hunter who kills the animal is the owner of the kill and makes the first division of meat. Meat sharing is apparently important in such societies to mitigate tensions, reduce the fear of hunger, and instill an attitude of mutual cooperation.

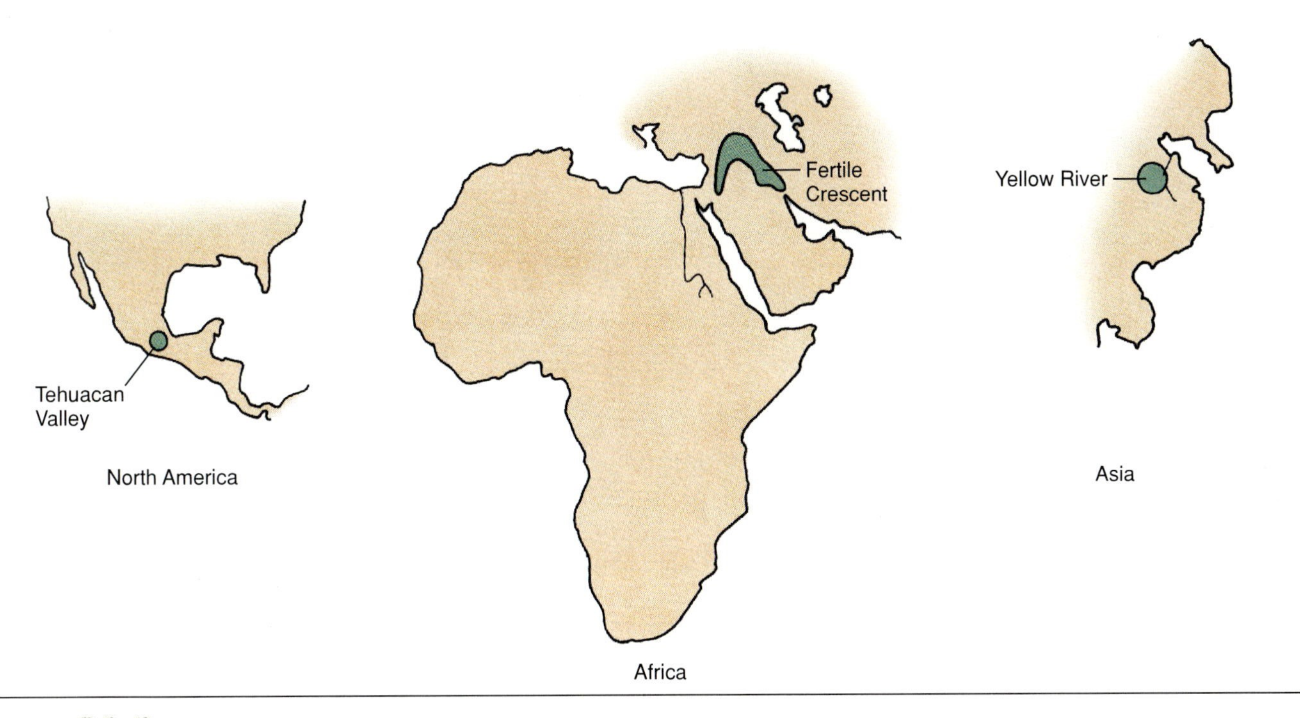

Figure 20.2
The beginnings of agriculture: the Fertile Crescent of the Tigris and Euphrates rivers of the Middle East, the Tehuacan Valley of Mexico, and the tropical area of the Yellow River of China.

It is interesting that these nomadic people, like those of Australia and the Eskimos, are totally committed to food sharing and gift giving. They do not collect artifacts as wealth, and personal possessions do not indicate status. A deemphasis of material wealth apparently helps them to survive in a harsh environment with a minimum of tension.

Although cultural anthropologists may argue that modern nomadic tribes do not duplicate the hunter-gatherers of ancestral *Homo sapiens,* there is reason to believe that many of the social activities are unchanged. The sense of resource preservation is uncanny. The nomads seem to know that conservation serves their best interests.

THE BEGINNINGS OF AGRICULTURE

Although much research remains to be done on the beginnings of agriculture, there is already a great deal of interesting knowledge in this area. The most popular theory is that agriculture originated earlier in temperate regions than it did in the tropics (figure 20.2). One idea to support this theory is that of *need:* there was no shortage of food and no need for shelter, clothing, or fire in the tropics, whereas all of these were concerns for those living in colder climates. More evidence for this theory is that the ancestors of many important domesticated crops are from the temperate regions. However, early agriculture in the tropics is difficult to document because the warm, moist conditions are not conducive to the preservation of archeological remains; thus we have more evidence from drier, colder regions on which to base our theories.

The Fertile Crescent

One such temperate area is near the Zagros Mountains of what is now Iraq. Located in the **Fertile Crescent** of the Middle East, formed by the Tigris and Euphrates rivers, the ancient city of Jarmo has been studied by archaeologists, who have documented the existence of agricultural activities there between 10,000 and 12,000 years ago. Jarmo is located in the foothills of the Zagros Mountains. Although it is hot and dry today, at the end of the last glacial period climates were cooler and wetter in that area, which made it ideal for agriculture.

Nomadic tribes migrated annually in the fall and winter out of the Zagros foothills to the nearby valleys, where forage and water availability lured the animals on which these people depended for meat and hides. These valleys also provided the plants needed for food, fiber, and fuel during the winter; they were not available at the higher elevations of the surrounding mountains.

In the spring, the animals slowly moved back toward the foothills of the mountains and spent the summer in these cooler elevations, where forage was plentiful for the animals as well as for the wandering nomads. This cycle was repeated year after year, and it is probable that the routes and temporary camps used by these nomadic people were the same on each trip. Each such camp had a designated trash dump area in which seeds and roots of the gathered plants were occasionally thrown. The composting of these dump areas produced rich nutrient conditions in which many of the plants sprouted and grew in greater densities than normally found in nature. As the people returned yearly to these areas, they must have gradually realized that using these plants near the campsite was more efficient than wandering far and wide gathering from wild populations. This realization could well have slowly evolved into experimentation by planting increased numbers of selected plants near the camp. Finally enough food could be grown in such a manner to last year-round, with only occasional hunting forays required to provide meat and hides.

Once these people limited their nomadic activities to one or two semipermanent annual camps and were increasingly dependent on cultivated plants, intensified agricultural activities developed fairly rapidly. The hazards that had been associated with a nomadic existence were reduced, while leisure time and population size increased as a result of this new sedentary lifestyle. This further intensified the need for adequate food supplies from cultivation. The domestication of animals followed, until finally villages such as Jarmo became firmly established.

In such villages the evolution from a nomadic existence to a stable one with a stored food supply allowed other advances. With a sedentary existence material goods could be accumulated, and potters, weavers, tanners, artisans, and scholars became important members of the community. Advanced civilizations rapidly evolved from such beginnings.

The Tehuacan Valley

We now know that agriculture developed independently in other areas of the world at a similar time in history. In the Tehuacan Valley of Mexico a parallel sequence of events led to an agricultural society dated by archaeologists also at 10,000 to 12,000 years ago.

The Tropics

An alternate theory of how agriculture may have first developed centers on the tropical regions of the world. Since these regions had ample supplies of water, year-round growing seasons, and warmth, there was less need to cultivate plants. Popularly termed the "genius" theory, this idea holds that since there were abundant food resources, free time was available for other pursuits; greater effort was applied to a consideration of planning, and subsequently the idea to concentrate some of the useful plants of the area close to the village soon developed. Because of the vegetative density of the tropics, this required clearing some land and planting seed or underground tubers and rhizomes. Since these areas would revegetate in a normal succession of plant species, a subsequent clearing of new areas every few years was necessary. This has been called the **slash and burn** technique, and because it allows the balance of nature to be reestablished, it has been a very successful small-scale agricultural practice in the tropics for thousands of years. How many thousands is not known because of the lack of preserved archeological remains to study and date.

The Yellow River of China runs through a tropical area in which agricultural activities are thought to have existed as long as 15,000 years ago. It is interesting that this area did not develop the complex societies that resulted from agricultural beginnings in the temperate areas of the Middle East and North America (Mexico).

(a)

(b)

Figure 20.3

(*a*) The coconut palm tree is a multipurpose plant thought to have been one of the earliest plants used by human populations. It is found worldwide in tropical climates and grows on sandy beaches, having been distributed by its buoyant fruit, which float readily in the ocean. (*b*) Palm trees are tall and have flexible trunks that bend and then return upright in the coastal winds.

EARLIEST CROP PLANTS

The period during which people first cultivated part or all of their plant food sources was almost certainly after the most recent glaciation. Cultivation arose independently in several different parts of the world. Until agriculture was used as the primary source of food supply in these early societies, the associated advances of civilization did not develop.

It is fairly certain that the first plants to be cultivated were those which had been gathered by these societies. We do know that every important civilization depended on cereal crops as the mainstay of their agricultural base. In addition, plants were more likely to have been cultivated if they had many uses or if they were abundant locally and easy to grow.

Multipurpose Plants

In the tropical regions the fruit of the coconut palm (*Cocos nucifera*) (figure 20.3) provides plentiful, nutritious,

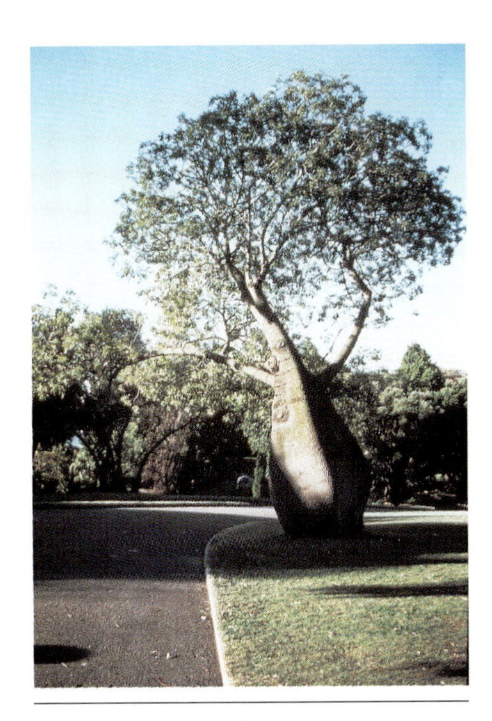

Figure 20.4
The baobab tree is a multipurpose plant native to Africa.

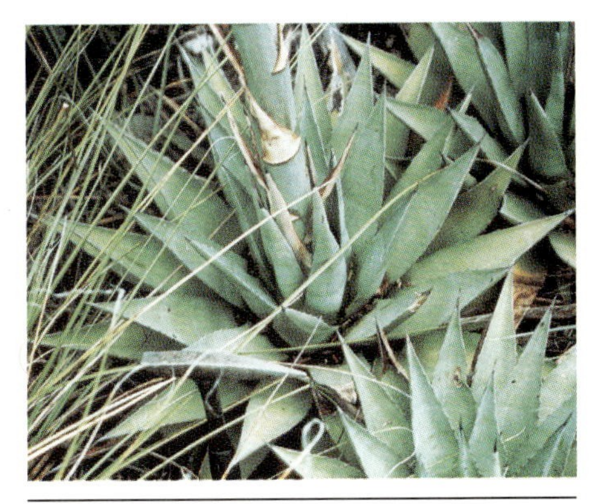

Figure 20.5
The *Agave*, or century plant, is still a highly used multipurpose plant in Mexico.

and bacteria-free "milk" and a solid flesh that can be eaten fresh or dried into copra, from which coconut oil can be extracted. The buoyant and watertight outer husks are useful containers, and the fiber of the husk can be made into rope and matting. Palm leaves are used for roof thatching on cottages, which are often constructed using palm trunks for the main supports. In addition, sap from the stem can be fermented into an alcoholic drink or evaporated to produce a sugar.

The mulberry tree (*Morus*) in China provides fruit for human consumption, leaves for silkworms to eat, and a beautiful yellow dye from the wood. In African savanna areas the baobab tree (*Adansonia digitata*) is also a **multipurpose plant** (figure 20.4). Its fruit contain seeds rich in oil, and the pulp of the fruit, high in vitamin C and tartaric acid, is made into a popular drink. The leaves have medicinal value, and the trunk provides fiber from which rope can be made. Even the trunks of old baobab trees are hollowed out for water storage during dry periods. The century plant (*Agave*) of Mexico yields fiber from its leaves and fluids from which several alcoholic drinks, including tequila, are made (figure 20.5). Locally, the "meat" of the developing flower stalk is a nutritious food source.

One of the best known of the multipurpose plants is the hemp plant (*Cannabis*). Requiring high levels of nitrogen, this plant might have originally grown, along with other weedy nitrophiles, in the nitrogen-rich rubbish piles of the nomadic camps. It was cultivated for fiber from the stems, oil from the seeds, and the medicinal properties of its leaves.

Root and Stem Crops

Because of the ease with which they may be cultivated and because of their high carbohydrate levels, plants with underground storage parts were probably early crop plants. Some of these had roots, and others stems, modified for carbohydrate storage, especially starch. Easily harvested by use of a digging stick, these edible underground parts could well have had their earliest agricultural beginnings in rubbish piles. Deliberate replanting of the leftover pieces of root or stem increased the already abundant food supply for these nomadic people. The use of such root crops in an early form of cultivation almost certainly existed in many different parts of the world.

The taro (*Colocasia*) from Asia and the similar tannia (*Xanthosma*) of the West Indies are both known to have been early root crops. They both have carbohydrate-rich corms, swollen underground stems with stem buds that can each produce a new plant. The Irish potato (*Solanum tuberosum*) is a modern crop tuber having buds (potato "eyes") that can be planted to produce new plants. Cassava (*Manihot ultissima*) is another tropical plant valued for its root; tapioca is made from this plant.

Although multipurpose plants and root crops had their place in the early development of plant cultivation, truly stable, nonnomadic, agriculturally dependent societies did not result from the exclusive use of such plants. The use of wild grass grains and the development of more highly productive domesticated strains were central to the ultimate development of agriculture in the great civilizations.

Cereals

As food plants, these members of the grass family have several desirable traits. Under cultivation they yield a large amount of grain per acre, and that grain—the single-seeded fruit—contains carbohydrates, minerals, fats, vitamins, and protein. The grains are compact and dry, which allows for long-term storage, or they can be ground into flour, which also stores well. Additionally, the grass stems (straw) can be woven or thatched into baskets, bedding, and housing.

Cereal plants can be encouraged to produce lateral shoots when the upper part of the plant stem is cut off. This process is called tillering and occurs when animals are grazed on young plants. Domestication of grazing animals not only provided these societies with food and skins, but it also increased the density of grain-producing cereal stems (tillers) for harvest. Cereals do well in plains areas near mountains of semiarid regions. According to archaeological records, millets (*Setaria, Echinochloa,* and *Panicum* species) were among the earliest cereal crops to have been cultivated in China. However, the crop around which advanced civilizations developed in China and Southeast Asia is rice (*Oryza*), a cereal plant of the west lowlands. In the Middle East wheat

Figure 20.6

A cross section of a chinampa shows how mud mixed with composting organic matter is put on top of the soil to increase fertility. Trees are planted along the edge of the chinampas and stakes are driven down below the canal bottom to stabilize the sides and prevent erosion.

(*Triticum*) was cultivated in the hilly regions of the Zagros Mountains, and was later introduced into the lower elevations of the river valleys after the development of irrigation techniques. Civilization had then spread from the mountainous highlands to the Tigris-Euphrates and Nile River valleys.

Maize (*Zea*) (Indian corn) is the cereal crop of the Americas, cultivated in the Tehuacan Valley of Mexico at least 8000 years ago. As in other areas of the world, sophisticated irrigation systems were developed that allowed cultivation to expand and yields to increase. By the time of Christ **chinampas** had been developed. These long, narrow strips of land bordered on three sides by irrigation canals yield several crops per year because there is no need to allow the land to lie fallow for part of each year. The rich muck of the canals is dredged each year and spread on the strips of land, thus replenishing the topsoil fertility. This maintains the canals while returning the rich nutrients to the soil (figure 20.6).

The chinampas system of irrigated agriculture was being practiced by the Mixtecs when the Aztecs conquered the region. The high productivity of this system was a major reason the Aztecs were able to dominate such a large region in such a short period of time. When the Spanish conquistadors arrived at Mexico City in 1519, they found the Aztec emperor was receiving 7000 tons of corn, 5000 tons of chilies, 4000 tons of beans, 3000 tons of cocoa, 2 million cotton cloaks, and several tons of gold, amber, and other valuables each year from his subjects. All this was possible because of advanced agricultural systems, which included the chinampas.

Other cereal crops that were cultivated include rye (*Secale*), oats (*Avena*), and sorghum (*Sorghum*) (figure 20.7). Rye was first developed as a secondary crop, growing as a weed among the primary crop species. Just as wheat was introduced from the Mediterranean region, so was rye; and since rye does better in colder climates than does wheat, it replaced wheat as the primary cereal crop in such areas and was grown even as far north as the Arctic Circle. Oats probably developed as a major crop plant in a similar way, because it is also tolerant of a wide variety of climates. Sorghum is known to have been cultivated in much of Africa, not only for its grain but also for the straw, which was used in the construction of walls and roofs of the village houses and for weaving baskets and sleeping mats.

The events discussed in this chapter occurred slowly over hundreds and even several thousands of years. The advent of agriculture, therefore, was truly a slow evolution from nomadic hunting and gathering to a stable society in which the components of civilization could develop.

Because of the much longer period that humans existed as small hunting and gathering groups, maybe as long as 2 million years, the relative suddenness of the beginnings of agriculture are often popularly referred to as the First Agricultural Revolution. Successive advances in agricultural technique and production have also been termed revolutions, and in fact a couple of such advances truly were revolutionary.

The need for continued agricultural advances stemmed from the combination of artificially selecting more productive strains of crop plants (but not necessarily more disease, climate, or insect resistant) and their continued intensification. These, combined with continued population growth, resulted in an exchange of crop plants around the world.

Figure 20.7

Wild and cultivated grasses, or cereal crop plants. Wild einkorn (A^1) and emmer (A^2) wheat with modern awn-less wheat (A^3). A^4, today's cultivated wheat (*Triticum aestivum*). Wild rice (B^1) and modern rice (B^2) are both *Oryza sativa*. Wild (C^1) and modern (C^2) barley are also the same species (*Horedum vulgaris*). Indian corn, or maize (D^1), compared with modern corn (*Zea mays*) (D^2).

TRADE ROUTES IN THE NEW WORLD

The discovery of America set into motion an unprecedented exchange of cultures and goods. Plants and animals were shipped between the Old and New Worlds, and some species managed to perform even better in their new environment than in their place of origin. As this exchange intensified, the capacity for food production expanded enormously. New **trade routes** developed throughout the world, and exotic plants and animals new to the "civilized" world became commonplace. The European settlers encountered a rather primitive agriculture among the American Indians, and they brought with them techniques that had led to agricultural advancements in densely populated Europe. Clearing of forests along the east coast of the United States, including the burning of debris and litter, resulted in fertile soils that supported the crops the immigrants had brought with them. Indians introduced them to a strange new crop called corn. The Europeans began to grow this nutritious new food and feed crop, and they gradually accepted its potential. (Even today, however, Europeans tend to find corn rather unpalatable and often refer to it as pig feed.)

Exotic plants from all parts of the world became commonplace although highly prized and expensive. Plant exploration became an essential part of government function, and many intriguing tales arose from the world travel of botanists and pseudobotanists. Coffee, tea, and spices were particularly prized in Europe, and exorbitant prices were paid for such commodities.

Mutiny on the Bounty

The attempted introduction of breadfruit (*Artocarpus communis*) into the tropical regions of the colonized world is one of the most interesting stories of plant introduction in all of history. Reports reaching Europe indicated that natives in the East Indies fed chiefly on the starchy breadfruit. This plant is sensitive to frost, but in the tropics it produces abundantly, some fruit weighing up to 10 kg. The bark of the tree is used for fiber, the wood can be used for shelter, the milky sap can be used for glue and caulking, and the seeds are high in protein. This seemed like an ideal plant for the British to introduce into the West Indies. In August of 1768, Sir Joseph Banks, Director of the Royal Botanic Gardens at Kew in London, set out on a trip around the world for scientific exploration, navigational charting, astronomy, and biological collecting. He sailed on the ship *Endeavour* with Captain James Cook and visited Tahiti, New Zealand, and Australia, where they ran aground. After a quick patch job, the ship limped to Jakarta, Indonesia, where nearly 50 of the ship's crew died of malaria and dysentery.

In a later exploration in 1787, Banks used his influence to have Captain William Bligh command a newly commissioned ship, the H.M.S. *Bounty,* and go to the East Indies to obtain breadfruit trees for the West Indies. Breadfruit was seen as an inexpensive staple food for the natives. The decision was made to load breadfruit seedlings at Tahiti, and accommodations were made to convert the deck into miniature greenhouses. Banks's letter to Bligh read: "The master and crew of the *Bounty* must not think it a grievance to give up the best accommodations." Captain Bligh was a tough disciplinarian.

During the time that the breadfruit seed were being germinated and potted up as young seedlings, the sailors were on extended shore leave. When these "softened" sailors reboarded and set sail, the plant seedlings received much better care and treatment than the sailors, causing a great deal of unhappiness. Only a few days out of Matavai Bay the crew mutinied, and Bligh, along with 17 loyal followers, was put overboard in a small boat. Somehow Bligh, the superb navigator, managed to get to the East Indies in just 47 days, a distance of 3600 miles. The mutineers threw the cargo of breadfruit seedlings overboard and sailed to an uncharted island near Tahiti. They sank the *Bounty* and lived out their self-imposed exile on Pitcairn Island.

In 1793, Captain Bligh commanded another ship, the *Providence,* and finally managed to deliver the breadfruit seedlings to Jamaica. A year later, George Washington heard of the success and wrote to acquire some plants to try at Mount Vernon. The tropical trees, of course, failed to survive.

✔ Concept Checks

1. What are the historical beginnings that led to human socialization?
2. Why is there reason to believe that agriculture originated in the temperate regions earlier than in the tropics?
3. What does *Mutiny on the Bounty* have to do with botany?

THE MECHANIZATION OF AGRICULTURE

Industrialization and the age of mechanization led to quantum leaps in the concentration of peoples; cities evolved rapidly along with new concepts in the production of material goods, communication, and transportation. This shift toward urbanization strained the existing food supplies and encouraged those in agriculture to find more efficient and productive means of supply to the cities (figure 20.8). This challenge was met with varying degrees of success, but eventually the size of towns and cities increased.

Figure 20.8
Rice paddy cultivation in Sri Lanka. Cultivation in many underdeveloped countries still follows the traditional methods.

Large cities developed in parts of the Old World thousands of years ago, and the Mayan civilization in Central and South America was highly structured in 5000 BC. Beans and squash were being cultivated in the New World 9000 years ago, and rice was cultivated in Thailand 12,000 years ago. By the time of the Industrial Revolution, agriculture was common throughout the world and established the basis for permanent settlement and civilization.

Food plants first domesticated—wheat, barley, rice, corn, and potatoes—remain today as staple crops. Only tomatoes and coffee among major food crops have been domesticated in the past 2000 years. Although there are as many as half a million species of plants in the world, only a few have been developed as major food crops. Farmers, traditionally reluctant to try new crops, have made up in quantity what they lacked in variety. Although surpluses were never large, farmers were able to produce more food than they required. Today, in industrialized nations all the food is produced by about 5% of the population.

FOOD COMMODITIES

Modern human diets depend heavily on the many forms of plant life. The grass seeds have given rise to cultivated cereals, most domestic animals are fed from cultivated crops (except for range animals), and fish continue to feed on aquatic plants of freshwater and the seas.

At least 90% of all human caloric intake is provided by commercially grown plants. Although meat continues to provide a large portion of the diet in countries that can afford it, the tendency is toward a greater consumption of plant products. Since energy losses increase in ascending trophic levels, it is possible to prevent these losses by eating "lower" on the chain, that is, closer to the producer organisms. In developing countries plant consumption has always been great (figure 20.9). In industrialized nations, however, the tendency toward meat consumption has long been associated with a higher standard of living. World food pressures now dictate even to the most developed nations that such eating patterns are no longer acceptable. Americans are rapidly learning that only a portion of what a cow, pig, or chicken consumes is converted into milk, meat, or eggs. A major portion of the feed consumed sustains metabolic function, bone, skin, and other body organs unimportant to humans. It is necessary to consider, however, that animals do convert a great deal of nonedible and nonharvestable plant material into important productivity.

At the present time three grain crops—wheat, rice, and corn—provide about 80% of the total human calorie consumption (along with potatoes, yams, and cassava). Although rich in carbohydrates, these crops are not particularly high in protein or fat. Diets must be supplemented with other foods to achieve some degree of nutritional balance. Vitamins and minerals are usually supplied through vegetables. Even though poor in protein, these six crops do provide considerable protein simply because they are consumed in large quantities. In addition, beans are an excellent source of protein, and peanut, soybean, and sunflower contain both protein and fat. Sugarcane and sugar beets were, until recently, chiefly a localized food source, but they have become major world crops as a result of better transportation (table 20.1).

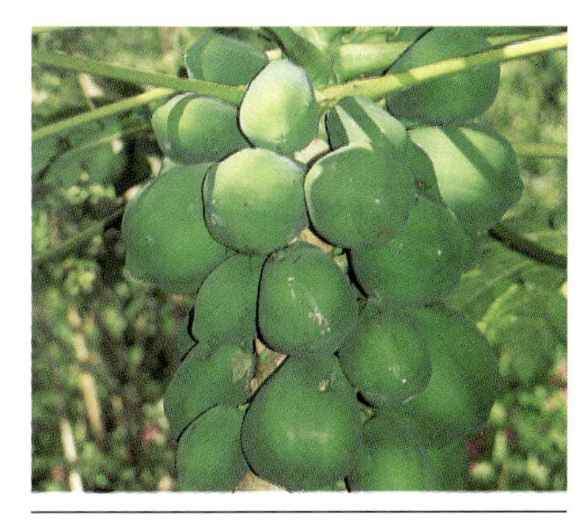

Figure 20.9
Papaya is an important tropical fruit in many regions, yet people from temperate areas often find its taste or consistency unacceptable. Human tastes change slowly, and many nutritious food sources are never accepted by a major segment of the population.

TABLE 20.1
Major World Food Plants

Grains	Wheat Rice Corn Sorghum Millet Barley Oats Rye
Tuber and Root Crops	Potato Sweet potato Yam Cassava (manioc)
Sugar Crops	Sugarcane Sugar beets
Protein Seeds	Beans Peas Soybeans Lentils
Oil Seeds	Olive Soybean Peanut Coconut Sunflower Corn
Fruit and Berries	Citrus Mango Banana Apple
Vegetables (including fruit eaten as vegetables)	Cabbage Squash Onion

A few other grain crops are important food sources. Barley, millet, sorghum, oats, and rye are consumed throughout the world. The world food consumption, however, is based on production of only a few species of the myriad possibilities available in the botanical world. It certainly is true that humans have tried and rejected many potential food plants because they were unpalatable, nonnutritious, or toxic. Others were too difficult to cultivate, harvest, transport, or store. On the other hand, human beings are exceedingly conservative when it comes to trying new foods. Most of us reject new foods because we don't like the look, smell, or feel. Nutritional value is seldom a consideration. Religious and social customs often prohibit certain foods, as do habits learned in early childhood. There are hundreds of nutritious food choices available to the human population that have not been adopted. "Experimentalists" try new foods or new methods of preparation everyday, but most discoveries fail to obtain general acceptance.

Specialty crops, such as strawberries, onions, and even mushrooms, are a recent phenomenon and contribute little to the total human diet; likewise, spices add to the palatability of foods but contribute little to human nutrition.

A major portion of the world's population can be divided into those who eat rice and those who eat wheat (figures 20.10 and 20.11). Rice is grown in the warmer parts of the world, generally in regions in which predictable monsoons sweep the countryside. In a year without the monsoons, severe famine can affect localized regions, although modern methods of food transportation, distribution, and storage have alleviated much of this problem. In Asia, rice comprises 75% to 85% of the entire diet. Rice has become a major agricultural crop in the Sacramento Valley of California and along the Gulf coast in Texas and Louisiana. Rice, however, has never become a major part of the American diet, and most American rice production is for export.

Wheat is the dominant cereal crop in most of the temperate or cooler regions of the world, and its production extends into the subtropics and higher altitudes of the tropics. Before the development of modern plant breeding, wheat had little cold tolerance and was confined to the warmer regions of the globe. During this century, however, geneticists have succeeded in breeding cold tolerance into the crop, and now it is grown in the most extreme latitudes.

Wheat and rye are considered the bread grains, because most production goes into flour. Although rye was a major crop a century ago in the United States, today almost all production is centered in Europe and the former Soviet Union. Regional food patterns change over time, and crop production must mirror these changes or fail economically.

The typical bread wheats are hexaploid (six sets of chromosomes in somatic cells), whereas the macaroni and noodle wheats of Europe and Asia are tetraploid (four sets of chromosomes). Wheat is grown both as a winter annual (planted in the fall) and as a spring annual (planted in the spring). The genetics of the cultivated varieties are different, the winter wheat containing genes for a great deal more cold tolerance. In recent years, livestock that graze on winter wheat have become a major source of income, sometimes exceeding the

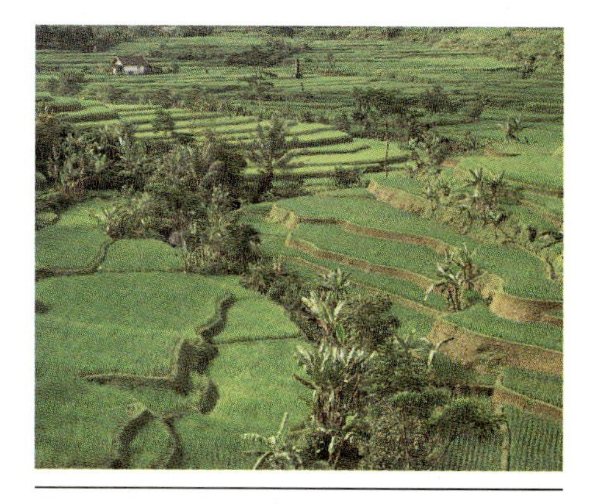

Figure 20.10
Almost half the world's population depends on rice as its primary food source. Shown are rice paddies in Indonesia.

(a)

(b)

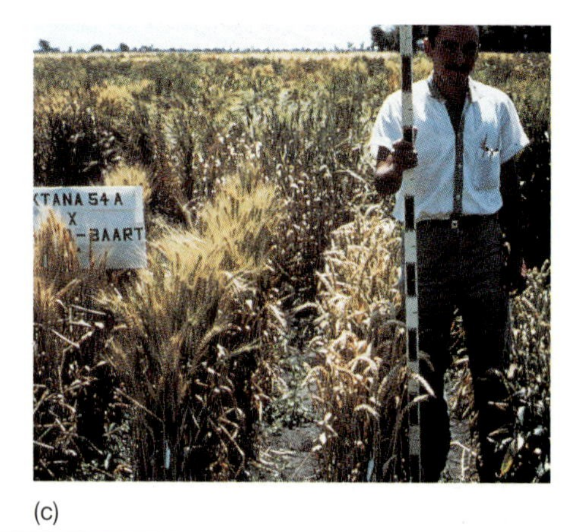

(c)

Figure 20.11
(*a*) Wheat is the dominant cereal crop in the temperate climates of the world. (*b*) This Mexican hacienda is surrounded by granaries once filled with wheat but now abandoned due to desertification. (*c*) New wheat varieties developed by the Rockefeller Foundation give hope for developing nations, including Mexico. This research led to the green revolution.

value of the grain crop. Winter wheat is an excellent source of green forage during the winter, and the grazing forces the plants to tiller (give off more stems) and produce more heads of grain.

In regions that are too warm and dry for the cool-season cereals, sorghum and millet are major food crops. Sorghum is particularly important in northern China, India, tropical Africa, and the southern United States (figure 20.12). Sorghum flour and millet flours are not considered good for baking, although they are eaten in many countries and yield a fermentation substrate for making beer. Sorghum consumption as food has never been high in the United States, and it is currently used as a feed for domestic livestock. The protein content of sorghum is higher than in most cereals, and because it can survive low rainfall, high temperatures, and alkaline soils, its popularity is increasing dramatically worldwide.

Corn was the food staple of the Incas, Mayans, and Aztecs, and today it remains a very important dietary component in the Americas. Other countries are beginning to accept corn as a food crop—slowly. It is well

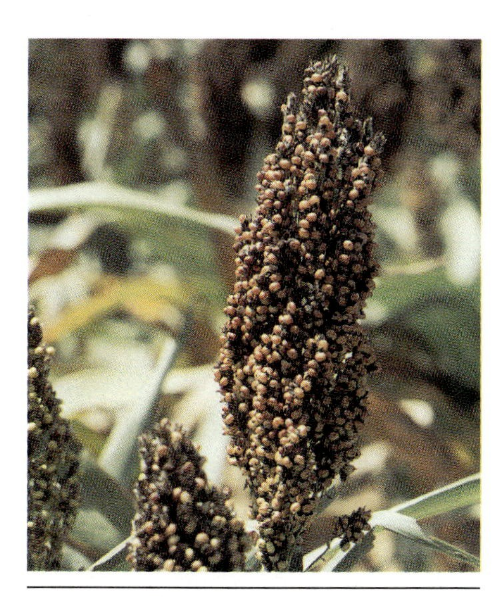

Figure 20.12
Sorghum is an important crop in the semiarid regions. Although often used as a human food source in many parts of the world, it has never gained acceptance in the United States. Instead, it is used domestically as livestock feed or exported for food.

Figure 20.13
Corn (*Zea mays*) is a cereal crop of the New World, long known to the civilization of Central and South America. Its acceptance in the Old World has been very slow.

adapted to warm, humid regions, and ideal production is achieved even in extreme latitudes during the warm summer. Corn requires more water than many parts of the world have to offer, and it is grown in those regions only with supplemental irrigation (figure 20.13).

Other grain crops contribute to total world consumption, although rice, wheat, and corn are certainly the most significant. Barley was formerly used extensively as a bread grain, but now about half the entire world production goes into the making of beer. Oats and buckwheat enjoy regional consumption but contribute little to total world caloric intake.

A synthetic grain crop has been introduced during the past few years. Plant breeders have hybridized wheat (*Triticum*) and rye (*Secale*) to form *Triticale,* which is reported to have the best characteristics of both parents, including a higher protein content than bread wheat and excellent baking properties. *Triticale* has been widely accepted in the former Soviet Union, but its success in the United States has been limited. The difference in quality and taste of *Triticale* bread does not seem to justify the development costs still attached to *Triticale* production. Additional protein content seems to be a minor consideration for the typical overfed American.

HISTORICAL PERSPECTIVE

The food situation on earth is now and always has been precarious. Approximately two-thirds of the world's population knows poverty and hunger (figure 20.14); in contrast, about 10% lives in relative affluence. The other 25% lives somewhere in between, never quite sure whether the supplies of food, shelter, and fuel will be adequate. The uncertainties are the same as they were for the first *Homo sapiens,* and the population is larger. Modern humans still depend largely on the weather and other interactions within the biosphere to ensure agricultural success. Insect and disease control is a major problem, and we have surprisingly little appreciation of how organisms interact in nature.

Throughout recorded history agricultural productivity has increased, but concomitant rises in population have always placed food supply at a dangerously close intersection with demand. Hunger is an ever present and prominent human condition. The stress of famine can induce strange behavior, such as hoarding, stealing, selling children, eating clay, diseased rodents, and bones, murder, suicide, and even cannibalism. Early accounts of the Roman famine of 436 BC and the Indian famine of 1291 AD report that thousands of people drowned

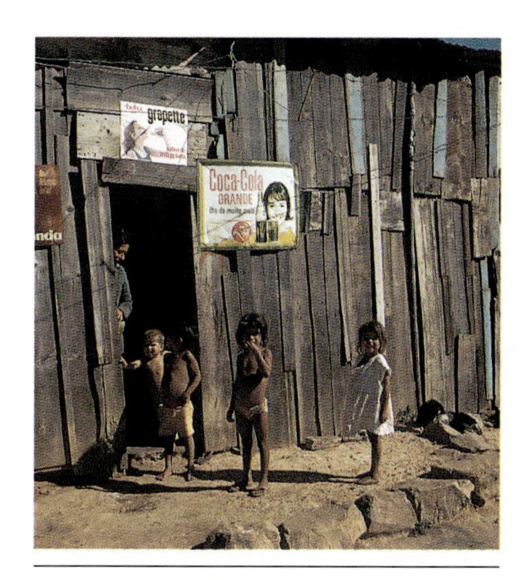

Figure 20.14
About two-thirds of the world's population lives with poverty, hunger, and disease.

themselves in rivers. Cannibalism is reported to have occurred in at least 15 famines in England, Scotland, Ireland, Italy, Egypt, and China. As recently as 1921, Russian cemeteries had to be guarded to ensure that hungry thieves did not steal recently buried bodies. Famines were often localized and involved only a few thousand people, but the great Indian famine of 1769 to 1770 cost the lives of approximately 10 million. Recent Indian famines claimed 1 million lives in 1866, 1.5 million in 1869, 5 million from 1876 to 1878, and 1 million in 1900. In China, between 9 and 13 million died in the famine years of 1877 to 1879. In Asia and on the Indian subcontinent, famine has been most often seen because of the unpredictable monsoon rains.

The **Irish potato famine** of 1846 and 1847 caused the death of 1.5 million people and prompted a massive emigration to the United States. Before the potato famine Ireland's population was 8 million; the current population is slightly more than 4 million, primarily a result of the long-term effects of the famine.

Through modern transportation and communication efforts we can anticipate famine and deliver food, supplies, and education materials to relieve its effects. On the other hand, massive aid is not always possible in other than localized situations. War activities often inhibit the flow of food to areas that need it most. Since World War II, for example, supply problems related to war caused the starvation of 2 to 4 million in the Bengal region of India. During 1969 to 1970, civil war in Biafra (eastern Nigeria) led to several hundred thousand deaths by starvation. More recently, periodic drought that sweeps the Sahel region of Africa continues to produce thousands of deaths. In spite of a worldwide campaign to send massive food and monetary aid to the region, there is little relief from the situation until the rains come, and by that time the ecological destruction is enormous.

Various industrialized nations have attempted to buffer the effects of malnutrition and starvation (figure 20.15). The United States contributes surplus food for approximately 100 million people per year. Beginning with the enactment of Public Law 480 in 1954, the United States began selling food to poor countries for payment in local currencies rather than in dollars and gold.

During the early 1960s American grain surpluses reached high levels, which were depleted after P.L. 480 came into existence. Land that had been idle was called back into production, and from 1966 to 1967 20% of the entire U.S. wheat crop was required to reduce famine in India. During the 1980s, American farmers were asked to reduce production because of temporary surpluses.

One of the social complications of a give-away food program is the false confidence instilled in the recipients concerning their future. Unless food is given out only in the case of a temporary disaster, countries (and individuals) tend to become reliant on the contribution and finally believe that the supply is inexhaustible. Many leaders believe that such false confidence merely results in further population increases and delays the day when food will not be available. Keeping humans alive at a subsistence level so that they may produce yet more offspring raises serious moral concerns about the future of such populations.

All these examples serve to point out the precarious position of world food supplies. Some sources contend that the biosphere has the ability to produce all the food needed for the foreseeable future. Others contend that the level of human population has already surpassed the ability of the world to provide food. The truth probably lies somewhere in between. New sources of inexpensive energy could improve food production greatly, but such energy sources seem many years away. While we wait, the world's population continues to double at an alarming rate.

Figure 20.15
Grain surpluses from the food-exporting nations feed the malnourished in many countries.

THE NEW AGRICULTURE

There is no question that agriculture has progressed steadily ever since the glimmer of domestication first touched the hunters and gatherers. Some periods of history have seen rapid progress; others have been slow. The tillage of rice paddies in Asia, for example, has been highly developed for thousands of years. Control of erosion by terracing has always been a standard practice. In some parts of the world, including Central America and Mexico, elaborate irrigation schemes ensured that water reached the crops at the right time.

In spite of these early advances, it was not until the Industrial Revolution that machinery became available for large-scale agriculture, which relieved a large portion of the population from the drudgeries of hand labor. The continued growth of mechanization has increased productivity in almost all crops. Some crops are still labor intensive (for example, pineapples), and mechanization has only partially alleviated the need for hand labor (figure 20.16). As recently as 1980,

(a)

(b)

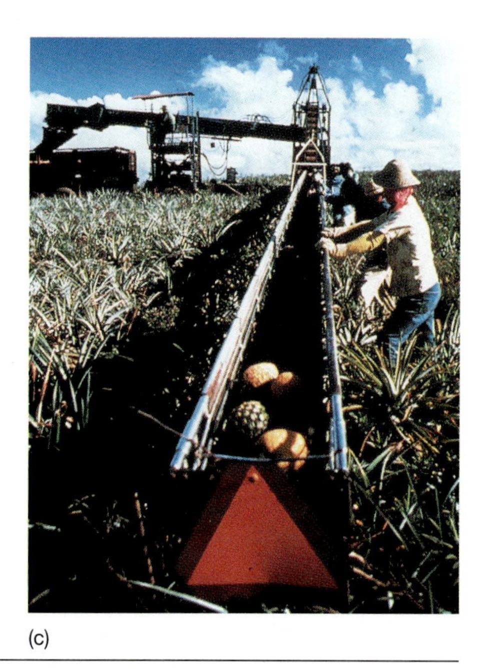
(c)

Figure 20.16
Crops such as pineapple are labor intensive, and a great deal of energy is devoted to production. (*a*) A field of young pineapple plants. (*b*) A mature fruit. (*c*) The harvesting operation in Oahu, Hawaii.

some portions of the labor movement in the United States have even advocated a slowdown in mechanization so that farm laborers will not be put out of work. This opposition has reached the point of filing legal restraining orders for the invention, development, and production of mechanized farm equipment. It would appear that American agriculture has done its job too well, and the consequence is social disagreement about priorities.

At the turn of the century, plant breeders and other agricultural and biological scientists began to accomplish impressive gains in crop yields. Productivity has accelerated during the past 80 years, and some yield increases have been truly remarkable. The "North American Breadbasket" was made possible through research and technology.

Basic Research

The elucidation of metabolic processes, including photosynthesis, respiration, and protein synthesis, is only one area of discovery that has led to increased productivity. An understanding of water and soils has improved conservation practices and irrigation methodology, and genetic research has led to a process of genotype selection to match specific crops with specific environments. Tremendous advances have been made in growth regulation with plant hormones, and the understanding of how diseases are transmitted has increased yields. Investigation of viruses and their transmission by insects has also been significant (figures 20.17 and 20.18).

Figure 20.17
Basic laboratory and greenhouse research provides a framework of understanding how plants function and how productivity can be increased.

Figure 20.18
Plants are sometimes grown in sterile-tissue culture so that all competition, including insects and disease, can be eliminated and environmental factors and nutrition can be studied individually.

Applied Research

Once basic concepts have been elucidated, the processes must be put to the test under field conditions (figure 20.19). Not all hypotheses born in a research laboratory survive the rigors of field testing. Some genotype screening programs may look excellent in laboratory/greenhouse studies, but in the field they sometimes fail because

Figure 20.19
Replicated field plots account for differences in soil type, nutrition, and other factors that can influence growth and development.

of susceptibility to insects or disease, low humidity, wind, soil chemistry, weed competition, and other factors. Only those gene sources that help the plant withstand the rigors of field competition have a chance of making a contribution to world agriculture.

Technology Transfer

Once innovative ideas have proven workable, the task of transferring the information to farmers is monumental. Many good ideas of the past have been filed away in a research paper and have never been put to use. Even with modern communication and impressive sales campaigns through the media, selling farmers on a new concept, new crop, or even a new variety of the same crop may be difficult. Tradition plays a major role in farming practices and procedures in all societies. Change comes slowly unless **technology transfer** concepts are applied judiciously (figures 20.20 and 20.21).

 *Concept Checks*

1. What are the most important cereal crops? Why do you think this is the case?
2. Why do you think that world food supplies are always on the verge of failing to meet requirements?
3. What is the difference between basic and applied research?

Figure 20.20
Cereal breeding, including hybrid corn, is accomplished in field plots like this one. This scientist is placing a paper bag over the tassel, or male flowers.

Perhaps the most impressive example of technology transfer in American agriculture is the **land-grant college system,** together with the state agricultural experiment stations and their accompanying agricultural extension service. This highly organized sector of the agricultural community managed to "sell" American farmers on innovative technology early in the twentieth century, using the demonstration plot to show that new methods and crops were superior to old ones. Farmers are impressed by success across the fence. Regional experiment stations have brought experimental farms close to every American farmer, and through the use of open houses, field days, and cooperative research they have been instrumental in communicating new farming practices. Other developed countries have similar systems with varying degrees of success. In recent years, extension systems have been tried in many developing countries, but lack of capital and poor communication have prevented large-scale success. Similar demonstration techniques have been used to sell products through home demonstration, communicate social and health services, and reach the industrial community through engineering experiment stations.

Figure 20.21
Aerial spraying of pesticides has been a major factor in increased productivity with high-technology agriculture. Application costs are often lower than with ground application, and there is no detrimental mechanical disturbance of the plants.

At the present time, mass communication makes every corner of the United States accessible. New varieties of crops are quickly recognized throughout the country, and fertilizers and feed are advertised widely. Seed companies, pesticide companies, and irrigation equipment companies spend a large portion of their budget on public relations and advertising. The campaign is unquestionably effective.

Sometimes farmers are eager to accept new ideas, but financial reserves and capital investment inhibit the progress of new technology. Lending institutions are often reluctant to take chances on innovative ideas, and the wheels of progress may move very slowly indeed. Capital investment is particularly critical in the developing countries, where lack of funds to buy seed or fertilizer may stand between poverty and a high level of productivity.

Agricultural Infrastructure

Much of the **agricultural infrastructure** is devoted to the processing, storing, packaging, and transporting of farm products. Perishability is a major factor in food production and delivery.

Perhaps some of the reasons for large-scale success of grain products are their slow perishability and ease of storage and transport. The industry is far better equipped for long-distance transport of a truckload of wheat than for a truckload of lettuce or strawberries.

Basic research has been instrumental in developing techniques to improve the processing and shipping of plant products. Consider, for example, the problem of maintaining fresh corn or pea quality. Harvesting affects sugar content, and the best produce is obtained if harvesting can be accomplished within a matter of hours after the highest sugar content is reached. The crop must be processed immediately because respiration rates are so high that even minor biochemical changes can affect flavor, texture, and overall quality.

Perhaps the most progress in storage techniques concerns apples. Recent advances in postharvest physiology have been so dramatic that a controlled atmosphere prolongs the storage life of apples up to an entire year with little deterioration of quality. Scientists have spent many years of effort in plant biochemistry, physiology, and genetics. The consequence of such storage and transportation techniques is a more stable market price and therefore a better pricing structure for both grower and consumer.

Even air freight is used to transport certain food products. In the United States some specialty crops are transported by air, but most shipment is by truck or rail. Some undeveloped countries have such poor road and rail systems that air is essentially the only effective means of transportation. In such countries many agricultural products are shipped in this manner, although the expense is very great.

THE GREEN REVOLUTION

The combination of scientific disciplines into teams (as in the postharvest storage of apples) has accelerated the research process. New varieties of rice and wheat introduced by research organizations have been readily accepted and in a few short years have led to tremendous increases in yield. An interdisciplinary team of researchers in Mexico, led by geneticist Dr. Norman Borlaug, has introduced several new varieties of wheat that far outproduce any of the existing varieties, provided that water and fertilizer are present in abundance. In parts of the world where these inputs are available, increases in productivity have been astounding. The program began in 1944, and within 20 years Mexico changed from a wheat importer to a wheat exporter. Since 1950, wheat production in Mexico has more than quadrupled. India and Pakistan, among other wheat-growing nations, have made similar gains. This dramatic change in worldwide productivity has included other cereal crops, primarily rice. The effort has come to be known as the **green revolution.** Dr. Borlaug won the Nobel Peace Prize in 1970, not so much for the technical aspects leading to the new varieties, but for his humanitarian efforts toward feeding a hungry world.

Despite the successes associated with the green revolution, critics have pointed out that the gains have been made only with tremendous energy inputs. The new varieties fail to perform in regions where irrigation or fertilizers are not available. Some have contended that only large corporate farms will be able to afford the massive inputs required to bring about the productivity, and small landowners will be forced out of business.

Few would question the success of the green revolution, and scientists surely will continue with the breeding of new, high-yielding varieties for specific crops in specific regions. At the same time, however, agricultural scientists recognize the need for improved production in those parts of the world where water and fertilizer are in short supply. Recently, similar efforts in breeding crops for arid regions and parts of the world without benefit of massive inputs of plant nutrients have begun. Breeding and selection for specific environmental conditions are sound ecological principles that allow the human interest to coexist with the environmental stability.

At the time of the 1973 OPEC oil embargo the developed countries were enjoying excellent productivity. Then came the crash: production of fertilizers from fossil fuels no longer looked so attractive in the face of skyrocketing prices, and the energy costs of pumping agricultural water became almost prohibitive. What had seemed like a dream of plenty was suddenly not cost effective, and the developed and developing countries alike were forced to reconsider the energy cost of all processes, including agriculture. Although the assumption had always been made that agriculture was energy efficient, it soon became apparent that advancing technology had led all of us so far afield from sound ecological management that serious environmental degradation was taking place. Important gains in yield were countered by surface erosion, buildup of soil salinity through irrigation with poor quality water, and buildup of pesticides in the soil and atmosphere.

The exploding world population had forced the hand of intensive agriculture to the point that maximizing productivity in a monoculture was the only acceptable way to maintain productivity. Even quality had to be perfect: the American consumer demanded and got blemish-free fruit and vegetables; the cost was tremendous application of insecticides, fungicides, bactericides, and herbicides. When the costs were added, it became clear that intensive, American-style agriculture was no longer a cost-effective method of production. The approach since the beginning of mechanized agriculture had been to select desirable crops and then tailor the environment accordingly. As long as water, fertilizer, pesticides, shading, supplemental light, enriched CO_2, heating, cooling, and other factors were inexpensive, only the genetic constraints determined the upper limits of crop production. Suddenly the cost of all these factors increased in direct proportion to the cost of fossil fuels. The farmer's price for agricultural commodities failed to keep abreast of inflation, and many farmers were forced out of agriculture.

The world population explosion is unabated, but productivity and food supplies are being carefully analyzed in cost/benefit considerations. Each bushel of wheat, rice, or corn is scrutinized to determine whether the cost justifies the product. Whether we like it or not, the cost of food production is tied closely to the costs of energy. This situation is complicated even more by talk of shifting prime food-producing land into land for energy crops, usually high-carbohydrate species that can be converted into alcohol through fermentation. To convert an acre of prime food-corn land into an acre of corn for alcohol ties the cost of farm products directly to the price of oil and gas. The farmers would then be in the business of producing energy, and additional farmland would be lost from production. This practice would appear to set a dangerous precedent, and various economists have warned against such an approach.

The Farmer's Bargaining Power

Whether a farmer can afford to maximize yields depends to a large extent on the price of a product. Although food prices have escalated dramatically throughout the world, most of that profit has been realized not by the farmer, but by the processor, packager, distributor, and transporter. Farmers traditionally have been so poorly organized as a political group that they fail to exert a major influence in the political arena or at the marketplace. They usually take the going offer, although modern storage facilities for some farm products allow the farmer to withhold from the market until a better price is offered. Producers of perishable products do not have this luxury, except through frozen foods, and for them the energy costs of withholding from the market are high.

The Limits of Productivity

The question is often asked whether farmers can continue indefinitely to produce as much food as the world requires. Throughout recorded history this has seldom been possible. Recall that the laws of thermodynamics predict that less energy comes out of a system than is put into it, that is, no system is perfectly efficient. The conversion of light energy into chemical energy is relatively inefficient.

The total sunlight available at any point on the earth is readily determined. Green plants capture only about 1% of that light energy on an annual basis, and much of that energy falls on land or water without green plants. Even if plants are properly spaced to maximize light trapping, the efficiency of conversion is low. Most of the energy is reradiated or lost to the system as heat. Thus, strictly from a thermodynamic point of view, the light energy available determines the upper limits of productivity. Seldom, however, in commercial agriculture is light the **limiting factor** in production. On a worldwide basis the limiting factors are unquestionably water and nutrition. Most of the world's so-called arable land is now in cultivation. The only land left is in the humid tropics, the desert, or mountainous regions too steep to cultivate safely. Although a great deal of land is available in the arid and semiarid parts of the world (somewhere between 25% and 33% of the entire land surface), water is the limiting factor in development.

Much of the North American breadbasket land is under irrigation (figure 20.22). In some cases irrigation comes from surface waters, and in others underground water must be pumped to the surface and then transported to the site of application. These pumps are usually located at the individual farms, and pumping may be done at very shallow depths, which requires very little energy, to hundreds of meters, which requires a great amount of energy. Many crops grown in arid regions require 2 to 5 acre-feet of water per year (1 acre-foot of water = 1 acre of water 1 foot deep, or approximately 325,000 gallons). In some regions this accounts for 90% of the entire water consumption, with all domestic and industrial uses requiring only 10%. Therefore irrigated agriculture comes under careful scrutiny in matters of water conservation.

You will recall from chapter 12 that transpiration accounts for about 99% of a plant's demand, and there are very few realistic suggestions for reducing that water loss. Anything applied to the plant to reduce water loss either seals off the stomata or causes them to close, and if the stomata are closed, no CO_2 can enter, and photosynthesis is greatly reduced. The correct approach appears to be to grow more water-efficient plants, those which produce more biomass while using less water. Sorghum, for example, is far more water efficient than corn if water is limiting. If water is not limiting, then corn is more efficient (figure 20.23).

The Future of Irrigation

The arid and semiarid parts of the western United States typify similar regions of the world. In some states **irrigation** has been developed to a high degree of sophistication with tremendous capital investment. In some cases irrigation water is transported hundreds of miles, as from the Feather River in northern California to the dry southern California deserts. A similar scheme has been proposed to transport water from the Arkansas River to the farmlands of western Texas and eastern New Mexico.

Quality is always a concern in the transport of water, because irrigation water always contains some salts. In some cases more than a ton of salt per

✔ *Concept Checks*

1. Why has the American agricultural extension system been so effective?
2. What is the "green revolution" and what do the critics have to say about its importance?
3. What ultimately determines the limit of plant productivity?

Figure 20.22

Sprinkler irrigation has allowed farming on steeper slopes than is often possible with furrow or flood irrigation. Although considerable water is lost to evaporation, new techniques and equipment apply the water closer to the ground.

acre-foot is solubilized in the water. As the water is stored and absorbed by the plant roots, the salt is left behind in the soil or the plant or in some cases leached beyond the root zone. Salt accumulated over a period of years is a major pollutant, and great efforts are made to drain the soil and leach the salts. The Imperial Valley in California is lined with a massive drainage system, and some experts predict that the agricultural system there cannot survive much longer. This region produces a great percentage of the nation's winter vegetables.

In the central part of the United States a massive underground aquifer holds water stored there for millions of years that is being pumped to the surface for irrigation of the Great Plains. The Ogallala Aquifer stretches from South Dakota to Texas and has allowed the high plains of Texas to contribute significantly to the nation's supply of food and cotton for export. Unfortunately, this aquifer is not being recharged, and the "fossil water" is being used up. Some areas have already had wells go dry, and others are becoming depleted. Depending on the locale, southern parts of the aquifer will no longer be a source of irrigation in the future. The suggestion has been made that interbasin transfer of water be accomplished from regions with a surplus of surface

Figure 20.23

Productivity is limited in many parts of the world because of a water deficit at some time during the growing season. Corn is relatively sensitive to water stress, and drought is a major factor in limiting corn production in dry regions.

water. Many proposals have been made for importing water from Canada, the Missouri River, the Arkansas River, and even from the mouth of the Mississippi River at New Orleans. While technically feasible, the costs of construction and maintenance of such a system are astronomical. The energy costs of pumping alone would be more than the states could afford. Even if such a system were to be approved by voters, legal ramifications would probably disallow construction. The Constitution of the United States relegates the water rights to the individual states, and it seems unlikely that one state would ever be willing to sell its water to another. The projected costs of water delivered to the farmer are absolutely prohibitive for the irrigation of crops.

The future of irrigated agriculture, then, seems dimmed by accurate cost/benefit analyses (figure 20.24). In those regions where water delivery is inexpensive, readily available, and dependable, irrigated agriculture is likely to persist. A recently introduced concept, deficit irrigation, implies that irrigation is to be used only in emergency situations to save a crop or to get over a period of particularly bad weather conditions. Although yields would not be maximized, decreased energy costs of production would offset the loss of yield, and the farmer would make just as much revenue.

In regions that have been irrigated but no longer have the water to do so, the future of agriculture will depend almost entirely on rainfall and the judicious selection of crops. A return to dryland agriculture is forecast for much of the Great Plains, where rainfall is sufficient to support drought-resistant grain crops and perhaps cotton (figure 20.25). In many cases revenue to the farmer is just as great as with irrigated agriculture, although the regional and individual farm productivity is decreased. Regions that are forced to switch from high-intensity, irrigated agriculture to low-intensity, dryland agriculture simply will not be able to provide as large a segment of the world's food supply.

Figure 20.24
The future of irrigated agriculture and expansion into the drier parts of the world are subject to careful scrutiny using a cost/benefit analysis. These large center-pivot sprinkler irrigation systems irrigate a circle of almost 65 hectares of land surface.

Figure 20.25
Marginal land for dryland agriculture, or land intended for irrigated agriculture but without a water supply, fails to support vegetation and often results in severe erosion.

Most of the world's arid and semiarid regions fall into desert or grassland biomes. If land has been broken out from a grassland biome, it is likely to support the original or similar vegetation (figure 20.26). If one decides to return the land back to its climax vegetation, reestablishment may be difficult. Under conditions of wind, high temperatures, and unpredictable rainfall, establishment of vegetation under dryland conditions is a difficult assignment. Even so, where dryland agriculture potential is marginal at best, farmers often consider returning the land to grassland for livestock or wildlife grazing. If a small amount of irrigation water is still available, the judicious use of that water for grassland establishment may be strongly considered.

The return to native vegetation is even more difficult in the deserts. The competition for water is so great that plants are far apart, and large root systems are essential for survival. When those plants are removed by farmers or ranchers, it is difficult to achieve reestablishment. Not only is water accumulation a problem, but nutrient cycling in the desert is particularly precarious. Individual plants are often referred to as islands of fertility; the nutrients simply cycle within the plant and in the soil directly under the plant. The desert soil between plants may be sterile, but nutrients abound around the plant. However, once that plant is removed, the nutrients are swept away by the wind or leached out of the root zone by thunderstorms. Once that nutrition has been lost, it is almost impossible to put it back again. The desert is indeed a delicate resource.

Sustainable Agriculture

Much has been said in recent years of the concept of **sustainable agriculture.** This production philosophy suggests that in order to preserve our precious resources for future generations, we must develop agricultural systems that will allow us to sustain productivity indefinitely, by minimizing the use of irrigation and chemical fertilizers, controlling soil erosion, and minimizing damage to the environment by

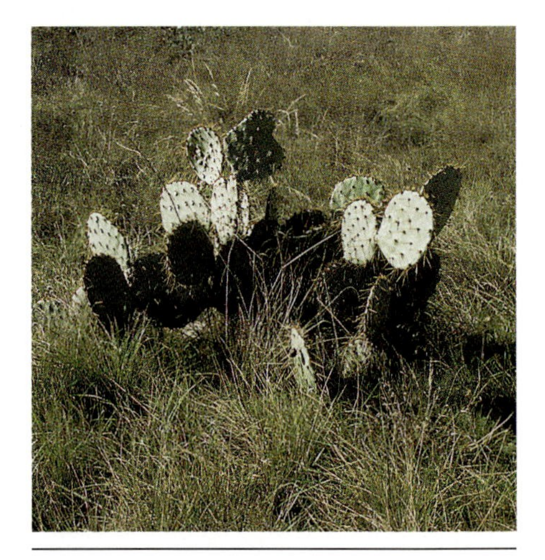

Figure 20.26
Prickly pear cactus is not necessarily indicative of a desert biome; it is merely an invader of a severely overgrazed grassland. Range management involves a stewardship of the land so that productivity is closely tied to the local ecology.

controlling the use of pesticides. All of this sounds good, but there are still questions about whether such a sustainable system has application to large-scale agriculture and whether in fact it can be achieved.

DEMOGRAPHY

Thomas Malthus published an essay in 1797 entitled *The Principle of Population* in which he put forth the hypothesis that biological species, including *Homo sapiens,* always have the ability to produce more offspring than can be accommodated. Natural systems keep population levels in check, primarily by limited food supply. No one is likely to be upset when a test tube of bacteria runs out of a nutrient source, or when a plague of grasshoppers runs out of forage plants, but the Malthusian concept strikes home vividly when there is a lack of human food. We have seen that famine and starvation are effective in controlling the size of the human population. The climatic problems involved in food production are compounded by social problems, including civil strife and war. War alone, however, is not nearly as effective in population control as famine. A nuclear war might rival starvation in actual numbers of deaths, but more than likely a nuclear war would be caused by problems with food supply. The message from Malthus—and we have no reason to believe that his hypothesis is not correct—is that if the human species fails to control its own population size, natural force will intervene and do it for us.

At the beginnings of agriculture some 12,000 to 18,000 years ago the human population was less than 5 million. At the time of Christ the world had 250 million inhabitants, and that number had grown to 500 million by 1650. In 1850 it had doubled again to 1 billion; and by 1930 it had doubled again in just 80 short years. By 1976 it had doubled again, this time in only 46 years, and the present rate of doubling takes approximately 39 years. Many factors kept early population levels under control, but the astounding leaps in doubling time are cause for alarm.

Countries that can least afford the population increases (the developing countries) are the ones with the most rapid rises. In Latin America, Africa, and Asia the annual number of births far exceeds the number of deaths, and the rate of increase is more than 3% per year in some regions. Throughout human history the birthrate has fluctuated considerably, but changes in the death rate usually have the greatest effect on population growth. Medical technology has greatly reduced the death rate and thus increased the rate of population growth.

In the developed countries, unlike the underdeveloped countries, population growth is declining because the birthrate is declining. This decline is often attributed to technological advancement and literacy rather than religious factors. In recent years some countries, such as Japan and China, have greatly influenced the rate of population growth through governmental decree and massive birth control campaigns. The availability and understanding of birth control devices and social acceptance of family planning are major factors in control. It appears that a stable food supply and lack of social upheaval are also important in reducing the desire for more children. Anticipation of insecurity in old age is often cited as a reason for large families.

The Population Reference Bureau has projected a world population at 7.022 billion by the year 2010 and a population of 8.378 billion by the year 2025. As long as the earth had adequate supplies or arable land, bringing that land into production to keep up with the population increases was a relatively simple matter. However, essentially all of the world's arable land is now in cultivation (except for the humid tropics).

TABLE 20.2

World Population Data (1994)

CONTINENT	POPULATION (MILLIONS)	DOUBLING TIME* (YEARS)	PROJECTION FOR 2010	TFR†	GROSS NATIONAL PRODUCT (PER CAPITA $U.S.)
Africa	700	24	1078	5.9	650
Asia	3392	41	4253	3.1	1820
North America	290	98	334	2.0	22,840
Latin America	470	35	584	3.2	2710
Europe	728	1025	738	1.6	11,990
Oceania	28	57	34	2.6	13,040
World	5607	43	7022	3.2	4340

Data from 1994 World Population Data Sheet, *Population Reference Bureau.*

Doubling time is the number of years required for population numbers to double.

†*TFR (total fertility rate) is the average number of children that would be born to each woman in a population if each were to live through her child-bearing years. A TFR of 2.1 to 2.5 is considered "replacement level" fertility.*

The problems confronting a move into the arid lands have already been discussed. The only other area with possibilities for expansion is in the humid tropics. The problems there are probably even greater than in desert agriculture; soil fertility, pest management, storage, and transportation are all major obstacles to agricultural development. In spite of these problems, the tropical rain forests are rapidly being developed, often without rational decision making.

Agriculture is forced to attempt to increase yields on land already in production. These efforts have led to the green revolution, in addition to all the problems associated with excessive resource usage and deterioration of the ecosystem. It would appear that the portion of the biosphere capable of supporting agriculture is already being taxed to the limit. What does the future hold?

HIGH-TECHNOLOGY ADVANCEMENTS

Recent advancements in molecular biology have led to the concept of **genetic engineering,** the process of combining portions of the genome of one organism with the genome of another. Thus desirable characteristics of totally unrelated organisms might be incorporated into another organism to impart some beneficial effect. Early results were achieved with microorganisms, but the future appears excellent for similar advancements with both plants and animals. Whereas genetic affinity usually limits the hybridization of organisms not closely related, genetic engineering holds forth the possibility of taking specific genes from one organism and combining them with the genes of another totally unrelated organism. Recent work has also produced a technique known as protoplast fusion. In this process the plant cell wall is enzymatically digested away, and the sticky cell membranes allow the protoplasts of the two species to combine. Occasionally the nuclei fuse, and the new organism has the characteristics of both parent cells.

It is still too early to say what effect genetic engineering and high-technology research will have on increasing productivity, but they do hold the promise of manipulating the genes of genetically superior plants to make them more photosynthetically efficient, have greater stress tolerance, and increase the efficiency of symbiotic nitrogen fixation. This last factor alone would be a major achievement, particularly if more desirable plants could take nitrogen from the atmosphere and reduce the dependence on commercial fertilizers made from fossil fuels. Nitrogen fixation, too, requires energy, and greater nitrogen fixation will result in a lower productivity of other organic materials.

Recent announcements in genetic engineering concern new varieties of crop plants that have been "transformed" by the insertion of foreign genes. Applications may be found in frost tolerance, insect and disease resistance, drought and salt resistance, and herbicide tolerance. When these new genetically-engineered plants come on line, they hold the promise of greater agricultural productivity while dealing with the problem of environmental contaminants.

THE FUTURE OF AGRICULTURE

Although the future demographic picture is bleak, if methods can be found to achieve population control, the future for *Homo sapiens* could be very bright. It might be possible under such conditions to achieve an ecological stability in which human needs balance with the rest of nature. A stable food supply could be the most important factor in the control of social unrest, which has plagued the world since the origin of our species.

> ✔ *Concept Checks*
>
> 1. Why can't we simply transport water over great distances to make certain that our level of agricultural production remains high?
> 2. What are "transformed" plants?

New Crops

Agriculture is gradually realizing that, to achieve ecological balance, we should abandon the idea of modifying the environment to fit the crop. Instead, we should select crops to fit specific environments. It was a mistake to take mesic crops to the desert and attempt to cultivate them there. The energy inputs were simply too great to justify the product. A recent effort involves the selection of new genotypes of existing crops that better fit the climate and soils for a region and the selection of crops never before considered for that particular region. In many cases new crops are actually old crops; for example, the grain amaranths being considered for semiarid regions were food crops thousands of years ago in Central and South America. In the past few years considerable effort has been given to the search for new species not used in technological agriculture. Although the following list is not exhaustive, it includes some of the recent prospects for food and forage crops.

1. **Grain amaranths** Several species of the Amaranthaceae (the pigweed family) have potential for small grain production under low rainfall, dryland conditions. The tiny seeds are high in protein and make an excellent flour and bread. These plants are well adapted to conditions of environmental stress, including the high temperatures and high light intensity found in desert regions.
2. **Saltbush** (*Atriplex*) Approximately 200 species of this genus are found throughout the hot, dry parts of the world (figure 20.27). The saltbushes are particularly adapted to saline water and soils and produce a high-protein forage similar to alfalfa. This potential feed for domestic livestock and wildlife could be grown on land considered too marginal for farming because of lack of rainfall or the presence of brackish water. Some species also have potential as biomass crops for fuel plantations in semiarid regions where land is not in competition with traditional agriculture.

3. **Ironweed** (*Kochia scoparia*) Related to the saltbushes of the Chenopodiaceae, this annual plant has been used as a high-protein livestock forage. It is a weedy invader of disturbed sites throughout the world and competes effectively under conditions of low rainfall, high temperatures, and high light intensity. *Kochia* is a prime example of a species that for years has been considered a curse to commercial agriculture because it was weedy and too competitive with agricultural crops.

4. **Winged bean** (*Psophocarpus tetragonolobus*) This plant has potential in the tropics primarily because it is a highly competitive legume that fixes nitrogen symbiotically and produces a high-protein food product.

5. **Buffalo gourd** (*Cucurbita foetidissima*) This plant is native to the deserts and grasslands of the southwestern United States (figure 20.28). It produces a large vine with baseball-sized fruit full of seeds rich in protein and fat. In addition, this herbaceous perennial survives the winter and periods of drought by storing water and nutrients in large underground tubers that are rich in starch. The seeds could be ground into meal, and the tubers could be extracted for food or industrial starch.

Figure 20.27
Atriplex canescens (four-wing saltbush) is native to western North America and appears to be a prime candidate for domestication. As a drought-resistant, salt-tolerant, perennial shrub it produces both high-protein forage and abundant biomass that may be used as a feedstock for energy production.

There are hundreds of potential food and forage plants available but never selected for agricultural production. Introducing a new crop is exceedingly difficult in today's society. The processes of (1) basic research, (2) applied research, and (3) technology transfer must be undertaken before there is any hope of breaking into commercialization. The infrastructure that surrounds each crop is unique to that particular species. Rice, for example, must have specialized planting, cultivating, and harvesting equipment; the processing, storage, and transportation are handled by people familiar with equipment peculiar to that crop. Even the marketing and acceptance are unique to certain parts of the world. If the world's rice production should suddenly double, marketing and acceptance would be a major problem throughout the Western cultures.

Even in the advanced American culture where we purport to try everything new, we are reluctant to add any new fruit and vegetables to our diet of some 15 or 20 standard species. Except for a few "trendy" new items (such as kiwifruit), Americans are suspicious of most new foods. By contrast, in Taiwan some 135 vegetable species appear at the local market, and the diversity is extremely important to the Chinese diet.

(a)

(b)

Figure 20.28
The buffalo gourd (*Cucurbita foetidissima*) is native to the southwestern United States and appears to be a choice candidate for a food and forage crop. (*a*) The large vines produce baseball-sized gourds with abundant high-protein seeds. (*b*) The underground storage tubers are rich in starch.

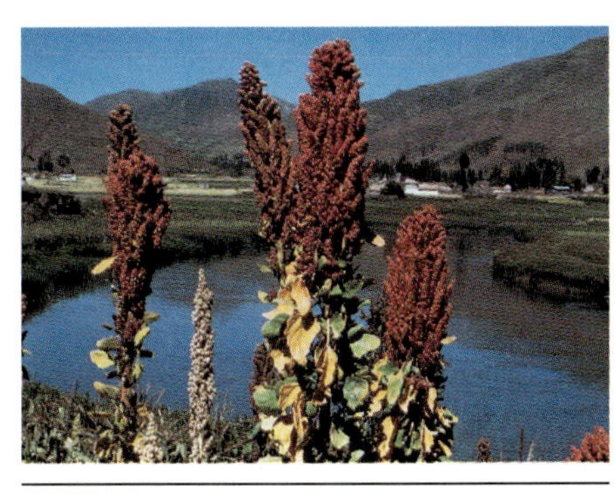

Figure 20.29
One of the most promising "new" food crops is the ancient Andean crop quinua (*Chenopodium quinoa*). This high-yield seed crop is grown at high altitudes and is readily acceptable in the local community.

New crops are sometimes suspect because they upset the traditional patterns of production and consumption. On the other hand, recent economic pressures have caused the world's agricultural producers to take a closer look at innovation, including new crops (figure 20.29). There may be, somewhere in the botanical world, an undiscovered or unrecognized species that has food potential even greater than rice, wheat, or corn. Although new crops are not a panacea for the world's food supply and population ills, they could be highly effective in alleviating hunger and famine in a world caught in its own ecological net.

Conclusions

No single factor will save *Homo sapiens*. We must use our best creative efforts to increase productivity within the ecosystem that surrounds us, while using the same ingenuity to keep our worldwide population at a safe and manageable level. Malthusian theory has worked through recorded history; there is now a race against time, to see whether humans are clever enough to save themselves from extinction.

Summary

1. Humans with upright posture probably first evolved in Africa, near the Olduvai Gorge in Tanzania. This posture freed the hands, which gave certain advantages in the struggle for survival.
2. Humans practiced hunting and gathering throughout the Paleolithic (Old Stone Age) until approximately 10,000 to 15,000 years ago. There are still some primitive tribes who exist today as nomadic hunters and gatherers, such as the !Kung of Africa.
3. Agriculture is documented for the Fertile Crescent region of what is now the Middle East to between 10,000 and 12,000 years ago. At a similar period agriculture was also developed in the Tehuacan Valley of Mexico. One theory explaining the development of domesticated plants is the "need" theory for temperate regions such as these two.
4. Another hypothesis for the origins of agriculture is the "genius theory" for tropical areas. Domestication of plants is believed to have occurred near the mouth of the Yellow River of China approximately 15,000 years ago.
5. The earliest crop plants were probably multipurpose plants such as palms, mulberry, and hemp. Root crops are undoubtedly another group of early domesticated plants because of their ease of propagation. "Root" crops also included underground stems (tubers) such as potato and the corms of taro and tannia.
6. The food basis of all advanced civilizations is a cereal crop such as rice, wheat, corn, barley, or sorghum. Irrigation systems developed in many early agricultural areas; the chinampas of Mexico are among the best known.
7. Increased travel ultimately resulted in crop introductions from one part of the world to another. One of the most unusual attempts at crop introduction involved the breadfruit transported by the H.M.S. *Bounty*.
8. Industrialization led to rapid increases in population concentrations, along with the production of material goods, communication, and transportation. Crops were domesticated between 12,000 and 18,000 years ago in many parts of the world, apparently in response to food pressures created by villages and towns.
9. The first domesticated food crops—wheat, barley, rice, corn, and potatoes—are still worldwide staples. Of major food crops (and there are only a few of them) only tomato and coffee have been domesticated in the past 2000 years. Even though variety is lacking in total number of species, farmers have been very successful in total productivity. In the industrialized nations about 5% of the population produces all the food consumed.
10. At the present time wheat, rice, and corn provide about 80% of the total human caloric intake. Most people receive enough carbohydrates, but protein and fat consumption is severely limited for much of the world.
11. Currently about two-thirds of the world's population suffers from poverty and hunger, 10% lives in relative affluence, and the other 25% lives somewhere in between, never being sure of adequate food and shelter. Throughout recorded history, agricultural productivity has increased but rises in population have always placed food supply at a dangerously low intersection with demand.
12. In recent years United States food production has enjoyed major surpluses, which have suppressed prices. Sometimes those surpluses are used for export and foreign aid, but political complications and the recipient's inability to pay for food often cause the producer nations to reduce production.
13. The exceptional productivity of the North American breadbasket reflects progress in basic and applied research, technology transfer, and the ingenuity of farmers. The land-grant college system and the agricultural extension system have been successful in providing information for farmers; similar plans have met with varying degrees of success in the developing countries.
14. The green revolution has brought high-yielding varieties of wheat and rice to a hungry world. These varieties are selected for maximum production under conditions of plentiful water and nutrition; some critics feel that the energy price for that productivity is too high.
15. Lack of bargaining power has forced farmers to settle for a less than satisfactory profit, even though food prices continue to rise. Most of those increases are demanded by the infrastructure that harvests, packages, transports, and markets the product.
16. Lack of sufficient arable land has caused concern about agriculture's ability to continue to feed a world that doubles in population every 40 years. Although desert land is available, the costs of water transport limit the potential for irrigation with current energy sources. The ecological and environmental problems associated with agriculture in the humid tropics are equally insurmountable.
17. Recent achievements in high technology, including the potential for genetic engineering, lends hope for increasing productivity to help feed the world's increasing population. This process must be coupled with population control if the human species hopes to survive.
18. New crops never before considered for agriculture might adapt to climates where traditional crops will not grow. Even though the genetic potential may be great, acceptance of a new crop depends on many factors, including the development of an infrastructure to support it. The human factor is likewise important in the acceptance process.
19. No single solution will save *Homo sapiens*. Our best creative efforts must be used to ensure survival within the constraints of the biosphere.

Key Terms

agricultural infrastructure 386
applied research 385
basic research 385
cereals 377
chinampas 378
demography 391
evolution 373
Fertile Crescent 375
genetic engineering 392
green revolution 387
hunting and gathering 373
Irish potato famine 384
irrigation 388
land-grant college system 386
limiting factor 388
multipurpose plants 377
sustainable agriculture 390
technology transfer 386
trade routes 379

Discussion Questions

1. From where did the earliest inhabitants of North America come?
2. Discuss the transition from hunting/gathering to sedentary agriculture.
3. Where are the earliest examples of agriculture?
4. Why were breadfruit worth all the trouble as an introduced crop?
5. What are the most important food crops? Why?
6. What are some of the major basic research questions in botany?
7. What is meant by technology transfer?
8. What was the green revolution and what do the critics have to say about it?
9. What are the ultimate limits on agricultural productivity?
10. What is meant by sustainable agriculture?

Suggested Readings

Anthony, K. R. M., J. Meadley, and G. Robbelen, eds. 1993. *New crops for temperate regions.* Chapman & Hall, New York.

Bettolo, G. B. M., ed. 1987. *Towards a second green revolution: From chemical to new biological technologies in agriculture in the tropics.* Elsevier Science Publishers, New York.

Bray, F. 1994. Agriculture for developing nations. *Sci. Am.* 271(1):30–37.

Campbell, B. M. S., and M. Overton, eds. 1991. *Land, labour, and livestock: Historical studies in European agricultural productivity.* St. Martin's Press, New York.

Conway, G. R., and E. B. Barbier. 1990. *After the green revolution: Sustainable agriculture for development.* Earthscan, London.

Crosson, P. E., and N. J. Rosenberg. 1989. Strategies for agriculture. *Sci. Am.* 261(3):128–135.

Farooq, A. L., M. Khan, and G. M. Buth. 1993. *Palaeoethnobotany: Plants and ancient man in Kashmir.* A. A. Balkema, Rotterdam.

Forbes, M. H., and L. J. Merrill, eds. 1986. *Global hunger: A look at the problem and potential solutions.* University of Evansville Press, Evansville.

Fukuoka, M. 1987. *The road back to nature: Regaining the paradise lost.* Japan Publications, Inc., New York.

Gasser, C. S., and R. T. Fraley. 1992. Transgenic crops. *Sci. Am.* 266(6):62–69.

Hancock, J. F. 1992. *Plant evolution and the origin of crop species.* Prentice-Hall, Englewood Cliffs.

Hawkes, J. G., ed. 1991. *Genetic conservation of world crop plants.* Academic Press, San Diego.

Hazell, P. B. R., and C. Ramasamy. 1991. *The green revolution reconsidered: The impact of high-yielding rice varieties in South India.* Johns Hopkins University Press, Baltimore.

Howell, J. H. 1987. Early farming in northwestern Europe. *Sci. Am.* 257(5):118–127.

Janick, J., and J. E. Simon, eds. 1993. *New crops.* Wiley, New York.

Lycett, G. W., and D. Grierson. 1990. *Genetic engineering of crop plants.* Butterworths, Boston.

Renfrew, J. M., ed. 1991. *New light on early farming: Recent developments in palaeoethnobotany.* Edinburgh University Press, Edinburgh.

Sauer, J. D. 1993. *Historical geography of crop plants: A select roster.* CRC Press, Boca Raton.

Smith, B. D. 1992. *Rivers of change: Essays on early agriculture in eastern North America.* Smithsonian Institution Press, Washington, D.C.

Strobel, G. A. 1991. Biological control of weeds. *Sci. Am.* 265(1):72–78.

Wills, W. H. 1988. *Early prehistoric agriculture in the American Southwest.* University of Washington Press, Seattle.

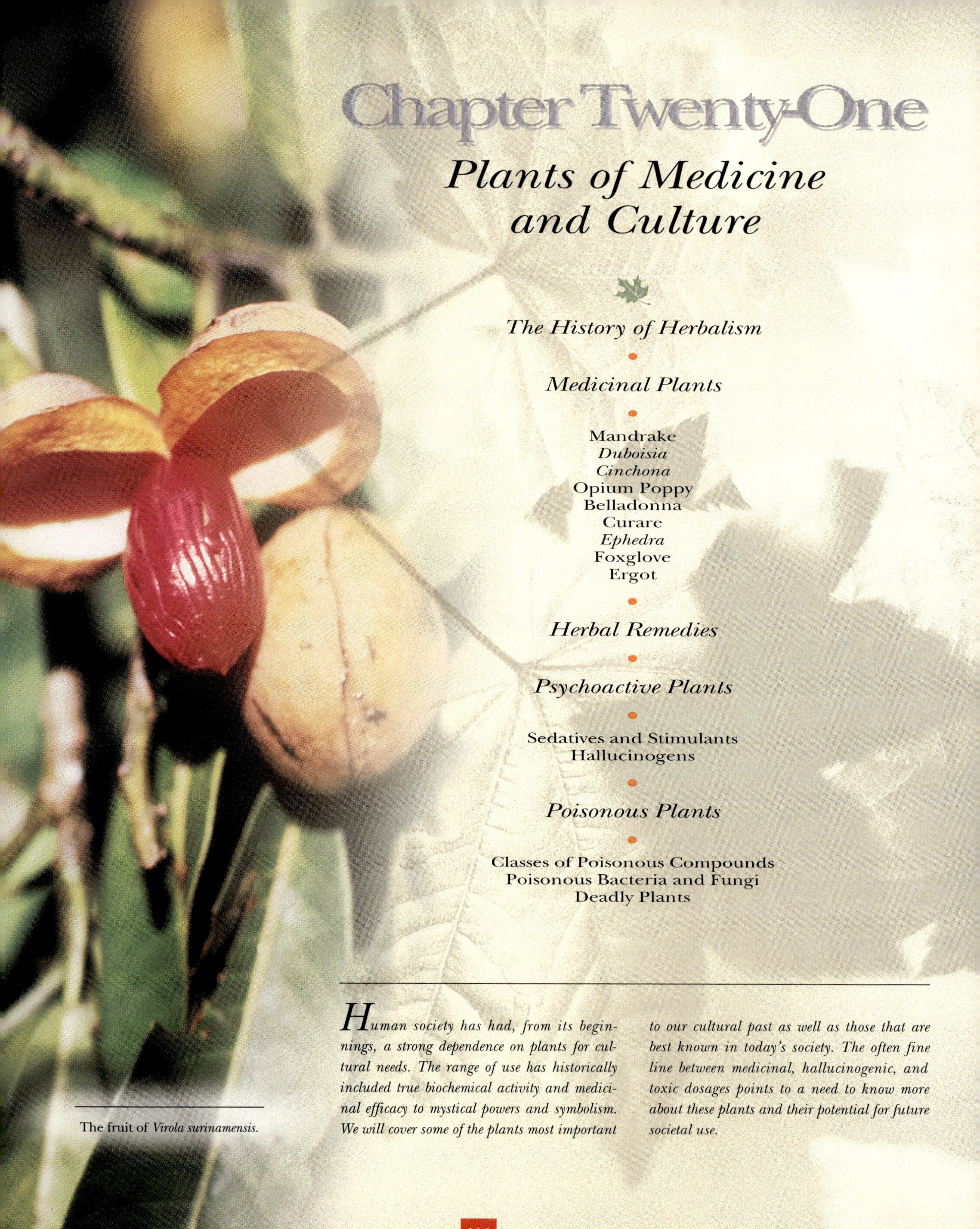

Chapter Twenty-One

Plants of Medicine and Culture

The History of Herbalism

Medicinal Plants

Mandrake
Duboisia
Cinchona
Opium Poppy
Belladonna
Curare
Ephedra
Foxglove
Ergot

Herbal Remedies

Psychoactive Plants

Sedatives and Stimulants
Hallucinogens

Poisonous Plants

Classes of Poisonous Compounds
Poisonous Bacteria and Fungi
Deadly Plants

The fruit of *Virola surinamensis.*

Human society has had, from its beginnings, a strong dependence on plants for cultural needs. The range of use has historically included true biochemical activity and medicinal efficacy to mystical powers and symbolism. We will cover some of the plants most important to our cultural past as well as those that are best known in today's society. The often fine line between medicinal, hallucinogenic, and toxic dosages points to a need to know more about these plants and their potential for future societal use.

If primitive humans found plants that eased pain, soothed stomach distress, or provided relief of any kind, it was undoubtedly an accidental discovery. Disease per se was not understood at that time; physical ailments were mystical and therefore were attributed to "bad spirits" or as punishment from displeased supernatural powers. Once plants that were effective in relieving human suffering were identified, knowledge of their healing properties was passed on through generations by word of mouth and later in written form.

THE HISTORY OF HERBALISM

The early history of botany was essentially the early history of medicinal plants, and **herbalism,** the use of plants for treating ailments, was the realm of the botanist. In the first century AD Dioscorides compiled detailed botanical and medicinal information on thousands of plants in his *De Materia Medica.* This book was to remain the authoritative reference on medicinal plants for over 1500 years. In the fifteenth to seventeenth centuries botany became more of a science, and the use of medicinal plants was finally a mixture of medical knowledge and ritual, not exclusively a magical practice. Herbalists were a combination of botanist, psychiatrist, and faith healer, but their knowledge of the effect plants had on the body was considerable. As more was learned about the true medicinal value of plants, less credence was given to those having only mystical powers. This was the beginning of a disenchantment with much of "traditional" medicine. In the twentieth century only a few medicinal plants of exceptional value and proven effectiveness have continued to be recognized.

Modern medicine is predominately composed of scientifically developed synthetic drugs, although about 40% of all prescribed drugs are natural substances or only semisynthetic. Medical schools generally offer only a single course (pharmacognosy) dealing with medicines directly from plant, animal, and mineral origins. This has been the case for only a little over a hundred years; throughout history, and even in many cultures today, the physician has been a person with excellent botanical knowledge.

The medicine men of the Native Americans, the witch doctors and herbalists of Africa and South America, and the shamans of Asia are trained practitioners who function not only in the realm of mystical and occult faith healing. Many years of training and apprenticeship are required before an individual is recognized as a medicine man or herbalist. Navajo medicine men, for example, learn about the biology and uses of over 200 different plants during their training. Their African and South American counterparts have comparable knowledge of the medicinal plants found in their parts of the world. The Miskitos of eastern Nicaragua, although occupying a very small geographical area, have medicinal uses for almost 100 different plants. Much as modern medicine often must rely on trained counselors, psychiatrists, and psychologists to treat symptoms involving the mind and the emotions of the patient, the herbalists and witch doctors have always considered treatment of the spirit an integral part of their craft. The modern Navajo Indians often blend the two worlds of medicine. The physician deals with most of their physical needs, including surgery, while the medicine man treats imbalances in body functions with his herbal teas and poultices and tends to the health of the mind and spirit.

Nowhere has a balance of traditional medicine and modern medicine been achieved with greater success or overall acceptance than in the People's Republic of China. Few cultures have a more thorough knowledge of herbal remedies. Along with acupuncture, this knowledge forms a solid base for their total health care. Both the traditionalists and the modern medical people accept the strengths of each other's expertise, knowledgeably blending them into an approach to treatment of the whole entity—the human body, spirit, and mind. **Holistic medicine** is beginning to gain acceptance by some Western physicians, and even the World Health Organization (WHO) now sponsors a program to promote and encourage traditional medicine throughout the developing countries of the world.

Biomedical scientists and other health experts are showing increased interest in the plants used by reputable herbalists, and biological conservationists are receiving increased support and acceptance of their efforts to prevent plant extinction, especially in the tropical parts of the world. It is estimated that some 20,000 species of flowering plants are threatened with extinction in the tropics because of clearing the forests for agriculture and development. It is not known how many plants may exist with as yet undiscovered potential for food, industrial, or medicinal use, and it will never be known if these plants become extinct. The renewed interest in medicinal plants is therefore a timely and desirable activity while both the plants and the lore about their medical use still exist.

MEDICINAL PLANTS

There is a degree of difficulty in categorizing some plants as medicinal plants because of the fine line that exists between medicinal, psychoactive, and poisonous dosages. For other plants, the distinction is clear. It should be noted, however, that essentially any chemical substance added to the body can be toxic or harmful in excessive amounts.

Mandrake (*Mandragora officinarum,* Solanaceae)

Although mandrake was written about more extensively than any of the early medicinal plants, its use was surrounded in myth and superstition. Appreciation of this plant goes back to several hundred years BC, and its use in ancient Rome is documented.

Mandrake the Mystical

Because it was so potent, and because of the lack of understanding of its properties, mandrake was considered a mystical plant. A mandrake root was a good luck charm, a talisman, endowing its possessor with good health, social prominence, sexual prowess, and great wealth. The owner of a mandrake root would often keep it in a velvet-lined box, feeding it and giving it special treatment. If the owner ever started having bad luck it was considered a sign that the talisman had turned against its owner. To get rid of such a powerful charm, however, was very difficult.

A popular belief of this time was the **doctrine of signatures,** which stated that the Creator placed certain items on earth for humans and identified their intended use by their shape. A kidney-shaped leaf was considered to be useful in treating the kidney; walnuts, with their furrowed ridges, were believed beneficial to the health of the brain or head; and liverworts were used to treat the liver. The shape of *Mandragora* frequently resembled the human body, with a thick, fleshy, two-branched or forked taproot and a short stem branching into the leafy part of the plant. It did not take too much imagination to conceive of this as a human form; thus *Mandragora* was believed to benefit the entire body (figure 1).

Biblical scholars think mandrake might have been offered to Christ on the cross to ease his pain. Shakespeare detailed important considerations in collecting the root: it must be accomplished without listening to its death shriek as it is pulled from the ground, lest the collector die of fright from the horror of the sound. Lucrezia Borgia used mandrake as a poison. Even in the nineteenth and twentieth centuries extracts from *Mandragora* root have been mixed with morphine to produce a twilight sleep to ease the pain of childbirth.

Figure 1
Mandrake (*Mandragora officinarum*). The thick and often forked taproot resembles a human body.

The active compounds are several alkaloids, primarily **hyoscyamine.** They may be extracted in great quantities from the root by soaking or boiling in wine. This extract was the first known effective anesthetic, sedative, and pain reliever.

Duboisia (Solanaceae)

In its native Australia this plant was used by aborigines during the same period of time that mandrake was used in Europe, the Middle East, and North Africa. Also containing hyoscyamine, the leaves were used to relieve pain. The hunters ran their prey into exhaustion, which often took several days. Once the kill was made, the hunters would chew *Duboisia* leaves to ease the fatigue, hunger, and thirst from their exertions. They also used it to stupefy emu and to poison fish.

Cinchona (Rubiaceae)

Native to tropical South America, this tree contains the alkaloids **quinine** and quinidine in its bark. Prepared commercially as early as 1823, quinine was used for over a hundred years to treat hundreds of thousands of malaria victims. After 1930 a synthetic, chloroquine, has been preferred for the treatment and suppression of malaria.

Cinchona was named by Linnaeus for the countess of Chinchone, wife of the viceroy of Peru. The countess was cured of a malarial fever in about 1638 by a tribal witch doctor (who the viceroy subsequently appointed as the royal physician). Following a visit to Peru, King Charles II of England and several members of the French and Spanish royal families contracted malaria and sent word to Peru. The viceroy, the countess, the witch doctor, and a support entourage set sail for Europe, taking with them an ample supply of the bark. After successful treatment of King Charles and the others, the drug became widely acclaimed in Europe, drawing the attention of Linnaeus, who named the plant and described it based on the witch doctor's descriptions and bark samples. Linnaeus later sent students to Peru to collect botanical specimens at the invitation of the countess.

Opium Poppy (*Papaver somniferum*, Papaveraceae)

This native Middle Eastern plant contains over 25 different alkaloids, including several that are among the most important pain relievers in human history (figure 21.1). The immature capsules yield a milky sap when cut. This sap dries to a gummy brown residue of pure **opium.** Known to the Sumerians as early as 4000 BC, opium was used

(a) (b)

Figure 21.1
Opium poppy (*Papaver somniferum*). (*a*) The beautiful flowers and (*b*) distinctive fruit. A similar shaped but much smaller fruit of *P. setegerum* is being held next to a capsule of *P. somniferum*.

Figure 21.2
Belladonna (*Atropa belladonna*) is an important medical plant because of the alkaloid atropine.

medicinally in ancient Greece, Rome, and China. During the Middle Ages **laudanum** (opium dissolved in wine) was widely used to relieve pain and produce a euphoric state.

By the nineteenth century "medicinal" opium use was widespread in Europe and North America, producing thousands of opium addicts, many of whom were prominent and wealthy. Opium use, however, was far more common and serious in China than anywhere else. There it was smoked in specially constructed pipes, addicting millions by the 1830s. Although opium was banned in China by the last Ming emperor (1628–1644), British and American clipper ships nevertheless smuggled tons of opium from India into China for almost two centuries until a crackdown on the smugglers was ordered by the Chinese emperor in 1838. This action resulted in the Opium Wars, which the British won, gaining Hong Kong Island as part of their victory. The Second Opium War occurred ten years later, followed finally by the Boxer Rebellion, during which the Chinese evicted all foreigners. Seldom has a single plant affected so many or resulted in such hostility.

Several powerful alkaloids can be extracted from opium; the most important are **codeine, morphine,** and **heroin.** All these compounds are analgesic (pain relieving), affecting the central nervous system. Codeine is the mildest of these, producing almost no euphoria but effectively relieving minor pain and functioning as a cough suppressant.

Morphine depresses the cerebral cortex, reducing brain arousal to pain and causing a euphoric feeling, thus eliminating anxiety. Morphine also depresses the respiratory and cough centers of the brain and impairs digestive action. Prolonged use of morphine results in physical tolerance, so gradually increasing dosages are necessary to achieve the same pain relief. Prolonged use also results in addiction, which at one time was common in wounded soldiers. As with any addictive narcotic compound, discontinued use results in physically painful withdrawal symptoms.

Heroin, a product manufactured from opium, is so powerful in its effects on the body and so dangerously addictive, it is not used medicinally even by prescription in the United States. It is available only through illegal means, so the price is high and the quality and purity are erratic. As with morphine, the body builds a tolerance to heroin and requires increased dosages to produce the same effect. Advanced addiction requires such large quantities that overdoses frequently occur. Thus from a single plant species, *Papaver somniferum,* some of the most widely used pain relievers and most dangerous and addictive drugs originate.

Belladonna (*Atropa belladonna,* Solanaceae)

Like *Mandragora* and a number of other plants in the Solanaceae family, **belladonna** contains several powerful alkaloids, including hyoscyamine, scopolamine, and atropine (figure 21.2). These alkaloids produce flushed skin and dilated pupils and, with lethal doses, delirium and respiratory failure. Atropine specifically blocks parasympathetic nervous system effects and is used as an antidote to nerve gas poisoning. Italian women used it in eyedrops to dilate the pupils and thus achieve a large, dark-eyed look, hence the name belladonna, Italian for "beautiful woman." Ophthalmologists take advantage of its pupil-dilating properties to facilitate examination of the interior of the eye.

Curare (*Chondodendron tomentosum,* Menispermaceae)

Chondodendron tomentosum is native to South America and produces the alkaloid tubocurarine, the principal ingredient in **curare.** Curare interferes with nerve impulses to the skeletal (voluntary) muscles, producing a reversible paralysis. This drug has been used as a muscle antispasmodic in the treatment of rabies and in spastic conditions. It can also induce muscle relaxation during surgery without anesthesia and helps control convulsions caused by poisons such as strychnine.

South American Indians boiled the bark and wood of these plants in combination with several species of *Strychnos* to make a thick, gummy extract with which they coated their arrow tips. The small arrows, capable of only shallow penetration, nevertheless could be depended on to paralyze the prey.

Ephedra (Ephedraceae)

Species of *Ephedra* (a gymnosperm) from India and China contain an alkaloid amine, **ephedrine,** that acts as a **bronchodilator.** Its mode of action is to relax the smooth muscles of the bronchial tubes, increasing their diameter and improving passage of air through them. Patients with bronchial asthma, chronic bronchitis, and emphysema are treated with this drug.

Several species of *Ephedra* also occur in western North America, but they have only small traces of ephedrine, so their effect as a bronchodilator is minimal. A tea is made from these plants, however, that is an effective and safe diuretic. One species, *E. antisyphilitica,* had a reputation for successfully treating syphilis; it is now known to be ineffective in this regard.

Foxglove (*Digitalis purpurea,* Scrophulariaceae)

Native to Europe, this exceptionally valuable medicinal plant contains several cardiac glycosides, including digitoxin (figure 21.3). Prior to the discovery of this drug, thousands of people died each year from congestive heart failure, the inability of the heart to pump blood at a sufficient pressure. This resulted in edema, or fluid buildup in tissues throughout the body, often causing such a swollen and distorted body that movement was difficult. The ailment was known commonly as **dropsy.** Treatment with digitalis slows and strengthens the heartbeat, which increases the volume of blood being circulated through the body. The filtering activity of the kidneys and the blood is more effective in removing wastes and preventing fluid buildups. It is estimated that, in the United States alone, over 3 million heart disease patients daily use digitoxin, digoxin, or one of the other cardiac glycosides found in this plant.

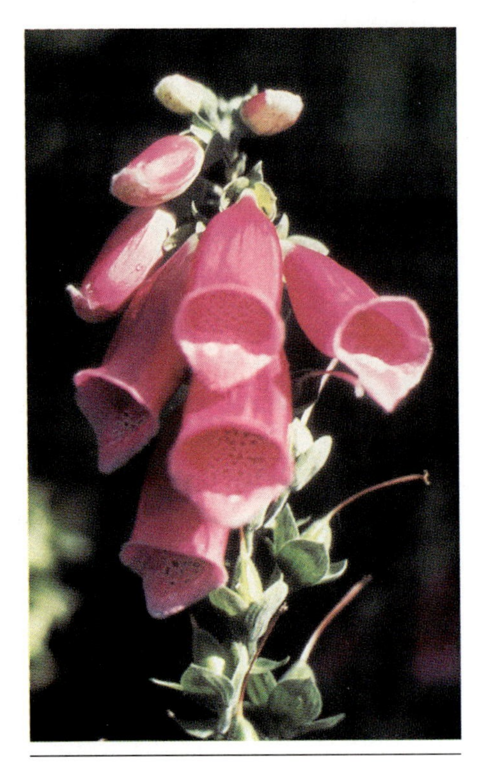

Figure 21.3
Foxglove (*Digitalis purpurea*) contains cardiac glycosides, including digitoxin, used in treating congestive heart failure.

Figure 21.4
Ergot is a fungus that infects many of the cereals, producing purplish-black sclerotia, which contain ergonovine.

Ergot (*Claviceps purpurea*)

This fungus infects wheat, rye, and other cereals, destroying the seed and producing purplish-black sclerotia (Ergots or ergot bodies). The sclerotia (figure 21.4) contain at least ten alkaloids, and one of these, ergonovine, has been an effective **oxytocic agent,** stimulating uterine contractions and speeding up labor. Ergonovine is more effective in this capacity than the animal hormone oxytocin, and it also reduces postpartum bleeding, has low toxicity, and acts rapidly. Ergot also has an interesting history as a hallucinogen.

✔ Concept Checks

1. Does herbalism have any scientific basis?
2. What was the doctrine of signatures?
3. Why is *Papaver somniferum* considered to be one of the most important sources of pain relievers in history?
4. What ailments do cardiac glycosides treat?

HERBAL REMEDIES

Besides the well-known and effective medicinal plants just discussed, there are hundreds of undiscovered and poorly understood herbal remedies. Fortunately, few of these plants have toxic or addictive properties (figures 21.5 to 21.9). Home remedies, folk medicine, herbalism, natural or holistic medicine, and all similar terms deal essentially with the same idea: the

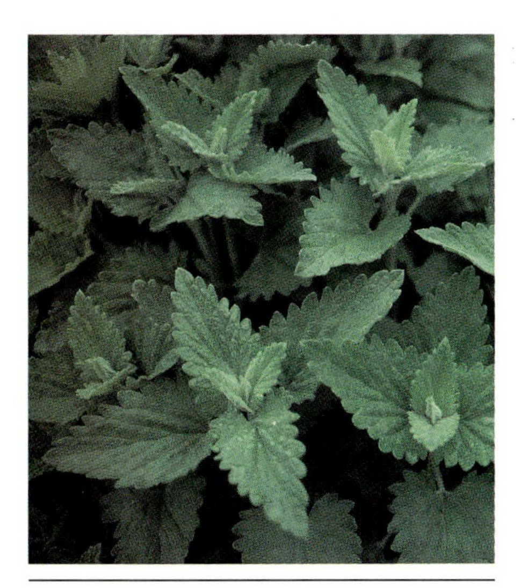

Figure 21.5
Catnip (*Nepeta cataria*) is a mild tranquilizer.

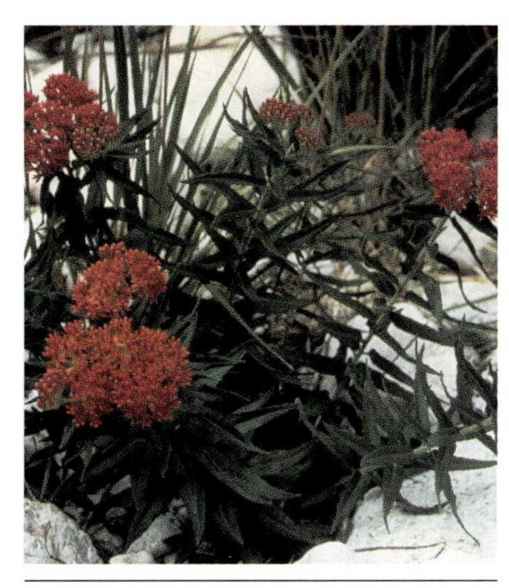

Figure 21.7
Pleurisy root (*Asclepias tuberosa*) increases perspiration and bronchial dilation.

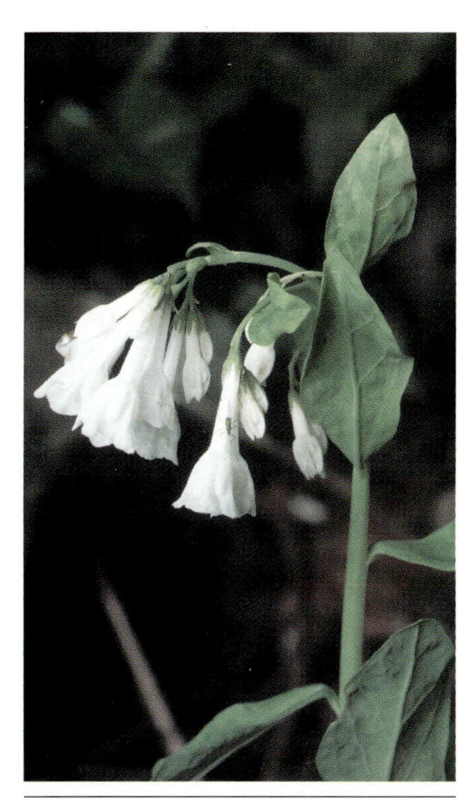

Figure 21.9
Comfrey (*Symphytum officinalis*) contains a mucilage that acts as a demulcent and tonic for the respiratory tract.

Figure 21.6
Mullein (*Verbascum thapsus*) contains compounds that ease coughing and asthma.

Figure 21.8
Plantain (*Plantago*) has been used to make a poultice for insect bites.

increased use of plant materials both as a means to better health and for the treatment of specific ailments. Many of the more popular herbal materials are available through health food stores, but if you are tempted to collect your own (and we do not recommend this), make sure the desired plant is correctly identified.

Table 21.1 lists a small percentage of the plants reported as having some medicinal value in the United States alone. There are many popular books detailing the correct collection, extraction, and use of these and many more herbal remedies. An intelligent combination of home remedies and health tonics with modern medical care is becoming a popular and widely accepted health care plan in the United States. Certainly, there is room for more scientific and biomedical study of many of these herbal cures and the plants used in their preparation.

PSYCHOACTIVE PLANTS

Psychoactive plants contain compounds that act on the central nervous system to produce a mind-altering state, visions, distortions of the senses, and changes in psychomotor ability. There are three categories of psychoactive compounds: stimulants, hallucinogens, and depressants. It is not uncommon, however, for all psychoactive compounds to be considered hallucinogenic because of the mind alteration produced. The drug culture of the 1960s and 1970s dubbed these experiences "trips" and called some of these compounds "mind expansion" or "psychedelic" drugs. Modern Western culture developed new terms, new uses, and some new synthetic drugs, but psychoactive compounds have been in use for thousands of years. Hallucinogens have

TABLE 21.1

Selected Herbal Remedy Plants of the United States

PLANT	PART USED	PURPORTED VALUE
Aloe (*Aloe vera*)	Leaves; mucilaginous juice	A topical pain reliever, promoting healing of burns and cuts
Alum root (*Huchera* spp.)	Root; chopped, 1 teaspoon boiled for 20 minutes	An astringent for treatment of stomach flu, ulcers, sore throat (gargle), and cuts and abrasions (promotes clotting)
Asparagus (*Asparagus officinalis*)	Root; chopped and steeped in boiling water to make a tea	A diuretic and laxative for treatment of gout and urate-related joint inflammations; helps prevent kidney stones
Birch (*Betula*)	Bark	An analgesic for headaches and arthritis; methyl salicylate is the active ingredient
Catnip (*Nepeta cataria*)	Entire plant	A mild tranquilizer; treatment for colic or teething in young children
Comfrey (*Symphytum officinalis*)	Root; extracted in water	Mucilage acts as a demulcent and tonic for the respiratory tract; also used as a stomach tonic and to soothe diarrhea and dysentery; may be toxic
Dandelion (*Taraxacum officinale*)	Leaves and flowers; steeped in boiling water to make a tea	Diuretic; treatment for kidney inflammations; as an aid to liver and spleen functions
Dogbane (*Apocynum androsaemifolium*)	Roots	Cardiac stimulant; may be toxic
Gentian (*Gentian* spp.)	Roots; chopped and steeped in hot water for 30 minutes	Considered an excellent stomach tonic for indigestion, heartburn, etc.
Mullein (*Verbascum thapsus*)	Large, basal leaves; washed, dried, chopped, steeped in water or smoked	Treatment of throat, chest, and lung maladies such as coughing, asthma, and respiratory tract infections
Oak (*Quercus* spp.)	Bark or small twigs; chopped and steeped in water to make a tea	An astringent for gum inflammations, sore throat gargle, intestinal tonic
Old man's beard (*Clematis* spp.)	Branches; chopped and steeped in boiling water to make a tea	Treatment of headaches, especially migraine; dilates veins but has a vasoconstricting effect on brain lining
Plantain (*Plantago major*)	Fresh leaves	Poultice; treatment for insect bites and skin abrasions; also useful for mild intestinal inflammations and hemorrhoids
Pleurisy root (*Asclepias tuberosa*)	Roots; chopped and boiled in water	Increases perspiration and bronchial dilation; treatment for pleurisy and mild pulmonary edema; may be toxic
Willow (*Salix* spp.)	Twigs or bark; boiled in water for 2 hours	General analgesic action; reduces inflammation of joints and membranes, eases headaches, lowers fevers, and reduces neuralgia; antiseptic wash or poultice for skin abrasions, eczema, or infected wounds

been a part of religious activities of many cultures, especially the more primitive ones, for a long time. Social use of such drugs has also been a regular part of many of the more advanced societies throughout history. Drug use has been acceptable in ceremonial and ritualistic context; however, it has varied from widely acceptable to forbidden and illegal in social settings.

Many of these compounds come from plants. Some of those considered to be hallucinogenic also have had medical applications in lighter dosages. Mandrake and belladonna in the Solanaceae fall into a medicinal, hallucinogenic, or poisonous category, depending on dosage. Others, while having no medicinal value per se, have been used to treat mental disorders and as pain relievers by altering the patient's level of consciousness. Still others are not hallucinogenic but poisonous. These are sometimes called pseudohallucinogens because they produce a delirium while acting on the system. The delirium makes the user very sick, and the aftereffect is painful. Nutmeg is a pseudohallucinogen, attractive for its availability and nominal cost. Several tablespoons, if it can be kept down, produce a painfully nauseating delirium from which the user comes down "hard" with severe and extended pain and toxic recovery.

Sedatives and Stimulants

Although incorrectly categorized by many as hallucinogenic, opium and opiate derivatives are actually **sedatives,** often producing a relaxed euphoric state but no true consciousness alteration. Reserpine, the indole alkaloid from *Rauwolfia* is also a depressant and lowers blood pressure (figure 21.10). Alcoholic beverages have a depressant action, reducing both mental and physical performance levels. Since alcoholic beverages all come from the fermentation process using plants as the sugar source, alcohol can be considered a plant product.

A drug that falls into the **stimulant** category is cocaine. Extracted from the leaves of *Erythroxylon coca,* a plant native to the eastern slopes of the Andes Mountains of South America, cocaine produces a feeling of euphoria, a lessening of fatigue, an absence of hunger, and increased energy (figure 21.11). Cocaine does not produce a hallucinogenic response, is not normally physiologically addictive, and the user does not require increased quantities due to body tolerance. It is, however, very psychologically addictive, and strong doses can produce feelings of paranoia and nervous insomnia. It also damages the mucosal tissues of the nose and throat, sometimes permanently destroying these sensitive areas. "Coke" addiction is one of the most serious drug problems of the 1980s and early 1990s.

Indians in the Andes Mountains chew the leaves, which enables incredible feats of endurance without food or rest. Coca chewing is common among the people of this region, as much a cultural phenomenon as an aid to extensive manual labor. It is reported that some of the early popularity of Coca Cola in the 1800s was because a small amount of cocaine was included in the recipe. Cocaine was banned as a component of the drink in the 1900s.

Hallucinogens

Hallucinogenic compounds, or **hallucinogens,** distort the senses, especially visual, and produce a departure from reality—hallucinations. The vascular plants that contain these compounds are all angiosperms, predominately dicots. There are no known hallucinogenic gymnosperms, ferns, bryophytes, or algae, but the fungi include a number of hallucinogenic taxa. Depending on the organism and the part that contains the active compound, one can eat the fresh or dried plant, chew the appropriate plant parts, concoct a beverage, inhale a powder, smoke the leaves, and even smear an oil-based ointment on the body to cause hallucinogenic effects.

Figure 21.10
Rauwolfia contains reserpine, an alkaloid that reduces hypertension and lowers blood pressure.

Figure 21.11
Coca leaves (*Erythroxylon coca*) contain cocaine.

Figure 21.12
Marijuana (*Cannabis*). (*a*) The tall, erect species also contains hemp fibers in the stem. (*b*) The typical palmately compound leaves with serrated, margined leaflets. (*c*) Pistillate flowers. (*d*) High concentrations of THC can be found in glandular hairs on the surfaces of bracts under the flowers.

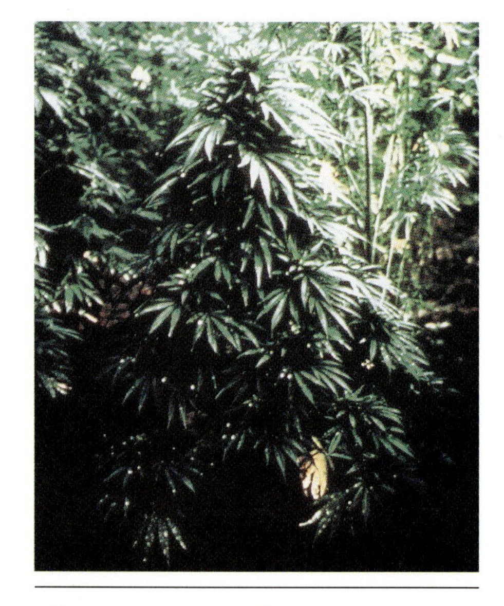

Figure 21.13
Cannabis indica is a low-growing species containing higher concentrations of THC than the more common *C. sativa.*

Marijuana

More than three species of *Cannabis* are recognized, depending on the legalistic or scientific (often mixed) point of view (figures 21.12 and 21.13). *Cannabis sativa* is the best known species, although *C. indica* contains a higher concentration of the active compound. *C. sativa* is an erect species with many fiber cells in the stem, a commercial source for making hemp rope, paper, and canvas. *C. indica* is a low-growing, highly branched species containing high levels of *trans*-tetrahydrocannabinolic acid, which converts on heating to **tetrahydrocannabinol (THC),** the compound that produces the hallucinogenic effects. THC is found in greatest concentrations in glandular hairs on the leaves, stems, and especially on the small bracts just below a pistillate flowering inflorescence. When collected from the glandular hairs separately, it is much more potent and is called hashish.

The physiological effects of THC on the body include an increased pulse rate, reddening of the eyes, and possibly a reduction in the internal fluid pressure of the eyeball. Marijuana is credited with enhancing visual perception, reducing muscular response time and coordination, altering time perception, increasing one's inner awareness, removing tension, and increasing sex drive.

THC is not an alkaloid and is not considered toxic or physiologically addictive. Its classification as a narcotic, therefore, is a legal category and not a chemical group. Psychological dependence, however, is not uncommon.

Arguments abound today: Is marijuana a dangerous drug or a mildly hallucinogenic compound less harmful than tobacco or alcohol? *Cannabis* is probably the most widely used and highly valued multipurpose plant in history. It has served as a source for fiber, as a medicine, and as a mild hallucinogen worldwide since it was introduced into North America in 1607. Only illegal as a crop plant since the late 1940s, over 63,000 tons of hemp (*C. sativa*) were raised in 1943 to provide the Navy with rope during the war. George Washington also raised hemp on his Virginia farm.

Ergot

Ergot itself is not the hallucinogen but is the spore-producing reproductive body of a fungus, *Claviceps purpurea,* that infects rye, wheat, and other cereal crops (see figure 21.4). Within the sacs are millions of tiny black spores containing several alkaloids, including ergonovine. Chemically known as **lysergic acid,** it is very similar structurally to a synthetic compound, D-lysergic acid diethylamide (LSD). Ergonovine produces smooth muscle and blood vessel constriction, which causes a burning sensation. Eating infected grain produces the disease that used to be known as St. Anthony's fire, which could result in gangrene of the ears, nose, and other body extremities from

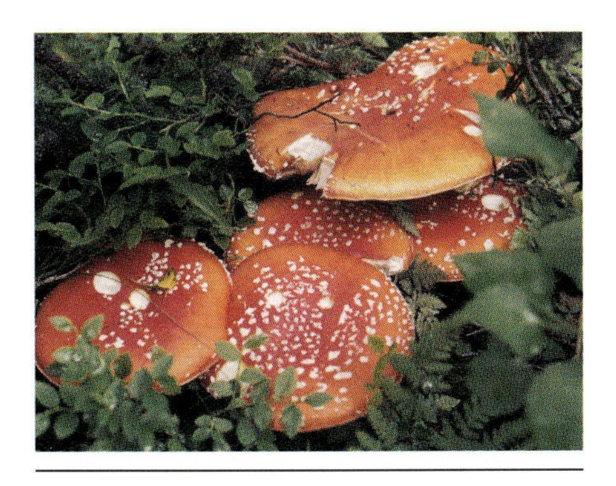

Figure 21.14
Fly agaric (*Amanita muscaria*), a beautiful but very poisonous mushroom with hallucinogenic properties.

lack of blood flow through the constricted vessels. This disease was thought to be caused by bad grain (the ergot sacs were mistaken for diseased grain), thus indirectly an affliction from the Devil. Victims would visit monasteries to repent and have the evil forces driven out. When fed with uninfected grain and given adequate freshwater and rest, most people would recover, but sometimes with the tip of their nose or ear or even hands or feet missing.

In addition to the burning sensation and danger of gangrene, the victims of St. Anthony's fire had hallucinations, an effect of the poisoning and the lysergic acid. Ergonovine (ergot) was not consumed purposefully as a hallucinogen, only accidentally. LSD, on the other hand, is a much more powerful synthetic relative and a popular modern hallucinogen.

Fly Agaric

The fly agaric mushroom is another hallucinogenic fungus (figure 21.14). Scientifically known as *Amanita muscaria,* these attractive mushrooms are large, bright yellow to orange-red with small white speckles on top. Their common name comes from the dead flies that are often found on the ground around them: In addition to being hallucinogenic, they are poisonous. Only the ones growing in the Old World are psychoactive, containing the alkaloid muscarine. This is the only known hallucinogenic compound that passes through the kidneys unaltered, and some cultures have been

Figure 21.15
Psilocybe, one of the "sacred mushrooms."

known to reuse the compound several times by drinking urine.

In larger doses muscarine produces blurred vision, sweating, lowered blood pressure, slow heartbeat, stomach pain, and breathing difficulty. These are the opposite symptoms of atropine poisoning, and fly agaric is used as an antidote to atropine overdoses.

Sacred Mushrooms

Several genera of mushrooms containing the alkaloids psilocin and psilocybin were used to produce hallucinogenic visions. The "mushroom cults" of modern times have most commonly used *Psilocybe mexicana* (figure 21.15), although *Conocybe* and *Stropharia* also contain significant levels of these alkaloids. These cults were originally a blend of Christian and pagan rituals, but today they are almost nonexistent, found only in a few villages in the northeastern mountains of Oaxaca, Mexico.

During the Mayan period in Mexico, mushroom use was common but officially available only to the priests. In contrast, almost anyone could use sacred mushrooms during the Aztec era. Some authorities believe that widespread mushroom usage was associated with the Aztec practice of offering human sacrifices by the hundreds to the gods. Current descendants, the Mazatec Indians of Oaxaca, believe that the mushrooms may be eaten only at night, and the activity is now a closed, family affair.

Peyote

Lophophora williamsii, a small, round, spineless cactus that grows flush with the ground, is found in the southwestern United States but grows primarily in Mexico. The hallucinogenic compound found in **peyote** is the alkaloid mescaline (figure 21.16). Peyote was used by the Aztecs and by the Comanches under the leadership of Chief Quanah Parker, who in 1875 was the last Indian chief to surrender his people to confinement on a reservation. It is now most closely associated with the religious rituals of North American Indians belonging to the Native American Church. Known for its nauseatingly bitter taste, peyote can be eaten fresh, or the dehydrated buttons can be chewed to achieve the desired effects. A drink can also be made from the buttons, but it is reportedly less effective in producing

(a)

(b)

Figure 21.16
(*a*) Peyote (*Lophophora williamsii*) is a spineless cactus that grows almost flush with the ground.
(*b*) When dried into "buttons," it does not lose any of its potency.

(a)

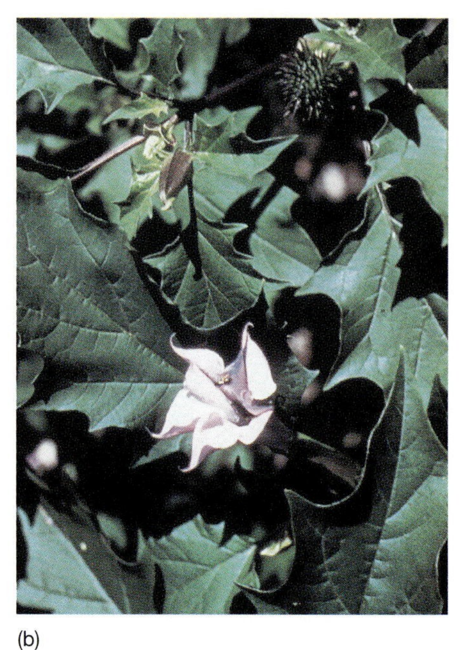

(b)

Figure 21.17
Henbane (*Hyoscyamus niger*) (*a*) is a hallucinogenic member of the Solanaceae, as is thornapple (*Datura stramonium*) (*b*).

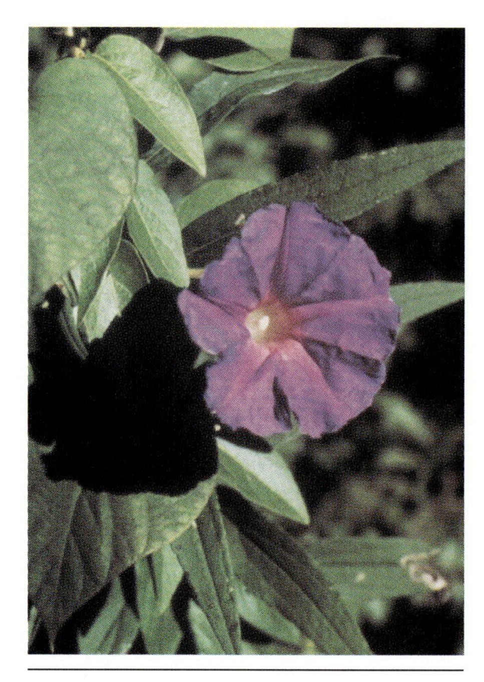

Figure 21.18
Morning glories (*Ipomaea*) were used by the Aztecs as a hallucinogen.

the vivid hallucinations. Bad trips are possible, as are spontaneous flashback recurrences.

Solanaceae: Henbane, Belladonna, Mandrake, and Thornapple

Henbane (*Hyoscyamus niger*), belladonna or deadly nightshade (*Atropa belladonna*), mandrake (*Mandragora officinarum*), and thornapple or jimsonweed (*Datura* spp.) are all members of the same plant family, the Solanaceae. They all contain a group of alkaloids including **scopolamine,** hyoscyamine, and **atropine** (hyoscyamine in the fresh plant changes to atropine when the plant is dried). These compounds have been used medically, hallucinogenically, and as toxins throughout history (figure 21.17).

In mild doses a feeling of timelessness and high spirits accompanies visual hallucinations and the sensation of flight. These plants were once a common ingredient in "witch's brew." When painted on the body with a straw broom, the alkaloids were absorbed through the skin and produced these reactions. The image of witches flying on broomsticks is a carryover from this ritual.

(a)

(b)

Figure 21.19
(*a*) Mescal bean (*Sophora secundiflora*) seeds contain hallucinogenic but poisonous compounds.
(*b*) Indians wore necklaces made from these seeds during their peyote rituals.

Other Hallucinogens

Other hallucinogenic plants include the South American *Virola* of the nutmeg family, which is administered to others by blowing a powder extracted from resins in the bark up the nose through a hollow snuff tube. Another South American plant, *Banisteriopsis,* yields an intoxicating beverage that produces a wide range of hallucinogenic responses. The Aztecs also used morning glories (*Ipomaea violacea* and *Rivea corymbosa*), which contain alkaloids similar to lysergic acid and its relatives (figure 21.18). Today's ornamental varieties often contain these compounds and are known by common names such as "pearly gates" and "heavenly blues." Mescal beans (*Sophora secundiflora*) were used by North American Indians before they began to use peyote. Peyote was preferred because of the high toxicity of mescal beans. Interestingly, these two plants were often found in identical habitats, and the Indians believed that the easily visible mescal bean plants were provided to lead them to the peyote cactus growing near them. They wore necklaces made of mescal beans during their peyote rituals (figure 21.19).

(a)

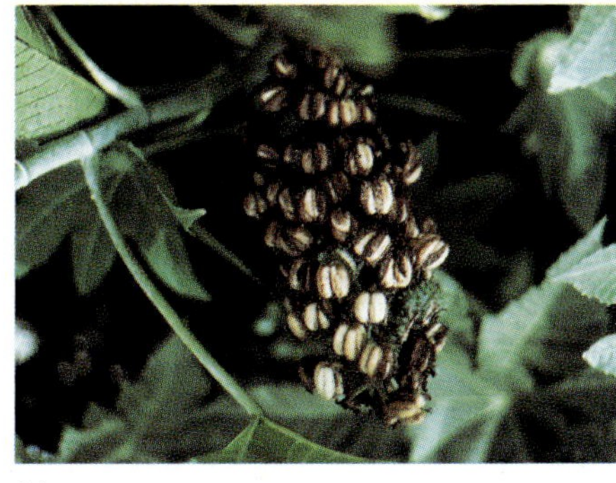
(b)

Figure 21.20
Castor bean plant (*Ricinus communis*). Although its vegetation is attractive enough for it to be a popular houseplant (*a*), and castor oil is a commercial product of its seeds (*b*), this plant also contains a deadly toxic alkaloid, ricin (not a component in the oil).

✔ *Concept Checks*

1. What are some herbal remedies for headaches?
2. What is a psychoactive plant?
3. What are some examples of hallucinogenic plants?
4. Why is the Solanaceae such an important plant family?

POISONOUS PLANTS

Of the approximately 500,000 plant species in the world, less than 1% of them are thought to be toxic to humans. In spite of this statistic, poison control centers in the United States find that plants consistently rank in the top three as sources of possible poisoning; for children under five, plants are the number one danger. This statistic is a recent phenomenon, resulting from significant increases in numbers of houseplants in homes, offices, and businesses during the 1970s and 1980s.

Historically, poisonous plants have been less a problem of accidental consumption than a purposefully ingested agent for medicinal or hallucinogenic purposes. The chemistry of plant toxicity is much better understood today, hence the reduced number of accidental poisonings resulting from incorrect dosage or application.

Classes of Poisonous Compounds

Internal Organ Poisons

These act primarily on the kidneys, liver, and stomach and include several distinct chemical types.

1. **Alkaloids** contain nitrogen and usually taste bitter. They are commonly found in several plant families, including the Solanaceae, Leguminosae, Papaveraceae, Rubiaceae, Apocynaceae, and Ranunculaceae.
2. **Glycosides**
   a. **Cyanogenic glycosides** include hydrocyanic acid (HCN), found in peach and cherry pits.
   b. **Cardiac glycosides** are the compounds found in *Digitalis*. They affect the heart.
3. **Oxalates** (oxalic acid) are found in rhubarb leaves and several other plants and produce severe stomach pain.
4. **Resins** are found in milkweed, laurels, and water hemlock and are especially potent toxins.
5. **Phytotoxins** are proteins, such as the active ingredients of castor bean plants (figure 21.20).

Nerve Poisons

These are mostly alkaloids. The toxins act on different parts of the nervous system.

1. Somatic nervous system toxins act on the striated skeletal or conscious control muscles. Curare is an example in this category.
2. Autonomic nervous system toxins affect the muscles of the heart. Examples are the alkaloids of the Solanaceae.
3. Central nervous system toxins, the opiate group, act on the brain and spinal cord.

Irritants

These compounds burn the skin, eyes, and throat, usually causing swelling, redness, welts, and even weeping lesions. Many are resins.

Mineral Poisons

Some plants accumulate unusually high concentrations of specific minerals found in the soil or in the air. Locoweed (*Astragalus mollissimus*) is a selenium accumulator. Sufficient ingestion by cattle or horses can cause "blind staggers" and even death. Lethal levels are difficult to consume, however, with horses requiring ingestion of over 500 pounds of plants over a 6-week period and cattle needing well over 2000 pounds over a 2- to 3-month period. Some roadside weeds accumulate lead from exhaust fumes and can produce lead poisoning in animals that feed on them.

Allergens

This is not actually a category of "poisonous" compounds because, unlike the preceding, allergens produce a toxic response only to those people (or other animals) who have a specific sensitivity. The true poisons produce a response in all individuals. Poison ivy (*Rhus toxicodendron*) and poison oak are probably the most common allergens (figure 21.21). Over 100 million people in the United States alone are sensitive to them.

Poisonous Bacteria and Fungi

Botulism is food poisoning caused by a powerful toxin produced by the bacterium *Clostridium botulinum*. It is an

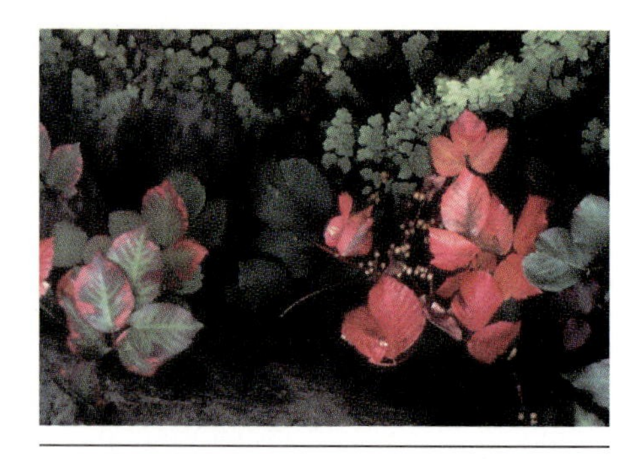

Figure 21.21
Poison ivy (*Rhus toxicodendron*) is probably the most widespread and serious plant allergen.

Figure 21.22
This *Amanita* (destroying angel) is related to *A. muscaria* (fly agaric) and *A. phalloides* (death angel).

anaerobic organism, sometimes infecting home- and commercially canned foods. It has been estimated that a single drop of pure toxin in the water supply of New York City could kill the entire metropolitan population. Since it is anaerobic, sealed cans and jars will build up carbon dioxide, causing the lids to puff outward, a sure sign of infected contents. Such cans should never be opened, since that would release the positive pressure and might spray botulism toxin all over the immediate area. Cooking the food at boiling temperatures for sufficient time will destroy the toxin.

Of the 15,000 or so species of fungi, approximately 75 are technically poisonous, but fewer than 10 are truly deadly. The most poisonous of these is the "death angel," *Amanita phalloides*, a close relative of fly agaric, *A. muscaria* (figure 21.22). The active compounds, **amanitin** and **phalloidin,** are internal organ poisons that attack the liver and kidneys, producing a painful death. It was once a popular means to eliminate political and other enemies while having it look like death by "natural causes." The symptoms vary according to dosage. A lethal amount ingested will not produce symptoms for 12 hours, whereas a relatively mild case results in symptoms almost immediately. The most trusted household servants during the Dark and Middle Ages became the cook and wine steward because of the ease with which someone could poison a meal. The practice of having an animal or slave taste food arose because of this danger, and the delayed symptoms of severe death angel poisoning were one of the reasons for its popularity among "professional poisoners."

Deadly Plants

The degree of toxicity among poisonous plants varies with the nature of the active compound and with its concentration in the plant. Some plants are only mildly poisonous and others deadly. Figure 21.23 and table 21.2 include some of the plants commonly reported to the Poison Control Center hospitals in the United States. This list includes most of the highly dangerous plants that one might encounter in home, yard, or garden but does not cover many wild poisonous plants. Several of the most lethal wild and domesticated plants are also included in the following discussion.

(a)

(b)

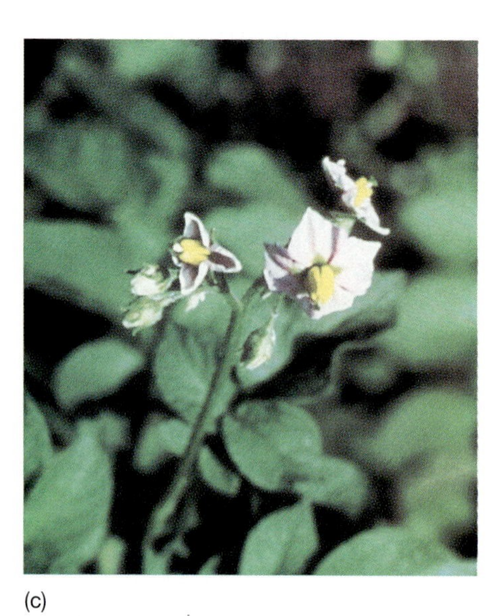

(c)

Figure 21.23
Some poisonous plants. (*a*) Mistletoe berries (*Phoradendron tomentosum*). (*b*) Daphne. (*c*) Irish potato (*Solanum tuberosum*).

TABLE 21.2

Common Poisonous House and Yard Plants

COMMON NAME	SCIENTIFIC NAME AND FAMILY	POISONOUS PARTS	TOXIC COMPOUND	SYMPTOMS
Houseplants				
Castor bean, rosary pea	*Ricinus communis* (Euphorbiaceae)	Seeds; must be chewed	Ricin, a lectin	Nausea, muscle spasms, purgation, convulsions, and death; eight seeds considered a lethal adult dose; a child could die from chewing one or two
Mistletoe	*Phoradendron* (Loranthaceae)	All parts, especially berries	Toxic amines; protein	Stomach upset and, in severe cases, death from inability to breathe
Dumb cane	*Dieffenbachia* (Araceae)	All parts	Calcium oxalate and proteolytic enzymes; asparagine	Burning sensation in mouth and throat, swelling of mouth and throat tissues; could lead to asphyxiation in small children
Caladium	*Caladium* (Araceae)	All parts, including bulbs	Calcium oxalate and/or irritant juices; asparagine	Burning sensation in mouth and throat, swelling of mouth and throat tissues; could lead to asphyxiation in small children
Elephant's ear	*Colocasia antiquorum* (Araceae)	All parts	Calcium oxalate and/or irritant juices; asparagine	Burning sensation in mouth and throat, swelling of mouth and throat tissues; could lead to asphyxiation in small children
Philodendrons	*Philodendron* and *Monstera* (Araceae)	All parts	Calcium oxalate and/or irritant juices; asparagine	Burning sensation in mouth and throat, swelling of mouth and throat tissues; could lead to asphyxiation in small children
Crotons	*Croton* (Euphorbiaceae)	Seeds, leaves, and stems	Croton oil	Powerful purgative action that can cause death on ingestion of small amounts
Poinsettia	*Euphorbia pulcherrima* (Euphorbiaceae)	All parts	Milky sap contains acrid irritants as complex esters related to the diterpene phoabol	Can cause severe irritation if ingested in large quantities; death unlikely but possible if the individual is allergic to the irritants
Crown of thorns	*Euphorbia milii* (Euphorbiaceae)	All parts	Milky sap contains acrid irritants as complex esters related to the diterpene phoabol	Can cause severe irritation if ingested in large quantities; death unlikely but possible; like poinsettia, it should not really be classified as a poisonous plant, only as an irritant or allergen
Yard and Garden Plants				
Daffodil, narcissus, jonquil	*Narcissus* (Liliaceae)	Bulbs	Calcium oxalate crystals	Nausea, vomiting, diarrhea; if eaten in quantity, could be fatal
Autumn crocus	*Colchicum autumnale* (Liliaceae)	Bulbs, flowers, seeds	Colchicine and other alkaloids	Burning in the throat and stomach; vomiting; weak, quick pulse; kidney failure; respiratory failure commonly resulting in death
Star-of-Bethlehem	*Ornithogalum umbellatum* (Liliaceae)	All parts, including bulbs	Cardiac glycosides	Nausea, gastroenteritis
Pokeweed, pokeberries	*Phytolacca americana* (Phytolaccaceae)	Roots, shoots, berries	Triterpene saponin, lectin-mitogen	Severe gastrointestinal disturbance, weakened pulse and respiration; can be fatal
Iris	*Iris* species (Iridaceae)	Leaves and rhizomes	A glycoside	Gastrointestinal disturbance; sometimes a burning in the mouth
Yew (English and Japanese)	*Taxus* species (Taxaceae)	Bark, leaves, seeds	Alkaloid, taxine	Fatal; taxine readily absorbed into the intestines; death is sudden

Continued

TABLE 21.2

Common Poisonous House and Yard Plants—cont'd.

COMMON NAME	SCIENTIFIC NAME AND FAMILY	POISONOUS PARTS	TOXIC COMPOUND	SYMPTOMS
Yard and Garden Plants—cont'd.				
Oleander	*Nerium oleander* (Apocynaceae)	All parts	Cardiac glycosides	Severe vomiting, bloody diarrhea, irregular heartbeat, drowsiness, unconsciousness, respiratory paralysis, and death; children have been reportedly poisoned by chewing a leaf and adults by eating hot dogs cooked on oleander stems used as skewers
Daphne	*Daphne mezerium* (Thymelaeaceae)	Berries	Coumarin glycosides and a diterpene, mezerein	Fatal; a few berries can kill a child; burning of the throat and stomach, internal bleeding, weakness, coma, and death
Rhododendron, laurel, azalea	*Rhododendron* species (Ericaceae)	All parts	Resins	Vomiting, weakness, and in extreme cases paralysis and death
Buckeye, horse chestnut	*Aesculus hippocastanum* (Hippocastanaceae)	Leaves, flowers, young sprouts, seeds, bark	Lactone glycoside, esculin	Vomiting, diarrhea, lack of coordination, paralysis. California Indians used extracts of buckeye as fish poison
Wisteria	*Wisteria* species (Leguminosae)	Seeds, pods	Lectin, mitogens	Mild to severe digestive upset
Jimsonweed, thornapple	*Datura stramonium* (Solanaceae)	All parts, especially the seed	Scopolamine, hyoscyamine alkaloids	Flushed skin, dilated pupils, dry mouth, delirium, and death due to respiratory failure
Woody nightshade, bittersweet	*Solanum nigrum* or *S. dulcamara* (Solanaceae)	Leaves, roots, berries	Solanine, alkamine aglycones	Burning in the throat, nausea, dizziness, dilated pupils, weakness, convulsions; can be fatal
Irish potato	*Solanum tuberosum* (Solanaceae)	When tubers are exposed to sunlight, the green tissue under the skin	Solanine, alkamine aglycones	Burning in the throat, nausea, dizziness, dilated pupils, weakness, convulsions; can be fatal
Jerusalem cherry	*Solanum pseudocapsicum* (Solanaceae)	Berries	Solanine, alkamine aglycones	Burning in the throat, nausea, dizziness, dilated pupils, weakness, convulsions; can be fatal
Elderberry	*Sambucus* species (Caprifoliaceae)	Stems, leaves, unripe berries, and especially roots	Alkaloids, cyanogenic glycosides, and calcium oxalates	Burning of mouth, throat, and stomach
Holly	*Ilex* species (Aquifoliaceae)	Berries	Toxic compound unknown	Not fatal, but purgative and emetic and most dangerous to children
English ivy	*Hedera helix* (Araliaceae)	Leaves, berries	Triterpene sapogenin, hederagenin	Vomiting, diarrhea, and nervous depression; can be serious in small children
Foxglove	*Digitalis purpurea* (Scrophulariaceae)	Leaves, seeds, flowers	Digitoxin, cardiac glycosides	Large amounts cause dangerously irregular heartbeat, mental confusion, and digestive upset; usually fatal
Lily of the valley	*Convallaria majalis* (Liliaceae)	Roots, leaves, flowers, fruit	Cardiac glycoside, convallatoxin	Large amounts cause dangerously irregular heartbeat, mental confusion, and digestive upset; usually fatal

Continued

TABLE 21.2

Common Poisonous House and Yard Plants—cont'd.

COMMON NAME	SCIENTIFIC NAME AND FAMILY	POISONOUS PARTS	TOXIC COMPOUND	SYMPTOMS
Yard and Garden Plants—cont'd.				
Monkshood	*Aconitum* species (Ranunculaceae)	All parts, especially the roots and leaves	Aconitine (alkaloid)	Numbness, paralysis of upper then lower extremities; death by respiratory paralysis; very poisonous
Rhubarb	*Rheum rhaponticum* (Polygonaceae)	Leaves (not the petioles, which are edible)	Oxalates, anthraquinone glycosides	Stomach pain, nausea, vomiting, and in serious cases convulsions, internal bleeding, coma, and death; very poisonous

Hemlock (Conium maculatum, *Umbelliferae*)

A herbaceous plant with leaves typical of the carrot family, **hemlock** contains the toxin coniine. It is not the same as the woody hemlock tree. This particular species is recognizable by the reddish purple spots on its stem (figure 21.24).

Socrates was poisoned with hemlock in 399 BC for crimes against the state. He was convicted by Athens politicians for "neglecting the gods whom the city worships" and "corrupting the youth of the city" with his ideas and unwillingness to support the politicians in power. There is evidence that the officials of Athens did not want him to die; they were afraid he would have more support as a martyr than alive. Socrates was given repeated chances to plead guilty and receive a light sentence, but he refused because he had done nothing wrong. He was even given a chance to escape, which he refused. Finally, the sentence was carried out, and he was given a cup of hemlock juice to drink, followed by a laudanum of wine and probably opium to ease the pain of his death.

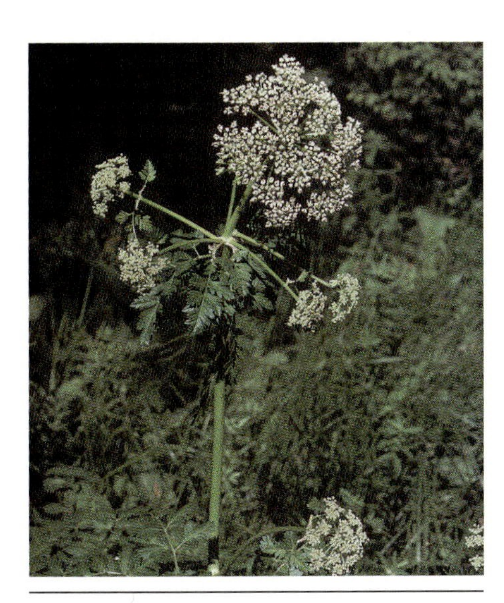

Figure 21.24
Hemlock (*Conium maculatum*) was the plant used to put Socrates to death.

Socrates insisted on having the stages of the hemlock poisoning accurately observed and recorded. He walked around until his legs began to grow heavy; then he lay down and had someone feel his feet and legs and report the progression of numbness and cold as it approached his heart. He died of respiratory failure.

Water Hemlock (Cicuta *spp., Umbelliferae*)

Also a herbaceous plant, water hemlock is found growing in wet habitats such as shallow streams and along the edges of ponds and lakes (figure 21.25). Its leaves are not fernlike, as are hemlock leaves; rather, they superficially resemble marijuana leaves. The hollow stems contain a yellow sap, which has the deadly cicutoxin in it. Especially concentrated where the stem and root join, this toxin is even more poisonous than hemlock juice. Death is a painful process involving severe convulsions and respiratory failure.

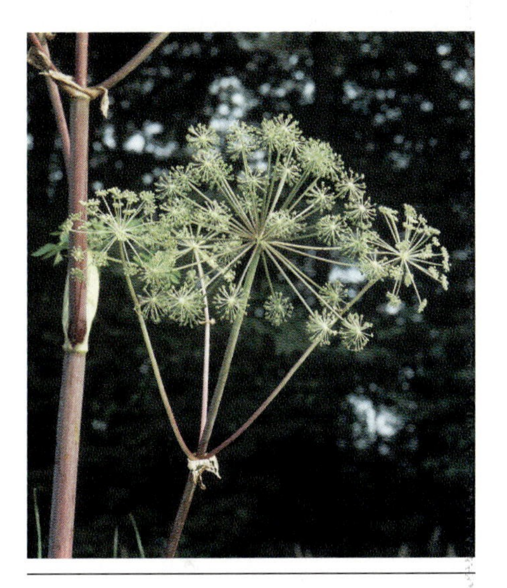

Figure 21.25
Water hemlock (*Cicuta*) is even more poisonous than its close relative hemlock.

Monkshood (Aconitum napellus, *Ranunculaceae*)

Also known as wolfsbane, this plant contains the alkaloid **aconitine,** which is a very powerful poison (figure 21.26). Used as an arrow poison by early stone-age cultures, this toxin is very fast acting, a feature that once also made it a favorite poison.

One of the most famous poisonings accredited to aconitine is that of the Roman emperor **Claudius,** who died suddenly in AD 54. Historians first blamed Julia Agrippina, Claudius's empress, of doing away with him so that her son by a previous marriage, Nero, could rule. It was later

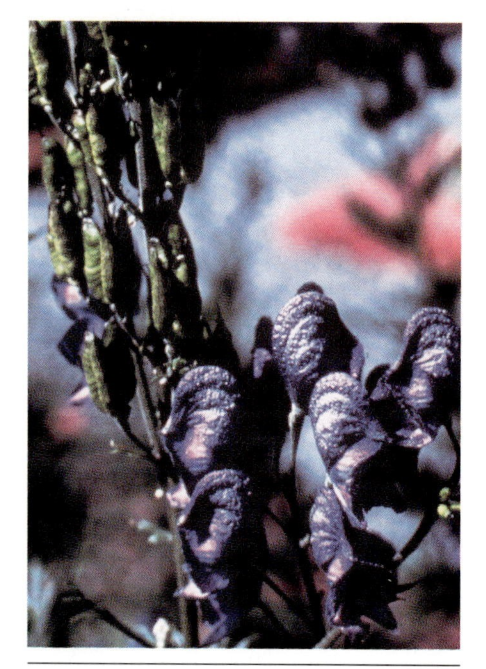

Figure 21.26
Monkshood (*Aconitum napellus*) contains a powerful toxin credited for causing the death of the Roman emperor Claudius.

Figure 21.27
Yew (*Taxus*) is very poisonous but also has excellent wood for bow making.

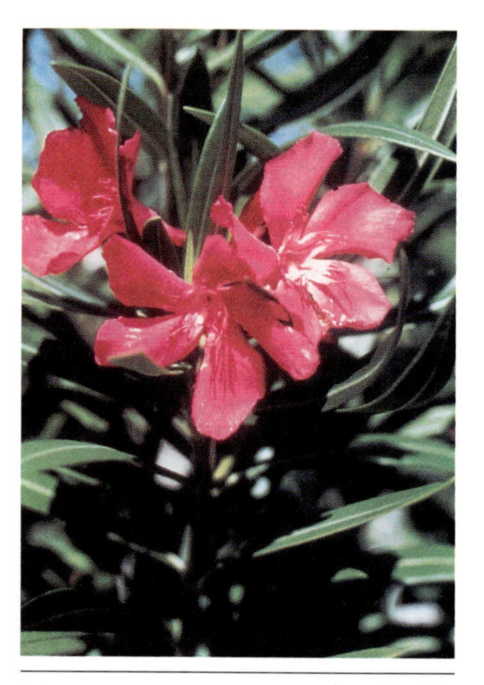

Figure 21.28
Oleander (*Nerium oleander*), a popular yard ornamental, is reportedly a very poisonous plant.

decided that Locusta was the poisoner, working with Stertinus Xenophon, Claudius's personal physician. Locusta also poisoned Britannicus, Claudius's son, so that Nero could rule alone rather than be co-emperor with Britannicus. Some reports have Claudius poisoned by death angel mushrooms, but the better-documented story is that he was fed only a small dose of these mushrooms to evoke immediate symptoms. A feather, previously dipped in aconitine, was used by the physician to tickle his throat to induce vomiting and save Claudius from the attempted mushroom poisoning. That he died anyway was probably accredited to heart failure from the close call, and murder was not officially suspected.

Monkshood is also credited by historians as being the "murder weapon" in the death of Pope Adrian VI and in an unsuccessful attempt on the prophet Mohammed's life. In spite of its well-known toxicity, this plant is attractive and is grown in gardens as an ornamental, especially in Scandinavian countries.

Yew (Taxus *spp.*, *Taxaceae*)

This plant was prized for the elasticity of the wood and particularly for its value in making the finest of bows (figure 21.27). Robin Hood used a yew wood bow because of its qualities. The alkaloid **taxine** is very poisonous and is found in highest quantities in Japanese and English yew (*T. cuspidata* and *T. baccata*). The word **toxon** means bow, and **toxin** means poisonous. The genus name *Taxus* is thought to be derived from some combination of these words.

Oleander (Nerium oleander, *Apocynaceae*)

Oleander is an extremely poisonous plant (figure 21.28). Reported deaths include that of a child who chewed on a single leaf. Unconfirmed reports attest to the danger of eating a hot dog cooked on an oleander stick. Many people value oleander as an attractive landscaping plant without being aware of its danger (or perhaps in spite of it).

Jimsonweed (Datura stramonium, *Solanaceae*)

Jimsonweed, along with many other members of this family, is highly toxic. Also known as **thornapple** because of the spiny fruit, and **moonflower** because of its night-blooming white flowers, the name jimsonweed stems from an incident that occurred in 1676. During Bacon's Rebellion a group of soldiers on a march near Jamestown, Virginia, added some *Datura* leaves to their stew. The result was a stupefied regiment that had to be replaced. The "Jamestown weed" in time became known as jimsonweed.

Tobacco (Nicotiana tabacum, *Solanaceae*)

Although tobacco is not normally listed as a poisonous plant, the presence of the alkaloid **nicotine** in the leaves qualifies it for inclusion in this section. A drop of pure nicotine is toxic enough to kill a medium-size dog in only seconds, and nicotine sulfate is one of the most effective insecticides in use (figure 21.29).

Figure 21.29
Tobacco (*Nicotiana*) contains the toxic and addictive alkaloid nicotine.

In addition to its toxicity, nicotine is a stimulant and a physiologically and psychologically addictive drug, complete with physical withdrawal symptoms. In most countries, however, it is legal and is not classified as a narcotic; all other addictive alkaloids are so classified in the United States. Finally, this toxic and addictive drug is known to be a significant contributing factor in a number of respiratory ailments and in throat and lung cancer. This is true even for "secondary smokers" (nonsmokers who breathe smoke from nearby smokers).

✔ Concept Checks

1. What are examples of the three classes of nerve poisons?
2. In what plant families are alkaloids commonly found?
3. Name ten poisonous house and yard plants.
4. How was Socrates put to death?

Summary

1. The early history of botany was the history of herbalism and medicinal plants. Modern medicine includes little "traditional medicine" in the Western world but a balanced approach in China. Some of the more primitive societies in Africa and South America also depend extensively on their medicine men even today.
2. Medical plants were often surrounded by mysticism and superstition even though many of them contained compounds that did affect the body physiologically. Some of the well-documented medicinal plants include mandrake, *Duboisia,* and belladonna, all in the Solanaceae. In addition, *Cinchona,* opium poppy, curare, *Rauwolfia, Ephedra,* foxglove, and ergot have medicinal properties.
3. Many herbal remedies have less dependable curative powers, but many people still depend on these plants for general well-being, minor pain relief, topical cuts and burns, internal organ disorders, and respiratory tract disorders.
4. Psychoactive plants include stimulants such as cocaine, depressants such as the opiates, and true hallucinogens such as marijuana, ergot, fly agaric, sacred mushrooms, peyote, and members of the Solanaceae.
5. Poisonous plants contain active ingredients in one of several classes of compounds. Internal organ poisons include the alkaloids and cardiac and cyanogenic glycosides. Also producing toxicity in the internal organs are plants containing oxalates, resins, and phytotoxins.
6. There are about 75 species of fungus that are poisonous and about 10 that are deadly. Members of the genus *Amanita* are among the most toxic of all mushrooms.
7. The deadly toxic plants include hemlock and water hemlock, monkshood, yew, oleander, jimsonweed, and tobacco. Many others are only slightly toxic.

Key Terms

aconitine 411
alkaloids 407
amanitin 408
atropine 406
belladonna 399
botulism 407
bronchodilator 400
Claudius 411
codeine 399
curare 399
doctrine of signatures 398
dropsy 400
ephedrine 400
hallucinogens 403
hemlock 411
herbalism 397
heroin 399
holistic medicine 397
hyoscyamine 398
laudanum 399
lysergic acid 404
morphine 399
nicotine 412
opium 398
oxytocic agent 400
peyote 405
phalloidin 408
psychoactive plants 402
quinine 398
sedatives 403
stimulant 403
taxine 412
THC (tetrahydrocannabinol) 404
thornapple 412
toxin 412

Discussion Questions

1. In today's world, could there be a place for herbalism and holistic medicine?
2. What is the future of new medicines from plants?
3. Why is experimentation with medicinal/hallucinogenic plant compounds dangerous?
4. Should nicotine be classified as an illegal narcotic drug or should others be made legal?

Suggested Readings

Duke, J. A., and R. Vasques, 1994. *Amazonian ethnobotanical dictionary.* CRC Press, Boca Raton.

Gold, M. S. 1989. *Marijuana.* Plenum Medical Book, New York.

Jacob, I., and W. Jacob, eds. 1993. *The healing past: Pharmaceuticals in the biblical and rabbinic world.* E. J. Brill, New York.

Kinghorn, A. D., and M. F. Balandrin, eds. 1992. *Human medicinal agents from plants.* American Chemical Society, Washington, D.C.

MacGregor, F. E., ed. 1993. *Coca and cocaine: An Andean perspective.* Greenwood Press, Westport.

Murphy, L., and A. Bartke, eds. 1992. *Marijuana/cannabinoids: Neurobiology and neurophysiology.* CRC Press, Boca Raton.

Musto, D. F. 1991. Opium, cocaine and marijuana in American history. *Sci. Am.* 265(1):40–47.

Painter, J. 1994. *Bolivia and coca: A study in dependency.* L. Reinner Publishers, Boulder.

Plotkin, M. J. 1993. *Tales of a shaman's apprentice: An ethnobotanist searches for new medicines in the Amazon rain forest.* Viking, New York.

Ratsch, C. 1992. *The dictionary of sacred and magical plants.* ABC-CLIO, Santa Barbara.

Schultes, R. E., and R. F. Raffauf. 1990. *The healing forest: Medicinal and toxic plants of the northwest Amazonia.* Dioscorides Press, Portland.

Tyler, V. E. 1992. *The honest herbal: A sensible guide to the use of herbs and related remedies.* 3rd ed. Pharmaceutical Products Press, New York.

United States Congress. 1994. *Medicinal uses of plants: Protection for plants under the Endangered Species Act.* Serial #103–74. U.S. Government Printing Office, Washington, D.C.

Weil, A., and W. Rosen. 1993. *From chocolate to morphine: Everything you need to know about mind-altering drugs.* Houghton Mifflin, Boston.

Weiss, R. D., S. M. Mirin, and R. L. Bartel. 1994. *Cocaine.* American Psychiatric Press, Washington, D.C.

Weissmann, G. 1991. Aspirin. *Sci. Am.* 264(1):84–91.

Wijesekera, R. O. B., ed. 1991. *The medicinal plant industry.* CRC Press, Boca Raton.

Chapter Twenty-Two

Our Precarious Habitat

Natural vs. Human Environment

Our Impact on Natural Ecosystems
Habitat Destruction and Ecosystem Imbalance

Conservation

Endangered Species
Biodiversity

Environmental Quality

Pollution

Air
Water
Soil
Noise

The Greenhouse Effect and Global Warming

The Ozone Layer

Acid Rain

Economic and Environmental Trade-Offs

Recycling and Clean-Up Efforts

Revegetation and Environmental Repair

Positive Solutions

A single fruit of the common dandelion sails on the wind.

This chapter pulls together the basic understanding of ecological interactions and ecosystem balances learned in chapters 2 and 3 with the basic physiological and genetic functions of plants covered in later chapters and addresses the role of human society in the natural biological world. What have our actions as a single species done to thousands of other species of plants and animals, and over what span of time? What damage has been done to our abiotic world, and what are the consequences? How have our actions now started affecting us as a species, as a society, as individuals? Can we live with pollution, global warming, acid rain, resource shortages, and economic consequences? Is there anything we can change?

Although the environmental problems and ecological imbalances are serious, awareness is rapidly changing to understanding, and when true understanding of the problems is widespread throughout all levels of human society, solutions are possible. Governmental regulation and resulting modifications in business practices will not be enough, however; individual action will be necessary. At the end of this chapter and throughout the book, each individual student should gain a clear understanding of the problems and an appreciation of their options as educated individuals; they should be among the leaders in changing what has been happening to our precarious habitat.

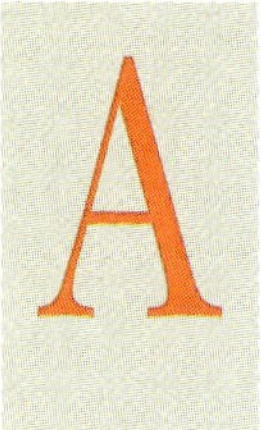

Are we humans *a part of* or *apart from* the earth's natural ecological balance? This is a question that more and more of us are asking. It is also a question whose answer requires objectivity and an understanding of how environmental interactions occur.

NATURAL VS. HUMAN ENVIRONMENT

The correct answer is *yes,* we are part of the natural environment because our activities affect the complex ecological interactions, and *no,* because we do not function in an interdependent way with the other species and their habitats. Because we use the same abiotic resources (water, light, soil, air) and because our use of those resources has an impact on the "natural" or wild plants and animals, we cannot separate ourselves from those ecological functions nor the other species involved in them.

On the other hand, we can (up to a point) control or influence the resources upon which our continued existence depends; we are not immediately subject to the same ecological rules of energy flow, resource cycling, population size, and carrying capacities that all other living organisms are. As a species we have learned to increase productivity and provide suitable habitat for many more humans than could otherwise survive as basic hunters and gatherers. Our intellectual capacities have resulted in a competitive advantage over other species and the ability to manipulate our environment and find resources unavailable to the rest of the biological world.

Therefore, we can evaluate how we humans affect the natural ecological balances of the biosphere from the perspective of an outsider, an agent of change that has an impact on other species and their habitats but not as part of those interactions. In this sense, we are not part of the natural world; we are apart from it. This has been the attitude that humans have taken for centuries, and not until relatively recently has there been widespread exception taken to this position.

As we learn more about how resource wasteful we have been and how rapidly we are running short of those resources, the more concerned and aware we have become. Today, most of the developed world and its better-educated populace are at least marginally aware of how dependent we are on the "natural" ecological balances that we have so seriously damaged. We are now starting to measure our future by the same ecological yardstick as all of the other interdependent biological components of the biosphere. Now we realize that we in fact are part of those interrelationships; *we are a part of our natural environment.*

Our Impact on Natural Ecosystems

Because of the obvious changes on the global environment that humans have made since the industrial revolution, one might make the assumption that our negative impact has been relatively recent. Even though the damage has not been globally visible until the last 20 or 30 years, studies have shown that in certain regions, human destruction of natural habitats began some 7000 years ago.

In the Mediterranean area, for example, domesticated plants and animals were common 6000 years ago. Serious deforestation and the replacement of native flora by introduced species in the Mediterranean region happened at least 2500 years ago. In the British Isles, **habitat destruction** during the past 3000 to 4000 years has resulted in the disappearance of nearly 90% of the natural forests and associated wildlife.

In North America, the 1 million square kilometers of Atlantic forest has been reduced to less than 7% of that total, and only about 1% is undisturbed. Our central prairies have been plowed or overgrazed into oblivion in the past 200 years from over 100 million acres to less than 1% of that total, scattered in small relictual patches. Our natural wetlands are also disappearing at an alarming rate.

How much damage humans cause to the environment depends on a combination of how many people there are plus how many resources each person consumes. In addition, how much damage occurs in producing these resources? No matter how careful we are in the extraction or production of the resources we consume, and no matter how much we

Figure 22.1
Forested land being cleared for agriculture.

modify our standard of living to conserve those resources, if the world's population continues to grow, we will continue to deplete earth's bounty faster than it can be replaced, and we will do so at the expense of environmental integrity.

Between 1950 and 1987, the world's population doubled from 2.4 billion to 5 billion. This is an increase in 37 years that matches the total human population increase from the time we first emerged as a species until 1950. United Nations' estimates are that by 2025 the global population will increase to 8.5 billion and ultimately 10 billion by the end of the twenty-first century. Most significantly, 95% of this growth will occur in developing countries. Even though the populations of developed countries consume far more resources per person than in developing countries, at the minimum each person requires food, shelter, and clothing. To provide these basics, forests will be cut, marginally productive land will be plowed and cultivated, and energy resources will be consumed (figure 22.1).

The disparity between the developed and developing countries is already great. Only about 25% of the earth's population reside in the wealthy countries, but they consume 80% of the commercial energy or about 32 barrels of crude oil per person per year.

Automobiles are one of the most significant users of our petroleum resources. In 1990, there were approximately 500 million cars in the world, each one burning an average of two gallons of fuel per day. This converts to about one-third of the world's daily oil production. In addition, the number of automobiles continues to increase at a rate faster than population growth. At current rates, there could be as many as 2 billion automobiles by the year 2025.

Habitat Destruction and Ecosystem Imbalance

Even though our own temperate forests, grasslands, and wetlands are represented only by very few intact examples of what used to be, more is heard about the loss of the world's tropical forests than any other ecosystem. There are two reasons for this. First, the tropical forests contain more than half of the world's known plant and animal species, with estimates that only 30% of the species actually housed in the tropics have even been identified and named. Thus the loss of such a huge amount of biodiversity is especially distressing.

The second reason for so much concern is that tropical soils are productive only while the forest is intact. Once cleared and kept clear for agricultural use, it is unlikely that a tropical forest could ever reestablish; they will be lost forever. The rate of forest clearing continues to accelerate in some countries because of their rapidly growing human populations. It is estimated that 11 million hectares (27 million acres) of tropical forests, are destroyed each year. At this rate, in 25 years tropical rain forests will disappear completely from nine countries, four in the Americas, three in Africa and two in Asia.

The tropical forests are cleared for firewood, for crops, and for sales of hardwoods (figure 22.2). The only way to prevent their ultimate destruction is either to halt the population growth in these countries or find other resources and ways of paying for them. Some efforts are being made in Costa Rica and other tropical countries to exploit the forest in a different way—as a tourist attraction. Revenue raised by protecting and sharing these natural areas can be used to purchase building materials, food, and fuel.

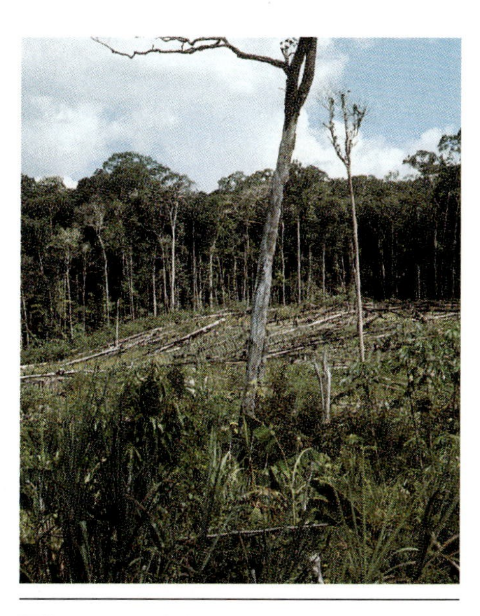

Figure 22.2
Tropical rain forests are being lost at an alarming rate worldwide, such as this one in Brazil.

The removal of our native plant communities is often followed by the introduction of exotic plant species or cultivars. Native range grasses have been replaced by more-productive hybrids and exotic selections. Many millions of acres of grassland and forest have been replaced by crop species for food and forage production. Even in our urban areas exotic introductions dominate our landscaped homes, businesses, parks, and roadsides. The loss of native plant communities removes the base of the food chain and reduces the habitat for the wild fauna of the region. The destruction of our native flora was not malicious nor were the consequences understood; however, those consequences are now known to be serious.

CONSERVATION

The pioneers who settled the United States saw before them an apparently endless vista of forests, grasslands, clear rivers, lakes, seashores—and opportunity. The natural resources seemed boundless, the biological diversity vast. The far-reaching extent of the resources probably precluded any consideration they might have had for conservation. Today, however, faced with shortages of freshwater, fossil fuels, clean air, food, and recreational sites, we wonder how they could have destroyed so much while settling America. A great deal is written about the need for conservation measures, but there is still not enough actually done. Although state and federal agencies have set aside some wilderness areas and nature preserves, much of the public still has a "parks are for the people" attitude. Too many people still cannot comprehend the expenditure of public tax dollars to establish tracts of land from which the public is barred. The pioneer spirit lives on: land is to be utilized, cleared, and made "productive" and "habitable."

The **conservation** movement began slowly in the 1940s and 1950s with relatively few champions, who were often regarded as extremists, doomsayers, and against progress. During the 1960s and 1970s laws were passed regulating air, water, and noise pollution; tracts of land were designated as preserves; and the public was slowly educated about the need for these measures. It was not until the late 1970s and 1980s, however, that this awareness has been based on understanding and the conservation movement supported by a large segment of the public (figure 22.3). With continued increases in public involvement, tax dollars, and legal controls, the next 20 years should hold actual gains and not just a slowing down of the rate of destruction. Such gains could ultimately result in a situation unique since hunting and gathering bands wandered the land: humans becoming in balance and in harmony with the land.

Endangered Species

Nothing more vividly symbolizes the impact of humans on the natural ecosystem balance than the number of plant and animal species that have become, or are in danger of becoming, **extinct,** even considering that the process of natural selection has produced millions of extinctions over the ages.

Figure 22.3
One of the more striking endangered plant species is found in Haleakala Crater on Maui, in the Haleakala National Park. The silver sword (*Argyroxiphium sandwicense*) grows in habitats that look almost like a moonscape.

The habitat destruction that has taken place is fairly recent. Between 1800 and 1850, only four plant species were known to have become extinct. Between 1851 and 1900, however, 41 species were lost to extinction and another 45 between 1901 and 1950. Today there are estimated to be some 20,000 species of vascular plants in the continental United States; 1200 of these are categorized as threatened, and approximately 750 species are in real danger of extinction (figure 22.4). These figures are based on information compiled by the Smithsonian Institution at the direction of the 1973 Endangered Species Act passed by the United States Congress. It is estimated that, since the Endangered Species Act was passed, as many as 100 plant species in the United States may have already become extinct. It would seem that the mere establishment of protective legislation is not sufficient to adequately preserve our natural biological resources. More preserves and wilderness areas need to be established, and a sincere effort at enforcing the laws needs to be made. Only in this way can theory be put into practice.

International efforts are even more inadequate. A 1940 "Convention Between the United States of America and Other American Republics" established an official document of agreement regarding nature protection and wildlife preservation in the western hemisphere. Unfortunately, some of the most massive habitat destruction and species extinctions are occurring in the overpopulated countries in the Americas as a result of massive land clearing for cultivation.

A 1973 "Convention on International Trade in Endangered Species of Wild Fauna and Flora" established a framework of cooperation among countries aimed at reducing the removal of threatened and endangered species and their sale in other countries. The intent of this convention was to reduce the effect that collectors have on rare plant and animal species. To date, these efforts have been minimal, but they show promise of having a positive impact. And yet, it is estimated that, worldwide, as many as two species become extinct every day! This is an alarmingly higher rate than the natural extinction process.

No single habitat type has been altered more significantly than the tropical rain forests. The tropics have an unusually delicate ecological balance. The specialized relationships of simple food webs result in a chain reaction. The removal of a link in the food chain is not an isolated event in such habitats; it can affect the survival of several other species as well.

The unfortunate response of some is: "so what?" There are plenty of other plant and animal species even if all of the endangered ones do become extinct. Of what importance is a snow leopard or silversword? They don't provide humans with food, medicine, or building supplies. Understanding the answer to this kind of question requires an appreciation of ecological

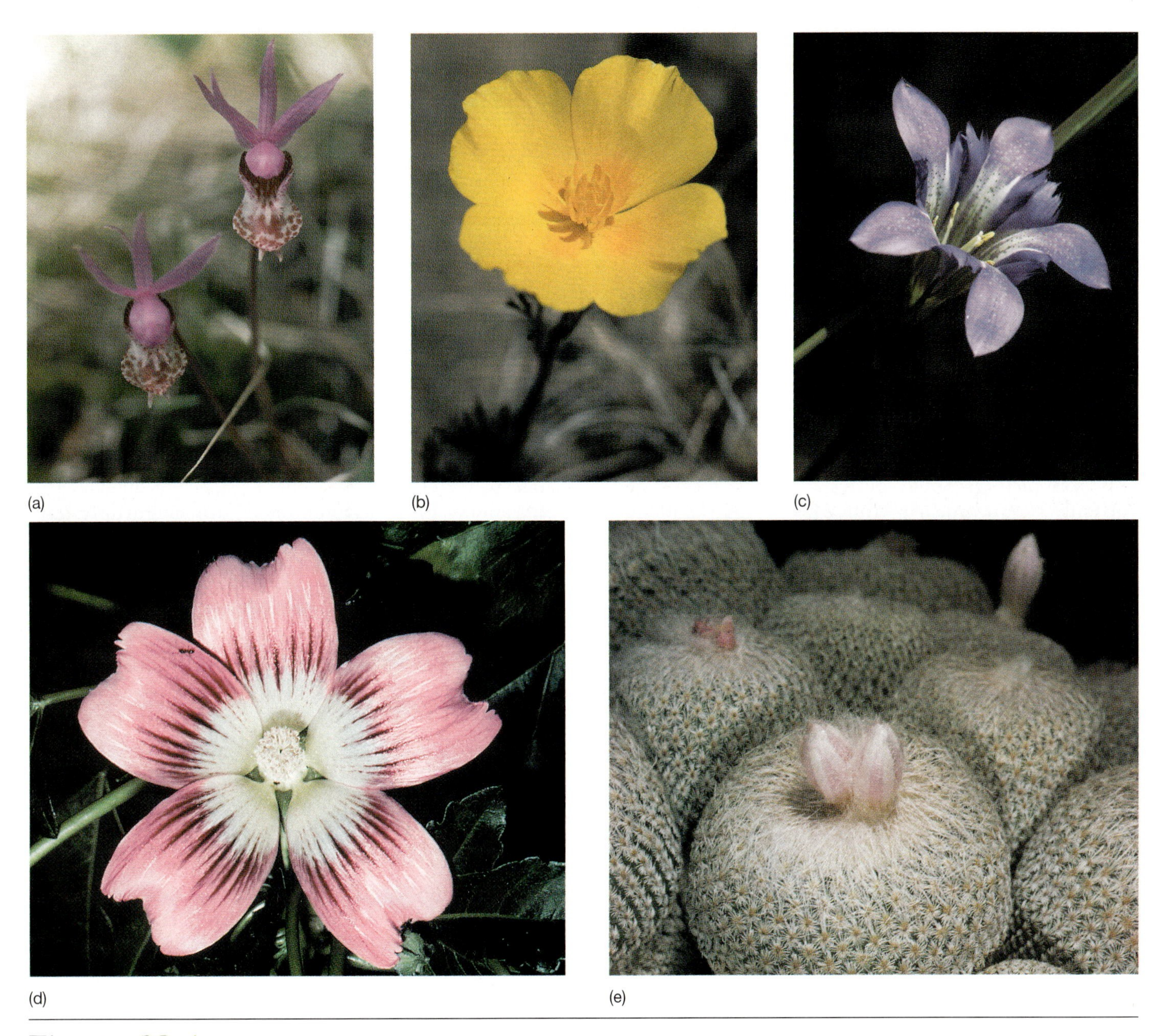

(a) (b) (c) (d) (e)

Figure 22.4

As environmental factors change naturally over time or as habitats are altered as a consequence of human activity small, relictual populations of plant species are often the result. These are several species that are considered threatened or endangered for extinction. (*a*) Fairy slipper. (*b*) *Eschschalzia ramasa.* (*c*) *Gentian.* (*d*) *Lavatera assurgentifolia.* (*e*) *Epithelantha bokei.*

balance. The loss of any given species may not be directly significant to human need, but that loss may affect the status of some other species in the chain that is important.

Another part of the answer concerns potential. So little is known about most species that humans do not use, that the rapid loss of species might include a potential crop plant, wonder drug source, or fuel reserve. Any unnecessary loss from the natural genetic pool stunts potential. In a world with massive shortages, this ought to be viewed as unacceptable. The propagation and preservation of rare and endangered plant species in botanical gardens and arboreta are significant steps in the fight against their extinction (figure 22.5). In these settings their biology can be studied, and they can then be reintroduced into appropriate protected habitats when available.

Finally, even if there is no applied use for a species, and even if it could be proven that its loss would not have an impact on other plants or animals, the loss of any aspect of the world's variety is reason enough to protect these species. It would be a great loss to future generations to know of certain plants and animals

(a)

(b)

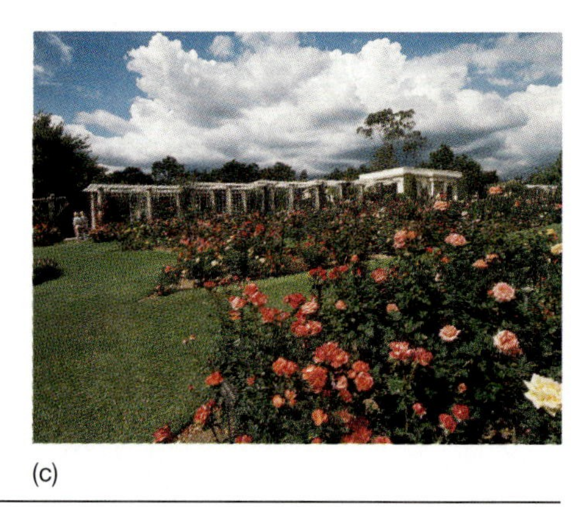
(c)

Figure 22.5
Botanical gardens, arboreta, and artificial-environment structures now house many of the world's plant species near extinction or known to be extinct in the wild. (*a*) The New York Botanic Garden, Bronx, NY; Enid Haupt Conservatory. (*b*) Missouri Botanical Garden, St. Louis; Japanese Gardens. (*c*) Huntington Botanical Gardens, San Marino, CA.

only from pictures. Some species are currently known primarily from zoos and botanical gardens, because they are already so scarce or already absent in the wild. The wholesale reduction in total habitat size for many species has already resulted in small relictual populations, sometimes protected by government decree, sometimes not.

The critical component in whether zoo protection, wilderness areas, legislation, and any other measures will significantly slow the rate of extinctions resulting from human activity is the awareness of the lay public. Until those who are actively destroying available natural habitats realize the consequences of their actions, there will be little progress. In underdeveloped and overpopulated countries the education of the lay person to these problems is especially difficult. The instinct for self-preservation is an immediate response to need; the significance of today's actions on the next generations is far too abstract to have much impact. The degree and the continuation of protection, therefore, are in the hands of the educated public. It is up to those of us who understand, care to speak out, and rationally support conservation measures.

Biodiversity

According to the National Biological Diversity Conservation and Environmental Research Act, biological diversity means "the full range of variety and variability within and among living organisms and the ecological complexes in which they occur and encompasses ecosystem or community diversity, species diversity, and genetic diversity."

The exact magnitude of **biodiversity** (biological diversity) on earth is unknown. The approximately 1.5 to 1.7 million formally named species are only a fraction of the number of plant and animal species in existence. In addition, the vast majority of the world's species are tropical, and this is the part of the earth in which the destruction of habitats and the concurrent loss of biodiversity is most serious. Tropical rain forests are being cut down at a 0.6% annual rate, which means that some 4.6 million hectares per year are being destroyed. Sadly, moist deciduous forests are being destroyed at a 1.0% annual rate or 6.1 million hectares per year.

This is the largest and most rapid loss of diversity since the massive natural extinctions at the end of the Paleozoic and Mesozoic eras, some 65 million years ago. Tropical forests occupy less than 7% of the earth's terrestrial surface but they are home to 50% of all plant and animal species. Another way of expressing human impact on earth's biodiversity is to point out that the rate of species loss through deforestation is approximately 10,000 times greater than the naturally occurring extinction rate that existed prior to the existence of humans. Past extinctions of this magnitude have changed the course of evolutionary history, the most recent one resulting in the end of the age of dinosaurs and the beginning of the mammals.

Why does the loss of biodiversity matter? Well, for one thing, we don't know what the ecological consequences will be and how we will be affected by these changes. More practically, only a very small percentage of the species with potential human economic importance have been used by humans. For example, only about 7000 plant species have ever been utilized by humans for food, and only a few of these have become staples of human life. There are many tens of thousands of edible species, many with proven superiority to those currently in use. Other plant species may be sources for future miracle drugs, cures for cancer or AIDS. If lost due to extinction before scientists can even fully assess their potential, human society will realize the greatest loss.

The saddest part of this loss is that it will affect our children and grandchildren far more than it will affect us. Because of our ability to alter natural processes, humans are the stewards of the planet, and we have the responsibility of maintaining its biodiversity. What we are doing, however, is heading towards the eradication of thousands of species in the span of a few hundred years—the blink of an eye in terms of an evolutionary time scale.

Figure 22.6
Still an all too common scene.

Figure 22.7
The dilemma of how to handle our "throw-away" society is reaching crisis proportions in some areas. Landfills are no longer the solution.

A translation from a "primitive" culture provides a perspective on our responsibility:

> Treat the Earth well,
> We did not inherit it
> from our parents,
> It is on loan to us
> from our children.

Concept Checks

1. Are humans part of the earth's natural ecosystem?
2. What is the basic cause of natural resource depletion in the past 200 years?
3. Why is the loss of the tropical rain forests such a concern?
4. What causes the extinction of a species?

ENVIRONMENTAL QUALITY

The most significant issue in human ecology during the past three decades is **overpopulation.** The degradation of our environment is a direct function of ever-increasing world population size. Resource shortages, pollution, and urban crowding all result from overpopulation. Until humans realize that the laws of natural balance do not apply only to all other species, but also to *Homo sapiens,* a rational solution to these problems is unlikely. Other solutions are possible but not desirable. Mass starvation, epidemics, or world war would effectively reduce the world's human population, but it is hoped that none of these will ever occur.

On the other hand, total dependence on continued technological advancements and vastly improved crop strains is not a realistic approach either. The quality of our environment will ultimately be determined by how many demands are made on it, and reduced demands can happen only when there is a reduced human population combined with more conservative resource consumption per person.

Fortunately, natural systems have amazing resiliency and buffering capacity. Even ecosystems that seem for all practical purposes to be "dead" have the ability to recover if given enough time for cleaning and restoration. Lake Erie, for example, had become so polluted with industrial wastes during the 1960s that almost all life in that once magnificent lake had died. Once the cleanup began, under strict controls, Lake Erie started to recapture its ecological balance.

The same may be true for the recovery from one of the world's most disastrous and costly oil spills, caused by the Exxon *Valdez* oil tanker that ran aground in Alaska. Clean-up costs have been enormous, but no one yet knows how long-term the effects on the environment will be in that region.

In an industrialized society the cost to environmental quality is great. Every convenience and time-saving gimmick has its price tag, and we have sometimes cut corners for which we pay dearly at a later time. The "throw-away" attitude has made some material goods obsolete and created mountains of garbage for which there are no landfills (figure 22.6). The intentions were valid in the beginning: paper plates meant less dishwashing (less labor, less water, less detergent, and fewer sanitation problems) and a boon to the fast-food industry. Plastic containers were convenient, readily disposable, lightweight, and required less labor and lower shipping costs, a benefit to the transportation industry. Disposable diapers were thought to be the panacea for modern parents looking for a solution to a huge laundry problem. Attention was not paid to the enormous tasks of waste disposal and cycling (figure 22.7). Degradation of plastic is slow or impossible, and burning leads to toxic fumes and

(a)

(b)

(c)

Figure 22.8
Water pollution (*a*) and (*b*) and air pollution (*c*) are the most noticeable of the environmental sacrifices modern industry, overpopulation, ignorance, and a poor land ethic have given us.

unsightly air pollution (figure 22.8*a, b,* and *c*). In spite of recent claims of biodegradable diapers and plastics, there is little evidence that such materials will really solve the solid waste problem. Even papers that are relatively biodegradable decompose very slowly when placed in a landfill with limited gas exchange. Air and water pollution are two of our most serious threats to the environment (figure 22.8).

POLLUTION

It is doubtful that anyone on this planet could question that pollution is a major problem for a civilized world. Even those who live in relatively pristine and unchanging environments (there are a precious few of them left) have visited or seen through the media the enormous problems associated with environmental pollution. **Pollution** is simply an overproduction of anything that becomes a societal problem because we have no way of getting rid of it. Pollution comes in many forms: air, water, soil, noise, people, etc.

Air

Air pollution has become so pervasive in many parts of the world that it is often an accepted way of life. Those who live in relatively clean air and who visit such regions, usually in large cities, are often horrified and wonder how the inhabitants can live there. Children who grow up in Los Angeles, or Mexico City, or Taipei, or Frankfurt, or Karachi have little to compare with the putrified air that they breathe (figure 22.9). Such severe pollution, often associated with gaseous emissions from automobiles, has become an accepted, standard by-product of an industrialized society. The burning of firewood and charcoal casts a gray/brown pall on many cities, as in New Delhi. Yet industrialization in that city causes a mixing of many sources of atmospheric gases, just as it does in other cities.

Figure 22.9
Smog in the large cities of the world, including New York City, is becoming a serious health concern.

One wonders just how much a society is willing to tolerate in order to achieve its goals. Is living in New York worth the environmental pollution? Is commuting from Connecticut to New York for two hours worth the advantage of avoiding the air pollution? Is the cost of additional transportation and the utilization of resources worth the commute?

In addition to our man-made pollution, much of the world is subjected to dust, pollen, mold spores, and other air pollutants that wreak havoc with the quality of life, particularly with those individuals susceptible to allergies, those wearing contact lenses, etc. In some regions, odors in the air become objectionable in the quality of life. It may be as simple as a cattle feedlot a few miles away, or it may be an industrial giant spewing out toxic fumes in a nearby industrial park.

Water

Throughout history, civilizations have risen and fallen based on many factors, but perhaps none so important as water. A plentiful supply of freshwater is always a plus in deciding where to "put down roots," and there can be no question that water access was a major determining factor in the beginnings of agriculture. One would have thought that such supply problems would have been a problem only in primitive societies, but such is not the case. Throughout the world, city planners worry as much about a constant access to a quality freshwater supply as any other factor. Even in relatively mesic regions, where water reservoirs are abundant and close by, there are many concerns today about short-term droughts and the rate at which water is consumed by a demanding public.

(a)

(b)

Figure 22.10
(*a*) Chelan Lake, Washington. (*b*) Canadian Lake, Ontario.

In the water cycle, water molecules at or near the earth's surface evaporate into the atmosphere and then return to a landmass or ocean, perhaps thousands of miles away. The amount of freshwater in the world is truly enormous—about 15 quadrillion gallons of water is stored within 0.5 mile of the soil crust within the continental United States alone. It is withdrawn on a daily basis at the rate of 89 billion gallons. Some of that water is recharged, but much of it is fossil water that has been stored in the rocks and porous material for millions of years. Recharge of fossil water is almost nonexistent due to the geologic formations that seal it off. About half the nation's population depends on groundwater, and the other half utilizes surface waters stored in rivers, lakes, and reservoirs (figure 22.10).

The quality of both surface and groundwaters varies enormously, and an exploding world population has put pressures on our water resources that we once never dreamed possible. First of all, agricultural irrigation utilizes about 67% of all groundwater in the United States (figure 22.11), public drinking water and other domestic purposes utilize about 14%, rural households and livestock use about 6%, and industrial uses account for the other 13%. In many irrigated regions of the west, irrigation accounts for greater than 90% of total water consumption.

Figure 22.11
Irrigation of our croplands uses vast amounts of water from underground aquifers.

What is the source of our water contamination? There are many natural sources, such as nitrogen, arsenate, metals, lead, and mercury. Leaching and evaporation leave behind many undesirable salt deposits, and saltwater intrusion in coastal regions continues to occur. As a lack of careful zoning or carelessness, intentional waste disposal such as septic tanks, cesspools, injection wells, and land applications of waste water and sludge are a major source of contamination in many regions. Many major cities of the world still dump raw sewage into nearby rivers, lakes, and coastal waters.

Humans also contribute to the deterioration of water quality by irrigation, pesticides and fertilizer use, animal feedlots, deicing salts, runoff from urban and rural areas, mining operations with associated tailing piles, and underground storage of toxic wastes.

The quality of our freshwater supply continues to diminish, despite many efforts to slow the pollution. In many parts of the industrialized world, bottled water, deionizers, and water softeners are a regular part of the household expense. Bottled drinking water is now an enormous industry and is likely to continue for the foreseeable future.

Soil

Publicity concerning the Love Canal fiasco and other toxic waste dumps around the country brings awareness to the problems that we associate with soil pollution. Although these events would appear to be relatively isolated, the truth of the matter is that soil pollution is a major problem throughout the world. Even in developing countries, the use of pesticides at alarming rates have brought about levels of pollution that were unheard of two decades ago. A particular herbicide or insecticide might seem to be ideal for a particular farm, but crop rotation may spell disaster for the seedlings of a new crop. It is not at all uncommon

Figure 22.12
Irrigated land ruined due to salt buildup in the soil.

to drive through the countryside and see fields of chlorotic crops with poor productivity, no doubt brought on by a soil contamination of pesticide to which the crop was not tolerant.

Salt contamination of soils is an insidious destroyer of crop productivity, because the decrease in yields come about very gradually as the salts continue to increase. In many semiarid regions, the irrigation water is accompanied by salt contamination that is left behind when the water is used for plant growth and development. Such waters may contain salts that are beneficial to the crop (fertilizers) and nonbeneficial salts that are toxic to plant growth. There may be direct effects on the toxic ions, but there are also osmotic effects, because the plant's ability to take up water is impeded. Such regions often have insufficient natural rainfall to leach the salts out of the root zone, and so the situation worsens year after year (figure 22.12).

All of these problems of soil pollution are compounded by poor management practices that contribute to soil erosion. Wind and water erosion may destroy in a matter of 24 hours or less a topsoil resource that nature has taken hundreds of years to develop.

Noise

One of the consequences of a technological society is noise pollution, the insidious villain that preys on all of our nerves, even when we fail to realize it. There are some noises in nature that we find pleasing—a waterfall, a summer rain, the wind whistling through the trees, the surf on the shore, the birds singing, the lonesome howl of a coyote, and the marvelous sounds of jungle animals.

But somewhere along the line, other noises crept into our daily routine, which are not so pleasant—jet aircraft engines, jackhammers and chain saws, automobile and truck noises along a freeway, a motorcycle on a quiet suburban street, and a myriad of other noises (figure 22.13). Even inside our houses, vacuum cleaners, air conditioners, electric mixers, and television create an atmosphere of tension that did not exist in earlier times. Perhaps some of the more threatening noises are those of telephones, office machines, typewriters, and computers that seem to control every business office in the world. Dot matrix printers and automatic typewriters can give a headache to anyone who is already under a great deal of stress.

No one really intended for all this noise pollution to happen, but it has, and it happened over a long period of time. We make attempts at zoning and noise suppression, but little is being done to treat this continuing menace. Voters and taxpayers can make their feelings known by refusing to accept housing near an airport or freeway, by refusing to buy appliances that are unusually noisy, and by voting to rid our government of individuals who are insensitive to noise pollution.

Figure 22.13
The noise level of New York City streets is compounded by the tall buildings "trapping" the noise.

THE GREENHOUSE EFFECT AND GLOBAL WARMING

Our atmosphere is a gaseous envelope that surrounds the earth, and it powers the physical climate system. When radiation produced by the sun reaches the earth's atmosphere, some of it is reflected back upward by clouds and dust, whereas the remainder of it reaches the land and water surface. Some of that radiation is absorbed by the soil or water surface, but some is reflected back to space by ice, snow, water, and other reflective surfaces, including those erected by humans, such as concrete streets and parking lots and the roofs of buildings. All of these radiation sources emanate from the sun, but the earth itself also emits infrared radiation in its cooling process. Part of this infrared radiation is trapped by certain atmospheric gases. **Carbon dioxide** does an excellent job of trapping infrared radiation, as does **methane** (CH_4), **nitrous oxide** (NO_2), and the man-made **chlorofluorocarbons** CFC-11 and CFC-12 (table 22.1).

The summer drought of 1988 throughout the Corn Belt of the United States caused a new concern about the potential of **global warming.** Many parts of the southwestern United States and semiarid and arid regions in other parts of the world are subjected to periodic droughts. These are rather predictable and accepted, but drought in the Corn Belt is another matter. Scientists have long known that certain gases

TABLE 22.1

Greenhouse Gases: Relative Effectiveness at Trapping Heat

CO_2	CH_4	NO_2	CFC-11	CFC-12
1	25	250	17,500	20,000

Reprinted with permission from One Earth, One Future: Our Changing Global Environment. *Copyright 1992 by the National Academy of Sciences. Courtesy of the National Academy Press, Washington D.C.*

produced by human activity will probably cause the earth's average temperature to increase within the lifetime of most people living today.

The decade of the 1980s was the warmest recorded in recent geologic history, but no one is quite certain of whether it represents a long-term trend or merely a blip in the statistical averages that make up climatic history. We do not know whether the 1988 drought was a signal of things to come, but we do know that corn production fell below consumption for the first time in history, water levels of the Mississippi River were so low that normal barge traffic was stranded for weeks, forest fires burned uncontrollably in the national parks, a superhurricane threatened the Gulf Coast, and floods in Bangladesh led to the deaths of more than 2000 people.

Trace gases such as water vapor, carbon dioxide, methane, nitrous oxide, chlorofluorocarbons, and tropospheric ozone create the so-called **greenhouse effect** by trapping heat at or near the earth's surface. This increase in temperature has a number of predictable effects, but perhaps others that have not yet become evident. The predictable effects are changes in precipitation patterns, a probable melting of the polar ice caps, which will raise the level of the oceans, and changes in the rate of photosynthesis in various parts of the world. Some regions may become more desertlike, whereas others may get increasing precipitation and cloudiness, both of which could reduce plant productivity and will certainly change species distribution and ecosystem balances.

After water vapor, carbon dioxide is the most abundant and effective greenhouse gas. It occurs naturally in the atmosphere, but the burning of fossil fuels, especially coal, releases enormous quantities into the atmosphere. Deforestation involves clearing of trees, subsequently to be burned or left to decay, and these processes also release carbon dioxide. The oceans act as a good buffer for carbon dioxide, and they absorb a great deal of it; plants, too, absorb much carbon dioxide during photosynthesis. The fear is, however, that the levels of carbon dioxide in the atmosphere will overwhelm these systems and begin to build to intolerable levels (figure 22.14).

Most scientists believe that the earth is already destined for an atmospheric temperature increase of 0.5° to 1.5°C, and many atmospheric scientists

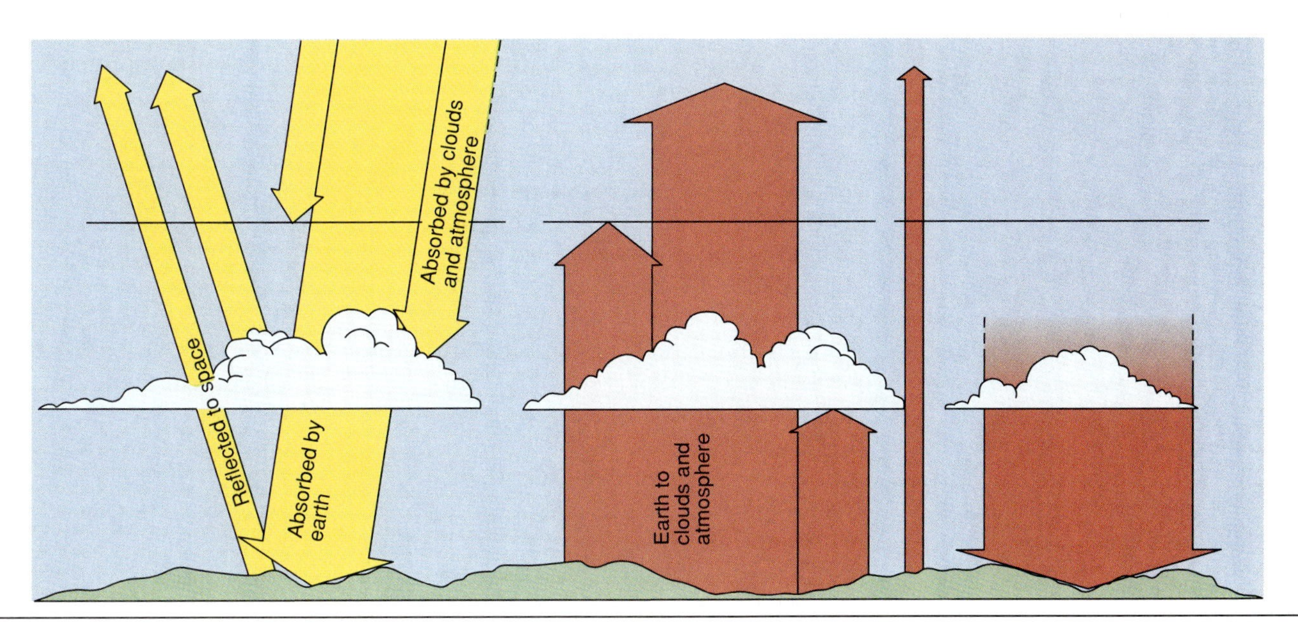

Figure 22.14
The buildup of atmospheric carbon dioxide helps trap heat near the earth's surface, resulting in the greenhouse effect.

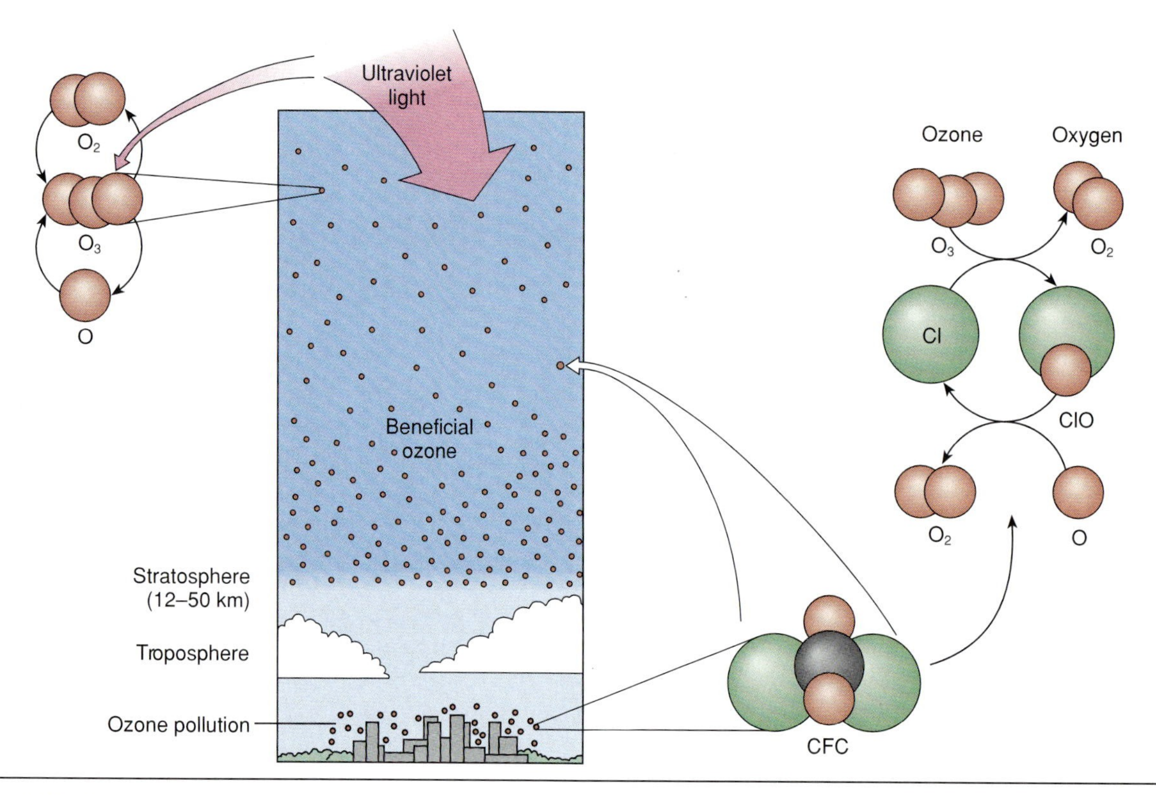

Figure 22.15
The protective ozone layer in the earth's stratosphere is being depleted by the presence of a number of halocarbon gases, especially chloroflourocarbon.

✔ Concept Checks

1. What is the basis for the concern about global warming?
2. What would be the consequences of a 3 or 4°C rise in temperature at the earth's surface?
3. What is the source of the heat attributed to the greenhouse effect?

believe that the average temperature could rise from 1°C to 5°C within the next 30 or 40 years. If the temperature rises 4°C, the earth will be warmer than at any time during the past 40 million years. During the last glaciation when most of North America was covered with ice, the average temperature of the earth was only about 5°C cooler than it is today.

Changes of the magnitude predicted would have disastrous effects on the survival of species; the rate of extinction would be increased manyfold over what it is at the present time.

THE OZONE LAYER

Earlier in this book we spoke of the atmospheric gases, of which oxygen comprises about 21%. Some oxygen molecules are energized by sunlight to split apart and recombine with regular oxygen atoms (O_2) to form O_3 or **ozone.** The rate at which ozone formation and ozone breakdown occurs is relatively constant under unchanging conditions. In the **troposphere** (below about 10 km) the levels of ozone are ordinarily quite low, whereas in the **stratosphere** (10 km to about 50 km) the concentration of ozone is quite high, reaching levels of about 9 parts per million (ppm) at about 38 km (figure 22.15). Little ozone occurs above the stratosphere because the air is so thin that few oxygen molecules have a chance to react with each other.

Stratospheric ozone acts as a powerful absorber of ultraviolet radiation and shields the earth's surface from these extremely high-energy and dangerous rays. The relatively long UV rays that do reach the earth's surface are so screened that it is usually safe to be exposed to the sun's rays in moderation. If something were to happen to the ozone layer in the stratosphere, the UV radiation could reach the earth's surface in quantities that would make it unsafe for normal outside activities without protective clothing. Unfortunately, that appears to be precisely what is happening.

A team of National Oceanographic and Atmospheric Administration scientists dispatched to the Antarctic in 1987 confirmed that the ozone layer over Antarctica had been reduced by more than 50% of values reported in 1979. At altitudes between 15 and 20 km, the depletion was as high as 95% of the baseline data from 1979, when measured during the first two weeks of October. For much of the year, the ozone layer is intact, but during the early spring of the southern hemisphere, severe ozone depletion occurs.

In addition to the gases described previously, we now believe that the ozone layer is being depleted by other

What Price Comfort?

Chlorofluorocarbons (CFCs) belong to a group of synthetic compounds found in refrigeration, insulation, foams, and other industrial products. Whenever these gases are released into the atmosphere (which they ultimately are), they release free chlorine, which then catalyzes the breakdown of ozone, the protective layer that shields our planet from ultraviolet radiation. Fluorocarbons are characterized by the ability to absorb heat. Therefore, even modest amounts of CFCs produce an enormous ability to trap heat at the earth's surface. Up to 20,000 times more effective than even carbon dioxide in absorptive capacity, there is no wonder that the world has finally come to the shocking reality that a few synthetic chemicals have the capability of producing so much damage to the environment.

Many nations of the world recognize the destruction caused by CFCs, and in September of 1987 came together to produce the "Montreal Protocol on Substances That Deplete the Ozone Layer." This agreement calls for a 50% reduction in CFC production from 1986 levels by the end of the century. CFC molecules are so stable that molecules being released today will remain to deplete ozone for a century or more. Researchers agree that CFC concentrations will continue to increase for 10 to 20 years after we stop releasing them. The answer would appear to be an even more restrictive policy, perhaps banning all CFC production as quickly as possible. Research is currently underway to find safe substitutes for these chemicals in our industrial products.

synthetic gases of the halocarbon group, in addition to the chlorofluorocarbons. One of these, CFC-113, is used as a solvent for cleaning electronic circuitry, and its atmospheric concentration is increasing about 11% annually. Even carbon tetrachloride (a cleaning fluid), methyl chloroform, and Halon 1301 and Halon 121 used in fire extinguishers, are major contributors to the ozone depletion. Bromine, a rather simple gas used in fumigants and fire extinguishers, is thought to cause 10 to 30% of the Antarctic ozone depletion.

Tropospheric ozone serves an entirely different function. Although it is the same molecule chemically, its position nearer the earth's surface causes it to act as a powerful absorber of heat, thus contributing to the greenhouse effect. Tropospheric ozone is not breaking down as fast as it once did, and pockets of it are causing heat buildup in specific regions, adding just one more piece of evidence that our earth is indeed increasing in temperature.

ACID RAIN

Many parts of the world are now plagued by a condition practically unheard of a few years ago. The term **"acid rain"** refers to acidic molecules that become suspended in the atmosphere and fall back to the earth's surface as rain, snow, or fog, or even as gases and dry particles. Whenever dry deposition occurs, the same chemical reactions occur as those involved in acid rain.

The source of most of the pollution is from the release of **sulfur oxide** into the atmosphere when coal is burned. So-called low-sulfur coal releases much less of the sulfur oxide, but virtually all coal and other fossil fuels release not only sulfur oxide, but nitrous oxides into the atmosphere. In some regions, automobile pollution from nitrous oxides is a major factor in the quality of life. Whenever these compounds are released into the air, they are transformed in a series of chemical reactions into sulfuric and nitric acids, and then into sulfate and nitrate. The rates of transformation are governed by sunlight, temperature, humidity, clouds, and the presence of other chemicals in the atmosphere. The acids dissolve in water droplets and eventually fall back to the earth in various forms of precipitation.

Unfortunately, since these products are airborne and at the mercy of wind patterns, acidic materials may be carried more than 1000 km. This situation creates some very difficult social and political problems, particularly when one political entity is accused of acidifying a region under another political entity. Acidified lakes in upstate New York are believed to be caused by midwestern power plants that burn coal. Acid rain in Scandinavian countries originates largely in Great Britain and central Europe, and about 50% of the acid rain falling in eastern Canada comes from the United States.

Acid rain is thought by some experts to be a major problem for both plants and animals in the regions affected. Acidified lakes in the northern United States and eastern Canada have been declared "dead" as a result of the loss of plant and animal life that once lived there. In most cases, the damage is not yet severe enough to kill many plants and animals, but the species diversity has been measurably reduced.

Forests and other vegetation may be damaged or killed, depending on the severity of the acidification process. In some cases, it appears that acid rain is yet another "environmental insult" with which plants and animals must cope. Thus, acid rain may merely compound the environmental problems created for a forest by drought, insects, disease, malnutrition, etc.

Even man-made buildings and monuments may be subject to the ravages of acid rain. The Environmental Pollution Administration estimates that the cost of building repair as a result of acid deposition in the United States is as high as $5 billion annually. Marble and limestone are particularly susceptible. The Acropolis in Athens and the Jefferson Memorial in Washington are showing

✔ Concept Checks

1. What difference does it make that there is a hole in the ozone layer over Antarctica?

2. What is the chemical difference in stratospheric ozone and tropospheric ozone? What are the differences in effect?

3. Why isn't acid rain a problem all over the world?

great signs of stress. Acid rain from Mexico's refineries may be causing severe damage to the Mayan ruins, and in southwestern Colorado acid rain fed by local power plants is damaging the ancient sandstone cities of the Ansazi Indians. There is increasing evidence that rusted steel in bridges and buildings is partially due to the acidified atmosphere.

Not everyone agrees on how serious a problem acid rain is; even though it is a major threat, there is some reason to be hopeful that acid rain can at least be slowed. New technologies and new laws controlling emissions have brought remarkable improvement in atmospheric conditions in some regions. Perhaps with additional social and political pressure, we can eventually add at least a partial success story to a society that seems bent on self-destruction in the name of progress.

ECONOMIC AND ENVIRONMENTAL TRADE-OFFS

A true democracy maximizes freedom of expression, thoughts, ideals, and the pursuit of personal goals. Citizens as a group decide what restrictions will and will not apply to the society. Modern technology has led industrialized, democratic nations into a trap: on the one hand, freedom of expression dictates to each to "do your own thing"; to some this includes polluting the environment in its many forms (litter, gases in the air, noise, pesticides). On the other hand, concerned citizens strive to impose restrictions on society as a whole to ensure a more acceptable quality of life for all. Voters and taxpayers find themselves lining up on one side of an issue at one time, and having to reverse themselves at another time. The determining factor is often economics. How much will it cost me (and society) to achieve this so-called improvement in the quality of life? Unfortunately, one person's quality of life is another person's pollution.

This freedom of expression in the developed countries has led to an environmental degradation unparalleled in history. Cleaning up the environment was simply considered too expensive for most of the paying customers. Factories spewed contaminants into the air and poured toxic wastes into rivers and streams, and increasing pressures for food forced modern agriculture to use more and more pesticides so that weeds, insects, and diseases could be kept under control.

The environmental awareness issue finally became so powerful in the 1970s that controls were implemented, and since then a great deal of progress has been made in improving the quality of air, water, and soil resources. Much remains to be done, and changes in political philosophy cause a roller-coaster effect for many years to come. Ultimately, however, individual responsibility must bring about the changes considered desirable by a majority of the society. Individuals must make these decisions as informed citizens, concerned taxpayers, and most importantly as members of the natural community, not exclusively as exploiters of its resources.

RECYCLING AND CLEAN-UP EFFORTS

With the true realization that natural resources were being used up at an unacceptable rate, the technology for **recycling** paper, aluminum, plastic, and glass improved by the end of the 1980s to such a level that recycling became cost effective. By the early 1990s, just over 50% of the aluminum cans produced in the United States came from recycled aluminum. In addition, all qualities of recycled paper from grocery sacks to glossy slick paper for color printing became available at competitive costs by 1991.

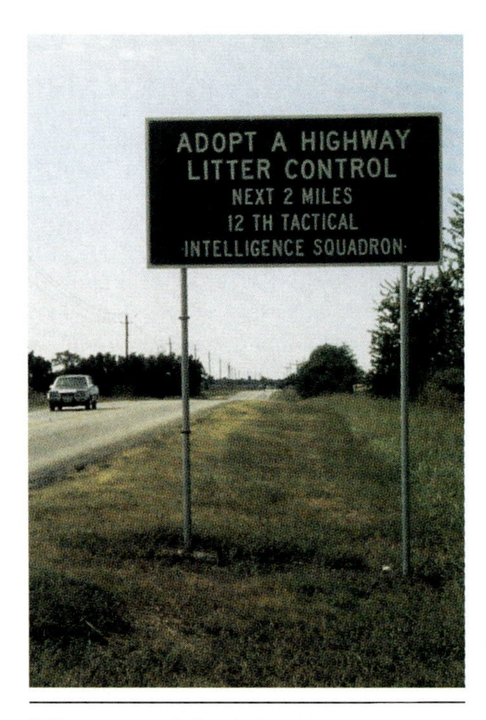

Figure 22.16
A segment of roadside in Texas "adopted" for litter pickup.

In the late 1980s, public pressure caused many fast-food chains to discontinue the use of styrofoam cups and food containers just as it had earlier forced more fuel-efficient American-produced automobiles during the oil shortage in the 1970s. Public education programs to reduce litter along our highways and in our public parks has been effective in cleaning up our visual world. Many states have developed "Adopt A Highway" programs for public litter pickup along highway rights-of-way (figure 22.16). Legal penalties have been stiffened for public littering, and enforcement has taken a higher priority than in the past. The number of clean air, clean water, and antivisual pollution organizations mushroomed in the 1980s and early 1990s at local, state, regional, and national levels, providing watchdog functions in support of antipollution ordinances and laws.

Public outcry over the Exxon *Valdez* oil spill in Alaska resulted in tighter controls on oil tankers and new technology for cleaning up such spills. In January of 1991, during the Persian Gulf War (Operation Desert Storm) the largest oil "spill" in history occurred when Iraq opened the valves at a major tanker loading station, dumping over 640 million gallons of Kuwaiti oil into the Persian Gulf as an act of **environmental terrorism.** Oil-eating bacteria and other technology was used to somewhat reduce the environmental damage, but this was an ecologically devastating event.

Public outcry to stop the war did not occur as a result of this massive oil slick, however; in fact, it was a military operation that was successful in stopping the flow of oil into the gulf, further galvanizing the United Nations coalition of countries in their resolve to end the war through military means, not stop the war to prevent further environmental damage.

A secondary act of environmental terrorism, coupled with military and economic goals, was the setting on fire or damaging of over 950 Kuwaiti oil wells, storage facilities, and refineries. Estimates set the amount of oil burned at 2.5 to 3 million barrels per day with fires reaching up to 500 feet in the sky and spreading black smoke over a distance of more than 1000 miles. Done near the end of the war when it was clear Iraq would soon be defeated, this act produced tons of airborne particulate matter and toxic fumes and darkened skies over large regions of the Persian Gulf area. Combined with the oil slick, these acts may result in the Persian Gulf War being the most environmentally destructive conflict in the history of warfare, even though the war lasted only three months.

The key to understanding what environmental problems exist is the same as it is for providing cleanup and corrective measures—education. The citizens of developed countries are more aware of what needs to be done than those of developing or third world countries. In addition, the vast majority of the goods and services that cause environmental imbalances are consumed by the developed countries, therefore it is their responsibility to conserve, reduce resource consumption, recycle, cleanup and modify their standard of living.

REVEGETATION AND ENVIRONMENTAL REPAIR

While there are myriad environmental problems and ecological imbalances, the single basis for many interrelated problems is the removal of the world's native flora. Although the destruction of the tropical rain forests is one of the most visible and rapidly occurring examples, just in North America there are already major ecosystem disasters. Our native hardwood forests have been reduced to only a fraction of what was originally here prior to settlement and clearing for lumber, croplands, pastures, and development. The once vast central plains have been described by ecologists as a nonfunctional ecosystem. Represented only by small, scattered remnant pieces of prairie land, which add up to only 1% of what was originally there, this land is now a vast farmland or modified pasture composed of introduced monocultures of exotic or hybrid grasses and other crop plants. Our wetlands, deserts, and other ecosystems are also severely damaged by removal of the native flora and the resulting destruction of habitat for the wild fauna that depends on these plants.

Compounding the problems caused by the removal of native plants is the subsequent introduction of exotic (introduced from their native range—whether from another continent or from elsewhere in the same continent) and cultivated (horticulturally selected—no longer native genotype) plants. As crops, pasture grasses, and landscaping plants, these introductions at best take up space that could be occupied by native plant species, and at worst they escape cultivation and spread in the wild, displacing native species (figure 22.17).

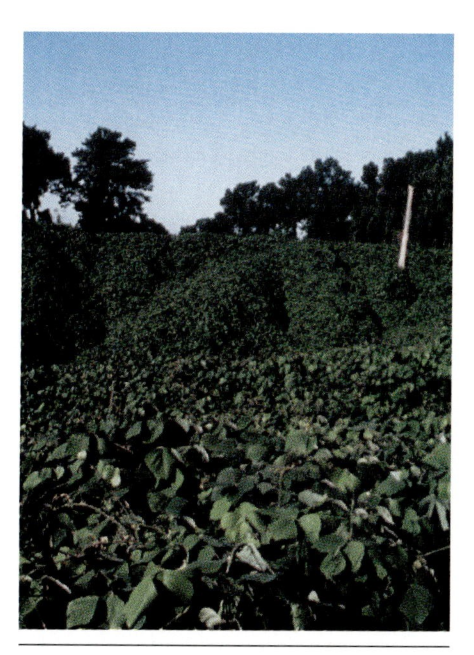

Figure 22.17
Throughout the southeastern United States, Kudzu (*Pueraria lobata*) has aggressively spread to choke out native vegetation, even killing shrubs and trees by blanketing them with dense growth.

POSITIVE SOLUTIONS

Fortunately, there is a rapidly growing interest in propagating and reestablishing native plants in a variety of settings. Abandoned farmland and pastureland is now more and more often being seeded with native grasses and other indigenous plants. Roadsides and other public lands are not only being managed to preserve what native flora is still there but are being seeded to reintroduce native species that were once present. Even federal legislation now exists to require state departments of transportation to spend one quarter of one percent of federal funds for highway rights-of-way revegetation on wildflower establishment (figure 22.18). In combination with new highway vegetation management guidelines that call for reduced mowing, reduced herbicide use, and higher mower settings, many states have made very pointed changes in their vegetation management practices that protect and encourage native species over introduced grasses and forbs.

Figure 22.18

Wisconsin roadside where native wildflower and grass reestablishment saves mowing costs while providing regional beauty and expanded habitat for native plants of the area.

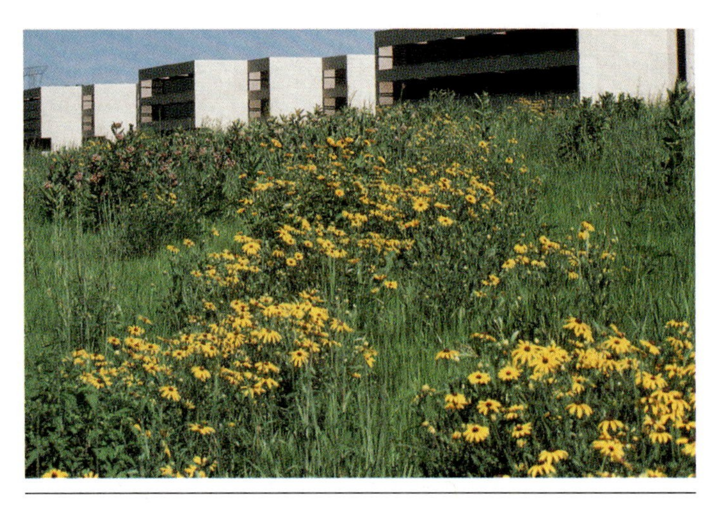

Figure 22.19

The use of native grasses and wildflowers by Prudential Insurance in Minnesota produced a beautiful reestablished prairie area that is far less expensive to maintain than traditional landscapes using exotic species.

The reasons for increased interest in the reestablishment and protection of native plant species in public land vary. While it would be nice to assume that there is always an ecological basis for such efforts, often there are other, stronger motives involved. For one thing, reduced roadside or public land maintenance costs can be dramatic. States such as Texas, Ohio, California, and Pennsylvania, which have large highway systems can literally save millions of dollars annually by reducing the number of mowings each growing season. In the mid-1980s when Texas instituted a new roadside vegetation management plan, which called for significantly reduced mowing, especially in rural areas, the highway department saved over 8 million dollars the first year of the program.

Another underlying reason for the increased interest in reestablishment and modified management programs is the resulting increase in seasonal wildflower displays. The spring and summer roadside wildflowers are even part of the tourist attraction for many states, as is the fall color in the northeastern states. The combination of maintenance cost savings, increased color and regional natural beauty and, of course, positive ecological action continues to cause an increased interest in the reestablishment of native flora.

This combination of positive factors carries over into the smaller and more formal, planned landscapes of businesses and homes. Once established, native plants used in landscapes require little or no supplemental water or fertilizer, no soil amendments, drastically reduced or no pesticide use, and reduced maintenance costs and time commitments (figure 22.19). How much savings depends on how formal or naturalistic the design and whether or not natives are being used exclusively or in combination with maintenance-intensive exotics.

Whatever the degree, increased use of native plants in our planned landscapes is a very positive economic, aesthetic, and especially, ecological activity. Many people even plan specific landscape themes using natives. Butterfly gardens, restored mini-ecosystems for increased bird and other wild fauna visitation, simulated meadows for seasonal color changes, and many other landscape design ideas can be developed with the use of native plant species. Of course, these plants should be obtained from propagated sources, not dug from the wild. The latter only causes a net loss in natural occurrence and total numbers, because some percentage do not transplant successfully. The goal is to increase the numbers of native plants on earth while reducing resource consumption that accompanies the maintenance of exotics. The use of regional native plants in our planned landscapes is an activity in which all individuals can participate, resulting in a dramatic, positive impact on our environment.

So, while ecological and environmental problems are very real and serious, human society, collectively, and individually, is beginning to respond effectively. As we work to reduce resource consumption and clean up pollution, we can also work toward environmental repair through the preservation of existing biodiversity and the reintroduction of indigenous native plants instead of the continued introduction of exotics.

Just as we treat our special friends with love and respect—we make an effort, we care—the land should also be treated like an important friend; it deserves the same effort, understanding, and respect as special people. *One person, one plant at a time, we can make a difference.*

Summary

1. Although able to modify our environment, often to its demise, humans are now understanding that we must learn to live in cooperation and in balance with the natural world.
2. Because of continued world population growth, it will be difficult to reduce wasteful consumption of natural resources; however, especially the developed countries must find ways to conserve more resources.
3. All ecosystems are impacted by human activity, but the tropical rain forests are the most rapidly disappearing habitats.
4. The removal of the native flora followed by the introduction of non-native exotic plant species is the basis for much of our ecological imbalance.
5. Conservation movements have resulted in endangered species legislation in the United States and even some international efforts to legally protect the environment. These efforts were made necessary by lack of environmental foresight in development of industrial products. The basic problem, however, is too large a world human population for the available resources.
6. The loss of local biodiversity is the greatest concern a given habitat faces.
7. Reduced environmental quality is a general phenomenon of overpopulation, and specific kinds of pollution are the most visible problems that affect us.
8. The global warming that results from the greenhouse effect may ultimately result in major changes in our earth's ecological stability.
9. Environmental awareness has led to improved recycling technologies, extensive clean-up efforts, and the reestablishment of native plant species in our planned landscapes.
10. Continued and increased efforts, especially by those members of the human population educated in the causes and significance of environmental problems, can result in a return to a biological world in balance, with all its members (including humans) existing as part of the world community, not apart from it.

Key Terms

acid rain 427
biodiversity 420
carbon dioxide 424
chlorofluorocarbons (CFCs) 424
conservation 418
environmental terrorism 429
extinction 418
global warming 424
greenhouse effect 425
habitat destruction 416
methane 424
nitrous oxide 424
overpopulation 421
ozone 426
pollution 422
recycling 428
stratosphere 426
sulfur oxide 427
troposphere 426

Discussion Questions

1. Are humans part of the natural world?
2. Why should we worry about the loss of biodiversity through extinction?
3. In what ways are *you* willing to modify your behavior and standards of living to close the gap on resource depletion?
4. How can the basis for these ecological problems, overpopulation, be solved?
5. Could you live without exotic landscaping and support the reestablishment of native, indigenous plant species in public and private planned landscapes?

Suggested Readings

Bolch, B. W., and H. Lyons. 1993. *Apocalypse not: Science, economics, and environmentalism.* CATO Institute, Washington, D.C.

Brown, L. R., C. Flavin, and S. Postel. 1991. *Saving the planet: How to shape an environmentally sustainable global economy.* W. W. Norton, New York.

Cagin, S., and P. Dray. 1993. *Between earth and sky: How CFCs changed our world and endangered the ozone layer.* Pantheon Books, New York.

Carson, P., and J. Moulden. 1991. *Green is gold: Business talking to business about the environmental revolution.* Harper Business, Toronto.

Clark, W. C. 1989. Managing planet earth. *Sci. Am.* 261(3):46–54.

Dunlap, R. E., and A. G. Mertig, eds. 1992. *American environmentalism: The U.S. environmental movement, 1970–1990.* Taylor & Francis, Philadelphia.

Foster, B. A. 1993. *The acid rain debate: Science and special interests in policy formation.* Iowa State University Press, Ames.

Maduro, R. A., and R. Schauerhammer. *The holes in the ozone scare: The scientific evidence that the sky isn't falling.* 21st Century Science Associates, Washington, D.C.

Morrison, M. 1993. *Fire in paradise: The Yellowstone fires and the politics of environmentalism.* Harper Collins Publishers, New York.

Ray, D. L. 1992. *Trashing the planet: How science can help us deal with acid rain, depletion of the ozone, and nuclear waste (among other things).* Harper Perennial, New York.

Repetto, R. 1990. Deforestation in the tropics. *Sci. Am.* 262(4):36–42.

Sachs, W., ed. 1993. *Global ecology: A new arena of political conflict.* Zed Books, Atlantic Highlands.

Simmons, I. G. 1993. *Interpreting nature: Cultural constructions of the environment.* Routledge, New York.

Chapter Twenty-Three

Now, Why Botany?

This final chapter is important to read because it brings together the information covered in the book in a very brief and personal way. As well-educated lay botanists, we all have the tools and the understanding of how important each of us are in the decision-making process and of how that process can and does affect us directly. We have both a responsibility and the knowledge that allows us to accept it.

Dodecatheon, the shooting star, with its characteristic reflexed petals.

"Draba"

Within a few weeks now Draba, the smallest flower that blows, will sprinkle every sandy place with small blooms.

He who hopes for spring with upturned eye never sees so small a thing as Draba. He who despairs of spring with downcast eye steps on it, unknowing. He who searches for spring with his knees in the mud finds it, in abundance.

Draba asks, and gets, but scant allowance of warmth and comfort; it subsists on the leavings of unwanted time and space. Botany books give it two or three lines, but never a plate or portrait. Sand too poor and sun too weak for bigger, better blooms are good enough for Draba. After all it is no spring flower, but only a postscript to a hope.

Draba plucks no heartstrings. Its perfume, if there is any, is lost in the gusty winds. Its color is plain white. Its leaves wear a sensible woolly coat. Nothing eats it; it is too small. No poets sing of it. Some botanist once gave it a Latin name, and then forgot it. Altogether it is of no importance—just a small creature that does a small job quickly and well.*

—Aldo Leopold

To the person lacking a background in botany, *Draba* is indeed as Leopold describes—small, plain, unimportant, and unnoticed (figure 23.1). Again, however, Leopold makes these comments tongue-in-cheek, for he knows that *Draba's* place in the botanical world is as important as larger, more visible, and more "useful" plants. That realization sets apart the botanically educated from the average citizen. When more of the world's people view themselves as natural members of the biotic community instead of users of it, then the future balance and stability of the community is secure, as is the future of each of its individual components. In a sense, then, humans are no more important than the lowly *Draba;* nor are we any less important. Now, it is possible to view the importance of the botanical world to the human world—they are one.

Figure 23.1
A small (15 to 20 cm) inconspicuous plant, *Draba* blooms early in the spring.

LAY BOTANISTS

Although few people become professional botanists, the simple awareness of plants and of their importance makes us all botanists in one sense; we are **"lay" botanists.** In addition to a realization of the applied worth of plants to human society, a basic knowledge of plant structure, function, reproduction, distribution, ecology, and cultivation allows us to better appreciate the total role of plants on earth. A plant's role should not be measured exclusively in terms of direct human utility, but also in terms of energy flow, oxygen production, erosion control, and overall natural ecological balance. Learning about the plant kingdom, therefore, is not purely an academic exercise, but rather it contributes to the present and future survival of our planet and the continuation of all the organisms on it, including humans.

The Basic Importance of Plants

The importance of plants to the existence of all other organisms on earth can be summarized by reviewing two aspects of photosynthesis: (1) *energy conversion* and (2) *oxygen production.* Plants form the base of the food chain, because they combine atmospheric carbon dioxide and water with sunlight energy to produce food energy in the form of carbohydrates. Without green plants, therefore, the animals of the world would have no source of food other than some single-celled aquatic organisms. The availability of plant tissues as a food source has resulted in the development of extensive and involved plant-animal and animal-animal interactions, as well as the evolution of complex forms of land-dwelling organisms.

As a by-product of this light to tissue energy-conversion process, oxygen is released into the atmosphere. With a few anaerobic exceptions, all organisms on earth require oxygen to carry on respiration, the process of producing from carbohydrates a form of energy capable of doing biological work. Additionally, the availability of oxygen ultimately resulted in sufficient atmospheric ozone accumulations to allow life on land to be screened from damaging ultraviolet radiation.

These two functions of plants, combined with their soil stabilization functions, their utility as nesting habitats, nutrient cycling, essential amino acid production, and many other interactions with other organisms, elucidate their central position in the balance of nature. Even lay botanists must acknowledge the magnitude of the importance of plants in nature.

(a)

(b)

(c)

Figure 23.2
"Natural areas" are popular as recreational sites because of their beauty, grandeur, and remoteness from urban settings. (*a*) A coniferous forest. (*b*) An island coastline. (*c*) The seashore.

Recreation and Aesthetics

Aldo Leopold began the foreword to *A Sand County Almanac* by stating, "There are some who can live without wild things, and some who cannot." More and more people are realizing the worth of wilderness areas and the relaxing beauty of nature. The ability to walk a quiet trail through the woods, to sit on a rock by a river, to notice the *Silphium*, the *Draba*, the first robin of spring—these are not to be taken for granted but are to be actively and consciously preserved. It is in such settings that the brightness of the stars and the contrasting darkness of the voids between them stimulate poetry; it is in such settings that people find peace of mind, relaxation, perspective. It is not without reason that we refer to "escaping" from the big city to pursue recreational activities in the forests, on the mountains, at the seashores—in natural areas (figure 23.2). Even within urban areas, parks can provide a measure of this feeling. Bringing plants into homes and offices is yet another example of the realization that natural surroundings relax and soothe.

Maybe it is instinctive or possibly even genetically controlled that humans want to spend time in natural settings. Also, many of us exist primarily in the concrete, plastic, and steel world we have built, and natural areas offer a change. Either way, millions visit the national parks and wilderness areas annually (figure 23.3). Too many of these visitors contribute to the decline of such areas by their thoughtless and destructive acts of littering and vandalism (figures 23.4 and 23.5). Fortunately, there are many more that follow a good **land ethic** by leaving these sites as they found them. When we enter these areas, we should do so as members of the natural environment, not as controllers or conquerers. This attitude comes from knowing and understanding what altering natural interactions can mean to the ecological balance. Lay botanists have such knowledge. They are and must be the **stewards** of the land, always.

An often unsuspected bonus of study is a more highly developed aesthetic appreciation. If beauty is in the

(a)

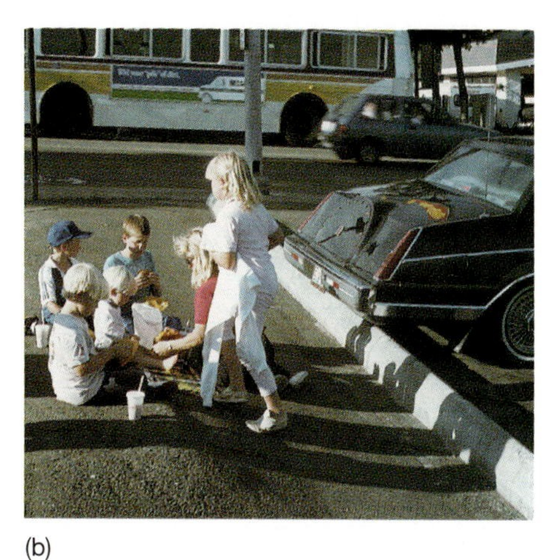
(b)

Figure 23.3
Crowded urban settings increase the popularity of natural recreational areas (*a*), which provide more scenic picnic spots than this sidewalk (*b*).

Figure 23.4
Unfortunately, the popularity of many recreational sites results in excessive crowding, reducing the appeal of such places and increasing litter and degradation of the area.

(a)

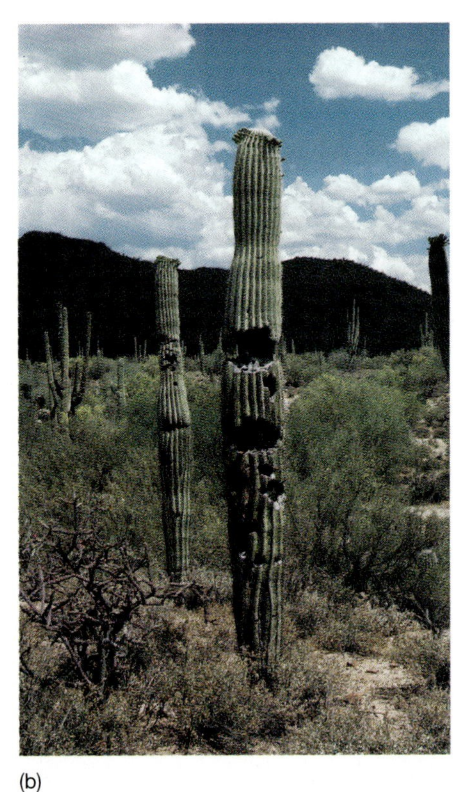
(b)

Figure 23.5
Vandalism reflects the destructive attitude of some of our nation's vacationers. (*a*) Carved initials on a tree trunk. (*b*) Shotgun destruction of a giant saguaro cactus in Arizona.

Figure 23.6
Fortunately, most visitors appreciate the beauty of the natural areas they visit and leave them as they found them, clean and beautiful.

Concept Checks

1. What ecological role do plants play in the biosphere, earth?
2. What practical uses have we humans made of the plant kingdom?
3. What kind of land ethic do we have?

eye of the beholder, then the better-trained eye has a greater potential for seeing beauty. It is understandable why trained naturalists are often the most emotionally involved in the areas they study. True love can develop on those trails through the woods, along an empty beach, or on a rock by the river (figure 23.6).

SOCIOPOLITICAL CONSIDERATIONS

The importance of plants to each individual should be clear. As sources for food, shelter, recreation, and industrials, plants obviously serve a purpose for each of us. To all organisms on earth, plants are essential. As the base of the food chain, as a producer of oxygen, and as components in the natural ecosystem balance, plants occupy an irreplaceable position. We should not isolate our newfound knowledge, however. The study of botany should improve our ability to function as a component of the biological world. We should now be wiser taxpayers and more thoughtful voters and citizens. Plants are as important in our daily activities as we humans are to each other.

It should always hold true that knowledge—in any discipline—results in an improved awareness of issues, problems, possible solutions, and personal responsibility for participation. Posing questions is a natural extension of the educational experience. Your knowledge of plants should be as much a part of your base of information for living as your actions are an impact on the botanical world around you. Directly and indirectly, both as an individual and as part of a collective society, you can have a positive influence on the world around you and thus on your niche within that world. Even though we question many policies and decisions about ecological issues, land management, increased pollution, and the like, there is a large portion of our world that is still beautiful, productive, clean, balanced, and that has enormous potential.

Your actions will serve both to solve problems and to ensure the continuation of the many "good" aspects of our world.

WORLD FOOD SUPPLIES AND POLITICAL INVOLVEMENT

One of the consequences of a world of "haves" and "have-nots" is a great deal of resentment toward those people with a relatively high standard of living,

(a)

(b)

Figure 23.7
The United States is one of the world's major wheat producers and exporters. A South Dakota wheat harvest (*a*) produces a bounty of grain (*b*).

sometimes manifested in quiet anger, sometimes in war. Humans demonstrate responsibility to the individual and family above all else. National pride, religious fervor, and charity pale when a family breadwinner is faced with starvation for his or her family. Psychologists tell us that the bottom line is self-preservation, and the individual will do almost anything to ensure that the family is fed. Too often in history zealous leaders have taken advantage of this human instinct to rally forces to battle, making shallow promises of protection and freedom from hunger. Most of the time, they have failed.

In the past, ignorance for many people really was bliss. Modern worldwide communication has allowed those in the underdeveloped countries to see and hear about a better life. Their demands are now being heard in the political arena, and local governments are under increasing pressure to provide a fair share of the world's wealth. While the wealthy countries of the world contain only about one-fourth of the world's population, they consume nearly 80% of the earth's commercial energy. The resulting utilization of natural resources and production of waste is proportionately out of balance.

One point is clear in the overall picture of global food supplies: the world is overly dependent on North America for its food resources (figure 23.7). This is not to say that other countries have totally failed in food production. In recent years India, for example, has made major advances in achieving self-sufficiency, at least in certain crops. Other nations have made specific gains in some areas, yet worldwide dependence on North America as the reliable food exporter remains constant.

Several reasons explain this growing dependence on North America. Increasing population pressures, particularly in the developing countries, have hastened the deterioration of food systems. The increased pressure on traditional farming systems to produce more and more has led to ecological degradation; once fallowed land (land cropped only once every two or three years so that moisture and nutrients can be stored) is now being called into constant production, depleting nutrients at a more rapid rate and often leading to severe erosion. In some parts of the world newly rich nations, particularly the oil-producing countries, are demanding a greater share of the world's food. The gap between the haves and many of the have-nots continues to widen.

These factors have resulted in North America having significant influence over world food supplies. The United States and Canada are in a position to dictate to the importing nations the terms on which food will be distributed. Serious moral and ethical questions arise from this potential political power: Does North America have the right to use food as a political force? Can food resources be used to dictate population control, religious expression, and personal philosophy? Should food ever be used as a weapon to beat a recalcitrant country into submission?

Economic sanctions were used during the last half of 1990 against Iraq by a coalition of nations under the guidance of the United Nations in an effort to force Iraq to withdraw from Kuwait. The result was that the Iraqi military received food and medical care while the civilians went without. In this case, the leadership of Iraq was willing to impose sacrifices unequally on the population, and a military solution to the situation was implemented instead of continuing economic sanctions.

There may not be any right or wrong answers to these questions, but they deserve informed thought. There may come another situation when such questions will again have to be addressed.

✔ Concept Checks

1. How do "have" countries influence "have-not" countries?
2. Have food supplies ever been used successfully as a political weapon?
3. How do individuals have any direct connection with these global ecological and political issues?

THE BOTANICAL WORLD

Obviously, from any perspective, the world is a human world. We can and do affect every corner of the global habitat in which we live and upon which we depend. Aesthetically, practically, and ecologically, we are involved in the total interactions of the natural world. True realization of how we are inescapably tied to our global actions has only come about in the very recent history of our existence as a biological part of that world. Only through that understanding and a desire to continue as part of the world community will we be able to modify our actions so that those interactions are sustainable ones. The future of the botanical world is in human hands; the future of human society rests in the continuation of that world (figure 23.8).

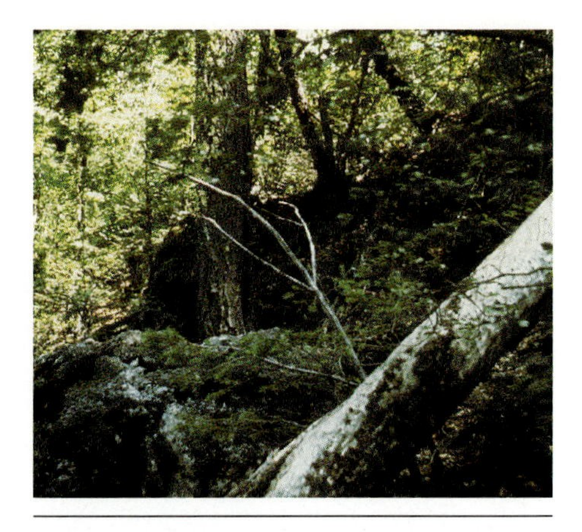

Figure 23.8
Proper management of our botanical world by an educated and concerned populace will maintain ecological balance, plant function, and pleasant natural areas.

Summary

1. All plants, no matter how large or small, are important in the total balance and functioning of the biological world. Obviously critical in photosynthesis, energy conversions, and oxygen production, plants form the base of the food chain.
2. Any person with a basic knowledge of plant structure and function is a lay botanist who can better appreciate the biological, ecological, aesthetic, and applied roles of plants.
3. Knowledge should help form a base for people to act more responsibly toward their natural environment and to participate more wisely in decision-making processes that affect their world. Many sociopolitical considerations are worldwide in scope, and yet rational evaluation of resource use can be a decision affecting each individual.
4. Modern agriculture now involves food surpluses in some countries, shortages in others. The United States is still a country with food surpluses, which could be used for political pressure.

Key Terms

Discussion Questions

1. Is each and every plant (or animal) species (let alone each individual organism) really all that important? Why and why not?
2. Can we establish a balance between our need to aesthetically experience the natural world with our need to utilize it for resources?
3. In another Persian Gulf War situation, should we impose economic sanctions only?

Suggested Readings

Asefa, S., ed. 1988. *World food and agriculture: Some problems and issues.* W. W. Upjohn Institute for Employment Research, Kalamazoo.

Barrett, E. C., and L. F. Curtis. 1992. *Introduction to environmental remote sensing.* 3rd ed. Chapman & Hall, New York.

Bohlen, J. T. 1993. *For the wild places: Profiles in conservation.* Island Press, Washington, D.C.

Burgan, R. E., and R. A. Hartford. 1993. *Monitoring vegetation greenness with satellite data.* U.S. Department of Agriculture, Forest Service, U.S. Government Printing Office, Washington, D.C.

Facklam, H., and M. Facklam. 1990. *Plants: Extinction or survival?* Enslow, Hillside, NJ.

Hedin, P. A., J. J. Menn, and R. M. Hollingworth, eds. 1994. *Natural and engineered pest management agents.* American Chemical Society, Washington, D.C.

Leopold, A. 1949. *A Sand County almanac.* Oxford University Press, Inc., New York.

Peterken, G. F. 1993. *Woodland conservation and management.* 2nd ed. Chapman & Hall, New York.

Scorer, R. S. 1990. *The satellite as microscope.* Halsted Press, New York.

U.S. Fish and Wildlife Service. 1993. *Why save endangered species?* U.S. Department of the Interior, U.S. Fish and Wildlife Service, Washington, D.C.

Whelan, T., ed. 1991. *Nature tourism: Managing for the environment.* Island Press, Washington, D.C.

Zalom, F. G., and W. E. Fry, eds. 1992. *Food, crop pests, and the environment: The need and potential for biologically intensive integrated pest management.* APS Press, St. Paul.

Appendix

TABLE 1

Metric Conversion Table

PHYSICAL PROPERTY	UNIT OF MEASUREMENT	NUMERICAL VALUE	SYMBOL	ENGLISH SYSTEM CONVERSION
Length	Meter		m	39.37 inches
	Kilometer	10^3 m	km	0.62137 mile
	Centimeter	10^{-2} m	cm	0.3937 inch
	Millimeter	10^{-3} m	mm	0.03937 inch
	Micrometer	10^{-6} m	µm	0.00003937 inch
	Nanometer	10^{-9} m	nm	0.00000003937 inch
Area	Hectare	10,000 m^2	ha	2.471 acres
Mass	Gram		gm	0.03527 ounce
	Kilogram	10^3 gm	kg	2.203 pounds
	Milligram	10^{-3} gm	mg	0.00003527 ounce
	Microgram	10^{-6} gm	µg	0.00000003527 inch
Volume (solid)	Cubic meter		m^3	35.314 cubic feet
	Cubic centimeter	10^{-6} m^3	cm^3	0.0610 cubic inch
	Cubic millimeter	10^{-9} m^3	mm^3	0.0000610 cubic inch
Volume (liquid)	Liter		l	1.06 quarts
	Milliliter	10^{-3} l	ml	0.034 fluid ounce
	Microliter	10^{-6} l	µl	0.000034 fluid ounce

Glossary

A

abiotic Pertaining to nonbiological factors.
abrasion The mechanical process of gradually breaking down a hard layer, as in a seed coat.
abscisic acid A plant hormone associated with dormancy, abscission of organs, and water stress.
abscission The detachment of leaves, flowers, or fruit from a plant, usually at a mechanically weak location, termed the *abscission zone.*
absorption The process of taking in, as uptake by roots.
accessory buds Those buds adjacent to a primary bud and usually smaller in size.
achene A small, dry, one-seeded indehiscent fruit; the pericarp is easily separated from the seed coat.
acid **H^+**(proton) donor; a substance that associates to release H^+ and thus cause the pH of the solution to be less than 7.0.
acid rain High concentrations of H^+ (protons) in rainfall which displaces many ions attached to soil particles; results in sulfuric and nitric acids, especially.
acidic Possessing a relatively large number of hydrogen ions; having a pH less than 7.0.
acridine dyes Organic pigment molecules that are capable of causing permanent genetic changes (mutations).
active transport The movement of ions or molecules against a concentration gradient using metabolic energy.
adaptation Conforming to a given set of environmental conditions.
adenine A nitrogen base found in both DNA and RNA.
adenosine diphosphate (ADP) The building block for ATP; by adding a terminal phosphate group and a large amount of energy, ATP can be formed.
adenosine triphosphate (ATP) The major source of chemical energy for biochemical reactions. The metabolic energy is stored primarily in the terminal phosphate ester linkage.
adhesion The attraction of unlike particles; water particles *adhere* to the surface of clays.
adventitious A structure arising at some location not usually expected, such as on a stem.
adventitous root A root produced by a stem or leaf rather than by a root.
aerate To supply with oxygen.
aerial Pertaining to being in the air, such as a root projecting from an aboveground stem.
aerobic respiration (Gr. *aero,* air + *bios,* life) respiration in the presence of oxygen.
agar (ah-ger) A complex polysaccharide made from red algae and used for preparing a semisolid substrate for growing microorganisms.
agent orange A herbicide used as a defoliant in the Vietnam War; composed of 2,4-D and 2,4,5-T.
aggregate fruit A fruit developing from numerous simple carpels from a single flower.
agricultural infrastructure The complexities of processing, storing, packaging, and transporting farm products.
aleurone layer (al-u-roan) A group of cells rich in protein granules and located as the outer layer of the endosperm of many grain seeds.
alga (pl. **algae**) A photosynthetic protist containing plastids.
algal bloom A proliferation of algae due to a nutrient-rich medium, usually resulting in a green scum on the water surface.
algin A polysaccharide derived from brown algae and used for many industrial processes.
alkaline Denoting substances that release hydroxyl (OH^-) ions into solution; *see* basic.
alkaloid A group of nitrogen-containing compounds having diverse structures; many alkaloids have medicinal, hallucinogenic, or toxic properties.
allele One of the two genes for a given trait at a specific locus on homologous chromosomes.
alternate Leaf arrangement in which there is only one leaf per node.
alternate host The alternate plant required to complete the life cycle of some microorganisms. For example, for *Puccinia graminis tritice,* wheat is the primary host and *Berberis vulgaris* is the alternate host.
alternation of generations The sequence of a diploid sporophyte plant producing haploid spores that develop into gametophyte plants (or stages). Gametophytes produce gametes, which fuse to form a sporophyte again.
amber Fossilized resin of ancient trees.
ambient Air temperature.
amino acid (ah-mean-o) An organic molecule including one or more amino ($—NH^2$) and acid (—COOH) groups; proteins are made up of these molecules.
amino group ($—NH^2$) A chemical part of a molecule that imparts basic properties to an amino acid.
α-amylase An enzyme that converts starch to sugars.
anaerobic Without oxygen.
anaerobic respiration The partial oxidation of pyruvate to lactic acid under anaerobic conditions.
anaphase The stage of nuclear division in which the chromosomes are pulled to opposite poles while attached to spindle fibers at the centromere.
androecium (Gr. *andros,* man + *oikos,* house) A collective term referring to the stamens within a flower.
androsterone An animal hormone not synthesized by plants.
angiosperm The group of plants characterized by having flowers as their sexual reproductive structures.

anisogamy The union of motile gametes of unequal size.
annual A plant that completes its life cycle during one growing season.
annual ring The secondary xylem produced during a single growing season.
annulus (L. *annulus,* a ring) In fern sporangia, a row of cells with both thin and thickened cell walls which facilitate the opening of the sporangia and the release of spores because of the unequal expansion and tension produced as they are moistened and dried.
anther The male reproductive organ enclosing and containing the pollen grains.
antheridiophore (Gr. *anthos,* flower + *phoros,* bearing) in some liverworts, a stalk that bears antheridia embedded on an elevated and expanded tip.
antheridium (an-thur-id-e-um) (pl. **antheridia**) The multicellular male sex structure of plants other than seed plants.
anthocyanins A group of water-soluble red to blue flavonoid pigments found in certain plants; especially important pigmentation in flower petals.
antibiosis The inhibition of growth of a microorganism by a substance produced by another microorganism.
antibiotic An organic molecule naturally produced by one microorganism that retards or prevents the growth of another organism.
anticodon loop The portion of a tRNA molecule responsible for the anticodon triplet, which pairs with the codon of mRNA.
antiparallel Opposite in direction, as in the structure of the two strands of DNA.
antipodal cell The three haploid cells at the end of the embryo sac away from the micropyle.
apex The tip of a structure; a leaf apex, for example, is the tip of the leaf.
apical dominance The phenomenon leading to controlled growth of lateral shoots; growth occurs primarily at the top of the plant.
apical meristem (L. *apex,* tip + Gr. *meristos,* divisible) A group of cells at the tip of the stem and root that give rise by cell division to the primary tissues and are ultimately responsible for the structural organization of the entire primary plant body.
apoplastic Pertaining to the movement of water in the free space of tissue; free space includes cell walls and intercellular spaces.
applied research Research conducted to address a specific problem for the purpose of application for productivity or commercial gain.
Archaea (archebacteria) (Gr. *arche,* beginning) Early prokaryotes, differing from bacteria in having unusual types of metabolism, membrane lipids, and amino acid and DNA and RNA base sequences. Inhabit extreme environments.
archegonium (ar-ke-go-ne-um) The multicellular female sex structure of plants other than seed plants.
ascocarp The dense, compact mass of hyphae constituting the fruiting body of the Ascomycetes.
Ascomycetes (Gr. *asko,* bag) A group of fungi whose spores are borne in an ascus or bag.
ascospore A spore produced within an ascus.
asexual Lacking sexual reproduction; vegetative reproduction.
aspartate Four-carbon amino acid found with malate to be the first products of the C_4.
athlete's foot A disease caused by imperfect fungi that flourish under warm, wet conditions. Usually dissappears if the feet are kept dry.
atmospheric pressure Ambient pressure created near the earth's surface by large air cells that circulate around the globe.
atom The basic unit of matter; the smallest complete unit of the elements, consisting of protons, neutrons, and electrons.
atomic number The number of protons within the nucleus of an atom, which determines the elemental properties of that atom.
atomic weight The weight of an atom determined by adding the number of protons and neutrons (the mass of electrons is usually considered to be negligible).
autotrophic (au-to-tro-fik) An organism that produces its own food by photosynthesis; green plants.
auxin A hormone that causes the bending response in *Avena* coleoptiles.
axil (Gr. *axilla,* armpit) The upper angle between a stem and a leaf.
axillary bud A bud, consisting of an apical meristem and leaf primordia, situated in the leaf axil.

B

β-carotene An important plant carotenoid and precursor of vitamin A.
bacillus (pl. **bacilli**) A rod-shaped bacterium.
backcross Crossing a hybrid offspring back to either parent.
bacteria (pl. **bacteria**) (Gr. *bakterion,* dim. of *baktron,* a staff) A prokaryote organism.
bacteriochlorophyll A type of chlorophyll found only in bacterial systems.
bacteriophage A virus that attacks bacterial cells.
bark Those portions of a woody plant stem or trunk exterior to the vascular cambium.
basal rosette Leaves clustered at ground level; can be alternate or whorled, but with extremely short internodes.
basalt A semiviscous layer of igneous rock underlying the granitic continental plates.
basic Possessing a large number of hydroxyl (OH^-) ions; a pH of more than 7.0.
basic research Research conducted to elucidate basic concepts alone with no direct application in mind.
basidiocarp The densely packed hyphae of the fruiting body in the Basidiomycetes.
Basidiomycetes (L. *basidium,* little pedestal) A group of fungi whose spores are borne in a basidium.
basidiospores (L. *basidium,* little pedestal + Gr. *spora,* seed) A haploid spore produced by a basidium in the Basidiomycetes.
basidium The club-shaped structure produced by the hyphae of Basidiomycetes and upon which sexual spores are produced.
basipetal Movement down the stem as with auxin moving from the shoot apex downward.
bedrock The solid rock underlying soil horizons A, B, and C.
benthic (ben-thik) Pertaining to the bottom region of an ocean, lake, or pond.

berry A fleshy, two- or multiple-carpeled ovary, each carpel having many seeds.
bicarbonate ion Carbonic acid ionizes to produce a bicarbonate ion, HCO_3^- and a proton, H^+ in rainfall.
biennial A plant that requires two growing seasons to complete its life cycle.
big bang The theory proposing that the entire universe was created at one time.
binomial The two names, genus and species, comprising the scientific name.
bioassay A quantitative assay of a particular substance using a portion of or an entire living organism.
biological clock An internal biological timing system that relates cyclic phenomena within the organism to the diurnal or annual clock.
biology The study of all living organisms.
bioluminescence The biological production of light using ATP as an energy source.
biomass The total amount of organic material produced; usually expressed on a dry weight basis.
biomes (bye-ohmes) Worldwide groupings of similar ecosystems.
biosphere The earth and all of its ecological interactions considered as a single system.
biotic Pertaining to the living part of the environment.
bisexual A flower having both stamens and pistil (both sexes); a perfect flower.
bivalent Two homologous chromosomes associated in a parallel fashion; formed during prophase I of meiosis.
blade The flattened portion of the leaf.
blue-green algae Prokaryote organisms with photosynthetic phycobilin pigments.
bog A wet, marshy region, such as a peat bog.
bomb calorimeter An instrument used for measuring the caloric content of any organic material; measures energy release in combusting organic materials to CO_2 and H_2O.
botany The study of plants.
bract A modified leaf or leaflike structure, usually much reduced in size.
branch roots Lateral roots arising from the pericycle of a larger root.
bryophytes (Gr. *bruon,* moss + *phyton,* plant) A group of nonvascular plants consisting of the hornworts, liverworts, and mosses.
bud An apical meristem and the appendages which surround and enclose it. Also refers to the asexual reproductive outgrowth in yeasts and some bacteria.
budding A method of asexual propagation characterized by placing a bud of one plant onto the stem of another plant.
bud scales Modified leaves surrounding and protecting a bud.
bud scale scar A scar or impression encircling the twig caused by the abscission of bud scales.
buffer Any substance that absorbs or releases protons (H^+) so that the pH of the solution remains stable, even when acid or base is added.
bulb An underground storage organ characterized by fleshy leaves attached to a stem base.
bulliform cell A large epidermal cell found on the upper surface of many grass leaves; turgor pressure in these cells controls the lateral rolling of the leaves during water stress.
bundle sheath cell Cells surrounding the vascular bundle in C_4 plants.
buttressing A stem modification in which the diameter is larger near the ground, giving additional mechanical support.

C

C_3 pathway The Calvin cycle of carbon fixation in which CO_2 is incorporated into ribulose 1,5-bisphosphate.
C_4 pathway Also called the Hatch-Slack pathway of photosynthesis; CO_2 is initially fixed in mesophyll cells into malate and aspartate.
callose A complex carbohydrate found in the sieve areas of sieve-tube elements; particularly abundant at the time of injury.
callus An undifferentiated group of cells formed as a response to wounding (as at the base of a stem) or in tissue culture.
calorie A unit of heat; one calorie is the amount of heat required to raise the temperature of 1 gm of water 1°C.
calyx Referring to all the sepals of a flower collectively.
CAM *see* Crassulacean acid metabolism.
Cambrian A geological period of the Paleozoic beginning about 590 million years ago and lasting about 85 million years, during which time many divisions of protists occurred in the oceans.
canopy The upper portion of a population of plants; the term is usually associated with forests and agricultural crops.
capillary pores Small spaces in the soil that become filled with a fluid (such as water) because of the adhesion of particles to the matrix (solid substrate) and the cohesion of the water molecules to themselves.
capsid The protein coat of a virus.
capsule (1) A simple fruit that develops from a compound ovary with two or more carpels; capsules dehisce in many ways. (2) The sporangium of a bryophyte.
carageenan A polysaccharide extracted from red algae and used for many industrial products.
carbohydrate An organic molecule consisting of a chain of carbon atoms, each having hydrogens (H^+) and hydroxyl (OH^-) groups attached in the basic pattern CH_2O; includes simple sugars and polysaccharides such as starch and cellulose.
carbon An element occurring native as the diamond and as graphite, forming a constituent of all organic compounds.
carbon dioxide A gas composed of one atom of carbon and two of oxygen and is necessary for photosynthesis.
carbon fixation The process by which CO_2 is incorporated into organic molecules.
carbon skeleton The central portion of any organic molecule consisting only of the carbon molecule.
carbonic acid Atmospheric CO_2 dissolved in water to yield H_2CO_3, a weak acid.
Carboniferous A geological period of the Paleozoic beginning approximately 360 million years ago and lasting about 85 million years. During this time forests appeared and became dominant, composed of arborescent club mosses, horsetails, and ferns. Their extinction resulted in the formation of coal deposits.
carboxyl group An acid group attached to a molecule; —COOH.

carnauba wax (car-now-bah) A high quality, hard industrial wax extracted from the carnauba palm tree.
carnivore An animal that feeds on other animals.
carotenoids Eight–isoprene unit terpenes synthesized by most plants.
carpel The reproductive unit of angiosperms composed of a placental surface and ovules. A component of the gynoecium.
carpellate A unisexual flower having carpels but no stamens.
carrying capacity The maximum number of organisms that can live in balance within the natural food supply of a given area.
caryopsis A dry, indehiscent one-seeded fruit in which the pericarp is united to the seed coat.
Casparian strip The suberized layer covering the radial and transverse walls of endodermal cells.
catalyst Any substance that causes a chemical reaction to proceed much faster than it would without the catalyst. In biochemical reactions enzymes are proteins that act as catalysts.
cell enlargement The expansion in volume of a cell typically as a result of increasing turgor pressure.
cellulase An enzyme that breaks down cellulose.
cellulose The primary structural carbohydrate of plant cells; composed of many glucose molecules.
cell wall The rigid outermost layer of the cells found in plants, some protists, and most bacteria. Found in plants composed principally of cellulose.
cementation Compacting and natural "gluing together" of sedimentary deposits to form sedimentary rocks.
Cenozoic (Gr. *kainos,* recent) A geological era beginning 65 million years ago and extending to the present, characterized by major adaptive radiation of specialized life forms and formation of biomes.
central bud The main bud in a cluster located at one position.
central cell Also known as the polar cell, this is the binucleated cell in the center of an embryo sac containing the two haploid polar nuclei.
centromere The point of attachment of a chromosome to the spindle fiber.
cereals Members of the grass family that yield a large amount of grain rich in carbohydrates, fats, proteins, and vitamins.
chaparral (shap-a-ral) A vegetative association typical of the Mediterranean region dominated by smaller, often thorny or roughly branched evergreen trees and shrubs and deciduous trees. Included in the temperate deciduous forest biome.
charophytes A group of green algae characterized by calcified filaments that have nodes and internodes.
chemical bond The force holding two atoms together; the force can result from the attraction of opposite charges (ionic bond), or from the sharing of one or more pairs of electrons (a covalent bond).
chemical element A substance that cannot be separated into different substances by ordinary chemical methods; *see* element.
chemotaxonomy Using the identification of groups of chemical compounds as genetic indicators in the establishment of taxonomic relationships.
chiasma (pl., chiasmata) The point of physical exchange of equal segments of adjacent nonsister chromatids during the bivalent stage of meiosis.
chinampas Long, narrow strips of land bordered on three sides by irrigation canals which can produce several crops per year.
chitin A hard polysaccharide found as a major component of the cell wall of most fungi and the exoskeleton of insects and crustaceans.
chlorenchyma Parenchyma tissue containing chloroplasts as in the leaf mesophyll.
chlorofluorocarbons (CFCs) Man-made gas compounds that enter the atmosphere and trap infrared radiation which increases global warming.
chlorophyll The pigment molecule responsible for trapping light energy in the primary events of photosynthesis.
Chlorophyta A division of protists commonly referred to as the green algae. May be unicellular, multicellular, or coenocytic.
chloroplasts Organelles found in cells of the aboveground portions of green plants; specialized for photosynthesis.
cholesterol An animal steroid found in some plants in low concentrations.
chromatid One replicate of a chromosome visible in prophase following DNA replication during interphase.
chromatin The dark-staining nuclear material present during interphase; includes the DNA and nuclear proteins.
chromatophore Discrete spherical bodies located in the membranes of prokaryotes that carry the enzymes and pigments important in photosynthesis.
chromoplasts Membrane-bound organelles specialized for carotenoid storage.
chromosome The microscopic strands within the nucleus of eukaryotic cells that carry DNA, which is responsible for inheritance.
Chrysophyta A division of protists consisting of golden-brown and yellow-green algae as well as the diatoms.
circulation cells Air movement in a circular pattern from the earth's surface up into the outer atmosphere and then back down again; produce high and low pressure systems and dictate precipitation patterns at different latitudes.
cladophyll A stem or branch that resembles a leaf.
clay The smallest soil particle size at less than .002 mm in diameter.
climacteric rise A point during the ripening process of certain fruit when the respiration rates rise to very high levels.
climatology The study of climates and the factors influencing them.
climax community The ultimate vegetative community that any given habitat can support; the final stage in ecological succession.
clone Genetically identical organisms.
coastal shelf The shallow region of the ocean surrounding a large landmass.
coccus (Gr. *kokkos,* berry or pit) Any of various spherical-shaped bacteria.
codon The three-nucleotide sequence of mRNA responsible for coding of a specific amino acid.
coelenterates (si-lent-tur-ates) Cnidaria; a group of aquatic animals represented by hydras, sea anemones, jellyfish, and corals.
coenocyte (see-nuh-site) Organisms possessing a large mass of cytoplasm with many nuclei.
coenzymes Various substances, including certain vitamins and heavy metals, required, in conjunction with a specific enzyme, to bring about a biochemical reaction.
coevolution The evolution of two species in concert, such that their survival and reproduction are mutually beneficial.

cofactor A nonprotein substance required by enzymes for proper function; they may be metallic ions or organic molecules called coenzymes.
cohesion The attraction of like particles; water molecules *cohere* to each other.
coleoptile The meristematic growing tip of a grass.
collenchyma The tissue type characterized by primary cell walls with thickened corners.
colonial A multicellular organism that produces a colony of cells, usually referring to *colonial* algae.
colonization The pioneer establishment of vegetation on a previously unvegetated area.
colonizer An organism that initiates the biological "conquest" of soil or rock; *see also* pioneer organism.
colony A group of cells all derived from a mother cell held together by an extracellular secretion, but not closely adhering to each other; there is no division of labor.
commensalism An interaction between two species in which one population is benefited but the other is not affected.
common name A regional name for well-known plants; in the language of the region, rather than in Latin, and not necessarily paralleling any scientific name.
community All living organisms sharing a given area.
companion cell The small cell adjacent to a sieve-tube element within phloem tissue.
compartmentalization The division of labor in living cells such that enzymes related to a particular function are packaged and separated from the other cell contents, usually by a membrane.
compensation point The condition in a living system in which the uptake of CO_2 equals the release of CO_2; that is, photosynthesis equals respiration.
competition Demand by two or more organisms for the same resources.
complementary strand Two polynucleotide chains in which the pairing of adenine is always with thymine (in DNA) or uracil (in RNA), and guanine is always paired with cytosine.
complete flower Having all four floral parts; sepals, petals, stamens, and pistil.
compost Partially decayed organic matter used in gardening and farming to enrich the soil and increase water holding capacity.
compound fruit A fruit consisting of several individual fruits held together (a multiple fruit) or in which separate carpels of a flower stay together (an aggregate fruit).
compound leaf A leaf composed of two or more completely independent blade units called leaflets.
compression A fossil formed when carbonized plant material is still present in the original shape but is greatly compressed and reduced in size by pressure.
compression wood The reaction wood produced along the lower side of leaning trees, straightening the trunk by expanding and pushing the tree upright.
concentration gradient The difference in concentration in two parts of a system.
cone A strobilus; the reproductive structure of a gymnosperm.
conidiophore Hyphae on which one or many conidia are produced.
conidiosporangium A structure that produces conidiospores.
conidiospore (pl. **conidia**) An asexual spore produced singly or in chains.
conifer Any of a group of plants that produce a strobilus or cone as a reproductive structure.
coniferous forest A community type dominated by cone-bearing gymnosperms mostly in the northern hemisphere.
conjugation A type of isogamy that occurs by the formation of a conjugation tube between cells of different mating strains, as in *Spirogyra.*
conservation To protect, conserve the natural world and all its components.
contractile root A specialized root, common in bulbs, that contracts, pulling the stem deeper into the ground.
convergent evolution Groups of unrelated organisms becoming similar in appearance because of evolutionary change in response to similar environments.
coral reef A hard, rocklike structure in shallow tropical waters. Structure is composed of plants and animals encrusted with calcium carbonate.
corepressor A substance that inhibits production of a particular enzyme.
cork The secondary tissue produced by cork cambium; the outer part of the periderm.
cork cambium The secondary cambium giving rise to cork tissue.
corm An enlarged underground vertical stem.
corolla Referring to all the petals of a flower collectively.
cortex Ground tissue located between the vascular bundles and epidermis of stems and roots.
cortisone An animal hormone not synthesized by plants.
cosmopolitan Worldwide in distribution.
cotyledon A seed leaf; the first leaf formed in a seed.
covalent bond A chemical bond between two atoms created as the result of sharing of electrons.
crassulacean acid metabolism (CAM) A type of carbon metabolism in which stomata open only at night, thus conserving water and allowing the leaves to take in CO_2. Photosynthesis is completed during the day, allowing a partitioning of reactions in time.
cristae The folded-membrane inner structure of mitochondria.
cross-pollinated Pertaining to a flower having pollen deposited on it from a different flower of another plant.
crossing-over The physical exchange of equal segments of adjacent nonsister chromatids during the bivalent stage of meiosis; formation of a chiasma.
crown The topmost portion of a plant.
curare A variety of poisonous plant extracts from the bark, roots, stems, and tendrils of several woody lianas, including *Chondodendron tomentosum.*
cuticle (kute-i-kuhl) The waxy coating of the epidermis on all aboveground parts of a plant.
cutin (kute-in) A lipid layer found in the outer walls of epidermal cells.
cuttings Portions of stems, usually consisting of two or three nodes, used for propagation by the production of adventitious roots.
cyanobacteria (Gr. *kyanos,* dark blue) Sometimes called "blue-green algae"; photosynthetic bacteria; important oxygen producers in the evolution of life on earth.

cycads Members of the Cycadophyta, which have unbranched stems and a terminal grouping of leathery, compound leaves.
cyclic photophosphorylation The formation of ATP by activation of PS I, but failure to activate PS II. Consequently ATP is formed, but NADP is not reduced.
cyclosis The spontaneous movement of cytoplasm and some organelles within the cell.
cytochrome A group of proteins involved in electron transfer in biological systems.
cytokinesis Cell division; the process accompanying nuclear division in which the cytoplasmic contents are divided between the daughter cells.
cytokinin A group of hormones that promote growth by stimulating cell division.
cytoplasm The viscous contents of a cell within the plasma membrane, but generally excluding the nucleus.
cytosine A nitrogen base found in both DNA and RNA.

D

dark reactions Reactions in photosynthesis not requiring light; the Calvin cycle.
day-neutral plant A plant whose flowering is not controlled by the length of day.
day length The number of hours that sunlight illuminates a given area on earth; dependent on the angle of the earth relative to the rays of the sun.
decarboxylation The removal of a single carbon atom as CO_2 from an organic molecule.
deciduous (de-sid-u-us) Referring to plants that lose all their leaves during the cool season; as opposed to evergreen plants.
decomposers Fungi and bacteria acting in nature to consume dead organic matter and return nitrogen and other elements to the soil.
deficit irrigation The concept of low levels of irrigation to achieve a moderate level of productivity; even though yields are not maximized, limited water supplies are conserved.
dehiscent Mature fruit that splits open to release the seed.
deletion mutant A mutation in which a base pair is deleted and shifts the sequence out of phase by one pair; also called a frame shift.
demography The tracking and reporting of population trends and changes over time for different countries and areas of the earth.
denature To change the configuration of a protein molecule such that it loses specificity and no longer functions as an enzyme.
dendrochronology The science of studying growth rings of trees to determine past conditions.
deoxyribonucleic acid (DNA) The double-stranded nucleic acid composed of adenine, guanine, cytosine, and thymine, in addition to phosphate and deoxyribose.
deoxyribose The five-carbon sugar contained in DNA.
deplasmolysis Water flowing into a plasmolized cell to correct the process that would cause cellular death from dehydration.
dermatophytes Certain imperfect fungi that cause skin diseases in humans; includes the organisms causing ringworm and athlete's foot.
desert An area characterized by low average annual rainfall (below 25 cm per year), high temperatures, low-growing scattered woody vegetation and succulents.
desertification (de-surt-i-fi-ka-shun) A process by which fragile, semiarid ecosystems lose productivity because of loss of plant cover, soil erosion, salinization, or waterlogging. Usually associated with human misuse.
desiccation (des-i-kay-shun) The process of drying.
determinate growth Pertaining to a leaf or stem that stops growing after differentiating into a terminal reproductive unit.
Deuteromycetes A miscellaneous assemblage of fungi also termed the imperfect fungi because the sexual reproductive features are either not known, not used, or have been reduced or lost; reproduction is asexual.
development The summation of all activities leading to changes in cells, tissues, organs, or organisms; a genetically controlled sequence.
developmental biology The study of how organisms, their cells, and their tissues achieve their final, predictable form and function.
Devonian A geological period of the Paleozoic era beginning about 400 million years ago. During this period major diversification of the early land plants occurred.
diatomaceous earth A powdery, soil-like material formed by the glass cell walls of dead diatoms deposited on the marine floor.
dichotomous (di-kot-o-mus) Pertaining to the division or forking of a single axis into two branches.
dicotyledon (dicot) The group of angiosperms that produce two cotyledons.
differentially permeable membrane A membrane that permits the passage of certain types of particles and inhibits the passage of others; also termed selectively permeable membrane.
differentiation The chemical and physical changes associated with the developmental process of an organism or cell.
diffuse-porous wood Secondary xylem characterized by the same-sized vessels and tracheids throughout the growing season so that growth rings are difficult or impossible to detect.
diffusion The random movement of particles from a region of high concentration to a region of low concentration.
digestion The process by which macromolecules are broken down into smaller molecules.
dihybrid cross A cross in which the inheritance of two characteristics is studied.
dihybrid cross Two traits being studied simultaneously in a single cross.
dioecious The condition of being unisexual with pistillate and staminate flowers on different plants.
diploid A nucleus with two sets of identical chromosomes; the sporophyte is diploid, having the 2n number of chromosomes.
dirt Displaced soil.
disaccharide A carbohydrate formed of two simple sugar molecules linked by a bond; sucrose is an example.
disulfide bond A linkage between the sulfur atoms of two different amino acids in a protein.
diurnal (dye-urn-al) Pertaining to daily cycles or events.
doctrine of signatures Early concept that the Creator placed certain items on earth for humans and identified their intended use by their shape.

dominant (1) The most prevalent species in a plant community. (2) The allele that has its trait expressed. (3) A gene that has its phenotypic expression appear in the offspring, regardless of the nature of its allelic partner.
dormant Having reduced metabolic and respiratory activity.
double bond A covalent bond sharing two pairs of electrons.
double fertilization The reproductive strategy in angiosperms in which two sperm are involved in the fusion with other nuclei.
double helix A helix composed of two molecules winding around each other, as in DNA.
dropsy An old term for congestive heart failure which produces edema and resulting decrease in manual dexterity, causing one to drop things.
drought (drout) An environmental condition in which precipitation is not sufficient to maximize biological productivity.
drupe A fleshy fruit with a one-carpeled ovary and only one seed. The endocarp is hard and stony, tightly enclosing the seed; the mesocarp is fleshy, and the exocarp is soft and thin.

E

earlywood The large, thin-walled xylem cells produced early in the growing season, which appear less dense than latewood.
ecological succession The sequential replacement of one vegetative community by another through a series of stages; succession ends when the climax community is established.
ecology The total interrelationships of all living organisms with each other and with the nonliving components of their environment.
economic sanctions Using the threat or action of withholding food, medical care, or other critical societal needs to force compliance with one's policies.
ecosystem The natural interrelationships among all the living organisms in a given area and with the environmental factors of that area. Ecosystems are self-sustaining, balanced, and self-perpetuating.
ectomycorrhizae Association of a fungus with a vascular plant root or rhizome system in which the fungus surrounds, but does not penetrate the living cells (protoplast) of the system.
egg The middle of three haploid cells at the micropylar end of the embryo sac; when fertilized, it will form the zygote.
elaters The spiral walls that aid in the release and dispersal of spores from the sporangia in horsetails (*Equisetum*) or the elongate, hygroscopic cells with spiral wall thickenings that aid in the release of spores in liverworts.
electromagnetic spectrum The entire radiation spectrum from the high-energy levels of cosmic rays to the low-energy levels of radiowaves; a small portion in the center provides wavelengths visible to humans.
electron The negatively charged particle that orbits around the atomic nucleus. The number of electrons is always equivalent to the number of protons.
electron transfer The process of energy transfer in biological systems, usually in small steps with only slight changes in energy levels.
electron transport system The passing of excited electrons through a series of molecules resulting in ATP production.
element A substance composed of a single kind of atom, the characteristic being determined by the number of protons in the nucleus.
embryo In plants, that portion of a seed that will form the growing seedling following germination; it has a radicle, apical meristem, and embryonic leaf or leaves.
embryo sac The mature megagametophyte; contains eight haploid nuclei.
embryonic axis The main root/shoot body of a seedling.
endangered species A living species that is in danger of becoming extinct because of small population sizes, poor reproduction, reduced available habitat, or a combination of these factors.
endergonic Pertaining to a reaction that requires energy input before it will occur; endergonic reactions never occur spontaneously.
endocarp The interior layer of the fruit wall.
endodermis The layer of cells directly outside the pericycle and inside the cortex of roots; a portion of the cell layer is suberized by the Casparian strip.
endogenous Originating within an organism, such as the origin of a lateral root.
endogenous factors Internal factors regulating an organism's growth and development.
endomycorrhizae Association of a fungus with a vascular plant root or rhizome system in which the fungus penetrates the living cells of the root.
endoplasmic reticulum The flattened membrane network running throughout the cytoplasm of cells; if ribosomes are attached, it is termed *rough endoplasmic reticulum;* if ribosomes are not present, the membrane is termed *smooth endoplasmic reticulum.*
endosperm A triploid nutritive tissue resulting from the fusion of a haploid sperm nucleus with the two haploid polar nuclei in the ovule of angiosperms.
endosymbiosis The theory that some cellular organelles arose by the incorporation of a prokaryote into the cytoplasm of a eukaryote.
entine The inner layer of a pollen grain shell.
entire A smooth leaf margin without teeth, lobes, or undulations.
entropy A physical concept describing the degree of orderliness in a system.
environment The external, physical factors affecting an organism's development and expression of its phenotype.
environmental insults Any factor in the physical environment that inhibits the growth and/or development of an organism.
environmental terrorism Conducting acts of terrorism through the destruction or pollution of the environment.
enzyme A protein that functions as a catalyst in biochemical reactions.
ephemeral (e-fem-ur-uhl) Temporary, such as vegetation that completes its life cycle in a short time.
epicotyl That portion of a seedling above the cotyledonary node.
epidermis The outermost layer of cells covering the entire plant body.
epinasty The unequal growth of petioles causing the leaf blade to curve downward.

epiphyte (ep-i-fight) A plant that grows on another plant as a support but derives no nutrition from the host.
epistasis When one gene has a masking effect on the expression of another, nonallelic gene, the former is said to be *epistatic.*
ergosterol The most common sterol; can be converted to vitamin D in the skin of animals exposed to sunlight.
ergot The spore-producing reproductive body of a fungus that infects grain crops; ergot contains lysergic acid and ergonovine.
essential oils Highly volatile and aromatic oils formed in glands or special cells by some plants; probably involved in pollinator attraction or repulsion of herbivores; used in perfumes, soaps, medicine, and food.
estrone An animal steroid found in some plants in low concentration.
estuary A marshy region where freshwater from a stream or river merges with salt water from the ocean.
ethanol Ethyl alcohol is the end product of anaerobic respiration (fermentation) in plants.
ethylene A plant hormone that promotes fruit ripening in addition to other physiological responses.
etiolation The abnormal elongation of stems caused by insufficient light or unbalanced hormonal relationships. *Etiolated* stems often lack chlorophyll.
Euglenophyta A group of unicellular protists, usually green or colorless with two flagella per cell.
eukaryotic Possessing a nucleus.
eutrophication (you-tro-fi-ka-shun) The natural process of dead organisms gradually filling a standing body of water (pond, lake) as *eutrophic* (nutrient-rich) cycles cause rapid population increases followed by crashes due to shortages of nutrients.
evapotranspiration The combined water loss from both leaf surfaces and from the soil surface; the sum of transpiration and evaporation.
evolution The process by which a lower, simpler form of life is changed through genetic mutations to a higher, more complicated form of life.
excitation Electron movement caused by light striking the chlorophyll molecule.
exergonic Pertaining to a reaction that gives off energy and occurs spontaneously.
exine The hard outer coat of a pollen grain.
exobiology The study of evidence relative to the possibility of life on other planets.
exocarp The outer layer of the mature ovary (fruit) wall.
exogenous factors External factors regulating an organism's growth and development.
exon The "sense" segments of mRNA that contain the actual genetic message for producing a given protein.
extinction The permanent removal of all individuals of a species from earth.

F

facultative anaerobes Microorganisms capable of switching pathways of respiration, depending on the presence or absence of oxygen.
family A taxonomic category composed of one or more related genera.
fascicular cambium The layer of cambium that develops between the xylem and phloem within a vascular bundle.
fats Organic molecules containing high levels of carbon and hydrogen, but little oxygen. Oils are merely fats in liquid state.
fatty acid A long chain hydrocarbon with little or no oxygen and terminating in an acid (—COOH) group.
fermentation The process of ethanol formation by partially oxidizing pyruvic acid; no oxygen is involved in this process.
Fertile Crescent Between the Tigris and Euphrates Rivers in the Middle East, a site of early agriculture.
fertilization The fusion of two haploid gametes, egg and sperm, producing a diploid zygote.
fertilizer burn When the concentration of solutes in the soil is too high due to over fertilization, plasmolysis at the roots occurs, damaging the plant.
fiber An elongated cell type found in many plants and associated as a support tissue of xylem or phloem; highly sclerified and usually dead at maturity.
fibrous roots A root system with many equally sized roots forming a mat, as in grasses. There is no primary taproot.
fibrous root system A root system in which roots are finely divided and lacking a main axis, common in monocotyledons.
fiddlehead The curled fern frond prior to unrolling and elongation; also known as a *crozier.*
field capacity The soil-water storage capacity; the saturated soil profile after gravitational percolation ceases to flow.
filament The elongated stalk of a stamen.
filamentous A slender or threadlike structure or organism, in the latter consisting of a chain of interconnected cells as in filamentous algae.
filial An offspring generation, for example, F_1, the first filial generation.
finite resources Resources that have a limit to their availability; not boundless.
fire ecology The study of the environmental effects of fire.
fission The most common reproduction in prokaryotes, characterized by a replication of the DNA and a splitting or pinching of the mother cell into two daughter cells.
flaccid Pertaining to a cell or tissue with less than full turgor pressure.
flagella Elongated appendages of certain cells used in locomotion.
flavin mononucleotide An electron acceptor in the electron transport scheme or aerobic respiration.
flavonoids A group of secondary compounds produced by plants and important in chemical identification of those plants; believed to be contained in petals that reflect ultraviolet patterns and thus an element in pollinator attraction.
flora All the plants of a given area.
florigen The hypothetical flowering hormone. This compound (or compounds) has never been identified chemically.
flower The reproductive structure of the anthophyta or angiosperms.
fluid mosaic Refers to the model for the structure of membranes, consisting of a bilayer of lipids in which globular proteins are embedded and can move laterally.
fluorescence Longer wavelength emissions which can result form excited electron energy, results in a glowing light.

follicle A fruit that develops from a single-carpeled ovary and splits down one side when mature.
food chain A sequence of organisms in which plants are the primary food source for herbivores, which are in turn the food source for carnivores, etc., until the top carnivore level is reached.
food web A community food chain depicting which species feed on each other and how many interrelationships are involved.
foot The basal portion of a moss sporophyte, embedded in the gametophyte.
fossil (L. *fossio,* a digging) Any evidence of pre-existing life which is preserved in the earth's crust. In addition to the remains of entire, or parts of, organisms, fossils also include indirect evidence such as animal tracks and footprints, as well as the impression of leaves.
fossil fuels Organic molecules derived from partially decayed plant and animal matter, produced primarily during the Carboniferous period; includes oil, gas, and coal.
fraction I protein Equivalent to RUBP carboxylase; the primary leaf protein in many green plants.
fragmentation A method of asexual reproduction by simply breaking into parts.
frame shift *See* deletion mutant; insertion mutant.
freshwater Water of relatively low salt concentration, as opposed to seawater or salt water.
frond The photosynthetic leaf blade of a fern.
fruit A mature ovary.
fruiting body The reproductive structure of certain fungi.
fucoxanthin An accessory brown pigment found in brown algae and some protists.
fungi Members of the kingdom Fungi, typically multicellular, eukaryotic organisms with absorptive nutrition.
funiculus The stalklike structure connecting an ovule to its placental surface within an ovary.

G

galactose A 6-carbon sugar.
β-galactosidase The enzyme responsible for the splitting of lactose into glucose and galactose.
gametangia Any cell or organ in which gametes are formed.
gamete A sex cell; the mature haploid reproductive cell of either sex.
gametophyte The haploid, gamete-producing plant in the alternation of generations; undergoes mitosis to produce the haploid gametes, which fuse to form the diploid zygote of the sporophyte.
gamma rays That portion of the sun's total range of radiation in which rays are shorter than x rays; below 0.1 nm in length.
gemmae cup (jem-me) A specialized structure on the gametophyte of certain liverworts that produces asexual plantlets (gemmae) capable of starting new gametophytes.
gene The unit of inheritance; a group of nucleotides on the DNA molecule responsible for the inheritance of a particular character.
gene activators Hormones that can turn a given gene on or off.
gene regulation The process by which genes are turned on and off to regulate growth and development of an organism.
gene synthesizer A machine that can produce a functional gene segment in a short period of time.
generative nucleus Produced by the haploid microspore nucleus of a pollen grain, the generative nucleus divides mitotically to form two sperm nuclei.
genetic engineering Modifying the genetic structure of one organism by splicing in selected genetic information from another organism.
genetic variability Produced by sexual reproduction, all organisms are genetically different from each other.
genome The entire genetic complement for a particular organism.
genotype The genetic composition of an organism.
genus A grouping of closely related species; the first word of a scientific binomial.
geotropism The bending responses of a plant to the forces of gravity.
germinate To resume growth and increase metabolic activity, as in a seed.
gibberellin A group of related compounds which cause single-gene dwarf mutants of corn and peas to elongate normally.
gill Slits in the underside of a mushroom cap from which basidiospores are produced.
Ginkgo A decidious gymnosperm tree (*Ginkgo biloba*) with distinctive fan-shaped leaves, native to eastern China.
girth Circumference, as of a tree trunk.
global warming Some experts contend our earth's annual average temperature is gradually increasing, partly due to the greenhouse effect.
globular protein A coiled polymer of amino acids forming a compact macromolecule.
glucose A 6-carbon monosaccharide (simple sugar); the primary substrate for respiration.
glycerol An organic molecule to which fatty acids are attached to form a fat.
glycogen The primary storage carbohydrate of animal cells.
glycolysis The series of reactions preceding anaerobic or aerobic respiration in which glucose is oxidized to pyruvic acid.
glycoprotein A macromolecule composed of a carbohydrate-protein complex.
glyoxysomes A subcellular microbody present in the cytoplasm of many oil seeds. Enzymes packaged in the glyoxysome convert lipids to carbohydrates during the germination process.
gnetophytes Taxa of the division Gnetophyta, a group of gymnosperms.
Golgi apparatus Organelles consisting of stacks of flattened membranes that function in packaging and synthesis of membranes and cell walls.
grafting A method of asexual propagation characterized by placing a shoot (the scion) onto the rootstock (the stock) of another plant.
gram-negative Denoting bacterial cells that do not stain with crystal violet and iodine.
gram-positive Denoting bacterial cells that are readily stained with crystal violet and iodine.
granite An igneous rock overlying most of the landmasses of the earth.
granum (pl. **grana**) Structures within chloroplasts, seen as green granules with the light microscope and as a series of stacked disk-shaped membranes with an electron microscope; the grana contain the chlorophylls and carotenoids and are the sites of the light reactions of photosynthesis.

grassland A community type composed of dense, turf-forming perennial grasses and herbaceous perennials.
gravel Inorganic soil particles larger than 2.0 mm in diameter.
green revolution New varieties of crop lasts developed by research which dramatically increased yields, resulting in what is called the green revolution.
greenhouse effect The climatic effect attributed to high levels of CO_2 in the atmosphere that traps incoming infrared radiation and raises the temperature of the earth.
ground meristem The basic or fundamental tissue of the apical meristem; dermal tissues surround the ground meristem, and the provascular strands are embedded in it.
ground parenchyma The basic ground tissue consisting of living parenchymal cells.
growth An irreversible increase in size or volume of a cell, tissue, organ, or organism.
growth ring A concentric layer of wood (secondary xylem) formed in one growing season through cell divisions of the vascular cambium. Also termed annual ring.
guanine A nitrogen base found in both DNA and RNA.
guard cell One of two epidermal cells associated with the stomatal apparatus.
guttation Root pressure forcing water out of the leaves, usually overnight when stomates are closed.
gymnosperm (Gr. *gymnos,* naked + *sperma,* seed) Seed-producing plants in which the seed is not enclosed in an ovary as in angiospems. Ginkgos, cycads, and conifers are examples.
gynoecium (Gr. *gynos,* women + *oikos,* house) The collective term referring to the carpels or ovule-bearing structures, within a flower.

H

habitat The biotic and abiotic components making up the home of all the organisms in a given region.
half-life The time required for half the radiation of a radioisotope to be emitted.
hallucinogen A compound that produces a mind-altering effect.
halophilic Pertaining to microorganisms that tolerate high concentrations of saline media.
haploid One set of chromosomes, the number of chromosomes a gamete would contribute to fertilization.
hardwood A term often used in reference to all woody dicots; more accurately, wood having a high specific gravity.
head Many individual flowers tightly compressed into the shape of a single large flower.
heartwood The wood found in the center of a tree trunk; often a darker color due to the accumulation of resins, oils, gums, and other metabolic by-products, which prevent water movement through this tissue.
hemicellulose A polysaccharide component of primary cell walls; similar to cellulose, but more easily degraded.
hemoglobin The blood protein responsible for the transport of oxygen throughout the body.
hemp *Cannabis sativa;* source for fibers used in making hemp rope.
herbaceous Without woody tissues; typical of most annuals and biennials; herbaceous perennials die back to the soil level each year.
herbalism The use of plants for treating ailments.
herbarium A collection of pressed, dried plant specimens mounted on sturdy (rag) paper and stored for reference and research.
herbicide A plant growth regulator that inhibits growth or kills a plant when applied at relatively low concentrations.
herbivore Animals that feed only on plants.
heredity The transmission of genetically controlled characters from parents to offspring through sexual reproduction.
heritable Capable of being inherited, such as a physical trait.
hesperidium A berry with a thick leathery peel (exocarp and mesocarp) and a fleshy endocarp arranged in sections.
heterocyst A specialized cell of certain cyanobacteria; usually larger and thick-walled, this cell excludes oxygen and allows for anaerobic nitrogenase activity.
heterosporous Producing two distinct types of spores.
heterotrophic An organism that obtains its food from other organisms.
heterotrophic theory A theory of the origin of life which proposes that the first organisms obtained nutrition from the spontaneous formation of organic molecules derived from primordial gases.
heterozygous A genotype for a given phenotypic expression containing a dominant and a recessive allele for that trait.
hexose A 6-carbon sugar.
Hill reaction The splitting of a molecule of water during the light reactions of photosynthesis; the photolysis of water.
histones Basic proteins constituting a portion of the nuclear material and functionally associated with DNA.
holdfast An organ at the base of macroalgae that attaches the stalk to a rocky surface.
holistic medicine The treatment of the whole entity, the human body, spirit, and mind—especially accepted in Eastern and primitive cultures.
homogeneous Consistent and similar in composition.
homologous pairs Two chromosomes that are morphologically and structurally identical and that pair during prophase of meiosis.
homologue One chromosome morphologically and structurally identical to another one in a somatic cell nucleus.
homosporous Producing only one kind of spore.
homozygous A genotype for a given phenotypic expression containing either two dominant or two recessive alleles for that trait.
horizons Soil zones from the surface downward that reveal visible horizontal layers of soil.
hormone An organic molecule synthesized by a plant that exerts, even in low concentrations, profound regulation of growth and/or development.
horticultural therapy The activities of propagating, growing, and planting plants is therapeutic for a variety of people including the elderly and mentally challenged.
humidity Water vapor in the atmosphere.
humus The organic portion of soil derived from partially decayed plant and animal matter.
hybrid An organism resulting from the fusion of gametes from different parental species.
hydrogen potential Same as pH; refers to the number of charged hydrogen atoms or protons present, acidity of the soil.

hydrolysis The splitting of a large molecule into smaller molecules by the addition of water.
hydrophilic The property of a substance that has a tendency to attract water.
hydrophobic The property of a substance that has a tendency to repel water.
hydroponics A system of growing plants in a liquid nutrient solution without soil.
hypha (pl. **hyphae**) The long slender thread of a vegetative fungus.
hypocotyl The shoot portion of a seedling below the cotyledonary node.
hypocotyl hook The hooked portion of a hypocotyl formed at germination that assists a dicot shoot in pulling itself through the soil crust.
hypothesis A proposed solution to a problem; a theory to explain unproven events or observations.

I

igneous Most of the molten mass of the earth, igneous material cooled to form the earth's crust.
igneous rock A rock of molten or volcanic origin.
imbibition The process of taking up water physically.
immunological Pertaining to the immune response, in which a protein antibody is synthesized by an organism to counteract some pathogenic factor.
impermeable Having the property of restricting the passage of substances.
incomplete dominance Referring to the phenotypic expression for a given trait demonstrating a blending of the genetic messages from the allele partners controlling that trait; no dominant allelic partner.
incomplete flower Lacking one or more of the four floral parts (sepals, petals, stamens, pistil).
indehiscent Denoting mature fruit that do not split open to release their seed.
independent assortment The random alignment of homologous chromosomes in meiosis.
indeterminate Pertaining to a stem that produces unrestricted vegetative growth; the stem does not terminate in flowering.
indigenous Native to, or originating in a particular area.
indoleacetic acid (IAA) The naturally occurring auxin.
inducer A compound that causes the induction of a particular enzyme.
inducible enzyme An enzyme present only when induced by some particular substrate.
indusium A thin membrane that covers the sori of many ferns; the membrane often breaks at the time of spore maturation.
inferior ovary Ovary attachment to a modified receptacle in which other floral parts are fused to the ovary.
inflorescence A flower cluster with a definite arrangement of individual flowers.
infrared That portion of the sun's total range of radiation having wavelengths immediately longer than the longest of the visible spectrum (red); between approximately 750 nm and 1 m in length.
inorganic A chemical compound without carbon as its skeleton atom.
insectivore A plant capable of deriving nutrition by digesting insects.
insectivorous plants Plants which actively or passively capture insects as an adaptation to acquire nitrogen. Examples include the venus flytrap, sundew, and pitcher plant.
insertion mutant A type of mutation in which an extra base pair is inserted and shifts the sequence out of phase by one pair; also called a frame shift.
integuments The two outer layers of an ovule that enclose the nucellus tissue within.
intercalary meristem A type of meristem present at the base of the blade and/or sheath of many monocots.
interfascicular cambium The layer of cambium that develops between vascular bundles and connects with the fascicular cambium to form the vascular cambium of woody tissues.
interkinesis The activities occurring between meiosis I and meiosis II; similar to interphase but without chromosome replication.
internode The stem distance between nodes.
interphase The nuclear condition between one mitosis and the next. The chromosomes are not visible, but intense metabolic activity is occurring.
intertidal zone The region between high and low tide.
intron The intragene segments of genetic gibberish that do not code for the production of a given protein; these segments are left out of the mRNA that leaves the nucleus to direct protein synthesis.
invertebrates Animals without backbones.
involutions Infolding of membranes to increase surface area.
ion An atom or molecule that has gained or lost an electron, causing the particle to become electrically charged.
ionize To split a molecule into two or more parts, each part becoming an electrically charged particle.
irregular symmetry Bilateral symmetry; having only one plane through which the structure (flower) could be cut to result in mirror image halves; zygomorphy.
irrigation The application of water to agricultural crops by transporting it to the crop through canals, pipes, and sprinklers.
island of fertility The region in a desert ecosystem directly beneath a tree or shrub. Leaf fall and accumulation of litter result in nutrient cycling directly around the plant, even though the bare soil between plants may be depleted of nutrients.
isogamy The union of gametes or gametangia of equal size.
isoprenes Five-carbon compounds that are the basic unit of terpenes.
isozyme Different chemical forms of the same enzyme; thought to be important in adaptation to environmental extremes.

K

kelp Brown algae, a macroalgae of temperate waters.
kinetic energy The energy resulting from the random movement of molecules.
knot Commonly seen as an irregular formation in a longitudinal section of wood formed when a branch becomes surrounded (buried) by formation of secondary xylem.
Kranz anatomy A specialized leaf anatomy found in C_4 plants in which the vascular bundle is surrounded by bundle sheath cells.

Krebs cycle The mitochondrial oxidation of pyruvic acid by condensing acetyl coenzyme A with oxaloacetic acid to form a series of 6-, 5-, and 4-carbon organic acids.

L

lactic acid The end product of anaerobic respiration in animals, causes muscle soreness.
lactose Milk sugar; a disaccharide composed of glucose and galactose.
land ethic A philosophy or attitude about land that provides for its protection or more careful management.
lateral meristem Meristems that give rise to secondary tissues, typically by the formation of radial files of cells and thereby increasing the diameter and circumference of an axis over time; the vascular cambium and cork cambium.
lateral root A root that arises from another, older root. Often termed secondary or branch roots.
laterite A type of tropical soil in which iron and aluminum oxides cause the soil structure to harden like concrete.
latewood The small, thick-walled xylem cells produced at the end of the growing season that appear as a dense ring of wood adjacent to the thin-walled earlywood.
latitude A geographical unit used to measure the distance from the equator toward either pole.
lattice A crystalline-like structure caused by the precise orientation of molecules in a solid or liquid.
lay botanist A nonscientist who understands botanical principles and appreciates the value of the plant world.
layering A method of asexual propagation in which portions of the stem are wounded and covered with a medium, usually soil, to stimulate the production of adventitious roots.
leaching The process of removal of ions or molecules by flushing with water.
leaf-area index A numerical index of the ratio of leaf area to ground area in a plant community.
leaf margin Refers to the edge of a leaf, the area between the apex and base.
leaf primordium (L. *primordium,* beginning) The first stage of leaf development; a small lateral protuburance formed by an apical shoot meristem that will expand to form a leaf.
leaf scars A scar left on a stem or twig when the leaf abscises.
leaf sheath The lower part of a blade or petiole that invests the stem more or less completely.
leaflet An individual blade unit of a compound leaf.
leeward The side away from the direction of a prevailing wind.
legume A fruit that develops form a single-carpeled ovary and splits along two sides when mature.
lenticel (L. *lenticella,* a small window) Isolated areas of loosely arranged cells in the cork surfaces of stems, roots, and fruit that allow interchange of gases between internal tissues and the atmosphere through the periderm.
leucoplasts Membrane-bound organelles specialized for starch storage.
liana A large, woody vine common to the tropical forests; it climbs the tall trees and often trails from the canopy.
lichen An organism composed of a symbiotic association of an ascomycete fungus with algal or cyanobacterial cells.
light The electromagnetic radiation produced by the sun that heats and illuminates the earth.
light intensity The strength of light rays; the degree of brightness dependent on the number of photons striking a given area at a point in time.
light reactions Light requiring parts of photosynthesis, photolysis and photophosphorylation.
lignification Impregnated with lignin.
lignified Impregnated with lignin, such as the secondary cell walls of woody plants.
lignin A complicated organic molecule found as an important constituent of many secondary cell walls; imparts strength and rigidity to the cellulose microfibrils.
limiting factor The climatic factor that would first curtail plant growth if unavailable.
limnetic zone The open-water zone of a lake or pond; extends to the depth of effective light penetration.
limnologist A scientist who studies freshwater biology.
linkage The occurrence of alleles for one trait on the same chromosomes as the alleles for a different trait.
lip cell Thin-walled cells that interrupt the annulus in fern sporangia.
lipid A fat or oil; composed of fatty acids and glycerol.
litter Dead organic material such as branches, tree trunks, and dry grass that accumulates on the floor of a forest. Litter acts as additional fuel in a forest fire, producing increased destruction.
littoral zone The shallow-water zone of a lake or pond; light penetrates to the bottom, and the area is occupied by rooted plants such as water lilies, rushes, and sedges.
liverwort A group of the division Bryophyta charcterized by a small, inconspicuous, liver-shaped thallus and that lives in moist environments.
loam A mixture of sand-, silt-, and clay-sized soil particles.
locus (pl., **loci**) The position that a given gene occupies on a chromosome.
long-day plant A plant that flowers when the length of day *exceeds* some critical value.
LSD Lysergic acid diethylamide; a synthetic hallucinogenic derivative of lysergic acid.
lumen The central cavity of a cell.
Lycophyta A division of "lower" (i.e., non-seed-producing) vascular plants, commonly referred to as club and spike mosses.
lysosome A membrane-bound organelle containing enzymes that break down proteins and other macromolecules.

M

macroalgae The multicellular brown, red, and green algae.
macroelements Essential elements needed for plant growth in relatively large amounts; C, H, O, P, K, N, S, Ca, Fe, Mg.
macrofibril An aggregation of microfibrils in the cell wall, visible with the light microscope.
macromolecule (Gr. *makros,* large + L. *moliculus,* a little mass) A very large molecule, generally used in reference to carbohydrates, lipids, proteins, and nucleic acids.
macropores Large pore spaces caused by invertebrates and larger animals, including reptiles and mammals, that permeate the soil. They drain water not held by capillarity.

malate Four-carbon organic acid found with aspartate to be the first products of the C_4 pathway of photosynthesis.
margin The edge of a flattened structure; in leaves, the lateral edge of the blade.
marginal meristem The meristem along the margin of a leaf primordium responsible with the formation and shape of the leaf.
marine Of or pertaining to the ocean.
marsh A wetland along the shallow margins of lakes and ponds and in low, poorly drained lands where water stands for several months of the year.
mass The fundamental unit of measurement equivalent to the weight of a substance when compared with the weight of hydrogen.
matric potential The water potential component caused by the attraction of water molecules to a hydrophilic matrix.
meadow An open clearing in a landscape in which low-growing, herbaceous plants dominate.
mean Occupying a median or middle position, halfway between the two extremes.
megagametogenesis The production of megagametes (large gametes) from megaspores in the ovules of angiosperms.
megagametophyte The gametophyte stage containing eight haploid nuclei within the embryo sac.
megaspore Haploid cell produced by meiosis in the ovules of angiosperms; a single megasporocyte produces four megaspores, only one of which remains functional.
megasporocyte Also known as the megaspore mother cell; this diploid cell undergoes meiosis to produce four haploid megaspores in the ovules of angiosperms.
megasporogenesis The process of megaspores (large spores) being produced via meiosis in the ovules of angiosperms.
meiosis The process of two sequential nuclear and cellular divisions resulting in four haploid cells from a single diploid cell; occurs in reproductive organs.
membrane In living organisms, a phospholipid bilayer impregnated with protein and certain other compounds; functions in partitioning of cellulose activities.
membrane selectivity See differentially permeable membrane.
meristem (Gr. *merizein,* to divide) The tissue or zone from which new cells are produced by cell division.
mesic (me-zic) Referring to a region that receives adequate precipitation to maintain biological productivity.
mesocarp The middle layer of the fruit wall, located between the exocarp and endocarp.
mesophyll Parenchymal tissue in the center of a leaf.
mesophyte A plant that grows in soils that contain moderate or intermediate amounts of moisture.
Mesozoic The geological era beginning about 250 million years ago. During this time gymnosperm forests, dinosaurs, mammals, birds, and flowering plants appeared. By the end of the Mesozoic, about 70 million years ago, dinosaurs became extinct.
messenger RNA (mRNA) A ribonucleic acid transcribed from the DNA template.
metabolic energy That energy obtained from ATP produced in metabolism.
metabolism The summation of all the biochemical events within a cell or organism; it includes both synthesis and breakdown of molecules.
metabolites Chemical substances required in metabolism.
metacentric Denoting a chromosome with a centromere located at the center.
metamorphic rock A type of rock, either granitic or sedimentary in origin, and structurally changed by high temperature and pressure.
metaphase The stage of nuclear division in which the chromosomes migrate to the center of the cell.
methane CH_4, natural gas.
methanogen A bacterium that obtains energy from CO_2 and H_2 and forms methane.
microfibril An elongated strand of cellulose molecules.
microgametogenesis The production of microgametes (small gametes) from microspores in the anthers of flowering plants.
microorganisms Single-celled living organisms that can be seen only with the aid of a microscope.
micropores Spaces between soil particles that hold water by means of capillary forces.
micropyle A small opening at one end of an ovule where the integuments come together.
microsporangium (pl. **microsporangia**) The structure (sporangium) in which microspores are produced.
microspore Haploid cells produced by meiosis in the anthers of angiosperms; four microspores are produced from a single microsporocyte.
microsporocyte Also known as the microspore mother cell, this diploid cell undergoes meiosis to produce four haploid microspores in the anthers of angiosperms.
microsporogenesis The process of microspores (small spores) being produced via meiosis in the anthers of angiosperms.
microsporophyll A leaflike structure giving rise to one or more microsporangia.
microtubules Tiny rodlike structures made of the protein tubulin and important in the synthesis of certain membranes.
middle lamella The cementing layer of pectic substances between primary cell walls.
midrib The central large vein of a leaf.
mimicry Pertaining to the resemblance of one organism to a totally unrelated organism; the similarity in color, form, or behavior may result in protection or some other advantage to the mimic.
Mississippian Part of the Carboniferous period, beginning around 360 million years ago.
mitochondrion (pl. **mitochondria**) A double membrane-bound cellular organelle associated with the Krebs cycle and electron transport aspects of respiration.
mitosis Nuclear division of somatic cells resulting in two genetically identical daughter nuclei.
modification A change, as in morphology, usually associated with a functional advantage.
mole One gram molecular weight of any substance; that is, the molecular weight of any substance in grams.
molecular weight The sum of all the atomic weights of a molecule.
molecules Groups of atoms that share electrons and therefore bond together.
monocarpic Denoting a plant that flowers only once.
monocot One of the two primary groups of angiosperms characterized by a single cotyledon, parallel venation of leaves, and floral parts in threes.
monoculture An agricultural system in which only one crop species is cultivated.

monoecious The condition of being unisexual with both pistillate and staminate flowers on the same plant.
monohybrid cross A cross in which the inheritance of only one character at a time is studied.
monosaccharide A sugar broken down to its simplest functional unit—a single saccharide.
monsoon Exceptionally heavy precipitation that occurs seasonally in certain parts of the world.
morphine A pain-relieving and addictive compound derived from the opium poppy (*Papaver somniferum*).
morphology The study of form and structure of living organisms.
moss A group of terrestrial, nonvascular plants; the dominant plant body is the gametophyte with the sporophyte embedded in it.
motility Ability to move.
mountain tundra That portion of tundra vegetation confined to alpine meadows. Low-growing grasses, sedges, and forbs with a very short growing season; permafrost is typical.
mucigel A slimy material secreted by and covering the root cap and root hairs.
mucilaginous Containing a mucilage, usually composed of mucopolysaccharides.
multigenic (polygenic) inheritance Inheritance in which the genetic control for a trait results in the phenotypic expression varying continuously.
multiple fruit Individual ovaries of many separate flowers clustered together.
mushroom Common name for a group of fungi (Basidiomycetes) that produce an aboveground reproductive structure.
mutagen Any substance capable of causing a mutation.
mutation Any permanent heritable change in the genetic code of the DNA.
mutualism An interrelationship in which both (or all) organisms benefit from their association.
mycelium A tangled mass of fungal hyphae.
mycologist (Gr. *mykes,* fungus) A person who studies fungi.
mycorrhizae (mi-kor-i-ze) Certain species of fungi growing in soil in a symbiotic relationship with a higher plant.

N

NAD (nicotinamide adenine dinucleotide) A coenzyme that acts as an electron acceptor; particularly important in respiration.
NADP (nicotinamide adenine dinucleotide phosphate) A coenzyme that acts as an electron acceptor; particularly important in photosynthesis.
natural law A theory that has been tested many thousands of times and found always to be true, e.g., the law of gravity.
nectar A sweet, syrupy exudate produced by some flowers as an attractant for pollinators.
netted venation A type of vein arrangement in leaves in which the profusely branching veins form an interconnecting network; also termed reticulate venation.
neutral Soil pH of 7, neither acid nor alkaline.
neutron Uncharged atomic particles of essentially the same mass as protons. In the most common stable atoms the number of neutrons is equivalent to the number of protons.
niche The specific and unique set of environmental conditions for a given species.
nitrogen base The basic component of nucleic acids; composed of a nitrogen-containing molecule of purine or pyrimidine.
nitrogen fixation The process by which nitrogen gas (N_2) in the atmosphere is reduced to ammonia by certain microorganisms.
nitrogenase The enzyme involved in the conversion of atmospheric nitrogen gas into ammonia.
nitrous oxide NO_2, an atmospheric gas that traps infrared radiation and also can form nitric acid in rain droplets, resulting in acid rain.
node The point on a stem at which leaves and buds are attached.
nodule A tumorlike growth on the roots of certain higher plants that encloses a population of nitrogen-fixing bacteria.
noncyclic photophosphorylation The formation of ATP utilizing the Z-scheme of photosynthesis.
nonseptate Lacking cross walls; commonly used in reference to certain fungal hyphae. Also termed aseptate.
nucellus The inner somatic tissue of an ovule.
nuclear envelope The membrane surrounding the nucleus in eukaryotic cells.
nuclear pores Pitted regions in the nuclear envelope through which processed mRNA migrates to the ribosomes.
nucleic acid A macromolecule composed of a polymer arrangement of nucleotides; DNA is double stranded, and RNA is single stranded.
nucleolar organizer An area on certain chromosomes associated with the formation of the nucleolus.
nucleolus (pl. **nucleoli**) Particularly dense region of DNA associated with certain chromosomes; interphase nuclei may exhibit one or many nucleoli.
nucleotide A portion of a nucleic acid molecule composed of a nitrogen base (purine or pyrimidine), a pentose sugar (ribose or deoxyribose), and a phosphate.
nucleus The central "kernel" or membrane-bound organelle of a eukaryotic cell; contains the chromosomes and most of the DNA of the cell.
nut A dry, indehiscent one-seeded fruit with a hard pericarp (the shell).
nutrients Food substances necessary for the sustenance of living organisms.
nyctinasty The "sleep movements" of leaves in response to change in turgor pressure of cells at the base of the petiole.

obligate anaerobe A microorganism that can survive only under anaerobic conditions.
observations Objectively measuring or recording the results of testing a theory; watching and evaluating.
ocean desert The concept of a region in the ocean in which lack of nutrients is reflected in a lack of plants and animals; a region of low biological productivity.
omnivore An organism that feeds on both plants and animals.
oogamy The union of a large nonmotile egg with a small motile or nonmotile male sperm.
oogonium (pl. **oogonia**) (Gr. *oion,* egg + *gonos,* offspring) In algae and fungi, a specialized egg-containing cell; a single-celled female gametangium, containing one or more eggs.

Oomycetes An assemblage of organisms, previously grouped with the fungi, but currently assigned to the kingdom Protista based in part on the presence of cellulose cell walls, rather than chitinous walls as found in the fungi.
operator A gene responsible for the activation and deactivation of the structural genes.
operon In gene regulation, the system consisting of an operator gene and one or more structural genes controlled by the operator.
opposite Leaf arrangement in which they are attached at a node directly across from one another on the stem.
orbital Discrete pathways or bands surrounding an atom through which electrons move.
Ordovician A geological period beginning around 500 million years ago, believed to be the time when plants may have begun to migrate onto land.
organ A structure composed of different tissues, such as root, stem, leaf, or reproductive (e.g., flower) parts.
organelle Subcellular particles that perform some particular function within the cell.
organic Living or once living material; compounds containing carbon formed by living organisms.
osmoregulator Any substance, either organic or inorganic, that functions to change the solute potential of a solution, and thereby controls the water relations of that solution by osmosis.
osmosis A special case of diffusion of water molecules across a selectively permeable membrane.
osmotic shock Alternating seawater and dilute freshwater in intertidal zones causes erratic osmotic balance in plants growing in these areas.
outcrosser A flower that must be cross-pollinated to successfully complete the reproductive process.
ovary The lower, enlarged portion of the female reproductive structure in a flower that gives rise to the fruit.
overpopulation Too many organisms (especially humans) for the available resources to support in a sustainable way.
ovule That portion of the female reproductive structure of a flower enclosed by an ovary; a developing embryo that gives rise to a seed.
ovuliferous scale In conifers, an axillary, scalelike shoot that bears one or more ovules. Each strobilus may contain one or many ovuliferous scales, depending on the species.
oxidation The loss of electrons by an atom.
oxidative phosphorylation The electron transport system associated with aerobic respiration and mitochondria. In the release of energy through a series of cytochromes, three molecules of ATP are made.
oxygen An important element existing as a gas when two atoms are bonded together as O_2.
ozone (O_3) A molecular form of oxygen in equilibrium with O_2. A layer of ozone in the stratosphere protects the earth from harmful ultraviolet radiation; used as a disinfectant.

P

paleontologist A person who studies fossils and all aspects of extinct life.
Paleozoic The geological era beginning 590 million years ago and ending about 250 million years ago. During this time plants invaded, established, and diversified on the land.
palisade The vertical photosynthetic cells below the upper epidermis in leaf tissue; these cells are a specialized parenchyma.
palmate An arrangement of leaflets (or lobes on a simple leaf), each originating from a common point, usually the axial end of a petiole.
parallel venation A type of venation in which the main veins of a leaf are parallel or nearly so, but converge at the apex and base.
parasitic Denoting an association where one living organism benefits at the expense of another.
parenchyma The tissue type characterized by simple, living cells with only primary cell walls.
parthenocarpy Fruit development without fertilization.
pathogen Any organism that causes a disease of another organism.
peat moss A relatively sterile, inert medium composed of partially decomposed plants of the genus *Sphagnum*. Exceptionally high water-holding capacity.
pectin The cementing substance of which the middle lamella is composed; primarily calcium pectate and pectic acid.
peduncle The stalk of a single flower or an inflorescence.
pelagic (pel-a-jik) Pertaining to the ocean.
pellicle A protein layer located just inside the plasma membrane in euglenoids.
penicillin An antibiotic produced by certain species of *Penicillium*. An ascomycete fungus.
Pennsylvanian A geological period of the Carboniferous beginning about 320 million years ago.
pentose A 5-carbon sugar.
PEP carboxylase The enzyme responsible for CO_2 fixation in the primary fixation of C_4 metabolism.
pepo A berry with the outer wall formed from receptacle tissue fused to the exocarp; the fleshy interior is mesocarp and endocarp.
peptide bond A chemical bond formed between the amino group of one amino acid and the carboxyl (acidic) group of an adjacent amino acid.
percolation The movement of water by gravitational pull down through soil particles.
perennial A plant that overwinters and continues to grow for many years. It may reproduce every year, or only on rare occasions.
perfect flower A flower having both stamens and carpels and therefore bisexual.
perianth Referring to all the sepals and petals collectively.
pericarp The fruit wall that develops from the ovary wall.
pericycle The layer of cells surrounding the xylem and phloem of roots and considered to be a part of the vascular cylinder.
periderm The protective tissue that replaces epidermis when it is sloughed off during secondary growth; includes cork, cork cambium (phellogen), and phelloderm.
permafrost The permanently frozen soil in polar regions.
permanent wilting point The soil moisture content at the point when a given plant's root system can no longer absorb water.
permeability A property of membranes allowing all substances to pass freely.
Permian The last geological period of the Paleozoic era beginning about 286 million years ago and ending 251 million years ago. The end of the Permian was a time of abrupt extinction in the diversity of flora and fauna, especially of sea life.
peroxisome A cellular microbody containing enzymes involved with photorespiration and photosynthesis.

petal The often showy flower component attached just inside the sepals; petals are usually colorful to attract pollinators.
petaloid A modified, flattened filament of a stamen that may resemble a petal.
petiole A stalklike portion of a leaf connecting the blade to the stem or branch.
petrifaction A fossil formed when plant parts are infiltrated or replaced by mineral substances such that the structure is preserved but the fossil is actually rock.
petroleum products Oil and the materials made from crude natural petroleum.
PGA 3-phosphoglyceric acid, the first compound produced in the Calvin cycle.
pH (1) The negative logarithm of the hydrogen ion concentration. (2) A numerical scale used to measure the acidity or basidity of any substance.
Phaeophyta The group of organisms commonly termed the brown algae, consisting of about 1500 species, including the kelps.
pharmaceuticals Medicinal drugs.
phellem Also termed cork; the outer, protective tissue of stems and roots composed of nonliving cells with suberized cell walls, formed centrifugally by the cork cambium or phellogen.
phelloderm A tissue laid down to the inside of the phellogen; the inner part of the periderm.
phellogen The cork cambium.
phenotype The physical features of an organism; the manifestation of the genotype as influenced by the environment.
phloem The food-conducting tissue of vascular plants, composed of sieve elements, various kinds of parenchyma cells, fibers, and sclereids.
phloem rays The part of the vascular ray which is located in the secondary phloem.
phospholipid A special kind of lipid molecule in which glycerol is attached to two fatty acids plus a phosphate group; important part of biological membranes.
photolysis The splitting of a molecule of water during the light reactions of photosynthesis; the Hill reaction.
photon An elementary particle of light; a discrete unit of light energy.
photooxidation The change in the structure of a molecule due to exposure to light; sometimes referred to as bleaching.
photoperiodism The system within organisms that causes certain events, including the onset of reproduction, to be related to the length of day.
photophosphorylation The formation of ATP utilizing light energy in photosynthesis.
photorespiration In C_3 plants, respiration rates increase in the light.
photosynthesis The production of carbohydrates by combining carbon dioxide and water in the presence of light energy; occurs in chlorophyll pigments of plants and releases oxygen as a by-product.
photosynthetic unit A group of associated chlorophyll molecules, including antenna molecules and a central chlorophyll *a* collector molecule.
photosystem I The second part of the Z-scheme in which a chlorophyll *a* molecule absorbs most effectively at 700 nm.
photosystem II The first part of the Z-scheme in which the chlorophyll *a* molecule absorbs most effectively at 680 nm.
phototropism The bending of a plant toward a unidirectional light source.
phycobilins Red and blue accessory pigments found in certain photosynthetic organisms.
phycobilosomes Groups of pigmented cells containing phycocyanin and phycoerythrin.
phycocyanin A blue photosynthetic pigment found in cyanobacteria and red algae.
phycoerythrin A red photosynthetic pigment found in cyanobacteria and red algae.
physiological limits The tolerance range of given species for temperature induced rates of metabolic activity.
phytochrome The protein pigment responsible for the phenomenon of photoperiodism.
phytoplankton The single-celled plants of aquatic biomes; the food source for zooplankton.
pigments Molecules that reflect and absorb light at particular wavelengths.
pinnate Denoting an arrangement of leaflets (or lobes on a simple leaf) along a main central unit.
pioneer organism An organism that first colonizes soil or bare rock.
pistil The female reproductive part of a flower, consisting of stigma, style, and ovary. Also termed the carpel.
pistillate Denoting a unisexual flower having a pistil but no stamens (=carpellate).
pith Parenchymal tissue in the center of a stem located interior to the vascular bundles.
pith ray The region or tissue located between vascular bundles in a stem and connecting the pith and cortex; also termed interfascicular region or medullary ray.
placenta The tissue to which an ovary is attached.
plant growth regulator Any molecule that exhibits hormone-like effects in a plant, whether synthesized or naturally occurring.
plasma membrane The cell membrane that holds the cytoplasm and lies against the cell wall (in plant and fungal cells).
plasmodesma (pl. **plasmodesmata**) Microscopic threads of cytoplasm that traverse cell walls and membranes.
plasmodium The naked body of cytoplasm with many nuclei in slime molds.
plasmolysis The osmotic removal of water from the cytoplasm and vacuole, causing the cytoplasm to pull away from the cell wall and clump in the center.
plasticity Cell wall "stretchability" induced by auxin.
plastids Any of a group of organelles including chloroplasts, chromoplasts, and leucoplasts.
pleiotropy The action of single genes that affects the expression of several traits.
plumule The shoot apex of a seedling, including embryonic leaves.
pneumatophore A specialized root that grows upward into the air from roots growing in the mud; aids in gaseous exchange.
point mutation A change in the genetic composition of a cell due to a change in a single triplet of the DNA template.
polar ice cap The portions of the globe close to the poles that are permanently covered with ice.
polar nuclei Two of the eight haploid nuclei of a megagametophyte that migrate, one from each end, to the middle of the embryo sac. These two nuclei fuse with a sperm nucleus during double fertilization.

polarity The establishment of poles or areas of specialization at opposite ends of a cell, tissue, organ, or organism. Polarity leads to the differentiation of roots from shoots.
pollen The collective term for pollen grains, the male gametophytes.
pollen grain The structure into which a haploid microspore develops; contains a halpoid tube nucleus and two haploid sperm nuclei at maturity.
pollination The transfer of pollen to a receptive surface; the stigma in angiosperms or the pollination droplet in most gymnosperms.
pollinator An organism that effects pollination.
pollution The overproduction of anything that becomes a societal problem because we have no way of getting rid of it, and where it is results in damage or an undesirable situation.
polycarpic Denoting a plant that flowers more than once during its lifetime.
polyembryony The development of more than one embryo.
polygenic Multiple genetic influences controlling a single trait, causes the trait to display continuous variation.
polymer A large molecule formed by the linkage of many smaller, identical molecules.
polynucleotide chains Attachment of one nucleotide to another in a linear fashion.
polypeptide Any number of amino acids linked by peptide bonds.
polyploid A term describing a cell with more than two sets (2n) of chromosomes.
polysaccharide A large molecule formed by joining sugar molecules end to end.
polysome A group of ribosomes functionally related and held together by a strand, presumably mRNA.
pome A fleshy fruit derived from a compound inferior ovary; the fleshy edible part is the ripened tissue surrounding the ovary (derived from receptacle and perianth tissue); the ovary matures into the core and contains the seed.
population All individuals of a given species in a given area.
pore space Space between soil particles.
porosity The relative amount of pore space versus soil particulate space in soil.
postharvest physiology The study of the storage and perishability of fruits following harvest.
potential energy Energy available to do work, but not yet expressed.
prairie An area of land dominated by grasses with occasional shrubby plants and small trees occurring where the grass cover is broken and with herbaceous perennials during certain seasons.
Precambrian The geological era beginning with the formation of the earth, about 4700 million years ago and extending until about 590 million years ago. During this time the origin of life, the origin eukaryotes, and multicellular life-forms appeared.
precipitation Moisture falling from the atmosphere in the form of rain, sleet, snow, or hail.
pressure potential The water force created by a real pressure against a membrane.
primary cell wall The cellulosic wall of all plant cells laid down at the time of mitosis and cytokinesis.
primary consumer An organism that consumes a producer organism as a food source; a herbivore.
primary growth Growth originating in the apical meristems of shoots and roots resulting in an increase in length of the axis.
primary pit fields Regions within the primary cell wall in which plasmodesmata traverse the cell wall.
primary succession Plant successional events occurring in a pristine or newly forming habitat.
primary tissue Any tissue derived from the apical meristem, either shoot or root.
primates Mammals such as apes, gorillas, and monkeys.
pristine Denoting a natural and undisturbed state.
producers Autotrophic organisms that synthesize organic molecules directly from CO_2 and H_2O.
productivity The amount of plant product resulting from a combination of water, nutrients and environmental factors.
profundal zone The deepest portion of a lake.
progametangia Swollen hyphae of fungi that fuse at the point of contact and eventually form gametangia.
progymnosperm (Gr. *pro,* before + *gymnos,* naked + *sperma,* seed) An extinct group of plants, the Progymnospermophyta, which included woody plants bearing large plannated braching systems; the prototypes of the gymnospems (e.g., *Archaeopteris*).
prokaryotes Organisms whose cells have no nucleus, including bacteria and cyanobacteria.
promoter A segment of DNA that controls a group of structural genes.
prop roots Adventitious roots arising on a stem above the ground and imparting some mechanical support to plants. The angled roots may provide for absorption of water and nutrients.
propagation The process of increasing in number.
prophase The first stage of nuclear division, characterized by the disappearance of the nuclear envelope and the appearance of the shortened chromosomes.
proplastids Membrane-bound particles that develop some internal structure; may subsequently develop into chloroplasts, chromoplasts, or leucoplasts.
protein A macromolecule composed of a given linear sequence of amino acids.
protein synthesis The assemblage of protein as determined by the nucleotide sequence of a messenger RNA and the assistance of transfer RNA aligning amino acids in the proper arrangement.
prothallial cell Sterile cells present in the pollen grain of gymnosperms, believed to represent the last remnant of the vegetative male gametophyte thallus, which was, ancestrally, free-living.
protists (Gr. *protos,* first) A member of the kingdom Protista, which includes unicellular eukaryotic organisms and some multicellular ones derived from them.
protoderm The dermal or outer tissue of an apical meristem that gives rise to the epidermis.
proton The positively charged atomic particle that determines the elemental characteristics of an atom.
protonema The early, filamentous growth of the gametophyte in bryophytes and ferns.
protoplast The living portion of the cell.
protoplast fusion The technique of enzymatically digesting the cell wall of two distinctly different cells, then treating the plasma membrane so that protoplasts of the two cells fuse. The resulting hybrid may be difficult or impossible to achieve with traditional plant breeding.
protoxylem The first xylem cells formed in the primary xylem.
protozoan Unicellular eukaryotic heterotrophs.

proviral state The condition of a host cell after having been transformed.
pruning The selective removal of parts of a plant, usually woody shrubs or trees.
Psilophyta A division of leafless, rootless, vascular, homosporous plants, as exemplified by the genus *Psilotum.*
psychoactive plants Plants that contain compounds that act on the human central nervous system to produce a mind-altering state.
Pterophyta The division of vascular plants commonly referred to as the ferns.
pubescence Having hairs or trichomes on the surface.
pulping The process of partially digesting and breaking up wood fibers to make paper.
pumice A lightweight white, yellow, or gray stone formed from volcanic glass; used in polishing and cleaning.
Punnett square A visual aid in determining the possible recombination frequencies for the existing gamete types of a cross.
purine A double-ring nitrogen base found in the nucleic acids, including adenine and guanine.
pyrimidine A single-ring nitrogen base found in the nucleic acids, including cytosine, thymine, and uracil.
Pyrrhophyta A division of protists commonly termed the dinoflagellates.
pyruvic acid The end product 3-carbon compound resulting from glycolysis of glucose which enters the Krebs cycle in aerobic respiration.

Q

quarks A fundamental constituent of matter generally found in protons and neutrons. To date, six quarks are known and termed; up, down, strange, charm, bottom, and top.

R

radial cut A longitudinal section cut along a radius.
radial micellation Reinforcement of the cell wall of guard cells in specific regions such that the cells curve outward when fully turgid.
radiation Radiant energy in the form of waves or particles.
radicle The primary root of a germinating seed.
radioisotope dating A technique for determining the age of fossil materials based on the rate of disintegration of the radioisotopes.
radioisotopes Unstable isotopes of certain elements that emit particles of radioactive decay.
rain-shadow desert A desert on the leeward side of a mountain range.
random motion Molecular movement that is always present but nondirectional.
rays Parenchymal cells found in secondary xylem and phloem in both woody angiosperms and conifers, which provide for lateral (radial) transport.
reaction wood Wood produced in response to a tree that has lost its vertical position, causing the tree to straighten.
receptacle The modified apex of a stem to which the floral parts are attached.
recessive gene Gene that is masked by its dominant allelic partner, having the recessive phenotype expressed only when both alleles for a given trait are recessive.
recognition surface The three-dimensional structure of a biological membrane surface that gives specificity due to macromolecules of various sizes and shapes that extend above the lipid bilayer.
recolonization Revegetation; the reestablishment of a natural community following a natural or unnatural event that removed the existing community.
recombinant DNA Genes produced by genetic engineering manipulations that contain genetic segments from other organisms.
recycling To reuse certain materials such as paper, aluminum, and plastic by reconstituting used products into new raw materials.
reduction A gain of electrons.
reduction division The first nuclear division of meiosis during which the paired homologues migrate to opposite poles, resulting in cells with a reduction from a diploid to a haploid number of chromosomes.
regular symmetry Radial symmetry; having two or more planes through which the structure (flower) could be cut to result in mirror-image halves; actinomorphy.
regulator gene A gene that inhibits the structural genes of an operon.
relative humidity A measure of the amount of water in the atmosphere.
repressible enzymes Enzymes whose production can be inhibited by a corepressor.
repressor A compound that binds to and controls the regulator in gene regulation.
resin canal Long intercellular spaces present in the longitudinal system of vascular cells of pine, spruce, larch, and Douglas fir. Adjacent parenchymal cells secrete the resin into the canals.
resins Complex carbohydrates synthesized by certain plants in glands, canals, or ducts; insoluble in water. Used in various industrial products, including paints and varnishes. They may aid wood in resistance of decay.
resonance a vibration of electrons from one atom to another within chlorophyll.
respiration The process of converting food energy in the form of glucose to ATP energy; occurs in the cells of all living organisms and releases carbon dioxide as a by-product. Aerobic respiration requires the presence of oxygen, but some organisms can respire anaerobically.
revegetate The natural or induced replacement of plants into a cleared area; the recurrence of the same plant community that existed prior to clearing.
rhizoid Rootlike absorptive structures on the underside of certain gametophytes.
rhizome A fleshy, horizontal underground stem.
Rhodophyta The division of protists commonly called the red algae.
Rhyniophyta An extinct division of early, vascular land plants characterized by their small stature, dichotomous branching, lack of leaves, roots, and possession of terminal sporangia (e.g., *Rhynia, Cooksonia*).
ribonucleic acid (RNA) The single-stranded nucleic acid composed of adenine, guanine, cytosine, uracil, phosphate, and ribose.

ribose A 5-carbon sugar important in RNA and many other compounds.
ribosomal RNA (rRNA) A ribonucleic acid involved in the formation of the ribosome.
ribosome The cellular organelle responsible for the translation part of protein synthesis.
rickettsia A very small and specialized pathogenic bacterium transmitted by ticks, mites, and fleas.
ring-porous wood Secondary xylem characterized by larger vessels and tracheids being produced early in the season (or following favorable growing conditions). Thus each term of growth activity is seen as a ring.
RNA polymerase The enzyme responsible for forming mRNA during transcription.
root cap A group of cells covering the root meristem that aid the root's penetration through soil.
root hairs Root epidermal cells that elongate, increasing the total absorptive surface area.
root pressure A positive pressure in the xylem due to a negative solute potential and closed stomata.
root system The portion of the plant formed from the growth of the radicle and its subsequent branching; generally the underground portion of the plant.
rootbound The condition of a restricted root system caused by excess growth in too small a volume of soil; sometimes called potbound.
RUBP Ribulose 1, 5-bisphosphate, the 5-carbon compound that combines with CO_2 in the Calvin cycle.
RUBP carboxylase The enzyme that fixes CO_2 in the Calvin cycle of photosynthesis.

S

salicylic acid (sal-ah-sill-ik) A compound with pain-relieving characteristics, found in willow bark and other plants; the basic ingredient of aspirin.
samara One- or two-seeded dry indehiscent fruit in which the ovary wall forms an outgrowth to form a wing.
sand Soil inorganic particles in the .02–2.0 mm size range.
sandstone A sedimentary rock composed of sand.
sapwood The functional secondary xylem found between the vascular cambium and the nonfunctional heartwood in the center of the trunk or branch.
savanna Found between the tropics and deserts, this vegetative community is composed of extensive grasslands with scattered deciduous trees.
scarification A mechanical or chemical degradation of a hard surface, such as seed coats, so that oxygen and water can penetrate the hard layers.
schizocarp Dry, indehiscent two- or multiple-carpeled ovary that splits at maturity into separate one-seeded sections that fall away.
science An objective discipline, an approach to understand what is and is not supportable.
scientific method The objective process of approaching a problem. Involves hypothesis establishment, testing and observing the results, reevaluation of the hypothesis in light of new knowledge, and retesting to seek repeatability and thus validity of the hypothesis.
scientific name A Latinized binomial (genus and species) unique to each identified organism.
scion The upper shoot portion of a plant placed onto a rootstock in the process of grafting.
sclereids Stone cells found in tissues varying from pear fruits to the hard shell of nuts.
sclerenchyma The tissue type characterized by thick, sclerified cell walls; includes both fibers and sclereids (stone cells).
seasonal Fluctuations in environmental factors such as temperature over an annual cycle.
seaweed A common name for various macroalgae.
secondary cell wall A cellulosic wall, often impregnated with lignin, laid down inside the primary cell wall of many woody species.
secondary compounds Organic molecules synthesized by certain species of plants and not thought to be directly involved in essential metabolism.
secondary consumer An organism that consumes a primary consumer; a carnivore.
secondary growth Growth derived from secondary or lateral meristems, the vascular and cork cambiums; secondary growth results in an increase in diameter or circumference; contrasted with primary growth, which results in an increase in length.
secondary phloem All phloem tissue formed by the vascular cambium in woody tissues.
secondary succession Revegetation of cleared land; return to previous community structure.
secondary xylem All xylem tissue formed by the vascular cambium in woody plants.
sedatives Compounds that produce a relaxed, euphoric state—opium and its derivatives, morphine and heroin, are sedatives.
sedentary Inactive.
sediment Particles that settle out during the transport of liquids, usually water.
sedimentary rock Sedimentary deposits cemented together.
seed A mature ovule within a fruit.
seed coat The protective layer that develops from the integuments around a maturing ovule.
seed ferns Fossil plants with fernlike foliage but producing seeds belonging to the division Pteridospermophyta.
seedling The embryonic product of the germination of a seed; the young shoot and root axis.
seed-scale complex The spirally arranged scales on a female strobilus in gymnosperms.
seive cell The organic solute-conducting cell of the phloem in gymnosperms.
selectively permeable membrane Living membranes that allow only certain elements to pass through.
self-incompatibility Condition of a flower that cannot successfully complete the reproductive process with pollen produced by its own stamens.
self-pollinating plant A plant that has its own pollen fall on its own stigma.
senescence The aging process, usually characterized by the loss of some functional capacity, including reproduction.
sepal The flower part attached outside the others, enclosing the flower when in bud.
septa Cross walls, as in the hyphae of certain fungi.
septate Pertaining to the possession of cross walls.

seral stage One stage of the communities in an ecological succession.
sessile Referring to leaf blades without petioles in which the blade is attached directly to the stem.
seta The stalk portion supporting the capsule of the moss sporophyte.
sexual Pertaining to the fusion of gametes; sexual reproduction.
shale A sedimentary rock composed of mud.
shard A piece of broken clay or ceramic pot placed over the drainage hole of a pot to prevent the loss of soil during watering.
sheath The base of a leaf in monocots, usually wrapping around the stem.
shoot system Collective term for the stem and its leaves; generally the aboveground portion of the plant.
short-day plant A plant that flowers when the length of day is shorter than some critical value.
shrub(by) A plant that is shrublike in habitat; usually a short, perennial plant without strong apical dominance.
sieve plate The part of the wall of sieve-tube members bearing one or more highly differentiated sieve areas or modified primary pit fields.
sieve pores Holes or modified plasmodesmata found within a sieve area either on a sieve plate or sieve cell.
sieve-tube member One of the cells of a sieve tube; found primarily in flowering plants and typically associated with a companion cell; also called sieve-tube element.
silent mutation A permanent genetic change, but one that is never expressed by the phenotype.
silique A simple fruit that develops from a two-carpeled ovary; at maturity the two halves fall away, leaving the seeds attached to the persistent central wall.
silt Soil inorganic particles in the .002–.02 size range.
Silurian The geological period beginning about 435 million years ago during which time the first land plants occurred.
simple fruit A fruit derived from the maturation of a single ovary.
simple leaf A leaf having a single blade portion; may be highly lobed or dissected.
sink A site of collection of metabolites, such as sugar; metabolic sinks may exist anywhere in the plant where organic solutes are being transported by the phloem and stored.
siphonogamy (Gr. *siphon,* pipe + *gamia,* marriage) A reproductive process in seed plants in which a pollen tube carries the sperm cells to the egg located within the integumented megasporangium.
slime mold Members of the Myxomycetes, characterized by a creeping, plasmodial stage.
sociobiology The study of the biological basis of human social behavior.
sociopolitical The awareness by society of how the political process can be influenced by well-informed individuals.
softwoods A term often used in reference to all coniferous gymnosperms; more accurately, wood having a low specific gravity.
soil Inorganic particles, organic material and pore spaces combine to constitute soil.
soil profile A vertical section of soil showing the zones of particle sizes from surface down to bedrock.
solubility The relative ability of a solute to be dissolved.
solute Any substance dissolved in a solvent.
solute potential The water potential component caused by the presence of solutes in water.
solvent The liquid matrix in which a solute is dissolved.
somatic cells All body cells other than sex cells, containing at least the two sets of chromosomes inherited by both parents.
sorus A region on the underside of a fertile fern frond where sporangia are concentrated.
SPAC Soil Plant Atmosphere Continuum referring to water movement in plants.
speciation (spe-see-a-shun) The evolutionary processes leading to the development of new species.
species A group of similar organisms that are capable of interbreeding with each other; can interbreed with members of a different species with only minimal success or not at all; the second word of a scientific binomial, always in Latin.
specific heat The amount of heat required to raise the temperature of 1 gm of any substance 1 degree Celsius.
sperm nuclei Each pollen grain produces two sperm nuclei, which effect double fertilization in angiosperms.
Sphenophyta The vascular plant division commonly termed the horsetails to which the genus *Equisetum* belongs.
spindle fibers The protein fibers formed during prophase of nuclear division; chromosomes attach to these fibers at the centromere.
spine A leaf or leaf part modified as a hard, sharp-pointed structure.
spirillum A spirally coiled bacterium.
spongy parenchyma A tissue composed of loosely packed, irregular parenchymatous cells containing chloroplasts; commonly found in leaves.
sporangiospore Any spore produced from a sporangium.
sporangium (pl. **sporangia**) The embryonic product of the germination of a seed; the young shoot and root axis.
spore Haploid structures produced by the sporophyte plant via meiosis that develop into the gametophyte.
sporophyll (Gr. *spora,* seed + *phyllon,* leaf) A leaf that bears sporangia.
sporophyte The diploid, spore-producing plant in the alternation of generations; undergoes meiosis to produce the haploid spores.
stamen The male reproductive structure in a flower, consisting of an anther supported on a filament.
staminate A unisexual flower having stamens but no pistil.
staminodium A sterile stamen; nonfunctional anthers and often with petaloid filaments.
starch A polysaccharide composed of glucose linkages; the primary storage carbohydrate of plants.
stele The central vascular cylinder of roots and stems of vascular plants.
sterols Complex alcohols (steroid alcohols) that are important in animals as hormones, coenzymes, and precursors for vitamin D.
stewards Protectors, careful managers of land and nature's resources.
stigma The apical portion of the pistil, the surface on which pollen lands and germinates.
stimulant A psychoactive category of compounds that produce euphoria and increased energy, such as cocaine.
stolon A horizontal aboveground stem.
stoma (pl. **stomata**) The epidermal complex consisting of two guard cells and the pore between them.

strata Layers of sedimentary rock, the oldest rocks occurring at the bottom.
stratosphere The earth's atmosphere from about 10 km to 50 km.
strobilus (pl. **strobili**) (Gr. *strobilos,* cone) A number of sporangia- or ovule-bearing stuctures (sporophylls, sporangiophores, scales, etc.) grouped together on an axis.
stroma The matrix between the grana in chloroplasts and site of the dark reactions of photosynthesis.
stromatolite A massive limestone deposit produced by cyanobacteria.
structural genes Those genes which are transcribed into proteins.
style The central, elongated portion of the pistil between the stigma and ovary.
suberin A lipid material found in the Casparian strip of the endodermis and in the cell walls of cork tissue.
submetacentric Pertaining to a chromosome with a centromere located between the center and one end of the chromosome.
substrate The beginning substance from which other molecules are synthesized.
succulent A plant with fleshy, spongy organs modified to hold water.
sucrose A disaccharide (glucose + fructose) found in many plants; the primary form in which sugar produced by photosynthesis is translocated.
sulfur oxide Produced from burning coal, this gas is a pollutant that can produce sulfuric acid in rain droplets.
superior ovary Attachment of the ovary to the receptacle above and free from the attachment of the other floral parts.
sustainable agriculture Agricultural systems that allow for continued productivity with minimal inputs of fertilizers and water.
symbiotic An association of two or more different organisms in which both benefit.
symplastic Pertaining to the movement of water and solutes through tissues by passing through biological membranes.
synangium Fused sporangia.
synapsis The coming together of homologous chromosomes to form bivalents during prophase I of meiosis.
synergid Two haploid cells on either side of the egg cell at the micropylar end of the embryo sac.

T

taiga (tie-ga) Areas of coniferous forests north of 50° latitude having extended cold and snow cover winter conditions.
tangential cut A longitudinal section cut at right angles to a radius.
tapetum The nutritive somatic tissue surrounding the microsporocyte.
taproot A primary root from which secondary roots originate.
taproot system A system of roots characterized by the presence of one dominant axis (the primary root) and several lateral, subordinate roots.
telocentric Pertaining to a chromosome with a centromere located at one end.
telome theory A theory that regards the primitive vascular plant as consisiting of upright, dichotomously branched axes, some of which bear terminal sporangia and from which, through a series of geometric, architectural modifications, the reproductive and vegetative organs of other vascular land plants evolved.
telophase The stage of nuclear division in which sets of chromosomes finally arrive at the poles of the dividing cell. The new nuclear envelope forms at this stage.
temperate Characterized by a mild or moderate temperature.
temperate deciduous forest Mostly in the northern hemisphere, a vegetative community composed of deciduous trees and understory species.
template The genetic message on the sense strand of DNA as determined by the sequence of nucleotides.
tendril A modified leaf or stem in which only a slender strand of tissue constitutes the entire structure.
tension wood The reaction wood produced along the upper side of leaning woody trees, straightening the trunk by contracting and "pulling" the tree upright.
terminal bud The meristematic tissue located at the tip of a stem.
terminal bud scale scars The scars left by the reduced leaves that enclose the apical meristem of a twig. The distance between successive terminal bud scale scars represents one year's growth.
terpenes A group of secondary compounds composed of two to many isoprene units in a chain or ring; sometimes categorized as hydrocarbons only, sometimes to include terpenoids.
terpenoids A term referring to all compounds composed of isoprene units.
terrarium A closed biological system in which plants and animals coexist without external inputs or discharges; H_2O, CO_2, O_2, and nutrients cycle in the closed system.
terrestrial Pertaining to the land.
testcross Crossing an organism having a dominant expression for a trait with an organism having a homozygous recessive genotype for that trait to determine the genotype of the organism expressing the dominant phenotype.
testing Application of theories to examine repeatability and accuracy, should have comparisons to known information.
thallus The plant body of organisms that is undifferentiated into roots, leaves, or stems.
theory An idea that is well supported by data (information) but not known to be universally true.
thermal stratification The layering of different temperatures of water or air caused by different densities, less dense floating on more dense layers.
thorn A modified stem terminating in a sharp point.
thorn forest The dry end of a savanna community; hotter and drier than a typical savanna, these areas are characterized by dense, thorny small trees.
thylakoid The lamellar structure of the grana of chloroplasts.
thymine A nitrogen base found in DNA.
tillering The production of lateral buds and shoots near the ground to result in a plant with several shoots instead of one; particularly important in grasses and grain crops.
tissue A group of cells, generally similar in structure and origin, that perform a common function.
tonoplast The membrane surrounding the vacuole.
top carnivore The consumer at the end of a food chain or web; a carnivore that ordinarily has no predator under those ecological conditions.
topography The surface condition of an area of land; relief features.
topsoil The uppermost layer of soil, highly weathered and often rich in organic matter and nutrients.
toxin A compound that is poisonous.

trace elements Microelements of plant nutrition, needed in small quantities such as boron, zinc, and copper.
tracheids Elongated, spindle-shaped cells that conduct water in the xylem; particularly important in gymnosperms.
tracheophytes Plants with a well-defined vascular system.
transcription The first stage of protein synthesis in which the template of DNA is transcribed into mRNA.
transfer RNA (tRNA) A small ribonucleic acid molecule involved in the transfer of specific amino acids to a protein being synthesized.
transformation The incorporation of viral nucleic acid into the nucleic acid of the host cell.
transforming factor A substance that can be passed from one cell to another and cause a permanent change in heredity.
transitional mutants A type of mutation in which a single purine-pyrimidine base pair is replaced by another.
translation The second stage of protein synthesis in which the codon of mRNA pairs with the anticodon of tRNA at the surface of the ribosome.
transpiration The evaporation of water from the surface of plant leaves and stems.
transpirational pull Water molecules being "pulled" up the xylem are triggered by transpirational water loss at the leaf surface.
transverse cut A section cut at right angles to the long axis. Also termed a cross section.
transversional mutant A type of mutation in which a purine-pyrimidine base pair is replaced by a pyrimidine-purine base pair.
trichome An epidermal protrusion such as a hair or scale.
Trimerophytophyta An extinct division of early vascular plants characterized by a monopodial axis, dichotomously branched lateral axes bearing terminal sporangia. Believed to have been the ancestor of the ferns and progymnosperms.
triplet A group of three nucleotides on a nucleic acid that codes for a particular amino acid.
trophic level (tro-fik) Feeding level; a step in the energy flow through a food chain.
tropical rain forest A vegetative community dominated by evergreen trees, having high rainfall, year-round growing seasons, and high species diversity.
tropics Equatorial regions with year-round warm weather.
troposphere The earth's atmosphere below about 10 km.
tube nucleus One division of the microscope nucleus in a pollen grain that is responsible for the formation of a pollen tube from the stigma through the style to the ovule.
tuber A horizontal, underground stem with a very enlarged tip.
tubulin The protein of which microtubules are composed.
tumor A spherical mass of cells in which cell divisions occur at random and often in an uncontrolled fashion.
tundra A terrestrial habitat having permafrost soils, low-growing scattered vegetation and low annual precipitation; found north of the arctic circle or at high mountain elevation.
turf The intertwined fibrous roots of grasses forming a mass with the soil just below ground level.
turgid Full, even distended with water taken in by osmosis.
turgor pressure The real pressure developed in living cells by pressing against a membrane.
turning over The mixing of thermal zones in a standing body of freshwater due to cold, more dense water "sinking" from the top through the other layers.
turpentine A solvent that includes two terpenes—camphor and pinene.

U

ultraviolet That portion of the sun's total range of radiation having wavelengths immediately shorter than the shortest of the visible spectrum (purple); between approximately 380 and 100 nm.
understory That vegetation which characterizes the lower level of plants in a forest; the vegetation below the canopy.
unisexual The condition of having either stamens or pistil, not both; an imperfect flower.
unit membrane The transmission electron microscopic interpretation of a biological membrane, consisting of two electron-dense lines separated by a translucent space.
uracil A nitrogen base found in RNA.

V

vacuole The aqueous cavity within the cytoplasm and surrounded by the tonoplast that stores low molecular weight ions and molecules.
variability Generally refers to genetic differences between all organisms that results in uniqueness.
vascular bundle Veins in herbaceous plants and in leaves of all plants, including the specialized cells of both xylem and phloem.
vascular cambium The lateral meristem characteristic of secondary growth; gives rise to secondary xylem and secondary phloem.
vascular plants A plant possessing xylem and phloem as conducting tissue.
vascular system The total of the tissues that function in the conduction of water and food by means of xylem and phloem in the plant body.
vector An agent of transfer, such as insects and other animals in pollination.
venation The pattern of vein development.
vermiculite Expanded mica used as a sterile medium for the rooting of cuttings; high water-holding capacity and relatively inert.
vesicles Membrane-bound particles pinched off by constriction of a membrane, as in the Golgi apparatus.
vessel The primary water-conducting cell system in the xylem of most angiosperms.
vessel member One of the cells comprising a vessel, characterized by the possession of a lumen and perforation plate shared with other vessel members above and below in the series. Also termed vessel elements.
viability The period of time an organism remains alive; often used to describe the length of time before a seed will fail to germinate.
viroid An infectious agent consisting of a single-stranded RNA molecule with no protein coat; produces diseases in plants.

virus A crystal-like particle (organism?) composed of a protein coat and a central core of DNA or RNA, but never both.
viscous Thick and highly dense.
visible spectrum That portion of all the sun's radiation that can be perceived as light by humans; between approximately 380 and 750 nm in wavelength.
viviporous Characterized by beginning embryo growth while still attached to the mother plant.
volatile oils Terpenes composed of two to four isoprene units; also known as essential oils, such as lemon and peppermint.
volatilization The process of evaporation; to pass into the atmosphere as a gas.

W

water potential The relative ability of water molecules to do work by interacting with each other.
water stress A shortage of water in plants resulting in decreased ability to function properly, because water affects all biochemical controls and is the medium for all chemical reactions.
wax A lipid material with considerable oxygen inserted in the molecule; high melting point and relatively impermeable to water.
weed Any plant growing in an area where it is not desired.
wetland Terrestrial areas having the presence of water, undrained hydric soils, and plants uniquely adapted to wet conditions.
whorled Three or more leaves attached at a node.
windward The side toward which the wind is blowing.

X

x rays That portion of the sun's total range of radiation which is immediately shorter than ultraviolet; between 0.1 and 100 nm.
x-ray crystallography A technique for studying molecular and atomic structure of a substance.
xeric (zee-rick) Characterized by a very dry environment.
xerophyte A plant adapted for growth under arid conditions.
xylem (Gr. *xylon,* wood) In vascular plants, the tissue that conducts water and minerals; xylem consists, in various plants, of tracheids, vessel members, fibers, and parenchyma.
xylem ray That part of a vascular ray which is located in the wood or secondary xylem. Consists of a radial file of parenchyma cells, Also termed wood ray.

Y

yeasts One of several species of ascomycete fungi in which an extensive mycelium is not produced, so that the oganism is usually unicellular.

Z

zone of differentiation An area within the plant, proximal to the apical meristems, where cells begin to assume their srtuctural and functional characters.
zooplankton (zo-o-plank-tun) The single-celled and small multicellular animals that feed on phytoplankton in aquatic ecosystems.
zoospore A motile asexual spore.
Zosterophyllophyta An extinct division of early vascular plants characterized by lateral sporangia and commonly spine-like emergences. Believed by evolutionary biologists to be the ancestor of the lycopods.
Z-scheme A diagrammatic representation of electron flow through PS II and PS I.
Zygomycetes Members of the fungal division Zygomycota, consisting of about 800 terrestrial species having primarily coenocytic hyphae. Bread mold is an example.
zygospore A thick-walled, resistant spore resulting from a zygote.
zygote A diploid cell formed by the fusion of two haploid gametes; the result of fertilization.

Credits

Photographs

part openers

1: © Rod Planck/Photo Researchers, Inc.; **2:** © Ed Reschke/Peter Arnold, Inc.; **3:** © Helmut Gritscher/Peter Arnold, Inc.; **4:** © J.H. Robinson/Photo Researchers, Inc.; **5:** © Dr. Jeremy Burgess/SPL/Photo Researchers, Inc.; **6:** © Nuridsany et Perennov/Photo Researchers, Inc.

chapter openers

1: Courtesy of Donald Myrick/Santa Barbara Botanical Garden; **2:** Courtesy of Desert Botanical Garden; **3:** © Ed Pembleton; **5:** Courtesy of J.R. Goodin; **8:** © Kjell Sandved; **9:** © Ed Pembleton; **10:** © Kjell Sandved; **11:** Courtesy of Santa Barbara Botanical Garden; **13:** Courtesy of K.F. Kenneally; **15:** Courtesy of J.R. Goodin; **17:** © Ed Pembleton; **18,19:** Courtesy of Colin Totterdell; **20:** © Kjell Sandved; **21:** © Loran Anderson; **22:** © Ed Pembleton; **23:** Courtesy of Santa Barbara Botanical Garden

chapter 1

1.1: Courtesy of J.R. Goodin; **1.2:** Courtesy of U.S. Forest Service; **1.3:** © Richard H. Gross/Biological Photography; **1.4:** © Richard Hutchings/Photo Researchers, Inc.; **1.5:** Courtesy of USDA; **1.6:** Courtesy of Conoco, Inc.; **1.7:** © Kjell Sandved; **1.8:** Courtesy of Noel Vietmeyer; **1.9:** Courtesy of USDA; **1.10:** © Kjell Sandved/Visuals Unlimited; **1.11:** Courtesy of USDA; **1.12:** Courtesy of Debi Hogan; **1.13, 1.14:** Courtesy of USDA

chapter 2

2.1: Courtesy of USDA; **2.10:** © Rich Buzzelli/Tom Stack & Associates; **2.12:** © F.C.F. Earney/Visuals Unlimited; **2.15:** © Cabisco/Visuals Unlimited; **2.16:** © W.M. Johnson/Visuals Unlimited; **2.19:** © Doug Sokell/Visuals Unlimited; **2.21:** © John Kaprielian/Photo Researchers, Inc.; **2.22A–C:** Courtesy of U.S. Forest Service; **2.23:** Courtesy of A.B. Way

chapter 3

3.4A: Courtesy of Mike Moulton; **3.4C:** © John D. Cunningham/Visuals Unlimited; **3.5A:** Courtesy of Mike Moulton; **3.5B:** © Kjell Sandved/Visuals Unlimited; **3.7:** © Steve McCutcheon/Visuals Unlimited; **3.10A:** Courtesy of Joseph R. McAuliff; **3.10B:** © Ed Pembleton; **3.11:** © Doug Sokell/Visuals Unlimited; **3.13B:** Courtesy of Carol Wittry; **3.13C:** © Richard Thom/Visuals Unlimited; **3.14:** © Jeff Greenberg/Photo Researchers, Inc.; **3.15A, B:** Courtesy of U.S. Forest Service; **3.17:** © A. Gurmankin/Visuals Unlimited; **3.19A:** © Ed Pembleton; **3.19B:** © Michael Gadomski/Photo Researchers, Inc.; **3.19C:** © George Ranalli/Photo Researchers, Inc.; **3.20:** © John D. Cunningham/Visuals Unlimited; **3.21:** © Stephen Krasemann/Photo Researchers, Inc.; **3.23:** © Link/Visuals Unlimited; **3.24:** © Rich Buzzelli/Tom Stack & Associates; **3.25:** Courtesy of U.S. Forest Service; **3.26:** © Don W. Fawcett/Visuals Unlimited; **3.28A,B:** Courtesy of R.W. Strandtmann; **3.28C, D:** Courtesy of Stan Shetler; **3.29A:** © Richard Thom/Visuals Unlimited; **3.29B:** © Martin G. Miller/Visuals Unlimited; **3.31B:** © Scott Blackman/Tom Stack & Associates; **3.33:** © Thomas Kitchin/Tom Stack & Associates; **3.35:** © Gregory K. Scott/Photo Researchers, Inc.

chapter 4

4.16: Courtesy of E. Frei and R.D. Preston, F.R.S.; **4.21, 4.24, 4.26, 4.28, 4.32:** Courtesy of Hilton Mollenhauer

chapter 5

5.15(all): Courtesy of Steve Combs

chapter 6

6.4B: © Biophoto Associates/Photo Researchers, Inc.; **6.9, 6.11:** Courtesy of Steve Combs; **6.12:** © Fred E. Hossler/Visuals Unlimited; **6.13:** © George Wilder/Visuals Unlimited;

chapter 7

7.3: © Lynwood M. Chace/Photo Researchers, Inc.; **7.8A, B, 7.9:** Courtesy of Steve Combs; **7.12:** © A.J. Cunningham/Visuals Unlimited; **7.13:** © Rudole Arndt/Visuals Unlimited; **7.15C:** © Stanley L. Flegler/Visuals Unlimited

chapter 8

8.3B: Courtesy of U.S. Forest Service; **8.4:** © John D. Cunningham/Visuals Unlimited; **8.5:** Courtesy of Steve Combs; **8.9A, B,** John D. Cunningham/Visuals Unlimited; **8.9C:** © Kjell Sandved/Visuals Unlimited; **8.10:** © John D. Cunningham/Visuals Unlimited; **8.11:** © Dick Poe/Visuals Unlimited; **8.16:** © John D. Cunningham/Visuals Unlimited; **8.17A:** © Ed Pembleton; **8.17B:** © John Trager/Visuals Unlimited; **8.25:** © Noah H. Poritz/Visuals Unlimited

chapter 9

9.1: © V.P. Weinland/Photo Researchers, Inc.; **9.7:** Courtesy of Steve Combs; **9.8:** From R.G.S. Bidwell, *Plant Physiology* 2nd ed. © 1979 Simon & Schuster; **9.9, 9.10:** Courtesy of Steve Combs; **Box 9.1A, B:** Courtesy of Kim Feisbie; **Box 9.1C:** Bridgeman/Art Resource, NY; **9.11:** © John D. Cunningham/Visuals Unlimited; **9.12:** © Ed Pembleton; **9.15A, B:** © Cabisco/Visuals Unlimited; **9.15C:** © John Gerlach/Visuals Unlimited; **9.15D:** © Kjell Sandved/Visuals Unlimited; **9.17:** © Jay B. Labov/Visuals Unlimited

chapter 10

10.1: © Link/Visuals Unlimited; **10.2:** © F.C.F. Eaxney/Visuals Unlimited; **10.3:** Courtesy of U.S. Forest Service; **10.4:** Courtesy of St. Regis Paper Co.; **10.5:** © Manfred Kage/Peter Arnold, Inc.; **10.6:** Courtesy of Steve Combs; **10.8:** Courtesy of U.S. Forest Service; **p. 156(right):** © Dale Jackson/Visuals Unlimited; **p. 157:** Courtesy of St. Regis Paper Co.; **10.10:** © D.J. Lambrecht/Visuals Unlimited; **10.13, 10.14, 10.15:** Courtesy of U.S. Forest Service; **10.16:** Courtesy of Steve Combs; **10.17:** © Matt Meadows/Peter Arnold, Inc.; **10.18:** © Kirtley-Perkins/Visuals Unlimited; **10.19A:** Courtesy of USDA; **10.19B:** Courtesy of U.S. Forest Service; **10.20:** Courtesy of St. Regis Paper Co; **10.21:** © Link/Visuals Unlimited; **10.22A:** © Loran Anderson; **10.23A:** Courtesy of David Magar; **10.23B, 10.24A,B:** Courtesy of Noel Vietmeyer; **10.24C:** Courtesy of D.M. Yermanos; **10.25:** Courtesy of California Agriculture; **10.26:** Courtesy of USDA

chapter 11

11.2A, B: Courtesy of Joan Nowicke, Smithsonian Institution; **11.2C:** Courtesy of John Skvarla; **11.8A:** © John D. Cunningham/Visuals Unlimited; **11.8B:** © William J. Weber/Visuals Unlimited; **11.9, 11.10:** Courtesy of U.S. Forest Service; **11.12A–D, 11.13A,B:** © Ed Pembleton; **11.13C:** Courtesy of Wayne Elisens; **11.13D–F, 11.15A, B:** Courtesy of Ed Pembleton; **11.16:** Courtesy of USDA; **11.19:** © William J. Weber/Visuals Unlimited; **11.22:** © Loran Anderson; **11.24C:** © Walt Anderson/Visuals Unlimited; **11.27A–F:** Courtesy of USDA

chapter 12

12.1: © Kevin & Betty Collins/Visuals Unlimited; **12.3,12.4:** Courtesy of USDA; **12.5A–C:** Courtesy of Mary Doohan; **12.10:** Courtesy of USDA; **12.16:** © Edward Hodgson/Visuals Unlimited; **12.22:** © Ed

Pembleton; **12.24:** Courtesy of Dr. Noel Vietmeyer, National Academy of Sciences; **12.28:** Courtesy of Steve Combs; **12.33:** © Arthur R. Hill/Visuals Unlimited

chapter 13

13.1: © David Matherly/Visuals Unlimited; **13.9B:** © C.G. Van Dyke/Visuals Unlimited

chapter 14

14.1: Courtesy of USDA; **14.10:** National Wildflower Research Center, Photo courtesy of B.H. Daniller; **14.11:** © Joe Eakes/Visuals Unlimited; **14.15:** © John D. Cunningham/Visuals Unlimited; **14.16:** © Nancy L. Cushing/Visuals Unlimited; **14.18:** © Robert E. Lyone/Visuals Unlimited; **14.25:** Courtesy of USDA; **14.28:** © P. Newman/Visuals Unlimited

chapter 15

15.6: © Cabisco/Visuals Unlimited

chapter 16

16.4B: © John D. Cunningham/Visuals Unlimited; **P. 288(fig A–C):** Courtesy of Dr. Sherwin Carlquist; **P. 288(fig D):** Courtesy of Dr. Bruce Baldwin

chapter 17

17.1: © Walt Anderson/Visuals Unlimited; **17.2:** © Dick Poe/Visuals Unlimited; **17.3:** © Nancy Cushing/Visuals Unlimited; **17.4:** Courtesy of Eileen Johnson; **17.7:** © T.E. Adams/Visuals Unlimited; **17.8:** © Cabisco/Visuals Unlimited; **17.9:** © Stanley Flegler/Visuals Unlimited; **17.10:** © William J. Weber/Visuals Unlimited; **17.11:** © Joe McDonald/Visuals Unlimited; **17.19:** Courtesy of Dan Stern

chapter 18

18.1A: Courtesy of Steve Combs; **18.1B:** © Stanley Flegler/Visuals Unlimited; **18.2:** © Patrick Eggleston/Visuals Unlimited; **18.5:** Courtesy of Steve Combs; **18.6:** © Cabisco/Visuals Unlimited; **18.7:** Courtesy of Dan Stern; **18.8:** © Cabisco/Visuals Unlimited; **18.9:** © Nancy Sefton/Photo Researchers, Inc.; **18.10:** © Kjell Sandved; **18.11:** Courtesy of Larry Roberts; **18.14:** © Ed Pembleton; **18.15:** Courtesy of Arthur Elliot; **18.17B:** © David M. Phillips/Visuals Unlimited; **18.20A:** © Ed Pembleton; **18.20B:** © Kjell Sandved; **18.20C–E:** © Ed Pembleton; **18.20F, G:** © Kjell Sandved; **18.20H, I:** © Ed Pembleton; **18.20J–N:** Courtesy of George Knaphus; **18.20O:** © Jane Thomas/Visuals Unlimited; **18.21A:** © D. Neumann/Visuals Unlimited; **18.21B:** Courtesy of Steve Combs; **18.22:** © A.H. Benton/Visuals Unlimited; **18.24A:** © Y. Ahmadjian/Visuals Unlimited; **18.24B:** © Kjell Sandved; **18.24C:** © Ed Pembleton; **18.24D:** © Sylvia Duran Sharnoff/Visuals Unlimited; **18.24E:** © Ed Pembleton; **18.25:** Courtesy of Dan Guravich; **18.26, 18.28:** © Ed Pembleton; **18.29:** © Kjell Sandved; **18.30:** Courtesy of Steve Combs; **18.31F:** © G. Shih-R. Kessel/Visuals Unlimited; **18.32:** © Steve & Sylvia Sharnoff/Visuals Unlimited

chapter 19

19.6B: © W. Ormerod/Visuals Unlimited; **19.7B:** © John D. Cunningham/Visuals Unlimited; **19.9,19.11A:** Courtesy of Tony Burgess; **19.12A:** © Wm. S. Ormerod, Jr./Visuals Unlimited; **19.13:** Courtesy of Chi-Chang Lee; **19.15:** © Ed Pembleton; **19.16A:** © Kjell Sandved; **19.16B:** Courtesy of Steve Combs; **19.20:** Courtesy of Santa Barbara Botanical Garden; **19.22:** © John D. Cunningham/Visuals Unlimited; **19.23:** © Walt Anderson/Visuals Unlimited; **19.24, 19.25:** © John D. Cunningham/Visuals Unlimited; **19.27A:** © Ed Pembleton; **19.27B:** © John D. Cunningham/Visuals Unlimited; **19.29:** © Ed Pembleton; **19.30:** © John D. Cunningham/Visuals Unlimited; **19.31A, B:** Courtesy of K. F. Kenneally; **19.32A:** Courtesy of Dieter Wilkin; **19.32B:** Santa Barbara Botanical Garden; **19.32C:** Courtesy of Dieter Wilkin; **19.33:** Courtesy of Dr. Sherwin Carlquist; **19.35:** © Forest W. Buchanan/Visuals Unlimited; **19.36A, B:** Courtesy of R.J. Newton; **19.37:** © Kjell Sandved; **19.38:** © Ed Pembleton; **19.39A–F:** © Kjell Sanved; **19.39G, H:** © Ed Pembleton; **19.39I–K:** © Kjell Sandved; **19.39L:** © Ed Pembleton; **19.39M,N:** © Kjell Sandved; **19.39O, P:** © Ed Pembleton; **19.40, 19.41, 19.42:** © Kjell Sandved; **19.43A–D:** Courtesy of USDA; **19.43E:** © Kjell Sandved; **19.43F:** © Richard Criley; **19.43G–I:** © Kjell Sandved; **19.44, 19.45:** © John Cunningham/Visuals Unlimited; **19.46:** © David Cavagnaro/Visuals Unlimited; **19.47A, B, 19.48:** Courtesy of John Neff; **19.49:** © Gregory G. Dimijian/Photo Researchers, Inc.; **19.50:** © Merlin D. Tuttle/Bat Conservation International; **19.52:** Courtesy of John Neff

chapter 20

20.1: Courtesy of Robert E. Murry, Jr., and The Center for the Study of Human Adaption; **20.5:** Courtesy of Tony L. Burgess; **20.8:** © Kjell Sandved; **20.9:** © Anthony Mercieca/Photo Researchers, Inc.; **20.10:** © Tim Hauf/Visuals Unlimited; **20.11A:** Courtesy of John Schmidt; **20.11B, C:** Courtesy of Norman Borlaug; **20.12:** © Garry D. McMichael/Photo Researchers, Inc.; **20.13:** © Michael P. Gadomski/Photo Researchers, Inc.; **20.14:** © Rapho/Photo Researchers, Inc.; **20.15:** AP/Wide World Photos; **20.16C, 20.17:** Courtesy of USDA; **20.19:** Courtesy of John Schmidt; **20.20:** Courtesy of USDA; **20.21, 20.22A:** © Ed Pembleton; **20.22B–D, 20.23:** Courtesy of USDA; **20.24:** © Science VU/Visuals Unlimited; **20.25:** Courtesy of USDA; **20.26:** Courtesy of B.H. Daniller; **20.28A, B:** Courtesy of Bill Bemis; **20.29:** Courtesy of Mario Tapia and International Development Research Center, Ottawa

chapter 21

21.1A: © Loran Anderson; **21.2:** © John Kaprielian/Photo Researchers, Inc.; **21.3, 21.4:** © Loran Anderson; **21.5:** © S.W. Carter/Photo Researchers, Inc.; **21.7:** Courtesy of Tony L. Burgess; **21.8, 21.9:** © William J. Weber/Visuals Unlimited; **21.10:** © Loran Anderson; **21.11:** © Kjell Sandved/Photo Researchers, Inc.; **21.12A:** Courtesy of Russell Strandtmann; **21.12B–D, 21.13:** © Loran Anderson; **21.14:** © George Loun/Visuals Unlimited; **21.15:** Courtesy of Arthur Elliot; **21.16A, B, 21.17A, B, 21.18, 21.9A, B, 21.21:** © Loran Anderson; **21.22:** © Michael P. Gadomski/Photo Researchers, Inc.; **21.23A–C:** © Loran Anderson; **21.24:** © Virginia P. Weinland/Photo Researchers, Inc.; **21.25:** © Mark Wright/Photo Researchers, Inc.; **21.26:** © Loran Anderson; **21.27:** © Derrick Ditchburn/Visuals Unlimited; **21.28:** © Loran Anderson; **21.29:** © Richard H. Gross/Biological Photography

chapter 22

22.1: © Steve McCutcheon/Visuals Unlimited; **22.2:** © R.L. Ciochon/Visuals Unlimited; **22.3:** Courtesy of James Conti; **22.4A:** © Don W. Fawcett/Visuals Unlimited; **22.4B:** © John Trager/Visuals Unlimited; **22.4C:** © Joe McDonald/Visuals Unlimited; **22.4D, E:** © John Trager/Visuals unlimited; **22.5A:** © George Loun/Visuals Unlimited; **22.5B:** Jack Jennings/Missouri Botanical Garden, St. Louis; **22.5C:** © Tim Hauf/Visuals Unlimited; **22.6:** © Frank Hanna/Visuals Unlimited; **22.7, 22.8A–C:** Courtesy of USDA; **22.9:** © John D. Cunningham/Visuals Unlimited; **22.10A:** © Link/Visuals Unlimited; **22.10B:** © David Addison/Visuals Unlimited; **22.11:** © Ed Pembleton; **22.12:** © John D. Cunningham/Visuals Unlimited; **22.13:** © Rafael Macia/Photo Researchers, Inc.; **22.16:** National Wildflower Research Center, photo by Katie McKinney; **22.17:** National Wildflower Research Center, photo by William D. Bransford; **22.18:** National Wildflower Research Center, photo by Darrel G. Morrison; **22.19:** National Wildflower Research Center, Prairie Restoration, Inc

chapter 23

23.2A: Courtesy of U.S. Forest Service; **23.2C:** Courtesy of USDA; **23.3A:** © Renee Lynn/Photo Researchers, Inc.; **23.3B:** © Ian Roberts/Visuals Unlimited; **23.5A:** © John D. Cunningham/Visuals Unlimited; **23.5B:** Courtesy of Joseph R. McAuliffe, Desert Botanical Garden, Phoenix; **23.6A:** Courtesy of Carol Wittry; **23.7A, B:** Courtesy of USDA

Index

Bold page numbers refer to terms that are set in bold. Italic page numbers refer to figures.